Water Pollution

Assessment and Management

Water Pollution

Assessment and Management

Arvind Kumar
G. Tripathi

2024

Daya Publishing House®
A Division of
Astral International Pvt. Ltd.
New Delhi – 110 002

First Published, 2004

Reprinted, 2024

ISBN: 978-81-7035-338-6 (HB)

Published by : **Daya Publishing House®**
A Division of
Astral International Pvt. Ltd.
– ISO 9001:2015 Certified Company –
4736/23, Ansari Road, Darya Ganj,
New Delhi-110 002
Ph. 011-43549197, 23278134
E-mail: info@astralint.com
Website: www.astralint.com

Printed at : **Replika Press Pvt. Ltd.**

Preface

Water is the elixir of life, without which no biota could survive in biosphere. An indisputable and inseparable bond exists between freshwater bodies and human beings. Human civilization originated, developed and thrived in places within easy reach of freshwater sources.

The role of water in nature is unique not only from the point of human consideration, but also for numerous organisms, which make the aquatic medium their abode. The stupendous increase in world's population resulting in spurt in urbanization, industrialization, agriculture etc. has put tremendous pressure on the limited resources thereby threatening the freshwater bodies with pollution.

With the rapid growth of population, abundant use of fertilizer in agriculture as well as untreated discharge of domestic sewage, the water bodies have become dustbin causing severe water pollution, a threat to the healthy existence of mankind. Considerable human interferences have added a pressure to the freshwater ecosystem. This has resulted in the degradation of physico-chemical as well as biological characteristics of water. The polluted state of natural water resources has led to a steady decline in aquatic productivity and it has also adversely affected the aquacultural strategies.

Thus, it seems to be imperative to know the quality of the waterbodies in order to conserve, manage and keep the freshwater free from pollution. Therefore, in order to assess and manage the freshwater ecosystem, we have ventured with humble notion to bring out such a comprehensive review volume with a hope that the environmental scientists and ecoplanners may undertake further investigation which are neglected hitherto.

In this book, all 40 chapters are written by authors who are acknowledged experts in their respective fields, with the intention to provide sufficient depth of the subjects to satisfy the needs at a level, which will be comprehensive and interesting. The book shall be useful to the students, teachers, scientists and researchers in the field of Environmental biotechnocrats and other people with similar interest.

Our special thanks and appreciation go to the scientists whose contributions have enriched this volume. We wish to express our sincere gratitude to Professor (Dr.) Indu Dhan, Hon'ble Vice Chancellor, S.K. University, Dumka who has been a source of constant inspiration. We are also grateful to Professor M.C. Dash, Vice Chancellor, Sambalpur University; Professor H.R. Singh, Pro-Vice Chancellor, University of Allahabad; Professor M.P. Singh, Pro-Vice Chancellor, V.B. University, Hazaribag; Professor N.C. Datta, Calcutta University; Professor Ajit Varma, J.N.U., New Delhi; Professor S.N. Kaul, NEERI; Professor Tanmoy Bhattacharya, Vidyasagar University; Professor S.K. Konar, Kalyani University; Professor P.C. Mishra, Sambalpur University; Professor Sudhendu Mandal, Viswa Bharati University; Professor P.P. Sood, Saurashtra University, Rajkot; Professor B.D. Joshi, G.K. University, Hardwar; Professor A.K. Mittal, B.H.U., Varanasi; Professor U.S. Bagde, Mumbai University; Professor D.B. Tembhare, Nagpur University; Professor B.N. Pandey, Magadh University; Professor J. Ojha, Bhagalpur University; Professor S.P. Roy, Bhagalpur University; Professor M. Raziuddin, V.B. University; Professor M.V. Subba Rao, Andhra University; Professor P.S. Murthy, Bangalore University; Professor P. Natrajan, Kerala University; Professor S.P. Hosmani, Mysore University; Professor R. Ramani Bai, Chennai University and Dr. M.P. Sinha, Ranchi University who have been a source of continuous encouragement and inspiration to me.

We are especially thankful to Dr. Sikandar Prasad Yadav, Principal, S.P. College, Dumka for his encouragement. We are indebted to respected Shri Tribhuvan Poddar of University Department of Zoology, Bhagalpur University, Bhagalpur for encouragement. We also acknowledge the incentives provided by one of my research scholars, Dr. Chandan Bohra, Head of Zoology, B.S.K. College, Barharwa for helping me actively in bringing out this book.

Last but not the least, we are also thankful to Mr. Anil Mittal, proprietor of Daya Publishing House, Delhi for publishing this book. Finally, we will always remain a debtor to all my well-wishers for their blessings, without which this book would not have come into existence.

Dumka **Professor (Dr.) Arvind Kumar & Dr. G. Tripathi**

Contents

1

Impact of Sewage Pollution on Primary Productivity of Wetland of Jharkhand (Santal Pargana), India

ARVIND KUMAR
Environmental Biology Research Unit, Post Graduate Department of Zoology, S.K. University, Dumka, Bihar

C. BOHRA
Pollution Research Laboratory, Department of Zoology, B.S.K. College, Barharwa, Jharkhand

Introduction

Sun is the ultimate source of life and is an eminent driving force of the biosphere. Its solar energy is trapped by chlorophyll bearing plants, which transform it into chemical energy. The energy that falls on the earth's surface as light is balanced by the energy that leaves the earth's surface as invisible heat radiation but the rate at which the energy is stored by green plants is called primary productivity and by the heterotrophs is called secondary productivity. The essence of life is nothing but the progression of such changes as growth, self duplication and the synthesis of complex relationship of matter which actually depends upon the biological productivity as well as energetics within the system. Thus, the primary productivity is the basis of whole metabolic cycle in natural aquatic ecosystems. The organic matter synthesized during the primary productivity by the primary producers is utilized by the consumers inhabiting the system.

All the materials synthesized by the producers, however, is not available to the consumers. Part of it is utilized by the producers themselves in their maintenance (respiration). Some part of it is wasted off and is used by the decomposers. The remainder is consumed by the herbivores, which transfers some amount to the carnivores. The accumulation of biomass in all other organisms except producers is referred as secondary production. The total rate of photosynthesis including the organic matter used up in respiration during the measurement period is called total photosynthesis or total assimilation or gross primary productivity, whereas the rate of storage of organic matter in producer tissues in excess of the respiratory use by the plants during the measurement period is called apparent photosynthesis or net assimilation or net primary productivity. Above all, the rate of storage of organic matter not used by the heterotrophs during the growing season is called net community productivity or this is actually the net primary production minus the heterotrophic consumption during the period of investigation.

The freshwater of the earth comprises less than 4 per cent of the total earth surface. Most of the surface of freshwater exists as ice and snow in glaciers and polar icecaps (Approximately 25.5×10^6 Km^3). Freshwater lakes and streams cover about 0.2 per cent and have an estimated area of about 2×10^6 Km^2. The mean net productivity is 200g $C/m^2/yr$. Freshwater marshes and swamps comprise almost the same area of about 2×10^6 Km^2, but it gives a comparatively much higher mean net productivity value *i.e.* 1500g $C/m^2/yr$.

Primary production studies are of paramount interest in understanding the effect of pollution on system's efficiency. High rates of production both in natural and artificial ecosystems occur when physico-chemical factors are favourable. But pollution of water in the long run leads to a reduction in primary production. Pollution also affects the production/respiration ratio, a proper level of which is very essential for the substenance of the system. In non-polluted water, the production usually exceeds respiration but in organically polluted system (sewage pollution) respiration exceeds production and no organic material is left available for the aquatic biota leading to system impairment.

Biological production is the key to the extent to which natural water resources may be utilized–for whatever purpose–and it is intended here to discuss ecological and biological production both in general and with particular reference to the impact of domestic sewage in freshwater ecosystems of Jharkhand. There are numerous factors affecting production, both positively and negatively, and such factors are subtly different from those influencing production in the typically more comprehensively studied temperate regions. By an understanding of these factors that affect and regulate production and determine the ecology of any aquatic system lotic or lentic, we can aim at a planned and sensible utilization of the regional resources. All too often a failure to understand has led, in the past, to environmental disasters. As a consequence mankind in the long term is the sufferer.

Production and its efficiency and a thorough understanding of ecology are of imperative importance in the tropical aquatic systems, where much of the supply of proteins comes from aquaculture. The whole yield of the aquatic ecosystems harvested by human beings in the form of aquatic life depends upon the primary production of the water bodies from which they come. Primary production of an aquatic system is the ability of water to support the production of organic materials from inorganic ones. Primary productivity is the path through which energy enters into its unidirectional path of flow through various trophic levels in the ecosystem. The radiant energy from solar system falling on the surface of water is trapped by microphytes (phytoplankton) through the process of photosynthesis and the radiant energy is then stored into the microphytes in form of organic matter. The harvest of predatory organisms, which occur at higher trophic levels, depends on the productions of microphytes through a series of transfers of materials–each step called trophic level and the whole relationship of eating and being eaten is called food chain. As only a fraction, say 10-20 per cent of the productivity of one trophic level can be transferred to the next; the proximity of harvested forms to the base of the chain is an important factor in determining the usable productivity or potential yield of the water concerned.

The Indian sub-continent comprises a predominantly tropical environment and because of generally stabilized climatic conditions throughout the central tropical region as a whole, with high temperature, high rainfall and bright illumination, the productivity of water of this region is more or less continuous and quite different from climatically induced variable rates of production found in temperate climates. Table 1.1 includes the data on primary production showing very clearly the ranges of microphytic production in different regions. It is clear that the shallow water bodies of oriental region are most productive in comparison to water bodies of different latitudes. In this region similar to the general features of tropical climate, however, production can reach levels high enough to result in dramatic surges in the species abundance so that algal bloom, particularly of the blue green algae *Microcystis aeruginosa* and *Oscillatoria rubens*, occur in freshwater with some higher value of nutrients. These, in turn, influence the abundance of secondary and tertiary consumers and hence the secondary and tertiary production. Thus, community structure can fluctuate widely and is brought about by the competitive interaction existing between opportunistic undergoing remarkable changes in population structure. Such a balance is also influenced by the generally longer breeding seasons of tropical species (Morton, 1978). This situation is exacerbated by the greater degree of adaptive radiation that has taken place in most of the principal aquatic families, resulting in a profusion of species, each exploiting a narrower niche in the environment as a whole.

Table 1.1: Range of Primary Production in Microphytic Communities of Shallow Freshwaters in Various Regions

Regions	*Sub Region*	*Mean Depth (Z)m*	*Gross Primary Production (Pg)mgC/m²/day*	*Annual Gross Primary Production gC/m²/Yr.*	*Authors*
Tropics	i) Tropic in General	2.5–5.0	158–15200	620–840	Goldman (1959), Kurasawa (1958), Ganf (1972), Imevbore *et al.* (1972), Burgis *et al.* (1973).
	ii) Oriental in Particular	(A) 0.9–3.0	1210–17550	—	Ganapati (1972), Ganapati and Sreenivasan (1972), Nasar & Datta Munshi (1975), Sarkar *et al.* (1977), Sinha (1988), Goral (1978).
		(B) 3.4–18.3	1601–3228		Ganapati and Pathak (1972), Ganapati and Sreenivasan (1972).
Temperate	i) American	1.0–20.0	60–13200	90–760.1	Steemann Nielsen (1955), Rodhe (1958), Wetzel (1966 a & b), Efford (1972), Wetzel (1975), Kelly *et al.* (1978).
	ii) European	1.2–14.0	0–11000	14–520	Nygaard (1955), Steemann Nielsen (1955), Rodhe (1958), Duthie and Kirton (1970), Steel *et al.* (1968), Morgan (1972), Moskalenko and Votinsev (1972),Wetzel (1975), Zutshi (1976).

In the tropics, in general as pointed out by Morton (1978) each species occupying a narrow niche might be considered more individually susceptible to the deleterious effects of man's activity in the aquatic environment. Alternatively, however, there is correspondingly greater variety of species possessing at least the potential for successful exploitation by man as a resource. Thus, it is now very clear that water body of this area possess an enormous potential for the successful exploitation of resources in one hand while on the other hand to the impacts of human activities–the pollutional one being the first and foremost.

Considerable work has been done on the primary productivity of an aquatic ecosystem. Odum (1956) studied the primary production in flowing waters. Eceles (1962) described the productivity of lake Nayasa and its possible significance in the nutrient cycle. Prasad and Nair (1962) have made a comparison of values of organic production obtained from oxygen and C^{14} methods. Sreenivasan (1963, 1964a, 1966, 1969 and 1972) has studied the role of primary productivity in fish culture. Tailang (1965) has also made a comparative study on problems of phytoplankton production and photosynthetic productivity in a tropical and temperate aquatic environment. Vijayaraghwan (1971b) observed seasonal variation in primary productivity in three tropical ponds. Prouse (1972) had made some observations primary and fish production in experimental fishponds of Malayasia. Thompson (1976) recorded the primary productivity of African wetlands with special reference to Okavanga Delta. Sinha and Sahai (1978) studied extensively the factors affecting growth and productivity of free floating microphytes in Ramgarh wetland Gorakhpur. Kaul *et al.* (1978) studied the distribution and primary productivity of macrophytes in some stagnant waterbodies of Kashmir. Gopal *et al.* (1978) also studied the eology and production in Indian freshwater ecosystems at primary producer level with emphasis on macrophytes. Rai *et al.* (1980) established interactions between a floating and submerged vegetation community in a fish pond of Bhagalpur with special reference to primary productivity. Laal (1981) studied the primary production and possibilities for utilization of swamps in north Bihar. Lipkin *et al.* (1986) observed the primary productivity of macrophytes with terminology approaches and a comparison of methods. Khan *et al.* (1988) estimated the primary production in Sheikha Zheel at Aligarh. Laal (1989) recorded seasonal variations in primary productivity and photosynthetic efficiency in two tropical wetlands. Rai and Sharma (1991) studied the energy transfer efficiency and primary productivity of tropical wetlands in north Bihar. Sen *et al.* (1992) has made a good observation on physico-chemistry, nutrient budget and the factors influencing primary production in a tropical lake of Hatia, Ranchi. Pati and Sahu (1993) studied primary production in Rengali wetland, Orissa. Das and Verma (1993) studied limnology of a paid and a lake at Rajnagar (Madhubani), Bihar. Gupta and Sharma (1994) studied the phytoplanktonic productivity of Devikund sagar village pond (Bikaner, northwest Rajasthan). Singh and Sinha (1994) observed the primary productivity and limnological profile of two ponds at Patna, Bihar. Sharma and Saxena (1994) made a good observation on studies on gross and net primary productivity of two bodies of water in the Indian desert. Vijay Kumar (1994) studied the seasonal variations in the primary productivity of a tropical pond. Pandey *et al.* (1995) observed the assessment of primary productivity of phytoplankton and macrophytes of Kawar lake wetland (Begusarai), Bihar. Agarwal *et al.* (1995) studied the comparatvie account of phytoplankton productivity in the Sagar lake and Military Engineering lake. Kumar (1995e) studied the estimation of productivity of odonata larva (Insecta) of a tribal pond of Santal Pargana, Bihar. Kumar (1996g) studied the impact of sewage pollution on chemistry and primary productivity of two freshwater bodies in Santal Pargana (Bihar). Florin am Montes (1998) observed the relevant scales to assess primary production of Mediterranean semi-arid Salt lake. Pandit (1999) studied the dynamics of primary productivity of certain ponds at Barharwa

(Santal Pargana), a tribal division of Bihar (India). Kumar (1999a) studied the impact of coal mining effluents on the primary productivity of water bodies and fish farming in and around the Chitra coal fields in Santal Pargana, Bihar. Mishra and Nayak (2000) observed the biomass productivity of energy plantation on Himalayan Foot Hill. Harikrishnan and Abdul Azis (2000) studied the primary production strategies in a freshwater temple tank in Kerala. Kumar (2001c) estimated the primary productivity of Tatloi thermal spring of Jharkhand (India).

However, no work has yet been done in and around the aquatic ecosystems of Barharwa except Pandit (1999). Thus, to fulfil this gap the present investigation is undertaken to assess the impact of sewage pollution on the primary productivity in two lentic water bodies of Barharwa.

Materials and Methods

Monthly estimation of primary productivity was made from January, 1999 to December, 2000 by light and dark bottle technique of Garder and Gran (1927). The water samples from both the ponds were collected and filled in two sets of initial light and dark (black painted) bottles. The dissolved oxygen contents of the initial bottles were determined immediately following the modified Winklers' method (APHA, 1989). The light and dark bottles were suspended at a depth of 0.5 m. with the help of nylon string at a regular interval of three hours during the day time from 6 am to 6 pm. After the expiry of the incubation periods, the suspended bottles were removed and the dissolved oxygen contents in each bottle were determined immediately. The calculated average hourly values were pooled in average monthly data.

Results

Monthly variation in gross production, net production, net production/gross production, respiration and percentage of gross production of two ponds from January, 1999 to December, 2000 were shown in Tables 1.2 to 1.5 and Figs. 1.2 to 1.8.

Table 1.2: Monthly Variations in Primary Productivity of Nathu Pond from Jan. to Dec., 1999

Period	*Production*		*Respiration*		
	GPP ($gC/m^2/day$)	*NPP ($gC/m^2/day$)*	*CR ($gC/m^2/day$)*	*Per cent of GPP*	*NPP Vs. GPP*
Jan.	2.97	0.80	2.17	73.06	0.27
Feb.	2.36	1.22	1.14	48.30	0.52
Mar.	3.51	2.16	1.35	38.46	0.62
Apr.	4.58	3.53	1.05	22.92	0.77
May	4.23	3.07	1.16	27.42	0.73
Jun.	4.19	1.75	2.44	58.23	0.42
Jul.	3.69	1.68	2.01	54.47	0.46
Aug.	2.60	0.69	1.91	73.46	0.27
Sep.	1.25	0.54	0.71	56.80	0.43
Oct.	2.45	1.21	1.24	50.61	0.49
Nov.	3.83	2.15	1.68	43.86	0.56
Dec.	2.89	1.25	1.64	56.75	0.43

Table 1.3: Monthly Variations in Primary Productivity of Munshi pond from Jan. to Dec., 1999

Period	*Production*		*Respiration*		
	GPP (gC/m²/day)	*NPP (gC/m²/day)*	*CR (gC/m²/day)*	*Per cent of GPP*	*NPP Vs. GPP*
Jan.	1.39	0.89	0.50	35.97	0.64
Feb.	2.19	1.42	0.77	35.16	0.65
Mar.	2.99	1.96	1.03	34.45	0.66
Apr.	3.89	2.85	1.04	26.74	0.73
May	3.64	2.12	1.52	41.76	0.58
Jun.	3.79	0.29	3.50	92.35	0.08
Jul.	3.65	0.88	2.77	75.89	0.24
Aug.	0.71	0.25	0.46	64.79	0.35
Sep.	1.07	0.33	0.74	69.16	0.31
Oct.	2.14	1.01	1.13	52.80	0.47
Nov.	3.49	1.05	2.44	69.91	0.30
Dec.	2.07	0.99	1.08	52.17	0.48

Table 1.4: Monthly Variations in Primary Productivity of Nathu pond from Jan. to Dec., 2000

Period	*Production*		*Respiration*		
	GPP (gC/m²/day)	*NPP (gC/m²/day)*	*CR (gC/m²/day)*	*Per cent of GPP*	*NPP Vs. GPP*
Jan.	3.07	0.92	2.15	70.03	0.30
Feb.	2.46	1.34	1.12	45.53	0.54
Mar.	3.61	2.28	1.33	36.34	0.63
Apr.	4.68	3.65	1.03	22.00	0.78
May	4.33	3.18	1.15	26.56	0.73
Jun.	4.29	1.78	2.51	58.51	0.41
Jul.	3.79	1.72	2.07	54.62	0.45
Aug.	2.70	0.72	1.98	73.33	0.27
Sep.	1.35	0.59	0.76	56.30	0.44
Oct.	2.55	1.25	1.30	50.98	0.49
Nov.	3.93	2.19	1.74	44.27	0.56
Dec.	2.99	1.31	1.68	56.19	0.44

Gross Production

The gross production ranged between 1.25g C/m²/day in September and 4.58g C/m²/day in April in Nathu pond whereas in Munshi pond, it varied between 0.71g C/m²/day in August and 3.89g C/m²/day in April, 1999 respectively. In the year 2000, it ranged between 1.35g C/m²/day in September and 4.68g C/m²/day in April in Nathu pond, whereas in Munshi pond, it varied between 0.77g C/m²/day in August and 3.95g C/m²/day in April.

Table 1.5: Monthly Variations in Primary Productivity of Munshi pond from Jan. to Dec., 2000

Period	*Production*		*Respiration*		
	GPP (gC/m²/day)	*NPP (gC/m²/day)*	*CR (gC/m²/day)*	*Per cent of GPP*	*NPP Vs. GPP*
Jan.	1.45	0.92	0.53	36.55	0.63
Feb.	2.24	1.45	0.79	35.27	0.65
Mar.	3.05	1.99	1.06	34.75	0.65
Apr.	3.95	2.88	1.07	27.09	0.73
May	3.69	2.15	1.54	41.73	0.58
Jun.	3.85	0.32	3.53	91.69	0.08
Jul.	3.71	0.91	2.80	75.47	0.25
Aug.	0.77	0.28	0.49	63.64	0.36
Sep.	1.13	0.36	0.77	68.14	0.32
Oct.	2.20	1.04	1.16	52.73	0.47
Nov.	3.55	1.08	2.47	69.58	0.30
Dec.	2.13	1.02	1.11	52.11	0.47

Net Production

The net production varied from 0.54g $C/m^2/day$ to 3.53g $C/m^2/day$ in September and April 1999 in Nathu pond whereas in Munshi pond, from 0.25g $C/m^2/day$ to 2.85g $C/m^2/day$ in August and April 1999 respectively. In the year 2000, the net production ranged between 0.59g $C/m^2/day$ and 3.65g $C/m^2/day$ in September and April respectively in Nathu pond whereas it varied from 0.28g $C/m^2/day$ to 2.88g $C/m^2/day$ in August and April in Munshi pond respectively.

NP/GP

The NP/GP value ranged between 0.27g $C/m^2/day$ to 0.77g $C/m^2/day$ in January-August and April 1999 in Nathu pond whereas in Munshi pond, it ranged from 0.08g $C/m^2/day$ to 0.73g $C/m^2/day$ in June and April 1999 respectively. In 2000, it varied from 0.27g $C/m^2/day$ to 0.78g $C/m^2/day$ in August and April in Nathu pond and from 0.08g $C/m^2/day$ to 0.73g $C/m^2/day$ in June and April in Munshi pond respectively.

Community Respiration

The community respiration value ranged between 0.71g $C/m^2/day$ and 2.44g $C/m^2/day$ in September and June in Nathu pond and 0.46g $C/m^2/day$ and 3.50g $C/m^2/day$ in August and June in Munshi pond in 1999, whereas in 2000, it ranged from 0.76g $C/m^2/day$ to 2.51g $C/m^2/day$ in September and June in Nathu pond and from 0.49g $C/m^2/day$ to 3.53g $C/m^2/day$ in August and June in Munshi pond.

The percentage value to gross production ranged between 22.92 per cent and 73.46 per cent and 22.00 per cent and 73.33 per cent in Nathu pond, while in Munshi pond, it varied from 26.74 per cent to 92.35 per cent and from 27.09 per cent to 91.69 per cent in 1999 and 2000 respectively.

Discussion

The primary productivity of any water body is primarily a function of nutrients and energy supply to the system. A large population under unfavourable conditions may have a low rate of production whereas a small population under a favourable condition may have high rate of production. The factors which mainly influence the primary production are the amount of available light, temperature and concentrations of essential nutrients.

So far as the nutrient is concerned its input to freshwater natural resources is clearly related to the use of catchments area, geological formation of the basin, input through effluents and also to its own standing crops.

Phosphorous and, in special sense, silica are generally regarded as the most critical nutrients for organic production in lakes (Rodhe, 1969; Schelsker and Stoermer, 1972; Schindler and Fee, 1974; Fee, 1978 and 1979) and phosphorous is the most implicated in lake eutrophication from human activities. In those studies in which annual productivity has been examined in relation to various physical and chemical factors, nitrogen, particular total (Mann, 1974), organic or Kjeldahl nitrogen (Herback, 1969) has been correlated with phosphorous.

More importantly, the direct nutrient pools in aquatic systems may be several steps removed from the free nutrient in solution. In tropical waters in general, most nutrients are incorporated into living tissue and are transformed by direct digestion/decomposition/absorption cycles (Henderson, 1978). A further complication is that excess nutrients are readily taken up by plant cells so that above certain critical levels, additional nutrients is taken up by the organisms into the intercellular pools without additional growth (Gerloff, 1969).

Thus as shown in the Fig. 1.1, when a nutrient is below critical concentration, any additions to the environment will be immediately taken up in growth with no change in concentration. But once

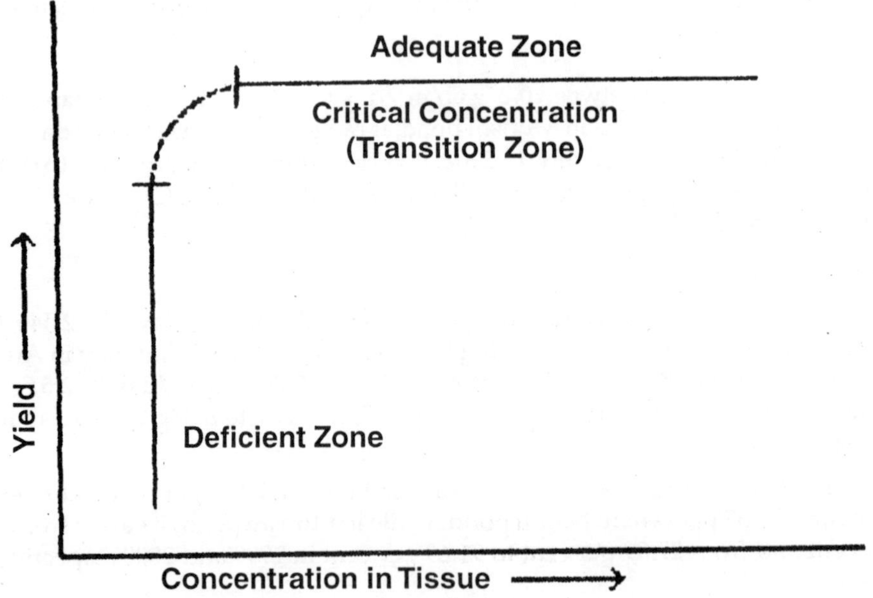

Fig. 1.1: Critical Concentration and Luxury Consumption (after Ulrich, 1961)

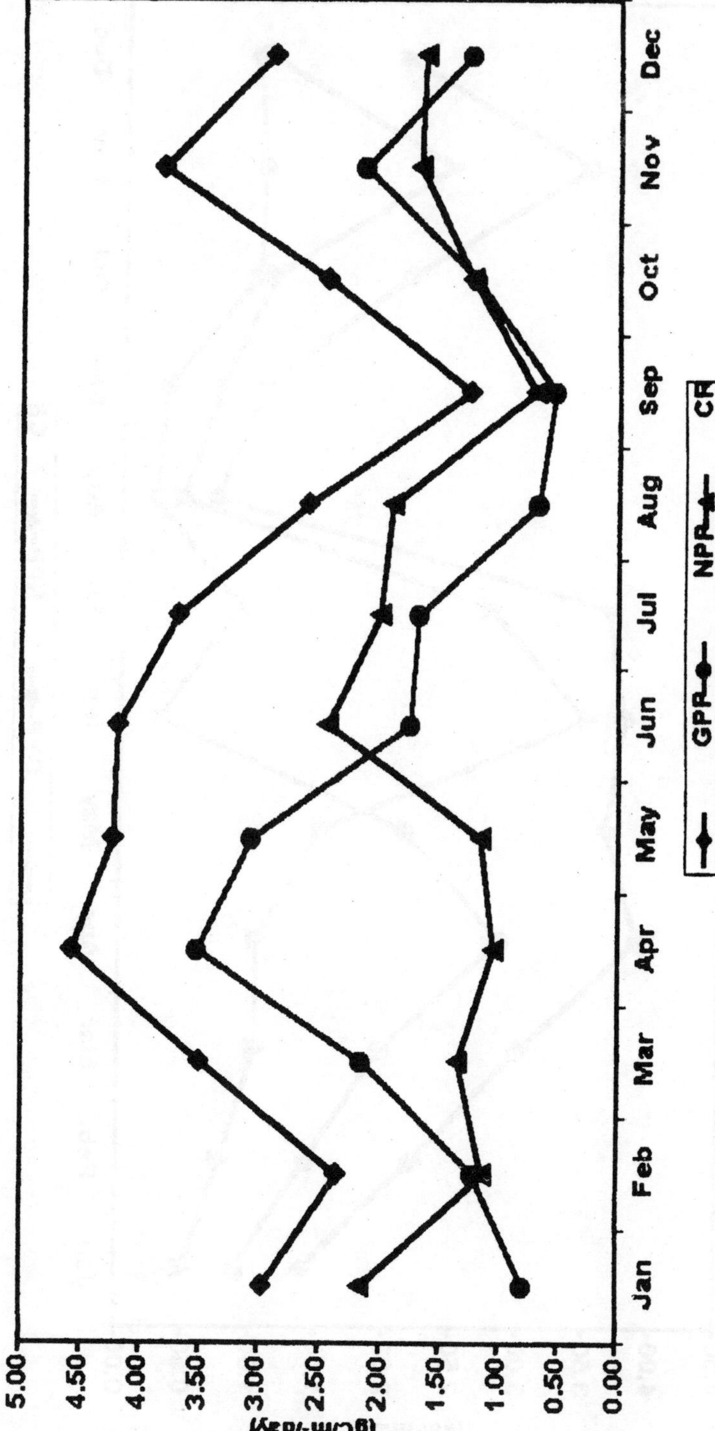

Fig. 1.2: Monthly Variations in GPP, NPP and Community Respiration in Nathu Pond During 1999

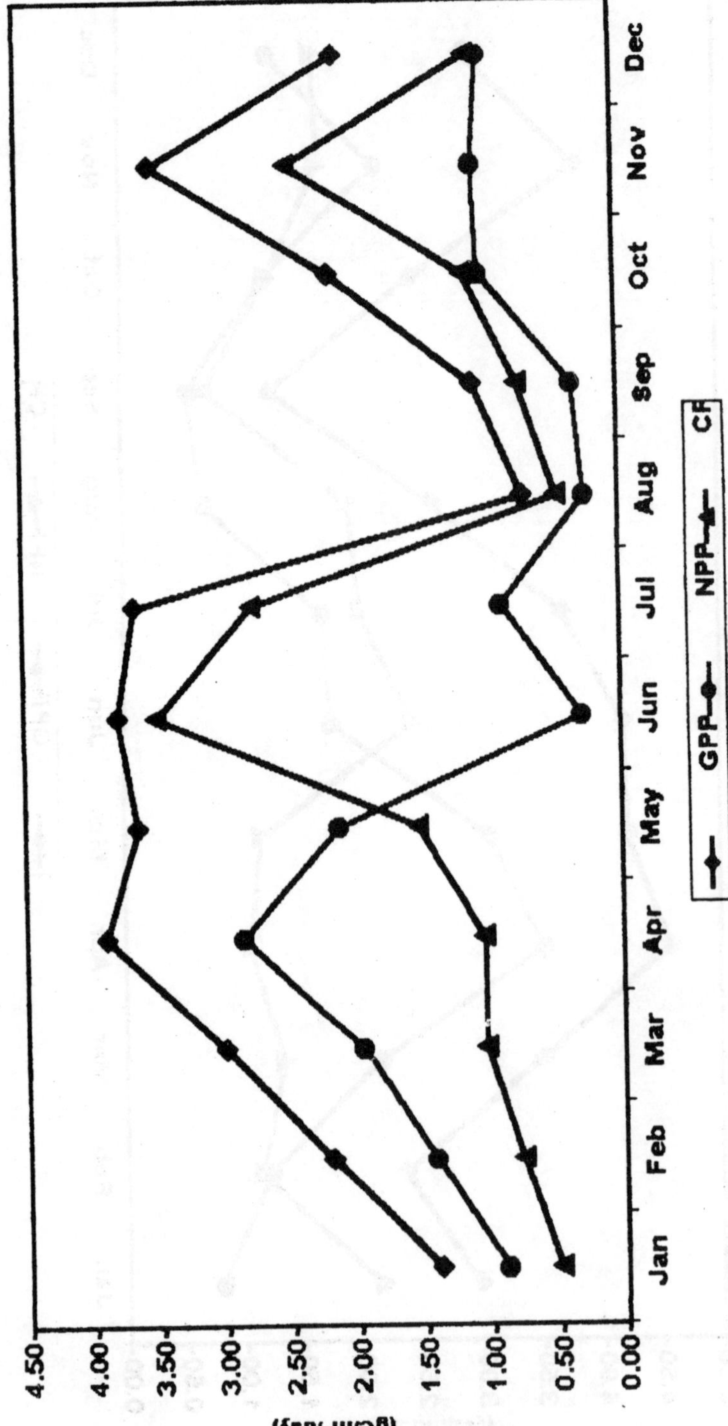

Fig. 1.3: Monthly Variations in GPP, NPP and Community Respiration in Munshi Pond During 1999

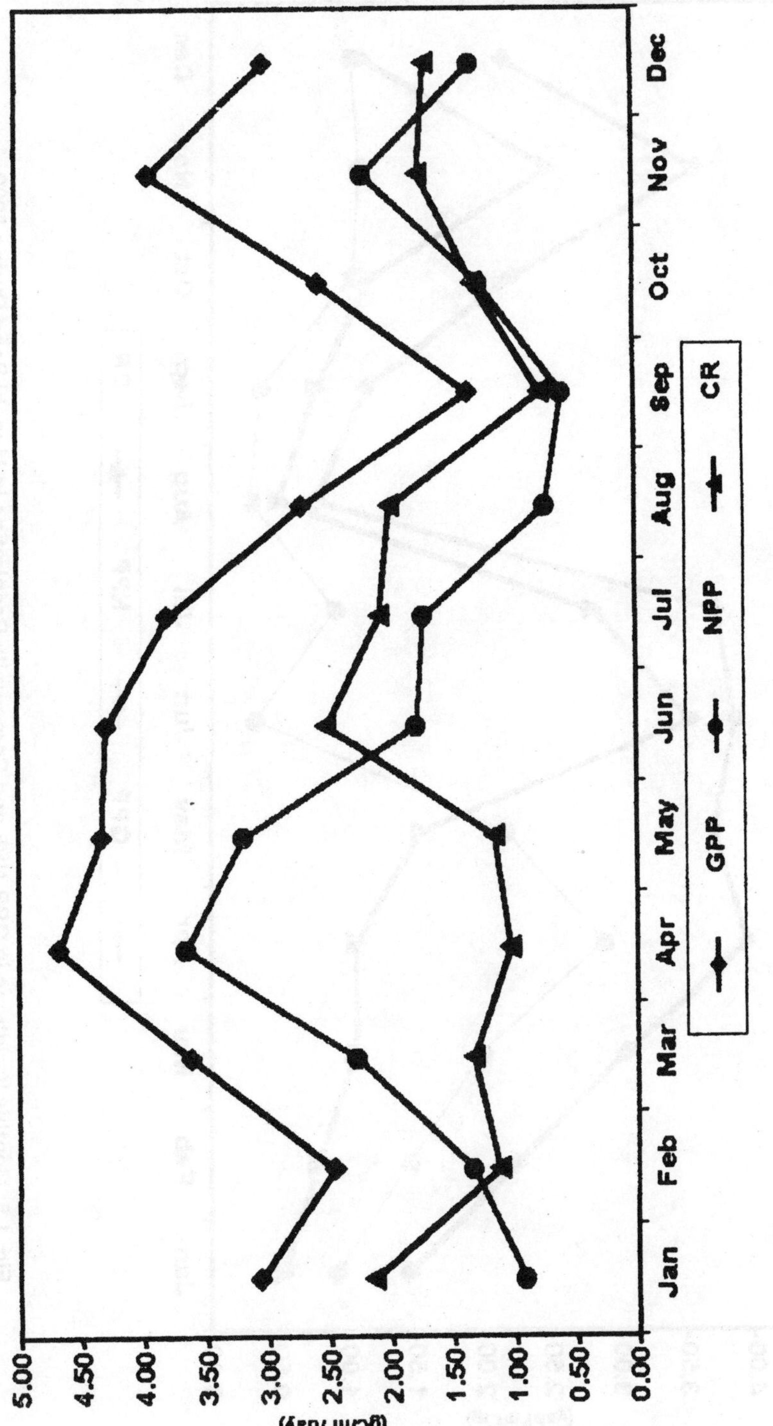

Fig. 1.4: Monthly Variations in GPP, NPP and Community Respiration in Nathu Pond During 2000

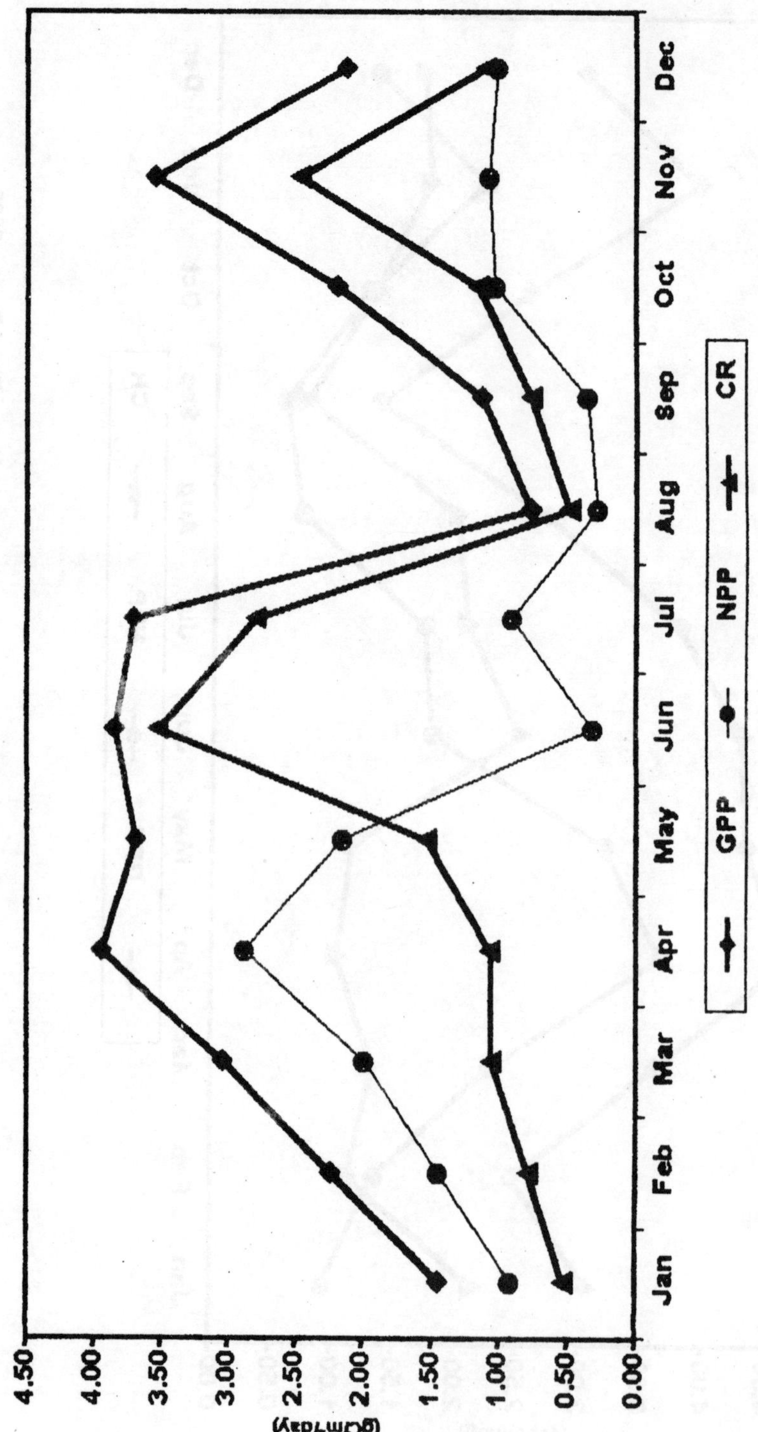

Fig. 1.5: Monthly Variations in GPP, NPP and Community Respiration in Munshi Pond During 2000

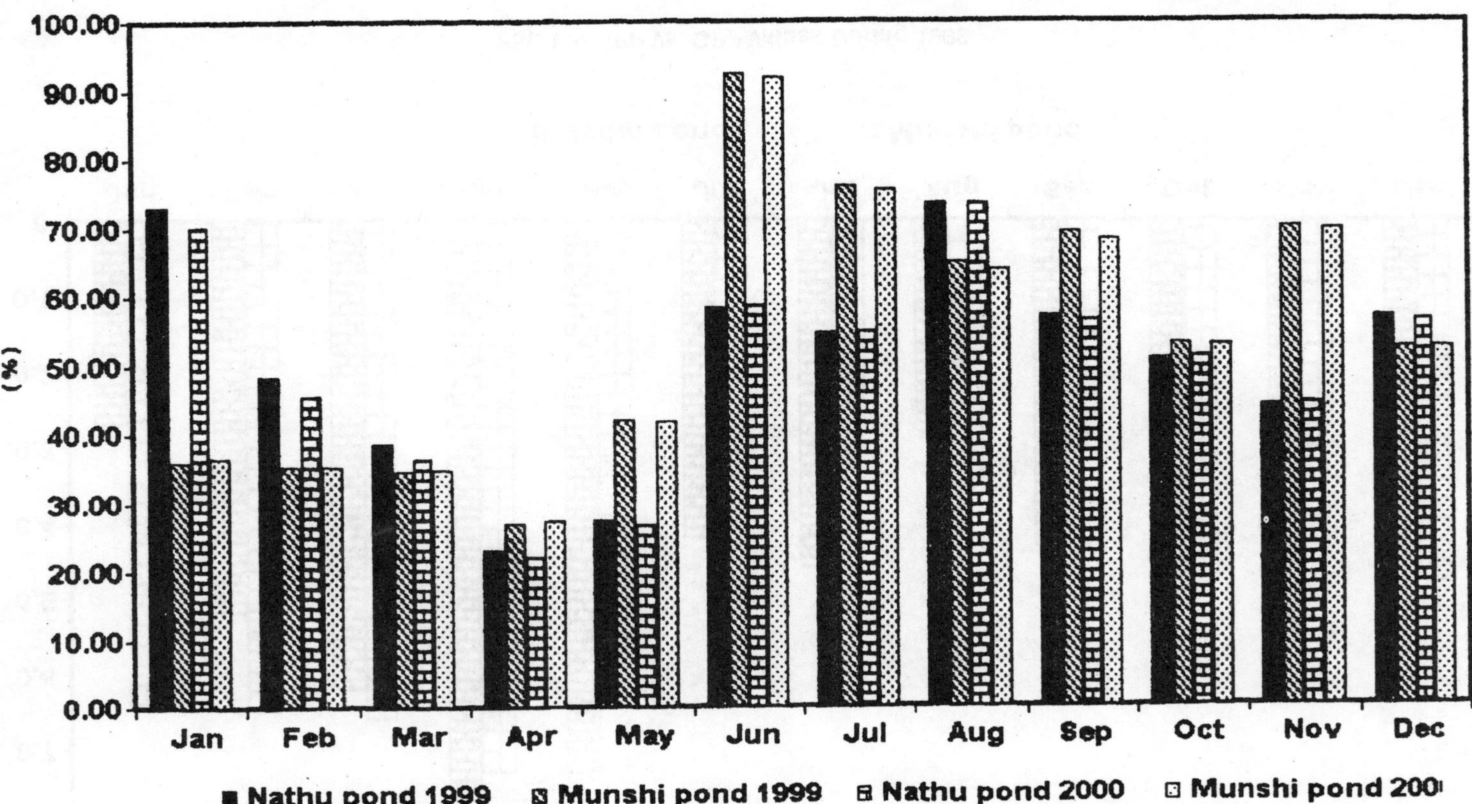

Fig. 1.6: Percentage of Gross Primary Productivity During 1999–2000

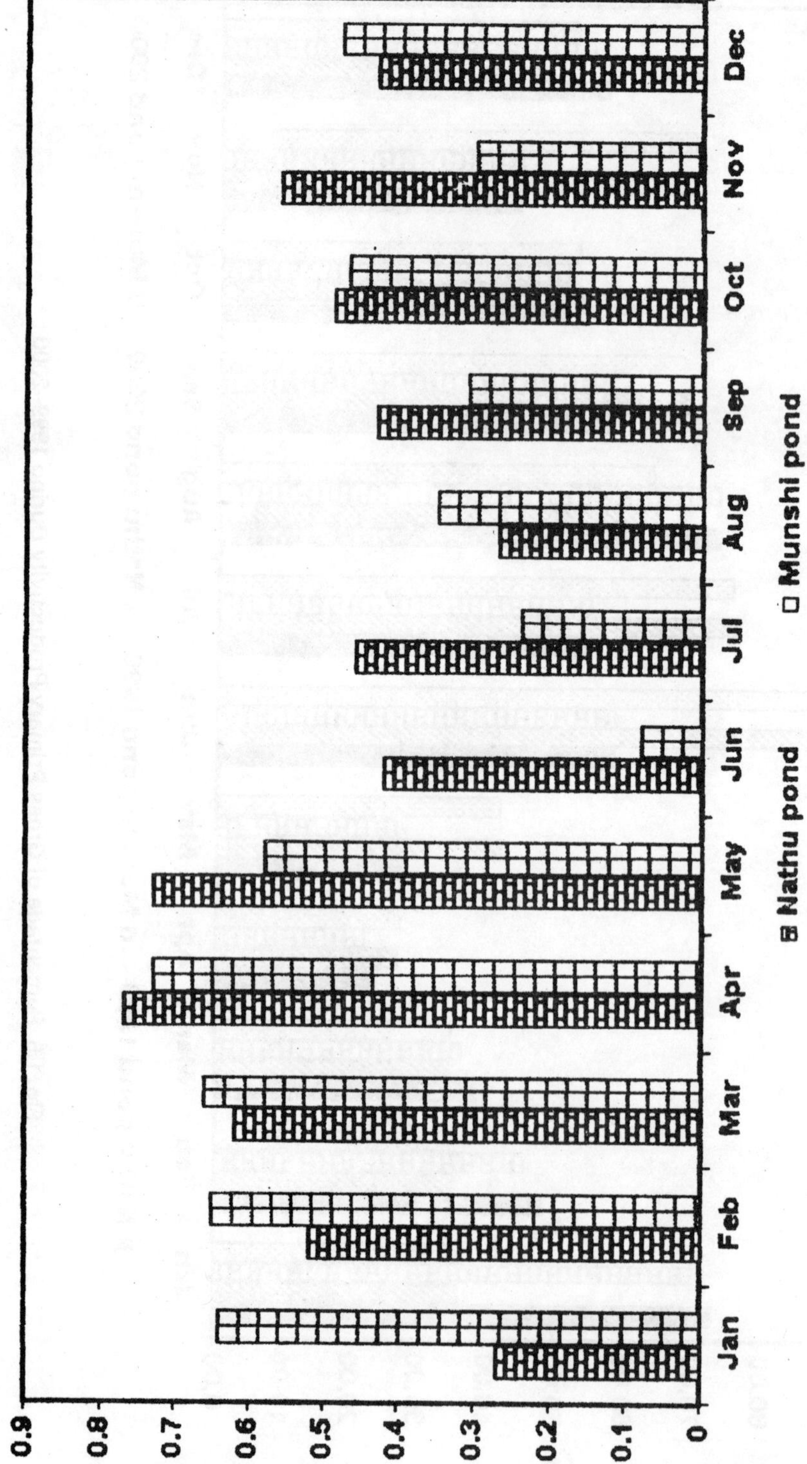

Fig. 1.7: NPP Vs. GPP Values During 1999

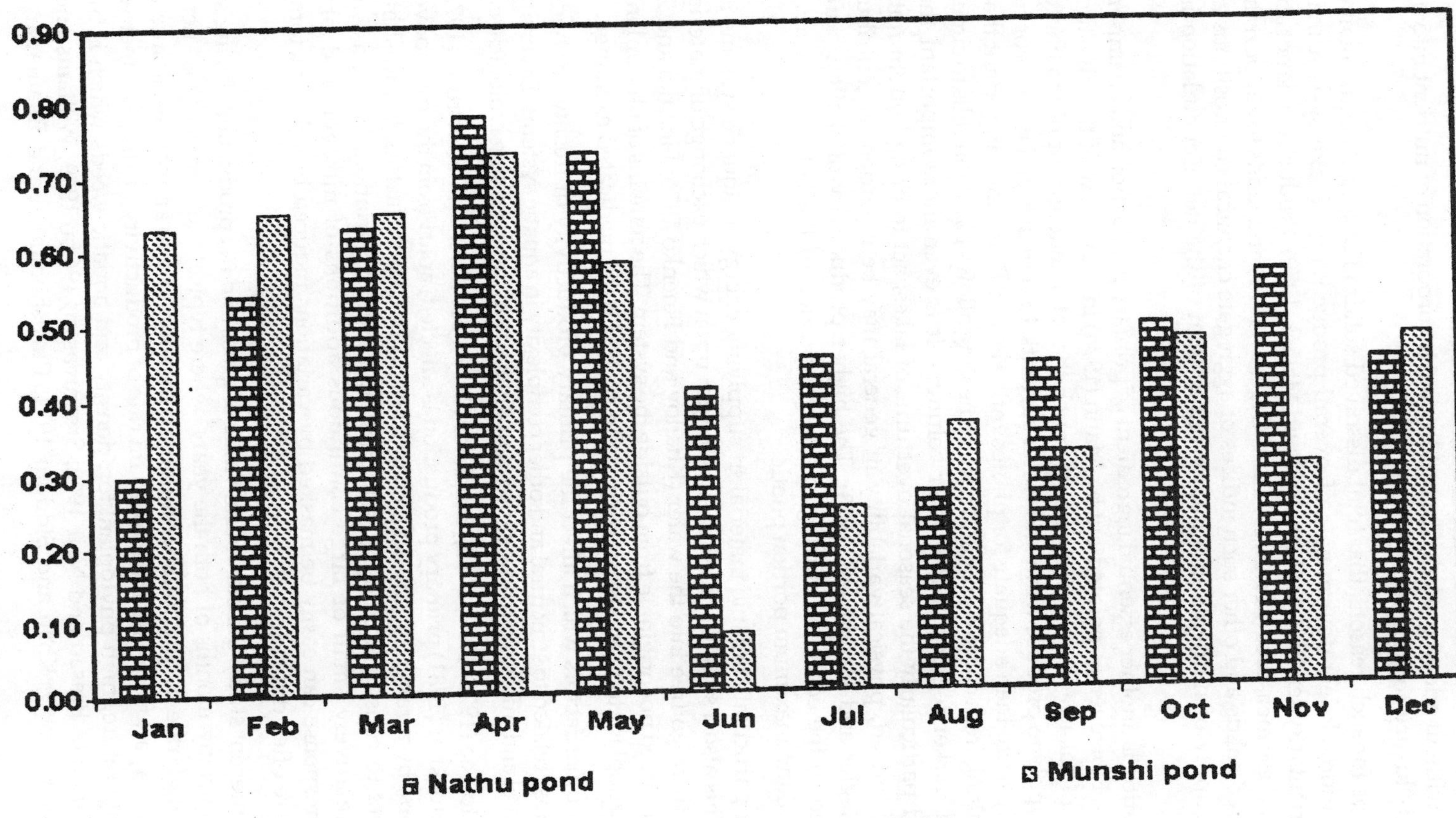

Fig. 1.8: NPP Vs. GPP Values during 2000

the critical concentration is reached, further addition of nutrients results in higher concentrations in organisms but without added growth. This additional incorporation of nutrient into internal stores has been called "luxury consumption."

It is perhaps for such reasons that total dissolved solids (TDS), or its equivalent conductivity, often appears more closely determinate of overall productivity, of lake system than any of the individual nutrients (Ryder, 1974; Jenkins, 1967 and Malack, 1979). Productivity is not always directly correlated with these measures (Mann, 1974 and Oglesby, 1979) and tends to support the hypothesis that the dissolved solids and other such indices of aggregate chemical compositions are correlated to the rate of supply of nutrients in a inflow rather than of the net accumulation in the lake in question.

Solar radiation is another essential ingredient of primary production and an important input to aquatic systems. However, as pointed out by Mann (1974) in his review of the IBP studies on lakes, the importance of solar radiation seems not so much that of an energy supply to photosynthesis as of the energy supply to produce wind and water currents. Except in high latitudes or turbid waters, light is generally more than adequate for photosynthesis. It is thus more important that the general turbulence resulting from current flow and surface heating allows plankton cells to circulate between the lighter and darker regions of the water columns. It is even more important that the water circulation, and particularly the seasonal overturns of lakes, act to bring bottom nutrients to the surface waters. Indeed, physical variability in general may be regarded as contributing to high productivity (Regier and Henderson, 1973). The higher productivity of shallow water bodies is probably related to the continual stirring of a larger portion of the bottom by waves as well as currents and making maximum portion photic.

The role of particulate organic material in supplying energy to aquatic systems is not simply that of source. It is also a suitable physical substratum upon which exchange of materials can occur without letting them go free into the water (Khailov and Finenko 1970; Finenko and Zaika, 1970). Detrital materials need not originate from outside the system. The remains of phytoplankton, rooted vegetation and zooplankton can also participate in such surface medicated exchange.

Among the other factors which affect the primary productivity of freshwater bodies the most important are those related to structural and other complexity in aquatic systems. Diversity in physical habitat, water level and inflow/outflow etc. lead to diversity in the overlying biological communities.

According to Fee (1978, 1979, 1980), Stockner (1979), Stockner and Shortreed (1978, 1979) and Shortreed and Stockner (1981) primary production is affected, apart from the two above mentioned important factors, by a range of physical, chemical and biological factors (*i.e.* flushing, isolation, etc.) each subject to considerable daily, seasonal, and annual variation. When a water body is perturbed by relatively small or large continuous additions of nitrogen and phosphorous, phytoplankton response can easily be masked by variations induced by other factors, and hence interpretations are often difficult.

During the present study, the maximum value of gross and net production during summer and minimum during rainy months of both the years, where typically observed gross net production data showed its high rate during summer months corresponding to an increase in intensity of light (28500-36800 lux). Apart from this, the maximum rate of productivity during summer was possibly due to high temperature, high phytoplankton density and aquatic weeds which is in agreement with the observation of Prasad and Nair 1962; Sreenivasan, 1964a, 1969; Williams and Murdoch, 1966; Spondniewska, 1969; Khan and Siddqui 1971; Purushottaman and Bhatnagar, 1976. High productivity during the summer months was also observed by Rajyalakshmi and Premswarup, 1975.

Lower rate of primary productivity during rainy season might be due to the cloudy days and to the addition of rainy and flood water coming through the catchment areas possessing a lot of organic sewage materials, which make the water bodies more dirty and turbid. Not only this, low temperature and light intensity (21000-23600 lux) also play an important role in declining productivity as insufficient light depletes photosynthesis activity.

Result shows that the effluent is rich in nutrients but primary productivity is not so high. Generally high contents of nutrient in a water body are attributed to a high rate and magnitude of primary productivity. But the present study shows that the primary productivity is not correspondingly high in spite of being rich in the supply of nutrients along the effluents. This might be due to the combined impact of physico-chemical characteristics of domestic sewage, which are quite unsuitable for good microphytic growth, and do not provide survival value to the planktonic communities. Rodhe (1969) also advocated that the enormous storage of nutrients is one of the reasons that help in an immediate or complete reversion of induced eutrophication. The community respiration values were also found lower in the present waterbody. This clearly reflects that the effluents receiving waterbody is possessing a very poor biota. The oxygen deficient nature of the effluent along with high BOD, low DO and other toxic substances might be the probable reasons for less number of plankton, which ultimately resulted into low respiratory processes.

Westlake (1966) suggested that annual net primary productivity is roughly 50 per cent of gross productivity in aquatic system. The aquatic macrophytes fix 40 per cent to 48 per cent carbon. Likens (1975) approximated value for respiration as 40 per cent of gross productivity.

Rivers range down to a minimum of < 1 mg carbon fixed per metre square per day because most of the rivers and streams have very little autochthonous production and depend mostly on autochthonous inputs. There is difficulty in studying the primary productivity of these flowing water systems. In fact a parcel of water at one location in a river is constantly changing as it moves downstream.

Arctic, Antarctic and Alpine lakes are less productive because of limitation in the growing season (covered by ice and snow during part of the year) and because of the seasonal delay or limitation in recharge of nutrients from the drainage system.

Rodhe (1969) summarized data for Danish lakes and Swedish lakes, all fertilized by domestic or industrial effluents which gave a maximum daily rate of 6000 mg C/m^2 for productivity. Lake Baikal which has largest volume of freshwater in the world, has an average phytoplankton productivity of about 310 mg C/m^2/day (Korzhov, 1963), an annual net productivity of 122.5g C/m^2 (Moskalenko, 1972) and is probably mesotrophic (Table 1.6).

Tropical lakes may be exceedingly productive if nutrients are plentiful. It has been known for a long time that very high productivities may be obtained in shallow tropical ponds and paddy fields. For the upper limit of the aquatic primary productivity in natural ecosystems, Wetzel (1966b) suggested that only in highly enriched situations it is likely to exceed 5g C/m^2/day.

In India the primary productivity of various freshwater systems (Table 1.7) ranges from very low values of 90-100g C/m^2/year in Kashmir Himalayan lakes to as high as 1130 to 3309g C/m^2/year in sewage doped eutrophicated swamps and Makhana swamps of north Bihar. Even the side spill over channel (Champanala) of the river Ganges is quite productive (876g C/m^2/year). The lakes with low fertility level had a phytoplankton dominance of diatoms and chlorophyceae, whereas in eutrophic lakes cyanophyceae predominate.

Table 1.6: Net Primary Productivity in Aquatic Ecosystems (from Likens, 1975)

Freshwater Systems	*mg C/m²/day*	*g C/m²/year*
Tropical lakes	100-7600	30-2500
Temperate lakes	3-3600	2-950
Arctic lakes	1-170	< 1-35
Antarctic lakes	1-35	1-10
Alpine lakes	1-450	< 1-100
Temperate streams	< 1-3000	< 1-650
Tropical streams	1-1000	1-1000
Oligotrophic lakes	50-300	—
Mesotrophic lakes	250-1000	—
Eutrophic lakes	600-8000	—
Dystrophic lakes	< 50-500	—
Ganga river	—	592*

* Bilgrami and Munshi 1982-85.

Table 1.7: Net Primary Productivity in Freshwater Lakes, Ponds and Streams of India

Freshwater System	*Net Primary Productivity (g C/m²/year)*		*Authors*
Southern India			
Ooty lake	1388		
Sandynalla lake	864		
Kodai lake	110		Ganapati (1972)
Yercand lake	1058		
Kashmir Himalayas			
Nilnag lake	90-100		Khan and Zutshi (1980)
Bihar			
Bhagalpur pond	1095		Nasar and Datta Munshi (1975)
Champanala side	785 (Phytoplankton)	876	Hassan (1982)
Spill over channels of Ganga river	91 (Macrophytes)		
Badua reservoir	1314		Verma and Datta Munshi (1983)
Ganga river	592		Bilgrami and Datta Munshi (1982-85)
Sewage doped swamps	156 (Phytoplankton) 974 (Macrophytes)	1130	Datta Munshi and Datta Munshi (1985)
Makhana swamps	172 (Phytoplankton) 3137 (Macrophytes)	3309	Datta Munshi and Datta Munshi (1985)

Table 1.8: Range of Primary Production of Microphytic Communities of Freshwaters in India

Sl. No.	*Type of Waterbody and Locality*	*Mean Depth (%)m*	*GPP mgC/m²/day* P_G	*NPP mgC/m²/day* P_S	*Respiration mgC/m²/day* R	*R as % of* P_G	*Photosynthetic efficiency %*	*Authors*
1.	Shallow lakes (Chilwa and Ramgarh, Gorakhpur)	1.25–1.9	36.5–405.0	19.8–210.0	16.6–195.0	45.4–48.1	—	Srivastava & Sahal (1976) Sinha (1969).
2.	Ponds/Fishponds/Temple Tanks/Sewage ponds (Fish pond) Ahmedabad Fishpond, Bhagalpur, Chetpat Fort Moat, Vellore and Chingleput, K.C. Kulam, Tarmaral Kulam, Ayyanakulam, Odathural tanks (all South India).	0.9–3.0	1210–17550	500.0–7795.9*	710.0–287.22**	(17.7–33.5) a (22.7–7.45) b	1.09–13.8	Ganapati (1972), Ganapati & Sreenivasan (1972), Sreenivasan (1972), Nasar & Datta Munshi (1975), Sarkar *et al.* (1977).
3.	Reservoirs (Man-made) Ajwa, Amaravathy, Stanley, Sathanur, Bhavanisagar Krishnagiri, Sandynulla, Ahmedabad and S. India.	3.4–18.3	1601–3228	1360.1–1640.2	964.8–1588.5	41.5–49.2	0.06–1.44	Ganapati (1972), Ganapati & Pathak (1972), Ganapati & Sreenivasan (1972), Sreenivasan (1972).
4.	Other lakes (South India), Kodal, Ooty, Yereaud	2.0–3.0	405–3802	—	—	—	0.17–1.55	Ganpati and Sreenivasan (1972), Sreenivasan (1972), Ganapati (1972).
5.	Temperate lakes (Kashmir, (Dal Anchar, Manasdal)							
	Autumn		28.0–306.6	7.2–194.0	15.8–112.5	—	—	
	Summer	0.6–3.0	26.9–979.0	16.1–529.0	13.6–450.0	—	—	Kaul (1977), Zutshi & Vass (1977).
	Spring		28.0–943.0	25.8–600.0	19.2–343.0	—	—	
6.	Rivers (Godawari, Rajamundry).	3.0–36	280–1232	175.0–542.0	105.0–690.0	43.8–62.4	0.1–0.58	Ganapati (1972).

* Including computed values.

a – Ponds other than fish pond, Bhagalpur.

b – Fish pond, Bhagalpur.

Gopal *et al.* (1978) while reviewing the studies on ecology and production in Indian freshwater ecosystems at primary producer level with emphasis on macrophytes have tried to generalize the microphytic primary productivity. According to their review, data on primary production of microphytes are available for a number of water bodies ranging from small fish pond and shallow lakes to large lakes and man made reservoirs and rivers. Ganapati and his associates have made extensive studies on freshwaters in south India and have recognized seven main types of freshwater bodies namely, reservoirs and lakes (perennial and periodic), ponds (including fish ponds), temple tanks, rivers, estuaries and sewage stabilisation lagoons (Ganapati, 1972; Ganapati and Pathak, 1972; Sreenivasan 1965a, 1968, 1972). Kaul and his students have investigated temple lakes in Kashmir (Kaul, 1977), Zutshi and Vass (1977) have studied Manasbal lake. But data on primary productivity of industrial effluent receiving freshwater ecosystems and mine drainage receiving systems are virtually nil.

The summarized data (Table 1.8) on normal water bodies or sewage fed systems or temple tanks which also have high organic load reflect that primary production ranges from about 36 mgC/m^2/day to above 17.5 gC/m^2/day. Studies on sewage fed ponds have shown even much higher production particularly at the time of algal blooms. In general, in comparison to deep lakes and big dams, the shallow lakes, fish ponds, temple tanks are more productive. Ganapati (1972) has presented a very good correlation between the depth of water body and gross primary production. The temperate lakes as in Kashmir are less productive than tropical and subtropical water bodies.

The higher productive values in Indian condition are associated with luxuriant algal growth particularly of *Microcystis aeruginosa* (Ganapati and Kulkarni, 1973; Talling *et al.*, 1973; Wassink, 1975). In spite of the generalization that tropical waters have more or less homogeneous production rates round the year as a matter of fact a good seasonal variation occurs corresponding to algal productive period and energy supply. Any factor which adversely affects microphytic production and nutrient and energy supply retards the production.

The sewage or other similar type of effluents which have high organic load are supportive to production while effluents toxic in nature, such as mine water and coal washery effluent, are suppressive to production by controlling the microphytic propagation. The net production, sometimes, for highly polluted water bodies has been found in negative values owing to high community respiration. Such situation indicates that the system is highly rich in biota and all the needs for survival and propagation are being met with the communities even after being the system polluted. The community respiration values vary from 1 to 2 per cent to more than 100 per cent, and similarly the photosynthetic efficiency varies from 0.1 per cent to more than 13 per cent in tropical Indian condition.

References

Agarwal, N.C., Bais, V.S. and Arasta, T. (1995). Comparative account of phytoplankton productivity in the Sagar lake and the Military Engineering lake. *J. Freshwater Biol.*, 7: 27-31.

APHA (1989). *Standard method for examination of water and wastewater*. American Public Health Association, Washington, D.C.

Bilgrami, K.S. and Datta Munshi, J.S. (1985). *Ecology of river Ganges: Impact of human activities and conservation of aquatic biota (Patna to Farakka)*. Final Technical Report (D.O En. Project). Bhagalpur University, Bhagalpur, p. 97.

Burgis, M.J., Darlington, J.P.E.C., Dunn, I.G., Gnaff, G.G., Gwahaba, J.J. and McGowan, L.M. (1973). *The biomass and distribution of organisms in lake George, Uganda.* Proc. Royal. Soc. London (B). 184: 271-298.

Das, H.N. and Verma, B.N. (1993). Limnology of a pond and a lake at Rajnagar (Madhubani), Bihar. *Biol. Journal,* 5: 41-46.

Datta Munshi, J.S. and Datta Munshi, J. (1985). Nature of freshwater resources of Bihar. In: *Recent Advance in Zoology.* (Eds. C.B.L. Srivastava and S.C. Goel). Rastogi Publ. Company, Meerut.

Duthie, H.C. and Kirton, W.L. (1970). Primary productivity and standing crops of phytoplankton in Bellwood reservoir. *Can. J. Bot.,* 48: 665-670.

Eceles, D.N. (1962). An internal wave in lake Nayasa and its possible significance in the nutrient cycle. *Nature,* 194: 832-833.

Efford, J.E. (1972). An interim review of the Marion lake project. In: *Problems of freshwaters,* (Eds. Z. Kajak and A. Hillbricht-Illkowska) P W N, Warsaw. pp. 89-109.

Fee, E.J. (1978). A procedure for improving estimates of *in situ* primary production at low irradiances with an incubator technique. *Verh. Internat. Verein Limnol.,* 19: 59-67.

Fee, E.J. (1979). A relation between lake morphometry and primary productivity and its use in interpreting whole-lake eutrophication experiments. *Limnol. Oceanogr.,* 24: 401-416.

Fee, E.J. (1980). Important factors for estimating annual phytoplankton production in the experimental lake area. *Can. J. Fish. Aquat. Sci.,* 32: 513-522.

Finenko, Z.Z. and Zalika, V.E. (1970). Particulate organic matter and its role in the productivity of the sea. In: *Marine food chains,* Oliver and Boyd (Ed. J.H Steele). Edinburgh, pp. 32-44.

Florin, M. and Montes, C. (1998). Which are the relevant scales to assess primary production of Mediterranean semi-arid salt lake. *International journal of ecology and environmental sciences,* 24: 161-177.

Ganapati, S.V. (1972). Organic production in seven types of aquatic ecosystem in India. In: *Tropical ecology with an emphasis on organic productivity* (Eds. P.M. Golley and E.B. Golley), Univ. Georgia, Athens, U.S.A., pp. 313-350.

Ganapati, S.V. and Kulkarni, P.D. (1973). *Primary production in the Siddanath temple tank at Baroda,* India. Supplementary paper submitted in the IBP-PP synthesis meeting, Aberystwyth.

Ganapati, S.V. and Pathak, C.H. (1972). Photosynthetic productivity in the Ajawa reservoir at Baroda, west India. In: *Productivity problems of freshwater* (Eds. Z. Kajak and A. Hillbricht Illikowska). PWN, Warsaw, pp. 725-731.

Ganapati, S.V. and Sreenivasan, A. (1972). Energy flow in aquatic ecosystem in India. In: *Productivity problems of freshwater* (Eds. Z. Kajak and A.Hillbricht Illikowska). PWN, Warsaw, pp. 457-475.

Garder, T. and Gran, H.H. (1927). Investigation of the production of plankton in the Osic Fjord. *J. Cons. Perm. Int. Explor. Mer.,* 42: 1-48.

Gerloff, G.C. (1969). Evaluating nutrient supplies for the growth of aquatic plants in natural waters. In: *Eutrophication: causes, consequences and corrective.* National Acad. Sci. U.S. Washington D.C., pp. 537-555.

Gnaf, C.G. (1972). The regulation of net primary production in Lake George: Uganda East-Africa. In: *Productivity problems of freshwater* (Eds. Z. Kajak and A. Hillbricht Illkowska). PWN. Warsaw, pp. 693-768.

Goldman, C.R. (1959). Primary productivity and limiting factors in three lakes of Alaska peninsula. *Ecol. Monograph*, 30: 207-230.

Gopal, B., Sharma, K.P. and Trivedy, R.K. (1978). Studies on ecology and production in Indian ecosystems at primary producer level with emphasis on macrophytes. In: *Glimpses of ecology* (Eds. J.S. Singh and B. Gopal). International Scientific Publ., Jaipur, pp. 349-376.

Gorai, A.C. (1978). *An ecological study of a freshwater tank in Dhanbad*. Ph.D. thesis (unpublished), Ranchi University, Ranchi.

Gupta, M. and Sharma, L. (1994). Phytoplanktonic productivity of Devkund Sagar Village pond (Bikaner, north-west Rajasthan). *Acta. Ecol.*, 16: 58-62.

Harikrishanan, K. and Abdul Aziz, P.K. (2000). Primary production studies in a freshwater temple tank in Kerala. *Indian J. Environ and Ecoplan.*, 3: 127-130.

Hassan, R. (1982). *Ecological investigation on right offtake side spill channel of the river Ganges at Bhagalpur*. Ph.D. Thesis, Bhagalpur University, Bhagalpur.

Henderson, H.F. (1978). The productivity of water bodies. In: *FAO/SIDA workshop on aquatic pollution in relation to protection of living resources*. Manila, Phillippines, 17 Jan.-27 Feb., 1977, pp. 267-287.

Herback, J. (1969). Relation of some biological productivity to fish production and the maintenance of U.S. reservoirs. In: *Proc. reservoir fisheries resources symposium*, Southern Division, American Fisheries Society, pp. 298-321.

Imevbore, A.M.A., Meszes, G. and Booszormenyi, Z. (1972). The primary productivity in a fish pond in Nigeria. In: *Productivity problems of freshwaters* (Eds. Z. Kajak and A. Hillbricht Illkowska). PWN. Warsaw, pp. 715-723.

Jenkins, R.M. (1967). The influence of some environmental factors on standing crop of fishes in U.S. reservoirs. In: *Proc. reservoir fisheries resource symposium*, S. Div. American Fisheries Soc., pp. 298-321.

Kaul, V. (1977). Limnological survey of Kashmir lakes with reference to trophic status and conservation. *Int. J. Ecol. Environ. Sci.*, 3: 29-44.

Kaul, V., Trishal, C.L. and Handoo, J.K. (1978). Distribution and production of macrophytes in some water bodies of Kashmir. In: *Glimpses of ecology*. (Eds. J.S. Singh and B. Gopal). Int. Sci. Publ., Jaipur, pp. 313-334.

Kelly, M.H., Fitzpatric, L.C. and Peasson, W.D. (1978). Phytoplankton dynamics, primary productivity and community metabolism in a north central Texas pond. *Hydrobiologia*, 50: 177-189.

Khailov, K.M. and Finenko, Z.Z. (1970). Organic macromolecular compounds dissolved in sea water and their inclusion in food chains. In: *Marine food chains* (Ed. J.H. Steele). Oliver and Boyd. Edinburgh, pp. 168-173.

Khan, A.A. and Siddiqui, A.Q. (1971). Primary production in a tropical fish pond at Aligarh, India. *Hydrobiologia*, 37: 447-456.

Khan, I.A., Khan, A.A. and Haque, N. (1988). Primary production in Sheikha jheel at Aligarh. *Environ. Ecol.*, 6: 858-862.

Khan, M.A. and Zutshi, D.P. (1980). Contribution to high altitude limnology of Himalayan system. I. Limnology and primary productivity of plant community of Nilnag lake. *Hydrobiol.*, 75: 103-112.

Kozhov, M. (1963). Lake Baikal and its life–*Monogr. Biol. II,* Verlag W. Junk, Den Hague.

Kumar, A. (1995e). Estimation of productivity of odonate larvae of a tribal pond of Santal Pargana (Bihar). *Environ. Ecol.*, 13: 313-315.

Kumar, A. (1996g). Impact of sewage pollution on chemistry and primary productivity of two freshwater bodies in Santal Pargana (Bihar). *Indian J. Ecol.*, 23: 86-92.

Kumar, A. (1999a). Impact of coal mining effluents on the primary productivity of water bodies and fish farming in and around Chitra coalfields in Santal Parganas (Bihar). In: *Environmental management in coal mining and thermal power plants* (Eds. P.C. Mishra and A. Nayak:). Technoscience Publications, Jaipur, pp. 188-196.

Kumar, A. (2001c). Estimation of primary production of Tatloi thermal spring of Jharkhand. *Indian J. Env. and Ecoplan.* (In press).

Kurasawa, H. (1958). Studies on biological production of fire pools III. Seasonal changes in the standing crop of phytoplankton. *Jap. J. Ecol.*, 8: 143-149.

Laal, A.K. (1981). Primary production and possibilities for utilization of swamps in north Bihar. Ph. D. thesis, Bhagalpur University, Bhagalpur.

Laal, A.K. (1989). Seasonal variation in primary productivity and photosynthetic efficiency in two tropical swamps. *J. Freshwater Biol.* 1: 55-59.

Likens, G.E. (1975). Primary production of inland aquatic ecosystems. In: *Primary productivity of the Biosphere* (Eds. H. Lieth and R.H. Whittaker). Springer, Verlag, pp. 185-202.

Lipkin, Y., Beer, S., Best, E.P.H., Kairesalo, T. and Salomen, K. (1986). Primary production of macrophytes: Terminology approaches and comparison of methods. *Aquat. Bot.*, 26: 129-142.

Malack, J.M. (1979). *Primary productivity and fish yield in tropical lakes.* Trans. Am. Fish. Soc.

Mann, K.H. (1974). Comparison of freshwater and marine systems: The direct and indirect effects of solar energy on primary and secondary production. In: *Proc. of the first int. cong. ecol. structure, function and management of ecosystems.* The Hague, Netherland, pp. 168-173.

Mishra, V.K. and Nayak, B.K. (2000). Biomass productivity of energy plantations on Himalayan Foot Hill. *Environ. and Ecol.*, 18: 13-16.

Morgan, N.C. (1972). Productivity studies at Loch Leven. (a shallow nutrient rich low land lake). In: *Productivity problems of freshwater*. (Eds. Z. Kajak and A. Hillbricht). PWN. Warsaw, pp. 207-226.

Morton, B. (1978). The ecology and productivity of south east Asian waters. In: *FAO/Resources,* Manila Philippines. 17 Jan.-27 Feb. 1977, pp. 288-331.

Moskalenko, B.K. (1972). Biological productive system of lake Baikal. *Verhandle Int. Ver. Limnol.*, 18: 568-573.

Moskalenko, B.K. and Vostinsev, K.K. (1972). Biological productivity and balance of substances and energy flow in lake Baikal. In: *Productivity problems in freshwater*. (Eds. Z. Kajak and A. Hillbricht Illkowska). PWN. Warsaw, pp. 227-243.

Nasar, S.A.K. and Datta Munshi, J.S. (1975). Studies on the primary production of a freshwater pond. *Jap. J. Ecol.*, 25: 21-23.

Nygaard, G. (1955). On productivity of fine Danish water Versh. *Int. Ver. Theor. Auglw. Lkimnol.*, 12: 123-133.

Odum, H.T. (1956). Primary production in flowing water. *Limnol. Oceanogor.*, 2: 85-97.

Oglesby, R.T. (1979). Community composition and yield of fish in relation to lake phytoplankton standing crop and production and to morphoedaphic factors. *J. Fish Res. Board*, Canada.

Pandey, K.N., Prakash, V. and Sharma, U.P. (1995). Assessment of primary productivity of phytoplankton and macrophytes of Kawar lake wetland (Begusarai) Bihar. *J. Freshwater Biol.*, 7: 237-239.

Pandit, B. (1999). Dynamics of primary productivity of certain ponds at Barharwa (Santal Pargana), a tribal division of Jharkhand, India. In: *Modern trends in environmental pollution and ecoplanning* (Ed. Kumar, A.) ABD publishers, Jaipur, pp. 97-99.

Pati, S. and Sahu, B.K. (1993). Studies on primary production in Rengali reservoir (Orissa). *J. Ecobiol.*, 5: 85-88.

Prasad, P.R and Nair, P.V.R. (1962). Studies on organic production. Gulf of Mannar *J. Mar. Biol. Assol. Ind.*, 5: 1-26.

Prouse, G.A. (1972). Some observations on primary and fish production in experimental fish ponds in Malaysia. *Malaysia Proc. IBP. Symp.*, 1972: 555- 561.

Purushottaman, A. and Bhatnagar, G.P. (1976). Primary production studies in Porto Novo waters, S. India. *Arch. Hydrobiol.*, 77: 37-50.

Rai, D.N. and Sharma, U.P. (1991). Energy transfer efficiency and primary productivity of tropical wetlands in north Bihar (India). *J. Freshwater Biol.*, 3: 89-97.

Rai, D.N., Verma, P.K. and Datta Munshi, J.S. (1980). Interaction between a floating (*Trapa bispinosa*) and submerged vegetation community in a fish pond of Bhagalpur pol. *Arch. Hydrobiol.*, 27: 137-142.

Rajyalakshmi, T. and Premswarup, T.V. (1975). Primary production in river Godavari, India. *J. Fish*, 22: 205-214.

Regier, H.A. and Henderson, H.A. (1973). Towards a broad ecological model of fish communities and fisheries. *Trans. Am. Fish Soc.*, 102: 56-72.

Rodhe, W. (1958). Primar produktion und Seetypen. Verh. Intenat. Ver. Limnol., 13: 121-141.

Rodhe, W. (1969). Crystallization of eutrophication concept in Northern Europe. In: *Eutrophication: causes, consequences and correctives*. National Academy of Sciences, Washington, D.C. pp. 50-64.

Ryder, R.A. (1974). The morphoedaphic index, a fish yield estimator: Review and evaluation. *J. Fish Res. Board* Can., 31: 663-688.

Sarkar, S.K., Gond, B.V. and Natrajan, A.V. (1977). A note on some distinctive limnological feature of Govindsagar reservoir. Himachal Pradesh. *Indian J. Animal Sci.*, 47: 435-439.

Schelsker, C.L. and Stoermer, E.F. (1972.). Phosphorous, silica and eutrophication of Lake Michigan. *Nutrients and Eutrophication*, 1: 17-171.

Schindler, D.W. and Fee, E.J. (1974). Experimental lake area whole lake experiments in eutrophication. *J. Fish Res. Board. Can.*, 31: 937-953.

Sen, N.S., Horo, P. and Mishra, V.K. (1992). Physico-chemistry, nutrient budget and the factors influencing primary production in a tropical lake. *J. Freshwater Biol.*, 4: 141-153.

Sharma, R. and Saxena, M.M. (1994). Studies on gross and net primary productivity of two bodies of water in the Indian desert. *Acta Ecol.*, 16: 125-128.

Shortreed, K.S. and Stockner, J.G. (1981). Limnological results from the 1979 British Columbia lake enrichment programme. *Can. Tech. Rep. Fish and Aquat. Sci.*, 9: 1-11.

Singh, M. and Sinha, R.K. (1994). Primary productivity and limnological profile of two ponds at Patna, Bihar. *J. Freshwater Biol.*, 6: 127-133.

Sinha, A.B. and Sahai, R. (1978). Studies on the factors affecting and the productivity of free floating macrophyte in Ramgarh lake, Gorakhpur. In: *Glimpses of ecology*. (Eds. J.S. Singh and B. Gopal). Int. Scientific Publ., Jaipur, pp. 377-382.

Sinha, M.P. (1988). Studies on polluted water ecosystem. Ph.D. Thesis (unpublished) of Ranchi University.

Spondniewska, I.T. (1969). Day-to-day variation in primary production of phytoplankton in Mikolajskie lake. *Ekologia Polska. Seria.*, A17: 503-514.

Sreenivasan, A. (1963). Primary production in three upland lakes of Madras state. *Curr. Sci.*, 32: 130-131.

Sreenivasan, A. (1964a). Limnology, primary production and fish production in a tropical pond. *Limnol. Oceanogr.*, 9: 391-396.

Sreenivasan, A. (1965a). Limnology of tropical impoundments. III. Limnology and productivity of Amravathy reservoir (Madras state), India. *Hydrobiologia*, 26: 501-506.

Sreenivasan, A. (1966). *Application of limnological and primary productivity studies in fish culture*. PAO world symp. warm water pond fish culture, Rome, 18-25 May, 1966, p. 12.

Sreenivasan, A. (1969). Primary production and fish yield in a tropical impoundment, Stanley reservoir, *Mettur Dam.*, 358: 125-130.

Sreenivasan, A. (1972). Energy transformation through primary productivity and fish production in some tropical fish water impoundments and ponds. In: *Productivity problems of freshwater*. (Eds. Z. Kazak and A. Hillbricht Illkowska) PWN, Warsaw, pp. 505-514.

Srivastava, V.C. and Sahai, R. (1976). Effects of water pollution on productivity and periodicity of phytoplankton of Chilka Lake. *Geobios*, 3: 187-190.

Steel, J.A.P., Duncan, A. and Andrew, T.H. (1968). The daily carbon gains and losses in the session of Queen Mary reservoir England, during early and mid 1968. In: *Productivity problems of freshwater* (Eds. Z. Kajak and A. Hillbricht-Illkowska) PWN. Warsaw, pp. 515-528.

Steemann-Nielsen, E. (1955). The production of organic matter by phytoplankton in a Danish lake receiving extra-ordinary great amounts of nutrient salts. *Hydrobiologia*, 7: 68-74.

Stockner, J.G. (1979). Lake fertilization as a means of enhancing sockeye Salmon population: The state of the art in the Pacific northwest. *Fish and Mar. Serv. Ech. Rep.*, 740: p. 14.

Stockner, J.G. and Shortreed, K.S. (1978). Limnological survey of 35 sockeye Salmon nursery lakes in British Columbia and the Yukon Territory. *Fish. Mar. Serv. Tech. Rep.*, 827: p. 47.

Stockner, J.K. and Shortreed, K.S. (1979). Limnological studies of 13 sockeye salmon nursery lakes in British Columbia. *Fish and Mar. Serv. Tech. Rep.*, 865: p. 125.

Tailang, J.F. (1965). Comparative problems of phytoplankton production and photosynthetic productivity in a tropical and temperate lake. In: *Primary productivity in aquatic environments.* (Ed. Goldsman, C.R.), Pallunza, Italy, pp. 399-424.

Talling, J.F., Wood, R.B., Prosser, M.V. and Baxter, R.M. (1973). The upper limit of photosynthetic productivity by phytoplankton. Evidences from Ethiopian Soda Lakes. *Freshwater Biology*, 3: 53-76.

Thompson, K. (1976). The primary productivity of African wetlands with particular references to Okavanga delta. *Okavanga delta symp. Gaborne*, 19: 67-79.

Ulrich, A. (1961). Plant analysis in sugar beet nutrients. In: *Plant analysis and fertiliser problems* (Vol. 8) (Ed. W. Reuther). Am. Inst. Biological Science. Washington D.C., pp. 190-211.

Verma, P.K. and Datta Munshi, J. (1983). Limnology of Badua Reservoir of Bhagalpur, Bihar. Proc. *Indian Nats. Sci. Acad.*, B. 49: 598-609.

Vijaya Kumar, K. (1994). Seasonal variation in the primary productivity of a tropical pond. *J. Ecobiol.*, 6: 207-211.

Vijayaraghvan, S. (1971b). Seasonal variation in primary production in three tropical ponds. *Hydrobiologia*, 38: 395-408

Wassink, E.C. (1975). Photosynthesis and periodicity in different environments. In: *Photosynthesis and productivity in different environments* (Ed. J.P. Copper).

Westlake, D.F. (1966). Some basic data for investigation of the productivity of aquatic macrophytes. In: *Primary productivity in aquatic environments* (Ed. C.R Goldman), pp. 229-259.

Wetzel, R.G. (1966a). Variation in productivity Goose and hypertrophic Sylvan Lakes. Indiana. Invest. *Indiana lakes streams*. 7: 147-184.

Wetzel, R.G. (1966b). Productivity and nutrient relationship in marl lakes of northern Indiana Verh. *Int. Ver. Limnol.*, 16: 321-332.

Wetzel, R.G. (1975). *Limnology*. W.B. Saunders Co., Philadelphia, U.S.A. p. 743.

Williams, R.B. and Murdoch, M.B. (1966). Phytoplankton production and chlorophyll concentration in the Beaufort Channel, North Carolina. *Limnol. Oceanogr.*, 11: 73-82.

Zutshi, D.P. (1976). Phytoplankton productivity, a algal dynamic and tropic status of lake. Mergozzo (Northern Italy) Mem. 1st. Ital. *Indrobiol.*, 33: 221-256.

Zutshi, D.P. and Vass, K.K. (1977). Estimates of the phytoplankton production in Manasbal lake, Kahsmir using C^{14} method. *Trop. Ecol.*, 18: 103-108.

2

Assessment and Management of Water Pollution: A Review

S. ANANTHI, P. UMA MAHESHWARI, K. USHA RANI, R. SARAVANAN & A. ARUN

P.G. Unit of Microbiology, Department of Zoology and Microbiology, Thiagarajar College, Madurai, Tamil Nadu – 608 002

ABSTRACT

Water pollution is a serious problem for the entire world. It threatens the health and well being of humans, plants, and animals. As the world become more industrial and smaller due to communications and trade, accidental and purposive hazardous dumping have contributed to the problem of sea pollution. All water pollution is dangerous to the health of living organisms, but sea and river pollution can be especially detrimental to the health of humans and animals. Rivers and seas are used as primary sources of potable water by populations all over the world. Another serious consequence of this pollution is the effect of this pollution on trade in the polluted areas. This paper examines cases, which reflect different causes of sea and river pollution, the seriousness of this pollution, the effect of this pollution on trade, and a possible global solution to this problem.

Keywords: Water Pollution.

Introduction

For the mankind, water quality is as important as water quantity. Water needs to be free of salinity, plants and animal waste, heavy metals, toxins and bacteria, to be safe to drink. Potable water is fresh unpolluted water that is safe to drink. Over 70 per cent of the earth's surface is covered by water but less than 3 per cent is freshwater, and not all of that is safe to drink.

Man also use water for cooling power plants, irrigating crops, bathing, watering lawns, washing cars and flushing sewage. Water scarcity and water pollution are becoming increasing problems in many areas of the world. Our water comes from rain fall, reservoirs, rivers and wells that tap into aquifers. Because of growing populations, there is a need to concern with water transport, storage, distribution, quality and quantity of water supplies.

Large-scale irrigation has become a way of meeting water needs on a third of all land used to raise food crops. The problem with irrigation is an increase in salination or water logging. These conditions result in stunted growth, decreased yields or in extreme cases, crop death. In many cases the water used for irrigation comes from wells. This can lead to depletion of underground aquifers. Some forms of irrigation require energy for pumps, which increases costs.

If adequate plant cover is not kept near bodies of water, rains often wash surface soil, salts and chemicals into the water. Deposits of silt cause sedimentation and cloudy water, making it difficult for fish to breathe and making the lake undesirable for swimming. Human wastes, insecticides, herbicides, chemicals, radioactive materials and heat pollute the water.

We may remember the Exxon veldez oil spill in Alaska and others more recently. One way to clean up spills is to use a bioengineered bacteria (USWFS, J&K, Hollingsworth, Sign warning of mercury contaminants at Felsenthal NWR) that eats up the oil spill. When no more oil is available as a food source, the bacteria self-destruct.

The addition of excess fertilizer to the soil or high concentrations of sewage has caused nitrogen compounds to be washed into lakes and streams, resulting in eutrophication. This causes overgrowth of aquatic plants and "algal boom". As they die, the plants are decomposed by the bacteria and other oxygen-using organisms in the water. This depletes the oxygen normally dissolved in the water. Eventually it can cause the deaths of fish and other aquatic organisms.

Health risks related to water quality, primarily due to agrochemicals, represent a growing problem. This is mainly due to the intensified production and the expansion of irrigated agriculture, often in association with food processing industries (FAO, 1993; Hespanhol, 1996).

Reuse is an important and natural method of managing drainage water. In order to develop the maximum benefit from a water supply and to help dispose of drainage waters, strategies for water reuse has evolved. In regions where irrigation water supplies are limited, drainage water can be used to supplement them.

Standing water and wet materials in the home can become a breeding ground for microorganisms, bacteria, mold and viruses. They can cause disease, trigger allergic reactions and continue to damage materials.

Marine Pollution

Marine pollution means the introduction by man, directly or indirectly, of substances or energy into the maritime area which results or is likely to result in hazards to human health, harm to living resources and marine ecosystems, damage to amenities or interference with other legitimate uses of the sea: Some potential pollutants such as oil and litter are highly visible. Others, such as toxic chemicals and radioactive material are invisible, hidden from view. The main types and sources of potential pollution are Sewage, Nutrients, Oil, Industrial wastes, Litter and Radioactivity.

Sources of Pollution

Table 2.1: Major Types of Water Pollutants and their Sources

Type of Pollutant	*Example*	*Major Sources*
Disease causing agents	Bacteria, viruses, protozoa, parasitic, worms	Poorly treated domestic, Sewage, animal wastes (agricultural)
Carbon compounds	Fecal waste (solid waste produced by animals)	Domestic sewage animal feed lots
	Oil	Machine waste spills pipe line breaks
	Chlorine compounds	Water purification Paper making plants
	Pesticides	Agriculture forestry
	Detergents plastic	Homes and Industries
Water-soluble acids and salts	Sulphuric and nitric acids	Mining, industrial wastes
	Salt	Irrigation, mining, de-icing of roads
Water-soluble poisons	Lead	Leaded gasoline, poisons, pesticides, smelting of lead
	Mercury	Industrial waste, fungicides
	Various herbicides and pesticides	Agriculture, forestry, mosquito control
Plant nutrients	Phosphates nitrates	Domestic sewage, industrial waste natural run off, agricultural run off, food processing industries
Sediment and suspended undissolved matter	Soil silt	Run off from agriculture forestry construction
Radio active substances	Uranium strontium	Natural sources (rocks, soils) Uranium, mining, nuclear weapons testing
Heat (from water used as a coolant)		Electrical power and industrial plants.

The Precautionary Principle

These concerns about various types pollution have been one of the factors that has led to the development of the precautionary principle. This means the Governments should take action to prevent pollution wherever there are reasonable grounds for concern that such pollution may occur (D.W. Connell, 1981).

Controlling Marine Pollution

Marine pollution is an international problem and cannot be tackled through piecemeal legislation on a country by country basis. Australia is a signatory to the International Maritime Organization's MARPOL 73/7 Convention Annex V, which prohibits the discharging of plastics anywhere at sea. Port reception facilities are supposed to be made available, but in many areas are not:

1. Cut down on the amount of plastic that you use. Reuse and recycle whatever you can.
2. Do not dispose of plastic or sanitary items down the toilet.
3. Never dispose of oil or oil based products down the drain.
4. Don't buy products with excess packaging that is not reusable, and let the storekeeper know why.

5. Don't dump litter or fishing gear from boats. Besides being illegal, it has a devastating effect on wildlife sort. Sort wastes on board for appropriate disposal or reuse on shore.
6. Contact your local port or marina and inquire whether they have adequate waste reception facilities. If they haven't globally your MP to have them installed.

Sources of Groundwater Contamination (Perimutter, N.M. & M. Lieber, 1970)

As water moves through the hydrologic cycle, the quality changes in response to differences in the physical, chemical and biological environments through which it passes. The changes may be either natural or non-influenced, in some cases they can be controlled. In other cases they cannot, but in most cases they can be managed in order to limit adverse water quality changes. Groundwater quality problems that originate on the Land Surface include infiltration of polluted surface water, land disposal of either solid or liquid wastes, stockpiles, dumps, disposal of sewage and water-treatment plant sludge, de-icing salt usage and storage, animal feedlots, fertilizers and pesticides, accidental spills, particulate matter from airborne sources. Groundwater quality problems that originate in the ground above the water table include septic tanks, cesspools and privies, holding ponds and lagoons, sanitary landfills, waste disposal in excavations, leakage from underground storage tanks, leakage from underground pipelines, artificial recharge, sumps and drywells, graveyards. Groundwater quality problems that originate in the ground below the water table includes waste disposal in well excavations, drainage wells and canals, well disposal of wastes, underground storage, secondary recovery, mines, exploratory wells, abandoned wells, water-supply wells, groundwater development.

Natural Controls on Groundwater Contamination

The first control includes the physical and chemical characteristics of the earth materials through which the liquid wastes flow. A major attenuating effect for many compounds is the unsaturated zone, which has been caved the "living filter".

The second major control includes the natural processes that tend to remove or degrade a contaminant as it flows through the subsurface from areas or points of recharge to zones or points of discharge. These processes include filtration, sorption, ion-exchange, dispersion, oxidation and microbial degradation, as well as dilution.

The third natural control relates to the hydraulics of the flow system through which the waste migrates, beginning with infiltration and ending with discharge.

The final control is the nature of the contaminant. This includes its physical, chemical and biological characteristics and particularly, its stability under varying conditions. The stability of the more common constituents and the heavy metals are fairly well known although more complex than commonly realized on the other hand, the stability of organic compounds, particularly synthetic organism compounds, has only recently come under close inspection and actually little is known of their degradation and mobility in the subsurface.

To a large extent, it is the aquifer frame work that controls the movement of groundwater and contaminants. Permeable zones are most often separated by layers of fine-grained material such as clay, shale or silt, which serve as confining beds. The geologic framework, in conjunction with surface topography, also exerts a major control on the configuration of the water table and the thickness of the unsaturated zone. Groundwater or interstitial velocity is controlled by the hydraulic conductivity, gradient, and effective porosity. Water movement through a permeable gravel with a gradient of 10 feet per mile averages about 60 feet per day, but in a clay with the same gradient and no secondary permeability the water movement would be only about 1 ft. in 30,000 years. In most aquifers groundwater velocity ranges from a few feet per day to a few feet per year.

Incidence of Diseases, Cases and Mortality and Health Risks (American Water Works Association, 1975) (Birley, 1995) (Cairncross and Feachem, 1983) (WHO, 1995)

Virtually all water pollutants are hazardous to humans as well as lesser species; sodium is implicated in cardiovascular disease, nitrates in blood disorders. Mercury and lead can cause nervous disorders. Some contaminants are carcinogens. DDT is toxic to humans and can alter chromosomes. PCBs cause liver and nerve damage, skin eruptions, vomiting, fever, diarrhoea, and fatal abnormalities. Along many shores, shellfish can no longer be taken because of contamination by DDT, sewage, or industrial wastes.

Dysentery, salmonellosis, cryptosporidium, and hepatitis are among the maladies transmitted by sewage in drinking and bathing water. In the United States, beaches along both coasts, riverbanks, and lake shores have been ruined for bathers by industrial wastes, municipal sewage, and medical waste. Water pollution is an even greater problem in the Third World, where millions of people obtain water for drinking and sanitation from unprotected streams and ponds that are contaminated with human waste. This type of contamination has been estimated to cause more than 3 million deaths annually from diarrhoea in Third World countries, most of them children.

Water-borne Excreta Related Infections

Most of the aquatic water or intermediate host habitats correspond to sites where humans use excess water for washing, bathing, swimming or fishing (Cairncross and Feachem, 1983) proposed a classification of water-borne diseases in order to more fully understand the effects of excreta disposal which includes Faecal-oral diseases (non-bacterial) *e.g.* Hepatitis A, Giardiasis, Faecal-oral (Bacterial) *e.g.* various diarrhoeas and dysenteries like cholera, E. coli diarrhoea or Salmonellosis, enteric faeces such as typhoid, soil transmitted helminthes *e.g.* Ascariasis, Trichuriasis, Hookworm, Beef and Pork tapeworms *e.g.*, Taeniasis, water based Helminthes *e.g.*, schistosomiasis, Excreta related insect vectors *e.g.*, Filariasis.

Health Risks and Chemical Pollution

Health risks related to water quality, primarily due to agrochemicals, represent a growing problem. This is mainly due to the intensified production and the expansion of irrigated agriculture, often in association with food processing industries (FAO, 1993; Hespanhol, 1996). In this context, both subsurface and open drains can have an impact on human health on 4 levels. They are drainage water reuse (Cairncross and Feachem, 1983), (Hespanhol, 1996), (FAO, 1993), drainage water treatment and disposal, application of insecticides or molluscicides during chemical vector campaigns, uncontrolled disposal of industrial liquid waste in drainage canals.

Control

Control of water related diseases always requires the interruption of transmission cycles, which include a number of quite different actors, man, animals or vectors. These all have different functions, react differently to environmental changes and have different capacities in either transmitting the disease or in resisting infection (WHO, 1982, BOS *et al.*, 1993). There are basically 3 ways to disrupt transmission:

1. By eliminating or reducing vector densities (Jobin, 1987; Oomen *et al.*, 1990; Fritsch, 1993)
2. By protecting the susceptible host through immunization, prophylactic drugs, reduced man-vector contacts using repellents, window screening, bed nets, or by installing adequate water supplies and sanitation facilities.
3. By reducing the reservoir of infection, by treating infected people or animals and by eliminating the pathogen in the host body.

The core of such an approach can be reduced to the formula, "TRANSMISSION AND MORBIDITY CONTOL". Transmission control contains chemical, biological and environmental management control of vectors, intermediate hosts, or any activity in the field of water supply, sanitation facilities and water treatment morbidity control includes chemotherapy or any medical control of the pathogen and treatment of diseases (Oomen *et al.*, 1990) visualized the environmental management component as a vehicle having hygiene education community participation as wheels rolling on the road of basic infrastructure.

Drainage Water Reuse

Reuse is an important and natural method of managing drainage water. In order to develop the maximum benefit from a water supply and to help dispose of drainage waters, strategies for water reuse have evolved. In regions where irrigation water supplies are limited, drainage water can be used to supplement them.

The major degradation factor of reused waters is the high concentration of ions. Waters with low ionic concentrations provide plants with an adequate supply of many of the essential nutrients needed for growth. Drainage and leaching of salts from the root zone are key factors in the management of salinity in agriculture. Another management factor is control of the range of salt tolerance expressed in crop species. Water reuse for agricultural crops has distinct economic incentives and a number of crops are known to be highly tolerant to salinity. However, as salinity increases in the irrigation water, there is a greater need to monitor and manage irrigation and drainage practices and to consider the sustainability of the system.

It is possible to safely reuse agricultural drainage water if the characteristics of the water, soil and the intended crop plants are known and can be economically managed.

Effects of Salinity on Crop Growth and Yield

Salinity in water or soil is an environmental factor that, in general, reduces crop growth. At relatively low salinity, especially among crop species such as cotton or the halophytic sugar beet, some salinity may actually improve crop production (Shannon *et al.*, 1994). This effect has been attributed in some instances to an improvement in water use efficiency of the plant (Letey, 1993). Sugar beet, sugarcane, dates, cotton and barley are among the most salt tolerant whereas beans, carrots, onions, strawberries and almonds are considered sensitive (Maas, 1986). Previous research has led to the development of large data bases on the salt tolerances of many crop species and varieties (Francois and Mass, 1978 and 1985; Mass, 1990).

Irrigation Systems and Scheduling

Traditionally, the goal of agriculture has been to maximize yield and profit, defined by the net difference between inputs and outputs. To maximize yields, soil salinity is reduced by leaching. The kind of management directed towards this goal is mediated by costs of water, drainage, applied nutrients and amendments to soils and waters. However, crops can be grown with saline waters provided with suitable irrigation and cropping strategies are used (Rhoades, 1988). Management needs to be more intensive and more precise methods should be used for water application and distribution. Water requirements needed for crop use and leaching should be accurately assessed and provided in a timely manner. Most crop water use coefficients have been developed for non-saline situations (Erie *et al.*, 1982) with saline water, growth and consumptive water use are reduced. However, preliminary evidence indicates that some crop water use coefficients may apply over a range of salinities because the water unused by the crop is needed to offset the increase in the leaching fraction requirement (Latey and Dinar, 1986). The greater the salinity of the irrigation water, the greater the need for adequate irrigation and drainage.

Sprinkling should be done at night or in the early morning when evaporation is less. Drip irrigation gives the greatest advantages when saline water is used.

Other Management Techniques

There are a number of cultivation and management methods which may offset yield loss when saline water is used for crop production. Some simple methods include: using more vigorous cultivars; using screens to select seed of larger size and higher seed weight; increasing population densities to offset smaller plant size and reductions in tiller number; and placing seed on sloped or modified seed-beds so that salts drawn to the surface by evaporation accumulate away from the seed line. Another way to move salts away from the seed or plant base is to irrigate in alternate furrows. An additional method of ameliorating the effects of moderately saline drainage water is to match fertilization requirements with the chemical composition of the water and soils. As drainage wastes often contain some of the essential fertilization requirements of plants, applied fertilizers can be reduced.

On the other hands, as ion interaction and competition reactions affect plant uptake, additions of calcium, potassium or some other element may be required for better crop growth. An important consideration is that high sodium concentrations in soils and waters may decrease soil permeability and prevent effective leaching of salts. The suitability of soils and waters is measured using a parameter called the sodium adsorption ratio (SAR).

Blending is the mixing of poor quality drainage water with good quality irrigation water. Provided that the blended water is sufficiently low in total salinity and toxic ions, this is the most economic and environmentally acceptable means of disposing of drainage water. This strategy can potentially increase the salinity of the groundwater over time. Often, separation of water of different qualities is the best strategy.

Egypt has an extensive drainage water reuse programme. Over 4000 million m^3 of agricultural drainage water, produced in the upper Nile Delta are reused to supplement irrigation water requirements in the lower delta. Pumping stations lift the better quality drainage water into irrigation canals, where mixing occurs. The Ministry of Public Works and Work Resources has an extensive reuse quantity and quality monitoring programme.

Reuse for Saline Agriculture and Forestry

Concept of Agriculture-Forestry and Solar Evaporators

The agriculture-forestry system and solar evaporator are still at the development stage (Tanji and Karajeh, 1992 and 1993; Jorgenson *et al.*, 1993). The concept compares favourably with other intermediate disposal technologies. It manages drainage water as a resource instead of directly disposing it into evaporation ponds, rivers or the ocean. Agriculture-forestry systems produce energy (biomass) while some other technologies for the treatment of drainage water have high energy demands.

By significantly reducing the final volume of non-usable drainage water, agriculture-forestry systems provide an opportunity for seasonal discharging of drainage water into rivers during high water flows.

Solar evaporators are preferable from the standpoint of wildlife safety as they provide for better bird control than do evaporation ponds. The trees create wildlife habitats, reduce air pollution, and enhance the overall environmental quality of fanning regions.

The agriculture-forestry method has been developed and tested in California for salt management on irrigated farmland. It has two objectives:

1. To utilize the drainage water as a resource to produce marketable crops.
2. To reduce the volume of discharged drainage water directly on the farm.

The agriculture-forestry system for the productive use of drainage water and its disposal into a solar evaporator has been developed in semi-arid conditions in California and other regions.

The main function of the trees is to use and evaporate large volumes of drainage water. This can be achieved not only through sequential reuse but also by uptake of the drainage water from shallow water tables, or by intercepting the flow of drainage water from upslope. The trees can be viewed as biological pumps. The tree biomass offers a number of marketing options ranging from electricity generation to the production of biofuels or biochemicals. Drainage water is further concentrated through the irrigation of halophytes that also have marketing potential as food or industrial crops. The salt concentration in the drainage water in each succeeding stage is increased, but the volume of water is reduced.

About 80-85 per cent of the initial volume of drainage water produced while growing salt tolerant crops. The remaining 15-20 per cent of drainage water, with increased concentration of salt, evaporates. The design of the solar evaporator contains of a levelled area lined with a plastic liner on which the crystallized salt is collected. The daily amount of drainage water discharged into the solar evaporator is correlated with daily evaporation rate to prevent water ponding.

The agriculture-forestry system operating in California for sequential reuse of water and salt-removal utilizes drainage water with an initial salt concentration of about 7000mg/litre, following the irrigation of salt-tolerant crops.

The system for sequential reuse of drainage water is also effectively reducing the level of Se. A typical Se concentration is 0.5mg/litre in drainage water applied to trees and 0.9mg/litre in the reduced volume of drainage water applied to halophytes. Consequently, the estimated quantity of Se is about 50 kg per 100,000m^3 of drainage water sequentially reused to irrigate halophytes. Selenium removal by trees is mainly concentrated in leaves, and the measured values (in terms of mg/kg) where 0.5–0.9 for eucalyputs, 0.6–1.0 for casuarinas, 2.6–3.6 for athel, and 0.6–1.8 for halophytes. Successful experiments were conducted on the transfer of Se through the harvesting halophytes and safely using this forage for cattle feeding (Forst, 1990).

Conclusion

Since water plays such a vital role in the life on earth, good quality water is a precious resource. Often water quality is more important than water quantity. The quality of the water affects the use we make of it, but the reverse is also true. Once we use the water, we affect its quality. The explosion in human population and industrial activities, and the rate at which new chemicals and products are being developed and used pose a global environmental threat. The natural decay processes in water bodies can no longer cope with these loads.

Water pollution has affected many people in many ways, more than they think. Contamination is very dangerous not only to animals but humans too. Oil pollution is a very unsafe problem in North America, especially in the U.S. Oil places a film on the water allowing no oxygen to get into the water. Many things are constantly being tried to stop serious water pollution such as oil spills et. but more needs to be done. There are many causes for water pollution, which is why each one has to be thoroughly considered in order to stop more regards to harming our living and breathing environment, in contrast, they act as if they are not concerned. Both the government and people should become very heavily involved in this significant issue that can someday put the earth's public health under stupendous risk. On the whole, it is strictly to all organisms' drawback if the water we use daily someday becomes as dangerous as gasoline to drink. This illustration represents the scene where waters enter lakes after being treated.

References

American Water Works Association (1975). Status of water-borne diseases in the U.S. and Canada. *Journal of American water works association*, 67(2): 95-98.

Birley (1995). *Reciclence of water borne disease.* www.msn.com

Bos *et al.* (1993). *Transmission and morbidity control.* www.google.com

Cairncross and Feachem (1983). *Classification of water borne diseases.* www.yahoo.com

Connell, D.W. (1981). *Water pollution causes and effects in Australia and New Zealand,* 2nd edition. University of Queensland Press, Site Lucia, (old), 203 pp.

Erie *et al.* (1982). *Irrigation system and scheduling.*

Espanhol, H. (1996). *Health risks and chemical pollution.* Chapter 7. Martin S. Fritsch. Swiss Federal Institute of Technology, Zurich, Switzerland.

FAO (1993). *Health issues related to drainage water management*. Chapter 7. Martin S. Fritsch Swiss Federal Institute of Technology, Zurich, Switzerland.

Francois and Maar (1978, 1985). *Drainage water reuse.* Chapter 4, USDA Salinity Laboratory, Riverside, California, USA.

Fritsch (1993). Control of water related diseases. Martin S. Fritsch, Swiss Federal Institute of Technology, Zurich, Switzerland.

Frost (1990). *Safely using forage for cattle feeding.* www.altavista.com

Jobin (1987). *Reducing vector densities*. www.msn.com.

Jorgenson *et al.* (1993). *Reuse for saline agriculture and forestry*. www.yahoo.com.

Latey (1993). *Improvement in water use efficiency of the plant*. www.google.com.

Latey and Dinar (1986). *Drainage water reuse: Effect of water salinity*. www.pollution.com.

Maas (1986). *Salt tolerances of many crop species and variation*. www.rediff.com.

Oomen *et al. Medical control of Pathogen and treatment of diseases*. www.about.com.

Pelmutter, N.M. and, M. Lieber (1970). Dispersal of plating wastes and sewage contaminants in groundwater and surface water, South Farmingdale Massapequa Area, Nassau country. New York. U.S. Geological survey water supply paper 1879-G.

Pelmutter, N.M., M. Lieber, and H.L. Frauenthal (1963). Chromium wastes in South Farmingdale, Nassau country, Long Island, New York. U.S Geological survey professional Paper 475-c, pp. c 179-c 184.

Rhoader (1988). *Irrigation and cropping strategies*. www.pubmed.com.

Shannon *et al.* (1994). Michael C. Shannon, USDA Salinity Laboratory, Riverside, California, USA. www.altavista.com.

Tanji and Karajeh (1992 and 1993). *Concept of agriculture forestry and solar evaporators*. www.science.com.

USWFS, Jond K. Hollingsworth. Sign Warning of mercury contaminants at Felsenthal NWR. Pollution; water pollution. www.msn.com.

WHO (1982). *Control of water borne diseases*. www.pollution.com.

WHO (World Health Organization) (1995). *Water borne excreta related infection*. World Health Organization, Geneva.

3

Quality of Water in Fruit and Vegetable Processing Industries and their Management

R. SARAVANA KUMAR, G. MANIMEGALAI, A. SOLAIMALAI & M. BASKAR
Krishi Vigyan Kendra, Tamil Nadu Agricultural University, Vriddhachalam

Introduction

Water is one of the important factors in activity of the fruit and vegetable processing industries. Water is needed in the processing plants for generating steam; washing and peeling of raw materials particularly fruits and vegetables as an ingredient in finished products; as a heat exchange medium in heating and cooling operations; for cleaning plant and equipments; and for protection against fire.

Different sources of water supplies may be classified as surface water (from lakes, streams and reservoirs) and subsurface water (from shallow and deep wells). The characteristics of water from these sources vary with rainfall, the nature of materials with which the water comes in contact, and the time of year.

The general characteristics of water from different sources are shown in Table 3.1.

Table 3.1: General Characteristics of Water from Different Sources

Type of Water	*General Characteristics*		
	Organic Matter	*Mineral Matter*	*Microbial Count*
Surface water	May be high	Ordinarily low	May be contaminated
Shallow well	Variable	Ordinarily low	Variable
Deep well	Usually low	Usually high	Usually low

The suitability of water for use in processing plants depends upon the following:

1. Physical properties, including colour, odour, flavour and turbidity.
2. Chemical properties, including dissolved solids and gases, pH and hardness. The presence of excessive amounts of hydrogen sulphide in ground and well waters, and phenols and other organic substances from industrial waste impart undesirable off- flavours.
3. Microbiological contaminants, including algae, and pathogenic and non-pathogenic organisms.

Most of the functions of fruit and vegetable processing plants call for water of a high degree of purity. An overall quality of water cannot be prescribed as so many specialized requirements prevail, so that characteristics that are objectionable for one use may not necessarily prove detrimental for others. In general, only potable water should be used in the preparation of any food intended for human consumption. Potable water is that water which contains no bacteria capable of causing human intestinal diseases and is aesthetically satisfactory for drinking purposes, *i.e.*, free from undesirable odours and flavours. The public health services of industrialized countries have laid down standards for potable water in terms of their physical, chemical and microbiological characteristics. These characteristics had determined through the physical examination and chemical and microbiological analysis of municipally treated water. Countries that do not have their own acceptable limitations have adopted these limits to safeguard the health of their populations with regard to potable water.

The water used in fruit and vegetable processing industries should be free from contamination and should be of potable quality in every respect. Its mineral content should be low, and it should be specially free from sulphate and iron salts. The water used in cannery should not be hard, as this toughens the product. In pickling, the carbonates and bicarbonates of calcium and magnesium combine with the lactic acid produced in the process, thereby lowering the acidity of the pickle and affecting its flavour. If the water used is excessively hard, scales are formed in the boiler and also a white crust is formed on the surface of the cans when they are cooled and dried after processing. Such water will, therefore, have to be softened suitably before use. When iron is present in the water, it combines with the tannins of fruits and vegetables and leads to their darkening. The water used should be colourless, odourless, tasteless, and bacteriologically pure. (Manorajan Kalia and Sangita Sood, 1996).

Table 3.2: Standards for Drinking/Potable Water

Sl. No.	*Water Characteristics*	*Recommended Maximum Limits (ppm)*	*Concentrations which Constitute Grounds for Rejection (ppm)*
(A)	**Physical Characteristics**		
1.	Turbidity (Silica scale)	< 10	–
2.	Colour (Cobalt scale)	<20	–
3.	Odour	–	–
4.	Taste	–	–
(B)	**Chemical Characteristics**		
5.	Hardness ($CaCO_3$)	<250	–
6.	Alkyl benzene sulphonate (detergent)	0.5	–
7.	Arsenic	0.01	0.05
8.	Barium	–	1.0
9.	Cadmium	–	0.01
10.	Carbon (Chloroform extract, exotic organic chemicals)	0.2	–
11.	Chloride	250.0	–
12.	Chromium	–	0.05
13.	Copper	1.0	–
14.	Cyanide	0.01	0.2
15.	Fluoride	1.7	2.2
16.	Iron	0.3	–
17.	Lead	–	0.05
18.	Manganese	0.05	–
19.	Nitrate	45.0	–
20.	Phenols	0.001	–
21.	Selenium	–	0.01
(C)	**Microbiological Characteristics**		
22.	Bacteriological examination *Escherichia coli communis*	Absent in less than 50 ml.	–

Water Quality for Various Uses in Fruit and Vegetable Processing Industry

General Purposes

Water used for general purposes in a fruit and vegetable processing industry, such as washing, processing and general sanitation. The total bacterial count of water considerably influences plant sanitation. The purpose of washing fruits and vegetables and cleaning equipment is to reduce contamination. Obviously, water used for this purpose should itself have a low bacterial count. The bacteriological–quality of water used throughout the plant should meet the standards required for drinking water. The tolerance limits and standards for water for general purposes are given below:

Table 3.3: Standards for Water for General Purposes

Water Characteristics	*Standard*	*Tolerance*
Colour	Should be quite colourless	–
Appearance	Clear, free from insoluble matter and appreciable deposit	–
Odour	–	–
Taste	–	–
Hardness ($CaCO_3$)	100–150	< 250 ppm
Chlorine	None–Trace	< 25 ppm
Free Ammonia	–	< 1.0 ppm
Albuminoid Ammonia	–	< 0.1 ppm
Heavy metals (Iron)	–	< 2.0 ppm
Nitrate nitrogen	–	Nil
Bacteriological examination *E. coli communis*	Absent in 100 ml	Absent in less than 50 ml

Canning of Fruits and Vegetables

The water used in cannery should not be hard, as this toughens the product. When very hard water is used for blanching vegetables some pecto-calcium and pecto-magnesium complexes are formed which starts the hardening of vegetable tissues. This process continues over the pasteurisation of the finished product. Soft water has negative consequences associated with minerals and hydrosoluble substances and losses during blanching of vegetables. When fruit is processed in sugar syrup, the use of hard water for the syrup preparation could induce the formation of pectin-sugar-acid gel facilitated by the medium pH and presence of calcium salts. The generally permitted level of hardness in water used for canning should not exceed 85 ppm as $CaCO_3$. Water used in a retort should be free from bicarbonate hardness, otherwise unsightly calcium carbonate deposits will develop on the outside of the filled cans or glass containers during processing. The sodium alkalinity of retort water should also be low, because high sodium alkalinity attacks can and gives them a spangled appearance.

Soft Fruit Drinks and Carbonated Beverages

Water from municipal supplies sometimes contains mineral residues and organic matter which makes it unsatisfactory for the preparation of carbonated beverages. For bottling carbonated beverages, it is necessary that the water used should be clear, colourless, odourless and free from water-borne organisms. Its alkalinity must be less than 50 ppm; its total dissolved solid content less than 500 ppm; and its iron or manganese content less than 0.1 ppm. Iron and manganese contents above this level may form a sediment in the beverage and contribute to a poor taste. If the water contains suspended matter, it does not carbonate readily and the beverage made from it rapidly becomes flat. Hardness further influences the degree of carbonation and in some cases, colour and clarity. Odours found in water alter the flavour of beverage often making them extremely disagreeable. Odours found in water are classified in a number of ways: they are indicated as "chemical" when coming from industrial wastes or chemical treatment plants; as 'fishy' when due to certain water-borne organisms; as 'septic' when from sewage contamination; as 'hydrocarbon' when from oil contamination as sulphurous when from hydrogen sulphide; and as 'chorine' when from the presence of fresh 'chlorine' in the water. For the taste to be acceptable water must be free from all these objectionable odours and must have no noticeable taste. Chlorinated water has a marked effect on the quality of citrus products, especially orange drinks. Alkalinity in excess of 50 ppm will neutralize sufficient acid in the beverage and destroy much of the zest or the tang, often making it extremely disagreeable. Neutralization of the acids also reduces the keeping quality of the product, because the concentration of acid aids in the preservation of beverages. Total dissolved solid exceeding 500 ppm will give the beverage a brackish taste.

Although water used in beverage manufacture may be potable and free from pathogenic organisms, it may contain micro-organisms which are capable of spoiling the product. The micro-organisms may give rise to growth which will cause deposits or opalescence and cause the product to spoil during storage. Carbonated beverage usually contain a flavouring extract, sugar, water and substances to produce foam. Some may even contain a fruit juice base. Algae and other water-borne organisms may find these produce favourable to their development and produce a off-taste and sediments. The products may frequently become cloudy and ropy; ropiness being caused by capsulated organisms. Carbonated beverages usually contain yeasts and bacteria. Yeasts frequently cause spoilage.

Most of the bacteria seem to remain dormant, but occasionally they may grow and cause spoilage. In addition, water used for beverage manufacture should not contain colloidal matter which, upon standing, may cause precipitation in the product.

Table 3.4: Standards of Water for Various Products

Water Characteristics	*Maximum Tolerance Limit (ppm)*		
	Canning	*Carbonated Beverages*	*Beverages*
Turbidity	1	1	5
Colour	–	10	–
Taste and odour	None	None	None
Iron and manganese	0.2	0.3	0.1
Sulphates (as SO_4)	–	250	–
Chlorides (as Cl)	–	250	–
Alkalinity (as $CaCO_3$)	–	30–85	50
Hardness (as $CaCO_3$)	85	–	150
Total dissolved solids	–	850	500
Organic matter	–	None	None
Fluorine	1.0	1.0	1.0

Fruit Acids and Pectin

In the manufacture of fruit acids and pectin, it is necessary to have water with a low mineral content. If possible, the ash content must be completely eliminated to avoid interference with the extraction of these products.

Citrus Fruit Packaging

When hard water is used for washing citrus fruit, the dust and dirt tend to cling to the fruit and give a dingy and unattractive appearance. Hard water thus needs to be softened. Soft water used in the washing of fruit has been found to loosen and remove the dirt, thus furnishing a clean and attractive product.

Bottle Washers, Keg Washers and Pasteurizers

Water used for these purposes should be of almost zero hardness. In pasteurization, waters which are high in bicarbonate hardness, form unsightly deposits on the outside of the bottles. This necessitates polishing operations later. Softening of this water eliminates these defects.

Boiler Feed Water

Water containing appreciable quantities of soluble salts, especially those of calcium and magnesium, is not suitable for boilers. As the steam formed evaporates, the soluble salts get concentrated and form compounds which are deposited in the equipment. Hard water forms scale or adherent sludge on the heat-transfer surfaces of the boiler, due to the deposition of insoluble carbonates and sulphates of calcium and magnesium, with or without silica. Scaling results in a loss of efficiency of heat transfer; a reduction in the capacity of the boiler and its eventual damage.

Table 3.5: General Standards of Water for Fruit and Vegetable Processing Plants

Water Characteristics	*Maximum Tolerance (ppm)*	*Typical Effect when in Excess of Maximum Tolerance*
Turbidity	1 to 10	Deposits on product and equipment
Colour	5 to 10	Possible discolouration from organic matter
Taste and Odour	Noticeable	Corresponding or intensified taste or odour in product
Iron and Manganese	0.2 to 0.3	Staining, discolouration off taste and possible growth of iron bacteria
Alkalinity	30 to 250	Neutralization of acid constituents, possible inhibition of bacterial action, deposits or white butts in ice
Hardness ($CaCO_3$)	10 to 250	Deposits, absorption by some products, precipitation with alkaline constituents
Total Dissolved Solids	850	Perceptible taste and possible inhibition of chemical reaction
Organic Matter	Generally any	Off taste, sediment, spoilage, undesirable reactions
Fluorine	1.0	Mottling of tooth enamel in children

A high solid content in feed water, which is alkaline and contains substances that promote foaming, causes these to be carried over in the steam. The steam generated is thus strongly alkaline and will corrode metals such as aluminium and tin. For this reason, maximum permissible limits are provided for boiler feed water and these vary with operating pressures. For operating pressures of 0 to 330 psig, the permissible limits of water solids are 3,500 ppm. For pressures between 1,000 to 1,500 psig, the permissible limits for water solids are 1,000 ppm.

It is important that the quality of water supplied to fruit and vegetable processing plants meets certain maximum tolerance limits prescribed as standard characteristics. Where such tolerance limits are exceeded, undesirable effects may show up in products and equipments (Srivastava and Sangeve Kumar, 1998).

Water for Cannery

In an average, nearly 5,000 litres of water is required for handling one tonne of fruit. The quantity however, depends upon the type of fruit and the product made. Water is extensively used in the preparation of syrups and brines and in several stage of processing such as washing, blanching, sterilizing, cooling, etc. Plenty of water is also needed to wash the machinery and equipment, floors, walls, drains, etc. For raising steam, the boiler also require sufficiently large quantities of water of the right type. A cannery must have plenty of water of good quality. It should be free from contamination and should be of potable quality in every respect. Its mineral content should be low, and it should be specially free from sulphates and iron salts.

The water used in a cannery should not be hard, as this toughens the product. In pickling, the carbonates and bicarbonates of lime and magnesium combine with the lactic acid produced in the process, thereby lowering the acidity of the pickle and affecting its flavour. If the water used is excessively hard, scales are formed in the boiler and also a white crust is formed on the surface of the cans when they are cooled and dried after processing. Such water will, therefore, have to be softened suitably before use. When iron is present in the water, it combines with the tannins of fruits and vegetables and leads to their darkening. The water used should be colourless, odourless, tasteless and bacteriologically pure. It should not show the presence of any faecal organism like *Bacillus coli.* It should not ordinarily have more than 20 parts hardness per 100 litres, *i.e.*, 0.096 grain per ounce. The upper limit is 50 parts hardness. Water containing 5 to 10 parts hardness per 100 litres is moderately soft, 10 to 15 parts slightly hard, 15 to 25 parts moderately hard and 25 to 30 parts and above, exceedingly hard. Organic carbon should not exceed 0.2 parts nor organic nitrogen 0.02 parts per 100,000 parts. Free ammonia should not exceed 0.05 parts per million. Water consuming 0.7 ppm or less of oxygen is very pure, while that consuming between 0.7 ppm and 2.1 ppm is of medium purity.

The natural sources of water are river, spring or deep well and sea. Rain water, although pure to start with, picks up oxygen, nitrogen, carbon dioxide, nitric acid, sulphur dioxide, etc. present in the atmosphere. It can be collected from clean rooftops, stored and filtered before use. River water contains several kinds of impurities of uncertain nature and should, therefore, be filtered and purified before use. Surface or pond water usually contains much organic matter and should not be used in a cannery. Deep well water is normally pure on account of its percolation through the earth for some considerable distance and thus become aerated, oxidized and filtered. In some cases, however, these waters may be highly hard, depending upon the locality. Besides, there may also be external contamination. They should, therefore, be tested before use.

Freezing Operations

While even less exact information is available in water uses in freezing operations than on canning operations, there is fairly close correlation between the two. Lamb and Havighorst (1983) described a plant processing 7½ tons of peas per hour, with the following quantities of water used as designated.

Table 3.6

	g.p.m.
Washing and Fluming	150
Froth Cleaner	10
Reel washer	120
Quality Grader	10
Elevators and Blanchers	20
Hydraulic Products Pump	120
Boiler Make up	10
Picking Belt	120

Without reuse, this amounted to 5,500 gal. of water per ton of peas. By reusing 50 per cent of the water through the washers, picking belts and product pump, the quantity was cut to 3,500 gal. per ton of peas. For cooling the compressed refrigerant, 1,100 g.p.m. of water were used on the freezing operation and 110 g.p.m. for maintaining room temperatures. The defrosters required 20 g.p.m. As abundant water was available at an initial temperature of 63° F it was used once with a temperature rise to 73° F and not recalculated through a tower or otherwise recooled.

Frozen concentrates have become a leading item in the frozen food field. The quantity of water employed depends upon the type of evaporator used. Evaporators of the type engineered by Cross use water only for condensing vapours from steam ejectors, for clean up, and small quantities for balancing the heat pump cycle by removing superheat from the refrigerant. The type of evaporator engineered by Kelly uses water as a heat exchange medium in evaporating and condensing water from the juice. Double effect recompression type concentrators use the largest quantities of water to condense vapours from the juice plus steam used to compress these vapours. A cooling tower is generally employed with this type of concentrator so make-up water is required only to remove entrained solids. Minimum water demand for the production of frozen concentrated orange juice has been estimated at 1,000 gal. of water per ton of fruit yielding 30 gal. of 42° Brix concentrate (Gridharilal *et al.*, 1986).

Dehydration

In the dehydration of fruits and vegetables, the quantities of water used in preparation and blanching compare closely with the quantities used in preparation for canning and freezing. Fruit prepared for sun-drying and dehydration, either as whole or cut, usually is not washed but some water is used in the hot water or lye-checking of grapes and prunes. The dried fruit during processing for marketing is washed and treated with hot water. The actual requirement for water in drying, dehydration and processing of dried fruits is not known. In the dehydration of vegetables considerably large quantities of water are used. This varies from 10,000 to 20,000 gal. per hour for a plant processing 100 tons per day of potatoes and sweet potatoes to 8,000 to 16,000 gal. per hour for carrots, beets, rutabagas and onions and to 2,400 to 4,000 gal. per hour for cabbage. This amounts to form a minimum of 600 gal. to a maximum of about 5,000 gal. per ton of raw material processed.

Fermentation and Beverages

In addition to alcoholic fermentations used in preparing wines and beer, lactic and acetic fermentations are extensively employed in preserving fruit and vegetable products. Juices are fermented to form vinegar. Cucumbers, cabbage and other vegetables are subjected to lactic acid fermentation for the production of pickles and kraut. In the production of vinegars, water is used for cleaning fruit before pressing and for cleaning lines, pumps, filters, tanks and other equipment to maintain sanitary conditions essential to controlled fermentation; also for steam used for packing pasteurised vinegar. In the production of distilled vinegar, the water used for diluting the vinegar stock should be of low solids, hardness and chloride content and particular care should be taken to avoid contamination with copper, iron, lead, zinc and tin salts which markedly reduce the efficiency of acetification in the generator process.

In the production of carbonated beverages assuredly sterile water, of low air content, free from odours of flavours and practically free of insoluble materials is needed. The production of such sparkling clear water, even from regular city drinking water, requires carefully controlled chemical coagulation, sand filtration, carbon purification, sterile filtration and deaeration. The laboratory control required in the production of such water is described by Kratz and Lewis (1990).

Bacteriology of Plant Water

All water that comes into contact with foods should meet the bacteriological standards for drinking water and preferably all freshwater at the plant should be that good. But this water also should be satisfactory from a bacteriological standpoint for use with the particular food being processed. A water supply may be adjudged potable yet be unsatisfactory for use with a food. Thus, for example, water containing appreciable numbers of psychrotrophs of the genera *Pseudomonas* or *Alcaligenes* might be unsatisfactory without treatment in a dairy plant making butter or cottage cheese. The slimy growth of iron bacteria in water supplies often leads to trouble in the food plant. More likely to be important is the chemical composition of the water, which must be suited to the use to be made of it. Thus hard water is undesirable in pea canning and in brewing; iron and manganese are bad in beet canning and in brewing; excessive organic matter may lead to off-flavours, etc.

Of special interest in canning factories is the bacteriology of the water in which the cans of processed foods are cooled after their heat treatment. If this water contains micro-organisms able to spoil the foods, it can enter defective cans through minute leaks and increase the percentage of cans of food spoiling during storage. Many canneries routinely chlorinate the cooling water to reduce or eliminate this trouble. The shortage of water in many food has necessitated reuse of part of the water and micro-organisms may build up in such reused water. Water employed for the final rinse of a food must be fresh and potable, out after use may be returned for soaking, first wash, or fluming, preferably after treatment with chlorine, chlorine dioxide, or a similar germicide.

In plant or continuous chlorinating beyond the break point (the point where the chlorine demand has been satisfied) to a residual of 5 to 7 ppm of chlorine is employed for continuous application to areas and equipment where slime bacteria may be a problem, *e.g.*, conveyors or belts, can coolers, product washers, and flumes. The chlorinated water may be applied as a spray, or parts of equipment may be immersed. When operations ease, chlorinated water may be applied to fillers, peelers, dicers, and similar equipment. Contaminated or polluted water lines are held filled with chlorinated water containing 50 to 100 ppm of chlorine for 12 to 48 hrs., the strength of chlorine and length of time depending on the extent of pollution. Ice used in contact with foods should meet the bacteriological requirements for potable water. Much work has been done on the incorporation of bacteriostatic or bactericidal chemicals in water and in ice to aid in food preservation. It has been noted previously that a chlortetracyline or oxytetracyline dip for dressed poultry had been approved, but approval was later revoked; however, these antibiotics may be incorporated in ice to be applied to fish and other seafood.

Recovery of Water from Processing Effluent

An often neglected source of water for food processing is waste liquid from antecedent operations in the same or adjacent plants. In a sense, most of the water obtained from surface sources is re-circulated or diluted and stabilized sewage, since waste from disposal systems of communities and factories enters, lakes and rivers from which other users draw their supplies. Sewage treatment is considered only insofar as it applies to recovery of water from contaminated effluent for reuse in food processing operations. Many procedures routinely used in treating sewage for discharge into streams, lakes and other public water courses are also applicable to the renovation of waste from food plants.

The processor located in a region with ample supplies of water at reasonable cost and having a disposal outlet for all of his waste has no reason for using such methods. However, some raw water supplies are actually more contaminated than process effluent and the number of these supplies is increasing constantly. Treatment of these effluents may be less costly than treatment of raw water. This is particularly true where the raw water is heavily contaminated with odorous organic substances or detergents, or is unusually high in salinity or hardness. Furthermore, the treatments required by local laws and regulations to be applied to process effluent before it can be dumped into streams may bring the liquid to a state of purity approaching that needed in the water supply. The cost of additional treatment which may be necessary can thus be brought within the acceptable ringe.

Industry consumes only about one-fiftieth of the water it withdraws. The rest is returned for downstream user. Clarke (1982) estimated that 80 per cent of returned water has been used merely for cooling. It is contaminated only by heat and the dissipation of heat through use of cooling towers and ponds is probably the most frequent example of water conservation by recycling. The other 20 per cent of returned water carries a great variety of industrial wastes and other impurities. Some of these substances are relatively easy to remove conventional techniques while others are extremely refractory and would require equipment of prohibitive cost for their extraction.

Bulk handling methods used in food manufacture include use of water dumps for receiving raw fruits and vegetables and hydraulic conveyors for moving raw and partially processed materials to and from the various operations. It has been shown that microbial counts for foods in the final container are influenced by the condition of the water in which they are dumped and conveyed. The extent of the permissible re-circulation in hydraulic conveying systems, therefore the water demand of these systems, is dependent in part on the sanitary standards for the food product. For all practical purposes, the sanitary condition of the product can be measured in terms of the number of bacteria in or on the food discharged from the system (Geyer 1983).

Liquid wastes from canneries may vary from 0.25 to 4 gal. per lb. of fruits or vegetables processed. Water used washing raw products may be high in inert solids (dirt) and contain insecticides or other chemicals. Biological methods of purification developed for treatment of domestic sewage are often inadequate for the stabilization of cannery wastes because of the poor balance of carbohydrates and nitrogenous compounds in the wastes (Hart and McGaukey, 1984). The heavy seasonal loads imposed on municipal plants can easily disrupt treatment procedures. Combinations of screening, flotation, heavy chlorination and chemical oxidants have been used to stabilize cannery effluent. Disposal methods frequently used by food plants for liquid waste include lagooning, spray disposal, digestion. In some cases, solids may be recovered to be sold as fertilizer or burned as fuel. Usually no thought is given to the recovery of the water present in these waste.

Probably recycling is more common than renovation of wastewater outlet from outside sources. Clarke (1982) classifies on site reuse systems as multiple recycle or cascade. Combinations of the two types are frequently used. A multiple recycle system consists of a number of parallel circuits carrying various grades of cooling waters and process water. Such an arrangement segregates wastes for simplified treatment and provides for variation in temperature, pressure and make-up water quality. One or more of the circuits may have integral purification stages for maintaining the necessary water characteristics. Usually, circuits are interconnected and have continuous blowdown drainage to limit retention time in a circuit and deliver the wastes to less critical systems or waste treatment facilities. If the blowdown waste passes to another process or cooling systems, the plant is a compound multi-cycle cascade type.

Cascade reuse systems depend on in-line integration of the water supply for two or more operators rather than differentiation of process and cooling waters. Water first enters the equipment or process requiring the cleanest or coldest supply and continues to successive operations where progressively higher temperatures and poorer water qualities can be tolerated. The effluent of one operation becomes the supply for the next until unacceptable temperature or contamination develops. In random reuse, one processor passes his effluent to another plant requiring a less pure supply.

Groundwater recharge is a method of maintaining the water table or at least of reducing the rate at which the table is being lowered or invaded by saline intrusion in areas of heavy withdrawal. In essence, this procedure takes advantage of natural purification processes for recycling water. The water is retained in the general area of withdrawal so that it can contribute to the total reserve available to users in the area whereas dumping the waste into a stream generally means the water is lost to that particular region. Obviously, groundwater recharge is not a simple matter of pumping the raw sewage into a hole drilled into the earth. A system for distributing the effluent over sufficient area to assure gradual percolation to the water table so that the material is not forced to the surface, proper treatment methods so that the natural purification steps can proceed rapidly, and isolation of withdrawal wells from the disposal area are the minimum necessary precautions.

Occasionally, some particularly objectionable waste substance which interferes with even limited reuse of the water can be recovered for sale or use. The recovery of tartrates from winery wastes is an example of a treatment step which may either pay for itself or yield a small profit.

Methods of Treatment

There is continuing effort among water technologists to develop new methods of waste treatment which will produce effluents of superior quality. The two sewage treatments which can be used are the trickling filter process and the activated sludge system (Weiner, 1983). In both of these procedures, waste is first passed through a screen of metal bars that removes all solid objects too large to be handled in subsequent operations. Usually, the second step is the passing of the liquid through a long, narrow tank called a grit chamber where large suspended particles settle out. If the content of such material is very low, as in some reliably homogeneous processing wastes, the grit chamber can be omitted.

The primary settling tank is a necessary part of the trickling filter process and is also used in most of the activated sludge systems. This is a large, usually open, basin in which the liquid waste is held until the larger suspended particles settle out and can be drawn off at the bottom as sludge. These concentrated impurities are sent to a digester which is a large cylindrical, sealed tank where they may be kept for about 30 days. Under the conditions of low oxygen tension existing in the digester, anaerobic bacteria metabolize most of the soluble materials, hydrolyze some of the insoluble matter, and convert part of the solids to gases such as methane and hydrogen sulphide. These actions change the sludge into a material of much greater biochemical stability. The precipitated material remaining in the digester is dried and disposed of in land fill, by burning, or as fertilizer while the effluent is drawn of and sent to the trickling filter.

The equipment which gives the trickling filter process its name is a large bed of rocks, crushed stone and gravel several feet thick. The discharge from the digester is distributed over the top of the filter bed by sprinklers which can be of several designs. The gravel or other material in the filter becomes coated with a film of aerobic bacteria which act upon the metabolizable soluble compounds in the liquid. These compounds are converted into carbon dioxide, water and cell components, as well as a few rather minor metabolic by-products. Since the cell components remain in the organisms attached to the gravel, they are effectively removed from the effluent which is then sufficiently pure to be discharged to a river or lake.

One of the advantages of the activated sludge system is that the amount of sludge going to the digester is considerably reduced so that the size of the digesters can be less than in the trickling filter method. The plant waste may, in the former process, be sent to a primary settling tank. This step may be omitted under certain conditions. In any case, the effluent next flows through a series of aeration tanks. Here it is agitated and oxygenated by air blown in from the sides and bottoms of the tanks. The conditions are extremely favourable for aerobic bacteria and they digest the wastes in a few hours. Sludge is removed in a settling tank and a small proportion of it-the activated sludge is–returned to the aeration tank where it assists in maintaining the microbiological activity at a high level. The rest of the sludge is sent to a digester for anaerobic stabilization. The effluent from the aeration tank is relatively pure but needs further treatment before it can be used in processing applications. The supplemental treatment advocated by Culp for efflluents encountered and under conditions existing in the test plants included chemical coagulation with 100 to 200 mg/l of alum and a coagulant aid, primary adsorption of alum floc and rapid filtration followed when necessary by secondary adsorption on activated carbon.

The separation bed which removes the precipitate and associated suspended material after addition of the flocculent is similar in outward appearance to a rapid sand filter, but the bed is arranged so that the waste first contacts the coarse material and then is passed through layers of increasingly finer granulation. Filtering in the direction of coarse to fine reduces surface plugging which would otherwise occur and makes it possible to use the entire depth of the bed for removal. The size of the filtration media is selected so that the raw waste will pass through the bed with no loss of particles until the flocculent are added.

The separation beds were washed similarly to rapid sand filters, with surface washing a necessary step. Washing consumed about four per cent of the treated water. Effluent from the separation beds operated in the preceding manner exceeded in clarity many drinking water supplies, with typical turbidity values being less than 0.5 standard units. Phosphates were greatly reduced (average values about 1.0 mg/l in the effluent) as a result of adsorption on the alum floc. Some of the alkyl benzene sulphonates were also removed and the BOD values were reduced about 50 per cent. These results are better than those to be expected from chemical precipitation, settling and rapid sand filtration. Part of this superiority is attributed to the very favourable conditions for adsorption in the separation bed.

The carbon filtration section of the plants consisted of a series of two or more columns about three feet in depth filled with granular activated carbon and passing fluid at five gallons per square foot per minute. The carbon filtration efficiency and economy were greatly enhanced by a filtration bed which removed materials that would otherwise quickly foul the carbon particles. The final effluent contained nothing that would identify it as a waste.

The alkyl benzene sulphonate concentration was below 0.05 mg/l, the phosphate content less than 0.5 mg/l, and the BOD less than ten milligrams per litre. Chlorinating could have been applied to render the water bacteriological pure. The pH of the system appears to have a strong influence on the effectiveness of the different treatment steps. Lea *et al.* (1984) indicated that phosphate adsorption on alum floc is optimal at pH 7.1 to 7.7. Some industrial contaminants evidently persist unchanged through all treatment processes and for long periods in natural water courses. These may lead to undesirable odour, flavour, or appearance effects in commercial food and beverage products. They may also disrupt the treatment processes by killing off the bacteria responsible for oxidation of other waste compounds or by interfering with coagulation and sedimentation.

Concentrated wastes removed by advanced waste treatment processes cannot be discarded into usable surface or ground supplies. Four approaches can be considered; (1) recovery of part of the concentrated waste for use as fertilizer, chemical raw materials, etc.; (2) disposal of the highly concentrated wastes into controlled waste sinks or into remote ocean areas; (3) subsurface disposal through injection into deep wells or sealed underground cavities; and (4) conversion of the contaminants to innocuous form by digestion, incineration, or wet oxidation.

Waste Treatment and Disposal

The food sanitation is concerned directly or indirectly with the adequate treatment and disposal of wastes from his industry. Solid and concentrated wastes ordinarily are kept separate from the watery wastes and may be used directly for food, feed, fertilizer, or other purpose, may first be used concentrated, dried, or fermented (*e.g.*, pea-vine silage), or may be carted away to available land as unusable waste. Care is taken to keep out of the wastewaters as much wasted liquid or solid food materials as possible, by taking precaution to avoid introduction into the watery wastes of drip, leakage, overflow, spillage, large residues in containers, foam, frozen-on food and food dust during the handling and processing of the food.

Wastes from food plants ordinarily contain a variety of organic compounds, which range from simple and readily oxidizable kinds to those which are complex and difficult to decompose. The strength food waste containing organic matter is expressed in terms of biochemical oxygen demand (BOD), which is the quantity of oxygen used by aerobic micro-organisms and reducing compounds in the stabilization of decomposable matter during a selected time at a certain temperature. A period of 5 days at 20°C is generally used and results are expressed as 5 day BOD. The BOD is determined by dilution of measured quantity of waste with water that has been saturated with oxygen and incubation of the mixture at 20°C, along with a control of dilution water alone. After 5 days, the residual oxygen in both control and test sample is measured by titration. The difference represents the oxygen-consuming capacity of the waste and is calculated to be expressed as parts per million of oxygen taken up by the waste. The strength of the waste in terms of pounds of BOD is calculated as follows:

$$\frac{\text{ppm 5–day BOD} \times \text{gallons of waste} \times 8.34}{1{,}000{,}000} = \text{pounds BOD}$$

This value can be converted to population equivalent (PE) by assuming that the domestic sewage of one person is equivalent to one-sixth of a pound of BOD per day.

Whenever appreciable amounts of wastes high in oxidizable organic matter (high BOD) are emptied into natural waters, such as streams, ponds, or lakes, the 7 to 8 ppm of free oxygen normally present in the waters is used up soon by oxidation processes carried out by aerobic or facultative micro-organisms. When the oxygen drops below 3 ppm the fish either leave or die, and when anaerobic conditions have been attained, hydrolysis, putrefaction, and fermentation by micro-organisms will follow, with the result that the body of water will become malodorous and cloudy and hence unsuited for recreational use and unfit for drinking and for use in the food plant. Wastes from a food plant to be emptied into a body of water must either be so greatly diluted by that water as to be innocuous or must be treated first to reduce the oxidizable compounds to a harmless level. Even the effluent from an efficiently operated sewage treatment system will encourage the growth of algae and higher aquatic plants in the water and make it less attractive for recreational purposes.

Preliminary treatments of food-plant wastes by chemical means may be employed, but most systems of treatment and disposal depend upon (1) screening out of large particles; (2) floating off of fatty and other floating materials; (3) sedimentation of as much of the remaining solids as practicable; (4) hydrolysis, fermentation, and putrefaction of complex organic compounds; and finally (5) oxidation of the remaining solids in the water to a point where they can enter a municipal sewage treatment and disposal system, a plant disposal system, a lake or stream, or soil. The completeness of oxidation required will depend upon the disposal to be made. Thus less oxidation might be required for feeding to a municipal system or for irrigating soil than for entering a stream or lake.

Chemical Treatment

In chemical pre-treatment, a chemical or mixture of chemicals is added to the sewage or waste so as to cause formation of a flocculent precipitate, which, in settling, carries with it much of the suspended and colloidal material, including bacteria. The effluent then is run into a body of water, onto soil, or into biological treatment system. The chemicals commonly used are soluble aluminium or iron salts, such as alum of ferrous sulphate, plus lime, giving a flocculent precipitate of aluminium or ferric hydroxide. Disposal of the sludge (seedlings) so obtained may be difficult.

Biological Treatment

The general biological methods for waste disposal and/or treatment include (1) *dilution*, by running wastewaters into a large body of water; (2) *irrigation*, in which wastewaters are sprayed onto fields of open-textured soil; (3) *lagooning*, by running the wastewaters into shallow artificial ponds (with or without other treatments); (4) use of *trickling filters*, made of crushed rock, coke, filter tile, etc.; (5) use of the *activated-sludge* method, in which wastewater is inoculated heavily with sludge from a previous run and is actively aerated in tanks; and (6) use of *anaerobic tanks* of various kinds, where settling, hydrolysis, putrefaction, fermentation take place, usually to be followed by some aerobic treatment.

The dilution method seldom is practicable because a sufficiently large or rapidly moving volume of water rarely is available or because the location is such that sewage decomposition cannot proceed without objections from nearby populations. Irrigation is increasing in popularity and is especially adaptable to use by plants located in rural areas and near open-textured soil. Lagooning has been used especially for seasonal wastes, as from canning factories. The wastes are decomposed slowly in these shallow ponds or lagoons until the liquid part can be run into a stream or other body of water during the rainy season or time of melting snow, when there is a good volume of water. Usually, sodium nitrate is added to reduce obnoxious odours. Sometimes the liquid is pumped from and returned to a lagoon, or it may be pumped from one lagoon to another in a series of lagoons. Trickling filters and activated sludge systems are probably the most effective of the systems listed, but they are expensive to run and require supervision by an expert. Anaerobic tanks yield an effluent that needs further treatment and should be either turned into a municipal system or given an aerobic treatment.

References

Clarke, F.E. (1982). Industrial reuse of water. *Industrial Engineering. Chem.*, 54(2): 18-27.

Geyer, J.C. (1983). *Water conservation.* National Research Council Publ., 942: 17-49.

Gridharilal, G.S., Siddappa and G.L. Tandon (1986). *Preservation of fruits and vegetables.* Publications and Information Division, ICAR, New Delhi.

Hart, S.A. and McGauhey, P.H. (1984). The management of wastes in the food producing and food processing industries. *Food Technol.*, 18: 432-438.

Kratz and Lewis (1990). The alcoholic fermentation of beverages. *J. Physiol.*, 36: 12-15.

Lamb and Havighorst (1983). Freezing of fruits and vegetables. *Science Cult.*, 28: 62-64.

Lea, W.L., Rohlich, G.A., and Kafz, W.J. (1984). Removal of phosphates from treated sewage. *Sewage and industrial wastes*, 26: 261-270.

Manorajan Kalia and Sangita Sood (1996). *Food preservation and processing*. Kalyani publishers, New Delhi.

McIntosh, O.H. and McGeotge, G.G. (1964). Year round lagoon operation. *Food processing*, 25(1): 82-86.

Srivastava, R.P. and Sangeve Kumar (1998). *Fruit and vegetable preservation: Principles and practices.* International book publication division, Luknow.

Weiner, D.L. (1983). *Understanding the world sewerage.* Can-Tex Industries Inc., Cannelton, Ind.

4

Wastewater Quality of Major Drains of Delhi Draining Wastewater to River Yamuna and Assessment of Water Quality of River Yamuna at Delhi Stretch

P.K. BEHERA & R.C. TRIVEDY

Central Pollution Control Board, East Arjun Nagar, Shahadra, Delhi

P.C. MISHRA

Department of Environmental Sciences, Sambalpur University, Jyoti Vihar – 768 019, Orissa

ABSTRACT

A detailed monitoring of the quality of wastewater discharged to river Yamuna from five major drains at Delhi as well as quality of river water at Delhi stretch was done over a period of two years during 1993-95. Twenty-eight parameters including heavy metals and pesticides content were selected for the study. Total coliform and faecal coliforms were also estimated. Based on few important water quality parameters, water quality index of river at sampling points were calculated which reveals that the river water beyond Wazirabad is unfit for bathing, swimming and even recreation.

Introduction

Yamuna River System

The Yamuna is the largest tributary of the river Ganga. The main stream of the river Yamuna–also called Kalindi in some reaches originates from the Yamunotri glacier near Badar Punch (30°58′ N, 78°27′ E) in the Mussoori range of the Lower Himalaya at an elevation of about 6320 metre above mean sea level (MSL) in the district of Uttar Kashi of Uttaranchal. After combining with tributaries Rishi Ganga, Kunta, Hanuman Ganga, Tons and Giri, Yamuna enters the plains at Tajewala. During its journey of nearly 1200 kms. from Tajewala to its confluence with the Ganga at Allahabad the Yamuna is joined by the other tributaries *viz.* Hindon, Chambal, Sindh, Betwa and Ken.

The total catchment area of the Yamuna river system is 3,45,848 square kilometre. It covers parts of Uttaranchal, Uttar Pradesh, Himachal Pradesh, Haryana, Rajasthan and Madhya Pradesh States and the entire Delhi. On percentage basis, contributions area-wise of the Yamuna are: U.P. 21.5, Himachal Pradesh 1.6, Haryana 6.1, Rajasthan 29.8, Madhya Pradesh 40.6 and Delhi 0.4. Sub-basin-wise, the Hindon contribution is 2, the Ken 7, the Sindh 7.5, the Betwa 13.9 and the Chambal 40.5. The balance 29.1 per cent is accounted for by direct drainage into the Yamuna or the smaller tributaries.

A number of medium and large scale irrigation projects, some based on storage and some on mere diversion, draw water from the Yamuna and its tributaries. While water drawn for municipal and industrial purposes is largely, if not entirely, returned to river system itself in the form of wastewater, that drawn for irrigation in mostly consumed in evapo-transpiration. At Tajewala itself, nearly entire flow (6025 million cubic metres) of Yamuna is diverted for irrigation through the Western Yamuna canal for Haryana and the Eastern Yamuna Canal for Uttar Pradesh. The other important water abstraction points are Wazirabad, Wazirabad-Okhla stretch, Okhla-Etwah stretch and Etawah.

In terms of economic activity the Malwa region in Madhya Pradesh (M.P.) and Braj Bhoomi in Uttar Pradesh (U.P.) and Haryana have for centuries been amongst the most prosperous agricultural areas. It has large industrial complexes at Kota, Gwalior, Indore, Nugda, Khetri etc. Paper mill at Saharanpur, a large urea factory at Panipat and oil refinery at Mathura, and the Ghaziabad-Delhi-Faridabad industrial complex developments are in the Yamuna basin. The water quality of Yamuna river in the Himalayan segment and the diluted segment is comparatively good. However, due to heavy abstraction and discharge of pollutants into the river system, there are critical segments which require pollution abatement measures to improve the water quality of the river. These segments with the causes of pollution are:

Wazirabad to Okhla

Domestic and industrial wastewater of Delhi.

Okhla to Vrindavan

Domestic wastewater from Delhi and industrial effluents from Saharanpur, Ghaziabad, Noida etc.

Vrindavan to Mathura

Domestic wastewater and industrial effluent from dyeing and printing industry of Vrindavan and Mathura.

Mathura to Etwaha

Domestic wastewater from Agra and Etawah.

Environmental Status of Delhi

Area

The National Capital Delhi is located on the bank of sacred river Yamuna. It has an area of about 1485 sq km. and lies between latitudes 20°14′15″ and 28°53′00″ (North) and longitudes 76°45′30″ and 77°21′30″ (East). It is situated at 250 metres above mean sea level. Delhi is surrounded on the north, south and west by the State of Haryana and on the east by the State of Uttar Pradesh.

Climate

The climate of Delhi is influenced by its inland position with the desert of Rajasthan to the west and southwest and Gangetic plains of U.P. to east. Extreme dryness with an intensely hot summer and cold winter are the characteristics of the climate. The year can be broadly divided into four seasons. The cold season starts in late November and extends to about the beginning of March. This is followed by the hot season which lasts till about the end of June when the monsoon arrives. The monsoon continues till the last week of the September. The two post-monsoon months, October and November continue, as transition period from monsoon to winter conditions. The climate of territory is semi-arid. The mean maximum temperature range from 27.6°C to 40.2°C. The mean minimum temperature range from 7.5°C to 22.2°C. The annual average rainfall is 754 mm.

Population

The population of Delhi has increased several folds after the partition (1947) making this ancient city as one of the largest cities of the world. This rapid urbanisation in turn enhanced the thrust on basic amenities to cope-up with the living standards. As per the census report of India, the population of Delhi was 9.42 million in 1991, registering an increase of 51.45 per cent in the decade 1981-1991. The population of Delhi in 1998 has been estimated to be 11.5 million, which includes 4 to 5 lakhs of floating population. Of the total population, 90 per cent is residing in urban areas and remaining 10 per cent in urban villages. It is about 13 million in year 2000.

Water Supply and Sewerage System

At present the Delhi Jal Board (DJB) is equipped to produce 2633 MLD of water and this water supply is further augmented by abstraction of groundwater for domestic consumption. Normally sewage generated is 70-80 per cent of total water supply and sewage generation works out to be 2107 in MLD. However the flow of wastewater in various drains is observed to be 2720 MLD thus confirming the use of groundwater for drinking and other purposes. The sewage treatment capacity was 1372 MLD during 1999. The average inflow at the Sewage Treatment Plants was about 953.4 MLD and accordingly about 1766 MLD of untreated sewage finds access into various drains and ultimately into the river.

Water Supply

The habitation of Delhi is in the following categories: (a) Unauthorised/regularised colonies. (b) Planned development/approved colonies (c) Re-settlement colonies (d) Urban villages (e) Unauthorised and non-regularised colonies (f) J.J. clusters. (g) Rural villages. Water supply has been provided to areas (a) to (d) and (g) above, subject to technical feasibility. Water supply is also being extended to unauthorised/non-regularised colonies by providing skeleton water supply either from the available distribution system nearby or through tube wells/deep bore hand pumps. In J.J.

clusters water supply is either through public water supply or deep bore hand pumps and in the absence of either of the above through water tankers, in the far flung areas.

Sewage System

Sewerage facilities have been extended to and being extended to all planned development approved colonies whether developed by DDA, MCD, NDMC, CPWD.

1. Out of 567 unauthorised colonies sewage facilities exist in 293 colonies and work is in progress by DJB in 65 colonies.
2. Out of 44 resettlement colonies, 32 have functional sewerage system and in another 4 work is in progress.
3. Out of 126 urban villages sewerage system exists in 88, work is in progress in another 2 urban villages. The sewerage system is yet to be provided to remaining 35 urban villages.
4. There are 219 rural villages whose population is estimated around 6 lakh, where drinking water supply has been provided. However sewerage system has not been provided in these rural areas as it is not technically feasible. The wastewater from rural belts find its way into storm waste drains.
5. There are 1017 unauthorised/non-regularised colonies which have no sewerage facilities.
6. As per policy in vogue, sewerage facilities are not to be provided in J.J. clusters due to meagre rate of water supply. As such sewage generated from such areas contribute to the flow in storm water drains ultimately out-falling into river Yamuna. The daily flow of wastewater currently received and handled at the seven sewage treatment plants and some oxidation ponds serving NCT Delhi is indicated in Table 4.1. Thus more than 60 per cent of the wastewater generated in NCT Delhi is reaching the river Yamuna via the sewers or open drains without receiving any treatment.

Table 4.1: Inflow at Sewage Treatment Plants

Sl. No.	*Location*	*Capacity in MGD*	*Daily Average Flow in MGD (Year 1999)*
1.	Okhla	124	104.35
2.	Keshopur	72	59.86
3.	Rithala	40	22.06
4.	Vasant Kunj	5.2	1.95
5.	Kondli	35.00	10.09
6.	Yamuna Vihar	10.00	1.66
7.	Coronation	10.00	7.61
8.	Oxidation ponds	6.00	2.47
	Total MGD	**302.20**	**210.05**

Industries and Industrial Wastewater

The growth of large industries is restricted due to introduction of planning restrictions in Delhi, but small scale industries have expanded rapidly. There was a sharp increase in industrial units from 18500 in 1961 to 93000 in 1993. Engineering, textiles, chemicals, electronics and electrical goods are the major categories of industries located in Delhi.

Delhi has 20 large, 25 medium and approximately 93000 small scale industries located at it's 28 planned industrial estates. Large and medium scale industries constitute less than 0.05 per cent of the total industries located in Delhi, but contribute 50 per cent of the total 300 million litres of wastewater generated every day (CPCB-1994). The large and medium scale industries have some facilities for effluent treatment and emission control but the small scale industries do not have such a set-up. The planned industrial estates in Delhi were mostly developed by the DDA and a few by CPWD. Though in a large number of estates, the effluent was originally out-falling into existing sewerage system but over the years the system has degenerated with the result that industrial effluents from most of the industrial estates are now flowing through open surface drains ultimately finding access into the river Yamuna. Some of the examples are Wazirpur Industrial Area, Rajasthan Udyog, Jhilmil Industrial Area etc.

Importance of River Yamuna to National Capital

Historically the city and the river have always remained inter-twined. The Yamuna originated at Yamunotri in the Himalayas and travels through the hilly areas of Himachal Pradesh and Uttar Pradesh. After entering plains it flows along the boarder of Haryana and Uttar Pradesh to enter the National Capital Territory of Delhi at Wazirabad. Within the Delhi territory the river flows from north to south. Although the NCT Delhi lies largely on the western bank of Yamuna, the eastern bank is fast developing and has attained significance economically. The main areas of old Delhi and New Delhi are on the right (or west) bank of the Yamuna, while the old Shahdra town is located on the left (or east) bank. The total length of the river from source to its confluence with the Ganga at Allahabad is 1376 kms. The length of the reach falling within the NCT Delhi is only 48 kms. At Wazirabad where the river enters the NCT Delhi, it is totally trapped for the city water supply. The river leaves the Territory of Delhi and enter into Uttar Pradesh at Okhla barrage. During its course through the city nineteen number (19 Nos.) of drains meet the river and carry into it sullage and other wastewater from various parts of city.

Methodology

Selection of Sampling Points

River Sampling Stations

The river Yamuna enters into National Capital Delhi at Wazirabad barrage and leaves at Okhla barrage. Thus the present study area covers approximately a length of 22 kilometers all over Delhi stretch, starting from upstream points at Wazirabad barrage to down-stream point at Okhla barrage. Selection of sampling sites has been done primarily on the basis of major drains joining to the river and also barrage and bridges existing on the river Yamuna at NCT Delhi stretch. Attempt was made to select the sites at cleanest point (where contamination of river water by drain water is negligible *i.e.* Wazirabad barrage and before and after joining of major drains). Thus the selected river monitoring stations and distance from Wazirabad barrage were *(i)* Wazirabad barrage 0 km., *(ii)* ISBT bridge 8 kms., *(iii)* ITO barrage 15 kms., *(iv)* Nizamuddin bridge 17 kms., and *(v)* Okhla barrage 22 kms.

Drain Sampling Points

Based on pollution load as well as flow of the drain water, only the major five (5 nos.) drains were selected for monthly collection of wastewater and sediment samples for the present investigation. The selected monitoring drains were *(i)* Najafgarh drain, *(ii)* Civil mill drain, *(iii)* Power house drain, *(iv)* Sen nursing home drain, and *(v)* Barpulla drain.

Water and Wastewater Sampling

From drains and river monitoring points samples were collected every month from December 1992 to January 1996. In case of river water, the individual grab sample were collected in triplicate from midstream in 2.5 litre plastic bottles with screw cap carefully after rinsing the sample bottles two or three times with samples. The record of every sampling collected was marked with an appropriate inscription on the container with marking pencil. Samples were collected from a depth of about 30 cm. from surface. After collection, samples were immediately transferred to the laboratory. The pH and dissolved oxygen were measured at the sampling point and for remaining parameters the water samples were immediately preserved. The suitable containers, time and chemical of preservation were selected as per APHA (1991). For pesticide analysis, separate brown glass bottles (one litre capacity) were used in sampling. The samples for heavy metals analysis were collected in polythene bottle (one litre capacity) containing concentrated HNO_3, (1 ml/bottle). For TKN, NH_3-N and nitrate samples were preserved separate bottle containing concentrated H_2SO_4, (2 ml/lt).

Analytical Methods for Estimation of Water and Wastewater

The water and wastewater samples collected from river and drains were analysed by the methods of APHA (1991) Trivedy and Goel (1984) and Saxena (1987). The following physico-chemical parameters were analysed:

1.	pH	2.	Conductivity	3.	Total Solids
4.	Suspended Solids	5.	Dissolved Solids	6.	DO (RW)
7.	Oxygen Absorbance	8.	COD	9.	BOD
10.	TKN	11.	Nitrate-Nitrogen	12.	NH_3-N
13.	Phosphate	14.	Total Alkalinity	15.	Total Sulphate
16.	Chloride	17.	Total Hardness	18.	Calcium
19.	Magnesium	20.	Sodium	21.	Potassium
22.	Total Coliforms (RW)	23.	Faecal Coliforms (RW)	24.	Iron
25.	Cadmium	26.	Chromium	27.	Copper
28.	Nickel	29.	Zinc	30.	Total BHC
31.	Total DDT	32.	Total Endosulfan		

(RW = River Water)

Before the actual analysis all the preserved samples were allowed to regain the room temperature, well shaken and filtered for individual tests. The final tabulated values represented the mean of 3 replicates.

Physico-chemical

The pH was recorded on a digital pH meter (Systronic make model No. 361). Conductivity was recorded on a conductivity meter (Systronic make No. 304). The observed values were expressed in µ mhos/cm. Total Solids was determined by evaporating a fixed volume of sample at 104°C in a hot air oven. Total Dissolved form a part of the total solids residue. In natural water, the TDS consist mainly of bicarbonates, carbonates, sulphates, chlorides, nitrates and possibly phosphates of calcium, magnesium, sodium and potassium with traces of iron, manganese and other substances. Total dissolved solids were determined by filtering the sample, drying the filtrate and determining its weight by difference. Total suspended solids (TSS) are the retained residue on a standard filter paper after filtration of a well mixed sample. The residue was dried at 103-105°C.Total suspended solids were determined by filtering the sample, drying the residue and determining its weight by difference. Dissolved Oxygen was determined immediately after collection of sample in the field by using modified Winkler's method. The samples were taken in to the ground glass stoppered 300 ml Corning make BOD bottles preventing any bubbles. 2 ml of saturated solution of Manganous sulphate and potassium iodide and potassium hydroxide solution were added after the settlement of the precipitate. It was dissolved in sulphuric acid and titrated against 0.025 N sodium thiosulphate solution. Biochemical Oxygen Demand refers to the quantity of oxygen required by bacteria and other micro-organisms in the biochemical degradation and transformation of organic matter (OM) under aerobic conditions. It is used as an index of organic pollution in water. More the oxidisable OM present in water, more the amount of oxygen required to degrade it biologically, hence more the BOD. BOD was evaluated by measuring the DO content of the sample before and after 5 days of incubation of 20°C in the dark. Preliminary dilution and aeration of samples were usually necessary to ensure that not all the oxygen is consumed during incubation. Chemical Oxygen Demand is the measure of oxygen required in oxidising the compounds present in water by means of chemical reactions involving oxidising substances such as potassium dichromate. Digestion of sample with $K_2Cr_2O_7$ and silver sulphate-sulphuric acid solution was done by closed reflux method using a Spectralab closed reflux condenser and the excess $K_2Cr_2O_7$ was titrated with standard ferrous ammonium sulphate (0.1 N) using 4 to 5 drops of ferroin indicator. Oxygen Absorbance was measured by acidifying the sample with sulphuric acid and adding potassium permanganate to absorb oxygen. The amount of oxygen absorbed by sample was calculated from the quantities of potassium permanganate added and the amount of left after the reaction by titrating with sodium-thiosulphate solution using starch as indicator. Total Kjeldahl Nitrogen was determined by Kjeldahl Distillation Method. The water/wastewater sample was digested with concentrated sulphuric acid (H_2SO_4), potassium sulphate (K_2SO_4) and mercuric sulphate ($HgSO_4$) as a catalyst. The digest was distilled with 40 per cent sodium hydroxide and the distillate was collected in 4 per cent boric acid with mixed indicator. The contents were then titrated with 0.01 N hydrochloric acid. Ammonical Nitrogen was determined by Nesslerization Method. The sample was adjusted to pH of 7.4 with phosphate buffer and distillated. The sample containing ammonia was treated with Nessler's reagent. The yellow brown colour produced was compared with that of standards. Nitrate Nitrogen was measured by Brucine Method. The sample containing nitrate ion reacts with Brucine-sulphanillic acid solution to form a yellow colour which was measured by spectrophotometer. Total Phosphate was estimated by Stannous Chloride Method. The sample was digested with concentrated HNO_3 and perchloric acid ($HClO_4$, 70-72 per cent), and neutralized with NaOH solution. The phosphate in neutralized and filtered sample reacted with ammonium molybdate and stannous chloride and developed blue colour was measured at 690 nm. Chloride was determined by Argentometric method involving titration of the sample against silver nitrate solution of 0.014 (N) strength using potassium chromate as an indicator. Sulphate was estimated by Turbidimetric method. The sample was treated with barium chloride and conditioning reagent. The sulphate ions were precipitated as barium sulphate crystals of uniform size in acid medium. Light absorbed by precipitate was measured by using a spectrophotometer at 420 nm. Alkalinity is the quantitative capacity of aqueous media to react with hydrogen ions. The alkalinity of natural water is normally due to the presence of bicarbonate, carbonate and hydroxide compounds of calcium, magnesium, sodium and potassium. Thus alkalinity may be expressed as total alkalinity or alkalinity due to industrial bases. Alkalinity was directly determined by titration with 0.02(N) H_2SO_4 using phenolphthalein and methyl orange indicator.

Hardness due to Calcium and Magnesium were determined by EDTA titrimetric method. For total hardness erichrome black-T was used as an indicator. For calcium hardness muroxide was used as an indicator. In this test, calcium and magnesium reacted with EDTA to form soluble complex and was indicated by the colour change of suitable indicators as mentioned above. Sodium and potassium in samples were measured at wavelengths of 589 and 769 nm respectively using a flame photometer. The sodium and potassium contents were expressed in mg/l after deducing the values from the standards curve. The pesticides namely DDT, BHC and endosulfan were extracted from water by liquid extraction methods and measured by Gas Chromatography (HP-5400, Hewlett Packard, USA). Heavy metals have the tendency to form complexes readily with the organic constituents. Thus, it is essential to destroy them by digestion with strong acids. After the preliminary acid digestion with strong acids, metals were determined by atomic absorption spectrometry. Acid eliminates interference from cyanides, nitrate, sulphide, sulphite, thiosulphate and thiocyanate. The heavy metals like cadmium, copper, iron, lead, nickel, chromium and zinc present in the water sample were analysed and determined by flame atomic absorption spectrophotometry by direct aspiration into air acetylene flame. 200 ml of well shaken water sample was digested with 2 ml of concentrated HNO_3 and evaporated gently to reduce it to 50 ml. The sample after filtration was stored in polythene bottles and analysed through an atomic absorption spectrophotometer (Model 703, Perkin and Elmer), provided with baseline corrections. The metal concentrations were calculated after appropriate blank correction and expressed in mg/l.

Bacteriological Examination of Water

Total coliforms and faecal coliforms were determined by using "Membrane filter technique for member of the coliforms group". The sampling of river water was done following standard methods (APHA, 1991) using sterilized top aluminium foil packed 300 ml capacity glass bottles. The samples were brought to the laboratory within eight hours of collection and were processed immediately. Taking sample volume of 10 ml, a serial dilution series was prepared in sterile buffered dilution water. A fixed volume of diluted sample was filtered with 0.45 μm in pore size sterile filter (Gelman, Millore and Whatman). For the filtration process, the sterilized filtration assembly (Paqualab System) was used. Immediately after filtration, the membrane filter was transferred by sterile forceps on to the solid M-FC agar (kept in petridishes), incubation was carried out by immersing the petridishes, sealed with lids and inverted inside incubation chamber at 44 ± 0.2°C for 24 hrs. After incubation, the faecal coliforms count was made on a filter with a 20-80 colony range. The faecal coliforms density was determined according to the following formula:

Faecal coliforms (CFU)/100 ml =100 × counted bluish colonies/Sample test volume in ml

CFU = colony forming unit

For total coliforms count, filtered membrane was transferred to solid M-Endo agar in petridishes and incubated at 37°C for 24 hrs. Total coliforms developed as darker metallic spleen shaped colonies were calculated as follows:

$$\text{Coliforms (CFU)/100ml} = \frac{100 \times \text{counted dark red metallic sheen colonies}}{\text{Total Sample test volume in ml}}$$

Assessment of Water Quality

The physico-chemical environment of water functions in one or more ways and influences the biotic components. Thus, it gives a picture of the environmental suitability of water for maintaining normal life. As such, the physico-chemical factor is important in elucidating some casual influences for the faunal composition and abundance of plankton, benthos and other aquatic life. Therefore, knowledge on physico-chemical characteristics of water is essential for proper utilization of riverine wealth (Kumar, 1997).

Monthly variation of physico-chemical factors in drain water and five river sampling points (averages of three year 1993-95) under observation has been shown in Tables. Two ways analysis of variance between stations and within monitoring months are given for all physico-chemical parameters. The microbial assessment of water, heavy metals and pesticides contents of the river and drain water are also presented in detail. The Water Quality Index of the river water has also been calculated.

Results and Discussion

Physico-chemical Characteristic of Drain and River Water

pH of Drain and River Water

pH is the term universally used to express the intensity of the acid or alkaline condition of a solution. It is the negative logarithm of the hydrogen-ion concentration in moles per litre. The pH of natural water, controlled by the carbon dioxide, bicarbonate equilibrium, usually ranges from 4.0 to 9.0. In majority of cases, natural waters exhibit slightly basic (pH >7) condition. It is due to the presence of bicarbonates and carbonates ions. Acidity affects the chemical and biological processes taking place in water in following manner: *(i)* Dissociation of organic and inorganic, molecules, thereby changing toxicity; *(ii)* Changing water solubility of compounds, also influencing toxicity; *(iii)* A lower pH may induce enhanced corrosion; *(iv)* Different biological species show optimum performance at different degrees of acidity, leading to shifts in species composition (D.de Zwart and Trivedy, 1994). pH of a raw water used for domestic purposes is important in that, it affects taste, corrosivity, efficiency of chlorination and treatment and processes such as coagulation etc. (Manivasakam, 1989).

Table DWM-1 reveals a pH range of 7.33 ± 0.28 (Nov.) to 7.5 ± 0.28 (Dec.) in Najafgarh drains, 7.2 ± 0.16 (Apr.) to 7.57 ± 0.17 (Jan.) Civil mill, 7.27 ± 0.12 (Aug.) to 7.57 ± 0.24 (Nov.) in Power house, 7.3 ± 0.05 (Feb.) to 7.63 ± 0.21 (Jul.) in Sen nursing home, 7.27 ± 0.19 (Aug.) to 7.5 ± 0.18 (May) in Barpulla and from 7.50 ± 0.30 (Dec.) to 8.18 ± 0.13 (Feb.) of Hindon river cut. Analysis of variance showed that significant difference exists in pH between different sampling drains ($F_1 = 18.86$, $P < 0.01$) as well as between different monitoring months ($F_2 = 2.85$, $P < 0.05$). It implies that the physico-chemical nature of effluents of different drains are not similar in nature.

Seasonal and annual variation of pH of the drains shows a minimum in summer (7.43 ± 0.17) and maximum in post-monsoon (7.45 ± 0.28) in Najafgarh drains, minimum in post-monsoon (7.35 ± 0.10) and maximum in summer (7.44 ± 0.19) in Civil mill, minimum in monsoon (7.33 ± 0.16) and maximum in post-monsoon (7.50 ± 0.19) in Power house, minimum in monsoon (7.37 ± 0.17) and maximum in winter (7.5 ± 0.23) in Barpulla drain, minimum in post-monsoon (7.52 ± 0.23) and maximum in winter (7.86 ± 0.26) in Hindon river cut. Annual averages of pH values were in order of Barpulla drain (7.49 ± 0.25) > Sen nursing home (7.43 ± 0.18) > Najafgarh (7.42 ± 0.21) > Power house (7.41 ± 0.19) > Civil mill (7.04 ± 0.18). Annual average pH values of Hindon river cut was higher (7.93 ± 0.32) in comparison to other five drains.

Table DWM-1: Monthly Variation of pH in Drain Water (Year 1993-1995)

Month	*NAJ-1*	*CIV-2*	*POW-3*	*SEN-4*	*BAR-5*	*HIN-6*
	Mean ± SD	*Mean ± SD*	*Mean ± SD*	*Mean ± SD*	*Mean ± SD*	*Mean ± SD*
Jan.	7.47 ± 0.17	7.40 ± 0.08	7.57 ± 0.24	7.47 ± 0.24	7.63 ± 0.26	7.84 ± 0.09
Feb.	7.33 ± 0.09	7.47 ± 0.05	7.28 ± 0.06	7.30 ± 0.00	7.50 ± 0.14	8.18 ± 0.13
Mar.	7.33 ± 0.12	7.57 ± 0.12	7.47 ± 0.12	7.37 ± 0.05	7.47 ± 0.31	8.12 ± 0.33
Apr.	7.37 ± 0.12	7.20 ± 0.26	7.33 ± 0.25	7.33 ± 0.09	7.53 ± 0.26	8.12 ± 0.34
May	7.60 ± 0.14	7.53 ± 0.05	7.47 ± 0.12	7.53 ± 0.12	7.75 ± 0.18	8.09 ± 0.20
Jun.	7.40 ± 0.16	7.47 ± 0.12	7.47 ± 0.09	7.33 ± 0.19	7.39 ± 0.31	8.13 ± 0.16
Jul.	7.47 ± 0.31	7.57 ± 0.19	7.40 ± 0.22	7.63 ± 0.21	7.43 ± 0.12	7.72 ± 0.28
Aug.	7.40 ± 0.28	7.27 ± 0.21	7.27 ± 0.12	7.47 ± 0.12	7.27 ± 0.19	8.01 ± 0.11
Sep.	7.37 ± 0.12	7.30 ± 0.08	7.33 ± 0.09	7.33 ± 0.05	7.40 ± 0.14	7.81 ± 0.12
Oct.	7.43 ± 0.09	7.33 ± 0.12	7.40 ± 0.14	7.40 ± 0.14	7.39 ± 0.08	7.95 ± 0.42
Nov.	7.33 ± 0.26	7.37 ± 0.19	7.57 ± 7.24	7.53 ± 0.21	7.60 ± 0.29	7.50 ± 0.30
Dec.	7.50 ± 0.28	7.33 ± 0.12	7.40 ± 0.08	7.40 ± 0.14	7.50 ± 0.22	7.71 ± 0.11

Two -way ANOVA Test

Significant difference in pH between monitoring drains, F_1=18.86 (P<0.01).
Significant difference in pH between monitoring months, F_2=2.85 (P< 0.05).

Table RWM-1 shows monthly averages (Mean ± SD) in pH variation at the five sampling stations along the river Yamuna at Delhi stretch. The pH ranged between 7.71 ± 0.26 (July) to 8.23 ± 0.12 (Feb.) at the site WAZ-1, 7.27 ± 0.19 (Feb.) to 7.63 ± 0.12 (Apr.) at ISB-2, 7.40 ± 0 (May) to 7.63 ± 0.21 (Oct.) at ITO, 7.30 ± 0.08 (Feb.) to 7.60 ± 0.16 (Sep) at the site NIZ-4 and 7.37 ± 0.17 (Feb.) to 7.87 ± 0.25 (Sep) at sampling station OKH-5. Analysis of variance shows a significant difference in pH between sampling stations (F_1=18.86, P<0.01) as well as seasons (F_2=0.85, P<0.05). This clearly signifies the effect of pollutants on amount of hydrogen ions along the Yamuna at Delhi stretch.

Seasonal and annual variation in pH at all sampling stations varied from 7.71 ± 0.24 (monsoon) to 8.16 ± 0.15 (winter) at WAZ-1, 7.44 ± 0.24 (winter) to 7.48 ± 0.18 (summer) at ISB-2, 7.37 ± 0.27 (winter) to 7.65 ± 0.17 (pre-monsoon) at site ITO-3, 7.37 ± 0.16 (winter) to 7.57 ± 0.14 (post-monsoon) at NIZ-4 and 7.51 ± 0.21 (winter) to 7.64 ± 0.24 (monsoon) at monitoring site OKH-5. The annual average pH values were in order of WAZ-1 (8.0) > OKH-5 (7.58) > ITO-3 (97.50) > ISB-2 (7.46) > NIZ-4 (7.46) in all river monitoring points.

In most of time the river water at Wazirabad barrage (unpolluted stretch) shows alkaline pH (7.7 to 8.20). Murugesan *et al.* (1994) reported an alkaline pH of water in river Tampraparni with minimum pH during August and maximum during May. A slight to moderate alkaline nature of most of the Indian river waters were reported by Vass *et al.* (1977), Rajesh (1981), Mitra (1982), Bhargava (1985), Venkateswarlu (1986) in river Jhelum, Datta *et al.* (1988) in Hoogly river, Sinha (1988) in river Damodar, Sharma *et al.* (1990), Sarwar and Refat (1991) in Doodhganga river, Isaraili and Ahmed (1993), Mishra (1999) in Brahmani river.

Table RWM-1: Monthly Variation of pH in River Water (Year 1993-1995)

Month	*WAZ-1*	*ISB-2*	*ITO-3*	*NIZ-4*	*OKH-5*
	Mean ± SD	*Mean ± SD*	*Mean ± SD*	*Mean ± SD*	*Mean ± SD*
Jan.	8.20 ± 0.16	7.47 ± 0.25	7.43 ± 0.37	7.37 ± 0.21	7.57 ± 0.09
Feb.	8.23 ± 0.12	7.27 ± 0.19	7.33 ± 0.09	7.30 ± 0.08	7.37 ± 0.12
Mar.	8.10 ± 0.08	7.50 ± 0.08	7.63 ± 0.12	7.30 ± 0.22	7.63 ± 0.17
Apr.	8.17 ± 0.12	7.63 ± 0.12	7.43 ± 0.05	7.43 ± 0.12	7.43 ± 0.05
May	8.17 ± 0.17	7.50 ± 0.08	7.40 ± 0.00	7.47 ± 0.09	7.57 ± 0.17
Jun.	7.95 ± 0.20	7.40 ± 0.14	7.47 ± 0.12	7.50 ± 0.08	7.50 ± 0.08
Jul.	7.71 ± 0.26	7.40 ± 0.14	7.47 ± 0.17	7.43 ± 0.09	7.50 ± 0.16
Aug.	7.76 ± 0.30	7.47 ± 0.12	7.53 ± 0.12	7.53 ± 0.09	7.57 ± 0.05
Sep.	7.76 ± 0.09	7.43 ± 0.26	7.53 ± 0.09	7.60 ± 0.16	7.87 ± 0.25
Oct.	7.83 ± 0.05	7.47 ± 0.05	7.63 ± 0.21	7.57 ± 0.09	7.63 ± 0.05
Nov.	8.03 ± 0.12	7.53 ± 0.12	7.67 ± 0.12	7.57 ± 0.17	7.63 ± 0.24
Dec.	8.13 ± 0.19	7.43 ± 0.33	7.43 ± 0.26	7.43 ± 0.12	7.63 ± 0.26

Two-way ANOVA Test

Significant difference in pH between river monitoring stations, F_1=18.86 (P<0.01).
Significant difference in pH between monitoring months, F_2=2.85 (P< 0.05).

In the present study, pH values were comparatively low at site ISB-2, ITO-3, NIZ-4 and OKH-5 in comparison to site WAZ-1, which might be due to regular additions of slightly acidic (domestic and industrial mixed) effluents from major and minor drains. Singh and Bhowmik (1985) also reported similar result in case of the river Ganga and Kumar (1999) in the river Mayurakshi.

Conductivity of Drain and River Water

Conductivity is a measure of ability of a water sample to conduct an electric current proportional to the ionic strength of the water. This mostly depends upon the nature of various dissolved ionised substances, their actual relative concentrations and temperature. The standard unit of electrical conductivity is seimence per metre (1 s/m = 10 milli-mhos/cms). In order to avoid the fractions of results in small decimal fractions, a smaller unit, the milli siemence per metre (1 MS = 10 micro-mhos/cm) is generally used.

Monthly average values of conductivity (μ mhos/cm) of five drains and one river cut (Table DWM-2) ranged from 666.7 ± 98.85 to 1332.3 ± 161.94 in the NAJ-1, 852.9 ± 59.16 to 1339 ± 379.12 in CIV-2, 808.3 ± 125.19 to 1330.3 ± 330.52 in POW-3, 1019.3 ± 28.9 to 2023.3 ± 968.5 in SEN-4, 1002 ± 363 to 1326 ± 62.36 in BAR-5 and from 396.3 ± 129.6 to 1185.3 ± 131.01 in HIN-6 (Hindon river cut). Analysis of variance shows a significant difference in conductivity between the five drains (F_1=19.2, P<0.01) as well as between monitoring months, (F_2=2.74, P<0.05). This clearly signifies that the pollution load (in term of load of minerals) of different drains was not same.

Seasonal average values of conductivity varied from 714.56 ± 108.74 (winter) to 994.89 ± 274.04 (monsoon) in NAJ-1, 899.44 ± 154.12 (winter) to 1110.107 ± 198.57 (post-monsoon) in CIV-2, 972.33 ± 163.98 (winter) to 1170.00 ± 208.19 (summer) in POW-3, 1303.89 ± 281.32 (monsoon) to 1468.00 ± 585.01 (winter) in SEN-4, 1087.78 ± 308.53 (monsoon) to 1265.58 ± 123.09 (summer) in BAR-5, and from 500.67 ± 280.46 (monsoon) to 1032.33 ± 186.02 (post-monsoon) in HIN-6 (Hindon river cut). Annual average values of conductivity (μ mhos/cm) were in order of 1453.33 > 1201.03 > 1048.7 > 1032.03 > 842.58 > 837.22 on SEN-4, BAR-5, POW-3, CIV-2 NAJ-1 and HIN-6 respectively.

Table DWM-2: Monthly Variation of Conductivity (μ mhos/cm) in Drain Water (Year 1993-1995)

Month	*NAJ-1*	*CIV-2*	*POW-3*	*SEN-4*	*BAR-5*	*HIN-6*
	Mean ± SD	*Mean ± SD*	*Mean ± SD*	*Mean ± SD*	*Mean ± SD*	*Mean ± SD*
Jan.	666.7 ± 98.9	917.3 ± 188.6	933.0 ± 75.0	1366.7 ± 378.1	1174.0 ± 55.7	1027.7 ± 123.0
Feb.	816.3 ± 82.2	884.6 ± 53.5	815.3 ± 168.0	1670.7 ± 905.9	1148.3 ± 74.4	1010.0 ± 99.0
Mar.	805.0 ± 107.22	1128.3 ± 106.4	1096.0 ± 53.8	2023.3 ± 968.5	1294.0 ± 99.3	802.3 ± 448.6
Apr.	862.7 ± 90.0	926.7 ± 649.0	1330.3 ± 330.5	1531.7 ± 331.0	1237.7 ± 128.4	863.7 ± 68.0
May	790.0 ± 28.3	1044.01 ± 179.2	1028.0 ± 17.2	1380.0 ± 269.2	1204.7 ± 146.4	1012.7 ± 106.6
Jun.	875.3 ± 80.0	1339.1 ± 379.1	1225.7 ± 80.6	1590.6 ± 289.7	1326.0 ± 62.4	1114.3 ± 177.7
Jul.	1332.3 ± 161.9	1037.0 ± 137.0	1113.3 ± 72.5	1527.3 ± 210.3	1142.7 ± 117.2	396.3 ± 129.01
Aug.	874.6 ± 27.1	1059.7 ± 265.3	395.0 ± 132.7	1365.0 ± 240.2	1002.0 ± 363.5	598.3 ± 237.3
Sep.	777.6 ± 151.1	852.7 ± 59.1	808.3 ± 125.1	1019.3 ± 28.9	1118.7 ± 358.3	507.3 ± 256.7
Oct.	753.3 ± 108.1	974.7 ± 74.4	1064.7 ± 61.1	1287.0 ± 89.9	1238.3 ± 51.7	636.7 ± 194.2
Nov.	785.3 ± 20.3	1198.3 ± 229.3	1058.7 ± 108.2	1356.7 ± 26.5	1270.7 ± 209.5	1185.3 ± 131.0
Dec.	771.7 ± 129.5	1022.3 ± 103.6	1140.0 ± 63.2	1321.7 ± 94.0	1255.3 ± 52.0	863.3 ± 69.4

Two-way ANOVA Test

Significant difference in conductivity between monitoring drains, F_1=20.2 (P<0.01).

Significant difference in conductivity between sampling months, F_2=2.73 (P< 0.05).

Monthly average values of conductivity in river water (Table RWM-2) were varied from 199.33 ± 18.21 (July) to 489 ± 313 (May) at WAZ-1, 325 ± 26.94 (Sept) to 767.67 ± 144.95 (March) at ISB-2, 271.33 ± 6.6 (Aug.) to 802.33 ± 26.66 (Feb.) at ITO-3, 297 ± 6.38 (Aug.) to 814.67 ± 38.9 (Feb.) 297 ± 6.38 (Aug.) to 814.67 ± 38.9 (Feb.) at NIZ-4 from 283.33 ± 65.62 (Aug.) to 660 ± 105.57 (Dec.) at OKH-5. Analysis of variance showed a significant difference in conductivity between the monitoring stations (F_1=20.20, P<0.01) and also in between the sampling months (F_2=2.73, P<0.05).

Seasonal conductivity values ranged from 234.11 ± 31.93 to 419.89 ± 105.76 at WAZ-1, 373.44 ± 58.91 to 675.33 ± 105.29 at ISB-2, 306.78 ± 50.42 to 790.11 ± 70.37 at ITO-3, 314.22 ± 57.47 to 796.70 ± 72.58 at NIZ-4, 308.33 ± 68.38 to 629.33 ± 80.65 at OKH-5. Annual averages were in order of 633.33 at NIZ-4 > 633.19 at ITO-3 > 595.11 at ISB-2 > 522.64 at OKH-5 355.0, at WAZ-1.

The winter season showed the highest seasonal average values at all the sampling stations followed by post-monsoon, summer and minimum in monsoon season. This could be associated with the lower water volume in the river during winter, coupled with wastewater from a number of drains entering the river Yamuna at Delhi stretch. There was no remarkable difference between the summer and post-monsoon values because almost the same volume of water was discharged from Wazirabad barrage, where as monsoon showed lower conductivity due of dilution. Imevbore (1978) and Adebisi (1980), Mishra (1999) stated the inverse relation of conductivity with water level which corroborates our findings.

Table RWM-2: Monthly Variation of Conductivity (µ mhos/cm) in River Water (Year 1993-95)

Month	*WAZ-1* *Mean ± SD*	*ISB-2* *Mean ± SD*	*ITO-3* *Mean ± SD*	*NIZ-4* *Mean ± SD*	*OKH-5* *Mean ± SD*
Jan.	442.67 ± 94.24	605.00 ± 71.92	768.00 ± 103.96	742.33 ± 81.94	577.33 ± 76.36
Feb.	435.67 ± 43.86	764.33 ± 36.75	802.33 ± 26.66	814.67 ± 38.90	656.00 ± 59.26
Mar.	437.33 ± 136.80	767.67 ± 144.95	780.33 ± 140.89	803.67 ± 155.24	642.33 ± 215.09
Apr.	300.00 ± 59.30	690.00 ± 155.13	661.3 ± 175.32	700.33 ± 174.90	465.67 ± 148.45
May	489.00 ± 312.96	654.33 ± 323.04	788.33 ± 149.56	763.33 ± 117.05	476.00 ± 66.10
Jun.	208.33 ± 27.74	667.33 ± 65.57	719.33 ± 129.84	650.67 ± 107.53	434.33 ± 49.1
Jul.	199.33 ± 18.21	242.33 ± 61.67	330.67 ± 56.10	336.00 ± 65.49	343.33 ± 65.62
Aug.	249.00 ± 24.54	371.00 ± 30.63	271.33 ± 6.60	297.00 ± 6.38	283.33 ± 22.88
Sep.	254.00 ± 17.28	325.00 ± 26.94	318.33 ± 49.75	309.67 ± 69.19	298.33 ± 85.14
Oct.	376.67 ± 91.77	565.67 ± 183.84	610.00 ± 128.89	620.67 ± 149.97	597.70 ± 85.83
Nov.	461.00 ± 138.74	681.00 ± 122.63	752.67 ± 138.12	767.67 ± 135.35	837.30 ± 132.55
Dec.	484.00 ± 11	625.67 ± 119.59	771.67 ± 90.40	794.00 ± 108.71	660.00 ± 105.57

Two-way ANOVA Test

Significant difference in conductivity between river monitoring stations, F_1=20.2 (P<0.01).
Significant difference in conductivity between monitoring months, F_2=2.73 (P< 0.05).

Total Solids of Drain and River Water

Solids refer to the matter that remains as residue upon evaporation after drying at a definite temperature. "Total solids" is the residue that includes both dissolved solids and suspended solids. The excess of TS creates imbalance for aquatic life.

Monthly variation (Mean ± SD) of TS content (mg/l) are given in Table DWM-3. The minimum and maximum TS content were 555.0 ± 9.9 (Apr.) and 986.33 ± 136.02 (July) in NAZ-1 drain, 652.3 ± 141.2 (Jul.) and 1016 ± 372.03 (Aug.) in CIV-2, 718.67 ± 293.21 (May) and 3542.0 ± 358.27 (Nov.) in POW-3, 598.33 ± 425.25 (Oct.) and 1198.0 ± 165.89 (Mar.) in SEN-4, 732.67 ± 52.25 (Jul.) to 879.33 ± 79.93 (May) in BAR-5 and from 585.33 ± 90.49 (Jul.) to 822.3 ± 67.61 (March) in HIN-6 (Hindon river cut). Analysis of variance shows a significant difference in TS content between five drains (F_1=22.9, P<0.01) as well as between sampling months (F_2=3.25, P<0.05).

Seasonal averages varied from as low as 563.83 ± 59.22 (pre-monsoon) to 825.89 ± 172.42 (monsoon) in CIV-2, 843.08 ± 138.71 (summer) to 2235.33 ± 2988.20 (post-monsoon) in POW-3, 1036.58 ± 170.61 (summer) to 1168.78 ± 214.49 (monsoon) in SEN-4, 743.56 ± 116.74 (monsoon) to 825.25 ± 87.23 (summer) in BAR-5 and 671.0 ± 94.39 (winter) to 746.50 ± 120.72 (post-monsoon) in HIN-6.

Annual averages in different drains were in the order of 1096.22 ± 1327.61 > 1072.92 ± 168.22 > 844.53 ± 213.85 > 803.53 ± 97.77 > 709.22 ± 115.9 > 655.89 ± 149.59 in POW-3, SEN -4, CIV-3, BAR-5, HIN-6 and NAJ-1 drain respectively.

Monthly averages (Mean ± SD) of total solids and the results of analysis of variance have been presented in Table RWM-3. The minimum and maximum TS contents were 137.67 ± 32.15 (Jan.) and 954 ± 481.08 (Aug.) at WAZ-l, 289.67 ± 177.90 (Apr.) 1270.33 ± 116.74 (Aug.) at ISB-2, 396.67 ± 74.7 (Oct.) and 1117.76 ± 51.8 (Aug.) at NIZ-4, 286.33 ± 58.54 (Apr.) and 1225.23 ± 225.41 (Aug.) at OKH-

5. Analysis of variance shows a significant difference between sampling stations (F_1=22.94, P<0.01) as well as between sampling months (F_2=2.8, P<0.05).

Table DWM-3: Monthly Variation in Total Solids (mg/l) of Drain Water (Year 1993-1995)

Month	NAJ-1 Mean ± SD	CIV-2 Mean ± SD	POW-3 Mean ± SD	SEN-4 Mean ± SD	BAR-5 Mean ± SD	HIN-6 Mean ± SD
Jan.	586.0 ± 100.4	938.0 ± 262.3	854.0 ± 384.6	1046.6 ± 33.9	846.6 ± 32.0	702.0 ± 101.5
Feb.	654.3 ± 42.0	834.0 ± 21.4	921.3 ± 495.9	1173.0 ± 186.1	839.3 ± 92.4	724.0 ± 37.5
Mar.	588.3 ± 31.7	897.0 ± 90.2	992.6 ± 555.7	1198.0 ± 185.8	879.3 ± 74.6	822.3 ± 67.6
Apr.	555.0 ± 9.9	984.6 ± 336.6	857.3 ± 467.6	1008.6 ± 128.1	841.3 ± 76.8	665.0 ± 65.8
May	556.3 ± 102.3	829.6 ± 136.6	718.6 ± 293.2	1049.6 ± 199.6	768.3 ± 79.9	749.0 ± 66.8
Jun.	628.0 ± 84.5	741.0 ± 131.8	803.6 ± 372.8	1058.00 ± 146.9	812.0 ± 77.2	741.6 ± 112.2
Jul.	986.3 ± 136.0	652.3 ± 141.2	789.3 ± 438.4	651.3 ± 467.3	732.6 ± 118.4	585.3 ± 90.4
Aug.	745.3 ± 162.8	1016.0 ± 372.0	950.6 ± 404.4	793.3 ± 597.8	741.3 ± 131.1	785.0 ± 55.6
Sep.	746.0 ± 74.2	813.3 ± 80.5	881.0 ± 310.1	781.6 ± 564.0	756.6 ± 961.7	645.6 ± 150.9
Oct.	590.6 ± 74.4	734.6 ± 77.9	834.3 ± 374.3	598.3 ± 425.2	791.3 ± 21.3	647.6 ± 114.5
Nov.	613.6 ± 40.4	881.3 ± 173.2	3542.0 ± 358.2	731.3 ± 517.4	808.3 ± 95.6	820.3 ± 76.5
Dec.	620.6 ± 131.1	839.3 ± 93.8	1000.6 ± 443.1	821.6 ± 579.5	825.0 ± 87.9	622.6 ± 63.3

Two-way ANOVA Test

Significant difference in total solids content between monitoring drains, F_1=22.97 (P<0.01).
Significant difference in total solids content between sampling months, F_2=3.25 (P< 0.05).

Table RWM-3: Monthly Variation of Total Solids (mg/l) in River Water (Year 1993- 1995)

Month	WAZ-1 Mean ± SD	ISB-2 Mean ± SD	ITO-3 Mean ± SD	NIZ-4 Mean ± SD	OKH-5 Mean ± SD
Jan.	409.0 ± 91.2	527.3 ± 7.6	546.0 ± 49.7	532.3 ± 30.2	410.0 ± 57.1
Feb.	254.3 ± 2.1	573.0 ± 65.0	527.0 ± 48.0	554.7 ± 24.6	474.3 ± 20.8
Mar.	300.0 ± 108.1	560.3 ± 103.0	508.3 ± 81.1	508.7 ± 80.7	432.0 ± 121.6
Apr.	224.0 ± 42.5	289.6 ± 177.9	393.0 ± 141.7	406.0 ± 117.1	286.3 ± 58.5
May	299.0 ± 156.5	455.3 ± 118.4	429.0 ± 84.7	507.3 ± 19.6	294.0 ± 10.0
Jun.	137.7 ± 32.1	483.6 ± 169.3	420.7 ± 32.7	406.0 ± 35.4	288.3 ± 16.8
Jul.	286.0 ± 59.5	707.6 ± 189.4	776.7 ± 411.5	522.3 ± 229.2	573.0 ± 291.7
Aug.	945.0 ± 481.0	1270.3 ± 116.7	1703.0 ± 496.3	1117.7 ± 51.8	1225.3 ± 225.4
Sep.	785.3 ± 358.1	1023.6 ± 482.6	1113.3 ± 576.3	782.3 ± 311.4	829.0 ± 456.0
Oct.	383.0 ± 142.3	500.3 ± 167.3	383.0 ± 61.6	396.7 ± 74.7	358.0 ± 38.6
Nov.	323.3 ± 70.7	456.3 ± 40.9	468.7 ± 93.6	491.3 ± 92.6	612.6 ± 179.2
Dec.	358.0 ± 65.0	497.3 ± 43.7	527.3 ± 63.9	512.3 ± 78.2	426.0 ± 54.8

Two-way ANOVA Test

Significant difference in total solids content between monitoring stations, F_1=22.94 (P<0.01).
Significant difference in total solids content between monitoring months, F_2=2.8 (P< 0.05).

Seasonal averages ranged from a minimum 240.17 ± 119.85 to the maximum 675 ± 448.91 at WAZ-l, 3743.44 ± 58.91 to 675.33 ± 105.29 at ISB-2, 306.78 ± 50.42 to 790.78 ± 72.58 at NIZ-4, 308 ± 68.38 to 629.33 ± 89.65 at OKH-5 (Okhla barrage). Exceptionally high values were recorded at all the stations of river in monsoon due to the top soil at the river bank and catchment area were carried by the stream downwards. Annual averages at different stations were in order of 649.6 > 612.08 > 561.47 > 517.42 > 392.8 at station ITO-3, ISB-2, NIZ-4, OKH-5 and WAZ-1 respectively.

Total Suspended Solids of Drain and River Water

No specific limit has been prescribed for suspended solids in water used for domestics and industrial purpose, but it indirectly controlled by the limits of turbidity (5 units). Disposal of sewage and industrial effluents contributes suspended matter to river and streams. The Indian Standards Institute has specified a maximum limit of 30 mg/l for suspended solids in wastewater dischargeable into water courses. Suspended solids are very objectionable in rivers for many reasons. Suspended solids containing much organic matter may cause putrefaction and consequently the stream may become devoid of dissolved oxygen.

The monthly averages (Mean ± SD) of total suspended solids and results of analysis of variance in drain water have been depicted in Table DWM-4. The TSS content varied from 128.3 ± 30.2 to 276.3 ± 175.7 in NAJ-1, 130 ± 160.3 to 499.3 ± 301.3 in CIV-2 167.0 ± 39.7 to 423.0 ± 148.3 in POW-3, 242.7 ± 70.1 to 526 ± 243.5 in SEN-4, 64.7 ± 28.2 to 194.3 ± 92.4 in BAR-5 and 73.3 ± 23.6 to 472.7 ± 76.4 in HIN-6 (Hindon river cut). Analysis of variance showed a significant difference in TSS content between sampling stations (F_1=20.7, P<0.01) as well as between monitoring months (F_2=2.53, P<0.05).

Table DWM-4: Monthly Variation of Suspended Solids (mg/l) in Drain Water (Year 1993-1995)

Month	*NAJ-1*	*CIV-2*	*POW-3*	*SEN-4*	*BAR-5*	*HIN-6*
	Mean ± SD	*Mean ± SD*	*Mean ± SD*	*Mean ± SD*	*Mean ± SD*	*Mean ± SD*
Jan.	167.0 ± 65.3	257.7 ± 74.2	219.7 ± 12.4	400.0 ± 107.4	76.7 ± 17.7	97.7 ± 38.9
Feb.	168.3 ± 31.7	268.7 ± 137.2	274.7 ± 42.2	511.7 ± 142.9	113.3 ± 65.6	104.7 ± 6.2
Mar.	153.0 ± 9.4	221.0 ± 119.1	282.7 ± 58.0	470.3 ± 185.4	87.7 ± 17.9	131.7 ± 2.8
Apr.	128.3 ± 30.2	360.3 ± 85.8	326.3 ± 86.9	248.3 ± 117.8	64.7 ± 28.2	141.7 ± 18.6
May	146.3 ± 80.5	298.7 ± 84.2	167.0 ± 39.7	242.7 ± 70.1	119.7 ± 35.2	126.0 ± 11.4
Jun.	209.7 ± 43.3	130.0 ± 60.3	199.3 ± 37.3	308.3 ± 77.7	194.3 ± 92.4	73.3 ± 23.8
Jul.	276.3 ± 75.7	145.3 ± 86.4	248.0 ± 76.7	279.3 ± 36.5	113.7 ± 29.8	316.7 ± 20.5
Aug.	210.3 ± 96.9	499.3 ± 301.6	423.0 ± 48.3	526.0 ± 243.5	189.0 ± 99.3	472.7 ± 76.4
Sep.	224.3 ± 48.3	275.0 ± 88.3	386.0 ± 67.3	517.3 ± 170.7	191.0 ± 103.3	309.3 ± 75.0
Oct.	165.3 ± 33.1	228.0 ± 43.3	242.7 ± 36.3	203.3 ± 69.7	98.3 ± 24.8	204.0 ± 8.6
Nov.	142.7 ± 46.6	200.7 ± 10.1	221.3 ± 22.5	222.3 ± 67.4	113.3 ± 26.1	114.3 ± 27.5
Dec.	145.7 ± 6.6	252.3 ± 24.1	303.0 ± 32.0	355.3 ± 68.0	112.7 ± 53.4	96.0 ± 33.5

Two-way ANOVA Test

Significant difference in suspended solids content between monitoring drains, F_1=20.76 (P<0.01).
Significant difference in suspended solids content between sampling months, F_2=2.53 (P<0.05).

Seasonal TSS content ranged from 141.5 ± 361 to 237 ± 67 in NAJ-1, 228.17 ± 50.03 to 306.65 ± 82 in CIV-2, 243.83 ± 119.84 to 352.33 ± 79.6 in POW-23, 317.0 ± 125.37 to 452.56 ± 140.5 in SEN-4, 87.11 ± 46.11 to 164.56 ± 63.8 in BAR-5, and 89.67 ± 30.90 to 366.22 ± 89.8 in HIN-6. Yearly averages of TSS in different drains were in order of SEN-4 (357.08) > POW-4 (274.47) > CIV-2 (261.42) > HIN-6 (179.75) > NAJ-1 (178.11) > BAR-5 (122.86).

Table RWM-4: Monthly Variation of Suspended Solids (mg/l) in River Water (Year 1993-95)

Month	*WAZ-1* *Mean ± SD*	*ISB-2* *Mean ± SD*	*ITO-3* *Mean ± SD*	*NIZ-4* *Mean ± SD*	*OKH-5* *Mean ± SD*
Jan.	38.3 ± 13.1	108.3 ± 36.8	56.7 ± 37.3	60.7 ± 18.0	44.3 ± 20.4
Feb.	35.6 ± 28.6	64.3 ± 12.0	66.3 ± 16.2	54.3 ± 23.8	70.0 ± 35.8
Mar.	35.0 ± 31.9	106.0 ± 47.1	50.0 ± 16.3	38.3 ± 6.2	40.0 ± 15.6
Apr.	31.7 ± 22.4	95.0 ± 8.2	37.3 ± 14.7	29.0 ± 4.2	29.7 ± 4.5
May	23.3 ± 12.0	54.3 ± 35.7	40.7 ± 18.8	53.3 ± 16.7	30.3 ± 7.6
Jun.	19.3 ± 13.3	128.3 ± 146.2	30.0 ± 4.2	33.7 ± 11.9	41.7 ± 26.2
Jul.	102.7 ± 3.8	464.7 ± 208.7	578.0 ± 437.1	300.0 ± 219.2	353.0 ± 314.6
Aug.	789.7 ± 540.9	790.0 ± 497.7	1536.3 ± 489.5	930.3 ± 60.8	1055.3 ± 210.0
Sep.	628.7 ± 372.0	806.7 ± 456.8	938.3 ± 599.0	523.0 ± 311.3	649.3 ± 495.6
Oct.	139.0 ± 166.3	186.0 ± 135.4	60.0 ± 12.7	56.3 ± 13.6	37.3 ± 22.4
Nov.	21.7 ± 14.4	63.0 ± 25.8	39:7 ± 19.7	32.3 ± 7.4	123.3 ± 141.9
Dec.	22.7 ± 11.6	55.3 ± 20.0	37.0 ± 20.8	24.3 ± 12.0	33.0 ± 10.2

Two-way ANOVA Test

Significant difference in suspended solids content between river monitoring stations, F_1=32.84 (P<0.01).

Significant difference in suspended solids content between monitoring months, F_2=2.17 (P<0.05).

Seasonal TSS contents ranged from a minimum 36.22 ± 19.29 (winter) to a maximum of 505.0 ± 465.98 (Monsoon) at WAZ-1, 75.56 ± 34.31 (Winter) to 687.11 ± 437.52 (Monsoon) at ISB-2, 39.5 ± 16.28 (Summer) to 1017.56 ± 647.58 (Monsoon) at ITO-3, 98.58 ± 14.23 (Summer) to 584.44 ± 343.01 (Monsoon) at NIZ-4, 35.42 ± 16.79 (Summer) to 685.89 ± 460.92 (Monsoon) at OKH-5. Annual averages at different stations were in order of 289.19 > 243.5 >208.94 > 177.97 > 159.31 at ITO-3, ISB-2, OKH-5, NIZ-4 and WAZ-1 respectively.

Exceptionally high values were recorded at all the monitoring stations of river in monsoon months because the top soil of sandy river bank and cultivated fields of catchment area were carried by the river stream down water. In summer, the TSS values were low which may be due settling of suspended matter on the river bed, because of lean flow.

Total Dissolved Solids of Drain and River Water

In natural water the dissolved solids exist mainly due to bicarbonates, carbonates, sulphates of calcium, magnesium, sodium and potassium with traces of iron, Manganese and other substances. The amount of dissolved solids present in water is a consideration for its suitability for domestic use. Presence of dissolved solids in industrial water is undesirable for many reasons. They form scales, cause foaming in boiler, accelerate corrosion and interfere with the colour and taste of the finished products. According to the 1962, USPHS, drinking water standards the permissible limit for total dissolved solids is 500 mg/l.

Monthly variations (Mean ± SD) of TDS content (mg/l) in river water are given in Table DWM-5. Monthly averages ranged from a minimum of 419 ± 73.27 to maximum of 535 ± 66.38 in NAZ-1, 507 ± 3.24 to 680.67 ± 119.4 in ISB-2, 495.0 ± 47.08 to 709.67 ± 91.51 in POW-3, 498.67 ± 148.95 to

832.00 ± 15.20 in SEN-4, 552.33 ± 199.74 to 791.67 ± 57.20 in BAR-5 and 266.00 ± 12.47 to 668.33 ± 276.71 in HIN-6 (Hindon river cut). Analysis of variance (Table DWM-5) shows highly significant difference in TDS content between different drains (F_1=19.1, P<0.01) as well as between sampling months (F_2=2.81, P<0.05).

Table DWM-5: Monthly Variation of Dissolved Solids (mg/l) in Drain Water (Year 1993-1995)

Month	NAJ-1 Mean ± SD	CIV-2 Mean ± SD	POW-3 Mean ± SD	SEN-4 Mean ± SD	BAR-5 Mean ± SD	HIN-6 Mean ± SD
Jan.	419.0 ± 73.3	679.7 ± 227.7	634.3 ± 84.7	695.7 ± 195.2	770.0 ± 15.3	604.3 ± 208.8
Feb.	486.0 ± 16.0	565.3 ± 47.6	646.7 ± 52.3	498.7 ± 148.9	726.0 ± 30.3	619.3 ± 253.0
Mar.	435.3 ± 39.3	676.0 ± 66.2	709.7 ± 91.5	656.1 ± 308.6	791.7 ± 57.3	684.7 ± 245.6
Apr.	426.7 ± 39.9	588.3 ± 178.4	708.0 ± 159.7	712.3 ± 35.9	693.0 ± 66.3	560.3 ± 210.2
May	430.0 ± 22.8	531.0 ± 72.8	551.7 ± 35.2	779.6 ± 158.9	348.7 ± 70.3	623.0 ± 218.4
Jun.	418.3 ± 44.8	611.0 ± 131.8	572.0 ± 38.0	716.0 ± 111.7	617.7 ± 15.7	668.3 ± 276.7
Jul.	710.0 ± 124.0	507.0 ± 73.2	550.3 ± 26.8	705.3 ± 95.8	619.0 ± 92.7	266.0 ± 12.5
Aug.	535.0 ± 66.4	516.7 ± 80.9	528.7 ± 30.7	744.0 ± 39.6	552.3 ± 199.7	312.3 ± 61.3
Sep.	488.3 ± 98.0	538.3 ± 39.9	495.0 ± 47.0	734.3 ± 49.6	569.0 ± 195.2	336.3 ± 120.5
Oct.	425.3 ± 42.0	515.7 ± 34.1	595.0 ± 78.0	733.7 ± 21.7	693.0 ± 8.2	443.7 ± 171.1
Nov.	467.7 ± 12.1	680.7 ± 119.4	647.3 ± 61.3	832.0 ± 15.3	695.0 ± 119.5	706.0 ± 308.7
Dec.	479.0 ± 136.0	587.0 ± 72.6	697.7 ± 65.7	810.7 ± 72.0	712.3 ± 37.2	530.0 ± 192.3

Two-way ANOVA Test

Significant difference in dissolved solids content between monitoring drains, F_1=19.15 (P<0.01).
Significant difference in dissolved solids content between sampling months. F_2=2.81 (P<0.05).

Table RWM-5: Monthly Variation of Dissolved Solids (mg/l) in River Water (Year 1993-95)

Month	WAZ-1 Mean ± SD	ISB-2 Mean ± SD	ITO-3 Mean ± SD	NIZ-4 Mean ± SD	OKH-5 Mean ± SD
Jan.	370.7 ± 99.9	419.0 ± 55.3	477.3 ± 65.4	471.7 ± 29.4	36.7 ± 42.2
Feb.	218.7 ± 28.8	508.7 ± 55.4	460.7 ± 32.7	500.3 ± 20.4	404.3 ± 55.7
Mar.	265.0 ± 76.2	454.3 ± 86.1	458.3 ± 76.7	476.3 ± 89.4	392.0 ± 106.5
Apr.	192.3 ± 25.6	371.7 ± 126.4	205.7 ± 138.8	377.0 ± 113.3	256.7 ± 54.4
May	275.7 ± 165.8	401.0 ± 116.4	388.3 ± 70.8	454.0 ± 34.8	263.7 ± 4.5
Jun.	118.3 ± 19.3	355.3 ± 58.9	309.7 ± 29.3	372.3 ± 25.2	246.7 ± 29.6
Jul.	183.3 ± 61.3	243.0 ± 97.3	198.7 ± 38.9	222.3 ± 68.2	220.0 ± 62.8
Aug.	164.3 ± 25.2	218.7 ± 102.3	166.7 ± 10.4	187.3 ± 16.5	170.0 ± 20.6
Sep.	156.7 ± 17.0	183.7 ± 122.1	175.0 ± 22.7	192.7 ± 37.1	179.7 ± 47.9
Oct.	244.0 ± 67.2	314.3 ± 101.3	323.0 ± 50.3	340.3 ± 82.7	322.3 ± 41.1
Nov.	301.7 ± 76.0	393.3 ± 19.7	429.0 ± 74.2	459.0 ± 93.5	489.3 ± 91.6
Dec.	335.3 ± 62.5	442.0 ± 29.8	490.3 ± 68.5	488.0 ± 78.4	393.0 ± 51.2

Two-way ANOVA Test

Significant difference in total solids content between river monitoring stations, F_1=20.38 (P<0.01).
Significant difference in total solids content between monitoring months, F_2=2.8 (P<0.05).

Seasonal TDS content ranged from 422.67 ± 47.5 to 577.78 ± 87.78 in NAJ-1, 520.67 ± 106.56 to 617.33 ± 50.03 in CIV-2, 524.67 ± 138.3 to 643.5 ± 57.75 in POW-3, 647.44 ± 178.64 to 791.83 ± 45.41 in SEN-4, 580.11 ± 105.6 to 729.67 ± 41.30 in BAR-5, 304.89 ± 163.4 to 634.67 ± 101.56 in HIN-6. Annual averages were in order of 718.53 (SEN-4) > 673.47 (BAR-5) > 611.36 (POW-3) > 583.06 (CIV-2) > 529.53 (HIN-6) > 467.72 (NAJ-1).

Table RWM-5 shows monthly averages (Mean ± SD) of total dissolved solids and the result of Analysis of variance in river water. Monthly averages varied from a minimum of 118.33 ± 19.29 (Jun.) to a maximum of 370.67 ± 99.87 (Jan.) at WAZ-1, 183.67 ± 122.06 (Sep.) to 508.67 ± 55.43 (Feb.) at ISB-2, 166.67 ± 10.4 (Aug.) to 490.33 ± 68.519 (Dec.) at ITO-3, 187.33 ± 16.52 (Aug.) to 500.33 ± 20.42 (Feb.) at NIZ-4, 170.00 ± 20.61 (Aug.) to 393.0 ± 51.23 (Dec.) at OKH-5 (Okhla barrage). Analysis of variance shows a significant difference in TDS between sampling stations (F_1=20.38, $P<0.01$) as well as between sampling months (F_2=2.80, $P<0.05$).

Seasonal values ranged between 168.11 ± 41.06 (Monsoon) to 305.89 ± 96.63 (Winter) at WAZ-1, 215.11 ± 33.57 (Monsoon) to 463.89 ± 64.17 (Winter) at ISB-2, 180.11 ± 29.92 (Monsoon) to 484.0 ± 52.69 (Winter) at ITO-3, 200.78 ± 48.35 (Monsoon) to 499.11 ± 34.65 (Winter) at NIZ-4, 189 ± 51.87 (Monsoon) to 405.83 ± 109.60 (Pre-Monsoon) at OKH-5.

All the downstream stations of Wazirabad barrage showed peak values during winter due to lean discharge of Wazirabad barrage and no dilution of drain effluents discharge to river Yamuna at Delhi stretch. The lower values during rainy season may be due to dilution of drain effluents by rain water. Yearly averages of TDS content were in order of 378.44 (NIZ-4) > 358.75 (ISB-2) > 346.97 (ITO-3) > 308.61 (OKH-5) > 235.5 (WAZ-1).

Dissolved Oxygen of River Water

The concentration of dissolved oxygen (DO) in natural and wastewater depends on physical, chemical and biological processes in the water. Since DO is essential for the aerobic degradation of organic pollutants, the measurement of DO is important in water pollution control and in wastewater treatment. The data on DO has been provided only for river water as DO could not be detected in drain water.

Table RWM-6 shows monthly averages and the result of analysis or variance for DO. The minimum and maximum values for DO were 6.07 ± 0.34 (Jul.) and 12.23 ± 1.26 (Mar.) at WAZ-1, 0.57 ± 0.42 (Mar.) and 5.97 ± 1.01 (Sep.) at ISB-2, 0.50 ± 0.41 (Mar.) and 6.03 ± 0.42 (Aug.) at ITO-3, 0.73 ± 0.25 (Feb.) and 5.4 ± 0.24 (Aug.) at NIZ-4, 0.54 ± 0.49 (Mar.) and 6.57 ± 0.66 (Sep.) at OKH-5. Analysis of variance showed a significant difference in DO between monitoring stations (F_1=21.25, $P<0.01$). However, no significant difference was observed between sampling months (F_2=1.69).

Seasonal DO values ranged from 6.42 ± 1.07 (Monsoon) to 10.39 ± 1.73 (Summer) at WAZ-1, 1.07 ± 1.14 (Summer) to 4.79 ± 1.62 (Monsoon) at ISB-2, 0.80 ± 0.47 (Winter) to 4.96 ± 1.1 (Monsoon) at ITO-3, 0.81 ± 0.44 (Winter) to 4.6 ± 1.25 (Monsoon) at NIZ-4, 2.83 ± 1.32 (Winter) to 4.97 ± 1.79 (Monsoon) at OKH-5.

Summer season showed the highest seasonal average at WAZ-1 (unpolluted stretch) may be attributed to maximum penetration of incident light energy due to high transparency of water resulting in higher photosynthetic activity and phytoplanktonic production of oxygen. The winter season showed the second highest seasonal average, which may be a function of water temperature (Datta *et al.*, 1988).

The decreased water temperature during winter season has a greater capacity to hold DO than warm water and probably led to a lower rate of respiration thereby allowing maximum DO in the winter (Welch, 1951). In case of four stations (*i.e.* down stream of Wazirabad barrage) the highest DO value was observed in monsoon season, may be due to dilution of drain effluents by rain water.

Table RWM-6: Monthly Variation of Dissolved Oxygen (mg/l) in River Water (Year 1993-95)

Month	*WAZ-1*	*ISB-2*	*ITO-3*	*NIZ-4*	*OKH-5*
	Mean ± SD	*Mean ± SD*	*Mean ± SD*	*Mean ± SD*	*Mean ± SD*
Jan.	10.9 ± 3.0	1.8 ± 0.6	0.8 ± 0.7	0.8 ± 0.7	2.9 ± 1.6
Feb.	10.3 ± 1.8	0.7 ± 0.6	0.9 ± 0.4	0.7 ± 0.2	2.1 ± 0.6
Mar.	12.2 ± 1.3	0.6 ± 0.4	0.5 ± 0.4	0.4 ± 0.3	0.6 ± 0.5
Apr.	9.9 ± 1.4	1.1 ± 0.5	1.4 ± 1.1	2.5 ± 1.7	3.7 ± 2.3
May	10.5 ± 1.1	2.3 ± 1.5	1.6 ± 0.7	1.3 ± 0.6	2.8 ± 1.6
Jun.	8.9 ± 1.2	0.3 ± 0.2	0.8 ± 0.5	2.1 ± 2.1	3.4 ± 0.4
Jul.	6.1 ± 0.3	3.1 ± 0.6	3.9 ± 0.1	3.3 ± 1.2	2.9 ± 1.0
Aug.	6.3 ± 1.3	5.3 ± 1.4	6.0 ± 0.4	5.4 ± 0.2	5.5 ± 0.9
Sep.	6.8 ± 1.2	6.0 ± 1.1	4.9 ± 1.1	5.1 ± 0.4	6.6 ± 0.7
Oct.	9.4 ± 1.4	3.1 ± 1.9	3.6 ± 2.0	3.0 ± 1.7	4.7 ± 1.7
Nov.	8.2 ± 0.7	1.7 ± 0.7	1.6 ± 0.8	1.0 ± 0.5	4.0 ± 2.5
Dec.	10.5 ± 2.2	2.2 ± 2.2	0.9 ± 0.5	0.9 ± 0.2	2.9 ± 1.9

Two-way ANOVA Test

Significant difference in DO content between river monitoring stations, F_1=21.25 ($P<0.01$).
No significant difference in DO content between monitoring months, F_2=1.69.

In flowing polluted water, the oxygen content varied with distance and depth, and there are more than one oxygen profiles. But generally on adding sewage or any organic wastes, DO first decreases and then slowly recovers (Kumar, 1997). Same type of trend was observed in the river Yamuna at Delhi stretch, *i.e.* DO sharply decreased at ISB-2 and some what recovered at monitoring station OKH-5 (Okhla barrage).

Yearly averages of DO were in order of 9.18 > 3.48 > 2.36 > 2.24 > 2.21 at the station WAZ-1, OKH-5, ISB-2, ITO-3, and NIZ-4 respectively. The present study on river Yamuna at Delhi stretch showed DO content below the tolerance limits set by CPCB (1978) for class-B water (3mg/l) at all the stations (expect WAZ-1), which indicates the unsuitability of water for outdoor bathing.

Oxygen Absorbance Value of Drain and River Water

This test determines the amount of oxygen absorbed by the sample from potassium permanganate in acid solution. The test gives some indication of the nature of oxidisable substances present in the sample. Though the test is an empirical one, it is useful as a quick means to find out the pollution potential of wastewater and is useful in laboratories. When it is not feasible to conduct Biochemical Oxygen Demand (BOD) test, due to interference by toxic substances, this test is of much use. It is also used to determine the McGowan strength of polluted waters, sewage and industrial effluents. (Manivasakam, 1989).

Table DWM-6 shows monthly averages (Mean ± SD) of oxygen absorbance and the results of analysis of variance. The minimum and maximum monthly averages of oxygen absorbance values (mg/l) were 17.58 ± 4.26 (Aug.) and 31.33 ± 18.53 (Jul.), in NAJ-1, 33.2 ± 9.34 (Mar.) and 63.73 ± 7.13 (Jan.) in CIV-2, 42.6 ± 19.95 (Sep.) and 64.93 ± 4.08 (Dec.) in POW -3, 48.27 ± 11.33 (Jul.) and 69.2 ± 3.12 (Dec.) in SEN-4, 19.93 ± 2.25 (Oct.) and 34.3 ± 18.19 (Feb.) in BAR-5, 2.87 ± 0.77 (Jul.) and 8.07 ± 0.88 (Apr.) in HIN-6. Analysis of variance showed a significant difference in OA between the different drains (F_1=19.2, P<0.01) and also between the sampling months (F_2=2.37, P<0.05).

Table DWM-6: Monthly Variation of Oxygen Absorbance (mg/l) in Drain Water (Year 1993-1995)

Month	*NAJ-1*	*CIV-2*	*POW-3*	*SEN-4*	*BAR-5*	*HIN-6*
	Mean ± SD	*Mean ± SD*	*Mean ± SD*	*Mean ± SD*	*Mean ± SD*	*Mean ± SD*
Jan.	25.1 ± 10.6	63.7 ± 7.1	53.6 ± 11.1	63.5 ± 5.2	32.6 ± 3.8	5.9 ± 0.7
Feb.	25.3 ± 6.0	56.1 ± 13.0	62.1 ± 3.0	67.1 ± 6.9	34.3 ± 18.2	7.1 ± 0.1
Mar.	22.4 ± 4.4	33.4 ± 13.5	57.7 ± 9.0	60.5 ± 8.1	25.5 ± 6.3	7.5 ± 4.1
Apr.	25.3 ± 4.7	59.1 ± 14.3	54.5 ± 10.5	53.7 ± 15.6	30.3 ± 12.0	8.1 ± 0.9
May	22.9 ± 7.0	51.5 ± 23.6	50.8 ± 3.3	62.9 ± 24.5	23.9 ± 3.1	4.0 ± 1.0
Jun.	20.5 ± 2.5	48.4 ± 19.5	47.7 ± 4.4	57.3 ± 8.8	29.7 ± 12.3	6.0 ± 0.5
Jul.	31.3 ± 18.5	33.2 ± 9.3	52.5 ± 6.2	48.3 ± 11.3	27.7 ± 10.8	2.9 ± 0.8
Aug.	18.3 ± 6.1	60.3 ± 21.4	52.1 ± 12.5	60.4 ± 25.7	19.3 ± 0.5	5.0 ± 2.5
Sep.	17.6 ± 4.4	51.6 ± 13.6	42.6 ± 20.0	68.4 ± 1.2	21.8 ± 1.9	3.4 ± 0.6
Oct.	18.7 ± 1.9	47.3 ± 4.1	58.5 ± 2.7	50.3 ± 8.3	19.9 ± 2.2	4.1 ± 1.2
Nov.	20.8 ± 0.7	52.7 ± 10.6	54.0 ± 4.3	60.7 ± 5.3	30.3 ± 6.7	4.4 ± 1.4
Dec.	21.1 ± 6.7	47.9 ± 6.7	64.9 ± 4.1	69.2 ± 3.1	28.0 ± 5.8	4.2 ± 1.4

Two-way ANOVA Test
Significant difference in oxygen absorbance content between monitoring drains, F_1=19.22 (P<0.01).
Significant difference in oxygen absorbance content between sampling months, F_2=2.81 (P<0.05).

Table RWM-7: Monthly Variation of Oxygen Absorbance (mg/l) in River Water (Year 1993-95)

Month	*WAZ-1*	*ISB-2*	*ITO-3*	*NIZ-4*	*OKH-5*
	Mean ± SD	*Mean ± SD*	*Mean ± SD*	*Mean ± SD*	*Mean ± SD*
Jan.	1.60 ± 0.33	10.40 ± 1.42	8.93 ± 2.85	9.60 ± 3.22	9.47 ± 3.49
Feb.	1.60 ± 0.33	10.13 ± 2.96	7.87 ± 2.45	6.93 ± 2.17	10.72 ± 4.58
Mar.	2.00 ± 0.86	13.07 ± 2.10	8.40 ± 1.50	8.80 ± 1.70	8.27 ± 0.38
Apr.	1.27 ± 0.09	9.00 ± 0.91	6.27 ± 1.32	7.73 ± 1.61	7.43 ± 2.78
May	1.60 ± 0.33	7.87 ± 2.74	6.73 ± 1.81	8.40 ± 1.50	6.73 ± 1.06
Jun.	1.87 ± 0.74	15.33 ± 9.89	8.67 ± 0.94	9.73 ± 0.68	6.67 ± 1.15
Jul.	2.33 ± 0.52	5.35 ± 0.41	5.37 ± 0.80	4.97 ± 1.27	4.07 ± 0.41
Aug.	4.53 ± 0.52	6.27 ± 0.52	5.93 ± 0.47	5.60 ± 0.33	5.73 ± 1.24
Sep.	4.13 ± 2.02	5.73 ± 2.10	6.87 ± 2.65	6.23 ± 2.12	6.40 ± 2.47
Oct.	1.87 ± 0.68	7.23 ± 3.63	5.87 ± 2.93	5.47 ± 1.80	4.80 ± 1.99
Nov.	1.27 ± 0.62	7.33 ± 1.89	5.60 ± 1.50	6.67 ± 1.15	8.80 ± 4.57
Dec.	1.43 ± 0.78	7.83 ± 2.77	6.03 ± 1.77	5.67 ± 0.96	6.80 ± 0.86

Two-way ANOVA Test
Significant difference in oxygen absorbance between river monitoring stations, F_1=20.17 (P<0.01).
Significant difference in oxygen absorbance between monitoring months, F_2=2.46 (P<0.05).

Seasonal and annual averages of oxygen absorbance varied from 18.53 ± 2.42 to 22.54 ± 5.26 in NAZ-1, 48.08 ± 20.44 to 56.71 ± 10.84 in CIV-2, 49.07 ± 14.77 to 58.09 ± 8.92 in POW-3, 58.62 ± 16.10 to 67.71 ± 6.32 in SEN-4, 22.93 ± 7.26 to 27.83 ± 12.18 in BAR-5 and 3.74 ± 1.8 to 6.42 ± 2.66 in HIN-6. Annual average oxygen absorbance values (mg/l) were 22.36, 50.43, 54.26, 60.18, 26.19 and 5.22 in NAJ-1, CIV-2, POW-3, SEN-4, BAR-5 and HIN-6 respectively.

Monthly average oxygen absorbance values (mg/l) ranged from 1.27 ± 0.09 (Apr.) to 4.53 ± 0.52 (Aug.) at WAZ-1, 7.23 ± 3.63 (Oct.) to 15.33 ± 9.89 (Jun.) at ISB-2, 5.37 ± 0.08 (Jul.) to 8.93 ± 2.83 (Jan.) at ITO-3, 4.97 ± 1.27 (Jul.) to 9.73 ± 0.68 (Jun.) at NIZ-4 and 4.07 ± 0.41 (Jul.) to 10.27 ± 4.58 (Feb.) at OKH-5 (Table RWM-7) in the river water. Analysis of variance showed a significant difference in OA between the sampling stations (F_1=20.17, P<0.01) and also between the sampling months (F_2=2.46, P<0.05).

Seasonal and annual averages of OA in Yamuna river at Delhi stretch ranged from 1.46 ± 0.46 (Winter) to 3.67 ± 1.57 (Monsoon) at WAZ-1, 5.84 ± 1.31 (Monsoon) to 11.32 ± 6.06 (Summer) at ISB.2, 5.73 ± 2.33 (Monsoon) to 7.64 ± 2.66 (Winter) at ITO-3, 5.60 ± 1.53 (Monsoon) to 7.7 ± 2.72 (Winter) at NIZ-4 and 5.4 ± 1.89 (Monsoon) to 8.64 ± 3.76 (Winter) at OKH.

It is observed that the peak values of OA were observed in summer season at all stations (except Wazirabad barrage). It may due to lean discharge of river water form Wazirabad barrage and also high concentration of oxidisable organic and inorganic pollutants present in drain effluents and discharged into river Yamuna. During the present investigation the oxygen absorbance values at Wazirabad barrage was found to be high during rainy season due to flushing of organic pollutants by rain water from river catchment areas. Annual averages of OA were in order of 8.81 >7.15 = 7.15 > 6.88 > 2.13 at station ISB-2, NIZ-4, OKH-5, ITO-3 and WAZ-1 respectively.

Chemical Oxygen Demand of Drain and River Water

The chemical oxygen demand determines the amount of oxygen required for chemical oxidation of organic matter using a strong chemical oxidant such as potassium dichromate (APHA, 1991). COD represents chemical oxidisable pollution load present in water and hence a reliable parameter for judging the quantum of organic pollution in the aquatic system.

Table DWM-7 presents the monthly average COD values of drain water. The COD ranged from 89.0 ± 20.8 to 147.3 ± 44.1 in drain NAJ-1, 284.0 ± 138.7 to 410.7 ± 196.0 in CIV-2, 278.7 ± 173.6 to 440 ± 71.2 in POW-3, 274.7 ± 69.8 to 465.3 ± 25.4 in SEN-4, 98.7 ± 23.6 to 205.0 ± 106.9 in BAR-5 and from 23.3 ± 9.6 to 35.0 ± 2.4 in Hundon river cut. Two way analysis of variance shows significant difference in COD values between the drains (F_1=19.5, P<0.01) as well as sampling months (F_2=2.33, P<0.05).

Seasonal averages of COD ranged from 122.67 ± 96.02 to 152.17 ± 66.27, 308.67 ± 68.91 to 373.0 ± 77.48, 323.11 ± 124.43 to 362.44 ± 64.22, 379.33 ± 66.59 to 449.78 ± 48.55, 136.0 ± 39.4 to 193.75 ± 47.48, 29.89 ± 22.14 to 34.11 ± 4.15 in Najafgarh, Civil mill, Power house, Sen nursing home, Barpulla drain and Hindon river cut respectively. Annual average of COD values (mg/l) were in order of 4.8028 > 349.61 > 336.31 > 166.92 > 133.89 > 31.92 in the drain SEN-4, POW-3, CIV-2, BAR-5, NAJ-1 and HIN-6 (Hindon cut) respectively. The maximum values of COD was noted in SEN-4 (Sen nursing home drain).

Monthly average COD values of river water (Table RWM-8) ranged from 5.33 ± 1.41 (Jun.) to 14.33 ± 6.94 (Aug.) at WAZ-1, 28.67 ± 15.86 (Sep.) to 65.33 ± 9.98 (Mar.) at ISB-2, 19.43 ± 5.73 (May.) to 49.33 ± 13.20 (Jan.) at, down stream station OKH-5 (Okhla barrage) observed. Analysis of variance showed a significant difference in COD values between different monitoring stations (F_1=20.45, P<0.01) and also between sampling months (F_2=2.31, P<0.05).

Table DWM-7: Monthly Variation of COD (mg/l) in Drain Water (Year 1993-1995)

Month	*NAJ-1* *Mean ± SD*	*CIV-2* *Mean ± SD*	*POW-3* *Mean ± SD*	*SEN-4* *Mean ± SD*	*BAR-5* *Mean ± SD*	*HIN-6*
Jan.	137.3 ± 81.1	365.3 ± 26.4	364.0 ± 76.8	433.3 ± 66.3	149.3 ± 22.2	35.0 ± 2.4
Feb.	147.3 ± 44.1	415.3 ± 40.7	387.3 ± 65.8	450.0 ± 35.6	205.0 ± 106.9	34.0 ± 6.2
Mar.	133.7 ± 24.5	309.3 ± 74.5	440.0 ± 71.2	465.3 ± 25.4	184.0 ± 50.9	35.7 ± 13.2
Apr.	137.3 ± 22.9	429.3 ± 133.5	371.0 ± 44.3	374.7 ± 54.4	199.0 ± 74.3	29.0 ± 0.8
May	117.3 ± 43.5	324.0 ± 99.7	310.0 ± 51.9	440.0 ± 56.4	174.7 ± 74.6	27.7 ± 1.2
Jun.	117.3 ± 13.2	318.7 ± 135.3	318.7 ± 31.0	410.0 ± 77.9	217.3 ± 90.1	22.7 ± 0.9
Jul.	188.3 ± 141.2	284.0 ± 138.7	325.2 ± 33.5	340.0 ± 86.4	186.7 ± 55.1	39.0 ± 34.0
Aug.	90.7 ± 28.5	410.7 ± 196.0	365.3 ± 106.8	414.7 ± 198.6	145.0 ± 21.6	27.3 ± 9.4
Sep.	89.0 ± 20.8	308.3 ± 74.4	278.7 ± 173.6	531.3 ± 70.7	132.7 ± 40.7	23.3 ± 9.6
Oct.	176.3 ± 79.7	280.0 ± 65.3	345.7 ± 15.0	274.7 ± 69.8	98.7 ± 23.6	38.3 ± 12.7
Nov.	141.3 ± 24.7	317.3 ± 65.4	322.7 ± 30.3	340.0 ± 28.3	146.7 ± 37.7	38.3 ± 2.1
Dec.	130.7 ± 43.4	273.3 ± 57.3	366.7 ± 27.8	425.3 ± 62.1	164.0 ± 35.3	32.7 ± 2.5

Two-way ANOVA Test

Significant difference in COD content between monitoring drains, F_1=19.31 (P<0.01).

Significant difference in COD content between sampling months, F_2=2.31 (P<0.05).

Seasonal and annual average values of COD ranged from 5.92 ± 1.93 (Winter) to 11.78 ± 5.43 (Monsoon) at WAZ-1, 33.78 ± 12.98 (Monsoon) to 55.89 ± 11.72 (Winter) at ISB-2, 28.67 ± 8.46 (Pre-Monsoon) to 45.56 ± 8.53 (Winter) at ITO-3, 34.67 ± 6.39 (Post-Monsoon) to 57.89 ± 22.97 (Winter) at NIZ-4 and 28.36 ± 10.34 (Summer) to 39.59 ± 10.75 (Winter) at OKH-5 (Okhla barrage) respectively.

Table RWM-8: Monthly Variation of COD (mg/l) in River Water (Year 1993-1995)

Month	*WAZ-1* *Mean ± SD*	*ISB-2* *Mean ± SD*	*ITO-3* *Mean ± SD*	*NIZ-4* *Mean ± SD*	*OKH-5* *Mean ± SD*
Jan.	8.00 ± 0.00	54.67 ± 15.43	53.33 ± 11.78	68.00 ± 29.03	49.33 ± 13.20
Feb.	6.67 ± 1.89	64.00 ± 3.27	42.67 ± 4.99	57.33 ± 13.20	38.67 ± 3.77
Mar.	5.33 ± 1.89	65.33 ± 9.98	36.00 ± 11.78	49.33 ± 9.98	34.67 ± 3.77
Apr.	8.00 ± 0.00	48.33 ± 8.50	31.33 ± 9.67	31.67 ± 2.87	30.00 ± 11.43
May	5.33 ± 1.89	34.33 ± 16.01	19.43 ± 5.73	31.33 ± 8.99	26.10 ± 10.16
Jun.	5.00 ± 1.41	81.33 ± 64.44	21.33 ± 5.66	41.33 ± 11.47	22.67 ± 9.98
Jul.	7.67 ± 0.47	32.67 ± 10.62	26.67 ± 12.08	35.67 ± 14.06	27.00 ± 11.52
Aug.	14.33 ± 6.94	40.00 ± 8.64	29.33 ± 4.99	33.33 ± 4.99	31.33 ± 5.73
Sep.	13.33 ± 3.77	28.67 ± 15.86	28.33 ± 10.21	37.33 ± 14.73	37.67 ± 9.53
Oct.	6.83 ± 2.01	35.67 ± 8.50	24.00 ± 8.22	32.00 ± 6.53	28.00 ± 5.66
Nov.	6.67 ± 3.77	33.33 ± 6.80	28.00 ± 8.64	37.33 ± 4.99	41.33 ± 24.51
Dec.	7.33 ± 2.49	40.33 ± 6.55	37.33 ± 8.64	35.00 ± 4.08	34.67 ± 3.77

Two-way ANOVA Test

Significant difference in COD content between river monitoring stations, F_1=20.45 (P<0.01).

Significant difference in COD content between monitoring months, F_2=2.32 (P<0.05).

In station WAZ-1 the higher values of COD in monsoon months was recorded perhaps due to organic and inorganic matter content reaching the water body through run-off water. The same trend was observed in Brahmani river Orissa as reported by Pradhan and Mohanty (1991). Annual averages of COD (mg/l) in the sampling sites were in the order of ISB-2 > NIZ-4 > ITO-3 > OKH-5 > WAZ-1.

In the present investigation, annual mean COD values of five sites revealed a minimum value in the site WAZ-1 (Wazirabad) showed the minimum value. There was sharp increase in COD at ISB-2 due to the entry of allochothonus organic and inorganic material from Najafgarh and Burari drain. At site ITO-3, the value again come down due to process of sedimentation by ITO barrage. The concentration of COD increased to some extent at site NIZ-4 due to fresh addition of drain water from Sen-nursing home. But at site OKH-5, the decrease values might be due to dilution (by Hindon cut), deoxygenation, reoxygenation, coagulation and settling to the river bed while travelling approximately six km from Nizamuddin bridge to Okhla barrage.

Biochemical Oxygen Demand of Drain and River Water

Biochemical oxygen demand (BOD) is defined as the amount of oxygen required by micro-organism to decompose biologically degradable organic matter in water under aerobic conditions. BOD is an essential parameter in water pollution control management and in the evaluation of the self-purification capacity of streams (Trivedy and Goel, 1984). It represents the biologically oxidisable organic loads present in water.

Table DWM-8 shows monthly averages (Mean ± SD) and results of analysis of variance of BOD at the five sampling drains. The BOD ranged from 51.0 ± 19.70 to 108.8 ± 86.03 in NAJ-1, 139.33 ± 48.95 to 245.7 ± 135.38 in CIV-2, 77 ± 28.08 to 244 ± 76.08 in POW-3, 142.60 ± 28.4 to 265.0 ± 64.11 in SEN-4, 53.53 ± 9.67 to 99.67 ± 31.34 in BAR-5, 3.67 ± 1.53 to 12.3 ± 10.0 in HIN-6 (Hindon river cut). Analysis of variance (Table DWM-8) shows significant difference in BOD content between sampling drains (F_1=19.5, $P<0.01$) as well as between sampling months F_2=2.22, $P<0.05$).

Seasonal and annual BOD averages (Mean ± SD) varied from 58.72 ± 16.24 (Summer) to 70.6 ± 50.56 (Monsoon) in NAJ-1, 158.85 ± 26.87 (Pre-Monsoon) to 216.4 ± 50.65 (Winter) in CIV-2, 175.33 ± 63.59 (Monsoon) to 217.5 ± 35.96 (Winter) in POW-3, 170.8 ± 48.23 (Pre-Monsoon) to 272.67 ± 44.38 (Winter) in SEN-4, 63.5 ± 11.22 (Pre-Monsoon) to 78.72 ± 27.32 (Winter) in BAR-5 and 5.0 ± 0.96 (Monsoon) to 7.06 ± 2.1 in HIN-6. The average annual values were in order of 225.24 > 195.94 > 188.2 > 75.16 > 65.77 > 6.83 in the sampling drains SEN-4, POW-3, CIV-2, BAR-5, NAJ-1 and HIN-6 respectively.

Monthly variation (Mean ± SD) in BOD content and the results of analysis of variance are given in Table RWM-9. The BOD varied from 1.93 ± 0.09 (April) to 5.57 ± 3.27 (Sep.) at WAZ-1, 11.6 ± 3.61 (Sep.) to 28.17 ± 3.97 (Mar.) at ISB-2, 8.7 ± 1.67 (Nov.) to 23.87 ± 7.31 (Jan.) at ITO-3, 12.07 ± 1.31 (Jul.) to 26.5 ± 8.78 (Jan.) at NIZ-4, 10.37 ± 4.88 (Oct.) to 23.7 ± 6.38 (Jan.) at OKH-5. Analysis of variance shows the significant difference in BOD content between the monitoring stations (F_1=20.2, $P<0.01$), as well as between the sampling months (F_2=2.31, $P<0.05$).

Table DWM-8: Monthly Variation of BOD (mg/l) in Drain Water (Year 1993-1995)

Month	*NAJ-1*	*CIV-2*	*POW-3*	*SEN-4*	*BAR-5*	*HIN-6*
	Mean ± SD	*Mean ± SD*	*Mean ± SD*	*Mean ± SD*	*Mean ± SD*	*Mean ± SD*
Jan.	63.4 ± 35.1	240.9 ± 84.3	199.0 ± 16.4	252.5 ± 22.3	67.3 ± 21.9	7.3 ± 1.2
Feb.	83.1 ± 12.9	225.0 ± 35.0	221.8 ± 67.5	263.0 ± 27.0	78.7 ± 37.7	8.3 ± 0.6
Mar.	61.8 ± 18.0	198.3 ± 42.5	244.0 ± 76.1	259.8 ± 54.4	78.3. ± 23.1	12.3 ± 10.0
Apr.	66.1 ± 12.2	245.7 ± 135.4	220.0 ± 70.0	178.5 ± 57.5	99.7 ± 31.3	9.7. ± 2.1
May	56.5 ± 31.3	184.2 ± 137.1	165.7 ± 25.3	265.0 ± 84.1	53.5 ± 9.7	4.7 ± 1.2
Jun.	50.5 ± 8.3	139.3 ± 49.0	170.0 ± 13.2	205.3 ± 71.3	97.7 ± 59.0	7.0 ± 1.0
Jul.	108.8 ± 86.0	149.7 ± 86.3	214.3 ± 67.7	200.5 ± 38.1	84.3 ± 65.6	3.7 ± 1.5
Aug.	52.0 ± 20.8	240.3 ± 132.1	174.7 ± 50.0	228.2 ± 144.1	72.5 ± 25.7	6.3 ± 3.5
Sep.	51.0 ± 19.7	160.7 ± 31.1	77.0 ± 78.1	255.0 ± 32.8	74.2 ± 30.3	5.0 ± 2.0
Oct.	62.1 ± 8.6	153.7 ± 6.2	161.7 ± 57.7	142.6 ± 28.4	54.0 ± 10.2	6.0 ± 2.0
Nov.	68.2 ± 4.1	162.1 ± 45.9	198.2 ± 21.3	202.3 ± 55.1	73.0 ± 2.0	6.0 ± 1.7
Dec.	65.8 ± 1.9	156.7 ± 23.6	225.0 ± 18.3	253.5 ± 120.8	87.8 ± 34.4	5.6 ± 0.5

Two-way ANOVA Test

Significant difference in BOD content between monitoring drains, F_1=19.59 (P<0.01).
Significant difference in BOD content between sampling months, F_2= 2.22 (P<0.05).

Table RWM-9: Monthly Variation of BOD (mg/l) in River Water (Year 1993-1995)

Month	*WAZ-1*	*ISB-2*	*ITO-3*	*NIZ-4*	*OKH-5*
	Mean ± SD	*Mean ± SD*	*Mean ± SD*	*Mean ± SD*	*Mean ± SD*
Jan.	2.60 ± 0.57	26.33 ± 2.62	23.87 ± 7.31	26.50 ± 8.78	23.70 ± 6.38
Feb.	2.43 ± 0.62	24.20 ± 3.35	20.57 ± 5.92	18.37 ± 3.16	18.33 ± 4.96
Mar.	3.07 ± 1.03	28.17 ± 3.97	16.60 ± 2.34	18.70 ± 1.07	15.77 ± 4.12
Apr.	1.93 ± 0.09	20.20 ± 2.97	13.97 ± 3.15	14.20 ± 2.47	13.77 ± 3.56
May	2.77 ± 0.65	18.97 ± 6.69	17.27 ± 2.58	14.87 ± 3.67	16.33 ± 2.05
Jun.	2.33 ± 0.81	30.60 ± 12.99	19.03 ± 5.72	15.33 ± 2.87	11.87 ± 3.43
Jul.	2.67 ± 1.70	12.13 ± 1.27	11.93 ± 0.57	12.43 ± 1.19	11.47 ± 1.87
Aug.	5.30 ± 1.84	11.87 ± 2.04	10.77 ± 1.28	12.07 ± 1.31	11.37 ± 2.25
Sep.	5.57 ± 3.27	11.60 ± 3.61	15.20 ± 5.87	15.53 ± 5.76	14.37 ± 5.14
Oct.	2.63 ± 0.74	16.73 ± 9.68	13.53 ± 8.92	13.03 ± 7.79	10.37 ± 4.88
Nov.	3.27 ± 1.24	13.53 ± 3.58	8.70 ± 1.67	15.33 ± 3.97	15.07 ± 12.33
Dec.	3.17 ± 0.90	18.47 ± 8.13	15.77 ± 1.78	15.17 ± 0.79	13.63 ± 2.67

Two-way ANOVA Test

Significant difference in BOD content between river monitoring stations, F_1=20.20 (P<0.01).
Significant difference in BOD content between monitoring months, F_2=2.31 (P<0.05).

Seasonal average values of BOD were varied from a minimum of 2.53 ± 0.71 (Winter) to 4.51 ± 2.72 (Monsoon) WAZ-1, 11.87 ± 2.51 (Monsoon) to 24.48 ± 9.19 (Summer) at ISB-2, 11.12 ± 6.86 (Pre-

Monsoon) to 20.18 ± 6.35 (Winter) at ITO-3, 13.34 ± 3.81 (Monsoon) to 20.67 ± 6.93 (Winter) at NIZ-4, 12.4 ± 3.69 (Monsoon) to 18.18 ± 6.64 (Winter) at OKH-5,

In the present investigation, the BOD values during monsoon and winter season showed higher and lower values respectively at unpolluted stretch of Yamuna river *i.e.* Wazirabad barrage. The low values of BOD in winter may be due to lesser quantity of total organic content in the water. Increase in organic matter load due to rain water may be the reason for higher BOD during rainy season at WAZ-1. In down-stream stations the trend of BOD content of river water was higher values during summer months followed by winter, seasonal and lower values during monsoon. The higher values during summer seasons may be due to lean water discharge from Wazirabad barrage and also the presence of higher concentration of biodegradable organic matter in effluents discharged into river at Delhi stretch.

In addition to this, BOD values also showed site-wise fluctuation. Annual averages were in order of 19.4 > 15.96 > 15.6 > 14.64 > 3.14 at ISB-2, NIZ-4, ITO-3, OKH-5 and WAZ-1, (Wazirabad barrage) respectively. The gradual decrease in BOD from station ISB-2 to OKH-5 supports gradual reduction due to effect of dilution and self-purification.

Total Kjeldahl Nitrogen of Drain and River Water

In water and wastewater, the forms of nitrogen of greatest interest are nitrate, nitrite, ammonia and organic nitrogen. All these compounds are inter-convertible and take part in the nitrogen cycle. Analytically, organic nitrogen and ammonia can be determined together and are referred to as "Kjeldahl Nitrogen".

Table DWM-9: Monthly Variation of TKN (mg/l) in Drain Water (Year 1993-1995)

Month	*NAJ-1* *Mean ± SD*	*CIV-2* *Mean ± SD*	*POW-3* *Mean ± SD*	*SEN-4* *Mean ± SD*	*BAR-5* *Mean ± SD*	*HIN-6* *Mean ± SD*
Jan.	29.20 ± 7.58	49.67 ± 15.33	42.33 ± 4.19	53.67 ± 3.09	37.33 ± 3.86	4.58 ± 1.01
Feb.	31.37 ± 5.22	47.13 ± 10.34	45.00 ± 8.29	55.83 ± 7.92	37.00 ± 2.16	14.05 ± 9.04
Mar.	26.00 ± 6.57	43.37 ± 5.85	50.53 ± 12.03	54.67 ± 12.50	38.00 ± 7.48	6.90 ± 4.13
Apr.	26.63 ± 3.63	51.23 ± 20.35	44.17 ± 14.90	42.67 ± 4.64	30.33 ± 3.68	7.66 ± 3.04
May	16.27 ± 15.00	48.77 ± 26.47	33.77 ± 1.52	52.00 ± 20.61	26.00 ± 4.24	5.57 ± 1.52
Jun.	25.33 ± 8.16	34.20 ± 6.24	31.10 ± 2.76	47.67 ± 14.52	39.50 ± 11.88	7.33 ± 3.40
Jul.	47.93 ± 34.62	37.30 ± 11.13	33.83 ± 0.85	38.33 ± 8.18	36.67 ± 8.06	1.86 ± 0.47
Aug.	24.00 ± 2.48	44.17 ± 16.28	34.77 ± 3.61	43.00 ± 16.33	40.33 ± 6.34	1.46 ± 0.32
Sep.	22.60 ± 5.24	36.17 ± 7.96	30.60 ± 7.03	49.33 ± 3.40	42.00 ± 6.48	2.06 ± 1.30
Oct.	26.77 ± 7.49	45.47 ± 8.87	41.27 ± 4.85	43.83 ± 4.37	42.33 ± 4.19	4.36 ± 2.68
Nov.	29.07 ± 3.44	40.10 ± 1.64	39.53 ± 9.15	53.73 ± 3.72	40.00 ± 6.98	3.89 ± 2.38
Dec.	27.83 ± 3.30	39.60 ± 7.11	44.93 ± 7.55	62.13 ± 9.83	41.20 ± 7.44	3.50 ± 1.23

Two-way ANOVA Test

Significant difference in TKN content between monitoring drains, F_1=19.33 (P<0.01).

Significant difference in TKN content between sampling months, F_2=2.76 (P<0.05).

Table DWM-9 is present as monthly average values (Mean ± SD) and the result of analysis of variance of TKN. The minimum and maximum average values were 16.27 ± 15 and 47.93 ± 35.62 in

NAJ-1, 34.20 ± 6.24 and 51.23 ± 20.35 in CIV-2, 30.6 ± 7.03 and 50.53 ± 12.03 in POW-3, 38.33 ± 8.19 and 62.13 ± 9.83 in SEN-4, 26.0 ± 4.24 and 42.33 ± 4.19 in BAR-5 and 1.46 ± 0.32 and 14.05 ± 9.04 in HIN-6, (Hindon cut). Analysis of variance shows that the variation in TKN between different sampling stations (F_1=19.3, P<0.01) as well as sampling months (F_2=2.76, P<0.05) were significant.

Seasonal and annual values in TKN of five drains (Mean mg/l ± SD) varied from a minimum of 23.56 ± 10.24 (Summer) to 29.97 ± 5.48 (Winter) in NAJ-1, 39.21 ± 12.78 (Monsoon) to 47.69 ± 12.47 (Winter) in CIV-2, 33.07 ± 4.92 (Monsoon) to 42.49 ± 8.53 (Pre-Monsoon) in POW-3, 43.56 ± 11.63 (Monsoon) to 58.17 ± 8.43 (Winter) in SEN-4, 33.46 ± 9.37 (Summer) to 40.94 ± 5.46 (Pre-Monsoon) in BAR-5, 1.79 ± 0.08 (Monsoon) to 7.10 ± 2.31 (Winter) in HIN-6 (Hindon cut). Annual average TKN values of five drains and one river cut were in the order of 49.74 (SEN-4) > 43.10 (CIV-2) > 39.32 (POW-3) > 37.56 (BAR-5) > 27.75 (NAJ-1) > 5.27 (HIN-6).

Monthly average (Mean ± SD) values and the result of analysis of variance of TKN (mg/l) in river water are tabulated in Table RWM-10. Monthly average TKN value varied between 0.79 ± 0.28 (Ju.l) to 1.62 ± 0.88 (Mar.) at WAZ-1, 4.27 ± 1.67 (Jul.) to 18.73 ± 5.23 (Jan.) at ISB-2, 3.75 ± 1.83 (Aug.) to 14.83 ± 6.35(Jan.) at ITO-3, 4.58 ± 2.38 (Aug.) to 16.70 ± 5.64 (Jan.) at NIZ-4, 3.7 ± 0.91 (Aug.) to 14.73 ± 3.81 (Jan.) at OKH-5. Analysis of variance showed significant difference in TKN, between different stations (F_1=21.05, P<0.01), as well as between monitoring months (F_2=2.27, P<0.05).

Table RWM-10: Monthly Variation of TKN (mg/l) in River Water (Year 1993-1995)

Month	*WAZ-1*	*ISB-2*	*ITO-3*	*NIZ-4*	*OKH-5*
	Mean ± SD	*Mean ± SD*	*Mean ± SD*	*Mean ± SD*	*Mean ± SD*
Jan.	1.39 ± 0.09	18.73 ± 5.23	14.83 ± 6.34	16.70 ± 5.64	14.73 ± 3.81
Feb.	1.56 ± 0.56	14.12 ± 2.79	11.26 ± 2.82	11.10 ± 1.10	10.40 ± 2.53
Mar.	1.62 ± 0.88	15.89 ± 1.68	9.85 ± 0.31	11.75 ± 1.66	11.08 ± 1.29
Apr.	1.35 ± 0.16	12.37 ± 1.97	7.52 ± 2.33	9.80 ± 2.33	9.96 ± i.20
May	1.12 ± 0.37	9.31 ± 2.62	10.57 ± 1.68	12.35 ± 3.68	12.53 ± 4.11
Jun.	0.84 ± 0.06	16.20 ± 11.08	10.28 ± 1.86	12.26 ± 3.43	6.95 ± 2.78
Jul.	0.79 ± 0.28	4.27 ± 1.67	4.20 ± 1.92	4.88 ± 2.78	4.60 ± 2.51
Aug.	1.45 ± 0.92	4.30 ± 1.77	3.75 ± 1.83	4.58 ± 2.39	3.70 ± 0.91
Sep.	1.46 ± 0.67	6.77 ± 2.35	11.63 ± 7.07	12.21 ± 8.08	10.72 ± 4.53
Oct.	1.45 ± 0.48	10.32 ± 6.23	8.31 ± 5.62	9.50 ± 7.70	8.84 ± 5.11
Nov.	1.20 ± 0.30	6.53 ± 1.65	7.11 ± 1.78	8.30 ± 1.81	10.11 ± 4.63
Dec.	1.35 ± 0.61	8.07 ± 2.40	11.94 ± 1.64	10.10 ± 1.93	8.27 ± 3.09

Two-way ANOVA Test

Significant difference in TKN content between river monitoring stations, F_1=21.05 (P<0.01).

Significant difference in TKN content between monitoring months, F_2=2.27 (P<0.05).

Seasonal and annual averages of TKN at five monitoring stations varied from 1.23 ± 0.56 (Summer) to 1.33 ± 0.42 (Winter) at WAZ-1, 5.11 ± 2.27 (Monsoon) to 14.65 ± 5.59 (Winter) at ISB-2, 6.53 ± 5.66 (Monsoon) to 12.71 ± 4.38 (Winter) at ITO-3, 7.22 ± 6.22 (Monsoon) to 13.07 ± 4.37 (Winter) at NIZ-4, 6.34 ± 4.35 (Monsoon) to 10.83 ± 4.32 (Winter) at OKH-5.

In the present study the maximum values of TKN were recorded during rainy season at Wazirabad barrage (unpolluted stretch) which might be due to influx of nitrogen rich flood water that brings large amount of autochthonous organic matter (loosely bound with soil). The other down stream stations showed peak values in summer months, which might be due to lean discharge of Wazirabad barrage, high evapo-transpiration rate and high concentration of drain effluents at polluted river stretch. The minimum values of TKN during monsoon at all down stream stations were obviously due to dilution.

Nitrate-Nitrogen of Drain and River Water

The nitrate is a stable product of aerobic oxidation of nitrogenous organic matter. Significant sources of nitrate are domestic sewage, effluent from nitrogenous fertilizer plants and agricultural runoff. Unpolluted waters contain usually small amounts of nitrate. A limit of 10 mg of nitrate-nitrogen per litre has been imposed on drinking water to prevent methaemoglobinemia in infants. In surface water, nitrate as nutrient is taken up by plants and converted in to cell protein. It is an essential nutrient for many photosynthetic autotrophs and in some cases has been identified as the growth limiting nutrients (APHA,1991).

Table DWM-10 presents monthly averages (Mean ± SD) and the results of analysis of variance of NO_3-N. The NO_3-N values varied from 2.43 ± 0.24 to 6.53 ± 0.47 in NAJ-1, 4.73 ± 1.36 to 8.87 ± 1.72 in CIV-2, 3.37 ± 0.56 to 9.33 ± 0.95 in POW-3, 3.93 ± 0.63 to 9.93 ± 2.70 in SEN-4, 2.57 ± 0.46 to 7.47 ± 1.81 in BAR-5, 1.25 ± 0.58 to 3.33 ± 1.17 in HIN-6. Results of analysis of variance shows a significant difference in NO_3-N content between monitoring drains, (F_1=19.7, P<0.01), as well as between monitoring months (F_2=2.75, P<0.05).

Table DWM-10: Monthly Variation of Nitrate-Nitrogen (mg/l) in Drain Water (Year 1993-1995)

Month	*NAJ-1*	*CIV-2*	*POW-3*	*SEN-4*	*BAR-5*	*HIN-6*
	Mean ± SD	*Mean ± SD*	*Mean ± SD*	*Mean ± SD*	*Mean ± SD*	*Mean ± SD*
Jan.	5.80 ± 0.00	8.87 ± 1.72	8.00 ± 1.07	9.47 ± 1.50	7.47 ± 1.81	2.09 ± 0.25
Feb.	6.53 ± 0.47	8.90 ± 0.99	8.57 ± 1.02	9.93 ± 2.70	6.87 ± 1.68	2.07 ± 0.37
Mar.	5.47 ± 0.47	8.23 ± 1.05	9.33 ± 0.95	8.50 ± 2.70	6.30 ± 0.49	1.76 ± 0.10
Apr.	5.57 ± 0.52	8.53 ± 2.69	7.80 ± 1.42	6.90 ± 1.28	5.20 ± 0.78	1.83 ± 0.59
May	6.40 ± 0.57	7.93 ± 2.78	7.23 ± 0.82	7.27 ± 3.15	4.63 ± 0.65	2.62 ± 0.17
Jun.	4.27 ± 0.24	6.77 ± 2.75	6.30 ± 1.02	7.33 ± 2.43	5.47 ± 1.25	3.33 ± 0.17
Jul.	2.43 ± 0.24	4.73 ± 1.36	3.37 ± 0.56	3.93 ± 0.63	2.57 ± 0.46	1.42 ± 0.49
Aug.	2.59 ± 0.44	4.90 ± 1.37	3.37 ± 0.60	4.27 ± 1.33	2.60 ± 0.16	1.25 ± 0.58
Sep.	3.97 ± 0.47	7.00 ± 2.67	4.57 ± 0.21	6.27 ± 1.07	4.80 ± 0.86	1.30 ± 0.33
Oct.	3.99 ± 0.13	7.77 ± 2.64	6.17 ± 0.45	6.30 ± 1.08	5.80 ± 0.41	2.04 ± 0.69
Nov.	4.78 ± 0.67	7.60 ± 1.91	6.63 ± 0.31	8.40 ± 1.10	6.43 ± 0.53	2.40 ± 0.54
Dec.	5.36 ± 0.37	7.80 ± 2.18	6.90 ± 0.65	8.83 ± 1.24	6.30 ± 1.08	2.11 ± 0.15

Two-way ANOVA Test

Significant difference in nitrate-nitrogen content between monitoring drains, F_1=19.77 (P<0.01).
Significant difference in nitrate-nitrogen content between sampling months, F_2=2.75 (P<0.05).

Seasonal averages shows variation with minimum of 3 ± 0.8 (Monsoon) to 6.10 ± 0.59 (Winter) in NAJ-1, 5.54 ± 2.16 (Monsoon) to 8.94 ± 1.79 (Winter) in NAJ-1, 5.54 ± 2.16 (Monsoon) to 8.94 ± 1.79 (Winter) in CIV-2, 3.77 ± 0.75 (Monsoon) to 8.03 ± 1.03 (Winter) in SEN-4, 3.32 ± 1.19 {Mon.) to 7.06

± 1.53 {Winter) in BAR-5, 1.32 ± 0.39 (Monsoon) to 2.38 ± 0.86 (Winter) in HIN-6. Yearly averages were in the order of 7.42 > 7.28 > 6.52 > 5.37 > 4.76 > 2.02 in drains of CIV-2, SEN-4, POW-3, BAR-5, NAJ-1 and HIN-6 respectively.

Monthly average values (Mean ± SD) of NO_3-N and the results of analysis of variance in river water are tabulated in Table RWM-11.

The values ranged from 0.84 ± 0.33 (May) to 1.38 ± 0.05 (Jan.) at WAZ-1, 0.95 ± 0.03 (Jul.) to 4.64 ± 2.09 (Jun.) at ISB-2, 0.71 ± 0.16 (Jul.) to 3.61 ± 1.44 (Jan.) at ITO-3, 1.11 ± 0.23 (Jul.) to 4.06 ± 1.46 (Jan.) at NIZ-4, 1.14 ± 0.22 (Aug.) to 3.47 ± 1.01 (Jan.) at OKH-5. Analysis of variance shows significant difference in NO_3-N between the monitoring stations (F_1=20.68, P<0.01) as well between sampling months (F_2=2.59, P<0.05).

Seasonal and annual averages of NO_3-N ranged from a minimum of 0.09 ± 0.17 (Monsoon) to a maximum of 1.31 ± 0.2 (Winter) at WAZ-1, 1.21 ± 0.54 (Monsoon) to 2.94 ± 0.64 (Winter) at ISB-2, 1.29 ± 0.80 (Monsoon) to 3.13 ± 0.93 (Winter) at ITO-3, 1.64 ± 0.88 (Monsoon) to 3.59 ± 0.96 (Winter) at NIZ-4, 1.5710.88 (Monsoon) to 2.80 ± 1.00 (Winter) at OKH-5.

Table RWM-11: Monthly Variation of Nitrate Nitrogen (mg/l) in River Water (Year 1993-1995)

Month	*WAZ-1*	*ISB-2*	*ITO-3*	*NIZ-4*	*OKH-5*
	Mean ± SD	*Mean ± SD*	*Mean ± SD*	*Mean ± SD*	*Mean ± SD*
Jan.	1.38 ± 0.05	3.34 ± 0.08	3.61 ± 1.44	4.06 ± 1.46	3.47 ± 0.81
Feb.	1.32 ± 0.17	3.12 ± 0.73	3.09 ± 0.30	3.53 ± 0.26	3.00 ± 0.14
Mar.	1.15 ± 0.25	3.27 ± 0.74	2.67 ± 0.30	3.11 ± 0.53	2.52 ± 0.45
Apr.	0.95 ± 0.10	2.6.7 ± 0.33	2.74 ± 0.31	3.67 ± 1.15	3.31 ± 0.63
May	0.84 ± 0.33	2.90 ± 0.22	3.12 ± 0.72	3.92 ± 0.72	3.57 ± 0.49
Jun.	0.99 ± 0.13	4.64 ± 2.09	3.38 ± 0.50	3.56 ± 0.59	2.92 ± 0.70
Jul.	0.76 ± 0.08	0.95 ± 0.03	0.71 ± 0.16	1.11 ± 0.23	1.49 ± 0.68
Aug.	0.99 ± 0.08	0.99 ± 0.08	0.85 ± 0.14	1.18 ± 0.33	1.14 ± 0.22
Sep.	0.96 ± 0.21	1.69 ± 0.72	2.30 ± 0.58	2.63 ± 0.83	2.09 ± 0.66
Oct.	0.97 ± 0.01	2.18 ± 0.77	2.20 ± 0.24	2.38 ± 0.39	2.25 ± 0.70
Nov.	1.28 ± 0.36	2.33 ± 0.34	2.24 ± 0.18	2.67 ± 0.19	2.43 ± 0.30
Dec.	1.19 ± 0.27	2.74 ± 0.49	2.73 ± 0.12	3.26 ± 0.37	2.47 ± 1.00

Two-way ANOVA Test

Significant difference in NO_3-N content between river monitoring stations, F_1=20.68 (P<0.01).
Significant difference in NO_3-N content between monitoring months, F_2=2.59 (P<0.05).

The peak values of NO_3-N were observed during winter season because of reduced water flow at Delhi stretch due to lean discharge of Wazirabad barrage and flow of drain effluents without dilution. The lowest values during the rainy season was due to dilution effect at down-stream stations.

Ajmal *et al.* (1985) reported minimum nitrate content in winter and maximum in summer in Kali Nadi. Sunder (1988) observed minimum during July and maximum in January. Shah (1988) reported the minimum in summer and maximum in winter months in river Jhelum. Lower values of nitrate has been reported in rainy season by Trivedy *et al.* (1990).

Annual averages were in order of 2.92 > 2.57 > 2.55 > 2.47 > 1.00 at monitoring station NIZ-4, ISB-2, OKH-5, ITO-3, and WAZ-1 respectively.

Ammonical-Nitrogen of Drain and River Water

The nitrogenous organic matter present in water is degraded biologically with the production of ammonia. The most important source of ammonia in natural water is the decomposition and ammonification of organic matter and also the sewage along with the excreta of same aquatic animals (Saxena, 1987). Presence of ammonia can be considered as the chemical evidence of organic pollution (Trivedy *et al.*, 1987; Ajmal and Razi-Ud-Din, 1988). Organically polluted waters has normally higher concentration of ammonia which is a product of ammonification of organic matter (Ellis *et al.*, 1946, Raybak and Sikarska, 1976).

Table DWM-11 shows monthly averages (Mean ± SD) and the results of analysis of variance of NH_3-N of different drains. The values (mg/l) ranged between a minimum of 13.97 ± 3.4 to a maximum of 26.53 ± 23.17 in NAJ-1, 20.10 ± 8.18 to 37.83 ± 21.76 in CIV-2, 22.57 ± 3.71 to 38.87 ± 5.33 in POW-3, 25.67 ± 4.71 to 49.33 ± 4.99 in SEN-4, 19.0 ± 4.85 to 32.67 ± 4.50 in BAR-5, 0.52 ± 0.30 to 4.49 ± 2.15 in HIN-6. Analysis of variance shows significant difference between sampling drains (F_1=19.42, P<0.01) as well as between monitoring months (F_2=2.68, P<0.05). Seasonal and annual average values of NH_3-N ranged from 16.85 ± 6.36 to 19.84 in NAJ-1, 23.47 ± 10.88 to 34.14 ± 10.0 in CIV-2, 24.81 ± 3.74 to 32.42 ± 6.7 in POW-3, 31.78 ± 8.9 to 44.61 ± 5.68 in SEN-4, 24.22 ± 9.54 to 32.0 ± 4.84 in BAR-5, 1.71 ± 0.6 to 5.63 ± 1.3 in HIN-6 respectively. Annual mean values of NH_3-N (mg/l) were 18.43, 29.04, 29.70, 38.67, 26.53, 26.53 and 4.1 in NAJ-1, CIV-2, POW-3, SEN-4, BAR-5 and HIN-6 respectively.

Table DWM-11: Monthly Variation of Ammonical-Nitrogen (mg/l) in Drain Water (Year 1993-1995)

Month	*NAJ-1*	*CIV-2*	*POW-3*	*SEN-4*	*BAR-5*	*HIN-6*
	Mean ± SD	*Mean ± SD*	*Mean ± SD*	*Mean ± SD*	*Mean ± SD*	*Mean ± SD*
Jan.	17.13 ± 2.74	33.10 ± 12.14	28.77 ± 6.27	43.17 ± 5.33	28.33 ± 1.25	2.95 ± 1.93
Feb.	22.10 ± 4.93	34.33 ± 6.76	33.00 ± 2.83	41.61 ± 3.09	25.33 ± 2.87	9.11 ± 6.30
Mar.	16.20 ± 3.96	30.17 ± 5.95	38.87 ± 5.33	43.43 ± 10.30	28.00 ± 6.68	3.15 ± 1.45
Apr.	20.17 ± 2.09	35.10 ± 16.21	34.47 ± 6.64	33.10 ± 2.02	22.33 ± 6.18	4.03 ± 0.60
May	14.40 ± 8.56	31.83 ± 21.76	29.47 ± 4.03	39.17 ± 16.88	19.20 ± 4.85	3.74 ± 1.44
Jun.	16.63 ± 7.13	22.27 ± 4.01	26.87 ± 2.29	31.67 ± 11.61	25.87 ± 10.57	4.49 ± 2.15
Jul.	26.53 ± 23.17	20.10 ± 8.18	27.20 ± 2.44	25.67 ± 4.71	19.00 ± 3.27	0.55 ± 0.08
Aug.	13.97 ± 3.40	24.27 ± 13.35	24.67 ± 3.40	30.33 ± 10.53	21.33 ± 4.99	0.63 ± 0.11
Sep.	15.80 ± 3.51	26.03 ± 9.55	22.57 ± 3.71	39.33 ± 2.87	32.33 ± 11.67	0.52 ± 0.30
Oct.	18.57 ± 5.30	30.20 ± 10.79	30.60 ± 2.41	36.00 ± 5.10	32.67 ± 4.50	1.36 ± 0.74
Nov.	20.13 ± 5.87	27.77 ± 5.12	29.13 ± 6.31	45.17 ± 1.55	31.67 ± 5.79	1.60 ± 0.95
Dec.	18.89 ± 0.94	27.33 ± 6.18	30.80 ± 7.59	49.33 ± 4.99	32.33 ± 10.66	1.89 ± 0.86

Two-way ANOVA Test

Significant difference in ammonical-nitrogen content between monitoring drains, F_1=19.42 (P<0.01).

Significant difference in ammonical-nitrogen content between sampling months, F_2=2.68 (P<0.05).

Monthly average (Mean ± SD) NH_3-N values of river stations and results of analysis of variance are tabulated in Table RWM-12. The values ranged from 0.82 ± 0.1 (Nov.) to 1.13 ± 0.55 (Mar.) at

WAZ-1, 1.64 ± 0.82 (Aug.) to 9.5 ± 2.56 (Jan.) at ISB-2, 1.45 ± 0.85 (Aug.) to 8.45 ± 3.49 (Jan.) at ITO-3, 1.33 ± 0.83 (Aug.) to 7.65 ± 2.51 (Jan.) at NIZ-4, 1.4 ± 0.85 (Aug.) to 7.23 ± 1.92 (Jan.) at OKH-5. Analysis of variance showed significant difference in NH_3-N content between different stations (F_1=22.14, P<0.01) as well as between monitoring months (F_2=2.40, P<0.05).

Seasonal values (Mean ± SD) ranged from a minimum of 0.8 ± 0.49 (Monsoon) to a maximum of 0.97 ± 0.35 (Pre-Monsoon) at WAZ-1, 1.78 ± 0.81 (Monsoon) to 7.42 ± 2.77 (Winter) at ISB-2, 2.70 ± 2.62 (Monsoon) to 5.93 ± 2.01 (Winter) at NIZ-4, 2.48 ± 1.24(Monsoon) to 5.48 ± 2.32 (Summer) at OKH.-5. Annual averages were in order of 5.05 > 4.68 > 4.63 > 0.90 at ITO- 3, ISB-2, OKH-5, and WAZ-1 river monitoring stations respectively.

Table RWM-12: Monthly Variation of Ammonical-Nitrogen (mg/l) in River Water (Year 1993-1995)

Month	*WAZ-1*	*ISB-2*	*ITO-3*	*NIZ-4*	*OKH-5*
	Mean ± SD	*Mean ± SD*	*Mean ± SD*	*Mean ± SD*	*Mean ± SD*
Jan.	0.82 ± 0.07	9.50 ± 2.56	8.48 ± 2.60	7.65 ± 2.51	7.23 ± 1.92
Feb.	1.10 ± 0.41	5.97 ± 0.97	6.49 ± 1.20	5.05 ± 1.19	4.48 ± 1.22
Mar.	1.13 ± 0.55	6.65 ± 1.79	5.15 ± 1.10	3.47 ± 0.53	4.79 ± 1.77
Apr.	1.02 ± 0.12	4.09 ± 0.81	4.32 ± 0.98	4.37 ± 0.68	5.08 ± 0.82
May	0.89 ± 0.28	5.02 ± 2.01	6.65 ± 1.36	7.12 ± 2.66	7.88 ± 1.89
Jun.	0.60 ± 0.15	7.58 ± 3.18	6.83 ± 0.70	7.20 ± 3.02	4.18 ± 2.47
Jul.	0.61 ± 0.25	1.69 ± 0.86	1.54 ± 0.40	1.50 ± 1.27	1.71 ± 1.20
Aug.	0.87 ± 0.53	1.64 ± 0.82	1.45 ± 0.20	1.33 ± 0.83	1.40 ± 0.85
Sep.	0.97 ± 0.55	2.02 ± 0.68	5.11 ± 1.30	4.42 ± 3.02	4.33 ± 1.73
Oct.	1.12 ± 0.44	3.88 ± 2.45	4.33 ± 1.00	4.55 ± 3.57	5.09 ± 2.56
Nov.	0.82 ± 0.10	3.28 ± 0.65	3.68 ± 0.90	3.64 ± 1.21	5.75 ± 2.91
Dec.	0.91 ± 0.34	4.85 ± 1.50	6.57 ± 1.30	5.27 ± 0.26	3.94 ± 1.45

Two-way ANOVA Test
Significant difference in NH_3-N content between river monitoring stations, F_1=22.14 (P<0.01).
Significant difference in NH_3-N content between monitoring months, F_2=2.4 (P<0.05).

Phosphate of Drain and River Water

Phosphorous is inevitable for all organisms as it is the chief constituent of ATP, the energy carrier acid present in nucleic and as phosphate group of nucleotides. In general, phosphorous is readily oxidised and occurs in the earth's rocks principally as orthophosphate. The phosphorous occurs in natural waters and in wastewater almost solely as phosphates rather than a free state. In water, phosphorous occurs both in inorganic and organic forms (as bound in the organic matter), only inorganic phosphorous (as orthophosphates) plays a dynamic role in an aquatic ecosystem. The most important anthropogenic sources of phosphate are the discharge of domestic sewage, detergents and agricultural run-off (Trivedy *et al.*, 1987). The amount of phosphorous in most freshwater bodies is very small and is subjected in seasonal and vertical variations (Michael, 1984). Phosphorous is essential to the growth of organisms that limits the primary productivity of a water body. The phosphate content of flowing water change slightly from time to time while that of stagnant waters it is more or less constant (Imevbore, 1978). The enrichment of this nutrient leads to the process of eutrophication.

The importance of phosphate in natural waters has been stressed by several authors (Juday *et al.*, 1928; Ohle, 1934; Phifer and Thompson, 1937; Cole, 1979; Dutta Munshi and Dutta Munshi, 1995). The total phosphate content is considered to be a nutrient of major importance in production process (Hutchinson, 1957 and Vollenwinder, 1968). The increased application of fertilizers, use of detergents and domestic sewages contribute significantly to the heavy loading of phosphorous in water (Golterman, 1975).

Table DWM-12 represents monthly average values and the result of analysis of variance of total phosphate present in the drain water in six monitoring drains. The average values ranged from a minimum of 2.12 ± 0.58 to a maximum of 4.78 ± 2.89 in NAJ-1, 3.93 ± 1.76 to 7.95 ± 1.28 in CIV-2, 3.39 ± 1.01 to 5.85 ± 0.18 in POW-3, 3.42 ± 1.17 to 5.97 ± 0.17 in SEN-4, 2.37 ± 0.31 to 4.62 ± 2.34 in BAR-5, 0.07 ± 0.05 to 1.19 ± 0.44 in HIN-6. Significant difference in PO_4-P content was found between five drains (F_1=19.43 P<0.01) as well as between monitoring months (F_2=2.69, P>0.05).

Table DWM-12: Monthly Variation of Total Phosphate (mg/l) in Drain Water (Year 1993-1995)

Month	*NAJ-1* *Mean ± SD*	*CIV-2* *Mean ± SD*	*POW-3* *Mean ± SD*	*SEN-4* *Mean ± SD*	*BAR-5* *Mean ± SD*	*HIN-6* *Mean ± SD*
Jan.	3.55 ± 1.63	7.95 ± 1.28	4.91 ± 0.27	5.47 ± 0.74	2.94 ± 0.24	1.19 ± 0.44
Feb.	3.93 ± 0.43	7.08 ± 0.64	5.34 ± 0.65	5.90 ± 0.71	2.96 ± 0.38	0.86 ± 0.85
Mar.	3.24 ± 0.95	6.28 ± 1.37	5.85 ± 0.18	5.97 ± 0.17	3.44 ± 0.85	0.45 ± 0.27
Apr.	3.25 ± 0.36	6.49 ± 0.49	5.59 ± 0.72	4.93 ± 1.42	3.11 ± 0.44	0.71 ± 0.32
May	3.37 ± 1.18	5.97 ± 1.24	4.06 ± 1.16	5.48 ± 0.87	2.37 ± 0.31	0.63 ± 0.69
Jun.	2.12 ± 0.58	5.03 ± 1.96	4.23 ± 1.33	5.13 ± 1.13	3.92 ± 2.09	0.94 ± 1.05
Jul.	4.78 ± 2.89	3.93 ± 1.76	3.69 ± 0.72	3.97 ± 0.65	3.63 ± 2.45	0.07 ± 0.05
Aug.	2.75 ± 0.56	6.46 ± 1.71	3.63 ± 0.45	4.33 ± 0.66	2.96 ± 0.75	0.07 ± 0.08
Sep.	2.53 ± 0.91	5.12 ± 0.98	3.39 ± 1.01	5.04 ± 0.96	2.97 ± 0.90	0.16 ± 0.17
Oct.	3.18 ± 0.85	5.31 ± 1.17	4.26 ± 1.61	3.42 ± 1.17	2.89 ± 0.77	0.19 ± 0.26
Nov.	3.71 ± 0.44	5.49 ± 2.33	5.03 ± 0.90	4.40 ± 0.59	3.39 ± 0.37	0.22 ± 0.17
Dec.	3.82 ± 0.24	5.66 ± 1.23	5.47 ± 0.60	5.20 ± 0.78	4.62 ± 2.34	0.19 ± 0.26

Two-way ANOVA Test

Significant difference in total phosphate content between monitoring drains, F_1=19.43 (P<0.01).
Significant difference in total phosphate content between sampling months, F_2=2.69 (P<0.05).

The seasonal average values of PO_4-P content in drain water ranged from a minimum of 2.99 ± 0.97 (Summer) to a maximum of 3.72 ± 1.02 (Winter) in NAJ-1, 5.17 ± 1.84 (Monsoon) to 7.25 ± 1.26 (Winter) in CIV-2, 3.57 ± 0.77 (Monsoon) to 5.29 ± 0.56 (Winter) in POW-3, 3.94 ± 0.97 (Pre-Monsoon) to 5.69 ± 0.62 (Winter) in SEN-4, 3.06 ± 0.63 (Pre-Monsoon) to 3.55 ± 1.57 (Winter) in BAR-5 and 0.49 ± 0.13 (Pre-Monsoon) to 1.06 ± 0.38 (Winter) in HIN-6. Annual averages were in order of 5.9 > 4.94 > 4.62 > 3.35 > 3.27 > 0.70 in CIV-2, SEN-4, POW-3, NAJ-1, BAR-5 drain and HIN-6 (Hindon river cut) respectively.

Table RWM-13: Monthly Variation of Phosphate (mg/l) in River Water (Year 1993-1995)

Month	*WAZ-1*	*ISB-2*	*ITO-3*	*NIZ-4*	*OKH-5*
	Mean ± SD	*Mean ± SD*	*Mean ± SD*	*Mean ± SD*	*Mean ± SD*
Jan.	0.17 ± 0.00	1.47 ± 0.32	1.82 ± 0.34	1.81 ± 0.41	1.30 ± 0.40
Feb.	0.14 ± 0.02	1.43 ± 0.18	1.58 ± 0.52	1.30 ± 0.33	1.42 ± 0.79
Mar.	0.16 ± 0.03	1.79 ± 0.13	1.26 ± 0.01	1.27 ± 0.18	1.54 ± 1.01
Apr.	0.17 ± 0.02	1.38 ± 0.24	1.20 ± 0.09	1.18 ± 0.13	1.18 ± 0.53
May	0.15 ± 0.04	1.27 ± 0.25	1.43 ± 0.31	1.24 ± 0.15	1.19 ± 0.36
Jun.	0.11 ± 0.05	1.81 ± 0.48	1.61 ± 0.29	1.12 ± 0.11	1.00 ± 0.16
Jul.	0.08 ± 0.03	0.89 ± 0.07	0.83 ± 0.09	0.82 ± 0.13	0.64 ± 0.22
Aug.	0.08 ± 0.03	0.84 ± 0.09	0.80 ± 0.07	0.82 ± 0.21	0.66 ± 0.14
Sep.	0.16 ± 0.14	0.96 ± 0.16	1.07 ± 0.30	1.11 ± 0.14	0.91 ± 0.11
Oct.	0.10 ± 0.01	1.42 ± 0.64	1.22 ± 0.55	1.17 ± 0.50	1.21 ± 0.35
Nov.	0.14 ± 0.05	1.15 ± 0.51	0.93 ± 0.55	1.28 ± 0.48	1.66 ± 0.70
Dec.	0.13 ± 0.06	1.03 ± 0.14	1.26 ± 0.15	1.44 ± 0.30	1.30 ± 0.74

Two-way ANOVA Test

Significant difference in phosphate content between river monitoring stations, F_1=19.81 ($P<0.01$).
Significant difference in phosphate content between monitoring months, F_2=2.20 ($P<0.05$).

Monthly averages (Mean ± SD) and the results of analysis of variance of phosphate in river water are tabulated in Table RWM-13. The values showed variation with a minimum of 0.08 ± 0.03 (Jul.) to a maximum of 0.17 ± 0.0 (Jan.) at WAZ-1, 0.84 ± 0.09 (Aug.) to 1.79 ± 0.13 (Mar.) at ISB-2, 0.08 ± 0.07 (Aug.) to 1.82 ± 0.34 (Jan.) at ITO-3, 0.82 ± 0.21 (Aug.) to 1.81 ± 0.41 (Jan.) at NIZ-4, and 0.66 ± 0.14 (Aug.) to 1.54 ± 1.01 (Jan.) at OKH-5. Analysis of variance showed a significant difference between the sampling stations (F_1=18.81, $P<0.01$) and also between monitoring months (F_2=2.2, $P<0.05$).

Seasonal as well as annual average values of river water showed fluctuation from 0.10 ± 0.09 (Monsoon) to 0.16 ± 0.02 (Winter) at WAZ-1, 0.90 ± 0.13 (Monsoon) to 1.57 ± 0.39 (Summer) at ISB-2, 0.90 ± 0.22 (Monsoon) to 1.55 ± 0.43 (Winter) at ITO-3, 0.94 ± 0.24 (Monsoon) to 1.54 ± 0.40 (Winter) at NIZ-4, 0.74 ± 0.20 (Monsoon) to 1.19 ± 0.57 (Winter) at OKH-5 respectively.

The phosphate content in polluted stretch were maximum during winter season, possibly due to lesser flow of river, *i.e.* the percentage contribution of drain water was much higher than the river water discharge from Wazirabad barrage. The reduced level of PO_4 in monsoon season might be due to dilution of drain water by rain water.

Annual averages were in order of 1.29 > 1.25 > 1.22 > 1.17 > 0.13 at monitoring stations ISB-2, ITO-3, NIZ-4, OKH-5 and WAZ-1 respectively.

Hutchinson (1957) has reported an increase of phosphate as a result of sewage contamination. Similar findings have also been reported by Shah (1988), Rana and Palria (1982), Ghose and Sharma (1989) and Kumar (1995a) in their studies of different river system.

Sulphate of Drain and River Water

Sulphate occurs naturally in water as result of leaching from gypsum and other common minerals. The sulphate is usually second to carbonate as the principal anion in freshwaters, although chloride sometime surpasses it. Free or elemental sulphur is inactive at normal temperature. This element can combine with both metals and non-metals to form many compounds.

Atmospheric sources of sulphate in water have increased with industrial activities. Coal combustion produces maximum SO_2 and copper smelting and paper manufacturing add to it. Through precipitation and run-off sulphate level of some freshwater becomes unusually high. The sulphate content of municipal water supplies is usually increased during clarification by alum. Monthly average values (Mean ± SD) of sulphate and the results of analysis of variance tabulated in Table DWM-13.

The average values varied between a minimum of 64.5 ± 2.67 to a maximum of 77.67 ± 2.05 in NAJ-1, 62.67 ± 11.12 to 91.0 ± 8.49 in CIV-2, 70.33 ± 15.52 to 111.67 ± 17.44 in POW-3, 88.33 ± 18.12 to 124.83 ± 21.14 in SEN-4, 91.67 ± 6.6 to 112.33 ± 11.09 in BAR-5, and 17.0 ± 5.72 to 33.0 ± 2.45 at HIN-6. Results of analysis of variance (five drains) shows significant difference between sampling drains (F_1=19.03, P<0.01) and also between sampling months (F_2=2.77, P<0.05).

Seasonal values ranged from 67.87 ± 10.95 (Winter) to 71.0 ± 15.94 (Monsoon) in NAZ-1, 68.67 ± 9.68 to 88.33 ± 10.22 (Summer) in CIV-2, 76.89 ± 14.76 (Monsoon) to 90.43 ± 18.44 (Pre-Monsoon) in POW-3, 90.22 ± 8.66 (Monsoon) to 111.0 ± 14.76 (Pre-Monsoon) in SEN-4, 96.56 ± 10.24 (Monsoon) to 112.14 ± 9.39 (Pre-Monsoon) in BAR-5, 20.44 ± 2.6 (Monsoon) to 32.22 ± 3.8 (Winter) in HIZ-6. Annual average SO_4 values of five drains and one river cut were in order of 104.81 > 102.63 >, 87.61 > 80.51 > 70.99 > 29.17 in BAR-5, SEN-4, POW-3, NAJ-1 and HIN-6 respectively.

Table DWM-13: Monthly Variation of Sulphate (mg/l) in Drain Water (Year 1993-1995)

Month	*NAJ-1* *Mean ± SD*	*CIV-2* *Mean ± SD*	*POW-3* *Mean ± SD*	*SEN-4* *Mean ± SD*	*BAR-5* *Mean ± SD*	*HIN-6* *Mean ± SD*
Jan	70.0 ± 8.8	83.0 ± 9.63	88.67 ± 8.18	105.67 ± 18.87	111.33 ± 3.40	32.00 ± 2.83
Feb	77.7 ± 5.2	76.3 ± 5.91	90.67 ± 3.68	86.33 ± 18.12	104.00 ± 6.38	33.00 ± 4.24
Mar	69.7 ± 2.1	91.00 ± 8.49	111.67 ± 17.44	93.33 ± 23.10	111.00 ± 6.48	30.67 ± 1.25
Apr	77.7 ± 2.1	87.00 ± 16.97	91.67 ± 6.18	113.00 ± 12.08	108.00 ± 14.24	34.00 ± 3.74
May	71.3 ± 8.5	89.00 ± 6.24	88.67 ± 10.34	124.83 ± 21.14	97.00 ± 4.55	31.33 ± 2.49
Jun	73.3 ± 6.8	85.67 ± 2.05	85.33 ± 10.21	106.67 ± 20.89	105.00 ± 12.96	32.00 ± 2.16
Jul	71.0 ± 20.1	67.33 ± 6.34	73.67 ± 13.07	83.33 ± 4.78	91.67 ± 6.60	23.67 ± 1.70
Aug	67.3 ± 17.8	62.67 ± 11.12	70.33 ± 15.52	88.33 ± 3.09	91.67 ± 3.09	20.67 ± 1.70
Sep	74.7 ± 4.0	76.00 ± 5.10	86.67 ± 5.79	99.0 ± 8.04	106.33 ± 10.87	17.00 ± 5.72
Oct	68.0 ± 5.7	79.83 ± 3.57	79.33 ± 17.00	98.67 ± 13.47	109.33 ± 7.36	30.00 ± 1.41
Nov	66.7 ± 2.6	86.00 ± 2.45	93.33 ± 13.10	117.67 ± 5.79	112.33 ± 11.09	33.00 ± 2.45
Dec	64.5 ± 11.1	81.67 ± 7.41	91.33 ± 18.80	112.67 ± 11.59	110.00 ± 8.16	32.67 ± 4.11

Two-way ANOVA Test
Significant difference in sulphate content between monitoring drains, F_1=19.03 (P<0.01).
Significant difference in sulphate content between sampling months, F_2=2.77 (P<0.05).

Table RWM-14 presents the monthly average (Mean ± SD) values and the results of analysis of variance of river water. The minimum and maximum sulphate content were 19.0 ± 0.82 (Jul.) and 36.33 ± 3.77 (Dec.) at WAZ-1, 21.0 ± 1.63 (Aug.) and 42.0 ± 2.94 (Feb.) at ISB-2, 25.0 ± 4.24 (Aug.) and

42.0 ± 8.49 (Mar.) at ITO-3, 26.33 ± 5.44 (Jul.) and 47.33 ± 4.19 (Nov.) at NIZ-4, 19.67 ± 1.25 (Aug.) and 54.38 ± 12.30 (Nov.) at OKH-5. Analysis of variance shows significant difference in sulphate content between monitoring stations (F_1=19.49, P<0.01) and also between sampling months (F_2=2.81, P<0.05).

Seasonal values varied between 20.0 ± 2.67 (Monsoon) to 28.78 ± 4.29 (Winter) at WAZ-1, 22.11 ± 4.31 (Monsoon) to 38.33 ± 7.96 (Summer) at ISB-2, 25.67 ± 4.47 (Monsoon) to 38.33 ± 2.81 (Pre-Monsoon) at ITO-3, 27.78 ± 4.94 (Monsoon) to 43.83 ± 5.46 (Pre-Monsoon) at NIZ-4, 23.6 ± 6.52 (Monsoon) to 46.73 ± 11.60 (Pre-Monsoon) at OKH-5. The concentration of sulphate in the river Yamuna at Delhi stretch showed some distinct fluctuation at different sites. Maximum concentration of sulphate was generally observed during summer and minimum during rainy season at Wazirabad barrage (non-polluted stretch). All down-stream stations showed peak values during winter season and lower values during monsoon season. Annual average values of SO_4 were 38.25 > 36.02 > 35.19 > 33.44 > 26.31 at NIZ-4, OKH-5, ITO-3, ISB-2 and WAZ-1 river monitoring points respectively.

Table RWM-14: Monthly Variation of Total Sulphate (mg/l) in River Water (Year 1993-1995)

Month	*WAZ-1*	*ISB-2*	*ITO-3*	*NIZ-4*	*OKH-5*
	Mean ± SD	*Mean ± SD*	*Mean ± SD*	*Mean ± SD*	*Mean ± SD*
Jan.	28.33 ± 0.47	33.67 ± 5.44	37.00 ± 8.52	37.33 ± 6.18	36.40 ± 5.38
Feb.	24.67 ± 0.94	42.00 ± 2.94	38.00 ± 2.83	40.67 ± 6.55	45.07 ± 6.54
Mar.	27.00 ± 4.32	40.00 ± 5.66	42.00 ± 8.49	43.67 ± 8.01	43.33 ± 9.29
Apr.	24.67 ± 4.50	39.67 ± 1.70	38.67 ± 6.60	42.67 ± 4.78	36.33 ± 6.94
May	28.33 ± 2.62	40.00 ± 6.68	39.67 ± 2.62	43.00 ± 5.72	35.07 ± 5.01
Jun.	29.33 ± 11.09	33.67 ± 12.04	36.67 ± 13.57	37.00 ± 13.14	34.50 ± 5.21
Jul.	19.00 ± 0.82	22.67 ± 6.02	23.67 ± 5.19	26.33 ± 5.44	28.07 ± 6.54
Aug.	20.00 ± 1.41	21.00 ± 1.63	25.00 ± 4.24	25.33 ± 4.11	19.67 ± 1.25
Sep.	21.00 ± 4.08	22.67 ± 3.86	28.33 ± 3.30	31.67 ± 1.89	23.07 ± 6.92
Oct.	27.33 ± 3.09	32.67 ± 2.49	36.33 ± 2.05	40.33 ± 4.19	39.07 ± 0.82
Nov.	29.67 ± 2.05	36.67 ± 3.09	40.33 ± 1.89	47.33 ± 4.19	54.38 ± 12.30
Dec.	36.33 ± 3.77	36.67 ± 5.91	36.67 ± 4.99	43.67 ± 5.31	37.30 ± 3.75

Two-way ANOVA Test

Significant difference in total sulphate content between river monitoring stations, F_1=19.49 (P<0.01).
Significant difference in total sulphate content between monitoring months, F_2=2.81 (P<0.05).

Alkalinity of Drain and River Water

Alkalinity commonly results from carbon dioxide and water attaching sedimentary carbonate rocks and dissolving out some of the carbonate to form bicarbonate solution. Each carbonate ion in solution takes up hydrogen ion to become bicarbonate ion. Further hydrolysis of bicarbonate or carbonate leads to the appearance of hydroxyl ions, which will increase the value of pH. Alkalinity measures the buffering capacity of water and the content of CO_2 in its various forms are involved in this carbonate bicarbonate-carbonic and buffering system. Alkalinity value provides guidance in applying proper dose of chemicals in water and wastewater treatment processes.

Table DWM-14, represents monthly average alkalinity values (Mean ± SD) and results of analysis of variance for five drains. The average maximum and minimum values (mg/l) were, 198.67 ± 8.82

and 362.67 ± 35.48, 256.0 ± 11.78 and 366.67 ± 80.16, 301.21 ± 36.71 and 350.36 ± 29.93, 288.0 ± 98.36 and 465.33 ± 195.97, 261.33 ± 98.00 and 415.33 ± 15.17, 117.33 ± 6.24 and 259.67 ± 12.36 in NAJ-1, CIV-2, POW-3, SEN-4, BAR-5 and HIN-6 respectively. Analysis of variance shows significant difference between sampling drains (F_1=19.1, P<0.01) as well as between sampling months (F_2=2.77, P<0.05).

Seasonal averages ranged from 216.17 ± 28.94 (Summer) to 256.78 ± 43.15 (Monsoon) in NAJ-1, 271.11 ± 15.06 (Monsoon) to 340.0 ± 73.82 (Pre-Monsoon) in CIV-2, 276.33 ± 39.19 (Monsoon) to 357.56 ± 28.15 (Winter) in POW-3, 306.33 ± 70.47 (Monsoon) to 398.44 ± 31.97 (Winter) in SEN-4, 302.44 ± 95.47 (Monsoon) to 402.22 ± 30.34 (Winter) in BAR-5, 121.67 ± 20.36 (Monsoon) to 259.11 ± 28.67 (Winter) in HIN-6. Annual averages were in order of 377.58 > 361.19 > 326.17 > 320.86 > 234.25 > 199.69 in SEN-4, BAR-5, POW-3, CIV-2, NAZ-1 and Hin-6 respectively.

Table DWM-14: Monthly Variation of Total Alkalinity (mg/l) in Drain Water (Year 1993-1995)

Month	*NAJ-1*	*CIV-2*	*POW-3*	*SEN-4*	*BAR-5*	*HIN-6*
	Mean ± SD	*Mean ± SD*	*Mean ± SD*	*Mean ± SD*	*Mean ± SD*	*Mean ± SD*
Jan.	242.7 ± 40.8	348.0 ± 17.0	313.7 ± 13.1	398.7 ± 16.4	413.3 ± 23.2	259.0 ± 12.4
Feb.	274.7 ± 15.43	315.7 ± 28.6	307.5 ± 39.7	369.3 ± 24.1	415.3 ± 15.2	258.7 ± 16.8
Mar.	219.3 ± 9.3	357.3 ± 49.1	307.9 ± 36.2	397.3 ± 11.5	394.7 ± 44.1	251.0 ± 8.5
Apr.	224.7 ± 28.5	366.7 ± 80.2	301.2 ± 36.7	390.7 ± 11.5	356.7 ± 66.1	205.0 ± 27.1
May	222.7 ± 44.4	333.3 ± 45.8	302.9 ± 59.3	376.0 ± 33.9	347.0 ± 50.2	193.0 ± 35.7
Jun.	198.7 ± 8.2	301.3 ± 21.0	307.2 ± 26.7	465.3 ± 196.0	345.3 ± 38.7	181.7 ± 8.1
Jul.	302.8 ± 36.0	282.7 ± 3.8	311.5 ± 43.2	288.0 ± 98.4	362.0 ± 41.2	118.7 ± 9.0
Aug.	232.0 ± 11.3	274.7 ± 12.4	309.5 ± 32.9	320.0 ± 41.0	361.3 ± 98.0	117.3 ± 6.2
Sep.	235.7 ± 31.6	276.0 ± 22.8	311.1 ± 25.8	311.0 ± 54.8	284.0 ± 102.3	129.0 ± 3.6
Oct.	206.7 ± 35.2	310.7 ± 30.2	328.6 ± 61.7	381.3 ± 32.7	396.0 ± 11.8	186.0 ± 38.0
Nov.	204.0 ± 11.8	361.3 ± 101.5	342.1 ± 34.9	396.0 ± 60.5	369.3 ± 47.5	237.3 ± 21.7
Dec.	248.0 ± 36.4	342.7 ± 16.4	350.4 ± 29.9	437.3 ± 19.7	389.3 ± 45.8	259.7 ± 8.5

Two-way ANOVA Test

Significant difference in total alkalinity content between monitoring drains, F_1=19.11 (P<0.01).
Significant difference in total alkalinity content between sampling months, F_2=2.77 (P<0.05).

Monthly average values (Mean ± SD) and the results of analysis of variance of alkalinity are tabulated in Table RWM-15. Monthly maximum and minimum values were 81.33 ± 6.6 (Aug.) and 161.0 ± 4.24 (Dec.) at WAZ-1, 104.67 ± 16.52 (Aug.) and 208.0 ± 26.13 (Mar.) at ISB-2, 90.0 ± 15.58 (Aug.) and 234.67 ± 43.12 at ITO-3, 86.67 ± 21.0 (Aug.) and 234.67 ± 16.44 (Feb.) at NIZ-4, 92.0 ± 20.40 (Aug.) and 244.0 ± 20.4 (Feb.) at OKH-5. Analysis of variance shows significant difference in total alkalinity between different monitoring points (F_1=19.98, P<0.01) and also between sampling months (F_2=2.78, P<0.05).

Seasonal values varied from a minimum of 85.78 ± 16.42 (Monsoon) to a maximum of 147.22 ± 13.53 (Winter) at WAZ-1, 108.56 ± 17.56 (Monsoon) to 2.4.56 ± 19.6 (Monsoon) at ISB-2, 116.0 ± 52.35 (Monsoon) to 228.0 ± 26.47 (Winter) at ITO-3, 98.56 ± 26.72 (Monsoon) to 231.56 ± 23.66 (Winter) at NIZ-4, 98.89 ± 20.85 (Monsoon) to 225.33 ± 19.60 (Winter) at OKH-5 respectively.

During the present investigation carbonate alkalinity was found to be higher during winter in all stations. This may be due to lean river flow and higher contribution from drain water at Delhi stretch and lower temperature bringing down the rate of decomposition of salts to minimum thereby increasing the alkalinity. The second peak in summer may be due to low water level and presence of excess free carbon dioxide produced in the process of decomposition. Ruttener (1953), Munawar (1970), Sunder (1988) and Kumar (1994) observed the accumulation of bicarbonate in summer due to increased rate of decomposition. Hegde *et al.* (1985) and Trivedy *et al.* (1990) also noticed the same observation. The winter peak values of total alkalinity have been observed by Munawar (1970), Bhatnagar and Sharma (1975), Bilgrami and Datta Munshi (1985). The minimum values of bicarbonate alkalinity were observed during monsoon months may be due to the dilution effect and condition less favourable for the photosynthetic conversion to carbonates. Similar results were also recorded by Rao (1971), Brajnandan *et al.* (1985) and Kumar (1995). They pointed out that the dilution plays a key role in lowering the alkalinity. Adebisi (1980) also explained that the alkalinity is inversely proportional to the water level.

Table RWM-15: Monthly Variation of Total Alkalinity (mg/l) in River Water (Year 1993-1995)

Month	*WAZ-1* Mean ± SD	*ISB-2* Mean ± SD	*ITO-3* Mean ± SD	*NIZ-4* Mean ± SD	*OKH-5* Mean ± SD
Jan.	141.33 ± 1.89	196.00 ± 17.28	234.67 ± 43.12	225.33 ± 34.62	224.00 ± 8.64
Feb.	141.33 ± 17.99	208.00 ± 26.13	2.25.33 ± 9.98	242.67 ± 16.44	244.00 ± 20.40
Mar.	126.67 ± 9.43	206.67 ± 15.43	218.67 ± 10.50	229.33 ± 21.25	213.33 ± 11.47
Apr.	113.00 ± 9.90	178.33 ± 12.76	187.33 ± 27.15	198.67 ± 34.92	171.30 ± 19.75
May	99.67 ± 5.91	108.00 ± 65.24	176.67 ± 25.94	199.33 ± 10.87	152.03 ± 14.97
Jun.	81.33 ± 6.80	166.00 ± 15.75	179.33 ± 26.91	183.33 ± 19.62	133.33 ± 15.69
Jul.	78.67 ± 10.50	116.67 ± 20.55	103.33 ± 11.81	99.33 ± 20.68	102.67 ± 13.20
Aug.	81.33 ± 6.60	104.67 ± 16.52	90.00 ± 15.58	86.67 ± 21.00	92.00 ± 20.40
Sep.	97.33 ± 21.25	104.33 ± 11.44	154.67 ± 74.21	109.67 ± 31.75	102.00 ± 25.35
Oct.	115.03 ± 22.73	163.33 ± 35.57	162.77 ± 34.92	164.00 ± 36.81	180.00 ± 40.13
Nov.	141.33 ± 11.47	193.33 ± 24.51	202.67 ± 25.37	213.33 ± 26.40	226.67 ± 49.89
Dec.	161.00 ± 4.24	206.33 ± 8.96	225.33 ± 9.43	228.00 ± 5.66	197.33 ± 12.36

Two-way ANOVA Test

Significant difference in alkalinity between river monitoring stations, F_1=19.98 (P<0.01).
Significant difference in alkalinity between monitoring months, F_2=2.78 (P<0.05).

Annual average values of total alkalinity at Delhi stretch were in order of 181.64 > 180.60 > 169.89 > 162.64 > 114.83 at NIZ-4, ITO-3, OKH-4, ISB-2 and WAZ-1 respectively.

The present observation also showed some extra load of alkalinity at site NIZ-4, ITO-3, OKH-4 and ISB-2 which might be due to the domestic and industrial effluents. Shah (1988) also recorded higher concentration of bicarbonate alkalinity in river Jhelum due to domestic sewage. Sreenivasan and Sunderaj, (1967), Munawar (1970) and Sreenivasan *et al.* (1974) have also reported similar findings.

Chloride of Drain and River Water

Contamination of water from domestic sewage can be monitored by chloride assays of the concerned water bodies. This is because human and animal excretions contains an average of 5g chloride/litre (Paramasivam and Sreenivasan, 1981). The chloride themselves do not pose any hazards as far as water quality is concerned but their concentration works out as a very good indicator of organic pollution (Trivedy *et al.*, 1990a). Chloride is influential in general due to osmotic salinity balance and ion exchange, but metabolic utilisation may not cause significant variations in the spatial and seasonal distribution with in an aquatic body. However, pollution sources of chloride *viz.* municipal wastewater, industrial sources and organic wastes can modify natural concentrations greatly (Ownbey and Kee, 1967). Chloride content also increase with the degree of eutrophication (Goel *et al.*, 1980, Sinha, 1986). Several workers like Zafar (1964) and Kumar (1995) described chloride as a pollution indicator. Thresh *et al.* (1944) also advocated that high chloride concentration are indicators of large amount of organic matter in the water which, have been further supported by Bhatnagar and Sharma (1975) and Adoni (1985).

Monthly average values (Mean ± SD) and the results of analysis of variance of chloride in drain water tabulated in Table DWM-15. The values fluctuated between 88 ± 16.33 and 182.67 ± 16.11, 88 ± 6.53 and 148.0 ± 33.15, 78.65 ± 9.98 and 136.0 ± 28.28, 128.33 ± 27.18 and 229.33 ± 78.27, 132 ± 60.78 and 192 ± 13.06, 29 ± 17.15 and 144 ± 11.05 in NAJ-1, CIV-2, POW-3, SEN-4, BAR-5, and HIN-6 respectively. Analysis of variance shows significant difference in chloride content between the drains (F_1=19.39, P<0.01) and also between monitoring months (F_2=2.7, P<0.05).

Table DWM-15: Monthly Variation of Chloride (mg/l) in Drain Water (Year 1993-1995)

Month	*NAJ-1*	*CIV-2*	*POW-3*	*SEN-4*	*BAR-5*	*HIN-6*
	Mean ± SD	*Mean ± SD*	*Mean ± SD*	*Mean ± SD*	*Mean ± SD*	*Mean ± SD*
Jan.	88.0 ± 16.3	100.0 ± 14.2	101.3 ± 24.7	145.3 ± 23.0	178.7 ± 6.8	137.0 ± 6.2
Feb.	98.7 ± 1.9	92.3 ± 23.3	97.3 ± 8.2	128.3 ± 27.2	156.0 ± 18.2	144.0 ± 11.1
Mar.	94.0 ± 12.3	133.3 ± 16.8	136.0 ± 28.3	173.3 ± 22.9	180.0 ± 21.4	101.5 ± 31.4
Apr.	108.0 ± 19.9	110.7 ± 10.0	108.0 ± 10.7	161.3 ± 3.8	130.7 ± 51.3	92.0 ± 25.7
May	101.3 ± 21.2	106.7 ± 11.5	101.3 ± 15.0	196.0 ± 6.6	182.0 ± 23.2	121.0 ± 22.1
Jun.	105.3 ± 18.0	118.0 ± 18.0	100.0 ± 16.3	229.3 ± 7.3	192.0 ± 13.1	117.0 ± 5.4
Jul.	182.7 ± 16.1	92.0 ± 13.1	84.3 ± 14.5	137.3 ± 24.5	132.0 ± 60.8	40.7 ± 9.0
Aug.	110.7 ± 11.5	89.3 ± 5.0	78.7 ± 10.0	149.3 ± 13.6	136.7 ± 51.0	29.0 ± 17.2
Sep.	112.7 ± 11.6	88.0 ± 6.5	88.7 ± 9.0	142.0 ± 10.7	137.7 ± 49.0	38.7 ± 22.1
Oct.	97.3 ± 11.5	97.3 ± 15.4	92.0 ± 17.0	149.3 ± 6.8	170.7 ± 10.0	51.0 ± 21.4
Nov.	105.3 ± 22.2	148.0 ± 33.2	124.0 ± 28.5	188.0 ± 5.7	176.0 ± 45.4	106.0 ± 15.3
Dec.	98.7 ± 34.6	116.0 ± 31.2	121.3 ± 29.3	176.0 ± 29.0	181.3 ± 26.6	100.7 ± 15.4

Two-way ANOVA Test

Significant difference in chloride content between monitoring drains, F_1=19.39 (P<0.01).
Significant difference in chloride content between sampling months, F_2=2.70 (P<0.05).

Seasonal values varied between a minimum of 86.0 ± 19.97 to a maximum of 135.33 ± 36.00 in NAJ-1, 95.89 ± 23.16 to 122.67 ± 36.2 in CIV-2, 83.89 ± 12.13 to 111.33 ± 23.75 in POW-3, 142.78 ± 26.96 to 168.67 ± 20.32 in SEN-4, 166.33 ± 19.62 to 173.33 ± 32.96 in BAR-5, 36.11 ± 5.3 to 131.89 ± 30.6 in HIN-6. Annual average values of chloride were in order of 340.0 > 296.67 > 294.53 > 268.08 > 205.0 > 185.36 in SEN-4, POW-3, BAR-5, NAJ-1 and HIN-6 respectively.

Monthly average (Mean ± SD) values and the results of analysis of variance for chloride in river water are tabulated in Table RWM-16. The chloride contents varied from 13.67 ± 1.7 (Jul.) to 45.33 ± 20.4 (Dec.) at WAZ-1, 27.33 ± 3.4 (Sep) to 105.33 ± 13.2 (Feb.) at ISB-2, 16.67 ± 3.4 (Aug.) to 109.33 ± 16.14 (Jan.) at ITO-3, 16.67 ± 4.71 (Aug.) to 109.33 ± 12.36 (Feb.) at NIZ-4, 18.00 ± 4.32 (Aug.) to 144.0 ± 69.13 (Nov.) at OKH-5 respectively. Significant difference in chloride content was found between the monitoring stations (F_1=21.93, P<0.01) and also between sampling months (F_2=2.81, P<0.05).

Seasonal averages showed a minimum of 15.78 ± 2.62 (Monsoon) to a maximum of 32.89 ± 25.0 (Winter) at WAZ-l 32.89 ± 9.57 (Monsoon) to 96.67 ± 12.1 (Winter) at ISB-2, 30.33 ± 20.9 (Monsoon) to 107.11 ± 12.56 (Winter) at ITO-3, 23.78 ± 8.13 (Monsoon) to 101.33 ± 15.10 (Pre-Monsoon) at OKH-5, respectively.

During the present investigation the chloride content showed minimum values during monsoon months. This may be certainly due to high dilution and high flow rate by rain water. Malviya *et al.* (1990) also observed a similar trend.

Table RWM-16: Monthly Variation of Chloride (mg/l) River Water (Year 1993-1995)

Month	*WAZ-1*	*ISB-2*	*ITO-3*	*NIZ-4*	*OKH-5*
	Mean ± SD	*Mean ± SD*	*Mean ± SD*	*Mean ± SD*	*Mean ± SD*
Jan.	37.3 ± 27.4	88.0 ± 6.5	109.3 ± 16.4	102.7 ± 11.5	54.7 ± 13.2
Feb.	24.0 ± 9.8	105.3 ± 13.2	105.3 ± 10.0	109.3 ± 12.4	76.0 ± 11.8
Mar.	38.7 ± 32.2	104.0 ± 34.6	105.3 ± 26.6	101.3 ± 29.3	74.7 ± 32.7
Apr.	20.7 ± 8.4	81.7 ± 35.4	77.0 ± 32.3	74.0 ± 28.6	51.3 ± 19.6
May	27.1 ± 14.5	98.0 ± 32.8	93.3 ± 24.9	98.7 ± 20.7	40.7 ± 13.9
Jun.	20.0 ± 14.2	85.3 ± 19.7	91.3 ± 18.9	88.7 ± 15.5	47.3 ± 3.4
Jul.	13.7 ± 1.7	43.3 ± 9.6	31.7 ± 9.0	29.3 ± 8.2	32.0 ± 5.7
Aug.	18.0 ± 2.8	28.0 ± 2.8	16.7 ± 3.4	16.7 ± 4.7	18.0 ± 4.3
Sep.	15.7 ± 0.5	27.3 ± 3.4	42.7 ± 29.6	25.3 ± 5.0	28.0 ± 11.8
Oct.	30.7 ± 13.2	62.7 ± 24.1	66.7 ± 21.8	65.3 ± 20.0	58.7 ± 7.5
Nov.	42.7 ± 27.2	85.3 ± 24.7	94.7 ± 24.9	100.0 ± 22.9	144.0 ± 69.1
Dec.	45.3 ± 26.4	95.0 ± 10.8	93.0 ± 29.0	104.0 ± 17.3	81.3 ± 21.3

Two-way ANOVA Test

Significant difference in chloride content between river monitoring stations, F_1=21.93 (P<0.0).

Significant difference in chloride content between monitoring months, F_2=2.58 (P<0.05).

The present study reveals that there are two peak values in chloride content *i.e.* winter and summer seasons. The peak value during winter might be due to reduced water flow (least discharge from Wazirabad barrage). Second peak values during summer may be attributed to reduced water flow, increases rate of evaporation and concentration of solutes. Similar results were reported by Rao (1979), Billore (1981), Oberoi (1987), Shrotriya (1987), Shrotriya and Dubey (1990) and Malviya *et al.* (1990).

In the present study, site-wise fluctuations of chloride content has also been noticed. The higher values of chloride were recorded at site ITO-3, which is much higher (77.25 ± 37.28) in comparison to rest of sampling sites including WAZ-1 (27.81 ± 6.29) site which might be due to continuous influx of much contaminated domestic sewage.

Munawar (1970) suggested that higher concentrations of chloride in water is an index of pollution of animal origin and there is a direct correlation between chloride concentration and pollution level.

Total Hardness of Drain and River Water

Cations that form insoluble compounds with soap contribute to what is termed as "hardness" of water. The soap consuming power of water is lowered by heavy metals and the alkaline earth metals. The two most common bivalent cations, calcium and magnesium of water accounts for hardness.

According the Ruttener (1953), the total hardness is the total amount of alkaline earths present without reference to the particular anion to which they are bound. Hard water contains large concentrations of alkaline earths derived from the drainage of calcareous deposits (Wetzel, 1975). Most inland waters are dilute solutions of calcium and sodium carbonates but variations contributed by lithology and climate are very wide. According to Rai (1974), the total hardness is an index of water quality and is of considerable significance in connection with the discharge of domestic and municipal sewage.

Table DWM-16 presents monthly average values (Mean ± SD) and analysis of variance results of hardness of drain water. The minimum and maximum average monthly values of hardness were 188.0 ± 29.49 and 246.33 ± 4.74 in NAJ-1, 223.0 ± 22.86 and 314.33 ± 109.78 in CIV-2, 243.33 ± 27.64 and 348.67 ± 36.16 in POW-3, 265.0 ± 93.85 and 435.33 ± 113.36 in SEN-4, 218.67 ± 77.46 and 341.67 ± 50.21 in BAR-5, 110.67 ± 8.99 and 245.0 ± 12.03 in HIN-6. Analysis of variance shows significant difference in hardness between five drains (F_1=19.39, P<0.01) and also between sampling months (F_2=2.76, P<0.05).

Table DWM-16: Monthly Variation of Hardness (mg/l) in Drain Water (Year 1993-1995)

Month	*NAJ-1*	*CIV-2*	*POW-3*	*SEN-4*	*BAR-5*	*HIN-6*
	Mean ± SD	*Mean ± SD*	*Mean ± SD*	*Mean ± SD*	*Mean ± SD*	*Mean ± SD*
Jan.	216.7 ± 45.7	298.7 ± 8.4	332.7 ± 12.7	360.3 ± 13.8	322.3 ± 10.7	245.0 ± 12.0
Feb.	246.3 ± 4.7	281.7 ± 17.2	322.0 ± 30.4	354.7 ± 55.8	335.7 ± 16.2	243.7 ± 15.1
Mar.	198.3 ± 8.5	282.3 ± 74.7	333.3 ± 36.9	374.3 ± 18.0	321.1 ± 23.0	227.3 ± 5.3
Apr.	191.3 ± 21.5	207.7 ± 75.2	302.7 ± 27.9	357.3 ± 19.3	274.3 ± 62.1	188.7 ± 18.0
May	194.3 ± 29.5	249.0 ± 65.7	293.3 ± 62.4	345.0 ± 24.8	260.7 ± 60.6	183.0 ± 36.7
Jun.	175.7 ± 10.3	258.7 ± 19.2	269.0 ± 36.5	435.3 ± 113.4	294.7 ± 40.9	175.3 ± 9.0
Jul.	239.7 ± 24.5	233.0 ± 16.3	267.0 ± 47.4	265.0 ± 93.9	263.0 ± 12.1	112.7 ± 10.5
Aug.	206.0 ± 11.0	232.0 ± 24.1	238.7 ± 37.1	288.3 ± 44.8	218.7 ± 77.5	110.7 ± 9.0
Sep.	203.3 ± 29.0	223.0 ± 22.9	234.3 ± 27.6	280.7 ± 56.8	259.0 ± 106.9	118.7 ± 7.4
Oct.	188.0 ± 29.4	267.0 ± 42.1	285.7 ± 64.8	326.0 ± 48.3	341.7 ± 31.2	163.0 ± 36.9
Nov.	186.0 ± 4.3	314.3 ± 109.8	332.3 ± 42.0	344.7 ± 53.6	314.0 ± 50.2	212.3 ± 25.3
Dec.	214.3 ± 36.8	299.7 ± 34.3	348.7 ± 36.2	349.0 ± 51.6	329.3 ± 42.3	244.0 ± 6.4

Two-way ANOVA Test

Significant difference in hardness content between monitoring drains, F_1=19.12 (P<0.01).

Significant difference in hardness content between sampling months, F_2=2.76 (P<0.05).

Seasonal minimum and maximum values of hardness were 187.0 ± 21.06 (Pre-Monsoon) and 217.33 ± 35.7 (Winter) in NAJ-1, 229.33 ± 21.32 (Monsoon) and 290.67 ± 86.43 (Pre-Monsoon) in CIV-2, 246.67 ± 40.89 (Monsoon) and 326.67 ± 26.33 (Winter) in POW-3, 278.0 ± 69.11 (Monsoon) and 351.22 ± 43.44 (Winter) in SEN-4, 246.89 ± 79.12 (Monsoon) and 327.83 ± 44.03 (Pre-Monsoon) in BAR-5, 114.0 ± 30.3 and 245.78 ± 31.12 in HIN-6. Annual average values of hardness were in order of 340.06 > 296.67 > 294.53 > 268.08 > 295.0 > 185.36 in SEN-4, POW-3, CIV-2, NAJ-1 and HIN-6 respectively.

Table RWM-17 shows monthly averages (Mean ± SD) of hardness (mg/l) in water of river Yamuna at five sampling stations. The range of hardness was 78.33 ± 10.5 (Jul.) to 179.0 ± 20.12 (Dec.) at WAZ-1, 111.0 ± 18.49 (Aug.) to 203.67 ± 64.01 (Nov.) at ISB-2, 97.33 ± 18.79 (Aug.) to 227.33 ± 71.26 (Nov.) at ITO-3. 93.67 ± 22.07 (Aug.) to 236.0 ± 57.15 (Dec.) at NIZ-4, 112.0 ± 30.74 (Aug.) to 259.33 ± 91.13 (Nov.) at OKH-5. Analysis of variance (Table RWM-17) showed significant difference in hardness (mg/l) between monitoring points (F_1=21.93, P<0.01) and also between sampling months (F_2=2.79, P<0.05).

Seasonal minimum and maximum values of hardness were 98.33 ± 43.20 (Monsoon) and 156.44 ± 24.06 (Winter); at WAZ-1, 117.67 ± 34.33 (Monsoon) and 195.22 ± 26.82 (Winter) at ISB-2, 120.0 ± 47.91 (Monsoon) and 184.22 ± 13.16 (Winter) at ITO-3, 107.33 ± 44.98 (Monsoon) and 208.33 ± 41.32 (Winter) at NIZ-4, 127.67 ± 48.26 (Winter) and 229.33 ± 68.09 (Winter) at OKH-5.

Table RWM-17: Monthly Variation of Total Hardness (mg/l) in River Water (Year 1993-1995)

Month	*WAZ-1* *Mean ± SD*	*ISB-2* *Mean ± SD*	*ITO-3* *Mean ± SD*	*NIZ-4* *Mean ± SD*	*OKH-5* *Mean ± SD*
Jan.	146.3 ± 14.0	169.7 ± 6.2	178.0 ± 15.6	186.3 ± 20.0	217.7 ± 72.5
Feb.	139.3 ± 13.1	198.7 ± 9.8	185.0 ± 0.8	201.3 ± 12.5	223.0 ± 75.2
Mar.	223.0 ± 10.2	182.0 ± 6.5	182.7 ± 5.2	196.0 ± 11.8	218.3 ± 51.4
Apr.	112.3 ± 6.6	157.3 ± 13.2	156.0 ± 11.3	168.0 ± 17.7	188.3 ± 48.1
May	96.3 ± 2.9	163.0 ± 13.1	161.7 ± 11.8	173.3 ± 8.7	168.3 ± 24.0
Jun.	83.3 ± 12.2	153.0 ± 5.7	158.0 ± 24.7	160.0 ± 18.7	151.0 ± 45.0
Jul.	78.3 ± 16.5	120.0 ± 9.1	104.3 ± 12.3	102.0 ± 21.4	136.0 ± 45.3
Aug.	102.7 ± 43.4	111.0 ± 18.5	97.3 ± 18.8	93.7 ± 22.0	112.0 ± 30.7
Sep.	114.0 ± 52.7	122.0 ± 55.2	158.3 ± 64.5	126.0 ± 67.4	135.0 ± 60.2
Oct.	136.0 ± 52.7	185.7 ± 52.7	186.7 ± 52.8	187.3 ± 68.1	216.7 ± 66.4
Nov.	162.3 ± 35.3	203.7 ± 64.0	227.3 ± 71.3	216.7 ± 79.7	259.3 ± 91.1
Dec.	179.0 ± 20.1	198.3 ± 38.6	184.0 ± 6.4	236.0 ± 57.2	237.3 ± 62.1

Two-way ANOVA Test

Significant difference in total hardness content between river monitoring stations, F_1=19.62 (P<0.01).

Significant difference in total hardness content between monitoring months, F_2=2.79 (P<0.05).

During the present investigation total hardness showed two peak values *i.e.* during winter and summer. The peak values during winter season might be due to lean discharge of river water from Wazirabad barrage and more contribution of drain water at Delhi stretch. The second peak values during summer might be due to higher temperature and low water level in the river as well as the addition of calcium and magnesium salts from detergents and soap used for bathing and cloth washing (through drain water). Its lower values were recorded during rainy season. This might be due to dilution on account of high rains. Ajmal and Razi-Ud-Din (1988) observed a higher hardness value in summer in river Hindon and Kali Nadi. Munawar (1970), Moss (1973), Sreenivasan *et al.* (1974), Singh (1979) and Patil (1982) found the same pattern during their studies which has been further supported by Bagde and Verma (1985). Annual average values of five monitoring stations were in the order of 188.58 > 170.56 > 164.94 > 163.69 > 122.75 at OKH-5, NIZ-4, ITO-3, ISB-2 and WAZ-1 respectively.

It was observed that the total hardness values increased sharply at ISB-2 from WAZ-1, due to input of domestic sewage through Najafgarh and Burari drain. The concentration of hardness gradually increased from station ISB-2, to OKH-5 which might be due to input of drain effluents joining to river Yamuna at Delhi stretch.

Calcium of Drain and River Water

Calcium is an essential element and human body requires approximately 0.7 to 2.0 g of calcium per day as a food element, the amount which cannot be supplied even by hard waters. Hardness are undesirable for household uses such as washing, bathing and laundering because of the consumption of more soap and other cleaning agents. Moreover calcium salts tend to cause incrustations on cooking utensil and water heaters.

Calcium is essential for normal plant growth and is desirable in water for irrigation. The presence of calcium in water is mainly due to its passage through or over deposits of lime stone, dolomite, gypsum and other gypsiferous materials (Manivasakam, 1989). The calcium content may range from zero to several hundred mg/l, depending on the source and treatment of water.

Table DWM-17, show monthly calcium average values in drain water and the results of ANOVA test for five monitoring drains. The values (mg/l) varied from a minimum of 44.2 ± 18.32 to a maximum of 68.07 ± 8.38 in NAJ-1, 72.0 ± 6.68 to 111.73 ± 39.65 in CIV-2, 68.00 ± 9.80 to 108.67 ± 16.05 in POW-3, 88.67 ± 13.70 to 130.87 ± 23.74 in SEN-4, 63.67 ± 26.16 to 84.33 ± 22.69 in BAR-5 and 34.0 ± 2.45 to 74.67 ± 8.06 in HIN-6. ANOVA (Table DWM-17) reveals significant differences in calcium content between the five sampling drains (F_1=19.11, $P<0.01$) and also between sampling months (F_2=2.66, $P<0.05$).

Seasonal mean values of calcium varied from 52.3 ± 14.1 (Summer) to 61.58 ± 8.8 (Winter) in NAJ-1, 74.44 ± 8.71 (Monsoon) to 102.7 ± 31.24 (Pre-Monsoon) in CIV-2, 74.44 ± 13.88 (Monsoon) to 104.78 ± 13.51 (Winter) in POW-3, 96.87 ± 33.81 (Monsoon) to 118.87 ± 14.5 (Summer) in SEN-4, 68.44 ± 23.08 (Monsoon) to 76.56 ± 8.96 (Winter) BAR-5, 36.0 ± 8.7 (Monsoon) to 75.0 ± 13.76 (Winter) in HIN-6. Yearly average values were in the order of 109.64 > 92.95 > 90.85 > 73.08 > 58.11 > 55.86 in SEN-4, POW-3, CIV-2, BAR-5, NAJ-1 drains and Hindon river cut respectively.

Table RWM-18 presents monthly average values (Mean ± SD) and the results of ANOVA test for calcium content (mg/l) in river water at five monitoring points. The values ranged from 20.67 ± 1.89 (Jul.) to 42.33 ± 4.92 (Nov.) at WAZ-1, 39.0 ± 7.29 (Jul.) to 52.33 ± 4.5 (Dec.) at ISB-2, 32.67 ± 2.49 (Jul.) to 56.67 ± 15.11 (Nov.) at ITO-3, 26.67 ± 4.11 (Aug.) to 64.67 ± 12.47 (Dec.) at NIZ-4, 32.33 ± 6.55 (Aug.) to 62.0 ± 17.66 (Dec.) at BAR-5. Result of ANOVA test shows significant difference in calcium content between monitoring stations (F_1=19.31, $P<0.01$) and also between sampling months (F_2=2.8 $P<0.05$).

Table DWM-17: Monthly Variation in Calcium (mg/l) of Drain Water Year (1993-1995)

Month	NAJ-1 Mean ± SD	CIV-2 Mean ± SD	POW-3 Mean ± SD	SEN-4 Mean ± SD	BAR-5 Mean ± SD	HIN-6 Mean ± SD
Jan.	60.0 ± 8.6	98.0 ± 5.8	106.0 ± 2.8	115.9 ± 2.9	73.6 ± 8.6	73.6 ± 8.9
Feb.	68.6 ± 8.3	95.5 ± 12.7	102.0 ± 18.5	110.0 ± 16.5	74.0 ± 7.7	74.6 ± 8.0
Mar.	44.2 ± 18.3	105.0 ± 8.2	108.6 ± 16.0	117.6 ± 4.0	71.0 ± 4.5	73.0 ± 5.8
Apr.	59.4 ± 12.3	94.0 ± 11.4	96.3 ± 14.3	113.9 ± 2.1	63.6 ± 18.1	62.6 ± 2.4
May	55.2 ± 13.2	86.0 ± 18.4	91.6 ± 22.8	113.0 ± 7.2	65.3 ± 23.8	56.3 ± 12.5
Jun.	50.4 ± 1.1	83.6 ± 4.6	90.0 ± 12.3	130.6 ± 23.7	79.6 ± 14.6	58.6 ± 6.6
Jul.	63.0 ± 5.4	72.0 ± 6.6	86.6 ± 13.1	107.9 ± 51.8	78.0 ± 4.3	34.3 ± 2.6
Aug.	46.3 ± 4.6	75.0 ± 11.5	68.0 ± 9.8	88.6 ± 13.7	63.6 ± 26.1	34.0 ± 2.4
Sep.	53.6 ± 9.7	76.3 ± 6.2	68.6 ± 9.4	94.0 ± 18.8	63.6 ± 27.5	39.6 ± 2.8
Oct.	51.6 ± 6.9	93.6 ± 14.7	83.6 ± 20.0	105.0 ± 14.1	78.3 ± 11.7	53.6 ± 11.5
Nov.	55.33 ± 6.6	111.7 ± 39.6	104.7 ± 13.5	107.0 ± 10.0	84.4 ± 22.6	64.3 ± 7.4
Dec.	62.4 ± 9.8	99.3 ± 16.4	109.0 ± 13.6	111.6 ± 12.2	81.6 ± 7.8	72.3 ± 6.9

Two-way ANOVA Test
Significant difference in calcium content between monitoring drains, F_1=19.11 (P<0.01).
Significant difference in calcium content between sampling months, F_2=2.65 (P<0.05).

Table RWM-18: Monthly Variation of Calcium (mg/l) in River Water Year (1993-1995)

Month	WAZ-1 Mean ± SD	ISB-2 Mean ± SD	ITO-3 Mean ± SD	NIZ-4 Mean ± SD	OKH-5 Mean ± SD
Jan.	40.0 ± 5.7	41.3 ± 1.2	50.7 ± 3.9	56.7 ± 6.6	48.3 ± 3.7
Feb.	41.0 ± 2.2	48.6 ± 6.7	53.0 ± 1.4	62.0 ± 3.5	54.3 ± 6.9
Mar.	37.7 ± 3.0	51.0 ± 5.4	55.3 ± 3.7	60.0 ± 3.6	58.7 ± 1.3
Apr.	36.3 ± 2.9	46.3 ± 2.4	52.0 ± 3.3	52.3 ± 5.6	52.7 ± 5.7
May	33.7 ± 0.9	47.0 ± 11.1	55.0 ± 2.5	50.3 ± 4.5	48.0 ± 6.5
Jun.	25.7 ± 2.4	44.0 ± 5.9	53.3 ± 6.1	50.7 ± 6.8	39.0 ± 5.0
Jul.	20.7 ± 1.9	39.0 ± 7.3	32.7 ± 2.5	29.3 ± 3.7	33.0 ± 4.5
Aug.	25.3 ± 6.9	40.6 ± 13.1	40.3 ± 16.0	26.7 ± 4.1	32.3 ± 6.6
Sep.	28.1 ± 10.0	41.3 ± 18.9	44.7 ± 16.2	36.0 ± 15.8	39.7 ± 19.5
Oct.	34.7 ± 9.8	44.6 ± 6.2	51.0 ± 13.1	50.0 ± 14.5	54.7 ± 16.9
Nov.	42.3 ± 4.9	44.3 ± 6.9	56.7 ± 15.2	58.7 ± 18.0	60.3 ± 20.9
Dec.	42.0 ± 0.8	52.3 ± 4.5	55.0 ± 5.1	64.7 ± 12.5	62.0 ± 17.7

Two-way ANOVA Test
Significant difference in calcium content between river monitoring stations, F_1=19.31 (P<0.01).
Significant difference in calcium content between monitoring months, F_2=2.8 (P<0.05).

Seasonal minimum and maximum calcium values were 24.67 ± 7.72 (Monsoon) and 41.33 ± 3.83 (Winter) at WAZ-1, 33.89 ± 6.12 (Monsoon) and 49.44 ± 5.96 (Winter) at ISB-2, 39.22 ± 14.13

(Monsoon) and 53.78 ± 3.88 (Winter) at ITO-3, 30.67 ± 10.40 (Monsoon) and 60.50 ± 8.75 (Winter) at NIZ-4, 35.0 ± 12.6 (Monsoon) and 55.89 ± 11.92 (Winter) at OKH-5. During present investigation calcium showed its peak values during winter due to lean discharge from Wazirabad barrage and high concentration of calcium in drain water. The lower values of calcium were observed during monsoon season due to dilution of drain water by rain water. Annual mean values (mg/l) of calcium at Delhi stretch were in order of 50.0 > 49.78 > 48.58 > 45.06 > 33.86 at ITO-3, NIZ-4, OKH-5, ISB-2 and WAZ-l respectively.

Magnesium of Drain and River Water

Magnesium is an essential element for human beings and is relatively non-toxic. However at higher concentrations, magnesium salts have a laxative effect particularly when it is present as magnesium sulphate. Magnesium salts act as cathartics and diuretics among animals as well as human beings. Magnesium in industrial water is undesirable since it is the major scale forming cation. Magnesium salts occur in significant concentration in natural waters. The industrial and commercial uses of magnesium salts are numerous and the occurrence of magnesium in industrial wastewater is common. Magnesium is essential to normal plant growth. Calcium and magnesium ions in irrigation water keep the soil permeable and in good tilt (Manivasakam, 1989).

Table DWM-18 represents monthly average values (Mean ± SD) and the results of ANOVA test of magnesium content (mg/l) of five drains. The minimum and maximum mean values of magnesium were 10.48 ± 1.84 to 21.57 ± 0.83 in NAJ-1, 7.81 ± 2.73 to 18.05 ± 3.11 in CIV-2, 10.67 ± 4.35 to 16.35 ± 4.41 in POW-3, 10.93 ± 3.62 to 28.48 ± 5.49 in SEN-4, 14.73 ± 3.89 to 34.73 ± 2.68 at BAR-5 and 4.83 ± 1.84 to 15.0 ± 1.0 in HIN-6. Analysis of variance (Table DWM-18) shows a significant difference in magnesium content between five monitoring drains (F_1=2.606, P<0.05) and also between monitoring months (F_2=2.65, P<0.05).

Table DWM-18: Monthly Variation in Magnesium (mg/l) of Drain Water (Year 1993-1995)

Month	*NAJ-1*	*CIV-2*	*POW-3*	*SEN-4*	*BAR-5*	*HIN-6*
	Mean ± SD	*Mean ± SD*	*Mean ± SD*	*Mean ± SD*	*Mean ± SD*	*Mean ± SD*
Jan.	15.8 ± 6.2	12.8 ± 3.0	16.3 ± 3.7	16.8 ± 1.8	29.9 ± 3.1	14.4 ± 3.2
Feb.	18.0 ± 3.9	10.3 ± 3.9	16.2 ± 4.7	18.1 ± 4.3	32.6 ± 3.8	15.0 ± 1.0
Mar.	13.4 ± 2.0	16.1 ± 7.3	14.2 ± 3.8	19.1 ± 5.4	34.7 ± 2.6	8.0 ± 3.5
Apr.	10.4 ± 1.8	18.0 ± 3.1	14.1 ± 5.7	18.1 ± 4.7	27.8 ± 4.1	11.3 ± 6.1
May	13.4 ± 6.7	16.1 ± 3.9	15.4 ± 9.8	15.1 ± 1.5	23.6 ± 0.4	10.0 ± 1.4
Jun.	11.9 ± 2.2	11.8 ± 1.8	10.6 ± 4.3	25.9 ± 13.0	23.3 ± 1.7	6.8 ± 1.6
Jul.	19.5 ± 2.2	13.0 ± 1.8	12.0 ± 5.2	28.4 ± 5.4	16.2 ± 0.3	6.5 ± 2.0
Aug.	21.5 ± 0.8	7.8 ± 2.7	16.5 ± 4.4	15.9 ± 4.2	14.5 ± 3.8	6.3 ± 1.7
Sep.	16.6 ± 1.8	8.3 ± 7.0	14.9 ± 5.6	10.9 ± 3.6	23.1 ± 11.2	4.8 ± 1.8
Oct.	14.2 ± 2.8	7.8 ± 1.3	18.2 ± 10.7	15.4 ± 3.8	31.6 ± 5.7	6.8 ± 3.1
Nov.	11.5 ± 3.7	8.4 ± 3.5	16.7 ± 4.5	18.4 ± 7.3	25.2 ± 3.3	9.8 ± 3.7
Dec.	14.0 ± 2.9	12.2 ± 2.2	18.2 ± 4.2	16.6 ± 5.7	30.0 ± 10.5	15.1 ± 2.7

Two-way ANOVA Test

Significant difference in magnesium content between monitoring drains, F_1=19.97 (P<0.01).
Significant difference in magnesium content between sampling months, F_2=2.65 (P<0.05).

The mean seasonal values of magnesium were varied from a minimum of 12.32 ± 4.16 (Summer) to maximum of 19.23 ± 3.00 (Monsoon) in NAJ-1, 9.73 ± 3.10 (Monsoon) to 14.50 ± 5.48 (Summer) in CIV-2, 13.92 ± 6.67 (Summer) to 17.45 ± 8.15 (Pre-Monsoon) in POW-3, 16.93 ± 6.08 (Pre-Monsoon) to 19.61 ± 8.48 (Summer) in SEN-4, 18.06 ± 7.80 (Monsoon) to 29.32 ± 5.10 (Winter) in BAR-5, and 5.89 ± 1.03 (Monsoon) to 23.04 ± 3.74 (Summer) in HIN-6 respectively. Annual mean values of magnesium were in order of 26.11 > 18.35 > 15.42 > 15.05 > 14.06 > on BAR-5, SEN-4, POW-3, NAJ-1, HIN-6 and CIV-2 respectively.

Table RWM-19 shows the monthly averages (Mean ± SD) and the results of ANOVA test for magnesium at five river monitoring points. Monthly values varied from 3.7 ± 1.5 (May) to 13 ± 6.5 (Nov.) at WAZ-1, 7.0 ± 2.9 (Aug.) to 17.0 ± 7.9 (Nov.) at ISB-2, 4.3 ± 1.2 (Aug.) to 18.3 ± 6.7 (Nov.) at ITO-3, 7.0 ± 2.4 (Aug.) to 16.5 ± 8.3 (Nov.) at NIZ-4 and 3.5 ± 1.1 (Aug.) to 20.0 ± 10.7 (Nov.) at OKH-5. Analysis of variance shows significant difference in magnesium content between monitoring stations (F_1=21.2, P<0.01) and also between monitoring months (F_2=2.9, P<0.05).

Table RWM-19: Monthly Variation of Magnesium (mg/l) in River Water Year (1993- 1995)

Month	*WAZ-1*	*ISB-2*	*ITO-3*	*NIZ-4*	*OKH-5*
	Mean ± SD	*Mean ± SD*	*Mean ± SD*	*Mean ± SD*	*Mean ± SD*
Jan.	10.0 ± 0.8	15.3 ± 0.9	12.0 ± 1.6	11.3 ± 1.9	12.0 ± 2.8
Feb.	8.7 ± 2.1	15.7 ± 1.2	12.7 ± 0.9	12.0 ± 1.6	13.0 ± 2.2
Mar.	6.7 ± 1.2	14.0 ± 1.4	11.0 ± 1.4	11.0 ± 0.8	8.0 ± 2.8
Apr.	7.0 ± 2.2	10.7 ± 0.9	6.3 ± 1.2	9.0 ± 1.4	6.0 ± 0.8
May	3.7 ± 0.5	8.3 ± 0.9	5.7 ± 1.2	10.7 ± 0.9	5.3 ± 0.5
Jun.	5.0 ± 1.6	9.0 ± 0.8	6.0 ± 2.2	7.8 ± 0.6	5.3 ± 0.5
Jul.	6.7 ± 3.1	8.7 ± 1.7	5.7 ± 1.2	8.3 ± 2.1	5.0 ± 1.4
Aug.	9.3 ± 6.1	7.0 ± 2.9	4.3 ± 1.2	7.0 ± 2.4	3.5 ± 1.1
Sep.	10.0 ± 7.8	10.0 ± 6.5	9.7 ± 5.3	8.7 ± 6.8	5.3 ± 4.0
Oct.	11.0 ± 6.2	15.0 ± 7.1	11.7 ± 5.9	14.7 ± 7.6	13.0 ± 5.7
Nov.	13.0 ± 6.5	17.0 ± 7.9	18.3 ± 11.1	16.5 ± 8.3	20.2 ± 10.7
Dec.	12.7 ± 4.1	11.3 ± 0.9	10.0 ± 3.6	16.0 ± 9.3	17.7 ± 11.0

Two-way ANOVA Test

Significant difference in magnesium content between river monitoring stations, F_1=21.2 (P<0.01).

Significant difference in magnesium content between monitoring months, F_2=2.90 (P<0.05).

The mean minimum and the mean maximum seasonal values were 5.58 ± 2.02 (Summer) and 12.0 ± 6.51 (Pre-Monsoon), 8.56 ± 4.4 (Monsoon) and 15.0 ± 9.49 (Pre-Monsoon), 8.0 ± 4.4 (Monsoon) and 15.6 ± 8.01 (Pre-Monsoon), 4.61 ± 2.66 (Monsoon) and 16.5 ± 9.25 (Pre-Monsoon) at WAZ-1, ISB-2, ITO-3 NIZ-4 and OKH-5, river monitoring station respectively.

During present investigation magnesium showed its peak value at Wazirabad during summer season might be due to lean flow of river, high temperature, increased rate of evaporation and concentration of solutes. Lower values during Monsoon might be due to dilution with rain water. At all downstream monitoring points *i.e.* ISB-2, ITO-3, NIZ-4 and OKH-5 magnesium showed peak values during post-monsoon might be due to lean discharge of water from Wazirabad barrage and higher concentration of drain water and lower values during monsoon might be due to dilution of drain water with rain water and huge water flow from upstream.

Annual average values (1993 to 1995) of magnesium content were in order of 11.83 > 11.09 > 9.51 > 9.44 > 8.64 at ISB-2, NIZ-4, OKH-4, ITO-3 and WAZ-1 respectively.

Sodium of Drain and River Water

Sodium is present in most natural waters from negligible to appreciable concentrations. The salt of sodium are highly soluble in water and impart softness to water. It has important considerations with regard to irrigation water only. Regarding drinking and industrial purposes it is of minor importance only. However it has to be mentioned that excessive amounts of sodium in drinking water is harmful to persons suffering from cardiac, renal and circulatory diseases. Sodium when present as sodium chloride in concentration more than 500 mg/l makes the water unpalatable and causes appetite disturbances. Soil permeability has also been found to be affected by a high sodium content (more then 60 per cent) in irrigation water (Rawat, 1978).

Domestic sewage is the important source of sodium of the freshwater. Sodium ions make their way in water through weathering of rocks. Sodium is said to be utilised by blue green algae in high concentrations (Gerloff *et al.*, 1952 and Provasoli, 1969). The threshold level of 400 mg/l is required for near optimal growth of several species (Kulshrestha *et al.*, 1992).

Table DWM-19 shows monthly mean values (Mean ± SD) and the results of ANOVA test of sodium content in drains discharging their effluents to river Yamuna at Delhi stretch.

Table DWM-19: Monthly Variation in Sodium (mg/l) of Drain Water (Year 1993-1995)

Month	*NAJ-1*	*CIV-2*	*POW-3*	*SEN-4*	*BAR-5*	*HIN-6*
	Mean ± SD	*Mean ± SD*	*Mean ± SD*	*Mean ± SD*	*Mean ± SD*	*Mean ± SD*
Jan.	115.3 ± 12.1	120.3 ± 8.6	124.4 ± 3.4	163.0 ± 18.4	168.0 ± 2.9	95.3 ± 2.5
Feb.	124.7 ± 8.7	115.0 ± 13.5	121.7 ± 23.2	162.7 ± 10.4	155.0 ± 2.8	94.0 ± 5.7
Mar.	123.3 ± 9.9	146.0 ± 20.5	135.4 ± 19.1	177.0 ± 9.4	169.7 ± 17.2	94.3 ± 5.9
Apr.	145.3 ± 12.3	134.0 ± 8.8	139.0 ± 25.9	178.0 ± 8.3	157.0 ± 20.8	83.7 ± 7.1
May	139.3 ± 19.0	139.3 ± 6.6	127.0 ± 19.6	171.3 ± 9.8	150.0 ± 21.4	84.7 ± 7.3
Jun.	144.0 ± 12.4	153.0 ± 8.6	137.7 ± 20.8	179.0 ± 5.4	163.2 ± 18.4	87.0 ± 9.9
Jul.	178.2 ± 23.8	106.7 ± 18.9	115.4 ± 11.7	159.7 ± 6.5	128.2 ± 7.9	62.3 ± 4.9
Aug.	128.4 ± 13.1	106.0 ± 20.1	69.0 ± 45.7	145.7 ± 14.8	107.3 ± 34.2	61.3 ± 0.9
Sep.	130.3 ± 24.3	103.7 ± 22.6	108.3 ± 27.1	132.7 ± 7.1	119.0 ± 43.5	66.3 ± 3.7
Oct.	131.0 ± 20.5	126.3 ± 1.2	121.7 ± 19.0	153.7 ± 3.4	133.0 ± 19.4	82.3 ± 3.3
Nov.	134.0 ± 13.8	163.7 ± 26.6	138.7 ± 6.1	172.3 ± 18.8	142.7 ± 36.4	91.3 ± 1.2
Dec.	130.0 ± 21.8	144.7 ± 18.1	140.0 ± 9.6	175.0 ± 15.1	159.0 ± 9.4	92.0 ± 4.3

Two-way ANOVA Test

Significant difference in sodium content between monitoring drains, F_1=19.13 ($P<0.01$).
Significant difference in sodium content between sampling months, F_2=2.83 ($P<0.05$).

Mean values varied from 115.3 ± 12.1 to 178.3 ± 23.8 in NAJ-1 103.7 ± 22.6 to 153.0 ± 8.6 in CIV-2, 69.0 ± 45.7 to 139.0 ± 25.9 in POW-3, 132.7 ± 71 to 179.0 ± 5.4 in SEN-4, 107.3 ± 34.2 to 169.7 ± 17.2

in BAR-5, 61.3 ± 0.9 to 95.3 ± 2.5 in HIN-6. Analysis of variance shows significant difference between monitoring drains (F_1=19.13, P<0.01) as well as between monitoring months (F_2=2.83, P<0.05).

Seasonal sodium values were varied between 118.67 ± 9.46 to 145.67 ± 30.14 in NAJ-1, 105.56 ± 20.62 to 145.0 ± 26.53 in CIV-2, 97.56 ± 37.48 to 128.56 ± 16.73 in POW-3, 146.0 ± 15.03 to 176.33 ± 8.91 in SEN-4, 118.22 ± 33.38 to 163.6 ± 21.6 in BAR-5, 63.33 ± 8.1 to 94.67 ± 9.23 in HIN-6. Annual mean values were in order of 164.17, 145.16, 132.56, 129.92, 123.17 and 82.89 in SEN-4, BAR-5, NAJ-1, CIV-2, POW-3 and HIN-6 respectively.

Table RWM-20 shows monthly average values of sodium at five monitoring points of river Yamuna and the results of ANOVA test. The sodium content of river water at different monitoring stations fluctuated from 19.9 ± 7.9 to 43.0 ± 10.0 at WAZ-1, 56.33 ± 2.87 to 95.0 ± 12.03 at ISB-2, 62.67 ± 14.29 to 112.0 ± 13.95 at ITO-3, 69.67 ± 11.90 to 121.0 ± 4.24 at NIZ-4 and 56.33 ± 6.13 to 115.67 ± 19.6 at OKH-5. The ANOVA test shows significant difference in sodium content between five river monitoring stations (F_1=19.38, P<0.01) and also between sampling months (F_2=2.44, P<0.05).

Table RWM-20: Monthly Variation of Sodium (mg/l) in River Water (Year 1993-1995)

Month	*WAZ-1*	*ISB-2*	*ITO-3*	*NIZ-4*	*OKH-5*
	Mean ± SD	*Mean ± SD*	*Mean ± SD*	*Mean ± SD*	*Mean ± SD*
Jan	38.7 ± 10.3	85.7 ± 5.4	110.7 ± 6.8	111.3 ± 7.4	82.3 ± 10.1
Feb	33.0 ± 14.5	93.0 ± 5.2	111.7 ± 14.1	121.0 ± 4.2	95.7 ± 4.8
Mar	34.0 ± 17.9	95.0 ± 12.0	19.3 ± 21.9	119.0 ± 11.1	99.0 ± 22.2
Apr	28.5 ± 7.9	91.7 ± 14.3	98.7 ± 11.6	113.0 ± 10.7	75.0 ± 10.4
May	28.7 ± 2.9	89.3 ± 21.9	112.0 ± 14.0	120.7 ± 8.6	77.0 ± 6.2
Jun	24.3 ± 5.4	89.0 ± 5.8	102.3 ± 25.3	109.0 ± 14.4	71.0 ± 5.1
Jul	19.9 ± 7.9	69.3 ± 7.4	72.0 ± 7.1	75.0 ± 19.2	62.3 ± 6.2
Aug	20.7 ± 9.2	64.0 ± 7.8	66.2 ± 13.1	69.7 ± 11.9	58.0 ± 6.1
Sep	23.1 ± 8.0	56.3 ± 2.9	62.7 ± 14.3	69.7 ± 18.6	56.3 ± 6.2
Oct	29.2 ± 9.2	76.0 ± 10.7	84.0 ± 9.3	94.7 ± 17.0	91.3 ± 9.0
Nov	38.5 ± 7.6	86.3 ± 8.4	98.7 ± 9.9	109.3 ± 20.2	115.7 ± 19.6
Dec	43.0 ± 10.0	88.7 ± 7.8	101.3 ± 0.9	111.0 ± 14.9	91.3 ± 0.9

Two-way ANOVA Test

Significant difference in sodium content between river monitoring stations, F_1=19.38 (P<0.01).
Significant difference in sodium content between monitoring months, F_2=2.44 (P<0.05).

The mean seasonal values varied from 21.33 ± 8.49 (Monsoon) to 37.33 ± 11.75 (Winter) at WAZ-1, 63.22 ± 8.34 (Monsoon) to 91.25 ± 14.83 (Summer) at ISB-2, 67.0 ± 12.53 (Monsoon) to 108.67 ± 9.94 (Winter) at ITO-3, 71.44 ± 17.03 (Monsoon) to 117.56 ± 6.7 (Winter) at NIZ-4, 58.89 ± 6.47 (Monsoon) to 103.5 ± 19.5 (Pre-Monsoon) at OKH-5. Seasonal lower values at all stations occurs during monsoon due to the effect of dilution. The peak in winter may be due to lean river flow (nil discharge from Wazirabad barrage) and microbial growth (Pradhan and Mohanty, 1991) mineralising the decomposed organic matter and releasing the nutrients to the sediment, increasing sodium in to the nutrient pool there by making more sodium to solubilise.

The sodium level in river of India has been reported by many workers. In river Bharalu, sodium content ranging from 2.0 to 27.0 mg/l reported by Lal and Bhattachary (1990) 32.7 mg/l in river

Cauvery (Somashekar, 1984), 9.3 mg/l in river Morar, Kalpi (Mishra and Saxena, 1991) from 2.5–3.6 mg/l in river Gelabil, Assam (Baruah *et al.*, 1993), 8.6-65.08 mg/l in river Kali (Krishnamurty and Bharati, 1995) 28.0–134.0 mg/l in river Pennar (Srikanth *et al.*, 1995).

The present study on river Yamuna at Delhi stretch showed the annual mean sodium values of 30.19 ± 12.3, 82.03 ± 15.87, 94.14 ± 22.26, 101.97 ± 23.50, 81.25 ± 20.34 at WAZ-1, ISB-2, ITO-3, NIZ-4 and OKH-5 respectively.

Potassium of Drain and River Water

Potassium is an essential nutritional element, but in excessive amounts it acts as a cathartic. Though potassium ranks seventh among the elements in the order of abundance, its concentration in most drinking water is trivial. Potassium in natural water is derived from weathering of rocks. The concentrations of potassium remain quite lower than the sodium, calcium and magnesium in aquatic ecosystems (Trivedy *et al.*, 1987). Although it is found in less quantity, it is also important in the ecology of blue-green algae (Provasoli, 1958 and Wetzel, 1966). Potassium has got a more or less similar chemistry like sodium and remains mostly in solution without undergoing any precipitation (Trivedy *et al.*, 1987).

Monthly averages (Mean ± SD) and the results of ANOVA test for potassium (mg/l) in the drain water are presented in Table DWM-20. The mean values ranged from 14.0 ± 0.82 to 28.33 ± 0.94 in NAJ-1, 11.0 ± 2.83 to 40.0 ± 21.28 in CIV-2, 12.67 ± 2.87 to 28.67 ± 3.2 in POW-3, 10.0 ± 1.41 to 22.0 ± 0.82 in SEN-4, 9.33 ± 2.36 to 20.33 ± 1.70 in BAR-5 and 7.67 ± 0.94 to 13.0 ± 0.82 in HIN-6. Result of ANOVA test shows a significant difference in potassium content between monitoring drains (F_1=20.36, P<0.01) and also between sampling months (F_2= 2.92, P<0.05).

Table DWM-20: Monthly Variation in Potassium (mg/l) of Drain Water (Year 1993-1995)

Month	*NAJ-1*	*CIV-2*	*POW-3*	*SEN-4*	*BAR-5*	*HIN-6*
	Mean ± SD	*Mean ± SD*	*Mean ± SD*	*Mean ± SD*	*Mean ± SD*	*Mean ± SD*
Jan.	14.0 ± 0.8	19.0 ± 2.1	26.6 ± 1.7	19.6 ± 4.1	20.3 ± 1.7	12.0 ± 1.4
Feb.	15.3 ± 1.7	20.2 ± 1.4	26.6 ± 2.6	10.0 ± 2.4	16.6 ± 2.4	11.3 ± 3.3
Mar.	15.0 ± 0.8	22.6 ± 4.0	28.6 ± 3.3	22.0 ± 2.9	19.0 ± 4.3	12.3 ± 2.3
Apr.	19.0 ± 3.7	24.6 ± 3.4	26.6 ± 4.7	22.0 ± 0.8	18.0 ± 4.0	9.0 ± 1.4
May	17.3 ± 1.7	24.0 ± 3.2	26.0 ± 2.5	21.0 ± 1.6	16.3 ± 4.5	8.3 ± 0.4
Jun.	20.0 ± 4.2	26.6 ± 1.4	26.0 ± 4.9	18.6 ± 1.2	17.3 ± 3.6	9.3 ± 2.0
Jul.	28.3 ± 0.9	11.6 ± 2.0	13.6 ± 2.6	13.3 ± 0.4	11.6 ± 1.2	7.3 ± 1.2
Aug.	19.0 ± 4.0	11.0 ± 2.8	12.6 ± 2.8	11.0 ± 0.8	9.3 ± 2.3	7.6 ± 0.9
Sep.	15.6 ± 4.5	11.0 ± 3.7	14.0 ± 5.3	10.0 ± 1.4	13.0 ± 5.7	7.6 ± 0.4
Oct.	16.0 ± 4.9	17.6 ± 0.4	15.2 ± 7.5	14.0 ± 2.5	13.6 ± 2.0	10.0 ± 0.0
Nov.	18.0 ± 2.1	25.6 ± 4.5	23.0 ± 7.2	16.6 ± 0.9	15.0 ± 4.5	11.6 ± 0.4
Dec.	16.6 ± 4.9	40.0 ± 21.2	24.3 ± 7.0	16.6 ± 1.2	17.6 ± 2.6	13.0 ± 0.8

Two-way ANOVA Test

Significant difference in potassium content between monitoring drains, F_1=20.36 (P<0.01).
Significant difference in potassium content between sampling months, F_2=2.92 (P<0.05).

Seasonal values varied from 14.33 ± 2.31 (Winter) to 21.0 ± 6.43 (Monsoon) in NAJ-1, 11.22 ± 2.97 (Monsoon) to 25.89 ± 15.86 (Winter) in CIV-2, 13.44 ± 4.14 (Monsoon) to 27.22 ± 2.48 (Winter) in

POW-3, 11.44 ± 1.71 (Monsoon) to 20.92 ± 2.29 (Summer) in SEN-4, 11.33 ± 3.94 (Monsoon) to 19.0 ± 2.45 (Winter) in BAR-5 and from 7.56 ± 1.2 (Monsoon) to 11.67 ± 1.8 (Winter) in HIN-6. Annual mean values (averages from year 1993 to 1995) of potassium were 17.86, 21.17, 21.97, 17.08, 15.67 and 9.97 in NAJ-1, CIV-2, POW-3, SEN-4, BAR-5 and HIN-6 respectively.

Table RWM-21 shows monthly averages (Mean ± SD) of potassium and the results of ANOVA test between five river monitoring points. The range of potassium content was 1.58 ± 0.59 (Aug.) to 6.07 ± 1.46 (Dec.) at WAZ-1, 3.67 ± 0.47 (Sep) to 8.0 ± 1.63 (May) at ISB-2, 4.0 ± 0.82 (Aug.) to 10.33 ± 1.7 (May) at ITO-3, 5.0 ± 0.82 (July) to 13.33 ± 1.41 (May) at NIZ-4, 3.0 ± 0.82 (Aug.) to 9.33 ± 0.94 (Nov.) at OKH-5 respectively. Analysis of variance showed the existence of significant difference in between the stations (F_1=20.21, P<0.01) as well as between sampling months (F_2=20.21, P<0.05), indicating that potassium content is influenced by seasons, drain water discharge and also river water discharge from Wazirabad barrage.

Table RWM-21: Monthly Variation of Potassium (mg/l) in River Water (Year 1993- 1995)

Month	*WAZ-1*	*ISB-2*	*ITO-3*	*NIZ-4*	*OKH-5*
	Mean ± SD	*Mean ± SD*	*Mean ± SD*	*Mean ± SD*	*Mean ± SD*
Jan.	3.37 ± 1.64	6.67 ± 0.47	9.00 ± 0.80	11.00 ± 0.82	6.67 ± 0.47
Feb.	3.83 ± 1.55	7.67 ± 0.47	8.67 ± 1.20	13.33 ± 1.70	7.67 ± 1.25
Mar.	3.80 ± 1.70	8.00 ± 1.41	9.67 ± 3.00	12.33 ± 2.05	8.67 ± 2.05
Apr.	3.17 ± 1.43	7.00 ± 1.63	8.00 ± 0.80	11.67 ± 1.89	7.67 ± 0.47
May	5.87 ± 2.64	8.00 ± 1.63	10.33 ± 1.77	13.00 ± 1.41	6.67 ± 0.47
Jun.	2.27 ± 1.32	7.67 ± 0.47	8.00 ± 3.56	11.67 ± 2.62	5.00 ± 0.00
Jul.	1.75 ± 0.35	5.00 ± 0.00	5.00 ± 0.80	5.00 ± 0.82	3.67 ± 0.47
Aug.	1.58 ± 0.59	4.33 ± 0.47	4.00 ± 0.82	5.33 ± 0.47	3.00 ± 0.82
Sep.	1.78 ± 0.31	3.67 ± 0.47	4.33 ± 0.47	5.33 ± 1.25	3.33 ± 0.47
Oct.	5.27 ± 2.53	6.67 ± 0.47	7.33 ± 0.47	8.33 ± 1.70	7.00 ± 1.41
Nov.	5.32 ± 1.69	7.00 ± 0.00	10.00 ± 2.16	10.67 ± 1.70	9.33 ± 0.94
Dec.	6.07 ± 1.46	7.67 ± 0.47	9.67 ± 1.25	10.67 ± 0.47	9.00 ± 0.00

Two-way ANOVA Test

Significant difference in potassium content between river monitoring stations, F_1=20.21 (P<0.01).

Significant difference in potassium content between monitoring months, F_2=2.53 (P<0.05).

Seasonal averages ranged from 1.71 ± 0.44 (Monsoon) to 5.29 ± 2.15 (Pre-Monsoon) at WAZ-1, 4.33 ± 0.67 (Monsoon) to 7.67 ± 1.43 (Summer) at ISB-2, 4.44 ± 0.67 (Monsoon) to 9.22 ± 1.13 at ITO-3, 5.22 ± 0.83 (Monsoon) to 12.17 ± 2.11 (Summer) at NIZ-4 and 3.33 ± 0.67 (Monsoon) to 8.17 ± 1.67 (Pre-Monsoon) at OKH-5. The high potassium level during winter might be due to lean discharge of river water from Wazirabad barrage and higher contribution of drain water at Delhi stretch. Another reason might be due to release of potassium from decomposed and mineralised organic matter during that period. The low potassium content in all stations during monsoon months marked the effect of dilution.

Annual average values of potassium were 3.67 ± 2.25, 6.61 ± 2.75, 7.83 ± 2.75, 9.86 ± 3.33 and 6.47 ± 2.33 at WAZ-1, ISB-2, ITO-3, NIZ-4 and OKH-5 respectively. The maximum concentration of K was observed at Nizamuddin bridge *i.e.* 9.86 (mg/l) and minimum was at Wazirabad barrage *i.e.* 3.67 (un-polluted stretch).

Heavy Metals in Water

Heavy metals occur at very low levels in almost all natural waters due to geo-chemical sources. Some of them (Fe, Cu, Zn, Ni, Co, Mn) are essential nutrients for aquatic life. At higher levels they are toxic. In our country majority of water bodies are invariably polluted by industrial and domestic wastes, which also contain metallic wastes. Therefore, monitoring in water courses become an essential part of water quality management. The measurements and characterisation of heavy metals are necessary to understand their distribution and pathways in the ecosystem and to assess the impact of their discharge into the environment (Ingole and Dhaktode, 1990).

Iron of Drain and River Water

Iron usually exists in natural water both in ferric and ferrous forms. Usually the ferric form is predominant in most of the natural water. Iron exists in the soluble state under reducing conditions. On exposure to air or on addition of oxidants soluble ferrous iron is oxidised to insoluble hydrated ferric oxides rapidly by dissolved oxygen at neutral pH (APHA, 1991). Under alkaline pH conditions of water, iron gets hydrolysed (Stumm and Lee, 1960) and forms insoluble hydroxide (Stumm and Morgan, 1981). The oxides and hydroxides of iron and manganese constitute significant sinks of heavy metals into aquatic system (Jenne, 1976). Ferric iron is only significantly soluble at pH less than 3.5. Iron was found harmful to the fishes even in low concentrations as ferric hydroxide, clogs up gills (Metelev *et al.*, 1971). At pH 6.5 to 7.5, 0.9 mg of iron per litre of water causes the death of fishes. The 1958 WHO international standards for drinking water specify a permissible limit of 0.3 mg/l and an excessive limit of 1.0 mg/l iron.

Monthly average values (Mean ± SD) of iron content in the drains and the result of ANOVA test are given in Table DWM-21. Monthly average values of iron of sampling drains ranged from a minimum of 12.1 ± 2.72 to a maximum of 27.89 ± 9.49 in NAJ-1, 3.34 ± 0.05 to 6.10 ± 0.15 in CIV-2, 3.33 ± 0.05 to 7.42 ± 0.56 in POW-3, 4.35 ± 0.51 to 8.09 ± 0.22 in SEN-4, 3.24 ± 1.06 to 4.07 ± 0.76 in BAR-5. Analysis of variance (Table DWM-21) shows significant difference in 'Fe' content (F_1=29.3, P<0.01) in between sampling drains as well as between sampling months (F_2=2.66, P<0.05).

Table DWM-21: Monthly Variation in Iron (mg/l) of Drain Water (Year 1993-1995)

Month	*NAJ-1*	*CIV-2*	*POW-3*	*SEN-4*	*BAR-5*
	Mean ± SD	*Mean ± SD*	*Mean ± SD*	*Mean ± SD*	*Mean ± SD*
Jan.	13.4 ± 2.1	3.4 ± 0.1	3.8 ± 0.5	4.3 ± 0.5	3.4 ± 0.1
Feb.	12.8 ± 0.6	3.3 ± 0.0	3.3 ± 0.0	4.4 ± 0.2	4.0 ± 0.7
Mar.	15.3 ± 2.3	4.4 ± 0.3	3.9 ± 0.0	4.6 ± 0.2	3.2 ± 1.0
Apr.	18.0 ± 1.7	5.2 ± 0.3	5.0 ± 0.0	4.9 ± 0.0	3.3 ± 0.5
May	27.8 ± 9.4	6.1 ± 0.1	7.4 ± 0.5	6.1 ± 0.2	3.4 ± 0.4
Jun.	17.0 ± 2.3	5.9 ± 0.1	8.5 ± 0.4	5.9 ± 0.8	3.8 ± 0.3
Jul.	13.8 ± 0.8	6.0 ± 0.3	8.6 ± 0.3	6.2 ± 0.1	3.8 ± 0.3
Aug.	13.6 ± 0.2	6.0 ± 0.1	8.7 ± 0.9	6.0 ± 0.2	3.8 ± 0.3
Sep.	12.1 ± 2.7	4.1 ± 0.2	9.0 ± 0.3	6.1 ± 0.2	3.9 ± 0.0
Oct.	17.6 ± 0.8	4.1 ± 0.1	8.0 ± 0.1	7.1 ± 0.2	4.0 ± 0.1
Nov.	24.5 ± 2.7	4.9 ± 0.1	7.3 ± 0.2	8.0 ± 0.2	3.7 ± 0.0
Dec.	22.0 ± 1.8	5.2 ± 0.1	7.3 ± 0.4	7.9 ± 0.9	3.8 ± 0.1

Two-way ANOVA Test

Significant difference in iron content between monitoring drains, F_1=29.3 (P<0.01).
Significant difference in iron content between sampling months, F_2=2.06 (P<0.05).

Seasonal values ranged from 13.20 ± 1.83 (Monsoon) to 22.58 ± 9.25 (Summer) in NAJ-1, 3.85 ± 0.69 (Winter) to 5.44 ± 0.70 (Summer) in CIV-2, 4.68 ± 1.6 (Winter) to 8.8 ± 0.61 (Monsoon) in POW-3, 5.19 ± 1.2 (Winter) to 6.15 ± 0.24 (Monsoon) in SEN-4, 3.46 ± 0.69 (Summer) to 3.88 ± 0.17 (Pre-Monsoon) in BAR-5. On basis of annual averages different drain points were arranged in order of NAJ-1 > POW-2 > SEN-4 > CIV-2 > BAR-5.

Table RWM-22 shows monthly average values (Mean ± SD) of iron (mg/l) at five river monitoring stations and the results of two way ANOVA test. The range of iron was 1.28 ± 0.02 (Jul.) to 2.78 ± 0.08 (Nov.) at WAZ-1, 1.64 ± 0.04 (Aug.) to 3.04 ± 0.19 (Dec.) at ISB-2, 2.34 ± 0.34 (May) to 3.66 ± 0.26 (Nov.) at ITO-3, 2.62 ± 0.37 (May) to 6.2 ± 0.335 (Nov.) at NIZ-4 and 2.98 ± 0.12 (Jun.) to 5.28 ± 0.35 (Nov.) at OKH-5. The data were subjected to two way ANOVA test. The test showed significant difference exist in iron content between different sampling stations (F_1=19.73, P<0.01), as well as between monitoring months (F_2=2.60, P<0.05).

Table RWM-22: Monthly Variation of Iron (mg/l) in River Water (Year 1994-1995)

Month	*WAZ-1* *Mean ± SD*	*ISB-2* *Mean ± SD*	*ITO-3* *Mean ± SD*	*NIZ-4* *Mean ± SD*	*OKH-5* *Mean ± SD*
Jan	1.695 ± 0.065	2.050 ± 0.050	2.445 ± 0.055	2.740 ± 0.010	3.200 ± 0.100
Feb	1.700 ± 0.000	2.065 ± 0.085	2.695 ± 0.015	2.735 ± 0.065	3.375 ± 0.115
Mar	1.905 ± 0.075	1.930 ± 0.070	2.755 ± 0.075	2.905 ± 0.045	3.755 ± 0.095
Apr	2.100 ± 0.120	2.295 ± 0.075	2.365 ± 0.065	2.730 ± 0.050	3.135 ± 0.145
May	2.325 ± 0.035	2.480 ± 0.020	2.340 ± 0.340	2.615 ± 0.365	3.375 ± 0.485
Jun	2.640 ± 0.220	2.890 ± 0.090	2.760 ± 0.040	3.040 ± 0.060	2.980 ± 0.120
Jul	1.280 ± 0.020	2.770 ± 0.080	2.920 ± 0.060	3.380 ± 0.020	3.635 ± 0.155
Aug	1.335 ± 0.045	1.640 ± 0.040	3.050 ± 0.070	3.295 ± 0.005	3.620 ± 0.060
Sep	1.725 ± 0.035	2.040 ± 0.060	3.455 ± 0.155	4.030 ± 0.070	3.805 ± 0.025
Oct	1.935 ± 0.045	2.055 ± 0.155	3.645 ± 0.215	5.025 ± 0.135	4.395 ± 0.195
Nov	2.780 ± 0.080	2.980 ± 0.120	3.655 ± 0.265	6.195 ± 0.335	5.280 ± 0.350
Dec	2.470 ± 0.490	3.040 ± 0.190	2.815 ± 0.165	4.860 ± 0.970	3.990 ± 0.310

Two-way ANOVA Test

Significant difference in iron content between river monitoring stations, F_1=19.73 (P<0.01).
Significant difference in iron content between monitoring months, F_2=2.66 (P<0.05).

Seasonal averages of iron content of five river monitoring stations varied from 1.45 ± 0.201 (Monsoon) to 2.36 ± 0.427 (Pre-Monsoon) at WAZ-1, 1.82 ± 0.178 mg/l (Monsoon) to 2.52 ± 0.483 (Pre-Monsoon) at ISB-2, 2.6 ± 0.27 (Summer) to 3.65 ± 0.241 (Pre-Monsoon) at ITO-3, 2.82 ± 0.249 (Winter) to 5.61 ± 0.638 (Pre-Monsoon) at NIZ-4, 3.3 ± 0.394 (Summer) to 4.84 ± 0.525 (Pre-Monsoon) at OKH-5. There were two peak seasonal values during post-monsoon and monsoon season which may be attributed to the enrichment of Fe brought from the catchment areas by rain water (Munawar, 1970; Rao, 1971; and Mahan, 1987). The high microbial growth in monsoon months (Pradhan *et al.*, 1991) might also be responsible for the high value of 'Fe' released due to mineralisation of the decomposed body. The lower values during summer and winter season may be attributed to the absence of external source of Fe from catchments areas. Average annual values were 1.99, 2.27, 2.9, 3.63, 3.7 at WAZ-1, ISB-2, ITO-3, NIZ-4 and OKH-5 respectively. The iron content of river water increased suddenly at ISB-2 (ISBT bridge), may be due to mixing of drain water from Najafgarh and Burai drain. The concentration of iron further increased at ITO-3, NIZ-4 and OKH-5 due to further contribution of drain water at Delhi stretch.

Cadmium of Drain and River Water

Cadmium is a bluish white metal. Cadmium is insoluble in water but soluble in acids. Cadmium salts are usually found in wastes from electroplating industries, textile printing, pigment works, lead mines and chemical industries. Cadmium is used as fungicides, insecticides, nematicides, polymerisation catalysts, pigments etc. and also as a contaminant of super phosphate fertilizers (Sittig, 1979). Cadmium has high toxic potential and minute quantities of cadmium are responsible for adverse renal arterial changes in human kidneys. Cadmium tends to concentrate in liver, kidneys, pancreas and thyroid of humans and animals. Once it enters the body it is likely to remain for long time (Nayak and Sawant, 1994; Forstner, 1977; Kudesia, 1980; Forstner and Wittmann,1981). Cadmium toxicity (50 mg/l) causes vomiting diarrhoea, abdominal pain, loss of bone, deformation, impaired kidneys functioning, impaired reproductive function, hypertension, tumour formation and teratogenic effects (Chatterjee, 1984). The WHO European Standard for cadmium in drinking water is 0.05 mg/l.

Monthly averages (Mean ± SD) of cadmium of five monitoring drains are given in Table DWM-22. Out of five drains monitored, only in Najafgarh drain Cd could be traced. The Cd content varied 0.016 ± 0.003 (Jan.) to 0.067 ± 0.002 mg/l (Nov). Seasonal values of cadmium for Najafgarh (DW-22) ranged from a minimum of 0.027 ± 0.012 (Winter) to a maximum of 0.064 ± 0.004 mg/l (Pre-Monsoon). Annual average value of cadmium was 0.038 ± 0.018 mg/l.

Table DWM-22: Monthly Variation in Cadmium (mg/l) of Drain Water (Year 1993-1995)

Month	*NAJ-1*	*CIV-2*	*POW-3*	*SEN-4*	*BAR-5*
	Mean ± SD	*Mean ± SD*	*Mean ± SD*	*Mean ± SD*	*Mean ± SD*
Jan.	0.016 ± 0.003	0.000 ± 0.000	0.000 ± 0.000	0.000 ± 0.000	0.000 ± 0.000
Feb.	0.024 ± 0.005	0.000 ± 0.000	0.000 ± 0.000	0.000 ± 0.000	0.000 ± 0.000
Mar.	0.022 ± 0.002	0.000 ± 0.000	0.000 ± 0.000	0.000 ± 0.000	0.000 ± 0.000
Apr.	0.028 ± 0.008	0.000 ± 0.000	0.000 ± 0.000	0.000 ± 0.000	0.000 ± 0.000
May	0.034 ± 0.006	0.000 ± 0.000	0.000 ± 0.000	0.000 ± 0.000	0.000 ± 0.000
Jun.	0.034 ± 0.008	0.000 ± 0.000	0.000 ± 0.000	0.000 ± 0.000	0.000 ± 0.000
Jul.	0.031 ± 0.001	0.000 ± 0.000	0.000 ± 0.000	0.000 ± 0.000	0.000 ± 0.000
Aug.	0.033 ± 0.004	0.000 ± 0.000	0.000 ± 0.000	0.000 ± 0.000	0.000 ± 0.000
Sep.	0.044 ± 0.001	0.000 ± 0.000	0.000 ± 0.000	0.000 ± 0.000	0.000 ± 0.000
Oct.	0.061 ± 0.002	0.000 ± 0.000	0.000 ± 0.000	0.000 ± 0.000	0.000 ± 0.000
Nov.	0.067 ± 0.003	0.000 ± 0.000	0.000 ± 0.000	0.000 ± 0.000	0.000 ± 0.000
Dec.	0.065 ± 0.016	0.000 ± 0.000	0.000 ± 0.000	0.000 ± 0.000	0.000 ± 0.000

Two-way ANOVA Test was not possible, due to inadequate data.

Monthly averages (Mean ± SD) of cadmium mg/l at five river monitoring points have been presented in Table RWM-23. Average values of Cd varied between 0.005 ± 0.0 to 0.017 ± 0.001 at WAZ-1, 0.0 to 0.08 ± 0.01 at ISB-2 and 0.006 ± 0.001 to 0.027 ± 0.001 at OKH-5 (Okhla barrage). The cadmium was not detected at ITO-3 and NIZ-4. Analysis of variance shows significant difference in cadmium content between five monitoring stations (F_1=20.52, P<0.01). However the test showed no significant difference in cadmium level between monitoring months (F_2=0.96).

Seasonal values varied with a minimum of 0.006 ± 0.002 (Monsoon) to a maximum of 0.014 ± 0.004 (Summer) at WAZ-1, 0.0 to 0.008 ± 0.004 (Pre-Monsoon) at ISB-2, and 0.009 ± 0.004 (Monsoon) to 0.002 ± 0.003 (Pre-Monsoon) at OKH-5. The peak values of cadmium were observed during summer due to lean river water flow and high concentration of drain effluents due to high rate of evaporations and low values during monsoon might be due to effect of dilution. Annual average values of cadmium were 0.01, 0.003 and 0.018 at WAZ-1, ISB-2 and OKH-5 respectively. The presence of cadmium at Wazirabad barrage (un-polluted stretch) might be due to plankton and nekton present in the water (Jain and Salman, 1995). The presence of cadmium at ISB-2 was due to drain effluents of Najafgarh drain and Cd at OKH-5 due to Hindon river cut, which carries industrial effluents.

Table RWM-23: Monthly Variation of Cadmium (mg/l) in River Water (Year 1994- 1995)

Month	*WAZ-1*	*ISB-2*	*ITO-3*	*NIZ-4*	*OKH-5*
	Mean ± SD	*Mean ± SD*	*Mean ± SD*	*Mean ± SD*	*Mean ± SD*
Jan.	0.005 ± 0.000	Nil	Nil	Nil	0.016 ± 0.001
Feb.	0.008 ± 0.001	Nil	Nil	Nil	0.018 ± 0.001
Mar.	0.009 ± 0.000	Nil	Nil	Nil	0.025 ± 0.001
Apr.	0.012 ± 0.002	Nil	Nil	Nil	0.021 ± 0.001
May	0.017 ± 0.001	Nil	Nil	Nil	0.017 ± 0.001
Jun.	0.018 ± 0.000	Nil	Nil	Nil	0.027 ± 0.001
Jul.	0.005 ± 0.000	0.003 ± 0.003	Nil	Nil	0.007 ± 0.001
Aug.	0.005 ± 0.001	0.003 ± 0.003	Nil	Nil	0.006 ± 0.000
Sep.	0.008 ± 0.001	0.005 ± 0.005	Nil	Nil	0.014 ± 0.002
Oct.	0.009 ± 0.001	0.008 ± 0.008	Nil	Nil	0.019 ± 0.001
Nov.	0.013 ± 0.002	0.007 ± 0.007	Nil	Nil	0.025 ± 0.001
Dec.	0.010 ± 0.002	0.010 ± 0.010	Nil	Nil	0.019 ± 0.001

Two-way ANOVA Test

Significant difference in cadmium content between river monitoring stations, F_1=21.60 (P<0.01).

No significant difference in cadmium content between monitoring months, F_2=0.96.

Chromium of Drain and River Water

Chromium salts impart colour to water. The salts of trivalent chromium are physiologically not harmful. It is reported that, the large doses of hexavalent chromium salts lead to corrosive effects in the intestinal tract and to nephritis (Manivasakam, 1989). It is one of the highly toxic metals responsible for various health hazards (Forstner, 1977; Kudesia, 1980; Forstner and Wittmann, 1981). Chromium is hazardous and create complication in the utilization of other nutrients by plants (Khan and Khan, 1983). Chromium compounds are used extensively in industrial processes such as metal pickling, electroplating, aluminium anodising, leather tanning and in the manufacture of paints, dyes, paper, explosives and ceramics. In addition chromates are frequently added to cooling water for corrosion control. The chromium content in natural waters are usually very small. Elevated concentrations can result from mining and industrial processes (Datar and Vashishth, 1990).

The harmful effects of water borne chromium are digestive track cancer and sometimes lung cancer (Chatterjee,1984). Banat *et al.* (1992) have attributed the chromium accumulation in the Rhine river sediment to industrial and domestic discharges. Chromium also discharged from electroplating units (Mathur *et al.*, 1987) galvanizing and tanning industries in river Rhine (Weijden and Middelburg, 1989). According to WHO International Standards for drinking water, the mandatory limit for hexavalent chromium is 0.05 mg/l.

Monthly averages (Mean ± SD) and the results of analysis of variance of chromium content (mg/l) in drain water are presented in Table DWM-23. Out of five monitoring drains the chromium was not found only in Civil mill drain water. Monthly averages values of chromium varied from a minimum of 0.308 ± 0.012 to a maximum of 0.825 ± 0.039 in NAJ-1, 0.081 ± 0.005 to 0.111 ± 0.025 in POW-3, 0.05 ± 0.004 to 0.01 ± 0.005 in SEN-4, 0.018 ± 0.002 to 0.073 ± 0.005 in BAR-5. Analysis of variance shows significant differences in chromium content between the monitoring stations (F_1=19.7, P<0.01). Difference between the monitoring months was not significant (F_2= 0.73).

Table DWM-23: Monthly Variation in Chromium (mg/l) of Drain Water (Year 1993-1995)

Month	*NAJ-1*	*CIV-2*	*POW-3*	*SEN-4*	*BAR-5*
	Mean ± SD	*Mean ± SD*	*Mean ± SD*	*Mean ± SD*	*Mean ± SD*
Jan.	0.30 ± 0.09	0.00 ± 0.00	0.08 ± 0.00	0.07 ± 0.01	0.04 ± 0.00
Feb.	0.38 ± 0.07	0.00 ± 0.00	0.08 ± 0.01	0.08 ± 0.00	0.04 ± 0.00
Mar.	0.59 ± 0.11	0.00 ± 0.00	0.08 ± 0.02	0.07 ± 0.00	0.05 ± 0.00
Apr.	0.59 ± 0.00	0.00 ± 0.00	0.16 ± 0.00	0.06 ± 0.00	0.04 ± 0.00
May	0.61 ± 0.01	0.00 ± 0.00	0.17 ± 0.01	0.08 ± 0.00	0.04 ± 0.00
Jun.	0.82 ± 0.03	0.00 ± 0.00	0.08 ± 0.01	0.09 ± 0.00	0.05 ± 0.00
Jul.	0.45 ± 0.04	0.00 ± 0.00	0.08 ± 0.01	0.05 ± 0.00	0.01 ± 0.00
Aug.	0.50 ± 0.01	0.00 ± 0.00	0.08 ± 0.00	0.05 ± 0.00	0.01 ± 0.00
Sep.	0.65 ± 0.06	0.00 ± 0.00	0.11 ± 0.02	0.07 ± 0.00	0.03 ± 0.00
Oct.	0.66 ± 0.02	0.00 ± 0.00	0.10 ± 0.01	0.08 ± 0.00	0.03 ± 0.00
Nov.	0.64 ± 0.12	0.00 ± 0.00	0.18 ± 0.00	0.09 ± 0.00	0.06 ± 0.00
Dec.	0.48 ± 0.16	0.00 ± 0.00	0.07 ± 0.00	0.10 ± 0.00	0.07 ± 0.00

Two-way ANOVA Test

Significant difference in chromium content between monitoring drains, F_1=19.7 (P<0.01).

No significant difference in chromium content between sampling months, F_2=0.73.

Seasonal average values ranged from 0.342 ± 0.058 (Winter) to 0.657 ± 0.089 (Pre-Monsoon) in NAJ-1, 0.081 ± 0.009 (Winter) to 0.126 ± 0.044 (Summer) in POW-3, 0.058 ± 0.010 (Monsoon) to 0.098 ± 0.014 (Post Monsoon) in SEN-5. Annual average values of chromium in four drains were 0.56 ± 0.15, 0.102 ± 0.034, 0.077 ± 0.017, 0.043 ± 0.017 in NAJ-1, POW-3, SEN-4 and BAR-5 respectively. The Najafgarh drain carries industrial effluents from Wazirpur industrial area and other industries functioning at its basin. The higher values of chromium observed in Najafgarh drain may be due to industrial water present in it.

Table RWM-24 presents for monthly averages (Mean ± SD) and the results of ANOVA test for chromium contents (mg/l) at five river monitoring stations. Though chromium was monitored at five stations (from Wazirabad to Okhla barrage) it was not detected at Wazirabad barrage (unpolluted stretch). The range of chromium content were 0.013 ± 0.001 (Jul.) to 0.030 ± 0.002 (Oct.) at ISB-2, 0.005 ± 0 (Jul.) to 0.025 ± 0.001 (Nov.) at ITO-3, 0.009 ± 0 (Jul.) to 0.037 ± 0.001 (Jan.) at NIZ-4, 0.007 ± 0.001 (Jul.) to 0.037 ± 0.002 (Jun.) at OKH-5. Analysis of variance (Table RWM-24) shows highly significant difference in chromium concentration in between sampling stations (F_1=20.52 P<0.01) and no significant difference in monitoring months (F_2=1.95).

Seasonal averages shows a minimum of 0.018 ± 0.005 (Monsoon) to a maximum of 0.028 ± 0.003 (Pre-Monsoon) at ISB-2, 0.007 ± 0.002 (Monsoon) to 0.019 ± 0.003 (Winter) at ITO-3, 0.011 ± 0.005 (Monsoon) to 0.0321 ± 0.003 (Winter) at NIZ-4, 0.015 ± 0.007 (Monsoon) to 0.032 ± 0.004 (Summer) at OKH-5. The higher concentration of chromium was observed during winter and summer season, may be due to lean river water flow (a little discharge from Wazirabad barrage) and higher percentage contribution of drain water comparison to river water at down stream river monitoring points. The monsoon months showed lowest values, indicating the effect of dilution. Yearly averages of chromium were in the order of 0.026 ± 0.008 > 0.024 ± 0.009 > 0.023 ± 0.006 > 0.015 ± 0.005 at OKH-5, NIZ-4, ISB-2 and ITO-3 respectively.

Table RWM-24: Monthly Variation of Chromium (mg/l) In River Water (Year 1994-1995)

Month	*WAZ-1*	*ISB-2*	*ITO-3*	*NIZ-4*	*OKH-5*
	Mean ± SD	*Mean ± SD*	*Mean ± SD*	*Mean ± SD*	*Mean ± SD*
Jan.	Nil	0.026 ± 0.002	0.017 ± 0.002	0.037 ± 0.001	0.027 ± 0.001
Feb.	Nil	0.023 ± 0.001	0.022 ± 0.001	0.031 ± 0.001	0.029 ± 0.002
Mar.	Nil	0.026 ± 0.002	0.019 ± 0.001	0.030 ± 0.002	0.028 ± 0.002
Apr.	Nil	0.019 ± 0.001	0.014 ± 0.003	0.025 ± 0.004	0.029 ± 0.001
May	Nil	0.024 ± 0.001	0.013 ± 0.003	0.020 ± 0.004	0.033 ± 0.003
Jun.	Nil	0.024 ± 0.005	0.022 ± 0.002	0.028 ± 0.002	0.037 ± 0.002
Jul.	Nil	0.013 ± 0.001	0.005 ± 0.000	0.009 ± 0.000	0.007 ± 0.001
Aug.	Nil	0.015 ± 0.001	0.006 ± 0.000	0.007 ± 0.001	0.016 ± 0.005
Sep.	Nil	0.025 ± 0.001	0.010 ± 0.002	0.018 ± 0.002	0.021 ± 0.002
Oct.	Nil	0.030 ± 0.002	0.016 ± 0.003	0.021 ± 0.002	0.026 ± 0.000
Nov.	Nil	0.026 ± 0.002	0.025 ± 0.001	0.037 ± 0.001	0.030 ± 0.000
Dec.	Nil	0.025 ± 0.000	0.020 ± 0.000	0.028 ± 0.002	0.034 ± 0.005

Two-way ANOVA Test

Significant difference in chromium content between river monitoring stations, F_1=20.52 (P<0.01).
No significant difference in chromium content between monitoring months. F_2=1.95.

Copper of Drain and River Water

Copper salts occur in natural surface water only in trace amounts up to about 0.05 mg/l. Copper occurs as native metals and in sulphide ores in nature. It is insoluble in water, but soluble in nitric acid and hot sulphuric acid. Metallic copper is used in all gauges of electric wire, coil and armature windings in electrical industries, made into casting, sheets, rods tubing and wire used in water and gas piping. It forms many important alloys. Aluminium, bronze, copper compounds are used as insecticides and algicides. They are also utilized in electroplating. Copper poisoning leads to toxic effects like ptyalism, nausia, hematuria, convulsions and death (Prasad, 1978). Copper salts are used in water supply systems to control biological growths in reservoirs and distribution pipes and to catalyse the oxidation of manganese. Corrosion of copper containing alloys in pipe fittings may introduce measurable amounts of copper into the water in pipe system. Effluents from electroplating plants constitute one of the important source of heavy metal pollution and normally contain objectionable quantities of copper, lead, zinc, cadmium, chromium, cyanides, oils and grease (Ramaswamy and Somashekar, 1982). However copper is essential to human and the adult daily requirement had been estimated to be 20 mg (APHA, 1991).

Table DWM-24, shows the monthly averages (Mean ± SD) of copper in drain water and the results of ANOVA test. The averages ranged from 0.176 ± 0.022 to 0.339 ± 0.029 in NAJ-1, 0.020 ± 0 to 0.143 ± 0.007 in CIV-2, 0.078 ± 0.002 to 0.285 ± 0.005 in POW-3, 0.054 ± 0.002 to 0.518 ± 0.002 in SEN-4, 0.017 ± 0.001 to 0.059 ± 0.003 in BAR-5.

Analysis of variance shows significant difference in Cu content between monitoring drains (F_1= 22.8, P<0.01) as well as between sampling months (F_2=2.07, P>0.05).

Seasonal and annual average values of copper (mg/l) varied between 0.193 ± 0.021 (Winter) to 0.318 ± 0.069 (Monsoon) in NAJ-1, 0.034 ± 0.017 (Monsoon) to 0.122 ± 0.033 (Winter) in CIV-2, 0.096 ± 0.025 (Monsoon) to 0.255 ± 0.030 (Summer) in SEN-4, 0.02 ± 0.003 (Monsoon) to 0.054 ± 0.008 (Summer) in BAR-5. Annual average copper values in drain water were 0.275 ± 0.068, 0.084 ± 0.036, 0.179 ± 0.074, 0.116 ± 0.013, 0.039 ± 0.016 at NAJ-1, CIV-2, POW-3, SEN-4 and BAR-5 respectively.

Table DWM-24: Monthly Variation in Copper (mg/l) of Drain Water (Year 1993-1995)

Month	*NAJ-1*	*CIV-2*	*POW-3*	*SEN-4*	*BAR-5*
	Mean ± SD	*Mean ± SD*	*Mean ± SD*	*Mean ± SD*	*Mean ± SD*
Jan.	0.176 ± 0.022	0.093 ± 0.002	0.185 ± 0.005	0.067 ± 0.003	0.049 ± 0.007
Feb.	0.198 ± 0.015	0.123 ± 0.038	0.255 ± 0.025	0.076 ± 0.003	0.059 ± 0.003
Mar.	0.228 ± 0.001	0.95 ± 0.005	0.270 ± 0.010	0.86 ± 0.001	0.046 ± 0.002
Apr.	0.246 ± 0.040	0.090 ± 0.000	0.255 ± 0.005	0.518 ± 0.002	0.056 ± 0.003
May	0.339 ± 0.029	0.090 ± 0.020	0.285 ± 0.005	0.099 ± 0.001	0.061 ± 0.007
Jun.	0.326 ± 0.010	0.085 ± 0.005	0.210 ± 0.020	0.109 ± 0.001	0.055 ± 0.007
Jul.	0.326 ± 0.070	0.025 ± 0.005	0.093 ± 0.002	0.054 ± 0.002	0.019 ± 0.001
Aug.	0.336 ± 0.091	0.020 ± 0.000	0.078 ± 0.002	0.074 ± 0.011	0.017 ± 0.001
Sep.	0.293 ± 0.005	0.058 ± 0.002	0.117 ± 0.034	0.080 ± 0.006	0.025 ± 0.001
Oct.	0.321 ± 0.014	0.090 ± 0.005	0.097 ± 0.000	0.080 ± 0.004	0.024 ± 0.06
Nov.	0.301 ± 0.015	0.095 ± 0.005	0.135 ± 0.015	0.077 ± 0.004	0.031 ± 0.006
Dec.	0.217 ± 0.021	0.143 ± 0.007	0.175 ± 0.005	0.074 ± 0.004	0.035 ± 0.005

Two-way ANOVA Test

Significant difference in copper content between monitoring drains, F_1=22.8 (P<0.01).
Significant difference in copper content between sampling months, F_2=2.07 (P<0.05).

Monthly average values (Mean ± SD) of copper and the results of ANOVA test for copper content in river monitoring stations are presented in Table RWM-25. The minimum/and maximum bimonthly averages (Mean ± SD) of copper were 0.014 ± 0.003 (Aug.) and 0.047 ± 0.005 (Dec.) at ISB-2, 0.01 ± 0 (Jul.) and 0.036 ± 0.002 (Oct.) at ITO-3, and 0.014 ± 0.002 (Aug.) and 0.044 ± 0.004 (Jan.) at NIZ-4 respectively. The copper was not observed at WAZ-1 (Wazirabad barrage) and OKH-5 (Okhla barrage). Analysis of variance test shows significant difference in copper content between different monitoring stations (F_1=20.33, P<0.01). However the difference between sampling months was not significant (F_2=1.33).

Table RWM-25: Monthly Variation of Copper (mg/l) in River Water (Year 1994-1995)

Month	*WAZ-1*	*ISB-2*	*ITO-3*	*NIZ-4*	*OKH-5*
	Mean ± SD	*Mean ± SD*	*Mean ± SD*	*Mean ± SD*	*Mean ± SD*
Jan.	Nil	0.028 ± 0.002	0.034 ± 0.002	0.044 ± 0.004	Nil
Feb.	Nil	0.038 ± 0.002	0.031 ± 0.001	0.041 ± 0.002	Nil
Mar.	Nil	0.036 ± 0.003	0.036 ± 0.003	0.037 ± 0.001	Nil
Apr.	Nil	0.040 ± 0.003	0.028 ± 0.002	0.028 ± 0.001	Nil
May	Nil	0.031 ± 0.001	0.032 ± 0.001	0.026 ± 0.002	Nil
Jun.	Nil	0.043 ± 0.003	0.029 ± 0.003	0.030 ± 0.003	Nil
Jul.	Nil	0.018 ± 0.001	0.010 ± 0.000	0.013 ± 0.000	Nil
Aug.	Nil	0.014 ± 0.002	0.010 ± 0.000	0.014 ± 0.000	Nil
Sep.	Nil	0.033 ± 0.003	0.016 ± 0.004	0.026 ± 0.002	Nil
Oct.	Nil	0.036 ± 0.002	0.024 ± 0.005	0.034 ± 0.002	Nil
Nov.	Nil	0.040 ± 0.002	0.021 ± 0.001	0.027 ± 0.001	Nil
Dec.	Nil	0.047 ± 0.005	0.023 ± 0.003	0.029 ± 0.004	Nil

Two-way ANOVA Test

Significant difference in copper content between river monitoring stations, F_1=20.33 (P< 0.01).
No significant difference in copper content between monitoring months, F_2=1.33.

Seasonal and annual average values in copper concentration varied from a minimum of 0.021 ± 0.009 (Monsoon) to a maximum of 0.038 ± 0.003 (Pre-Monsoon) at ISB-2, 0.012 ± 0.004 (Monsoon) to 0.031 ± 0.003 (Winter) at ITO-3, 0.017 ± 0.006 (Monsoon) to 0.039 ± 0.006 (Winter) at NIZ-4. The peak concentration of copper observed during winter season might be due to lean river water discharge from Wazirabad barrage and higher percentage contribution of drain water at river monitoring points. The low values during monsoon was the effects of dilution by rain waters. Annual average copper values were in order of 0.033 ± 0.01 > 0.029 ± 0.009 > 0.024 ± 0.009 at ISB-2, NIZ-4 and ITO-3 respectively. Absence of copper at OKH-5 might be due to dilution of copper containing drain effluents.

Zinc of Drain and River Water

Zinc has no known adverse physiological effects upon human beings. In fact it is an essential and beneficial element in human nutrition. From aesthetic considerations however, high concentrations (above 5 mg/l) of zinc of domestic water are undesirable. A bitter astringent taste is produced by zinc. Concentration of zinc above 30 mg/l give water milky appearance and causes a greasy film on boiling. It is now reported that even 5 mg/l can cause a greasy film on boiling. Zinc is commonly found in small quantities on domestic water supplies and industrial waters due to corrosion of galvanized iron and brass in condensing, cooling and distributing systems. Zinc may also enter into a water supply through the discharge of industrial effluent such as galvanising water, zinc plating wastes, viscose rayon wastes and rubber factory waste (Manivasakam, 1989). A relatively higher concentration of zinc may be toxic to aquatic life (Kothandaraman and Viswanathan, 1986).

Table DWM-25 represents for bimonthly average values (Mean ± SD) and the results of ANOVA test for zinc concentration in five monitoring drains. The minimum and maximum monthly average values were 0.320 ± 0.001 and 1.2 ± 0.122 in NAJ-1, 0.560 ± 0.050 and 1.086 ± 0.1 in CIV-2, 0.745 ± 0.185 and 1.675 ± 0.025 in POW-3, 0.425 ± 0.035 and 1.115 ± 0.135 in SEN-4, 0.30 ± 0.02 and 0.745 ± 0.065 in BAR-5 drain respectively. Analysis of variance shows significant difference in zinc concentration between sampling stations (F_1=20.4, P<0.01) as well as between sampling months (F_2=2.66, P<0.05).

Seasonal and annual average values varied between 0.424 ± 0.113 (Summer) to 0.982 ± 0.076 (Pre-Monsoon) in NAJ-1, 0.628 ± 0.132 (Monsoon) to 1.01 ± 0.204 (Pre-Monsoon) in CIV-2, 0.955 ± 0.210 (Monsoon) to 1.273 ± 0.130 (Pre-Monsoon) in POW-3, 0.585 ± 0.239 (Winter) to 0.960 ± 0.097 (Pre-Monsoon) in SEN-4, 0.140 ± 0.134 (Winter) to 0.175 ± 0.044 (Pre-Monsoon) in BAR-5 respectively. Annual averages of zinc content were in order of 1.050 ± 0.337 > 0.836 ± 0.194 > 0.688 ± 0.224 > 0.593 ± 0.138 > 0.558 ± 0.152 in POW-3, CIV-2, SEN-4, NAJ-1 and BAR-5 drain respectively.

Table RWM-26 shows the monthly averages (Mean ± SD) and the results of analysis of variance for zinc content in river water at five monitoring points. The minimum and maximum average values of zinc were 0.125 ± 0.015 (April) and 0.445 ± 0.035 (Nov.) at WAZ-1, 0.295 ± 0.005 (Dec.) and 0.405 ± 0.095 (Jun.) at ISB-2, 0.195 ± 0.005 (Feb.) and 0.470 ± 0.04 (Nov.) at ITO-3, 0.270 ± 0.01 (Jan.) and 0.570 ± 0.05 (Oct.) at NIZ-4 and 0.27 ± 0.01 (Feb.) and 0.620 ± 0.06 (Jul.) at OKH-5. Analysis of variance shows the significant difference in zinc content between different river monitoring stations (F_1=19.81, P<0.01), as well as between the monitoring months (F_2=2.91, P<0.05).

Table DWM-25: Monthly Variation in Zinc (mg/l) of Drain Water (Year 1994-1995)

Month	*NAJ-1*	*CIV-2*	*POW-3*	*SEN-4*	*BAR-5*
	Mean ± SD	*Mean ± SD*	*Mean ± SD*	*Mean ± SD*	*Mean ± SD*
Jan.	0.358 ± 0.032	0.766 ± 0.027	0.745 ± 0.335	0.425 ± 0.105	0.300 ± 0.020
Feb.	0.320 ± 0.001	0.825 ± 0.029	0.790 ± 0.200	0.425 ± 0.035	0.370 ± 0.010
Mar.	0.344 ± 0.106	0.790 ± 0.142	0.745 ± 0.185	0.570 ± 0.010	0.435 ± 0.045
Apr.	0.593 ± 0.033	0.704 ± 0.031	0.820 ± 0.140	0.52 ± 0.040	0.525 ± 0.105
May	0.378 ± 0.011	0.861 ± 0.025	1.040 ± 0.060	0.585 ± 0.085	0.525 ± 0.035
Jun.	0.383 ± 0.014	0.841 ± 0.096	1.370 ± 0.240	0.710 ± 0.070	0.435 ± 0.055
Jul.	0.62 ± 0.032	0.560 ± 0.050	0.795 ± 0.065	0.605 ± 0.025	0.595 ± 0.055
Aug.	0.414 ± 0.054	0.655 ± 0.025	0.890 ± 0.090	0.645 ± 0.035	0.630 ± 0.050
Sep.	0.710 ± 0.050	0.830 ± 0.100	1.180 ± 0.200	0.730 ± 0.050	0.745 ± 0.065
Oct.	0.923 ± 0.045	0.855 ± 0.036	1.230 ± 0.130	0.905 ± 0.075	0.725 ± 0.055
Nov.	1.042 ± 0.048	1.165 ± 0.185	1.315 ± 0.115	1.015 ± 0.085	0.705 ± 0.025
Dec.	1.286 ± 0.122	1.180 ± 0.100	1.675 ± 0.025	1.115 ± 0.135	0.705 ± 0.015

Two-way ANOVA Test

Significant difference in zinc content between monitoring drains, F_1=20.41 (P<0.01).

Significant difference in zinc content between sampling months, F_2=2.66 (P<0.05).

Seasonal and annual averages varied from 0.250 ± 0.033 (Summer) to 0.418 ± 0.038 (Pre-Monsoon) at WAZ-1, 0.283 ± 0.058 (Monsoon) to 0.478 ± 0.064 (Pre-Monsoon) at ISB-2, 0.237 ± 0.041 (Winter) to 0.453 ± 0.046 (Pre-Monsoon) at ITO-3, 0.295 ± 0.026 (Winter) to 0.490 ± 0.094 (Pre-Monsoon) at NIZ-4, 0.30 ± 0.033 (Winter) to 0.583 ± 0.055 (Monsoon) at BAR-5. Peak values of zinc were observed during monsoon and post-monsoon seasons might be due to solubilization of zinc ions present in sediment of river bank by rain water. The lower values of zinc during winter might be due to precipitation of this metal from the water to the sediment due to lean river flow and low temperature (Jain and Salman, 1995). Annual average values (mg/l) were in order of 0.412 > 0.380 > 0.353 > 0.330 > 0.328 at OKH-5, NIZ-4, ISB-2, ITO-3 and WAZ-1 respectively.

Table RWM-26: Monthly Variation of Zinc (mg/l) in River Water Year (1994-1995)

Month	*WAZ-1*	*ISB-2*	*ITO-3*	*NIZ-4*	*OKH-5*
	Mean ± SD	*Mean ± SD*	*Mean ± SD*	*Mean ± SD*	*Mean ± SD*
Jan.	0.290 ± 0.010	0.345 ± 0.055	0.245 ± 0.045	0.270 ± 0.010	0.290 ± 0.010
Feb.	0.285 ± 0.015	0.380 ± 0.080	0.195 ± 0.005	0.305 ± 0.025	0.270 ± 0.010
Mar.	0.265 ± 0.025	0.430 ± 0.070	0.255 ± 0.015	0.300 ± 0.010	0.325 ± 0.005
Apr.	0.215 ± 0.015	0.330 ± 0.070	0.305 ± 0.025	0.320 ± 0.040	0.335 ± 0.055
May	0.275 ± 0.015	0.330 ± 0.130	0.295 ± 0.015	0.410 ± 0.050	0.320 ± 0.020
Jun.	0.245 ± 0.035	0.405 ± 0.095	0.385 ± 0.005	0.410 ± 0.020	0.390 ± 0.010
Jul.	0.335 ± 0.015	0.325 ± 0.065	0.390 ± 0.010	0.425 ± 0.035	0.620 ± 0.060
Aug.	0.355 ± 0.005	0.355 ± 0.155	0.345 ± 0.015	0.360 ± 0.020	0.585 ± 0.025
Sep.	0.410 ± 0.020	0.305 ± 0.005	0.370 ± 0.010	0.440 ± 0.020	0.545 ± 0.045
Oct.	0.390 ± 0.010	0.335 ± 0.055	0.435 ± 0.045	0.570 ± 0.050	0.435 ± 0.025
Nov.	0.445 ± 0.035	0.385 ± 0.125	0.470 ± 0.040	0.410 ± 0.050	0.435 ± 0.055
Dec.	0.430 ± 0.070	0.295 ± 0.005	0.270 ± 0.010	0.340 ± 0.050	0.390 ± 0.070

Two-way ANOVA Test

Significant difference in zinc content between river monitoring stations, F_1=19.81 (P<0.01).

Significant difference in zinc content between monitoring months, F_2=2.91 (P<0.05).

Nickel of Drain and River Water

Nickel does not occur in natural water. Metal plating wastes when discharged in to the water courses contribute nickel. Increased use of stainless steel and other nickel containing alloy in water systems also contribute nickel. Nickel compounds are found in many ores and minerals. It is relatively non-toxic elements at lower level, but higher amount (more than 30 mg) may cause changes in muscle, brain, lungs, liver, kidney and can also cause cancer (Ramteke and Moghe, 1986). Nickel is an important non-ferrous metal closely related to iron and cobalt (Carbson, 1956). As an alloying metal it is used in steel, separately or with chrome, or with zinc or simultaneously with zinc and copper or with titanium etc. for making different industrial and strategic products (Smith *et al.*, 1961). Nickel is insoluble in water and soluble in acid. It forms alloys with copper, manganese, zinc, chromium, molybdenum etc. Stainless steel is the most widely used nickel alloy (Sittig, 1979).

Table DWM-26 shows the monthly averages (Mean ± SD) and the results of analysis of variance for nickel content in monitoring drains. The minimum and maximum average values of nickel (mg/l) were 0.221 ± 0.002 and 0.594 ± 0.326 in NAJ-1, 0.02 ± 0.002 and 0.071 ± 0.002 in CIV-2, 0.055 ± 0.005 and 0.180 ± 0.010 in POW-3, 0.043 ± 0.002 and 0.089 ± 0.004 in SEN-4, 0.019 ± 0.001 and 0.06 ± 0.007 in BAR-5. Analysis of variance results shows that, difference in nickel concentration in between monitoring drains (F_1=20.8, P<0.01) but no significant difference was observed between sampling months (F_2=1.33).

Table DWM-26: Monthly Variation in Nickel (mg/1) of Drain Water (Year 1994-1995)

Month	*NAJ-1*	*CIV-2*	*POW-3*	*SEN-4*	*BAR-5*
	Mean ± SD	*Mean ± SD*	*Mean ± SD*	*Mean ± SD*	*Mean ± SD*
Jan.	0.221 ± 0.002	0.045 ± 0.000	0.105 ± 0.005	0.080 ± 0.004	0.03 ± 0.005
Feb.	0.270 ± 0.013	0.051 ± 0.007	0.155 ± 0.015	0.072 ± 0.001	0.035 ± 0.001
Mar.	0.300 ± 0.004	0.061 ± 0.002	0.180 ± 0.010	0.063 ± 0.018	0.054 ± 0.004
Apr.	0.68 ± 0.002	0.072 ± 0.002	0.155 ± 0.035	0.085 ± 0.001	0.060 ± 0.007
May	0.221 ± 0.002	0.045 ± 0.000	0.105 ± 0.045	0.080 ± 0.008	0.049 ± 0.013
Jun.	0.306 ± 0.037	0.064 ± 0.001	0.135 ± 0.030	0.069 ± 0.009	0.43 ± 0.003
Jul.	0.217 ± 0.006	0.020 ± 0.002	0.055 ± 0.005	0.043 ± 0.002	0.019 ± 0.001
Aug.	0.594 ± 0.326	0.026 ± 0.000	0.055 ± 0.005	0.050 ± 0.006	0.20 ± 0.004
Sep.	0.511 ± 0.478	0.042 ± 0.003	0.078 ± 0.006	0.073 ± 0.005	0.029 ± 0.001
Oct.	0.326 ± 0.030	0.058 ± 0.002	0.082 ± 0.004	0.087 ± 0.001	0.40 ± 0.001
Nov.	0.328 ± 0.018	0.071 ± 0.002	0.097 ± 0.002	0.080 ± 0.004	0.048 ± 0.005
Dec.	0.273 ± 0.005	0.069 ± 0.004	0.122 ± 0.013	0.089 ± 0.004	0.044 ± 0.005

Two-way ANOVA Test

Significant difference in nickel content between monitoring drains. F_1=20.8 (P<0.01).
No significant difference in nickel content between sampling months, F_2=1.33.

Annual as well as seasonal average values of nickel varied between 0.253 ± 0.024 (Winter) to 0.440 ± 0.016 (Monsoon) in NAJ-1, 0.029 ± 0.01 (Monsoon) to 0.067 ± 0.005 (Summer) in CIV-2, 0.063 ± 0.012 (Monsoon) to 0.150 ± 0.038 (Summer) in POW-3, 0.055 ± 0.014 (Monsoon) to 0.84 ± 0.005 (Pre-Monsoon) in Sen-4, 0.023 ± 0.005 (Monsoon) to 0.051 ± 0.010 (Summer) in BAR-5 drain. Annual average values were in order of 0.327 ± 0.199 > 0.112 ± 0.044 > 0.072 ± 0.015 > 0.054 ± 0.017 > 0.039 ± 0.013 at NIZ-1, POW-1, SEN-4, CIV-2 and BAR-5 respectively.

Table RWM-27 represents the monthly averages and the results of analysis of variance test for the nickel content at different drain monitoring stations. Monthly averages varied from a minimum of 0.031 ± 0.005 (Sep) to a maximum of 0.061 ± 0.020 (Nov.) at WAZ-1, 0.016 ± 0.002 (Aug.) to 0.071 ± 0.002 (Jan.) at ISB-2, 0.01 ± 0 (Aug.) to 0.057 ± 0.006 (Nov.) at ITO-3, 0.016 ± 0.003 (Aug.) to 0.081 ± 0.002 (Nov.) at NIZ-4, 0.020 ± 0.002 (Aug.) to 0.070 ± 0.001 (Nov.) at OKH-5. The ANOVA test shows significant difference in nickel content between monitoring points (F_1=21.41, P<0.01) as well as between monitoring months (F_2=2.96, P<0.05).

Seasonal and annual average values are depicted in Fig. RW 27. Seasonal values varied from a minimum of 0.031 ± 0.007 (Monsoon) to a maximum of 0.066 ± 0.015 (Pre-Monsoon) at WAZ-1, 0.026 ± 0.011 (Monsoon) to 0.049 ± 0.009 (Pre-Monsoon) at ITO-3, 0.020 ± 0.005 (Monsoon) to 0.064 ± 0.004 (Winter) at NIZ-4, 0.027 ± 0.008 (Monsoon) to 0.065 ± 0.005 (Pre-Monsoon) at OKH-5. Peak values of nickel were observed during post-monsoon season might be due to lean discharge of river water from Wazirabad barrage and higher quantities of drain water contribution at river monitoring points. Yearly averages (mg/l) of nickel were in order of 0.052 > 0.048 > 0.045 > 0.044 > 0.032 at ISB-2, WAZ-1, NIZ-4, OKH-5 and ITO-3 respectively.

Table RWM-27: Monthly Variation of Nickel (mg/l) in River Water Year (1994-1995)

Month	*WAZ-1*	*ISB-2*	*ITO-3*	*NIZ-4*	*OKH-5*
	Mean ± SD	*Mean ± SD*	*Mean ± SD*	*Mean ± SD*	*Mean ± SD*
Jan.	0.055 ± 0.025	0.049 ± 0.001	0.025 ± 0.004	0.070 ± 0.016	0.031 ± 0.002
Feb.	0.041 ± 0.002	0.064 ± 0.004	0.028 ± 0.002	0.080 ± 0.022	0.034 ± 0.004
Mar.	0.040 ± 0.011	0.054 ± 0.006	0.027 ± 0.003	0.081 ± 0.002	0.036 ± 0.007
Apr.	0.051 ± 0.010	0.063 ± 0.005	0.037 ± 0.001	0.062 ± 0.009	0.056 ± 0.007
May	0.058 ± 0.009	0.060 ± 0.001	0.038 ± 0.004	0.051 ± 0.002	0.062 ± 0.001
Jun.	0.058 ± 0.010	0.071 ± 0.002	0.049 ± 0.002	0.035 ± 0.010	0.059 ± 0.003
Jul.	0.047 ± 0.026	0.041 ± 0.002	0.015 ± 0.005	0.018 ± 0.019	0.023 ± 0.003
Aug.	0.042 ± 0.014	0.016 ± 0.002	0.010 ± 0.000	0.016 ± 0.003	0.020 ± 0.002
Sep.	0.031 ± 0.005	0.041 ± 0.002	0.025 ± 0.001	0.027 ± 0.006	0.038 ± 0.003
Oct.	0.041 ± 0.019	0.067 ± 0.004	0.041 ± 0.002	0.035 ± 0.009	0.061 ± 0.002
Nov.	0.061 ± 0.020	0.070 ± 0.002	0.057 ± 0.006	0.034 ± 0.008	0.070 ± 0.001
Dec.	0.048 ± 0.005	0.053 ± 0.003	0.031 ± 0.001	0.036 ± 0.002	0.043 ± 0.003

Two-way ANOVA Test

Significant difference in nickel content between river monitoring stations, F_1=21.41 (P<0.01).

Significant difference in nickel content between monitoring months, F_2=2.96 (P<0.05).

Lead of Drain and River Water

Lead is not found in natural water supply. It may be due to the corrosion reactions and wastes contamination. Pollution due to lead is mainly attributed to the industrial effluents, coal combustion products and petroleum washes from various mechanical workshops garages, because lead is used as antiknocking compounds in petroleum (Katariya, 1994) and also from domestic sewage discharge in the city (Hydn, 1961; Harrison and Laxen, 1984). Automobile exhaust also adds considerable amount of lead particulates to the river water (Pophali *et al.*, 1990) so also the traffic of diesel and petrol vehicles (Agarwal *et al.*, 1978) and lead dust fallout. Lead is a minor element in the earth crust but the significant contribution of lead is from the atmosphere. Lead is classified as highly toxic element in nature (Forstner, 1977; Kudesia, 1980; Forstner and Wiltmann, 1981). Lead in high doses has been recognized as a cumulative general metabolic poison (Suess, 1982).

Table DWM-27 shows the monthly averages (Mean ± SD) of lead content in drain water of Najafgarh. The lead was not found in Civil mill, Sen-nursing home, Power house and Barpulla drains. The maximum concentration of lead was observed in the month of November *i.e.* 0.09 ± 0.004 (mg/l) and minimum in July *i.e.* 0.033 ± 0.005 (mg/l) in Najafgarh drain. Annual and seasonal mean values are depicted in Table DW26. Seasonal values of lead varied from 0.038 ± 0.008 (Monsoon) to 0.075 ± 0.015 (Pre-Monsoon) in Najafgarh drain. Annual mean values (Table DW26) was 0.061 ± 0.02 in drain water of Najafgarh.

The lead was not detected in river water of Yamuna between Wazirabad barrage and Okhla barrage.

Table DWM-27: Monthly Variation in Lead (mg/l) of Drain Water (Year 1994-1995)

Month	NAJ-1 Mean ± SD	CIV-2 Mean ± SD	POW-3 Mean ± SD	SEN-4 Mean ± SD	BAR-5 Mean ± SD
Jan.	0.068 ± 0.005	0.000 ± 0.000	0.000 ± 0.000	0.000 ± 0.000	0.000 ± 0.000
Feb.	0.073 ± 0.005	0.000 ± 0.000	0.000 ± 0.000	0.000 ± 0.000	0.000 ± 0.000
Mar.	0.085 ± 0.001	0.000 ± 0.000	0.000 ± 0.000	0.000 ± 0.000	0.000 ± 0.000
Apr.	0.088 ± 0.008	0.000 ± 0.000	0.000 ± 0.000	0.000 ± 0.000	0.000 ± 0.000
May	0.053 ± 0.015	0.000 ± 0.000	0.000 ± 0.000	0.000 ± 0.000	0.000 ± 0.000
Jun.	0.042 ± 0.004	0.000 ± 0.000	0.000 ± 0.000	0.000 ± 0.000	0.000 ± 0.000
Jul.	0.033 ± 0.005	0.000 ± 0.000	0.000 ± 0.000	0.000 ± 0.000	0.000 ± 0.000
Aug.	0.035 ± 0.007	0.000 ± 0.000	0.000 ± 0.000	0.000 ± 0.000	0.000 ± 0.000
Sep.	0.045 ± 0.005	0.000 ± 0.000	0.000 ± 0.000	0.000 ± 0.000	0.000 ± 0.000
Oct.	0.061 ± 0.004	0.000 ± 0.000	0.000 ± 0.000	0.000 ± 0.000	0.000 ± 0.000
Nov.	0.090 ± 0.004	0.000 ± 0.000	0.000 ± 0.000	0.000 ± 0.000	0.000 ± 0.000
Dec.	0.067 ± 0.002	0.000 ± 0.000	0.000 ± 0.000	0.000 ± 0.000	0.000 ± 0.000

Two-way ANOVA Test was not possible, due to inadequate data.

Pesticides in Drain and River Water

BHC (HCH) of Drain and River Water

The BHC which is HCH (Hexachloro Cyclohexane) is generally included in the class of organochlorine compounds but they are actually isomers prepared from chlorination of benzene and are saturated cylic hydrocarbons. BHC are available in four forms, alpha-BHC, beta-BHC, gama-BHC and delta-BHC depending upon the position of six chlorine atoms. Mehrotra (1985) has given an exhaustive review on the use of HCH and its environmental effects in India. HCH was first introduced in 1949 to control insect pests of the agriculture. Its use in public health started in 1960. The large scale use of HCH has led to its presence in different components of the environment.

In India, HCH residues have been reported from soil (0.025 to 124 ppm), water (0.02 to 6.61 ppm), cereals and animal feed (0.01 to 3.52 ppm), vegetables oil and oil seeds (0.31 to 0.58 ppm), milk and milks products (0.35 to 82 ppm), meat and poultry (0.04 to 1.522 ppm), wildlife as high up to 13.98 and 22.487 ppm in human tissues (Kulshrestha, 1989). Dikshith *et al.*, 1990, reported a total HCH residues of 4.654 (range 1.576 to 7.747) in wells, 6.130 (range 4.320 to 8.718) in hand-pump and 9.941 (range 6.221 to 15.58) in ponds.

Monthly average values (Mean ± SD) and the results of ANOVA test for T-BHC (mg/l) content of drain water are provided in Table DWM 28. Monthly average values varied from 430.00 ± 220.0 to 6650.0 ± 2300 in NAJ-1, 1600 ± 100 to 4390 ± 410 in CIV-2, 365.0 ± 550 to 1620 ± 320 in POW-3, 455 ± 5 to 2160 ± 220 in SEN-4 and 400 ± 10 to 2429.0 ± 431.0 in BAR-5. Analysis of variance shows the existence of significant difference in T-BHC content between monitoring drains (F_1=26.31, P<0.01) as well as between monitoring months (F_2=2.86, P<0.05).

Table DWM-28: Monthly Variation in Total BHC (mg/l) of Drain Water (Year 1994-1995)

Month	*NAJ-1*	*CIV-2*	*POW-3*	*SEN-4*	*BAR-5*
	Mean ± SD	*Mean ± SD*	*Mean ± SD*	*Mean ± SD*	*Mean ± SD*
Jan.	6650.00 ± 2300.00	2660.00 ± 20.00	925.00 ± 65.00	1585.00 ± 55.00	1045.00 ± 55.00
Feb.	5272.50 ± 592.50	2620.00 ± 220.00	1070.00 ± 170.00	1230.00 ± 90.00	1490.00 ± 300.00
Mar.	4830.00 ± 600.00	3280.00 ± 300.00	1420.00 ± 460.00	935.00 ± 45.00	1680.00 ± 200.00
Apr.	2740.00 ± 390.00	3109.00 ± 71.00	1110.00 ± 30.00	1090.00 ± 130.00	1725.00 ± 35.00
May	1870.00 ± 30.00	3666.50 ± 313.50	1310.00 ± 320.00	1510.00 ± 270.00	2125.00 ± 35.00
Jun.	935.00 ± 145.00	4390.00 ± 410.00	1620.00 ± 330.00	2025.00 ± 65.00	2429.00 ± 431.00
Jul.	780.00 ± 200.00	1815.00 ± 135.00	435.00 ± 95.00	455.00 ± 5.00	410.00 ± 50.00
Aug.	3550.00 ± 3010.00	1600.00 ± 100.00	365.00 ± 55.00	520.00 ± 10.00	400.00 ± 10.00
Sep.	430.00 ± 220.00	1840.00 ± 50.00	610.00 ± 70.00	1010.00 ± 70.00	810.00 ± 130.00
Oct.	730.00 ± 250.00	2295.00 ± 165.00	834.50 ± 54.50	1925.00 ± 255.00	779.00 ± 219.00
Nov.	1319.00 ± 679.00	2325.00 ± 235.00	905.00 ± 45.00	2160.00 ± 220.00	925.00 ± 235.00
Dec.	2435.00 ± 1415.00	2035.00 ± 95.00	925.00 ± 85.00	2015.00 ± 135.00	820.00 ± 170.00

Two-way ANOVA Test

Significant difference in BHC content between monitoring drains, F_1=26.31 ($P<0.01$).
Significant difference in BHC content between sampling months, F_2=2.86 ($P<0.05$).

Seasonal and annual average values of T-BHC (mg/l) in five monitoring drains varied from a minimum of 1024.5 ± 590.34 (Pre-Monsoon) to a maximum of 5104.17 ± 1932.96 (Winter) in NAJ-1, 1751.67 ± 147.81 (Monsoon) to 3611.38 ± 577.26 (Summer) in CIV-2, 470.0 ± 127.54 (Monsoon) to 1365.0 ± 374.13 (Summer) in POW-3, 661.67 ± 251.09 (Monsoon) to 2042.50 ± 265.55 (Pre-Monsoon) in SEN-4, 540.0 ± 207.28 (Monsoon) to 1145.0 ± 311.01 (Winter) in BAR-5. Annual mean values of BHC were 2628.46, 2636.29, 960.8, 1371.7 and 1219.8 in NAJ-1, CIV-2, POW-3, SEN-4 and BAR-5 respectively. The higher concentration BHC was observed in Najafgarh drain and may be due to contamination of drain water by agricultural run-off.

Monthly average values (Mean ± SD) and the results of analysis of variance for T-BHC content of five river monitoring points are tabulated in Table RWM-28. The bimonthly average values (mg/l) varied from 0 to 203 ± 7.0 at WAZ-1, 149 ± 49.0 to 600.0 ± 160.0 at ISB-2, 62.0 ± 9.0 to 640 ± 20 at ITO-3, 94 ± 4.0 to 760 ± 20.0 at NIZ-4, 119.5 ± 23.5 to 1275.0 ± 95 at OKH-5. Analysis of variance shows significant difference in T-BHC content between monitoring stations (F_1=29.69, $P<0.01$) and also between monitoring months (F_2=2.78, $P<0.05$).

Seasonal and annual average values of T-BHC (mg/l) concentration of river varied from 0 (Winter) to 154.5 ± 56.72 (Monsoon) at WAZ-1, 103.67 ± 31.4 (Monsoon) to 550.0 ± 101.98 (Winter) at ISB-2, 72.16 ± 12.84 (Monsoon) to 521.25 ± 115.8 (Summer) at ITO-3, 105.67 ± 11.74 (Monsoon) to 573.75 ± 138.65 (Summer) at NIZ-4, 127.5 ± 21.45 (Pre-Monsoon) to 1052.75 ± 190.06 (Summer) at OKH-5. The maximum content of T-BHC recorded during summer and/or winter season in all down stream stations of Wazirabad barrage might be due to lower water discharge from Wazirabad barrage and higher contribution of drain water at monitoring points. The lower values during monsoon was due to dilution. At Wazirabad barrage the peak values of T-BHC were observed during monsoon, might be due to washing out of agricultural lands of basin areas, by rain water. Annual average values of T-BHC were in order of 530.40 > 386.29 > 361.58 > 308.08 > 51.83 > at OKH-5, ISB-2, NIZ-4, ITO-3 and WAZ-1 respectively.

Table RWM-28: Monthly Variation of Total BHC (mg/l) in River Water (Year 1994- 1995)

Month	*WAZ-1*	*ISB-2*	*ITO-3*	*NIZ-4*	*OKH-5*
	Mean ± SD	*Mean ± SD*	*Mean ± SD*	*Mean ± SD*	*Mean ± SD*
Jan.	Nil	594.0 ± 26.0	378.0 ± 72.0	495.0 ± 35.0	304.5 ± 15.5
Feb.	Nil	581.5 ± 18.5	424.0 ± 26.0	495.0 ± 15.0	675.0 ± 185.0
Mar.	Nil	506.0 ± 4.0	475.0 ± 65.0	625.0 ± 45.0	799.0 ± 31.0
Apr.	Nil	334.0 ± 23.0	600.0 ± 10.0	760.0 ± 20.0	1028.0 ± 68.0
May	Nil	336.5 ± 226.5	640.0 ± 20.0	490.0 ± 70.0	1275.0 ± 95.0
Jun.	45.0 ± 45.0	331.0 ± 171.0	370.0 ± 60.0	420.0 ± 40.0	1109.0 ± 111.0
Jul.	203.0 ± 7.0	149.0 ± 49.0	66.5 ± 1.5	108.0 ± 12.0	323.0 ± 37.0
Aug.	184.5 ± 4.5	248.0 ± 138.0	62.0 ± 9.0	94.0 ± 4.0	277.5 ± 12.5
Sep.	76.0 ± 13.0	300.0 ± 140.0	88.0 ± 5.0	115.0 ± 5.0	172.5 ± 16.5
Oct.	83.5 ± 5.5	439.0 ± 241.0	128.0 ± 18.0	154.0 ± 9.0	135.5 ± 15.5
Nov.	30.0 ± 30.0	478.0 ± 92.0	179.0 ± 11.0	225.0 ± 35.0	119.5 ± 23.5
Dec.	0.0 ± 00.0	600.0 ± 160.0	286.5 ± 23.5	358.0 ± 62.0	147.0 ± 39.5

Two-way ANOVA Test

Significant difference in BHC content between river monitoring stations, $F_1=29.69$ ($P<0.01$).
Significant difference in BHC content between monitoring months, $F_2=2.78$ ($P<0.05$).

Total-DDT of Drain Water and River Water

The insecticidal property of DDT was discovered by Muller and his workers in 1939. It is an organochlorine insecticide which is a white crystalline solid, tasteless and almost odourless. DDT is a non-systemic stomach and contact insecticide of high persistence on solid surfaces. It is non-phytotoxic and a suspected carcinogen.

During the last decade, many scientific reports have shown the presence of persistent insecticides in man and the environment (Durham, 1965; Kraybill, 1969; USDHEW, 1969). The residues of DDT have been reported in several life forms, food, air and water. Even the mother's milk has been found to be contaminated with these chemicals (IARC, 1974). The man and animals are exposed to these chemicals either directly or indirectly over their entire life time mainly through the food chain. Food is reported to be responsible for more than 90 per cent of the total body intake of these chemicals in general population (Kraybill, 1969; Campbell, 1965). DDT has a strong potential for food chain biomagnification. Concentration of DDT in surface water depends on the soil as well as rains. DDT is highly toxic aquatic invertebrates at concentrations as low as 0.3 (mg/l). Toxic effects include impairment in reproduction and development, cardiovascular modifications and neurological changes.

Table DWM-29 shows monthly averages (Mean ± SD) and the results of analysis of variance of T-DDT (mg/l) content in five monitoring drains. The minimum and maximum values of T-DDT were 202.5 ± 17.5 and 910 ± 24 in NAJ-1, 96.5 ± 1.5 and 385 ± 25 in CIV-2, 0 and 300 ± 20 in POW-3, 30.0 ± 5 and 264.0 ± 96.0 in SEN-4, 103.0 ± 5 and 1230.0 ± 320.0 in BAR-5. Analysis of variance shows significant difference in T-DDT content between monitoring drains (F_1=28.8, P<0.01) as well as between monitoring months (F_2=2.16, P<0.05).

Table DWM-29: Monthly Variation in Total DDT (mg/l) of Drain Water (Year 1994-1995)

Month	*NAJ-1*	*CIV-2*	*POW-3*	*SEN-4*	*BAR-5*
	Mean ± SD	*Mean ± SD*	*Mean ± SD*	*Mean ± SD*	*Mean ± SD*
Jan.	320.00 ± 70.00	179.50 ± 3.50	190.00 ± 10.00	91.50 ± 18.50	248.00 ± 38.00
Feb.	485.00 ± 95.00	175.50 ± 14.50	104.00 ± 16.00	67.50 ± 8.50	455.00 ± 25.00
Mar.	490.0 ± 20.00	253.00 ± 43.00	61.50 ± 3.50	93.00 ± 7.00	840.00 ± 80.00
Apr.	700.00 ± 20.00	385.00 ± 25.00	0.00 ± 0.00	85.00 ± 11.00	765.00 ± 95.00
May	910.00 ± 240.00	720.00 ± 260.00	0.00 ± 0.00	75.50 ± 7.50	1130.00 ± 150.00
Jun.	900.00 ± 80.00	100.00 ± 10.00	0.00 ± 0.00	87.00 ± 11.00	1230.00 ± 320.00
Jul.	202.00 ± 17.00	96.50 ± 1.50	0.00 ± 0.00	30.00 ± 5.00	103.00 ± 5.00
Aug.	205.0 ± 25.50	108.00 ± 12.00	0.00 ± 0.00	36.50 ± 6.50	110.00 ± 0.00
Sep.	413.00 ± 117.00	198.00 ± 12.00	59.00 ± 9.00	66.50 ± 11.50	249.00 ± 41.00
Oct.	720.00 ± 360.00	215.00 ± 25.00	139.50 ± 29.50	140.00 ± 20.00	285.00 ± 65.00
Nov.	570.00 ± 60.00	200.00 ± 10.00	215.00 ± 25.00	245.00 ± 35.00	254.00 ± 56.00
Dec.	385.00 ± 125.00	174.00 ± 4.00	300.00 ± 20.00	264.00 ± 96.00	235.00 ± 45.00

Two-way ANOVA Test

Significant difference in DDT content between monitoring drains, F_1=28.8 (P<0.01).
Significant difference in DDT content between sampling months, F_2=2.16 (P<0.05).

Seasonal and annual mean of T-DDT varied from a minimum of 23.56 ± 10.22 (Summer) to a maximum of 23.56 ± 8.30 (Monsoon) in NAJ-1, 39.21 ± 12.78 (Monsoon) to 47.69 ± 12.47 (Winter) in CIV-2, 33.07 ± 4.9 (Monsoon) to 42.49 ± 8.53 (Pre-Monsoon) in POW-3, 43.56 ± 11.63 (Monsoon) to 58.17 ± 8.43 (Winter) in SEN-4, 33.46 ± 9.32 (Summer) to 40.94 ± 5.46 (Pre-Monsoon) in BAR-5. Annual mean values of DDT in NAJ-1, CIV-2, POW-3, SEN-4 and BAR-5 (five drains) were 27.75, 43.10, 39.32, 49.74 and 37.56 mg/l respectively.

Table RWM-29 shows the monthly average values (Mean ± SD) of river water and the results of Analysis of variance. The T-DDT content varied from a minimum of '0' to a maximum of 200 ± 10 (May) at WAZ-1, 0 to 114 ± 8.5 (Jun.) at ISB -2, 0 to 152.0 ± 13.0 (Oct.) at NIZ-4 and 0 to 66.0 ± 13.0 (Oct.) at OKH-5. Two way of analysis variance shows a significant difference in T-DDT content between monitoring stations (F_1=70.96, P<0.01) and also between sampling months (F_2=50.06, P<0.05).

The minimum and maximum values of T-DDT at five river monitoring stations were 52.33 ± 21.28 (Monsoon) and 170.5 ± 73.91 (Pre-Monsoon) at WAZ-1, 11.33 ± 25.34 (Monsoon) and 98.0 ± 23.89 (Summer) in ISB-2, 0 and 73.67 ± 24.67 (Monsoon) at ITO-3, 0 and 137.17 ± 193.25 (Pre-Monsoon) at SEN-4, 0 and 365.25 ± 68.04 (Pre-Monsoon) at OKH-5. Peak values of T-DDT were observed during monsoon and post-monsoon season might be due to washout of agricultural fields by rain water. Annual average values of T-DDT were in order of 80.667 > 64.2 > 55.04 > 33.29 > 12.37 at WAZ-1, ISB-2, NIZ-4, ITO-3 and OKH -5 respectively.

T-DDT was observed in higher concentration at Wazirabad barrage which might be due to contamination of river water by pesticides used in agricultural purpose in the basin area of Yamuna river.

Table RWM-29: Monthly Variation of Total DDT (mg/l) in River Water (Year 1994-1995)

Month	WAZ-1 Mean ± SD	ISB-2 Mean ± SD	ITO-3 Mean ± SD	NIZ-4 Mean ± SD	OKH-5 Mean ± SD
Jan.	Nil	62.5 ± 9.2	Nil	Nil	Nil
Feb.	Nil	74.5 ± 7.8	Nil	Nil	Nil
Mar.	Nil	70.0 ± 42.4	Nil	Nil	Nil
Apr.	Nil	95.0 ± 7.1	Nil	Nil	Nil
May	200.0 ± 10.0	113.0 ± 9.9	Nil	Nil	Nil
Jun.	165.0 ± 5.0	114.0 ± 8.5	Nil	Nil	Nil
Jul.	40.0 ± 20.0	Nil	61.5 ± 21.5	115.0 ± 25.0	Nil
Aug.	43.0 ± 15.0	Nil	56.5 ± 3.5	113.0 ± 17.0	Nil
Sep.	74.0 ± 5.0	34.0 ± 8.5	103.5 ± 7.0	183.5 ± 14.5	51.0 ± 13.0
Oct.	103.0 ± 7.0	92.0 ± 6.4	152.0 ± 13.0	195.0 ± 15.0	66.0 ± 13.0
Nov.	238.0 ± 42.0	48.5 ± 21.5	26.5 ± 26.5	54.0 ± 54.0	31.5 ± 31.5
Dec.	105.0 ± 5.0	51.0 ± 4.2	Nil	Nil	Nil

Two-way ANOVA Test

Significant difference in DDT content between river monitoring stations, F_1=70.96 ($P<0.01$).
Significant difference in DDT content between monitoring months, F_2=50.06 ($P<0.05$).

Table DWM-30: Monthly Variation in Total Endosulfan (mg/l) of Drain Water (Year 1994-1995)

Month	NAJ-1 Mean ± SD	CIV-2 Mean ± SD	POW-3 Mean ± SD	SEN-4 Mean ± SD	BAR-5 Mean ± SD
Jan.	280.50 ± 12.50	244.00 ± 34.00	139.50 ± 139.50	3.50 ± 71.50	335.00 ± 25.00
Feb.	254.00 ± 56.00	303.00 ± 7.00	153.00 ± 153.00	5.00 ± 53.00	650.00 ± 170.00
Mar.	320.00 ± 50.00	349.00 ± 51.00	126.50 ± 126.50	3.50 ± 58.50	620.00 ± 40.00
Apr.	445.00 ± 85.00	370.00 ± 10.00	152.00 ± 152.00	4.00 ± 75.50	790.00 ± 30.00
May	521.50 ± 58.00	460.00 ± 20.00	110.00 ± 110.00	0.00 ± 65.50	1095.00 ± 15.00
Jun.	493.00 ± 97.00	490.00 ± 100.00	105.00 ± 105.00	75.00 ± 76.00	1220.00 ± 240.00
Jul.	137.50 ± 7.50	108.00 ± 12.00	70.50 ± 70.50	2.50 ± 36.50	69.50 ± 30.50
Aug.	125.00 ± 5.00	104.50 ± 4.50	60.50 ± 60.50	4.50 ± 39.00	58.00 ± 10.00
Sep.	364.00 ± 116.0	219.00 ± 21.00	92.50 ± 92.50	3.50 ± 71.00	223.00 ± 37.00
Oct.	474.00 ± 94.00	280.00 ± 40.00	112.00 ± 112.00	16.00 ± 77.50	265.00 ± 35.00
Nov.	370.00 ± 50.00	291.50 ± 18.5	105.00 ± 105.00	5.00 ± 84.00	355.00 ± 5.00
Dec.	233.50 ± 23.50	249.50 ± 3.50	109.50 ± 109.50	18.50 ± 78.00	299.00 ± 9.00

Two-way ANOVA Test

Significant difference in endosulfan content between monitoring drains, F_1=28.8 ($P<0.01$).
Significant difference in endosulfan content between sampling months, F_2=2.16 ($P<0.05$).

Endosulfan of Drain and River Water

Endosulfan is a common member of organochlorinated hydrocarbons. There is ample evidence to show that endosulfan like many other chlorinated hydrocarbons used in agricultural programmes, is lethal to a variety of aquatic and terrestrial organisms even in very small doses. Sediment and water samples from twenty six fish ponds of West Bengal were analysed for the detection of pesticides residue by TLC and at all of these ponds DDT, endosulfan, BHC, methyl parathion residues were detected (Joshi, 1987).

Monthly averages (Mean ± SD) and the results of analysis of variance for endosulfan content (mg/l) in monitoring drains are shown in Table DWM-30. The average values ranged from 125 ± 5 to 521.5 ± 58.50 in NAJ-1, 104.5 ± 4.5 to 490.0 ± 100 in CIV-2, 60.2 ± 4.5 to 153.0 ± 5 in POW-3, 36.5 ± 6.5 to 84.0 ± 2.0 in SEN-4, 58.0 ± 10.0 to 1220 ± 240.0 in BAR-5. Analysis of variance shows the significant difference in T-Endosulfan content between monitoring drains (F_1=26.8, P<0.01) and also between monitoring months (F_2=2.12, P<0.05).

The minimum and maximum total endosulfan values in five drains were 208.83 ± 128.75 (Monsoon) and 422.0 ± 91.5 (Pre-Monsoon) in NAJ-1, 143.83 ± 55.04 (Monsoon) and 285.75 (Pre-Monsoon) in CIV-2, 74.5 ± 1384 (Monsoon) and 137.17 ± 15.32 (Winter) in POW-3, 48.83 ± 16.2 (Monsoon) and 80.75 ± 3.7 (Pre-Monsoon) in SEN-4, 116.83 ± 80.4 (Monsoon) and 931.25 ± 268.0 (Summer) in BAR-5. Annual mean values of total endosulfan of five drains were in order of 498.29 > 334.83 > 289.04 > 111.3 > 65.5 in BAR-5, NAJ-1, CIV-2, POW-3 and SEN-4.

The minimum and maximum total endosulfan values in five drains were 208.83 ± 128.75 (Monsoon) and 422.0 ± 91.5 (Pre-Monsoon) in NAJ-1, 143.83 ± 55.04 (Monsoon) and 285.75 (Pre-Monsoon) in CIV-2, 74.5 ± 1384 (Monsoon) and 137.17 ± 15.32 (Winter) in POW-3, 48.83 ± 16.2 (Monsoon) and 80.75 ± 3.7 (Pre-Monsoon) in SEN-4, 116.83 ± 80.4 (Monsoon) and 931.25 ± 268.0 (Summer) in BAR-5. Annual mean values of total endosulfan of five drains were in order of 498.29 > 334.83 > 289.04 > 111.3 > 65.5 in BAR-5, NAJ-1, CIV-2, POW-3 and SEN-4.

Table RWM-30: Monthly Variation of Total Endosulfan (mg/l) in River Water Year (1994-1995)

Month	*WAZ-1* *Mean ± SD*	*ISB-2* *Mean ± SD*	*ITO-3* *Mean ± SD*	*NIZ-4* *Mean ± SD*	*OKH-5* *Mean ± SD*
Jan.	0.00 ± 0.00	52.0 ± 5.7	Nil	Nil	Nil
Feb.	0.00 ± 0.00	69.5 ± 37.5	Nil	Nil	Nil
Mar.	0.00 ± 0.00	47.5 ± 10.6	Nil	Nil	Nil
Apr.	10.0 ± 10.0	80.0 ± 28.3	Nil	Nil	Nil
May	27.5 ± 2.5	100.5 ± 27.6	Nil	Nil	32.5 ± 6.5
Jun.	3.5 ± 17.5	89.5 ± 26.2	Nil	Nil	39.0 ± 6.0
Jul.	23.0 ± 7.0	Nil	Nil	64.5 ± 4.5	26.5 ± 26.5
Aug.	31.0 ± 5.0	Nil	Nil	79.5 ± 6.5	0.00 ± 0.00
Sep.	43.0 ± 0.0	20.0 ± 28.3	62.0 ± 6.00	113.0 ± 7.0	169.5 ± 9.5
Oct	34.5 ± 34.5	77.0 ± 29.7	69.5 ± 6.50	147.5 ± 21.5	305.5 ± 19.5
Nov.	22.5 ± 22.5	61.75 ± 24.7	149.0 ± 19.00	239.0 ± 41.0	421.0 ± 47.0
Dec.	Nil	44.0 ± 21.2	100.0 ± 10.00	204.5 ± 41.5	155.0 ± 155.0

Two-way ANOVA Test

Significant difference in endosulfan content between river monitoring stations, F_1=52.83 (P<0.01).
No significant difference in endosulfan content between monitoring months, F_2=1.53.

The monthly average values (Mean ± SD) and the results of analysis of variance of total endosulfan are presented in Table RWM-30. Monthly values varied from a miniinum of 0 to a maximum of 43.0 ± 5.6 (Sept.) at WAZ-1, 0 to 100.50 ± 27.6 (May) at ISB-2, 0 to 149.0 ± 19.0 (Nov.) at ITO-3, 0 to 239.0 ± 41.0 (Nov.) at NIZ.- 4 and 0 to 421.0 ± 47.0 (Nov.) at OKH-5. Analysis of variance shows a significant difference in endosulfan content (mg/l) between monitoring stations (F_1=52.83, P<0.01). No significant difference was observed between monitoring months (F_2=1.53).

The minimum and maximum endosulfan values were 0 (Winter) and 32.33 ± 9.6 (Monsoon) at WAZ-1, 6.67 ± 3.8 (Monsoon) and 53.5 ± 8.56 (Pre-Monsoon) at ISB-2, a (Summer) and 109.25 ± 42.21 (Pre-Monsoon) at ITO-3, 0 (Summer) and 193.25 ± 56.26 (Pre-Monsoon) at NIZ-4, 0 (Winter) and 363.25 ± 68 (Pre-Monsoon) at OKH-5. The maximum values of endosulfan were observed during post-monsoon might be due to contamination of river by washing of agricultural lands. Annual average values at endosulfan were in the order of 95.75 > 70.67 > 39.58 > 31.70 > 18.92 at OKH-5, NIZ-4, ISB-2, ITO-3 and WAZ-1 respectively.

Microbial Assessment of River Water

To assess the quality of water in terms of pathogenic and parasite organisms indicating a health risk for the human population, the use of an indicator system is considered to be indispensable, due to the vast variety of these organisms. The coliform bacteria, especially the thermo-tolerant types are considered to be a convenient indicator for the exposure of a water body to the faeces of warm blooded organisms, including man. Since human water-borne diseases are mainly transferred by exposure to faecal matter, the health risk can be assessed by measuring the presence of these bacteria (D.de Zwart and Trivedy, 1994).

Total Coliforms

The coliforms bacteria are discharged from the human intestine and their presence indicates the possibility of the presence of phathogenic organisms. The coliform group comprises all of the facultative and aerobic gram-negative, non-spore forming rod shaped bacteria. The coliform bacteria include the genera *Escherichia, Citrobacter, Enterobacter* and *Klebsiella*. The *Escherichia coli* are entirely of human origin but their exclusive estimation is difficult and hence the entire coliforms are used as indicator. (Trivedy and Goel, 1984). The membrane filtration (MF) method which gives a direct count of total coliforms was used in river water monitoring.

Monthly average number of total coliforms (Mean ± SD) in river water and the results of analysis of variance are presented in Table RWM-31. Average total coliforms at five river monitoring points varied from a minimum of 613 ± 147 (Dec.) to a maximum of 17267 ± 20318 (Jul.) at WAZ-1, 43633 ± 114398 (May) to 405333 ± 188378 (Mar.) at ISB-2, 169667 ± 107370 (May) to 396000 ± 78254 (Feb.) at ITO-3, 163667 ± 103384 (Jun.) to 377667 ± 69264 (Jan.) at NIZ-4, 223333 ± 33993 (April) to 392667 ± 8994 (Jan.) at OKH-5. Analysis of variance shows a significant difference in total coliforms between river monitoring stations (F_1=20, P<0.01) as well as between monitoring months (F_2=2.07, P<0.05).

Seasonal and annual means of total coliforms varied from 895.0 ± 391.0 (Summer) to 12611.0 ± 15504.6 (Monsoon) at WAZ-1, 178291.7 ± 152604.3 (Summer) to 333666.7 ± 27624 (Winter) at ISB-2, 219500.0 ± 99181.2 (Summer) to 346222.2 ± 74011.1 (Winter) at ITO-3, 235000 ± 64172.9 (Summer) to 353444.4 ± 73798.7 (Winter) at NIZ-4, 279333.3 ± 74402.81 (Monsoon) to 339888.9 ± 56396 (Summer) at OKH-5.

Table RWM-31: Monthly Variation of Total Coliforms (CFU/100 ml) in River Water (Year 1994-1995)

Month	WAZ-1 Mean ± SD	ISB-2 Mean ± SD	ITO-3 Mean ± SD	NIZ-4 Mean ± SD	OKH-5 Mean ± SD
Jan.	1200 ± 668	355000 ± 71447	362667 ± 37959	377667 ± 69264	392667 ± 8994
Feb.	933 ± 189	332667 ± 130956	396000 ± 68254	344333 ± 10209	348667 ± 42468
Mar.	1033 ± 544	405333 ± 188370	318667 ± 29273	345000 ± 40208	316667 ± 41899
Apr.	1200 ± 216	215333 ± 2625	219333 ± 28767	256667 ± 9428	223333 ± 33993
May	733 ± 94	43633 ± 114398	169667 ± 107370	174667 ± 102818	326667 ± 46428
Jun.	613 ± 211	48867 ± 131671	170333 ± 106772	163667 ± 103384	322200 ± 9356
Jul.	17267 ± 20318	255000 ± 63068	346667 ± 18856	312000 ± 55450	255000 ± 49497
Aug.	14200 ± 15416	282667 ± 30070	282333 ± 16049	336667 ± 41899	231667 ± 70985
Sep.	6367 ± 2745	279333 ± 58545	309333 ± 98256	315667 ± 56073	351333 ± 32714
Oct.	2100 ± 1042	306000 ± 40343	319667 ± 109393	320000 ± 112250	366000 ± 80862
Nov.	900 ± 82	275000 ± 55126	325000 ± 21213	313333 ± 47140	313333 ± 91094
Dec.	613 ± 147	323333 ± 43205	303333 ± 78457	275000 ± 109621	255000 ± 39370

Two-way ANOVA Test

Significant difference in number of total coliform between monitoring drains, F_1=20.00 (P<0.01).
Significant difference in number of total coliform between sampling months, F_2=2.06 (P<0.05).

The maximum number total coliforms were detected during monsoon season at Wazirabad which might be due to wash-out of faecal matter and over flowing of sewage drains by rain water. The peak coliforms numbers were observed during summer season at all down stream stations of Wazirabad might be due to lean discharge of river water from Wazirabad barrage and higher concentration of drain effluents discharged into the river Yamuna.

Annual mean of total coliforms observed were in the order of 308527 > 294555 > 293583 > 260180 > 3930 at OKH-5, ITO-3, NIZ-4, ISB-2, WAZ-1 respectively.

According to the "Designated Best-Use Criteria (CPC8-1978)" the limit of number of total coliforms is 500/100 ml for "Class-B" type water. From the monitoring data it was observed that, in all river monitoring points the total coliforms numbers were more than the required limit.

Faecal Coliforms

Monthly average faecal coliforms (Mean ± SD) in river water and the results of analysis of variance are depicted in Table RWM-32. Monthly average faecal coliforms number ranged from 93 ± 90 (May) to 15000 ± 17678 (Jul.) at WAZ-1, 27433 ± 1819 (Jun.) to 280000 ± 16330 (Jan.) at ISB-2, 64967 ± 29973 (Jun.) to 304000 ± 41215 (Jan.) at ITO-3, 6000 ± 22760 (Jan.) to 308333 ± 76630 (Jan.) at NIZ-4, 113333 ± 61283 (Jun.) to 343333 ± 32998 (Jan.) at OKH-5. Analysis of variance shows a significant difference in faecal coliforms between river monitoring stations (F_1=20.94, P<0.01). However, no significance difference was observed between sampling months (F_2=2.06).

Seasonal minimum and maximum values of five river monitoring points were 334 ± 164 (Summer) and 9889 ± 13245 (Monsoon) at WAZ-1, 100233 ± 81010 (Summer) and 270889 ± 20295 (Winter) at ISB-2, 133500 ± 73307 (Summer) and 2890000 ± 76262 (Winter) at NIZ-4, 150167 ± 67133

(Summer) and 291667 ± 85716 (Winter) at OKH-5. Annual mean numbers were in the order of 223417 > 221750 > 213297 > 199494 > 2818 at NIZ-4, OKH-5, ITO-3, ISB-2, WAZ-1 respectively.

Table RWM-32: Monthly Variation of Faecal Coliforms (CFU/100 ml) in River Water (Year 1994-1995)

Month	*WAZ-1*	*ISB-2*	*ITO-3*	*NIZ-4*	*OKH-5*
	Mean ± SD	*Mean ± SD*	*Mean ± SD*	*Mean ± SD*	*Mean ± SD*
Jan.	627 ± 339	280000 ± 16330	304000 ± 41215	308333 ± 76630	343333 ± 32998
Feb.	433 ± 94	274333 ± 19431	299000 ± 61822	272667 ± 37250	220000 ± 142361
Mar.	400 ± 216	217667 ± 12815	266333 ± 41153	226000 ± 25140	150667 ± 93243
Apr.	433 ± 125	127000 ± 34322	159667 ± 16337	175000 ± 7071	166667 ± 9428
May	293 ± 90	28833 ± 850	69267 ± 30279	72000 ± 31284	170000 ± 58878
Jun.	210 ± 83	27433 ± 1819	64967 ± 29973	61000 ± 22760	113333 ± 61283
Jul.	15000 ± 17678	211667 ± 41700	211667 ± 34238	246667 ± 46428	213333 ± 59067
Aug.	10900 ± 12092	248333 ± 13123	241667 ± 20138	280000 ± 40825	209333 ± 65429
Sep.	3767 ± 1725	240333 ± 42664	268333 ± 84097	273333 ± 38586	282667 ± 19345
Oct.	1150 ± 618	251667 ± 45522	213333 ± 23570	273333 ± 65997	326667 ± 67987
Nov.	367 ± 170	238333 ± 39651	176667 ± 54365	236667 ± 52493	256667 ± 87305
Dec.	243 ± 154	248333 ± 10274	284667 ± 58909	256000 ± 86595	208333 ± 44033

Two-way ANOVA Test

Significant difference in number of faecal coliform between monitoring drains, F_1=20.94 (P<0.01).

Significant difference in number of faecal coliform between sampling months, F_2=2.06 (P<0.05).

Water Quality Index

Water Quality Index (WQI) may be defined as a rating reflecting the composite influence on the overall quality of a number of quality characteristics or water quality parameter (Tiwari and Ali,1988). It reflects the collective influence of various physico-chemical and biological criteria of water (Sahu *et al.*, 1991). Hence WQI may be used as one of the most effective ways to communicate information on water quality trends to the policy makers and general public (Sinha *et al.*,1990a). Although WQI is not absolutely objective, it is more efficient method of monitoring the trends in water quality than those based on existing classifications. Therefore Water Quality Index (WQI) enables reduction of large amount of data to a single reproduction index, that provides distinct managerial and operational advantages (Patnaik and Mohapatra, 1994). Basically and essentially the WQI concept is meant for use in the management of natural streams. A number of attempts have been made to develop Combined Water Quality Index (CWQI) for municipal sewage, but they have met only with limited success. That is the reason why in the present work, all the river monitoring stations of the Yamuna river at Delhi stretch are calculated, season-wise as well as year-wise (from year 1993-95).

Considerable efforts have also been made in the few last decade to develop comprehensive mathematical management of water quality in river basin. The work on water quality modelling for the development of Water Quality Index (WQI) has been undertaken by various investigators. Horton (1965) reported an index number system for rating water quality and Harkins (1974) published a paper with objective study of Water Quality Index.

Due to the manifold uses of water, the quality of water required for each use differs significantly from others. This creates problem for formulation of WQI, because high value of a parameter may be desirable for some purpose whereas it may be indifferent or even detrimental for another. For example, a high value of DO is essential for aquatic life, but it is only of marginal value in drinking water supply and highly undesirable in boiled feed water. This situation poses a number of question like whether DO should be included as a parameter in formulating water quality index and, if included, should it affect the WQI positively or negatively and what weight should be assigned to it? Deninger and Maciunas (1971) suggested that the WQI developed with a specific use orientation does not seem to rate water quality levels in a manner markedly different from the rating made by a general non-specific use of oriented index. Thus water of a certain quality retains that relative water quality rating regardless of use for which it is being considered. Instead of developing a number of indices for various uses of water, it appears to be more meaningful to further develop and refine a sensitive and general water quality index. On the other hand, if one is faced with the task of judging the sensitivity of an index, the knowledge of what use the water is to be put to make the work of determining the relevant parameters much easier.

Water Quality Parameters

The most basic use of water is for drinking and personal hygiene. But the Central Pollution Control Board (1978-79) has classified the river and coastal water bodies into five major classes (*i.e.* A to E) based on Designated- Best-Use criteria (Appendix-I). Based on this classification the desired water quality of the river Yamuna at Delhi stretch is classified as "Class-B" (*i.e.* outdoor bathing, swimming and recreation). So for the present study the use of river water for outdoor bathing, swimming and recreation purpose has been considered.

The following water quality parameters have been selected for calculating the water quality index:

1. pH, 2. Dissolved Oxygen, 3. Biochemical Oxygen Demand, 4. Total Coliforms, 5. Total Dissolved Solids, 6. Total Suspended Solids, 7. Ammonical-Nitrogen.

The first four water quality parameters have been selected from CPCB Designated-Best-Use Standards for out-door bathing (Class-B) and last three parameters from NIEM (Network for Industrial Environmental Management, 1989).

Standard and Weightage

The "Standards" (permissible values of various pollutants) for "Class-B" (outdoor bathing, organised) are shown in the Table W-1.

Table W-I: CPCB Designed-Best-Use Standards for "Class-B" and Unit Weights of Water Quality Parameters

Sl.No.	Water Quality Parameters	Standard	Unit weight
1.	pH	6.5–8.5	0.0040
2.	DO	5 mg/l	0.200
3.	BOD	4 mg/l	0.333
4.	MPN/100 ml	500	0.002
5.	TDS	500 mg/l	0.0002
6.	TSS	100 mg/l	0.01
7.	NH_3-N	1.24 mg/l	0.08

It is obvious that the more harmful a given pollutant is the smaller in its permissible value or standard (Vs) recommended for out-door bathing "Class-B". So the proportional to the recommended standards for the corresponding parameters,

$$i.e.\ Wi = K/Si \qquad \text{(WQ. 1.1)}$$

Where,

Wi = Unit weight for the 'I' th parameters Pi (l= 1.2, ..., N) and
K = Constant of proportionality and
Si = Standard for 'I' th parameter.

For the sake of simplicity we assume that K=1. For pH, we have assumed the same unit weight as that for chloride in both cases.

Calculation of WQI

The WQI, reproduced into a formula for calculating the index involves a series of judgments which ultimately will determine the form and utility of the index. For that out of six different methods mentioned below (Botten *et al.*, 1978) of combined water quality rating curves and associated weightings, one method may be selected by the parameters are to be combined.

1. Unweighted Arithmetic Index
2. Weighted Arithmetic Index
3. Unweighted Solway Index
4. Weighted Solway Index
5. Unweighted Geometric Index and
6. Weighted Geometric Index.

For the present work, weighted arithmetic index has been chosen for calculation of WQIs.

Calculation of water quality index consists of two fundamental steps (Ott, 1978),

1. calculation of the quality ratings (or sub indices) for each the water quality parameters used in the index and
2. aggregation of these sub-indices into the overall index.

The present study involves a very simple method (Tiwari *et al.*, 1986) for calculating the water quality index following the two stages mentioned above.

Let there be N water quality parameters, Pi (I = 1, 2, ..., N) to be taken into account for calculating the WQI. Then the quality rating (or sub index) qi corresponding to the ith parameter Pi is a number reflecting the relative value of this parameter in the polluted water with respect to its "standard" or permissible value. One of the simplest relations to calculate qi is

$$qi = 100\,[(Vo - Vio)/(Vs - Vio)] \qquad \text{(WQ. 1.2)}$$

Where,

Vo = The measured value of the 'I'th parameter in the polluted water
Vio = The ideal value of this parameters in pure water
Vs = The standard for the "I" th parameter.

Since in general the ideal Vio = O for the outdoor bathing "Class- B", for most parameters *e.g.* WQ.1.2 assumes the simple form.

$$qi = 100\ (Vo/Vs) \qquad \text{.................................(WQ. 1.3)}$$

Equations WQ. 1.2 and WQ. 1.3 ensure that qi = O, if the 'I'th parameter is totally absent in the polluted water qi = 100, if the amount of this parameters is just equal to its permissible value Si for the drinking water except the following 2 exceptions.

1. For pH the ideal value is 7.0 (for natural water) for "Class-B" (out door bathing purpose) and the permissible value ranges from 6.5 to 8.5 so *e.g.* WQ. 1.2 for this case may be written as

$$qpH = 100\ (V_{pH}-6.5)/(8.5-6.5) \qquad \text{.................................(WQ. 1.4)}$$

Where V_{pH} is the observed value of pH and the sign of difference (–) indicates the numerical value of the difference, ignoring its algebric sign.

2. For DO, the ideal value may be taken as 14.6 mg/l (the solubility of oxygen in pure water at 0°C). Since the standard for "Class-B" Water quality is 5.0 mg/l *e.g.* WQ.1.2 reduces to

$$q_{DO} = 100\ [(V_{DO} - 14.6)/(5\text{-}14.6)] \qquad \text{.................................(WQ. 1.5)}$$

Where V_{DO} is the observed value of DO.

The overall water quality index (WQI) may be calculated by aggregating the quality ratings (or sub indices) qi linearly. Thus it may be written.

$$WQI = \frac{(\sum_{i=1}^{n} q_i w_i)}{(\sum_{i=1}^{n} w_i)} \qquad \text{.................................(WQ. 1.6)}$$

Table W-2: Sample Calculation of Water Quality Index (WQI) for River Yamuna at Wazirabad Barrage Up Stream Water During Winter Season of Year 1995 for "Class-B" Water Quality

Parameters (Pi)	*Observed Value (Vo)*	*Unit (Weight) (Wi)*	*Quality Rating (qi)*	*(q_iW_i)*
pH	8.13	0.004	75.33	0.301
DO	9.13	0.200	56.98	11.396
BOD	3.43	0.333	114.33	38.07
TDS	3.50	0.002	70	0.14
TSS	38.67	0.01	38.67	0.387
NH_3-N	1.02	0.80	82.25	65.8
MPN/100 ml	513	0.002	100.5	0.0201

*All values except pH are in mg/l

$$WQI = \frac{\sum q_i W_i}{\sum W_i} = \frac{116.114}{1.3492} = 86.06$$

Results and Discussion

A sample calculation of WQI using equation WQ. 1.3 to WQ. 1.6 for all river monitoring stations for (Class-B) bathing, swimming and recreation purpose during winter season of year 1965 is shown in the Table WQ. 3. Similarly WQIs for all seasons as well as annual WQIs from year 1993 to 1995 have been calculated and presented in Table WQ-1 and Fig. WQI-1.

The seasonal WQIs shows a minimum of 61.73 and a maximum of 72.89 in the year 1993, 63.21 and 99.87 in 1994, 82.19 and 105.20 in 1995, at Wazirabad barrage (up stream water). The annual WQIs of this station were 68.16, 74.75 and 93.85 for year 1993, 1994 and 1995 respectively. This implies that the Wazirabad barrage (reference station) satisfies the criteria for "Class-B" water (bathing swimming and recreation purpose) because all values were well below 100, the permissible limit for "Class-B" (except rainy season during year 1995).

The seasonal WQIs (year 1993-95) ranged from a minimum of 143 to a maximum of 700.18 at ISBT bridge, 171.18 to 627.30 at ITO barrage, 171.24 to 539.36 at Nizamuddin bridge and 156.57 to 554.87 at (Okhla barrage).

Maximum values of WQIs were recorded at all the stations during winter (except Wazirabad). The reason may be that during winter the water discharge from Wazirabad barrage was lean. So that the drain effluents discharged to river Yamuna flowed without dilution at Delhi stretch. The lowest values of WQIs were obtained during the rainy season at all the stations (except Wazirabad barrage) due to the effect of dilution of pollutants by rainwater. The annual WQIs of river monitoring station; Wazirabad barrage, ISBT bridge, ITO barrage, Nizamuddin bridge and Okhla barrage were 78.92, 399.8, 384.6, 384.6, 372.4 and 362.2 respectively.

In the present investigation it was observed that the WQIs of all river monitoring stations (except Wazirabad barrage) were well above the permissible limit through out the year. The above finding clearly shows that the river water at Delhi stretch (except Wazirabad) is totally unfit for bathing, swimming and recreation purpose (Class-B).

Water quality models proved to be powerful tools in water resources management as they can incorporate the complexity of the relevant processes in the water body into a utilitarian form for management consideration. Through the use of model both diagnostic and predictive capability is provided, diagnostic in the sense of permitting the identification and isolation of specific factors affecting the water quality and predictive in permitting evaluation of future effects of proposed changes in the water body, its inputs or its uses (Singh and Anandh, 1996).

Thus, the water quality index itself does not give the final answer to the problem of water quality management. It can only represent the standard of quality of water and can point out where the trouble is and what utmost action can be taken for the largest improvement in the quality of a given body of water (Mishra, 1999).

So far as the criteria for "Class-B" (bathing, swimming and recreation) for Yamuna river at Delhi stretch are concerned it is totally unsuitable for bathing, swimming and recreation purpose. The WQIs values were so high that the water quality of Yamuna river from down stream of Wazirabad barrage is not even fit for "Class-D" (propagation of wild life and fisheries).

Acknowledgements

PKB is grateful to authorities of Central Pollution Control Board, New Delhi and Sambalpur University for providing facilities to undertake the work.

References

Agarwal, V.K., Raj, K.P.S., Panchal, D.I., Balaraman, R. (1978). Heavy metal content in the water and sediments of Sursagar lake of Baroda (Gujarat). *Intern. J. Environment Studies*, 12: 245-250.

Ajmal, A. and Razi-Ud-Din (1988). Studies on the pollution of Hindon river and Kali Nadi (India). In: *Ecology and pollution of Indian rivers* (Ed. Trivedy, R. K.), Ashish Publishing House, New Delhi, pp. 87-112.

Ajmal, M., Razi-Ud-Din and Khan, A. (1985). Physicochemical aspects of pollution in Kali Nadi. *IAWPC Tech. Annual*, 12: 106-114.

Ali, M. and T.N. (1988a). Multiple regression techniques for rapid monitoring of water quality, *J. IWWA*. 2: 237-240.

APHA (1991). *Standard methods for the examination of water and wastewater, 18th Ed.*, APHA-AWWA-WPCF, American Publis. Health Association, Washington, D.C.

Bagde, U.S. and Verma, A.K. (1985). Limnological studies of J.N.U. Lake, New Delhi. Proc. Nat Symp. Pure and Applied Limnology (Ed. Adoni, A. D.). *Bull. Bot. Soc.* Sagar, 32: 16-23.

Banat, K., Forstner, U. and Muller, G. (1992). Schwermetall-annereich-Erungen in den sedimentn wichtiger Flusse in Bereich der Bundesrepublik Deutschland, line Bestandsaufnahme. *Internal rep.* Lab., Sedimentsforsch, Heidelderg.

Baruah, A.K., Sharma, R.N. and Borah, G.C. (1993). Impact of sugar mill and distillery effluents on water quality of river Gelabil, Assam. *Indian J. Environ. Hlth.*, 35: 288-293.

Bhargava, D.S. (1985). Water quality variation and control technology of Yamuna river. *Env. Pollut.*, 37 (Series B): 355-376.

Bhatnagar, G.P. and Sharma, G.P. (1975). Physico-chemical features of sewage polluted lower lake, Bhopal, Environ-agents and their biological effect. *Pro of the Inter. Symp. Int.* 5(4): 212-223.

Bilgrami, K.S. and Datta Munshi, J.S. (1985). *Ecology of river Ganges: Impact of human activities and conversation of aquatic biota* (Patna to Farakka), Final Technical Report submitted to D.O. En, Allied Press, Bhagalpur (India), p. 97.

Billore, D.K. (1981). *Ecological studies on Pichhola lake, Udaipur*. Ph.D. Thesis, Dept. of Botany, School of Basic Sciences and Humanity, University of Udaipur, Udaipur.

Brajnandan, P., Jatly, Y.C. and Singh, A. K. (1985). Periodicity and interrelationship of physico-chemical. factors in ponds. Proc. Nat. Symp. Pure and Appl. Limnology (Ed. Adani, A. D.) *Bull. Bat. Soc.*, Sagar, 32: 1-11.

Campbell, J.E., Richardson, L.A. and Schafer, M.L. (1965). Insecticide residues in the human diet. *Arch. Environ. Hlth.*, 10: 831.

Chatterjee, C.C. (1984). *Human physiology*. Medical Allied Agency, Calcutta.

Cole, G.A. (1979). *A text book of limnology, 2nd Ed.* C.V. Mosby Company, Site Louis, p. 426.

CPCB (1993-94). *Pollution statistics of Delhi*. Issued by Public Relation Cell for mass awareness, Central Pollution Control Board, Delhi.

D.de.Zwart and Trivedi, R.C. (1994). *Manual on integrated water quality evaluation*. Report No. 802023003, Joint Publication of RIVM and Central Pollution Control Board, East Arjun Nagar, Delhi, India.

Datar, M.D. and Vashishtha, R.P. (1990). Investigation of heavy metals in water and silt sediments of Betwa river, *IJEP*, 10: 666-672.

Datar, M.D., Jain, K.C. and Vashishta, R.P. (1990). Distribution of heavy metals in Betwa river water. In: Trivedy, R.K. (Ed.) *River Pollution in India*, Ashish Publishing House, New Delhi (India), 179-187.

Datta Munshi, J. and Datta Munshi, J.S. (1995). *Fundamentals of freshwater biology*, Narendra Publishing House, New Delhi, p. 222.

Deniniger, R.A. and Maciunas J.J. (1971). *A water quality index for public water supplies*. Department of Environmental and Industrial Health University of Michigan Ann Arber Michigan (U.S.A).

Dikshith, T.S.S., Raizada, R.B., Kumar, S.N., Srivastava, M.K., Kulshrestha, S.K. and Adholia, U.N. (1990). Residues of DDT and HCH in major sources of drinking water in Bhopal, India. *Bull. Environ. Contam. Toxicol.*, 45: 389-393.

Durham, W.F. (1965). *Effects of pesticides on man research in pesticides*. Academic Press Inc., N.Y., p. 63.

Ellis, N.M., Westfalk, B.A. and Ellis, M.D. (1946). Determination of water quality for fish and wildlife service. *U.S. Interior Res. Rept.*, 9: 121-122.

Forstner, U. (1977). *Metal concentrations in metal pollution in freshwater sediments natural background and cultural effects, Interactions between sediments and freshwater*. Goltorman, H.L. (Ed.) Washington/ The Hague, Pudoc/Junk, B.V. Pub., pp. 94-103.

Forstner, U. and Wittman, G.T.W. (1981). *Metal pollution in the aquatic environment*, Springer-Verlag, Berlin.

Ghose, N.C. and Sharma, C. B. (1989). *Pollution of Ganga river*. Ashish Publishing House, New Delhi.

Golterman, H.L. (1975). Physicological Limnology. Elsevier Scientific Publ. Co., N.Y. p. 489. *Indian J. Environ. Hlth.*, 21: 131-142.

Harkins, R.D. (1974). An objective water quality index. *J. Water Poll. Cont. Fed.*, 3: 589-591.

Harrison, A.D. (1963). Hydrobiological studies on the Vall river in the Verrenigung area, Part II, The chemistry, bacteriology and invertebrates of the bottom muds. *Hydrobiologia*, 21: 66-89.

Harrison, R.M. and Laxen, D.P.H. (1984). *Lead in atmosphere, in lead pollution: causes and control*. Chapman and Hall Ltd., London, p. 10.

Hegde, G.R., Bengale, U.D. and Bharati, S.G. (1985). Diurnal variation of physico-chemical factors and phytoplankton in a pond and a lake near Dharwad, Karnataka State, India. *Comp. Physiol. Ecol.*, 10: 245-250.

Horton, R.K. (1965). An index-number system for rating water quality. *J. Water Poll. Cont. Fed.*, 3: 300-305.

Hutchinson, G.E. (1957). *A treatise on limnology, Vol. 1: Geography, Physics and Chemistry*. Wiley, N.Y., p. 1015.

Hydn, H.S. (1961). Distribution of uranium and other metals in crude oils U.S., *Goel. Surv. Bull.*, 1100-3: 72-89.

IARC (1974). Monograph on evaluation of carcinogenic risk of chemicals to man some organochlorine pesticide pollution. *Baioteku (Biotech)*, 3: 589.

Imevbore, A.M.A (1978). The chemistry of river Niger in the Kainj reservoir Area. *Arch. Hydrobiol.*, 67: 412-431.

Ingole, S.A. and Dhaktode, S.S. (1990). Pollution studies of Amba river estuary: A comparative study. In: Trivedy, R.K. (Ed.), *River Pollution in India*, Ashish Publishing House, New Delhi (India), pp. 237-253.

Israili, A.W. (1991). Occurrence of heavy metals in Ganga river water and sediments of Western Uttar Pradesh. *Poll. Res.*, 10: 103-109.

Israili, A.W. and Ahmad, M.S. (1993). Chemical characteristics of river Yamuna from Dehradun to Agra. *Indian J. Environ. Hlth.*, 35: 199-204.

Jackson, M.L. (1964). *Chemistry of the soil*. Bear, F.E. (Ed.), 2nd Ed., Reinhold, New York, Chapter 2.

Jain, S. and Salman, S. (1995). Heavy metal concentration in a highly eutrophic lake sediments and overlying water. *Poll. Res.*, 14: 471-476.

Jenne, E.A. (1976). *Trace elements sorption by sediments and soil sites and processes*. Symposium on molybdenum, Marcel Dekker, New York, 2: 425-553.

Joshi, H.C. (1985). Pesticides residue monitoring in the Bhagirath Hoogly stretch of Ganga river system. *Proc. Nat. and Sen. Poll. Control of Env. Manage.*, NEERI Nagpur W.O., 123-124.

Joshi, K.M. (1982). Heavy metal pollution in India status and impact. *Proceeding of the seminar on Status and impact of heavy metal pollution in India*. Centre for Environ. Studies, Anna Univ., Madras.

Juday, C., Birge, E.A., Kemmerer, J.J. and Robinson, R.J. (1928). Phosphorous content of lake waters of Northern Wisconsin. Trans. Wisconsin, *Acad. Sci. Arts and Lett.*, 23: 233-248.

Katariya, H.C. (1994). An evaluation of water quality of Kaliasot river. *IJEP*, 14: 690-694.

Khan, S. and Khan, N.N. (1983). Influence of lead and cadmium on the growth and nutrient concentration of tomoto (*Lycopersicum esculentum*) and egg-plant (*Solanum melongena*). *Plant and Soil*, 74: 387-394.

Kothandaraman, H. and Viswanathan, S. (1986). Coocum, a perspective study, In river Coocum, Let it be a resource, Univ. of Madras, 4: 1-4.10.

Kraybill, H.F. (1969). Significance of pesticide residues in foods relations to total environmental stress. *Canad. Med. Ass. J.* 100: 204.

Krishnamurthy, S.R. and Bharati, S.G.(1995). Evaluation of water pollution in the river Kali near Dandeli (North Kanara district), Karnataka State, India. *Poll. Res.*, 14: 93-98.

Kudesia, V.P. (1989). Physico-chemical properties of Ganga river at Farrukhabad. *Indian J. Environ. Agric.*, 3: 81-82.

Kulshrestha, S.K. (1986-89). *Monitoring and surveillance of residues of organochlorine pesticides and heavy metals in fishes from selected environmental components of Madhya Pradesh and Rajasthan*. Technical Report (19/85/86/RE) Motilal Vigyan Mahavidyalaya, Bhopal.

Kumar, A. (1994). Studies on pollution in the Mayurakshi in south Bihar. *J. Environ. Pollut.*, 2(1): 21-26.

Kumar, A. (1994a). The effect of sewage pollution on the ecology of river Mayurakshi in Bihar, Abs. No. 40. Symp. on Management of ecosystem and sustainable utilization of biological resources at Vellyani (Trivendrum), 7- 9 Sept. 1994.

Kumar, A. (1995). Studies on pollution in the river Mayurakshi in south Bihar. *J. Environ. Pollut.*, 2(3): 137-141.

Kumar, A. (1997). *Limnobiotic studies of River Mayurakshi in Santal Praganas, Bihar*. A Ph.D. thesis submitted to S.K. University, Dumka (Bihar).

Kumar, A. and Sharma, R.K. (1999). Physico-chemical characteristics of river Sengar in district-Etawah (U.P.). *Ind. J. of Environment. Sci.* 3(2): 201-203.

Lal, P.C. and Bhattacharya, K.G. (1990). Trace metal status of the Bharalu river. *IJEP*, 10: 747-751.

Malviya, S., Yadav, M. and Palharya, J.P. (1990). Aspect of water pollution in the river Narmada at Hoshangabad (MP). In: Trivedy, R.K. (Ed.); *River Pollution in India*, Ashish Publishing House, New Delhi (India), pp. 59-70.

Manivaskam, N. (1989). *Physico-chemical examination of water, sewage and industrial effluents*. Pragati Prakashan, Meerut, p. 380.

Mathur, A., Sharma, Y.C., Rupainwar, D.C., Murthi, R.C. and Chandra, (Mrs.) S. (1987). A study of river Ganga at Varanasi with special emphasis on heavy metal pollution. *Poll. Res.*, 6: 37-44.

Mehrotra, K.N. (1985). Use of DDT and BHC and its environmental effects in India, *Proc. Indian North Sci. Acad.*, 51(5): 581-595.

Metelev, V.V., Kanev, A.I. and Dzasokhova, N.G. (1971). *Vodnaya Toksikologya*. Kolos Publishers, Moscow.

Michael, P. (1984). *Ecological methods for field and laboratory Investigations*. Tata MaGraw Hill Publishing Co. Ltd., New Delhi (India).

Mishra, S. (1999). *Quantification of heavy metals in Brahmani river ecosystem at Rourkela and its modelling*. A Ph.D. thesis submitted to Sambalpur University (Orissa).

Mishra, S.R. and Saxena, D.N. (1991). Pollution ecology with reference to physico-chemical characteristics of Morar (Kalpi) river, Gwalior (MP). In: Shastree, N.K. (Ed.) *Current trends in limnology, Vol. 1*, Narendra Publishing House, Delhi (India), pp. 159-184.

Mitra, A.K. (1982). Chemical characteristics of surface water at selected gauging stations in the river Godavari, Krishna and Tungabhadra. *Indian J. Environ. Hlth.*, 24: 165-179.

Mohan, K.S. and Reddy, N.S. (1987). A study of cations in two freshwater lakes of Hyderabad, India. *Poll. Res.*, 6: 5-8.

Moss, B. (1973). The influence of environmental factors on the distribution of freshwater algae: An experimental study II, Role of pH and the carbon dioxide bicarbonate system. *J. Ecol.*, 61: 157-177.

Munawar, M. (1970). Limnological studies on freshwater ponds of Hyderabad, India. II. *Hydrobiologia*, 31: 101-128, 39(3): 307-320.

Murugesan, A.G., Hameed, A.K.M.S.A. and Sukumaran, N. (1994). Water quality profile of the perennial river Tambraparni. *IJEP*, 14: 567-572.

Nayak, M.S. and Sawant, A.D. (1994). Seasonal variation in heavy metal concentration in water of lakes supplying water to Bombay city. *Journal of Indian Water Works Assoc.*, 14: 241-243.

Ohle, W. (1934). Chemisch and physikalishe unterruch ungienhor dentscher seen. *Arch. Hydrobiol*, 26: 386-464.

Patil, P. (1982). *An ecological study of the algal flora of the Lakes of Bhopal*. Ph. D. thesis, Bhopal University, Bhopal: 310.

Phifer, L.D. and Tompson, T.G. (1937). Seasonal variation in the surface water of San, J. Van. Channel-I during the five year period, January, 31st to December, 30th. 1935. *J. Mar. Res.*, 34-53.

Pophali, S., Siddiqui, S. and Khan, L.H. (1990). Sources and distribution of heavy metals in the abiotic components of a polluted urban stream in Bhopal. *IJEP*, 10: 600-603.

Pradhan, S. and Mohanty, R.K. (1991). Occurrence of microbes in Steel Plant effluents. *IJEP*, 11: 201-205.

Prasad, A.S. (1978). *Trace elements and iron in human metabolism*. Plenum Press, New York, USA.

Provasoli, L. (1958). Nutrition and ecology of protozoans. *Rev. Microbiol.*, 12: 278-308.

Rai, H. (1974). Limnological studies on the river Yamuna at Delhi, India. 1. Relation between the chemistry and the state of pollution in the river Yamuna. *Arch. Hydrobiol*, 73(3): 369-393.

Rajesh, K. (1981). *Some aspects of limnological studies in river Jhelum*. Ph. D. Thesis, Kashmir University, J&K, 420.

Ramaswamy, S.N. and Somashekar, R.K. (1982). Ecological studies on algal of electroplating wastes. *Phykos*, 21: 8-12.

Ramteke, D.S. and Moghe, C.A. (1986). *Manual on water and wastewater analysis*. NEERI, Nagpur (India).

Rao, K.L. (1979). *India's water wealth, its assessment, uses and projections*. Orient, Longman, Delhi, p. 210.

Rao, V.S. (1971). An ecological study of three freshwater ponds of Hyderabad, India. *Hydrobiologia*, 38: 213-223.

Raybak, J.J. and Sikorska, V. (1976). Environment. In: Pieczynska, E. (Ed.) *Selected problems of lakes, Littoral ecology*. Univ. of Warsaw.

Sahu, B.K., Panda, R.B., Sinha, B.K. and Nayak, A. (1991). Water quality index of the river Brahmani at Rourkela industrial complex of Orissa. *J. Ecotoxico. Env. Monit.*, 1: 169-175.

Sarwar, S.G. and Rifat, M. (1991). Abiotic features of a freshwater lotic ecosystem of Kashmir (Doodhganga river). *Poll. Res.*, 10: 69-73.

Screenivasan, A. and Sunderaj, R. (1967). Effects of certain wastes on the water quality fisheries of rivers Cauvery and Bhavani. *Environ. Hlth.*, 9: 13-21.

Shah, A.R. (1988). Physico-chemical aspects of pollution in river Jhelum (Kashmir) during 1981-83. In: Trivedy, R.K. (Ed.) *Ecology and pollution of Indian rivers*, Ashish Publishing House, New Delhi (India), pp. 163-207.

Sharma, R.C., Gausain, O.P. and Juyal, C.P. (1990). Ecology of high altitude river Bhilanga of Garhwal Himalaya. In: Trivedy, R.K. (Ed.) *River Pollution in India*, Ashish Publishing House, New Delhi (India), pp. 11-30.

5

Management and Reclamation of Water for Silk Reeling

C. DORESWAMY & RAMAKRISHNA NAIKA

Assistant Professor in Sericulture, Sericulture College, Chintamani

The reeling performance and quality of raw silk depends upon the combination of factors such as raw material, water quality, type of machinery, cocoon processing technology and skill of the reeler. Attempt has been made to improve the quality of cocoons and efficiency of reeling machine. The skill of the reeler has been transformed from time immemorial, but not much importance has been given to the quality of reeling water in the country and unaware of the influence of reeling water on raw silk quality.

Water is important to the silk industry due to its excellent solvent properties and being an ideal carrier of heat. Since, it is directly used in cooking, reeling, preparation of by-products and also for boiler for generation of steam required for stifling, cooking and reeling basin. About 850-1000 litres of water is required to manufacture 1 kilogram of raw silk. The quality of water used in the reeling industry is of great significance as it has a direct bearing on the economy of reeling on one hand and the quality of raw silk on the other hand (Natheshwar Sah, 1983).

The total availability of water in the nature:

97 per cent in the form of salt water

2 per cent in the form of ice and snows

1 per cent in the form of freshwater

In the native form, a waxy material called sericin–a low molecular weight protein component embeds silk fibres. The interaction of water and cocoon to unwind raw silk yarn is mainly to loosen the cocoon structure and swell the sericin for easy extraction of silk filament. The water of desired quality if not used might deteriorate the quality of silk on the one hand and may not loosen the cocoon structure there by leading to poor extraction of silk (Halliyal *et a1.*, 1987).

Water requirement for the cooking process are different from reeling because the cooking process is carried out at higher temperature and also it demands swelling and softening of sericin. The reeling process requires this state of sericin to continue during the entire process of unwinding. This calls for different quality of water for reeling, which is carried out at lower temperature compared to cooking. The chemical properties of water used for cooking and reeling plays a major role in the process of swelling and dissolution of sericin (Bhandiwad *et a1.*, 1988).

Sources of Water Available for Silk Reeling

Pond Water and River Water

These are more turbid in nature and they contain lot of physical and chemical impurities known to effect the colour and luster of the silk.

Rain Water

It mainly contains sodium, potassium, calcium, magnesium, sulphur and chlorides and pH is very low due to absorption of carbon dioxide from the atmosphere.

Well Water

Though it looks clean, it has many dissolved impurities as it is in continuous contact with soil and rocks and contains calcium, magnesium and sodium etc.

Halliyal *et al.* (1987) classified the water available for silk reeling as rainwater and sub-surface water, which are collected into streams, ponds, lakes and rivers. Sub-soil water, which has percolated a small distance into the ground. It is collected from shallow wells or surface springs and deep well water, which have usually percolated through several strata. Apart from this mode of collection, storage also affects the quality of water.

Kim (1989) reported that the sources of water available for silk reeling are rainwater, well water, river water, lake water and sea water as existing in nature. The main sources of filature water in the present time are water from wells, rivers and lakes. As properties of water are classified according to the source of supply, it is advisable to know the properties of water according to the source of supply before selecting the best water for use in the filature.

The reelers use water from different sources like open well, bore well, river, ponds for reeling. The water from these sources will have different pH values and total hardness. The above qualities vary depending on weather conditions and also composition of the soil (Sonwalkar *et al.*, 1991).

Quantity of Water Required for Filature Operation

In any filature much water is used for cocoon cooking, silk reeling and re-reeling, in addition to the boiler. About 850-1000 tonnes of water is required to manufacture a tonne of raw silk in semi-automatic machine. Halliyal *et al.* (1987) stated that the silk industry is one of the sector in mechanical processing namely reeling alone consumes 850-1000 litres of water with semi-automatic machine or automatic reeling machine and 250-300 litres of water for cottage basin to produce one kilograms of raw silk.

Bhandiwad (1988) reported that about 50-250 litres of water would be required for producing one kilogram of raw silk based on the type of machinery used. It is reported that on an average about 200 litres of water is required to produce one kg of raw silk (Sundaramurthy and Narasingh, 1992).

Kim (1989) it can be presumed that a water requirement for the production approximately 850 tonne. It would be safer to place the water needs at 1100 to 1300 tons for every tonne of raw silk. The filature industry can be categorized as one of the heavy water consuming sector.

Table 5.1: Standard Quality of Reeling Water*

Sl.No.	*Properties*	*Standard Concentration*	*Permissible Range*
1.	Colour	Colourless	–
2.	Smell	No smell	–
3.	Suspension and sediment	No	–
4.	pH before boiling	7.00	6.80–7.40
5.	pH after boiling	8.60–9.00	8.40–9.40
6.	Conductivity (M. mhos)	100	30–300
7.	Hardness (ppm)	45.00	35–90
8.	Calcium (ppm)	10.00	5–50
9.	Magnesium (ppm)	10.00	5–50
10.	Chloride (ppm)	<50.00	–
11.	Iron (ppm)	0.10	0.10–0.80
12.	Manganese (ppm)	0.00	0.10
13.	Sulphur (ppm)	–	–
14.	Carbonates (ppm)	6.00	0–20
15.	Bicarbonates (ppm)	6.00	0–20

* Set by Bhandiwad *et al.*, 1989 and Kim, 1989.

Impurities Present in the Water

Kanwar and Chopra (1976) analysed the chemical constituents present in the water and stated that the total solids, soluble salts, carbonate, bicarbonate, sulphate, calcium, magnesium, sodium, potassium, boron, nitrate, nitrite, silica, phosphate and fluorides were present. Hence, it is frequently necessary to ascertain the quality of water so as to gauge the possible effects of the water.

The various impurities present in the water are suspended matter, dissolved matter and solid dissolved matter. Suspended matter includes sands, clay, iron, aluminium and living organisms, sewage and oil. Dissolved matter includes gases like oxygen, nitrogen, carbon dioxide and hydrogen sulphides and solid dissolved matter includes salts of calcium, magnesium salts of sodium, potassium, organic and inorganic salts of iron, salts of manganese and aluminium, salts of silica and other organic compounds. Hard water contains calcium, magnesium, bicarbonates, sulphates and chlorides of calcium and magnesium (Natheshwar Sah, 1983a).

The natural water available locally would be the type of water used for the silk reeling and such water may contain one or more of the impurities like suspended solids, dissolved mineral salts, organic matter, colouring matter, dissolved gases, micro organisms and colloidal substances were present (Halliyal *et al.*, 1987).

Shamachary (1987) reported that unsuitable water contains some physical impurities like suspended in the water and chemical impurities like salts of the elements of calcium, magnesium are dissolved and makes the water hard which will not produce much lather with soap. A hard material called calcium sulphate deposits layer by layer over the inner surface and becomes narrower and dissolved chemical elements like sodium, bromide and silicates change the acidity and alkalinity of the water.

The properties of water is dynamic in nature due to environment factors like geological factors, carbon dioxide content in the soil, decomposition of organic matter and weather conditions like temperature, rain and snow etc.

Effect of Impurities on Reeling Performance and Raw Silk Quality

Calcium Sulphate

Silk filament becomes hard and slightly reddish colour. Tenacity and elongation becomes reduced.

Calcium Hardness ($CaCl_2$)

Silk becomes light brown, coarser and reduces the strength and luster will be affected.

Calcium Carbonate

Reduces the strength of the silk.

Sodium Bicarbonate

Solubility of sericin is greater in the presence of Sodium bicarbonate than calcium bicarbonate.

Magnesium Sulphate

This gives light brown colour to silk.

Iron

End picking is difficult, silk becomes brown colour, higher loss of Gum/Sericin and silk readily absorbs iron and manganese present in water.

Alkaline pH

As the silk is pertinacious in fibre, high pH hydrolysis the peptide linkage of fibroin to some extent, cocoon become soft and get damaged during brushing and filament attached to each other which does not unwind easily at the time of reeling and suffers from a defect known as plastering (Kim, 1989).

Chloride

Tenacity and elongation is reduced.

Acidic pH

The solubility of sericin is least when the pH of water is acidic. Acidity of water has a negative correlation with sericin solubility, which decrease with increase in water acidity.

Permanent Hardness

It inhibits swelling and solubility of sericin. It may not loosen the cocoon structure thereby leading to poor extraction of silk (Natheshwar Sah, 1983).

Reclamation of Unsuitable Water

If water from the natural sources is good enough to be used in the industry without any treatment, it would be very useful for the development of the industry. It is however observed that water from natural sources exhibit wide variations in the quality. Important methods of treating the water are broadly classified into four.

1. Filteration
2. Lime-soda
3. Ion-exchange
4. Sequestering

Filtration

The impurities in water may be either dissolved or suspended. Water can be purified by filtration before using in the filature reeling and also in boilers. Allowing the water to pass through a bed of a medium like sand or anthracite or the mixture of both usually does this. The different methods include

1. Filter bed of mixed materials
2. Copping of sand filters
3. Gravity filters
4. Pressure filters
5. Precoat filters

It is a common practice that to add aluminium sulphate in the form of Alum to the water, so that the suspended impurities settle at the bottom of the tank thereby improving the working efficiency. It is also a common practice that in some of the traditional silk reeling areas, the reelers use the seeds of the plant *Strychnos potatorum* L. in order to get similar effect of alum. Charka reelers are using this seed extensively. This seed is also being used extensively in paper and pulp industry to clear muddy water before discharging out (Doreswamy *et al.*, 1993).

Lime Soda Process

Water softening is carried out using a mixture of calcium oxide and sodium carbonate. The precipitated hardness component is allowed to settle and the clear soft water is collected. By this method, the soft water contains some amount of soda ash leading to the alkaline pH. This method requires neutralization with acid to bring down the pH to the required standards. This process is good for very hard water, which yields very low ppm hardness.

Ion-exchange Process

This process is widely used in the desalination, water purification and softening the water. A complex compound of sodium-aluminium-silicate known as Zeolite is used as an ion-exchange resin. The great advantage of the above process is that it yields water of zero hardness. Normally, total hardness is removed through H^+ type cation resin.

Sequestering Method

This method involves the use of Sequestering agents, which reacts with the metallic ions in such a way that they become part of a complex anion. These agents soften the water to make it suitable for reeling. The most common sequestering agents employed in textile processing are Sodium hexa meta phosphate, Nitrilotriacetic acid, Diethylene triamine penta acetic acid and Ethylene diamine tetra acetic acid.

References

Bhandiwad, R., Halliyal, V.G. Shankar, A.G. and Mangannavar, D.R. (1998). Effect of water on silk reeling industry. *The Indian Text. J.*, 99: 116-120.

Doreswamy, C., Chinnaswamy, K.P., Venkatesh, M. and Nageshchandra, B.K. (1993). Chillada seeds (*Strychnos potatorum* L.): A promising water clearing agent for silk reeling. *Indian Silk*, 31: 41-42.

Halliyal, V.G., Shankar, A.G., Mangannavar, D.R., Vijayeendra, M.K. and Nadiger, G.S. (1987). Studies on the effect of quality of water on reeling performance and raw silk quality. *Proc. of the seminar on prospects and problems of sericulture in India*, pp. 231-237.

Kanwar, J.S. and Chopra, S.L. (1975). *Analysis of irrigation water: Analytical agricultural chemistry*, Kalyani Publishers, New Delhi, pp. 306-330.

Kim, B.H. (1999). *Filature water engineering*. Associated Business Center Ltd., Colombo, Sri Lanka, p. 223.

Natheshwar, Sah (1983). The water and silk industry part I and II. *Indian Silk*, 21(4&5): 3-13.

6

Quality Assessment of Water from Fish Processing Industries and Their Pollution Management

R. SARAVANAKUMAR & A. SOLAIMALAI

Research Associate, Krishi Vigyan Kendra, Tamil Nadu Agricultural University, Vriddhachalam

G. MANIMEGALAI

Dean, Home Science College and Research Institute, Tamil Nadu Agricultural University, Madurai

M. BASKAR

Research Associate, A.D. Agricultural College and Research Institute
Tamil Nadu Agricultural University, Trichy

Introduction

India has a commendable natural water resources comprising of rivers, tanks, reservoirs, seas and ocean. The freshwater resources of India have been estimated to be 5.7 million ha. The total length of the.Indian coast is 8,041 km having about 3,726 fishing villages and the total fisher folk population is 59,59,144. The continental shelf is about 0.5 million sq. km. with an Exclusive Economic Zone of 0.02 million sq. km. There are 46,918 mechanized and 2,22,933 non-mechanized fishing vessels actively engaged in marine fish capture.

India stood in the third place during 1998, next to China and Japan with the total fish production of 5.24 million tonnes. India's fish production is approximately 5 per cent in the total world production. However, considering the potential resources, it can produce up to 10.0 million tonnes per annum both from capture and culture. Of the total marine fish production, about 6 per cent has been used for the processing and export. India export trade has fetched Rs. 5,100 crores during 1998-99. We have a greater scope for further increasing the export through enhanced production of fish and shrimp by aquaculture.

Fish Processing Industries in India

Fish processing is one of the most important industries in India. Fish processing industry includes processing plants, fish meal plants, fish canning plants and other supporting industries like ice plants, cold storage and pealing sheds. The details of the industries state-wise and category-wise are furnished in Table 6.1.

Table 6.1: Details of the Fish Processing Industries in India

Sl.No.	States	FP	IP	FMP	CS	PS	CP
1.	Gujarat	41	7	8	43	45	–
2.	Maharashtra	39	9	4	37	21	–
3.	Goa	7	6	–	10	5	3
4.	Karnataka	14	26	3	26	16	7
5.	Kerala	121	53	–	167	729	3
6.	Tamil Nadu	47	30	–	66	23	–
7.	Andhra Pradesh	51	9	–	52	48	–
8.	Orissa	21	7	–	20	15	–
9.	West Bengal	35	2	–	29	1	–
	Total	**376**	**149**	**15**	**450**	**903**	**13**

FP: Freezing Plants; IP: Ice Plants; FMP: Fish Meal Plants; CP: Canning Plants; CS: Cold Storage; PS: Peeling Sheds.

Fish Catch

The Total Fish Catch of the world is of the order of 70 million tonnes per year. India has a coast line of about 5,600 km and the continental shelf bordering the Indian coast has an area of about 2.6 million sq. km. in which infinite varieties of fish are found. In addition, the rivers, reservoirs, lakes and canals in the country have an area of about 1.13 million sq. km., which afford a wide variety of habitats of fishes. In 1981, the catch of fish was 2.6 million tonnes. About 151,000 tonnes of fish are exported per year; the total value of exports of fish and fish products amount to over 300 crores per annum. More than 80 per cent of our exports is frozen shrimp. The consumption of fish in India is about 2.5 kg per capita per year, which is perhaps the lowest in the world. This may be due to the fact that fish is eaten mostly by people residing in coastal areas. With increased fish catch by the use of modern techniques, the development of fish processing and better methods of storage and transport, there is likely to be an increase in the consumption of fish.

Types of Fish

It is estimated that there are about 25,000 species of fish living today, varying in size from very small fish to very large sharks. They are second to insects in the largeness of population and outnumber all other vertebrate (back-boned animals), mammals, birds, reptiles and amphibians put together. About 250 different species of fishes are used for edible purposes. Edible fishes are classified into two major categories based on their anatomical differences. Fishes having vertebrae with fin appendages, which help balance and steer the body in its movements through water, are termed as Finfish. The skin of these fishes is covered with protective scales and is abundantly coated with mucous. Fishes without a skeleton but covered with some type of hard shell are known as shellfish.

The shellfish are of two types; the molluscs and crustaceans. The former have soft unregimented body while the latter have segmented bodies. The molluscs are either partially or wholly enclosed in a hard shell that is largely of mineral composition. These include oysters, clams, scallops and mussels. The crustaceans are covered with crust like shells and they include lobsters, crabs, shrimps and crayfish. Edible shellfishes are mainly salt water fishes. The kinds of finfish available for food are more than those of shellfish and the types differ in different locations. They are found both in salt water and freshwater. Their flavour and quality depend partly on the water in which they are grown. Saltwater fishes have a more distinctive flavour than freshwater fishes. The common sea or marine fishes available in the country are all types of shark, salmon, mackerel, Bombay duck, ribbon fish, pomfret, sole, tuna, etc. Fishes from freshwater include sardine, mullet, catfish, perch, pearlespot, etc.

Fishes are also classified based on their fat content as lean (less than 2 per cent fat), medium (2 to 5 per cent) and fat (more than 5 per cent) fishes. Bombay duck, halibut, cod, perch and sole are examples of lean fish. Fat fish include salmon, sardine, mackerel, tuna, etc. In fish with more than 5 per cent fat in the edible flesh, the flesh is usually more highly pigmented than that of the low-fat varieties which are generally white fleshed.

Composition and Nutritive Value

The composition of fish varies considerably, amongst other factors, due to variation in fat content. The fat content of fish varies from 0.1 per cent to about 25 per cent. Many factors like the season, sex and stage of maturity determine the fat content of fish. In all cases, the liver and viscera (the pyloric caecum and mesenteria) constitute the location of fat depot. However, fats are also encountered in muscle tissue, mutt and roe (mass of eggs of fish). The principal component of most fish oil are glycerides of fatty acids which make up 95 per cent of the oil. Fish oils differ remarkably from vegetable oils in containing a great variety of fatty acids, especially highly unsaturated fatty acids having from four to six unsaturated linkages. The fatty-acid composition of seawater fish oil and freshwater fish oil varies, the latter containing smaller amounts of C-20 and C-22 unsaturated acids than marine fish oils, but a greater amount of C-18 unsaturated acids. Fish fat also contains cholesterol, lecithin, waxes and fatty alcohols. The fat from fish eggs is especially high in lecithin (up to 20 per cent). Fish is an excellent source of protein. The protein content of finfish is, on an average, about 20 per cent. The muscles of some lean fish contain exceptionally high levels of protein when compared to the best meat. As a whole, the protein content of shellfish is lower than that of finfish.

Fish contains some glycogen, shellfish containing a higher proportion; some shellfish have 3 to 5 per cent glycogen. The sweet taste of various shellfish is due to glucose formed from the glycogen by enzymic action. Fish is a good source of minerals. It contains more minerals than other types of meats. Fish meat is a good source of copper, sulphur and phosphorous. In general, saltwater fishes contain more iron than freshwater fishes, although Indian species of the latter group show high values. Marine fish is a dependable source of iodine. The iodine content of marine fish is nearly 30 times that of freshwater fish. Its content in marine fish varies from 0.01 to 0.5 mg per 100 g of fish meat. Iodine is essential for the prevention of goitre. Fish oils are the richest known sources of vitamins A and D. However, there are extreme differences in the vitamin reserves of different species. Fish flesh is a fair source of thiamine, being somewhat poorer in the vitamin than most meats. In some raw fish the enzyme thiaminase is present, and this destroys the vitamin and makes it unavailable if fish is held in a raw state or consumed raw. The riboflavin content in fish flesh is as much as in beef. Red flesh has a higher content of this vitamin than white muscle. Both seawater and freshwater fish are good sources of niacin. The vitamin C content of fish is very low and does not have any nutritional significance.

Systematic investigations of the nutritive value of fish have shown that it is equal to that of meat. Fish proteins are also easily digested. The amino acid pattern of fish protein closely approximates that found in any other class of animal protein. Generally, fish meat has lysine and histidine at a high level and occasionally, methionine and tryptophan not the low side of mammalian meat. Because of their high nutritive value, fish protein concentrates (FPC) have been developed from inexpensive and surplus fishes to combat malnutrition of people around the world.

Fish Spoilage

Microbiological

Flesh of healthy live fish is bacteriologically sterile. However, there are many types of bacteria on the surface skin, gills and the digestive tract of living fish. Generally, spoilage takes place shortly after the death of the fish when bacteria attack all the constituents of the tissues. Since these bacteria live on cold-blooded fish at a rather low ocean temperature, they are well adapted to cold and continue to grow even under common refrigeration conditions.

Physiological

Bacterial spoilage of fish does not begin until the fish has gone into rigor and passed out of it. For rigor mortis to set in, the glycogen in fish muscle must be converted into lactic acid accompanied by a decrease in pH. However, fish struggle when caught and use up virtually all of the glycogen in their muscles. When killed there is little glycogen left to be converted into lactic acid to slow bacterial growth. Rigor mortis in fish is comparatively shorter when compared to other animals which are rested before slaughter to build up glycogen reserves.

Chemical

Fish as taken from water have virtually little or no odour. Yet, virtually all fish products have a fishy odour and this is an evidence of some deterioration. The fishy odour is due to the production of trimethylamine by the action of bacteria on phospholipids and chorine present in fish. Also, the unsaturated fish fats undergo oxidation and become rancid, which also contributes to the smell of deteriorated fish.

Factors Influence Kind and Rate of Spoilage

The kind and rate of spoilage of fish vary with a number of factors:

The Kind of Fish

The various kinds of fish differ considerably in their perishability. Thus some flat fish spoil more readily than round fish because they pass through rigor mortis more rapidly, but a flat fish like the halibut keeps longer because of the low pH (5–5) of its flesh. Certain fatty fish deteriorate rapidly because of oxidation of the unsaturated fats of their oils. Fishes high in trimethylamine oxide soon yield appreciable amounts of the "stale-fishy" trimethylamine.

The Condition of the Fish When Caught

Fish that are exhausted as the result of struggling, lack of oxygen and excessive handling spoil more rapidly than those brought in with less ado, probably because of the exhaustion of glycogen and hence smaller drop in pH of the flesh. "Freedy" fish, *i.e.*, those full of food when caught are more perishable than those with an empty intestinal tract.

The Kind and Extent of Contamination of the Fish Flesh with Bacteria

These may come from mud, water, handlers and the exterior slime and intestinal content of the fish and are supposed to enter the gills of the fish, from which they pass through the vascular system and thus invade the flesh, or to penetrate the intestinal tract and thus enter the body cavity. Even then, growth probably is localized for the most part, but the products of bacterial decomposition penetrate the flesh fairly rapidly by diffusion. In general, the greater the load of bacteria on the fish, the more rapid the spoilage. This contamination may take place in the net (mud), in the fishing boat, on the docks, or, later, in the plants. Fish in the round, *i.e.*, not gutted, have not had the flesh contaminated with intestinal organisms, but it may become odorous because of decay of food in the gut and diffusion of decomposition products into the flesh. This process is hastened by the digestive enzymes attacking and perforating the gut wall and the belly wall and viscera, which in themselves have a high rate of autolysis. Gutting the fish on the boat spreads intestinal and surface-slime bacteria over the flesh, but thorough washing will remove most of the organisms and adequate chilling will inhibit the growth of those left. Any damage to skin or mucous membranes will harm the keeping quality of the product.

Temperature

Chilling the fish is the most commonly use method for preventing or delaying bacterial growth and hence spoilage until the fish is used or is otherwise processed. The cooling should be as rapid as possible 0 to –1°C, and this low temperature should be maintained. Obviously, the warmer the temperature, the shorter the storage life of the fish. Prompt and rapid freezing of the fish is still more effective in its preservation.

Bacteria Causing Spoilage

The bacteria most often involved in the spoilage of fish are part of the natural flora of the external slime of fishes and their intestinal contents. The predominant kinds of bacteria causing spoilage vary with the temperatures at which the fish are held, but at the chilling temperature usually employed, species of *Pseudomonas* are most likely to predominate, with *Achromobacter* and *Flavobacterium* species next in order of importance. Appearing less often, and then at higher temperatures, are bacteria of the genera *Micrococcus* and *Bacillus*. Reports in the literature list other genera as having species involved in fish spoilage, such as *Escherichia, Proteus, Serratia, Sarcine* and *Clostridium*. Most of these would grow only at ordinary atmospheric temperatures and probably would do little at chilling temperatures.

Normally pseudomonads increase in numbers on chilled fish during holding, achromobacters decrease and flavobacteria increases temporarily and then decrease. The bacteria grow first on the surfaces and later penetrate the flesh. Fish have a high content of non-protein nitrogen and autolytic changes caused by their enzymes increase the supply of nitrogenous foods (*e.g.*, amino acids and amines) and glucose for bacterial growth. From these compounds the bacteria make trimethylamine, ammonia, amines (*e.g.*, putrescence and cadaverine), lower fatty acids, aldehydes and eventually hydrogen and other sulfides, mercaptans and indole, which products are indicative of putrefaction.

A musty or muddy odour and taste of fish has been attributed to the growth of *Streptomyces* species in the mud at the bottom of the body of water and the absorption of the flavour by the fish.

As has been indicated, discolorations of the fish may occur during spoilage; yellow to greenish-yellow colours caused by *Pseudomonas fluorescens*. Yellow micrococci and others; red or pink colours from growth of *Sarcina, Micrococcus*, or *Bacillus* species, or by molds or yeasts; and a chocolate-brown colour by an asporogenous yeast. Pathogens parasitizing the fish may produce discoloration or lesions.

Preservation

Of all the flesh foods, fish is the most susceptible to autolysis, oxidation and hydrolysis of fats and microbial spoilage. Its preservation therefore involves prompt treatment by preservative methods, and often these methods are rigorous compared with those used on meats. When fish are gathered far from the processing plant, preservative methods must be applied even on the fishing boat. Evisceration should be done promptly to stop active digestive enzymes in the gut. Advantages gained by gutting may be offset by a possible delay in rapid cooling of the fish. Rigor mortis is especially important in the preservation of fish, for it retards post mortem autolysis and bacterial decomposition. Therefore any procedure that lengthens rigor mortis lengthens keeping time. It is longer if the fish have had less mucular activity before death and have not been handled roughly and bruised during catching and later processing and is longer in some kinds of fish than in other. Reducing the holding temperature will lengthen the period.

Aseptic methods to reduce the contamination of seafood are difficult to apply, but some of the gross contamination before processing can be avoided by general cleaning and sanitization of boats, decks, holds, bins, or other containers and processing equipment in the plant and by use of ice of good bacteriological quality. The removal of soil from contaminating surfaces and from the fish by adequate cleaning methods, including effective detergent solutions, helps greatly to reduce the microbial load on the fish. The removal of organisms is difficult, but the fact that most of the contamination is on the outer surface of the fish and other seafood permits the removal of many of the micro-organisms by washing off slime and dirt.

Use of Heat

Live crabs are cooked in retorts at temperatures up to 121°C to facilitate removing the meat from the shell. Hand picking and hand packing of the meat is common. Processing times and temperatures for canned crabmeat range from 85.6 to 87.2°C for 92 to 150 min (Dickerson and Berry, 1974). These processes are considered pasteurization and cans are preserved by refrigeration. Some seafoods, *e.g.*, oysters, are "canned" by packing into cans or jars and are not heat-processed but are preserved by refrigeration.

Most canned seafoods, however, are heat-processed so that they are sterile, or at least "commercially sterile." Like meats, seafoods are low-acid foods and for the most part have a slow rate of heat penetration and hence are difficult to heat process. Also some kinds of seafoods soften considerably or even fall apart when sterilization in the can is attempted. The process varies with the product being cannned and the size and shape of the container. In general, the heat processes are more severe than those used for meats, but some special products are lightly processed. Canning practices carried out in the FAO code of practice (FAO, 1973) minimize health hazards arising from canned seafoods.

Use of Low Temperatures

It is only after death of the fish or other sea animal that autolysis gets under way, with softening and production of off-flavour, and microbial growth becomes uncontrolled: as has been stated these changes are delayed by rigor mortis. Oyster in their shells, for example, will not decompose so long as they remain alive, and life is lengthened by chilling storage of shell oyster. Carp seined from Middle Western Lakes have been kept alive and hence in good condition by shipment in tanks to the Newyork Market. "Feedy" fish, that is, those stuffed with food, seem to decompose faster than normal fish.

Chilling

Because fish flesh autolyzes and the fats become oxidized at temperatures above freezing-rapidly at summer temperatures and more slowly as the temperatures is dropped towards freezing–preservation by chilling temperatures is, at best, temporary. When fish or other seafood is obtained at some distance from the receiving plant, the necessity for chilling on the boat depends upon the kind of fish, whether it is dressed there or not, and the atmospheric temperature. In general, small fish are more perishable than large ones; and dressed fish autolyze more slowly than whole fish but are spoiled more readily by bacteria. When outside temperatures are warm and distances of transportation are great, it becomes necessary to chill the fish and related foods on the fishing boat by packing in crushed ice or mechanical refrigeration in order to slow down autolysis and microbial growth until the products are marketed or are processed for longer preservation. The time allowable for holding in ice or in chilling storage will vary considerably with kind of fish or other seafood but will not be long in most instances. In general, chilling storage on shore is useful only when retail markets are near at hand and turnover is rapid. Otherwise, some other method of preservation is applied, such as freezing, drying, salting, smoking, or canning, or combinations of these methods.

Freezing

Most of the modern methods of freezing foods initially were developed for freezing fish. In olden days, ice with added salt was employed. With the advent of mechanical refrigeration, sharp freezing was employed and the fish were "glazed", *i.e.*, a layer of ice was frozen around the outside. Whole fish, especially the larger ones, usually are sharp-frozen in air or in a salt brine. Quick freezing is applied to wrapped fillets or steaks, although whole smaller fish may be so frozen. Like meats, quick-frozen fish may thaw to more like their original condition than fish frozen more slowly. During storage the fats of frozen fish are subject to hydrolysis and oxidation. Fatty fish deteriorate more rapidly than lean ones, probably because of more hydrolysis.

Decapitated raw shrimp are frozen and glazed and some cooked shrimp are frozen. Other seafood's preserved by freezing include scallop's, clams, oysters, spiny lobster tails and cooked crab and lobster meat. Most of these products are packaged before freezing. As with meats, freezing kills some but not all of the micro-organisms present, growth will take place after thawing if time permits. Fish carry a flora of psychrotrophic bacteria, most of which survive freezing and are ready to grow on thawing, *e.g.*, *Pseudomonas*, *Alcaligenes* and *Flavobacterium* species. Spores of type E *Clostridium botulinum* will survive freezing and storage and may grow and produce toxin when temperature reach 3.3°C or above. Frozen raw seafoods contain few enterococci, coliforms, or staphylcocci. Numbers of these organisms may be increased in breading and battering operations. Precooking reduces only coliforms to any extent.

Use of Irradiation

Preservation of fish by ultraviolet rays been tried but not put into practice. Experiments have indicated that gamma or cathode irradiation of some kinds of fish may be successful.

Preservation by Drying

The dry-salting of fish or immersion in brine constitutes a method of drying, in that moisture is removed or tied up. Oxidation of fish oils is not retarded and may cause deterioration. Salting of fish is being done to a lesser extent but still is used widely throughout the world. Salt cod is prepared by a combination of salting and air drying. The flesh is then removed from bones and skin. Sun drying of fish, either of small fish or of strips of flesh, is not practised extensively in the United States. Part of the preservative effect of smoking is the result of the drying of the fish.

Use of Preservatives

The salting or marination of fish by dry salt or in brine is effective not only because of the drying effect, mentioned in the preceding section, but also because of the effect of the sodium chloride as a chemical preservative. This method is used to a considerable extent in many countries of the world. The chemical and bacteriological qualities of the salt are important, for impurities such as calcium and magnesium salts may hinder the penetration of the sodium chloride and halophilic or salt-tolerant bacteria that are introduced may cause discolourations of the fish. Because of the great perishability of fish, investigators have tried numerous chemicals as preservatives, either applied directly to the fish or incorporated in the ice used in chilling them.

Preservatives Used on Fish

In the intensive search for chemical preservatives that could be applied directly or as dips to round fish or fillets, a large number of chemicals have been tried, ranging from those which most control agencies would approve to those whose use would be questionable. Sodium chloride, an acceptable preservative, has been discussed.

Fish may be dry-salted so to contain 4 to 5 per cent of salt. The salt contributes halophiles which may discolour the fish (*e.g.*, a red colour from *Serriatia salinaria**). Species of *Micrococcus** usually grow on the fish and there is a decrease in *Flavobacterium, Alcaligenes, Pseudomonas* and others. Curing of fish may be "mild," *i.e.*, with light salting, or may be in heavy brine or with solid salt and may be followed by smoking. Benzoic acid and benzoates have been only moderately successful as preservatives. Sodium and potassium nitrites and nitrates have been reported to lengthen the keeping time and are permitted in some countries. Sorbic acid has been found to delay spoilage of smoked or salted fish. Boric acid has been used in Europe with some improvement in the keeping quality. Other chemicals for which claims of success have been made, but whose use is contraindicated, include formaldehyde, hypochlorites, hydrogen peroxide, sulphur dioxide, undecylenic acid, capric acid, P-oxybenzoic acid, and chloroform.

Antibiotics also have been tried experimentally, usually in a dip or in ice. Of those tested, chlorteracycline and oxytetracycline seemed best and now their use is permitted. Chloramphenicol is fairly effective and penicillin, streptomycin, subtilin are poor or useless. Storage of fish in an atmosphere containing over 20 per cent of carbon dioxide has been found to lengthen the keeping time, but this method has not come into commercial use.

Pickling of fish may mean salting or acidification with vinegar, wine or sour cream. Herring is treated in various ways; salted, spiced, acidified. Various combination of these treatments, coupled with an airtight container, preserve the fish although refrigeration also must be employed for some products.

Formerly, fish was smoked primarily for its preservation and the smoking was heavy, but now that canning, chilling and freezing are available to lengthen the keeping time, much of the smoking of fish is primarily for flavour and hence is light. The smoke treatment and other preservative methods combined with it vary with the kind of fish, size of pieces, keeping time desired. Fish to be smoked usually are eviscerated and decapitated but may be in the round, split, or cut into pieces. Commonly, salting, light or heavy, precedes smoking and serves not only to flavour the fish but also to improve its keeping quality by reducing the moisture content. Drying may be aided by air currents. The smoking may be done at comparatively low temperatures (26.7° to 37.8°C), or at high temperatures like 63° to 92°C, which result in partial cooking of the fish.

Microbiology of Fish Brines

Numbers of bacteria in fish-curing brines vary with the concentration of salt, temperature of the brine, the kind and amount of contamination from the fish introduced and the duration of use of the brine and range from 10,000 to 10 million bacteria per millilitre. Salt concentrations usually are between 18 per cent and saturation but may be lower, especially after fish are introduced. The higher the temperature of the brine, the more salt necessary to prevent its spoilage. Contamination comes from the fish, which ordinarily introduce chiefly species of *Pseudomonas*, *Alcaligenes* and *Flavobacterium*, from ice, which introduces these genera plus *Corynebacterium* and cocci and from mechanically introduced sources, *e.g.*, dust, which add cocci. On continued use of the brine, numbers of organisms increase, because of addition from successive lots of fish and because of growth of salt-tolerant bacteria such as the micrococci. As the brine ages there is a decrease in numbers and an increase chiefly in corynebacteria in low-salt brines and in micrococci in high-salt brines.

Preservatives Incorporated in Ice

So-called "germicidal ices" are prepared by adding a chemical preservative to water before freezing. These ices are "eutectic" when the added chemical is uniformly distributed throughout, as with sodium chloride, or "noneutectic", when distribution is not uniform, as with sodium benzoate. Noneutectic ice is finely crushed for use on fish so as to get the chemical evenly spread in it.

Many investigators have sought the ideal chemical to be incorporated in ice for icing fish and have tested with some success a large number of chemicals, including hypochlorites, chloramines, benzoic acid and benzoates, colloidal silver, hydrogen peroxide, ozone, sodium nitrite, sulfonamides, antibiotics, propionates, levulinic acid and many others. The purpose of the application of preservative chemicals to fish either directly or as dips or germicidal ices is to kill or inhibit micro-organism on the surface of the fish where at first they are most numerous and active.

Antioxidants

Fats and oils of many kinds of fish, especially the fatter ones, such as herring, mackerel, mullet, salmon, are composed to a great extent of unsaturated fatty acids, and hence are subject to oxidative changes producing oxidative rancidity and sometimes undesirable alterations in colour. To counteract these undesirable changes, antioxidants may be applied as dips, coatings, glazes or gases. Good results have been reported with nordihydroguaiaretic acid, ethyl gallate, ascorbic acid, and other compounds and with storage in carbon dioxide.

Products

Noodles

Supapun-Brillantcs (1992) prepared fish noodles using Indian Carp (*Labeo rohita*) and other ingredients. Whole fish fillets were treated in 3 ways, minced by mechanical mincer minced manually, washed twice with iced water, drained and minced manually. After mixing with the other ingredients each fish dough was passed through the rollers, left to rest and then passed through a cutter. The fish noodles obtained were then steamed for 5 min. dried under the sun and packed in polyethylene bags. The chemical analysis and organoleptic assessment showed no significant differences in moisture content, TBA and water activity during storage. Organoleptic tests showed that fish noodles made from fillets with the dark meat removed were the most acceptable.

Surimi

Femeena-Hassan and Saleena-Mathew (1999) evaluated physico-chemical, microbiological and sensory properties of washed fish mince prepared from different species for their suitability for product development. Sensory scores were found to vary between species, indicating that separate procedures should be adopted for each species and incorporated into standardized methodology for Surimi preparation in commercial establishments.

Cutlet

Yaligar *et al.* (1993) reported the preparation and frozen storage characteristics of value added product–Fish cutlet. A ready-to-cook product, fish cutlet was prepared from an under tulized fish *Nemipterus japonicus*. The product was frozen at –28°C and further stored at –20°C. The biochemical, microbiological and organoleptical quality changes were studied during the storage period of 26 weeks. The study revealed that the product was acceptable even after 26 weeks of storage at –20°C indicating possibility for commercialization of the product.

Pickle

Chattopadhyay *et al.* (1986) prepared low cost freshwater fish pickle using juice from cheap abundantly available citrus fruit (*C. reticulata*). The product was shelf stable at RT for six months.

Balls

Pruthiarenum *et al.* (1994) investigated marine catch and freshwater fish to determine (1) the relationship between the freshness index of the raw material and functional properties of the meat; (2) the effect of leaching, addition of polyphosphate, setting time and species on fish ball quality and; (3) the quality of fish balls from a mixture of minced by-catch and commercial species. Results showed that storage life of by-catch species in ice was 9-15 days and freshwater fish was 16-18 days. The cohesiveness of fish balls from by-catch was acceptable for 10-12 days while tilapia was satisfactory even after 23 days and common carp 36 days. Good quality fish balls could be made from meat leached in diluted salt solution and heat treated at 40°C for 20 min.

Sausages

Xiong *et al.* (1999) evaluated the factors influencing the quality of freshwater fish sausages. Effects of pH (2.0-9.2), NaCl concentration (0.8–6 per cent) storage time (0–12 months) and concentration (5–15 per cent) and processing procedures on the quality of sausages made from freshwater fish were examined. Gel strength of ground fish flesh was highest (158.3–162.1 g/cm^2) at pH 7.2–6.8 and increased with increased NaCl concentration (optimum 2–3 per cent) and decreased with storage duration. With respect to starch, the best quality sausage was obtained with 10 per cent potato starch. The formula was optimized as (taking ground fish as 100): water 50, starch 15, NaCl 4, sugar 3.5 and sodium glutamate 0.85. For processing procedure the key points were grinding at 0°–10°C for 10–15 min and the sequence for adding supplements was NaCl, starch, water and others.

Freshwater Fish Waste Utilization

Docampo and lezcano (1994) reported the Vitamins A and B_1, iron and calcium contents of some important wastes from Cuban fishery industry. Using these wastes, way formulated two products, namel sausages from common carp (*Cyprinus carn*) silver carp (*Hypo phhalmichthys melitrala*) and canned dark meat from bonito (*Katsutov pelamis*). Taste evaluation and protein contents of both products were also determined. The results obtained demonstrated the presence of appreciable quantities of nutrients indicating their potential for use as dietary supplements to satisfy the essential nutri requirements in the human diet. Taste evaluations of the products showed high quality Xu-Shungan (1996) reported on the calcium powder from freshwater fish bones as an important nutrient essential for human body. Calcium deficiency gives rise to hypocalcimia symptom or even more serious calcium nutrition diseases. Calcium powder of the bone was made by heating, saponification and degreasing, degumming etc. The biological utilization and clinical trials showed that this powder had good absorbable calcium which was much better than common, calcium powder for therapeutic effects. Suryanarayana Rao *et al.* (1978) studied the utilization of fishery wastes for the production of microbiological media. The preparation of peptones as microbiological media from fish muscle and fish waste yielded promising results. One of the samples prepared from the heads and viscera of the water fish was found to be equal to the bactopeptone in growth promotion of organisms with stringent requirements such as *Lactobacillus arabinosus* and *streptococcus faecalis*. Papain treatment at 40°C for 4th hr. followed by vacuum drying gave a pale yellow low hygroscopic product containing 204 of amino N. A method, which relies on autolysis could be adopted for certain type of wastes such as material waste having proteolytic activity.

Ahmed and Mahendrakar (1995) studied the effect of different levels of molasses and salt on acid production and volume of fermenting mass during ensiling of tropical freshwater fish viscera. Viscera from free water fish, constituting 5 –11 per cent body weight consisted of 67 per cent water, 10 per cent protein and 3 per cent minerals.

Ahmed *et al.* (1996) described the change in microbial population during fermentation of tropical freshwater fish viscera (FV). Freshwater fish viscera was homogenised, mixed with 10 per cent (w/w FV) molasses and 0, 2 or 4 per cent salt and allowed to ferment at ambient temperature (26 ± 2°C) under micro-aerophilic conditions. The results revealed a reduction in total viable count and the number of spores, coliforms, *Escherichia coli, staphylococci* and *enterococci* and increase in yeasts and moulds as well as lactic acid bacteria during fermentation. Inclusion of either 0.5I per cent propionic acid, 0.3 per cent calcium proprionate or 0.1 per cent sorbic acid suppressed the growth of yeasts and moulds with propionic acid being the most effective. The study indicated that a microbiologically stable product could be prepared by ensiling fish viscera with 10 per cent molasses and 0.5 per cent propionic acid. Ristic *et al.* (1990) evaluated the quantity and quality of by-products obtained during processing of freshwater fish. The yield and characteristics of by products from the edible freshwater fish *Aristichuthys nobils* were determined. The mean by product yield was 40-90 per cent of total body weight. The results showed that head meat had high nutritional value and may be used as an edible by product. The other by products were considered to be inedible.

In the light of foregoing literature, it is concluded that there are not many reports on the value-added products from freshwater fish. The only reports available are on fish balls, patties, noodles and shelf stable products from freshwater fish. There is an urgent need for value addition to the fresh after fish through processing and extension of shelf life. It will go a long way in the development of freshwater fish industry and success of blue revolution.

Water Uses in Fish Processing Industry

There is few food industries that use no water at all, and many use very large quantities. Current emphasis on water conservation has seen the development of schemes designed to use less water and recover and recycle it wherever process conditions permit. These measures have become especially critical in areas in which water has become very scarce and/or very expensive.

Formulation

This type of water becomes a constituent of the fish processing. It should be of highest purity and potability. An obvious example of formulation water use is in the fish processing industry. Here, potable water is usually mixed directly with fishes. Of course, such water must meet the highest standards of potability even if the process calls for a subsequent pasteurizing step. If the 'process requirements are such that chlorinated water is unsatisfactory, special sterilizing filters may be used. Whatever the needs, inlet pipes to make-up or formulation tanks must be constructed so that an open air gap exists, which will thwart accidental backflow of the solution into the water supply system. The size of this gap is usually specified by local health code, but it should not be less than 4 in. These tanks should be covered, and hatches and ports in them should be designed to prevent the entry of foreign materials. The decision as to whether or not to filter incoming or outgoing formulation water depends on the needs of specific process involved. At the very least, constituent water should be strained to prevent the entry of large foreign objects. In cases in which a tank is batch filled, it often is a good practice to discard the initial of the water added. This procedure ensures that water that has entered the tank in fresh and that the tank has been rinsed before use. Whenever such a tank is filled, the water stream should be checked for clarity and colour. This is especially important if maintenance or construction work is occurring (or recently has occurred) on the water supply lines, or if the lines have remained unused for extended periods of time.

Cleaning

The use of water for cleaning involves washing plant structures and surfaces, equipment, raw materials, and in some cases, the finished product. This procedure may be required in some processes but not in others. The use of water to clean an inherently dry process line can lead to serious microbiological problems unless proper precautions for drying are taken.

Such cleaning can be for purposes of sanitation or for removing undesired colours or flavoured residues from a process stream. Depending on the process, raw materials also may require cleaning. This use of water is one of the few instances in which chlorination is of little benefit because the heavy "loads" of organic matter that are present deplete active chlorine very rapidly. Consequently, active residues of this compound are almost impossible to maintain, and in any event, are non-essential at this point in the process. After the initial removal of soil and dirt, the product may be sprayed with water as it moves over a conveyor or through a tumbling, washer reel. Parker and Litchfield (1962) have noted that spray nozzles should be located as close possible to the product in order to maximize pressure effects, the sprays should not overlap, water pressure should be optimized, and the nozzles should be free of obstructions. Generally, the greater the water pressure applied, the greater will be the cleaning efficiency. Less water is usually required in reels and tumblers than in other cleaning devices to produce a given level of cleaning.

Conveying

Water is used for conveying principally in the fish processing industries. Wherever practical, conveying water is reused after filtration or sedimentation. In many situations, cleaning and conveying occur at the same time.

Fluming is one method of transporting product to the process area. The raw product may be stored in large impoundment or bin that has troughs in the cement floor. When a heavy stream of water is directed at the product, it is washed into and through the trough to the process area. As mentioned, this water often is recycled.

Food products also can be conveyed within the plant by means of' "Pumping" in water. The diameter of pipes used for this purpose depends on the size of the product to be conveyed. Stainless steel or acrylic resin pipes are recommended for this type of conveying. Aluminium pipes or tubes also have been used for this purpose.

One of the principal difficulties with transporting foods from one plant location to another by water pumping is the tendency for product to contact the sides of the conveying tube. Residues of the product may thus adhere to the tubing's inside wall and become a focus for microbial growth. Such accumulations, when and if they form, can be minimized by periodically pumping a "pig', which is nothing more than a resilient plastic sponge of a type similar to that used to scrub ordinary household pots and pans. Of course, care must be taken to assure that "pigs" do not inadvertently enter the process stream. Depending on the rate of residue build up, such pipes may be disassembled periodically and scrubbed thoroughly by hand to remove accumulations. Process shutdowns on weekends are excellent times for this task.

Cooling

Water or water solutions are used to cool a wide variety of foods. In fact, in most industries, more water is used for cooling than all other uses combined. As attempts to conserve water have gained momentum, the installation of closed-loop cooling systems, involving a mechanism such as a cooling tower for the exchange of heat, has become fairly common. Whenever such closed heat exchange systems are used, they must be checked periodically to make sure that leaks or the potential for leaks do not exist in the system. This is normally done in several ways or combinations of ways. The best method is to look for pressure drops in the system. Another way to detect leaks is to measure water volume depletion. A third way is to place a substance in the cooling water such as a dye which can be instrumentally or visually detected in the food.

The canning industry relies on cooling water to terminate the thermal process and establish a partial vacuum within cans. When seam leaks occur, cooling water, along with whatever contaminants it might contain, enters the can. Micro-organisms that might be present thus find their way into the food and cause spoilage. The total aerobic bacterial counts of can cooling water were found by Graves *et al.* (1977) to vary from <1 to 210,000/ml; however, food-borne disease organisms were rarely observed. Normally, the numbers of organisms are inversely related to the level of residual chlorine in the cooling water. Usually 2-3 ppm will maintain bactericidal levels in most can-cooling systems including recirculating types, which employ cooling towers in the "loop". Such systems, however, because they seem to be especially susceptible to the build-up of bacterial and algal slimes, may have to be "burned out" with periodic treatment using very high concentrations of chlorine. If this procedure is an excellent oxidizing agent and can quickly deteriorate the wood surfaces of cooling towers, metals in cooling vessels and the cans themselves. Normally, "burn outs" are scheduled for down periods and are quickly followed by a rinse (with chlorine-free water) and discharge period.

Water is also used to cool airstreams that are impinged on a product requiring cooling. Air is blown through a fine spray of water, and the temperature of the air drops as a result of the evaporative cooling of the fine water droplets in the spray. This water is about 70-90 per cent recycled, the amount of freshwater added being dependent on the size of the droplets, temperature of the incoming air, and product. Occasionally, evaporative coolers develop microbiological problems; algal growth if the system is exposed to sunlight, and bacterial slimes if the water is not properly chlorinated (usually 0.5–1.0 ppm residual chlorine is sufficient). On the other hand, even very small amounts of chlorine may endanger the flavour of some products. In these situations a high rate of dilution with new water must be substituted for chlorine, in addition to frequent cleanouts. It is good to remember that high microbial levels in the water of an evaporative cooler may impinge directly on the product itself and can quickly produce a high-count product.

Steam

Steam is used in the fish processing industry to heat, peel, humidify, and clean. As a result, the water from which steam is generated is considered to be a food ingredient and must meet regulatory requirements for potable water. This creates problems in some situations because the boiler water from which steam is generated may contain corrosion inhibitors and water-conditioning compounds. Such materials are regarded as food additives by regulatory authorities and their use in process steam generation is controlled accordingly.

The use of steam to clean and sanitize plant and equipment surface is a common practice in fish processing industries. Presumably, this is an effective way to remove fatty, oily residues from some processes in which the use of hot water containing a detergent is impractical or ineffective. On the other hand, steam cleaning produces a very high-humidity environment, which can cause problems in other areas of the plant. In addition, steam generation is a costly and energy-intensive process. As with other sanitation-related tasks, the benefits of steam cleaning must be weighed against costs.

Water Quality

Microbiological Quality

There are two general groupings of micro-organisms that might be found in water: non-pathogens and pathogens. The former group includes organisms that potentially can form off-flavours, odour, and slimes, and can clog filtration systems. Although these problems can be annoying to deal with, they pose little threat to the public health.

The contamination of water by pathogenic micro-organisms, however, is quite another matter; the consumption of such water may result in outbreaks of disease in which hundreds and even thousands are made ill. Historically, water-borne outbreaks have been caused by gram-negative bacteria such as *Salmonella, Shigella,* and *Vibrio* species, which produce typhoid fever, dysentery, and cholera. These organisms enter water supplies most commonly from faecal source such as septic tanks, cesspools, or even leaking sewage lines.

Bryan (1977) has listed (Table 6.2) a number of circumstance that must be present to produce illness in individuals consuming food that has become contaminated by infected water. Obviously, the disease risk can be great if food is prepared with or from water containing pathogens, especially if a terminal pasteurization step is not present.

Table 6.2: Circumstances Required to Produce Illness from Wastewater-Contaminated Foods

1.	Infectious or toxic agents must be present and must contaminate sewerage or drainage systems.
2.	Agents remain viable during wastewater treatment.
3.	Wastewater effluent must contact food.
4.	Agents must survive during contact and contaminate the food.
5.	Agents must survive food process conditions in sufficient numbers to produce diseases symptoms.

In addition to water-borne diseases of bacterial origin, viruses and parasites also can be serious threats. The principal viruses that cause problems in water are hepatitis A and Norwalk agent. Chlorination normally inactivates these viruses.

Despite the best efforts of the food-plant sanitarian to control and even eliminate micro-organisms from the plant water supply, eventually the question probably will be asked, "How many bacterial are too many?" Standards for the microbiological purity of water are primarily set by local health authorities, and the sanitarian should be well versed in these limits. Water analyses should be performed only by laboratories certified by the respective state governments. Normally, no more than one coliform organism per 100 ml of water should be expected, and positive samples should be rare indeed. If, however, the incidence of sample positive for coliforms begins to increase fairly abruptly, investigation and corrective actions are indicated.

One other potential source of water contamination in the food environment deserves mentions: water softening, or deionizing, devices. Often this source of micro-organisms is first suspected when water has an off-odour or flavour. Total plate counts will be extremely high, and in fact, the water may even be turbid as a result of the high numbers of bacterial which it contains, Despite this, tests for faecal contamination of coliforms will be negative. Although the micro-organisms mostly apt to grow in deionizers are non-pathogenic, their presence is undesirable, and every effort should be made to prevent their growth. Frequent back flushing of the resin bed, plus replacement when the bacterial counts of exiting water begin to rise, are the most satisfactory means of controlling contamination. The use of antibacterial agents normally is not recommended because these materials could enter the water and, even at very low concentrations, could constitute an illegal food additive. Charcoal filters also may constitute an excellent source of bacterial in plants in which these devices are used.

Chemical Quality

Chemicals in water most often are a problem when cations such as calcium and magnesium accumulate. This is more apt to occur, when wells are the source of water as opposed to surface water supplies. Although these ions can cause colour, texture, and even flavour problems in certain foods, they are seldom a public health concern.

Likewise, while harmful chemicals have caused individuals who have consumed water containing them, reports of transmittal of such chemicals to fish during processing in sufficient concentration to cause food-borne illness are rare. However, such incidents do occur occasionally and when they do, the food must be withdrawn immediately from commercial channels. Most often, this type of chemical contamination occurs when there is an accidental cross-connection between pipes bearing chemically contaminated water and those used to carry potable water. Contamination from wastewater, as stated above, is primarily a microbiological threat, but such water also may certain high levels of toxic chemicals and minerals.

The greatest difficulty, as noted by Camp and Meserve (1974), is recognizing that chemical contamination of water supplies is occurring, because the spectrum of potential contaminants is so great and their recognition difficult. A knowledge of which impurities could enter the water supply is of some help in this regard, but often the food processor must rely on the odour, colour, and flavour of water reaching his plant to determine its acceptability for processing his product. Whether or not one can reasonably expect that water be free of all chemicals is a difficult question indeed. Certainly it should not contain substances known to be deleterious to living things if it is to be used in (or in contact with) food.

Radiation

As nuclear power installations have become more prevalent over the past 10 or 15 years, considerable concern has arisen over the potential for radioactive contamination of surface waters used for cooling these installations. Many such streams are now routinely monitored for radioactivity by public health authorities. Presumably, adequate warnings of excessive radiation emanating from water downstream from nuclear plants would be review these warning systems periodically and assure themselves that adequate safeguards are being provided.

Water Treatment

Chlorination

The principal means of ensuring the absence of disease bacteria and viruses from potable water supplies is through the addition of chlorine. Simple automated chlorination systems that can meet preset chlorine demands are commercially available. Residual chlorine levels are checked periodically to ensure that chlorine concentrations are sufficient to meet the various needs throughout the plant. Chlorine gas, sodium chloride-based generators, and sodium hydrochloride solutions are used as sources of chlorine.

The actual amount of chlorine that must be added to water to meet these concentrations depends on the chlorine demand of the system. This demand depends on such factors as pH, temperature, the concentration and type of impurities that might be present.

Chlorinated Hydrocarbons

The contamination problems currently besetting potable water supplies in the United States are many and varied. There is, however, one aspect of this subject of which food sanitarians should be aware. This pertains to the formation of potentially carcinogenic chlorinated organic compounds, which may occur in potable water supplies as a result of water chlorination.

This regulation basically places a maximum limit of 0.10 mg/litre total trihalomethanes and establishes monitoring and reporting requirements. Charcoal filters can be used to attain these levels; however, filters themselves can become a source of contaminating bacteria. In addition, residual chlorine levels can expected to drop markedly after such treatment, with the result that the water could become a public health risk.

Deionization

Hard water is especially front wells and springs where calcium and magnesium salts (primarily carbonates) are leached from the subsurface soil into groundwater. Their removal may be required to prevent textural and colour changes in certain foods, to improve the effectiveness of bottle washing operations, and to prevent the accumulation of mineral deposits within pipes and equipment. These requirements are only peripherally related to food sanitation, and so the deionization of food-plant water supplies will not be discussed further, except to note that microbiological problems can arise from growth in deionizing columns.

Coagulation and Filtration

Water, especially surface water is seldom suitable for food use as it arrives at the plant or water treatment facility. Depending on the types of anticipated impurities, coagulates such as aluminium to iron sulphate are added to the water to form flocs, which entangle contaminants and can be filtered from the water. Coagulation aids such as activated silica also may be employed. Following coagulation, the water is settled and filtered through a bed of fine sand. Flow rates through these filters may be as high as 3 gal/min for each square ft. of filter surface. Organic materials that can contribute to off-odours and flavours are removed, when necessary, by use of activated carbon filters or by the additions of carbon pellets during the coagulation step. An aeration step preceding coagulation also may be helpful in oxidizing organic impurities to odourless, colourless, and tasteless constituents.

References

Ahmed, J., Mahendrakar, N.S. (1995). Effect of different levels of molasses and salt on acid production and volume of fermenting mass during ensiling of tropical freshwater fish viscera. *J. Food Sci. Technol.*, 32(2): 115-118.

Ahmed, J., Ramesh, B.S., Mahendrakar, N.S. (1986). Changes in microbial population during fermentation of tropical freshwater fish viscera. *J. Appl. Bacteriol.*, 80(2): 153-156.

Bryan, F.L. (1977). Disease transmitted by foods contaminated by wastewater. *J. Food Prot.*, 40: 45.56.

Camp, T.R. and Meserve, R.L. (1974). *Water and its impurities 2nd edn.* Dowden Hutchinson and Ross Inc., Stroudsburg, Pennsylvania.

Chattopadhyay, A.K., Bandyopadhyay, J.K., Bhattacharyya, S.K. (1986). *Low cost freshwater pickle.* Fish Technol. Soc. Fish Technol., Cochin, 171-173.

Docampo, S., Lezcano, M. (1994). Processing of value-added products for human consumption from fishery industry wastes. Working papers presented at the third expert consultation on fishery product technology in Latin America–Porlamar, Venezuela, 21-25, March 1994, 16-19.

Graves, R.R., Leisniewski, R.S. and Lake, D.E. (1997). Bacteriological quality of cannery cooling water. *J. Food Sci.*, 42: 1280-1285.

Parker, M.E. and Litchified, J.H. (1962). *Food plant sanitation.* Reinhold, New York.

Pruthiarenun, R., Yamprayoon, J., Suwansakornkul, P., Kiatkungwalkrai, P., Suwanrangsi, S. (1994). Utilization of fish by-catch for fish ball manufacture. Symp on spoilage of tropical fish and product development processing. Royal Melbourne Institute of Technology, Melbourne, Australia. 23-26 Oct. 1984, pp. 428-449.

Ristic, M., Kormanjos, S., Milic, M., Curcic, R., Petrovic, N. (1990). Evaluation of quantity and quality of by-products obtained during processing of freshwater fish. *Technologija-Mesa*, 31(5): 184-189.

Supapun-Brillantes (1992). Fish noodles using Indian carp. *Asean Food J.* 7(3): 137- 140.

Suryanarayana Rao, S.V., Saraswati, C.R., Dwarkanath, C.T. (1978). Studies on the utilization of fishery wastes for the production of microbiological media. *Proc. International Publications of Fishery Council*, 18(3): 364-369.

Xiong, G.Q., Ye, L.X., Zhang, H. (1999). Factors influencing the quality of freshwater fish sausages. *Food Industry*, 1: 39-40.

Xu, Shungan (1996). Calcium powder from freshwater fish bone. *J. Shanghai Fish using Shanghai Shuichan Doxue Xuchao*, 5(4): 216.

Yaligar, M.D., Hirematt, G.G., Basavakumar, K.V., Shetty, T.S. (1993). Studies on the preparation and frozen storage characteristics of value added product–fish cutlet. *Punjab Fish Bull.*, 27(1): 21-24.

7

Management of Water Pollution in Ponds Used in Trapa-cum-Fish Culture Practices

SHIVESH PRÁTAP SINGH
Department of Zoology, Government P.G. (Autonomous) College Satna, Madhya Pradesh
SURENDRA GUPTA
Fisheries Department, Satna, Madhya Pradesh

ABSTRACT

The study were conducted in two ponds stocked with six species of carps, namely *Catla catla, Labio rohita, Cirrhinus mrigala, Cyprinus carpio, Hypopthalmichthys molitrix* and *Ptenopharyngdon idella* and trapa (*Trapa bisponia*) was lodged. The abundance peak of planktonic organisms were achieved during overall spectrum of hydrobiological parameters reflected that trapa-fish integration is not eco-friendly due to use of pesticide and tobacco as a medicine of trapa culture. Gross production of fish was affected.

Keywords: Trapa-cum-fish culture, Water pollution

Introduction

Aquaculture has made rapid strides in India in the three decades. Efforts are presently being made directly towards aqua-farming in embanked pond. The Catla, Rohu, Mrigal, Silver carp, Grass carp and Common carp are considered to be the best culturable varieties as they are the quickest growing fishes in terms of flesh return, non-predatory and domesticated in nature and best compatible species for their short-spectrum omni-harbivorous feeding habits in the composite and integrated fish culture.

Aquaculture plays an important role in state economy of Madhya Pradesh. It provides livelihood to thousands of people directly or indirectly. The FFDA of Satna district have brought under production 3015.1 ha. of water area of these fish farmers and have been imparted in fish culture.

The total areas of private and govt. sector ponds are 3829.357 ha in Satna district of M.P. However only 58.93 per cent is used perfunctorily for fish culture. Remaining water area is unutilized or is used only for trapa (*Trapa bisponia*) or singhara (water nut). The right of pond fishery development is vested in the department of fisheries. The ponds are given by panchayats on lease to groups of fisherman or individual persons from scheduled castes, scheduled tribes or to co-operatives in order of priority. However, only a Machua caste fisherman practice trapa-cum-fish culture.

During the research two perennial ponds (Banjarha and Matehna pond) of Satna district have been selected for detailed investigations.

Material and Methods

The selected ponds are Banjarha and Matehna ponds situated in Sohawal block of Raghurajnagar tehsil of Satna district. Banjarha pond (Pond A) is surrounded by agriculture fields and open area. The total area of the Banjarha pond is 5.068 ha. Whereas Matehna pond (Pond B) is located about 15 km. from Satna and situated at Matehna village, which is near Madhavgarh and 2 km. from Satna-Rewa road. The total area of Pond B is 4.160 ha (Table 7.1).

Table 7.1: Location and Morphometry of the Banjarha and Matehna Pond

Parameters	*Banjarha Pond (Pond A)*	*Matehna Pond (Pond B)*
Location	At Satna town	At Matehna village near Madhavgarh
Land khasara No.	265/2	349
Total land area (ha)	5.068	4.160
H.S.W. (ha)	5.000	4.000
L.S.W. (ha)	1.000	1.000
Productive water area (ha)	3.000	2.500
Catchment area (ha)	40.000	35.000
Mean Depth (feet)	8.0	6.5
Water source	Rain water and tubewell	Rain water and Tamas river water
Water use	Fish and trapa culture, Animal and domestic uses	Fish and trapa culture, domestic, Animal uses and sometimes P.H.E. supply through this pond
Selected for Central fisheries	Not selected	Selected

In the present investigation a regular uninterrupted monthly sampling from March 2000 to February 2001 was conducted to cover all the three seasons. In order to estimate the quantitative values of physico-chemical parameters, the water samples were taken to laboratory and were analyzed by using standard methods APHA (1985), Trivedy and Goel (1986).

The phytoplankton was systematically identified by using compound research microscope with the help of standard works. Edmondson, 1959; Prescott, 1962 and APHA, 1995 etc.

Whereas zooplankton was identified with the help of information given in standard works (Edmondson, 1959 and Topani, 1980 etc.).

Results and Discussion

Water Quality Characteristics

The water temperature of Banjarha (Pond A) and Matehna (Pond B) pond appear to show almost similar monthly trends. The lowest water temperature of 16.5°C (Pond A) and 18.0°C (Pond B) in January 2001 were recorded and highest of 39.0°C (Pond A) and 38.0°C (Pond B) in June 2000 (Table 7.2). The resultant water temperature of Banjarha and Matehna ponds appear to show monthly trends.

Table 7.2: Physico-chemical of Banjarha and Matehna Ponds of Satna During 2000-01

Parameters	*Banjarha Pond (A)*		*Matehna Pond (B)*	
	Minimum	*Maximum*	*Minimum*	*Maximum*
Water temp (°C)	16.5	39.0	18.0	38.0
Transparency (cm)	8.0	26.0	9.0	30.0
pH	5.5	8.1	6.3	7.4
Dissolved Oxygen (mg/l)	5.4	8.9	5.5	8.4
BOD (mg/l)	3.10	5.20	3.08	5.35
Total Alkalinity	80.0	260.0	160.0	260.0
Total Hardness (mg/l)	238.0	252.0	250.0	260.0
Calcium (mg/l)	221.0	234.0	229.0	240.0
Magnesium (mg/l)	16.0	22.0	16.0	25.0
Chloride (mg/l)	144.0	152.0	57.0	62.0
Phosphate (mg/l)	0.21	0.35	0.13	0.22
Ammonia (mg/l)	0.40	0.60	0.10	0.50
Plankton (mg/l)	1210	6853	1560	6820
Fish productivity (Year)	3300 (1999-00)	1300 (1995-96)	2400 (2000-01)	1200 (1998-99)

The values of water temperature observed from present ponds have shown more or less similar pattern and fluctuations to the observations of Saha *et al.* (1979), Prasad *et al.* (1985), Saha and Pandit (1986), Sharma and Jain (2000), Singh and Gupta (1996), Singh *et al.* (2002) and Mishra and Patel (2002).

In the present study seasonal change in light transparency was observed. Secchi transparency (cm) was observed at present water bodies, which ranged in Pond A from 8 to 26 cm. and Pond B from 9 to 30 cm. with the maximum in December and minimum in July to August of it changed in both ponds (Table 7.2). According to Beeton (1958) Secchi transparency not only represent the percentage of incident light but also the degree of suspended solids and turbidity. The high transparency was noted in summer and moderate in monsoon and winter. Zutshi (1985) have also noted the same trend and changes in water transparency at certain freshwater bodies. According to Welch (1935) and Wetzel (1975) the high amount of Ca, Mg and chlorides reduces water transparency and give mild turbid colour to water.

The pH values registered at different sampling stations which ranged from 5.5 to 8.1 in Pond A and from 6.3 to 7.4 (Pond B) in Table 7.2. The pH below 7 indicates that the sample water were slightly acidic which may be due to the wide spread use of pesticide, tobacco and various fertilizers in the nearby agricultural lands. Pearsall (1930) and Zafar (1966) observed that pH of the water appears to be dependent upon the relative quantities of calcium, carbonates and bicarbonates. The water trends to be more alkaline when it possesses carbonates. It is much less alkaline when it possesses larger quantities of bicarbonates, carbon dioxide and calcium.

A number of workers like Zafar (1966), Rao (1972), Quadri and Yousuf (1980), Prasad (1988) and Nayak (1993) have documented a positive relation of pH with alkalinity same positive relationship also registered during present investigation. The high photosynthesis and higher consumption of CO_2 provide its storage in water. It is derived from bicarbonates and then carbonates are formed, that results an increase in pH.

The annual cycle of dissolved oxygen in Pond A and Pond B appear to be closely related with water temperature and turbidity. The mean values of dissolved oxygen for present ranged from 5.4 to 8.9 mg/l in Pond A and from 5.5 to 8.4 mg/l in Pond B. The lower values were recorded during summer and maximum during winter in both ponds. Biological Oxygen Demand values varied between 3.10 to 5.20 mg/l at Pond A and 3.08 to 5.35 mg/l at Pond B with the minimum in December and maximum in June in each pond in Table 7.2.

According to Jackson (1961) the lake of high trophic status ordinarily has a range of high alkalinity of more than 50 reveals. The water bodies were noted to had alkalinity range from 80.0 to 260 mg/l at Pond A and 160.0 mg/l to 270.0 mg/l at Pond B (Table 7.2). In both ponds highest and lowest concentration of total alkalinity at present water bodies were noted during winter and during summer seasons respectively. The alkalinity concentration (except summer) of both the ponds is suggestive of the high productivity levels of the ponds. Adverse effect on the water quality of the Pond A was recorded due to the use of tobacco and pesticide for trapa culture.

According to Shrivastava (2001), lime convert waste biomass into animal feeds, *e.g.* solid waste, sewage sludge, manure and agriculture residues with lime converted into H_2, ethanol and acetic acid.

Waste Biomass
↓ Lime
Lime Treated Biomass
(Animal feed)
↓ Fermentation
$[RCOO]_2$ Ca
H+ ↓
Carboxylic acid + Ketone
↓ [H]
Alcohol

The increase in alkalinity was recorded during winter season. Analogous variation in alkalinity was reported by Goal *et al.* (1985). It is evident from present study that the ponds had a range of total hardness from 238.00 to 252.00 mg/l at Pond A and 250.00 to 262.00 mg/l at Pond B. Calcium ranged 221.00 to 234.00 mg/l at Pond A and 229.00 to 240.00 mg/l at Pond B. The value have been registered to be minimum in March (Pond A) and July (Pond B), and maximum in April (Pond A) and June (Pond B). The magnesium contents of Banjarha and Matehna ponds appear to show almost similar maximum and minimum values. Magnesium contents ranged between 16.00 to 22.00 mg/l at Pond A and 16.00 to 25.00 mg/l at Pond B. Chloride contents ranged between 144.00 to 152.00 mg/l (Pond A) and 57.00 to 62.00 mg/l (Pond B) with the minimum concentration during March (Pond A) and July (Pond B). The maximum concentration noted during December (Pond A) and May (Pond B). Low chloride 4-10 ppm only indicates the purity of water free from pollution (Sreenivasan, 1965). Our results show the concentration of chloride are much higher in Banjarha Pond throughout the year, indicating the water being polluted one.

Gonzalves and Joshi (1946) took chlorides into account in the tank water and according to them, the chloride concentration increase in the summer when the water level is low. In the present investigation too, the chloride concentration showed a regular periodicity, being higher in the warmer months when there is high rate of evaporation. Rao (1972) found direct correlation between chloride concentration and pollution level. Phosphate had been recorded from 0.21 mg/l to 0.35 mg/l at Pond A and 0.13 mg/l to 0.22 mg/l at Pond B. The level of Phosphate showed decreasing trend during winter and post-monsoon season due to rapid utilization of there vital nutrients by the macrophytes and phytoplankton. Similar result in this system was also found by Sinha (1995).

The positive relation of ammonia nitrogen and plankton indicates that its concentration in present water bodies is not lethal and is found under normal, tolerant limit. The present water was noted to have a range of ammonia from 0.40 to 0.60 mg/l (Pond A) and from 0.10 to 0.45 mg/l (Pond B). The minimum value of ammonia was observed in Pond A in month of March and Pond B in month November increasing trends had been observed maximum in Pond A during Summer month and in Pond B during 1st week of the July. The entire data of these parameters had shown significant changes with seasons (Table 7.2).

Plankton Characteristics

The qualitative and quantitative estimation of plankton made in present study indicates apparently that Chlorophyceae, Cyanophyceae, Bacillariophyceae and Euglenophyceae have constituted the bulk of phytoplankton in present water. The study ponds were found to be very rich in the phytoplankton and zooplankton populations. The rich population of plankton in Pond A and Pond B can be attributed due to the high concentration fertilizer. High population of phytoplankton population is a characteristic feature of eutrophicated systems (Sharma, 1991). The planktons were maximum recorded (6853, 6265, 6270, 6820 in both ponds during 1999-00 to 2000-01) in the month of June, while the minimum plankton population was recorded in winter (Table 7.2).

The temperature is considered important in controlling the fluctuation of plankton but Chacko and Krishna Murthy (1954) could not establish any co-relation between water temperature and plankton variation. The higher values of oxygen are associated with the rise of phytoplankton population (Bhatt and Negi, 1985).

The phytoplankton community is comprised of three major groups *viz.* Cyanophyceae, Chlorophyceae and Bacillariophyceae. The Cyanophyceae population was comprised by *Oscellatoria calybea*, *Nostoc muscorum*, *Anabena spiroides* etc. The Cholorophycean members were mainly represented by *Pediastrum* sp., *Sceriedesmus* sp., *Scenedesmus* sp., *Ulothrix* sp., *Cladophora* sp., *Chatophora* sp., *Oedogonium* sp., *Spirogyra* sp., *Cosmarium* etc. The Bacillariophyceae was dominated by *Navicula* sp. in both ponds.

Protozoa, Rotifera, Copepoda and Cladocera represented the zooplankton community. The *Amoeba* sp., *Euglypha* sp., *Paramecium* sp., *Verticella* sp. etc. protozoan were recorded. Certain common species of Rotifers were *Branchionus* sp., *Keratella* sp., *Gastropus* sp. etc. *Cyclop virdis* and *Mesocyclops* represented copepods. *Cladoceran* included *Daphnia* sp. and *Moina* sp. in both the ponds.

If the total plankton in the water bodies under study are taken into account then it was found that they are present in appreciable numbers in both water bodies. The highest and least number is observed in Pond A.

Several physico-chemical and biological factors acts simultaneously to influence the plankton fluctuation. There exists a definite correlation between phytoplankton and zooplankton (Verma and Munshi, 1987).

Fish Culture

The fish landing of Banjarha and Matehna ponds have shown a consistency during the study period, the value of fish caught have increased significantly, mainly attributed to quality of catch.

In the present study it is seen in Pond A that fish stocking upto 23,333 fry/ha/yr gave a yield of 866 kg/ha/yr and in Pond B 20,000 fry/ha/yr gave a fish yield of 2,400 kg/ha/yr during 2000-01. A conclusion that can be drawn from the present study is that 26.92 fry required producing one kg of fish at Pond A whenever in 8.33 fry in Pond B. More numbers of fry were however noted to have been stocked of 1,20,000 at Pond A in 1999-00 and stocked of 50,000 at Pond B in 1999-2000 and 2000-01. Initially, but the yield decreased later due to overstocking (Tables 7.3 to 7.8).

Table 7.3: Details of Seed Stocked and Fish Production in Banjarha Pond During 1991-92 to 2000-01

Year	Total Seed Stocked Standard Fry (Nos.)	Seed/Stocked Standard Fry (Nos./ha/yr)	Major Carps		No. of Fry/Yielding One Kg Fish Production
			Total Fish Production (kg)	Fish Production ha/yr (kg)	
1991-92	20000	6666	3900	1300	5.13
1992-93	15000	5000	4140	1380	3.62
1993-94	25000	8333	4260	1420	5.87
1994-95	21000	7000	4500	1500	4.67
1995-96	30000	10000	3900	1300	7.69
1996-97	20000	6666	5100	1700	3.92
1997-98	50000	16666	4800	1600	10.41
1998-99	60000	20000	6000	2000	10.00
1999-2000	120000	40000	9900	3300	12.12
2000-01	70000	23333	2600	866	26.92
Total	311000	143664	49100	16366	6.33

It is also concluded that the stocking of ponds with fingerlings in place of fry is beneficial and provides higher production. The fish production increased gradually with the increase of stocking density. However, after the saturation points, *i.e.* very high or over-stocking, the fish production decreased. At least 50 per cent of the total food of carp in fishponds must come from the pond itself. Just as land crops can usually be improved by the application of fertilizers, even when the pond is already productive, also can fish. The overall fish productivity of Pond A without using pesticide and tobacco for trapa culture (*Trapa bispinosa*) and without fertilizers and supplementary feeding was considerably good.

Table 7.4: Details of Seed Stocked and Fish Production in Matehna Pond During 1991-92 to 2000-01

Year	Total Seed Stocked Standard Fry (Nos.)	Seed/Stocked Standard Fry (Nos./ha/yr)	Major Carps		No. of Fry/Yielding One Kg Fish Production
			Total Fish Production (kg)	Fish Production ha/yr (kg)	
1991-92	12000	4800	3600	1440	3.33
1992-93	15000	6000	4200	1680	3.33
1993-94	10000	4000	3800	1520	2.63
1994-95	20000	8000	4100	1640	4.88
1995-96	18000	7200	3600	1440	5.00
1996-97	15000	6000	3800	1520	3.95
1997-98	25000	10000	3200	1280	7.81
1998-99	30000	12000	3000	1200	10.00
1999-2000	50000	20000	5800	2320	8.62
2000-01	50000	20000	6000	2400	8.33
Total	245000	98000	37500	16440	5.96

Table 7.5: Details of Stocking, Harvesting and Fish Production in Banjarha Pond During 1999-2000

Details	Trapa	Catla	Rohu	Mrigal	Silver carp	Grass carp	Common carp
Species ratio %	–	20	30	15	15	5	15
Nos. stocked	–	24000	36000	18000	18000	6000	18000
Initial average Size (mm)	–	101.2	102.7	103.7	129.2	78	91.7
Initial average Weight (gm)	–	9.5	8.8	9.3	20.4	3.6	7.5
Final average weight (gm)	–	950	750	600	980	1200	700
Total weight harvested (kg)	2500	2000	1800	1400	1200	2000	1500

Table 7.6: Details of Stocking, Harvesting and Fish Production in Matehna Pond During 1999-2000

Details	Trapa	Catla	Rohu	Mrigal	Silver carp	Grass carp	Common carp
Species ratio %	–	35	20	15	–	10	15
Nos. stocked	–	17500	10000	7500	–	5000	7500
Initial average Size (mm)	–	47.4	48.3	70.1	–	74.2	73.6
Initial average Weight (gm)	–	1.5	1.4	4.9	–	2.8	4.2
Final average weight (gm)	–	800	600	400	–	1500	800
Total weight harvested (kg)	2000	1700	900	600	–	1800	800

Table 7.7: Details of Stocking, Harvesting and Fish Production in Banjarha Pond During 2000-2001

Details	*Trapa*	*Catla*	*Rohu*	*Mrigal*	*Silver carp*	*Grass carp*	*Common carp*
Species ratio %	–	40	30	20	–	–	10
Nos. stocked	–	28000	21000	14000	–	–	7000
Initial average Size (mm)	–	101.2	102.7	101.6	–	–	73.6
Initial average Weight (gm)	–	9.5	8.8	8.6	–	–	4.2
Final average weight (gm)	–	400	300	200	–	–	300
Total weight harvested (kg)	2500	900	700	400	–	–	600

Table 7.8: Details of Stocking, Harvesting and Fish Production in Matehna Pond During 2000-2001

Details	*Trapa*	*Catla*	*Rohu*	*Mrigal*	*Silver carp*	*Grass carp*	*Common carp*
Species ratio %	–	40	20	15	–	10	15
Nos. stocked	–	20000	1000	7500	–	5000	7500
Initial average Size (mm)	–	47.4	4.31	50.3	–	76.3	51.2
Initial average Weight (gm)	–	1.5	1.4	1.7	–	3.1	1.8
Final average weight (gm)	–	850	600	400	–	1500	750
Total weight harvested (kg)	2400	1800	800	700	–	1900	900

Both study ponds are highly productive in terms of both primary production and fish yield potential. In the absence of natural breeding (except minor and local major carp) and stocking of major carps, fish seed can be enhanced only through regular stocking. The results indicate that with stocking of fish seed, fish production irregular increased in present study period. A direct relationship between stocking and fish yield rate in reservoir has been observed (Piska, 2000) Sugunan (1997), Velayudhan and Sidhathan (1998) reported a fish production in reservoir.

Piska (2000) reported that stocking of surface, columnar and bottom dwelling fishes and grass carp were 30–35 per cent, 15–20 per cent , 45 per cent and 5–10 per cent of the total respectively in culture ponds. During the present study, the species composition of fish seed stocked showed variations. The stocking rate was 40 per cent catla, 30 per cent rohu, 20 per cent mrigal and 10 per cent common carp at Pond A and 40 per cent catla, 20 per cent rohu, 15 per cent mrigal, 10 per cent grass carp, and 15 per cent common carp at Pond B, during 2000-01.

The production of major carps from the Pond A ranged from 1300 to 3300 kg/ha/yr during the 1991-92 and 1999-00 and from the Pond B ranged from 1200 to 2400 kg/ha/yr during the period 1998-99 and 2000-01. This yield gradually increased upto 1999-00. Thereafter a fall was observed in the following one year. The overall major carp production in the Satna district in the 1993-94 to 2000-01 were 17465.056 MT with an average of 1902.34 kg/ha/yr. The fish yield 2020 kg/ha/yr during 2000-01,thereafter decline was observed in 1994-95 and 1999-2000.

Maximum amount of fish production was observed in case of catla when compared to other major carp. Catla yield during 1999-2000 to 2000-01, which formed total production 2000, 1700, 900, 1800 kg at both ponds. Continue Banjarha and Matehna pond during 1999-2000 to 2000-01. Catla was followed by Rohu 1800, 900, 700, 800 kg common carp 1500, 800, 600, 900 kg, with 1400, 600,

400, 700 kg Mrigal 2000, 1800, 1900 kg Grass carp and Silver carp 1200 kg. Maximum overall weight of catla production was 2000 kg at Pond A and 1700 kg at Pond B during 1999-2000. Production of over 866 kg/ha (Pond A) and 2400 kg/ha (Pond B) during 2000-01 per annum was observed. The maximum rohu yield observed 3300 kg/ha/yr (Pond A) during 1999-2000 and 2400 kg/ha/yr (Pond B) during 2000-01.

Annual fish production ranging from 3446 to 5370 kg/ha has been reported (Sharma *et al.*, 1997; Sharma and Das, 1988). In freshwater pond less than one ha, through composite culture of Indian exotic carps with supplementary feed and fertilizers, production as high as 3000-5800 kg/ha and 7000-9000 kg/ha/yr were achieved by CIFRI in different parts of the country (Jhingran, 1976). In large ponds (1.48–2.15 ha) at Krishnanagar (W.B.), under composite fish culture with different species combinations of Indian and exotic carps and by providing supplementary feed and fertilizers, CIFRI achieved gross and net production ranging from 2654–4290 and 2514–4143 kg/ha/yr, respectively. With species combination of catla, rohu, mrigal, silver carp and common carp. Accordingly, with judicious management, particularly when hydrological conditions of Pond B are favourable in comparison with Pond A during 2000-01 for productivity of carp fishes. A reason was responsible for disturbance of hydrological condition of Pond A, causes of Epizootic Ulcerative Syndrome (ESU) disease, use of tobacco and pesticide for trapa culture, spray of oil emulsion mixed with synthetic detergent (in fish seed nurseries, which are in Pond A). The fish disease, Epizootic Ulcerative Syndrome (ESU) has been causing considerable damage to the inland fisheries of Asia-Pacific region for last two decades. Efforts to control the disease have met with varying degrees of success by application of lime and salt (Sinha, 1991 and Bhaumik *et al.*, 1993). Potassium permanganate and ash (Das and Das, 1993) have been tried as curative agents.

Thus, it may be concluded that the both ponds are certainly suitable for carps and on the basis of eco-based planning and judicious management is possible. Synthetic detergents have become one of the important contributory substances that create pollution in natural water systems. Toxicological impacts of synthetic detergents have been well documented by many workers (Abel, 1974 and Sparugue, 1971). Use of fertilizers in both ponds that effect some times changed concentration of inorganic concentration confirm by Alikunhi (1957), Banerjee *et al.* (1969), Sharma (1985, 1993) Jayapala, (1986) Sharma and Saini (1990, 1991) Saini and Sharma (1992). A widely used phoskill organo phosphate pesticide in the trapa culture of Vindhya and Satna region. Its chemical composition Monocrotophos a. i. 36 per cent w/w, other related ingredients 64 per cent w/w. is recommended to control insects pests of trapa.

The annual data of water temperature indicate clearly a significant with pH, transparency, Dissolved Oxygen, Biological Oxygen Demand, phosphate, ammonia and plankton at Pond A and pH, transparency, Dissolved Oxygen Biological oxygen demand, total alkalinity, total hardness, magnesium, ammonia and plankton at Pond B (Tables 7.9 and 7.10).

The values of pH have been calculated to be positive correlation with dissolved oxygen, total alkalinity, magnesium and plankton at Pond A and total alkalinity, total hardness, magnesium and plankton at Pond B. However an insignificant relation were noticed with transparency, biological oxygen demand, total hardness, calcium, chloride, phosphate and ammonia (Pond A) and transparency, dissolved oxygen, biological oxygen demand, calcium, chloride, phosphate and ammonia (Pond B). Similarly pH had very strong direct relationship with alkalinity ($r = 0.7493$ in summer) in Pond A and total alkalinity ($r = 0.8716$ in summer) in Pond B. A insignificant correlation was noticed between transparency and total alkalinity, total hardness, calcium, magnesium and ammonia (Pond A) and total hardness, total alkalinity, calcium, magnesium and chloride (Pond B), while it does show notable relation to other physico-chemical parameters at both ponds. Strong correlation has been shown BOD (r = –0.9632 in monsoon) in Pond A and ammonia (r = 0.8609 in monsoon) in Pond B.

Table 7.9: Correlation Co-efficient (r) with among Annual Data of Physico-chemical parameters of Banjarha Pond (Pond A) During February 2000 to March 2001

	Water Temp.	pH	Transp.	DO	BOD	Total Alkali.	Total Hard.	Ca	Mg	Cl	PO_4	NH_3	Plankton
Water Temp.													
pH	-0.5566*												
Transp.	-0.6163*	0.4751											
DO	-0.9004*	0.5918*	0.7302*										
BOD	0.8265*	-0.4446	0.7479*	-0.9431*									
Total Alkali.	-0.5299	0.9263*	0.2359	0.5545*	0.3997								
Total Hard.	0.0316	0.1209	-0.3448	-0.1675	0.2059	-0.0202							
Ca	0.1059	0.5026	-0.3941	-0.1325	0.3163	-0.1498	0.8601*						
Mg	-0.2888	0.6065*	0.1369	0.3153	0.2492	0.4355	0.4391	-0.3062					
Cl	-0.2465	0.4563	0.5571*	0.0344	0.5101	0.6439*	-0.1231	0.2511	0.3128				
PO_4	0.5537*	-0.5145	0.8643*	-0.5893*	0.5577*	-0.5461*	0.3305	-0.1971	-0.4181	-0.4061			
NH_3	0.6711*	-0.3412	-0.3404	-0.7739*	0.5235	-0.3582	-0.2831	-0.1593	-0.1728	-0.1611	0.0153		
Plankton	0.9079*	0.9013*	0.7515*	0.9069*	0.9100*	-0.5804*	0.9341*	0.8449*	0.9485*	0.9100*	0.9083*	0.9181*	

* Significant value at 0.05 level

A significant correlation observed between dissolve oxygen and biological oxygen demand, total alkalinity phosphate, ammonia and plankton (Pond A) and biological oxygen demand, total hardness, magnesium, ammonia and plankton (Pond B). On the other hand, there is strong inverse relationship for mostly negative significance in season at both ponds except positively some parameters. The biological oxygen demand shown a significant correlation between phosphate and plankton at Pond A and total alkalinity, total hardness, magnesium, ammonia and plankton at Pond B while other parameters does show insignificant in both ponds. BOD showed strong correlation with maximum parameters of both ponds during seasonally and annually.

The values of total alkalinity have revealed that present ponds had significant variation to annually and seasonally had shown a positive relation with chloride, phosphate and plankton at Pond A and total hardness and magnesium at Pond B. The values of total hardness recorded at present water bodies varied significantly only with calcium and plankton (Pond A) and magnesium ammonia and plankton (Pond B) respective of annually. It does not show any notable correlation with other parameters. In Pond A no significant relation was observed with magnesium to ammonia but significant with plankton. In Pond B a positive and significant relation between calcium and plankton was calculated.

Table 7.10: Correlation Co-efficient (r) with among Annual Data of Physico-chemical parameters of Matehna Pond (Pond B) During March 2000 to February 2001

	Water Temp.	pH	Transp.	DO	BOD	Total Alkali.	Total Hard.	Ca	Mg	Cl	PO_4	NH_3	Plankton
Water Temp.													
pH	-0.5391*												
Transp.	-0.6196*	0.0042											
DO	0.9695*	0.5166	0.5977*										
BOD	0.9647*	-0.4953	0.6770*	-0.9579*									
Total Alkali.	-0.5928*	0.9291*	0.0573	0.5261	0.5422*								
Total Hard.	0.7474*	0.7605*	-0.2687	0.7104*	0.6517*	-0.7494*							
Ca	0.3809	-0.1449	0.1631	-0.3946	0.4091	-0.2934	0.0806						
Mg	0.7028*	0.8110*	-0.3770	0.6937*	0.7019*	-0.8575*	0.8445*	0.1307					
Cl	0.2385	0.1101	0.2087	0.0774	-0.1263	-0.0437	0.2893	0.2055	-0.0065				
PO_4	0.3415	-0.1749	0.7786*	-0.3402	0.3405	-0.1741	0.2917	-0.4737	0.4685	-0.0822			
NH_3	0.7591*	-0.4476	0.7223*	0.7709*	0.7681*	-0.4979	0.7170*	0.0172	0.8234*	-0.1540	0.6968*		
Plankton	0.9018*	0.8832*	0.6774*	0.8712	0.9080*	-0.0774	0.7611*	0.7722*	0.8740*	0.8808*	0.9029*	0.9019*	

* Significant va1ue at 0.05 level

A significant correlation calculated between magnesium and plankton (Pond A) and ammonia and plankton (Pond B). An insignificant correlation was noticed between magnesium and chloride, phosphate, ammonia (Pond A) and chloride, phosphate (Pond B). Strong correlation has been shown ammonia and phosphate (r = 0.8584, r = 0.8388 in summer) in Pond A and chloride (r = 0.8451 in summer) in Pond B. The significant values of chloride with plankton at both ponds. It does not show any notable correlation with other parameters. The values of phosphate recorded at present water bodies varied significantly with plankton (Pond, A) and ammonia and plankton (pond B). The values of ammonia have been calculated to be correlation positively with plankton in both ponds.

By applying student *t* test between fish production and water quality it was noted that transparency, pH, dissolved oxygen, biological oxygen demand, total hardness, calcium, phosphate and plankton at Pond A and water temperature, transparency, biological oxygen demand, total alkalinity, total hardness, magnesium, phosphate, ammonia and plankton at Pond B were found highly significant. However the production of fish was found high in Pond B. A significant relation with fish production and between water temperature, transparency, pH, biological oxygen demand, Total alkalinity, phosphate, plankton at Pond A and water temperature transparency, dissolved oxygen, biological oxygen demand, phosphate, ammonia and plankton at Pond B (Tables 7.11 and 7.12).

The specific growth rate (SGR) of fishes in both Banjarha (Pond A) and Matehna (Pond B) are apparent that SGR of *C. catla, L. rohita, C. idella, C. carpio,* and *C. mrigala* were higher in Matehna Pond, whereas *C. catla, L. rohita, C. mrigala, C. idella, C. carpio* except *H. molitrx* (was not culture) had lower SGR in Banjarha Pond (Table 7.13).

Table 7.11: Correlation Co-efficient between Water Quality and Fish Productivity in Banjarha Pond (A) during March 2000 to February 2001

Sl.No.	*Water Quality*	*Correlation Co-efficient with Fish Production*	*t value*
1.	Water temperature	–0.7628933335*	1.2907646
2.	Transparency	0.826542522*	4.870237*
3.	pH	0.582320833*	2.3756909*
4.	DO	–0.226552331	3.17861*
5.	BOD	–0.752094964*	4.280301*
6.	Total Alkalinity	0.521509983*	2.027146
7.	Total hardness	–0.114083624	3.6296383*
8.	Calcium	–0.295109397	04.4100523*
9.	Magnesium	0.502408141	2.2289056
10.	Chloride	0.462528492	1.7296756
11.	Phosphate	–0.693937928*	5.3010957*
12.	Ammonia	–0.098294067	1.7734655
13.	Plankton	–0.612674962*	3.5837843*

* Significant value at 0.05 level

Table 7.12: Correlation Co-efficient between Water Quality and Fish Productivity in Matehna Pond (B) during March 2000 to February 2001

Sl.No.	*Water Quality*	*Correlation Co-efficient with Fish Production*	*t value*
1.	Water temperature	–0.720385555*	5.551766
2.	Transparency	0.788384634*	4.2504052*
3.	pH	0.300059015	1.0432556
4.	DO	0.683977066*	0.9375924
5.	BOD	–0.770918395*	2.9120613*
6.	Total Alkalinity	0.446326578	3.3166254*
7.	Total hardness	–0.378458309	5.0127653*
8.	Calcium	0.15898874	0.5340993
9.	Magnesium	–0.29083783	3.1565562*
10.	Chloride	0.272812436	1.0727213
11.	Phosphate	–0.675594292*	3.3909221*
12.	Ammonia	–0.586143659*	2.3994052*
13.	Plankton	–0.545783459*	–10.414639

Table 7.13: Comparative Specific Growth Rate (SGR) of Fishes during 2000-01

Fish Species	*SGR per cent*	
	Banjarha Pond	*Matehna Pond*
C. catla	0.54	0.91
L. rohita	0.50	0.85
C. mrigala	0.45	0.79
H. molitrix	–	–
C. idella	–	0.89
C. carpio	0.45	0.88

Studies on physico-chemical parameters of water plankton and fish production have indicated that Matehna Pond is a highly productive in comparison of Banjarha Pond because Banjarha (Pond A) is more polluted than Matehna (Pond B).

References

Abel, P.D. (1974). Toxicity of synthetic detergents to fish and aquatic invertebrates. *Fish. Biol.*, 6: 279-298.

Alikunhi, K.H. (1959). Fish culture in India. *Farm Bulletin, ICAR*, 20: 144.

APHA (1985-95). *Standard method of water and wastewater*. APHA Washington.

Banerjee, R.K., Singh, G.S. and Ray, P. (1969). Some observation on the use of poultry manure as fertilizer in rearing major carp fry. *Indian J. Fish.*, 16(1-2): 29-34.

Beeton, A.M. (1958). Relationship between Sacchi disc readings and light penetration in lake Huron. *Trans. Amer. Fish, Sac.*, 87: 73-79.

Bhatt, S.D. and Negi, Usha (1985). Physico-chemical features and phytoplankton population in a subtropical pond. *Camp. Physical. Ecol.*, 10(2): 85-88.

Bhaumik, U., Pandit, P.K. and Chatterjee, J.G. (1993). Impact of Epizootic ulcerative syndrome on the fish yield, consumption and trade in West Bengal. *J. Inland Fish Soc. India*, (23): 45-51.

Das, M.K. and Das, R.K. (1993). A review of the fish disease Epizootic ulcerative syndrome in India. *Environ. Ecol.*, 2: 134-145.

Edmondson, W.T. (1959). *Ward and Whipple's freshwater biology, 2nd ed.* John Wiley and Sons Inc., New York, 1248 pp.

Goel, P.K., Trivedy, R.K. and Bhave, S.V. (1985). Studies on limnology of a few freshwater bodies in South Western Maharashtra. *Indian J. Environ. Prot.*, 5: 19-25.

Gonzalves, E.A. and Joshi, D.B. (1946). Freshwater algae near Bombay I. The seasonal succession of the algae in a tank of Bandra. *J. Bombay. Nat. Hist. Soc.*, 46: 154-176.

Jackson, D.F. (1961). Comparative studies on phytoplankton photosynthesis in relation to total alkalinity. *Ver. Limnol.*, 14: 125-133.

Nayak, T.R. (1993). Phytoplankton and certain physico-chemicals condition, in freshwater lake "Matyatal" Panna (M.P.) *India. Proc. Acad. Environ. Biol.*, 2(2): 179-185.

Nayak, T.R. and Khare, Bhawana (1993). Plankton as indicator of eutrophication in shallow water Lakes of Panna (M.P.) India. *Proc. Acad. Environ. Biol.*, 2(1): 69-75.

Pearsall, W.H. (1930). Phytoplankton in English lakes I. The proportion in the water of some dissolved substance of biological importance. *J. Ecol.*, 18: 306-320.

Piska, R.S. (2000). Impact of stocking densities of major carp seed on fish production in a minor reservoir. *Fishing Chimes*, 20(9): 38-41.

Prasad, Braj Nandan, Jaitly, Y.C. and Singh, Yashpal (1985). Periodicity and interrelationships of physico-chemical factors in ponds. *Proc. Sym Pure and Appl. Limnology*, (Ed.) Adoni, A.D. Bull. Bot. Soc. Sagar, 32: 1-11.

Prasad, Sheela and Srivastava, B.B.L. (1988). Relative sterility of diacrisia sik and spodoptera mauruta boisd to Bisalir, *Geo.*, 15: 245-249.

Prescott, G.W. (1962). *Algae of the Western Great lakes area.* W.M.C. Brown Co. Dubuque, Iowa.

Qadri, M.Y. and Yusuf, A.R. (1980). Limnological studies on lake Manipur sar. I the bitope. *Geobios*, 7: 117-119.

Rao, R.S. (1972). Botanical survey work in Arunachal Pradesh in Arunachal research bulletin independence Silver Jubilee volume (Ed. B. Shaatri) Directorate of Information and Pubic Relation, Arunachal Pradesh, Shillong, pp. 49-76.

Saha, G.N., Sahgal, L., Mitra, E. and Nandy, H.C. (1979). Studies on seasonal and diurnal variation in physico-chemical and biological conditions of perennial freshwater pond. *J. Inland Fish. Soc. India*, 3: 79-102.

Saha, L.C. and Pandit, B. (1986). Comparative limnology of Bhagalpur ponds. *Comp. Phyiol. Ecol.*, 1(14). 213-216.

Schnabel, Z.E. (1938). The estimation of the total fish population of a lake. *American Mathematics Monthly*, 45(6): 348-352.

Sharma, A.P. (1991). Evaluation of ecosystem characteristics and trophic status with reference to fishery potential of Kumaon lakes. In: *Ecology of mountain waters* (S.D. Bhatt and R.K. Pandey Eds.), Ashish Publishing House, New Delhi, pp. 149-179.

Sharma, A.P., Singh, U.P., Chauhan, R.S. and Singh, V.K. (1997). Integrated duck-cum-fish culture in tarai region of Uttar Pradesh. Proc. International Conference on Ecological Agriculture Toward Sustainable Development, Chandigarh, pp. 285-292.

Sharma, B.K. (1985). The Indian species of the genus Brachionus (Eurotatoria : Monogonoria : Brachionidae). *Hydrobiologia*, 104: 31-39.

Sharma, B.K. and Das, M.K. (1988). Integrated fish-livestock-crop fanning system. *Proc. First Indian Fisheries Forum*, Mangalore, pp. 27-30.

Sharma, D. and Jain, R. (2000). Physico-chemical analysis of Gopalpura tank of Guna district (M.P.). *Ecol. Env. and Cons.*, 6(4). 441-445.

Sharma, O.P. and Saini, V.P. (1990). Mustard oil cake: A potential source for the production of zooplankton. *J. Hydrobiol.*, 6(1): 1-6.

Sharma, O.P. and Saini, V.P. (1991). Rotifer (micro-zooplankton) production through NPK fertilization. *Indian J. Fish*, 38(4): 242-244.

Sharma, O.P. and Saini, V.P. (1992). Evaluation of pig manure fertilization in relation to zooplankton production and water quality. *J. Ecobiol.*, 4(1): 27-31.

Shrivastava, K.P. (2001). Chemical approach of pollution prevention green chemistry. *Res. J. Chem. Environ.*, 5(2): 77-79.

Singh, S.P. and Gupta, S.K. (1999). Viability of integrated fish culture in Satna with special references to Sakaria tank, Satna. In: *Environmental pollution management* (Ed. V.S. Bais), Creative Publishers, Sagar, pp. 199-201.

Singh, Shivesh Pratap, Singh, R. and Pathak, D. (2002). Hydrobiological studies of two ponds of Satna. *Eco. and Cons.*, 2: 289-292.

Sinha, A.K. (1995). *Benthic ecology of Kawar lake Begusarai*, Bihar. Ph.D. Thesis, Bhagalpur University, Bhagalpur.

Sinha, J.P. (1991). Spread of EUS in fishes of Bihar. *Fishing Chimes*, 11: 60-61.

Sparugue, J.B. (1971). The measurement of pollutant toxicity to fish III. Sub-lethal effects and "safe" concentrations. *Wat. Res.*, 5: 245-266.

Sreenivasan, A. (1965). Limnology of tropical impoundments III. Limnology and productivity of Amaravathi reservoir. *Hydrobiologia.*, 26: 501-516.

Sugunan, V.V. (1997). *Reservoir Fisheries of India*. Daya Publishing House, Delhi.

Topani, J.T. (1980). *Freshwater animals of India: An ecological approach*. Oxford and IBH Publishing Co., New Delhi.

Trivedy, R.K. and Goel, P.K. (1984). *Chemical and biological methods for water pollution studies*. Environmental Publications, Karad, pp. 210.

Velayudhan, T.D. and Sidhathan, T. (1998). Influence of stocking of fish yield in Chulliar reservoir and its impact on rural economy. *Fish. Chimes*, 18(4). 17-20.

Verma, P.K. and Datta Munshi, J. (1987). Plankton community structure of Budua reservoir Bhagalpur (Bihar). *Trop. Eco.*, 28: 200-207.

Welch, P.S. (1935). *Limnology (1st ed)*. Mc., N.Y. and London.

Welch, P.S. (1948). *Limnological methods*. McGraw-Hill Book Co., New York, pp. 381.

Wetzel, R.G. (1975). *Limnology*. W.B. Saunders Co., Philadelphia, pp. 743.

Zafar, A.R. (1961). On the ecology of algae in certain fish ponds of Hyderabad, India I. Physico-chemical complexes. *Hydrobiologia*, 23: 179-195.

Zafar, A.R. (1966). Limnology of Hussain Sagar Lake, Hyderabad, India. *Phykos*, 5(1-2): 115-126.

Zutshi, D.P. (1985). The Himalayan lake ecosystem. In: (Ed. Singh, J. S.) *Environment regeneration in Himalayan: Concepts and strategies*, pp. 325-338.

8

Pollution Potential Studies of Groundwater Around Two Mining Areas in the Western Part of Sundargarh District, Orissa

S.K. DASH
Department of Geology, Government College, Sundargarh, Orissa
H.K. SAHOO
Department of Earth Science, Sambalpur University, Burla, Orissa

ABSTRACT

Mining activity at Sargipali (Pb-Zn Mines), and Tikilipara (Coal Mines) areas in the Tangarpali and Hemgir Blocks respectively in the western part of Sundargarh District, Orissa is an ongoing process which necessitates the study of the effects of such mining practices on the groundwater regime of the above areas. Heavy metal analyses of the groundwater samples from a number of dugwells imply that at present the groundwater of the above areas is not polluted with heavy metals and can be safely used for different purposes, but care must be taken as there is always a possibility of deterioration of the surface and groundwater quality if proper management measures are not taken for the wastewaters from the mines and concentration plants. To know the pollution potentials of the above two areas, an attempt has been made to calculate the Pollution Potential Index (PPI)/DRASTIC Index (DI) using Drastic system of modelling, which indicates that groundwater (shallow depth) of Sargipali area (DI–123) is more susceptible to pollution than the Tikilipara area (DI-104). Therefore the wastewater from the mines and the effluents from the concentration plant at Sargipali mining area needs to be managed properly and monitored regularly so as to save the undergroundwater from contamination by various heavy metals and other toxicants. But nevertheless both the areas need proper attention due to their higher values of Pollution Potential Indices.

Keywords: *Pollution Potential, Hydrogeologic settings, Environmental model, Groundwater contamination, Drastic Index.*

Introduction

During the last decade groundwater quality has emerged as one of the most important environmental issues. The chemical composition of groundwater is related to the environmental conditions in general and geological, pedological and topographical conditions in particular, as well as the influent nature of the surface water. Groundwater comes in contact with various minerals which are soluble in it in varying degrees. The dissolved minerals determine the usefulness of water for different purposes and presence of some ions beyond certain limits make it polluted. Because restoration of groundwater quality is such a formidable and cost prohibitive task, great emphasis is being placed upon protection of the resource.

Various studies on groundwater pollution clearly show that the problem varies spatially and not all the regions are equally vulnerable. Effective management strategies therefore need to be targeted so that limited personnel, funds and technology can be focused upon those areas most threatened so as to provide the greatest benefit for the investment (Merchant *et al.*, 1987). Perhaps, because it is so critical for public agencies to assess and map groundwater pollution hazard, spatial model designed to evaluate groundwater vulnerability to contamination have been more widely implemented in Drastic than any other single type of groundwater related model. Drastic has been used with exceptional frequency, developed by the National Water Well Association (NWWA) in collaboration with the U.S. Environmental Protection Agency (EPA), provides a systematic, standardized and widely applicable method for assessing and mapping groundwater pollution potential (Aller *et al.*, 1987). In this system one can utilize the available hydrogeologic information from various sources.

In the western part of Sundargarh district, only two mining areas (Sargapali and Tikilipara) are operative which need attention from groundwater point of view. With a view to establish the effects of mining and waste disposal on the groundwater regime of the areas under study, the present study was initiated. The areas under study lies towards the western part of the Sundargarh district and falls in the S.I.T.S. Nos. 64N/12 and 64N/16. Geologically the Sargapali area forms a part of the Gangpur Super Group of rocks and the Tikilipara area Lower Gondwana Super Group of rocks and both the areas fall under the Tribal Sub-Plan. The areas have good network of roads and are connected directly to the district headquarters. The areas under study are characterized by extreme climate with very hot (around 47°C) summer and very cold (around 4°C) winter showing large variation in mean daily temperature. About 85 per cent of the annual rainfall is received during the monsoon months. The region is formed extensively of hills and valleys exhibiting an undulating topography. The area is an agricultural dependent one. The surface water does not satisfy the growing needs of agriculture which further accelerates the use of groundwater in the region.

Materials and Methods

A detailed systematic hydro-geological investigation of the area was conducted for evaluating, the geological framework of the area. Data of 14 wells were utilised for analyzing the groundwater conditions. Geo-chemical parameters along with their heavy metal concentrations were determined for all the well waters according to standard methods. In the present study, Drastic system has been used for evaluating the groundwater pollution potential of the areas. The hydro-geological setting forms the basic framework of the Drastic model which is a composite description of all the major geologic and hydrogeologic factors affecting and controlling the groundwater movement into, through and out of an area. All the parameters using a relative ranking scheme (Aller *et al.*, 1987) are given below:

Parameters	*Weight*
Depth to Water (D)	5
Net Recharge (R)	4
Aquifer Media (A)	3
Soil Media (S)	2
Topography (T)	1
Impact of Vadose Zone (I)	5
Hydraulic Conductivity of the Aquifer (C)	3

These factors have been arranged so as to form the acronym Drastic. A numerical ranking system has been used in order to assess the groundwater pollution potential in hydrogeologic setting using the Drastic factors. It contains three significant parts *viz.* Weights, Ranges and Ratings. Each of the Drastic parameters has been assigned a weight (ranging from 1 to 5) to describe relative importance in the pollution processes as compared to other parameters. Similarly for the ranges of each parameter, different ratings (ranging from 1 to 10) have been assigned. Once all the factors have been assigned a rate, each rate is multiplied by the assigned weight and the resultant numbers are summed up to get the Drastic Index. The index value computed by the model is considered a relative indication of pollution potential.

Results and Discussion

The Pb-Zn mine at Sargipalli is a typical under ground mine extracting mainly lead and the important ore is galena. The trace elements associated are Zn, Cu, Ag, As, Fe etc. The wastewater from the concentration plant is diverted to the tailing dam (settling tank) which is finally discharged to Ichha nallah. The concentrations of different metallic ions in the wastewater of tailing dam and the groundwater samples imply that all the contents are within the permissible limits recommended by WHO (1984) & MINAS (1993) which are given in Table 8.1.

Table 8.1: WHO (1984) and MINAS (1993) Drinking Water Standards (For Heavy Metals)

Element	*WHO (1984)*		*MINAS (1993)*	
	HDL	*MPL*	*HDL*	*MPL*
Pb	50	–	–	5
Zn	–	–	–	1000
Cu	50	1500	–	1000
Fe	500	100	–	–
Ag	–	–	–	–
As	50	500	30	150
Ni	–	–	–	100
Se	0	10	10	10

All values expressed in ppb; HDL – Highest Desirable Limit, MPL – Maximum Permissible Limit.

Therefore, at present the groundwater of the surrounding areas of the Pb-Zn mining is safe to be used for drinking purpose. But obviously the ores and host rocks possess strong potentiality to produce metal toxicants directly or indirectly which may pollute the groundwater accordingly. As it is an under ground mine there is always possibility of direct contamination of groundwater from such toxicants. Care must always be taken for the As content of groundwater in this area, because sulphide minerals are generally associated with As and it is also adsorbed onto clay colloids bound to organic matter to form water soluble complexes with Al, Fe, Ca and Mg (Ganje & Rains, 1982). Also over extraction of groundwater during summer should be checked because it enhances the liberation of As into water column (Ramesh *et al.*, 1995). Arsenic becomes toxic when its concentration reaches 50 ppb in water used for drinking purposes (Table 8.1). It accumulates in the biota and other inorganic materials because of its low mobilisation factor of 3.3 and soon becomes poisonous. Grantham and Jones (1978) suggested that As may cause neurological damage when water contaminated with As greater than 100 ppb is consumed.

The present practice of letting out the acid mine-water and the liquid waste of the concentration plant to the Ichha nallah through the settling tank contributes surface water pollution to certain extent. The adverse impact or such pollution on the health of the inhabitants, local faunal and aquatic life, flora and vegetation may apparently be marginal, but it is yet to be properly assessed. In addition to these, the long exposure to atmosphere of old non-sealable stocks of lead concentrates heaped near the concentration plant and also the mine dumps and mine debris used as refilling materials may contribute to indirect pollution of the groundwater from the leaching solutions in future. Of course HZL is actively engaged in monitoring the groundwater quality around the area, and accordingly the effluent standards are being maintained. In contrast to the underground mining for lead ores, the coal mine at Tikilipara (near Gopalpur) is open cast one. Coal is found to be associated with the Barakars and Karaharbaris of Lower Gondwana, consisting of medium to coarse sand stones, shales and fire clay. Production of coal in the area only started recently during 1996. Disturbing coal strata during mining operations accelerates the natural deterioration of water quality by increasing the surface area of reactive minerals exposed to atmosphere, hydrosphere and biosphere (Powell, 1988) which ultimately affects the groundwater of the area. Normally iron sulphides (Pyrite and Marcasite) are found associated with coal deposits. The other trace metals associated with coal deposits are Mercury, Molybdenum, Lead Zinc, Arsenic, Cadmium, Boron, Antimony and Selenium. Keeping the above facts in view water samples from open wells from the surrounding areas were collected and analysed for some of their heavy metal contents (Table 8.2). From the analysis results is observed that all the contents are within the permissible limits recommended by WHO (1984) and MINAS (1993) which are given in Table 8.1.

But it is a fact that the groundwaters of Sardega, Tikilipara and Bankabahal have inferior taste due to relatively more iron content. Therefore, care must be taken to use the water for domestic supply as it may stain cooking utensils, sink basins, cloth washers and clothing. Further such water may promote the formation of iron bacteria. Though the groundwater from shallow aquifers in this area is within the potable range, care must be taken to monitor the wells regularly for the quality of groundwater, as there is every possibility of its contamination in such coal mining areas, because the heavy metals get enriched many times with respect to their average composition in the crust, as described by Taylor (1964) and NAPC (1980) (Table 8.3). From the figures (Table 8.3), it can be expected that in future the groundwater as well as the surface water may be contaminated with As and Se contents, if proper management procedures are not adopted for the mine drainage and the stock piles in the area.

Table 8.2: Heavy Metal Concentrations in the Groundwater Samples of Selected Localities in the Study Area (in ppm)

Sample No.	*Location*	*Pb*	*Zn*	*Cu*	*Fe*	*Ag*	*As*	*Ni*	*Se*
P_1	Gopalpur	*	0.8	*	0.2	*	*	*	*
P_2	Sargega	*	0.75	*	0.4	*	*	*	*
P_3	Tikilipara	*	1.0	*	0.5	*	*	*	*
P_4	Siarmal	0.1	1.2	*	0.3	*	*	*	*
P_5	Kulda	0.2	1.35	*	0.2	*	*	*	*
P_6	Bankabahal	*	1.5	*	0.5	*	*	*	*
P_7	Jupranga	*	0.85	*	0.3	*	*	*	*
P_8	Mahikani	0.01	0.03	0.07	0.2	0.01	*	*	*
P_9	Barbhanga	0.02	0.04	0.08	0.2	*	*	*	*
P_{10}	Bharatpur	*	0.08	0.1	0.3	0.02	*	*	*
P_{11}	Lokdega	0.02	0.02	0.07	0.3	*	*	*	*
P_{12}	Sargipalli	0.01	0.04	0.05	0.2	0.01	*	*	*
P_{13}	Sribhubanpur	*	0.02	0.1	0.1	*	*	*	*
P_{14}	Budelkani	0.02	0.02	0.04	0.1	*	*	*	*

* Not detected
Detection limit-0.001 ppm

Table 8.3: Enrichment Factors of Trace Elements in Coal with respect to the Average Composition of the Earth's Crust

Element	*Enrichment Factor*
Hg	2.25
Mo	2.00
Pb	1.28
As	8.30
Cd	6.50
B	5.00
Sb	5.00
Se	82.00

The generation and quality of wastewater depend on a variety of factors including the type of mine and mining process, amount and type of pyrite in the over burden rock, coal and other ore bodies, time of exposure, characteristics of over burden, techniques used for mining and water usage during extraction of coal and characteristics of coal particularly its hydrogen to carbon ratio and inherent moisture content. Generally mining operations result in generation of four different wastewater streams, such as *(i)* highly acidic, *(ii)* partially oxidised and/or neutralised or less acidic, *(iii)* oxidised and neutralised and/or alkaline and *(iv)* neutralised but not oxidised without any acidity (Pandey, 1997). Therefore proper management procedures are to be adopted so as to save the surface and groundwater resources of the area from being contaminated because once the groundwater source gets polluted, the effect may persist for decades or longer. The reclamation of surface water is easier than that of groundwater, hence prevention of groundwater pollution due to any cause is important.

In the present study, Drastic Index has been evaluated for both the mining areas, *i.e.* Sargipalli and Tiklipara taking into account the ranges and ratings of different parameters outlined by Aller *et al.* (1987) (Tables 8.4 and 8.5) and the Drastic or pollution potential index has been classified using the following expression:

$$D.I = Dr.Dw + Rr.Rw + Ar.Aw + Sr.Sw + Tr.Tw + Ir.Iw + Cr.Cw$$

Where,

r = rating of the parameters and
w = weight of the parameters

Table 8.4: Drastic Chart for the Sargipalli Area (Shallow Water Table Condition)

Parameter	*Range*	*Weight*	*Rating*	*Number*
Depth to water table	6–8	5	7	35
Net Recharge	83 mm	4	3	12
Aquifer media	Granite Gneiss	3	4	12
Soil media	Red and Yellow Sandy Soil	2	6	12
Topography	2–6 per cent	1	9	9
Impact of Vadose zone	Mica-Schist, Boulder	5	5	25
Hydraulic conductivity	3.5–5.0 m/day	3	6	18
				Drastic Index–123

Table 8.5: Drastic Chart for the Tikilipara Area (Shallow Water Table Condition)

Parameter	*Range*	*Weight*	*Rating*	*Number*
Depth to water table	10-14	5	5	25
Net Recharge	102 mm	4	6	24
Aquifer media	Sand Stone	3	6	18
Soil media	Sandy Soil	2	6	12
Topography	2-6 %	1	9	9
Impact of Vadose zone	Shale, Clay etc.	5	2	10
Hydraulic conductivity	0.001-0.1 m/day	6	1	6
				Drastic Index–104

The index value computed by the model is considered a relative indication of pollution potential. Higher scores indicate greater vulnerability. The index must, however be interpreted within a specific hydro-geologic setting, a combination of all the major geologic and hydro-geologic factors which affect and control groundwater movement into, through and out of an area, a mapable unit with common hydro-geologic characteristics and as a consequence common vulnerability to contamination (Aller *et al.*, 1987). Such a modelling has been applied in different areas like Vellar Basin in Tamil Nadu (Jayakumar, 1996) and Rewa in M.P. (Mishra *et al.*, 1996) and accordingly the affected areas have been taken care of from pollution point of view.

From the above discussions now it is clear that though both the mining areas represent medium pollution potential index at present it may be enhanced due to change in the parameters in future.

But nevertheless groundwater of Sargipalli area is more susceptible towards pollution than that of Tikilipara area and therefore proper management of the wastewater from the mine and the concentration plant is to be done so as to check the expansion of pollution plumes up to the shallow water table.

It is also felt that prediction and evaluation of pollution of the subsurface environment and planning remedial actions at existing sites may be useful for sitting and designing now land based waste treatment or disposal facilities. Though it is thought that most models used to make such predictions are based on the assumption that the system behaves ideally, but a variety of factors such as the model and pollution transport parameters and the geometric uncertainty, which introduce uncertainty into the model prediction. Therefore it is suggested that other uncertainty analysis techniques such as Monte-Carlo Analysis Modelling for groundwater pollution may be applied in such areas to know the transport mechanisms and concentrations of the pollutants in a more deterministic way (Bobba *et al.*, 1995).

References

Aller, L., Lehr, J.H., Petty, R., and Bennet, T. (1987). Drastic: A standardised system to evaluate groundwater pollution potential using hydro-geologic setting. *Jour. Geol. Soc. of India,* 29: 23-37.

Bobba, A.G., Singh, V.P., and Bengtsson, L. (1995). Application to uncertainty analysis to groundwater pollution modelling. *Env. Geol.*, 25: 89-95.

Ganje, T.J. and Rains, D.W. (1982). *Methods of soil analysis, Part 2: Chemical and microbiological properties.* Agronomy Monograph, No. 9 (2nd ed.), pp. 385-402.

Grantham, D.A. and Jones, J.F. (1978). Reply to comments on Arsenic contamination of water wells in Nova Scotia. *J. Amer. Water Well. Assoc.*, 38: 707.

Jaykumar, R. (1996). Evaluation of groundwater pollution potential using DRASTIC model: A case study from South India. *Indian J. Environ. Hlth.*, 38(4): 225-232.

Merchant, J.W., Whitemore, D.O., Whistler; J.L., Mc. Elwee, C.D. and Wood, J.J. (1987). *Groundwater pollution hazard assessment: A GIS approach.* Proceeding of the International Geographic Information Systems (IGIS) Symposium; Assoc. Am. Geographers, Washington. D.C., (3): 103-105.

MINAS (Minimum National Standards) (1993). Central Pollution Control Board, Govt. of India.

Mishra, R. and Richaria, L.K. (1996). Pollution potential in limestone aquifers. *Indian J. Environ. Hlth.*, 38(4): 256-260.

NAP (National Academy Press) (1980). *Trace element geochemistry of coal resources development related environmental quality and health,* Washington, D.C.

Pandey, R.A. (1997). Management options for water pollution control in coal industry: An overview. *Jour. IAEM,* 24: 34-44.

Powell, J.D. (1988). Origin and influence of coal mine drainage on streams of United States. *Environ. Geol. Water. Sci.*, 146: 209-218.

Ramesh, R., Shivkumar, K., Eswarwmorthi, S. and Purvaja, G.R. (1995). Migration and contamination of major and trace elements in groundwater of Madras City, India. *Environmental Geol.* (Springer-Verlag), 25: 126-136.

Taylor, S.R. (1964). Abundance of chemical elements in the continental crust: A new table. *Geochim. et. Cosmochim. Acta.*, 28: 1273-1285.

WHO (World Health Organisation) (1984). *Guidelines for drinking water quality Vol. 1: Recommendations*, Geneva, Switzerland, WHO, 130 p.

9

Studies on Selected Major Elements and Nutrients in Rushikulya Estuary (East Coast of India)

TAPAN RANI MAHAPATRO

Department of Marine Sciences, Berhampur University, Berhampur, Orissa

ABSTRACT

Like chlorinity the major elements (Ca, Mg, B and F) in general, exhibited an increasing trend from riverine to estuarine end. Major elements showed higher concentrations observed during summer, whereas lower during monsoon. This indicated the predominance of these elements in seawater. Positive correlations observed between chlorinity and said major elements as well as among the said major elements. Negative correlations observed between the major elements and nutrients. The vertical gradient for chlorinity and major elements also showed increasing trend. The nutrients (PO_4-P, SiO_4-Si, NO_3-N and NH_4-N) showed decreasing trend from riverine to estuarine end. Which indicated their dominance in river water. The nutrients showed negative correlation with chlorinity and major elements. The negative value may be due to the river runoff/terrestrial input to the estuary. The positive relations among nutrients indicated their addition to the estuary from same source. Whereas the negative correlation with chlorinity suggested that the freshwater is the main source of addition of nutrients. The vertical gradient for the nutrients showed decreasing trend.

Keywords: Chlorinity, Positive, Negative, Correlation, Major element, Nutrient.

Introduction

Estuaries are the environments that play an important role in the transfer of products of continental weathering to the ocean. These are unique water systems as they are transitional regions between fresh river water and saline coastal water. These are the regions of high biological productivity and are often subjected to high degree of pollution. These are also important regions of the coastal ecosystem since their enclosed nature often protects them from extreme winds and waves providing a rich source of nourishment. As a complex system the estuary receives huge quantity of dissolved chemical inputs from a number of sources as runoffs through the course of river (Das *et al.*, 1997). Input of the same from the sea itself cannot be ruled out. The distribution of chemical constituents in the estuarine environment is greatly regulated by physical, chemical and biological processes. The estuarine environment plays a great role in controlling the net fluvial transport and fate of the chemical constituents. Hence studies on the behaviour and distribution of dissolved constituents such as major elements and nutrients are highly important for assessing river inputs into the ocean.

Nutrients are the limiting factors for the growth of plants. Nutrient distribution is principally determined by mixing processes between freshwater of high nutrient and sea water of low nutrient contents. Nutrients are functionally involved in the processes of living organisms. Human activities within river catchments particularly clearing, agriculture and sewage production increase the nutrient content of runoff entering the receiving water body. The incorporation of the nutrients into the cells, tissues and extra cellular structures of living organisms and the excretions and regenerations of these elements in solution will impose additional mechanisms of addition, removal and transport on the geophysical and geo-chemical processes which influence the concentrations of the nutrients and other major elements. The primary agencies for the biological removal of inorganic nitrogen, phosphorous and silicon from solution in estuarine water are unicellular algae of the phytoplankton. Most of the phytoplankton has developed in the brackish layer due to nutrient enrichment (Legovic *et al.*, 1994). The site of these biologically activated mechanisms will depend upon the ecology of the organisms responsible and this will set up quite different patterns of distribution of the elements from those produced by a biological processes. Therefore, biological activity will tend to modify the movements towards geophysical and geo-chemical equilibrium. These effects will also be apparent in the distribution of any element, which is incorporated adventitiously into living organisms in sufficiently large quantities relative to its concentrations in estuarine water (Uthup, 1996). In terms of nutrient limitation of phytoplankton growth, it is generally assumed that growth in the later stages of bloom, in estuaries is either nitrogen or phosphorous limited. This comes from the fact that, in most rivers, phytoplankton growth is phosphorous limited and in the sea, growth is nitrogen limited (Legovic *et al.*, 1994). Hence in estuary the phytoplankton growth would be more in presence of both nitrogen and phosphorous. Nutrients regulate biological processes by limiting or enhancing organic production in almost all types of aquatic environments. The study of nutrients in the dissolved forms would help in understanding the potential availability of life supporting elements in any particular region of the aquatic system (Uthup, 1996). River waters, though relatively less uniform in their composition of major constituents, are still important as principal suppliers to the seas through estuaries. The transport of these constituents, however, depends largely on the geological nature of the drainage area. Concentration levels of major constituents, in general, do not exhibit appreciable changes and they bear a constant ratio with chlorinity in open seas. However, rapid changes in the behaviour of major constituents do occur in estuaries due to changes in physico-chemical parameters such as land runoff, precipitation and evaporation as a result of temporal changes. In addition, the highly productive features of estuaries alter the concentration levels of some bio-geochemically important elements such as fluorine, boron, calcium and magnesium (Barnes and Green, 1970).

It forms shallow and well mixed estuary wherein the influence of tide decreases sharply towards upstream. The estuary remains freshwater dominated during monsoon while the seawater dominated during pre-monsoon and summer, when the influx of freshwater from riverine source is very less/ negligible. The estuary is highly productive. A number of reports have been published on major elements and nutrients from time to time in this estuary (Tripathy *et al.*, 1990; Gouda and Panigrahy, 1992-95; Panigrahy *et al.*, 1999; Mahapatro *et al.*, 2001; Mahapatro and Padhy, 2001a and Mahapatro and Padhy, 2001b). However, no effort has been made so far to trace the path source of major elements and nutrients and to understand the impact of land drainage on the water of Rushikulya estuary. The present communication is an attempt to find out the correlation factor among these parameters and chlorinity.

Materials and Methods

The climate of this region is semi-arid type. Basing on the meteorological condition of the area, five different seasons namely monsoon (July-Sept), post-monsoon (Oct- Nov), winter (Dec-Jan), pre-monsoon (Feb-Apr) and summer (May-June) are experienced. The tides of this estuary are mixed and semi-diurnal type. The sampling was undertaken (Dec 1999–Nov 2000) at seven selected stations (Fig. 9.1), which cover the entire estuary, during the seasons mentioned above. The surface water samples were collected using polythene bucket, while bottom water samples were collected 1.5 m below the surface using shallow water sampler. The samples were filtered by 0.45 μm filter paper. Before filling the acid cleaned polythene bottles were rinsed 2/3 times with the respective station water. These filled polythene bottles with leak-proof stoppers were preserved in a refrigerator till analyses, which was completed within a month from the date of collection. Generally, samples preserved in refrigerator gives no change in result (for major elements) even after 1 month of storage. The nutrients were analysed within 24 hours of collection.

Chlorinity determinations were made using argentometric titration procedure (Grasshoff, 1976). Nitrate was determined after its reduction to nitrite followed by photometric method based on the formation of an azodye, which was proposed by Bendschneider and Robinson (1952) and later modified and described in the Manual of Methods in Aquatic Environment Research (1972) was followed for the fabrication and operation of cadmium reductor column. The photometric determination of ortho-phosphate in seawater was based on its formation of phosphomolybdenum complex with acidified molybdate reagent, which on reduction with ascorbic acid results in the formation of a highly coloured molybdenum blue compound. The method adopted in the present investigation was a modified procedure of Murphy and Riley (1962) as given by Grasshoff (1976). Phenol-hypochlorite method (Solorzano, 1969) was adopted for the determination of ammonium in seawater. The determination of dissolved silicon was based on the formation of a yellow silicomolybdic acid complex when an acidified sample was treated with ammonium molybdate solution. This on reduction with ascorbic acid yields intensely coloured molybdenum blue complex. The method of Koroleff (1971) was employed in the present investigation. Calcium and magnesium were estimated by a complexometric titration. In this method (Culkin and Cox, 1966) total concentration of calcium and magnesium was first estimated in an aliquot of sample by titrating with standard EGTA solution at pH 10 using NH_4OH/NH_4Cl buffer and Eriochrome Black-T (EBT) as indicator. At the end point the colour changes from wine red to purple blue. Calcium was then estimated by titrating another (same) aliquot of sample with standard EGTA solution at pH 12 using

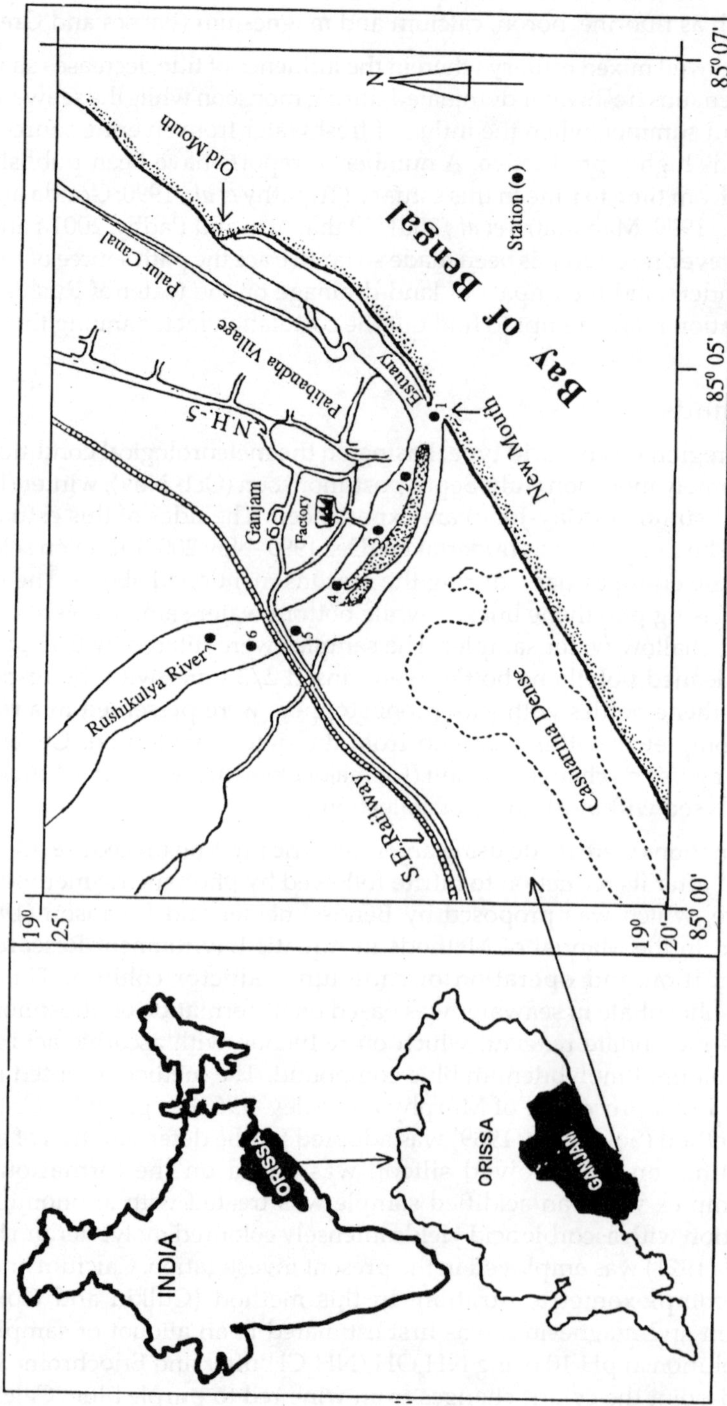

Fig. 9.1: Location Map of Rushikulya Estuary Showing Sampling Stations.

a buffer (sodium tetra borate, NH_4Cl and NaOH) in presence of Zincon as indicator. At the end point the blue colour of the solution changes to orange-red (Ringbom *et al.*, 1958). The interference of strontium and magnesium in the titration of calcium using EGTA is negligible in riverine and estuarine waters. The amount of magnesium present in the aliquot of sample was computed from the difference in the two titres. Fluoride was estimated by alizarine complexone procedure (Greenhalgh and Riley, 1961). Boron was estimated spectrophotometrically by curcumin method (APHA, 1975). The mean, standard error (SE), correlation coefficient and percentage of coefficient of variation (CV) were calculated by MS-Office–2000, Excel Package. Due to very less water at the upstream stations during pre-monsoon and summer, the surface and bottom samples could not be collected.

Results and Discussion

Chlorinity

Chlorinity is considered as the parameter in studying the mixing processes and intrusion of saline waters into estuaries by tidal action. The mixing and diffusion phenomena occurring in estuaries are largely influenced by chlorinity distribution pattern. Wide fluctuations in chlorinity are generally observed in estuaries.

The chlorinity concentrations for surface water ranged from 3.371–8.464, 7.799–12.145, 10.180–14.375, 1.063–6.117 and 2.552–7910 per cent during winter, pre-monsoon, summer, monsoon and post-monsoon respectively. The minimum and maximum surface water chlorinity concentrations were 1.063 and 14.375 per cent (mean 6.674 ± 0.608 per cent) respectively. The percentage of CV for surface water chlorinity was 11.074. The bottom water chlorinity concentrations during winter, pre-monsoon, summer, monsoon and post-monsoon were 3.886–9.438, 8.165–12.444, 10.440–14.486, 1.268–6.803 and 2.895–8.779 per cent respectively. The minimum and maximum bottom water chlorinity concentrations were 1.268 and 14.486 per cent (mean 7.117 ± 0.602 per cent) respectively. The percentage of CV for it was 10.864.

The surface water chlorinity progressively diminished from the estuary mouth at Site 1 and attained the lowest value at Site 7 on the riverine end. The bottom water, chlorinity also exhibited the same trend from estuarine to riverine end but maintained higher concentrations than surface one throughout [Fig. 9.2(a)]. The higher concentrations during all seasons at Site 1 may be due to the dominance of saline water at this point. The lower concentrations during all seasons at Site 7 may be due to the dominance of freshwater over saline water. Similar observations were reported elsewhere (Naik, 1978; Nair *et al.*, 1983; Narvekar *et al.*, 1983; Narvekar and Zinge, 1987; Palanichamy and Balasubramanian, 1989; Das *et al.*, 1997). Minimum chlorinity was observed during monsoon. The minimum values during this season may be due to the efflux of more freshwater and land drainage system. Similar chlorinity distribution in this estuary was observed by earlier workers (Mahapatro and Padhy, 2001 b).

Nitrate (NO_3-N)

It is the final oxidant product of nitrogen compounds in saline water and is considered to be the only thermodynamically stable oxidant level of nitrogen in the presence of oxygen in saline water (Sillen, 1961). In the present study surface water nitrate concentration ranged from 2.237–4.428, 1.158–1.976, 0.752–0.910, 2.719–4.492 and 2.589–3.878 µg at l^{-1} during winter, pre-monsoon, summer, monsoon and post-monsoon respectively. The minimum and maximum concentrations for the surface water were 0.752 and 4.492 µg at l^{-1} (mean 2.647 ± 0.208 µg at l^{-1}) respectively. The calculated percentage of CV for the same was 2.032. The bottom water nitrate concentration ranged from 2.153–

4.073, 1.151–1.784, 0.752–0.897, 2.719–4.079 and 2.598–3.280 μg at 1^{-1} during winter, pre-monsoon, summer, monsoon and post-monsoon respectively. The minimum and maximum concentrations for the bottom water were 0.736 and 4.280 μg at 1^{-1} (mean 2.541 ± 0.199 μg at 1^{-1}) respectively. The calculated percentage of CV for it was 1.186.

In the present study NO_3-N concentrations for both surface and bottom water exhibited a clear increasing trend from estuarine to riverine end (Site 1 to Site 7) during all the seasons. The bottom water NO_3-N concentration was less than the surface one at all the stations [Fig. 9.2(b)]. It can be attributed to the outflow of freshwater containing higher concentrations into the sea than the intrusion of seawater at bottom containing low concentration into the estuaries. Highest peak was observed during monsoon at Site 7, whereas lowest peak during summer at Site 1. The highest peak may be due to the cumulative impact of rainfall and river discharge experienced at this station. The distribution of NO_3-N at all stations maintained the order monsoon > post-monsoon > winter > pre-monsoon > summer, overall all these variations are highly significant. Heavy rain and consequent land drainage cause high values during monsoon. Similar studies on concentrations of NO_3-N associated with monsoon rain have been reported from several areas (Sankaranarayanan and Qasim, 1969; Solarzano and Ehrlich, 1975; Mackey and Leatherland, 1976; De Sousa *et al.*, 1981; Nair *et al.*, 1983, 1984 and Chandran and Ramamoorthi, 1984). Further, very low concentrations of NO_3–N observed during summer. This could be attributed to the rapid utilization of NO_3–N by phytoplankton. Phytoplankton production along the East Coast of India is normally high during pre-monsoon and summer (Ganapati and Rao, 1958) as was observed by me and thus phytoplankton assimilation could be taken as a major factor controlling NO_3-N distribution.

Both surface and bottom water NO_3-N were negatively correlated with chlorinity and major elements (B, F, Ca and Mg), whereas positively correlated with other nutrients (NH_4-N, SiO_4-Si and PO_4-P) [Tables 9.1(a) and (b)]. All the correlations were highly significant. The negative correlation indicated that freshwater input is primarily responsible for the nutrient economy in estuarine water and the positive correlation indicated the common source of occurrence of nutrients. Similar reports were putforth for the coastal water of Orissa (Panigrahy *et al.*, 1999; Padmavathi and Satyanarayana, 1999). The negative correlation with chlorinity is also indicated that low saline water is favourable for biotic population and conforms the views of Dehadrai (1970) and Dehadrai and Bhargava (1972). This states that biological productivity was maximum during pre-monsoon and summer. So, the nitrate loss can be attributed entirely to the biological utilization. It is in agreement with the view of De Sousa (1983).

Phosphate (PO_4-P)

The presence of PO_4-P in an estuary can be taken as an index of potential fertility of the ecosystem as a whole (Redfield, 1934). Phosphorous occurs in natural waters solely as PO_4-P. The PO_4-P rich detritus falls and is decomposed by bacteria. The presence of phosphates in large quantities in freshwater indicates pollution through sewage and industrial waste (Bandela *et al.*, 1999).

The surface water PO_4-P concentrations ranged from 0.579–0.763, 0.328–0.386, 0.257–0.297, 0.705–0.836 and 0.672–0.801 μg at 1^{-1} during winter, pre-monsoon, summer, monsoon and post-monsoon respectively. The minimum and maximum concentrations for the same were 0.257 and 0.836 μg at 1^{-1} (mean 0.610 ± 0.036 μg at 1^{-1}) respectively. The calculated percentage of CV for it was 0.039. The bottom water PO_4-P concentrations ranged from 0.633–0.782, 0.334–0.396, 0.261–0.297, 0.690–0.824 and 0.661–0.782 μg at 1^{-1} during winter, pre-monsoon, summer, monsoon and post-monsoon respectively. The minimum and maximum concentrations of the same were 0.261 and 0.824 μg at l^{-1} (mean 0.589 ± 0.034 μg at l^{-1}) respectively. The calculated percentage of CV for it was 0.035.

It showed decreasing trend from riverine to estuarine end (Site 1 to Site 7). The bottom water PO_4-P concentrations were higher than surface one at all the stations during the period of study [Fig. 9.2(c)], which indicates its depletion at surface due to utilization by phytoplankton and regeneration in the water column at bottom. The PO_4-P concentrations were observed in the order monsoon > post-monsoon > winter > pre-monsoon > summer. Very low concentrations during the pre-monsoon and summer picked up with the advent of the winter. Highest concentration was observed during monsoon at Site 7, whereas lowest concentration was observed during summer at Site 7. Similar report has been reported by Satpathy (1996) in the coastal waters of Kalpakkam. PO_4-P as the seasonal variations have been found to be significant it can be stated that the impact of rainfall and river discharge on PO_4-P changes is prominent. According to Nair *et al.* (1983) rainfall resulting in heavy river discharge leads to transportation of sediments in a big way. Thus an increase in the value of PO_4-P was noticed with the onset of monsoon rains. In addition, anthropogenic inputs of super phosphates applied to the fields as fertilizers, sewage and domestic waste (which contains alkali phosphate used in households as detergent). Similar result was observed by Das *et al.* (1997). The higher concentration at Site 7 during monsoon may be due to freshwater influx. Similar observations were made for the same estuary by several workers (Gouda and Panigrahy, 1995 and Mahapatro *et al.*, 2001)

Both surface and bottom PO_4-P showed highly significant values with the studied elements. With nutrients it showed positive correlation, whereas with major elements and chlorinity it showed negative correlation [Tables 9.1(a) & (b)]. It may be due to desorption of PO_4-P in high chlorinity. Similar reports has been made in Godavari estuary (Padmavati and Satyanarayana, 1999).

Table 9.1(a): Correlation Analysis of Surface Water Chlorinity, Nutrients and Major Elements in Rushikulya Estuary

	Chlorinity	*PO_4-P*	*SiO_4-Si*	*NO_3-N*	*NH_4-N*	*Fluoride*	*Calcium*	*Magnesium*	*Boron*
Chlorinity	1								
PO_4-P	−0.94	1							
SiO_4-Si	−0.98	0.94	1						
NO_3-N	−0.95	0.90	0.97	1					
NH_4-N	−0.95	0.99	0.96	0.94	1				
Fluoride	0.97	−0.97	−0.97	−0.95	−0.98	1			
Calcium	0.97	−0.96	−0.97	−0.93	−0.97	0.97	1		
Magnesium	0.98	−0.97	−0.96	−0.92	−0.98	0.97	0.98	1	
Boron	0.99	−0.97	−0.98	−0.94	−0.97	0.99	0.99	0.99	1

Silicate (SiO_4-Si)

The SiO_4-Si showed significant variations during the study period. During winter, pre-monsoon, summer, monsoon and post-monsoon the surface water concentrations were ranged from 21.352–31.018, 15.761–22.127, 14.589–17.986, 25.129–33.287 and 23.754–30.227 μg at l^{-1} respectively. The minimum and maximum surface water concentrations were 14.589 and 33.287 μg at l^{-1}

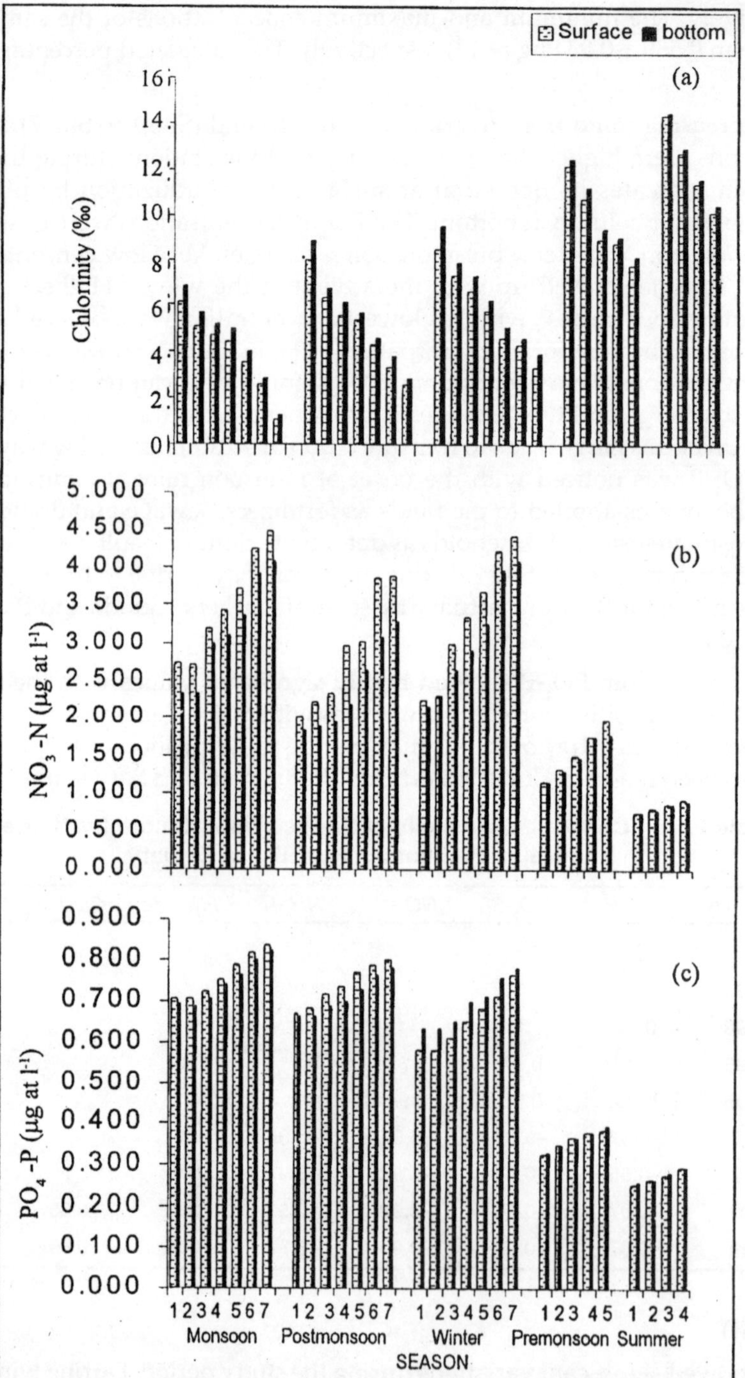

Fig. 9.2(a): Seasonal Variations of Chlorinity in Rushikulya Estuary.

Fig. 9.2(b): Seasonal Variations of NO_3-N in Rushikulya Estuary.

Fig. 9.2(c): Seasonal Variations of PO_4-P in Rushikulya Estuary.

(mean 24.194 ± 0.949 µg at l^{-1}) respectively. The calculated percentage of CV was 0.016. During winter, pre-monsoon, summer, monsoon and post-monsoon the bottom water concentrations were ranged from 19.341–29.895, 15.747–22.004, 14.581–17.985, 23.683–31.643 and 22.789–28.541 µg at l^{-1} respectively. The minimum and maximum concentrations of it were 14.581 and 31.643 µg at l^{-1} (mean 23.052 ± 0.851 µg at l^{-1}) respectively. The percentage of CV for it was 0.746.

Table 9.1(b): Correlation Analysis of Bottom Water Chlorinity, Nutrients and Major Elements in Rushikulya Estuary

	Chlorinity	*PO_4-P*	*SiO_4-Si*	*NO_3-N*	*NH_4-N*	*Fluoride*	*Calcium*	*Magnesium*	*Boron*
Chlorinity	1								
PO_4-P	–0.92	1							
SiO_4-Si	–0.97	0.90	1						
NO_3-N	–0.92	0.90	0.95	1					
NH_4-N	–0.95	0.99	0.93	0.91	1				
Fluoride	0.97	–0.96	–0.93	–0.90	–0.98	1			
Calcium	0.98	–0.94	–0.95	–0.90	–0.96	0.98	1		
Magnesium	0.98	–0.97	–0.94	–0.90	–0.97	0.98	0.98	1	
Boron	0.99	–0.96	–0.95	–0.92	–0.97	0.99	0.99	0.99	1

The minimum peak was observed during summer at Site 1 and maximum during monsoon at Site 7. The lowest concentrations in the surface waters during summer attained highest concentrations during monsoon at all the stations presumably due to the river water that brings in additional quantities of the nutrients to it. Similar observation was reported by Nair *et al.* (1983) in Ashtamudi estuary. During all the seasons increasing trend from estuarine to riverine end was noticed. It may be to the dominance of riverine water over sea water which carries wide spectrum of finely divided siliceous materials. Similar reports have been putforth by several workers (Gouda and Panigrahy, 1992; Padmavati and Satyanarayana, 1999 and Mahapatro and Padhy, 2001a). The minimum concentrations observed during summer may be due to its irvolvement in biological process and adsorption onto suspended sediments. Similar reports were published by Anirudhan and Nambisan (1990) in Cochin estuary. The maximum concentrations observed during monsoon may be due to high influx of freshwater during this season. Many of the siliceous materials have been transported to the sea by rivers or by wind (Riley and Chester, 1971). Its concentration during the study period appeared in the order monsoon > post-monsoon > winter > pre-monsoon > summer. The SiO_4-Si exhibited relatively high concentration at surface when compared with bottom during all the seasons [(Fig. 9.2(d)]. It may be due to the same reason as mentioned for NO_3-N.

Both surface and bottom SiO_4-Si concentrations were showed highly significant correlation [Tables 9.1(a) and (b)]. Positive correlations were observed among all the nutrients. It showed negative correlations with chlorinity and major elements. It may be due to the same reason as stated for NO_3-N.

Ammonium (NH_4-N)

The surface water NH_4-N concentrations during winter, pre-monsoon, summer, monsoon and post-monsoon were varied from 2.418–3.436, 1.006–1.976, 0.801–0.927, 3.093–4.027 and 2.834–3.618

μg at l^{-1} respectively. The minimum and maximum concentrations were 0.801 and 4.027 μg at l^{-1} (mean 2.546 ± 0.188 μg at l^{-1}) respectively. The percentage of CV for it was 1.060. Similarly, the bottom water NH_4-N concentrations during winter, pre-monsoon, summer, monsoon and post-monsoon were varied from 2.193–3.115, 0.993–1.501, 0.797–0.924, 2.730–3.861 and 2.517–3.367 μg at l^{-1} respectively. The minimum and maximum concentrations of it were 0.797 and 3.861 μg at l^{-1} (mean 2.336 ± 0.164 μg at l^{-1}) respectively. The percentage of CV for it was 0.812.

The NH_4-N (for both surface and bottom) exhibited an increasing trend from estuarine to riverine end. The surface water NH_4-N concentration showed higher value when compared with bottom one during all the seasons [Fig. 9.2(e)]. It may be due to its predominant freshwater origin in the study area. The NH_4-N concentrations during the study period observed in the order monsoon > post-monsoon > winter > pre-monsoon > summer. High concentrations were observed during monsoon may be due to the high influx of freshwater during this season as compared with other seasons. The combined effects of precipitation, river runoff and land drainage during monsoon may be another reason for high concentration during this season. Low concentrations during pre-monsoon and summer may be due to its utilization by phytoplankton. Similar observations were made by several workers (Sarala Devi *et al.*, 1983; Verlencar, 1987; Pradhan, 2000 and Acharya, 2001). Like other nutrients it also showed positive significant correlation. Positive correlations were observed, [Tables 9.1 (a) and (b)] among nutrients, whereas negative correlations were observed with chlorinity and major elements.

Fluoride

It has long been considered as an acute pollutant to natural environment because of the ability of plants and marine organisms to accumulate it (Ares, 1978) and because of its detrimental effects upto aquatic biota, which can be affected by concentrations as low as 1.5 mg/l (Wright and Davison, 1975). According to Carpenter (1969) about 80-90 per cent of fluoride in river water is cyclic derived from sea salts transported through rain. He also suggested that the contribution of dissolved fluoride by weathering and leaching of rocks is very small since most of the fluoride mobilized during weathering is carried in suspension still bound in detrital minerals.

The surface water fluoride concentrations during winter, pre-monsoon, summer, monsoon and post-monsoon varied from 0.35–0.63, 0.74–0.89, 0.83–0.98, 0.25–0.54 and 0.29–0.61 mg/l respectively. The minimum and maximum concentration of the same were 0.25 and 0.98 mg/l (mean 0.592 ± 0.037 mg/l) respectively. The calculated percentage of CV for it was 0.41. The bottom water fluoride concentrations during winter, pre-monsoon, summer, monsoon and post-monsoon varied from 0.49–0.69, 0.78–0.91, 0.83–0.99, 0.32–0.59 and 0.33–0.67 mg/l respectively. The minimum and maximum concentrations of the same were 0.32 and 0.99 mg/l (mean 0.640 ± 0.034 mg/l) respectively. The percentage of CV of it was 0.034.

Maximum fluoride concentration was observed during summer at Site 1, whereas minimum during monsoon at Site 7. It showed a clear increasing trend from riverine to estuarine end. The surface water concentration was less than bottom at every station during all the seasons of study [(Fig. 9.2(f)]. Whereas during pre-monsoon and summer the bottom and surface water concentrations showed very less or no difference. It may be due to the partial stratification of bottom water during these seasons. With the onset of monsoon its concentration dropped abruptly and remained low during the entire season. Highest concentrations were noticed in both surface and bottom waters

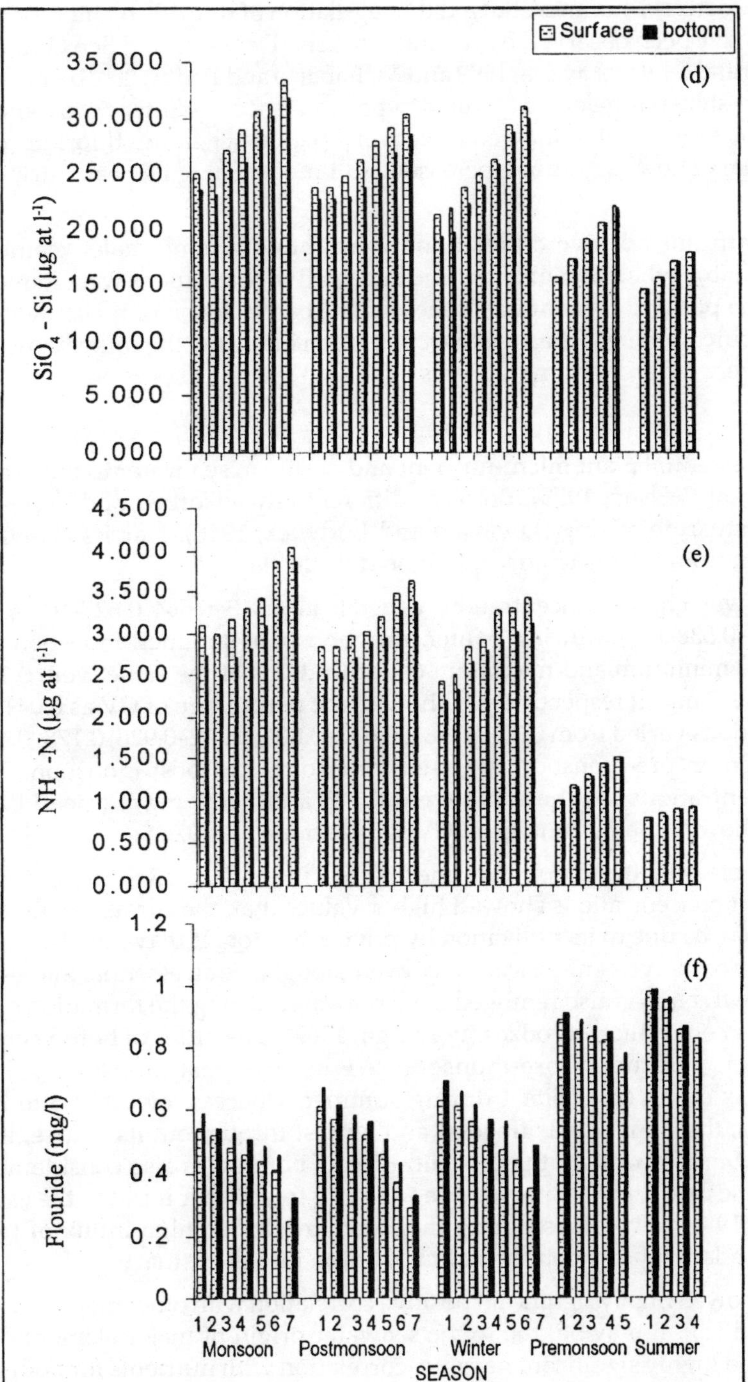

Fig. 9.2(d): Seasonal Variations of SiO_4-Si in Rushikulya Estuary.

Fig. 9.2(e): Seasonal Variations of NH_4-N in Rushikulya Estuary.

Fig. 9.2(f): Seasonal Variations of Fluoride in Rushikulya Estuary.

during pre-monsoon and summer showed the regulation of fluoride by highest salt water intrusion. Similar reports have been observed by several workers (De Sousa and Sen Gupta, 1988; Sarma *et al.*, 1993; Padmavati and Satyanarayana, 1999 and Mahapatro and Padhy, 2001b). Its concentration during the study period showed the order summer > pre-monsoon > winter > post-monsoon > monsoon. The post-monsoon and winter showed a recovery period when the fluoride concentration at any point in the estuary showed gradual increase with time reaching its peak during pre-monsoon and summer.

Highly significant positive correlations were found among major elements and chlorinity [Tables 9.1(a) and (b)], whereas highly significant negative correlations were found between nutrients and fluoride. The positive correlations indicated the contribution of major elements to the estuary from the same source, whereas the negative correlation indicated the source of occurrence of fluoride and chlorinity were different from nutrients.

Boron

Boron is an essential plant micro-nutrient and occurs in sea water to the extent of 4.5 mg/kg at salinity 35 per cent (Wilson, 1975). Boron is differentially adsorbed and deposited on suspended matter during estuarine mixing (Levinson and Ludwick, 1966). Harriss (1969) has attributed the removal of boron from waters to adsorption on detrital clays.

The surface water boron concentrations varied from 0.313–0.564, 0.627–0.796, 0.741–0.922, 0.122–0.429 and 0.219–0.538 mg/l during winter, pre-monsoon, summer, monsoon and post-monsoon respectively. The minimum and maximum concentrations of the same were 0.122 and 0.922 mg/l (mean 0.498 ± 0.037 mg/l) respectively. It showed the percentage of CV as 0.041. The bottom water boron concentrations varied from 0.356–0.581, 0.632–0.804, 0.742–0.928, 0.178–0.452 and 0.279–0.567 mg/l during winter, pre-monsoon, summer, monsoon and post-monsoon. The minimum and maximum concentrations for the same were 0.178 and 0.928 mg/l (mean 0.525 ± 0.034 mg/l) respectively. It showed the percentage of CV for the same was 0.035.

It showed increasing trend from riverine to estuarine end (Site 7 to Site 1) during all the seasons. The bottom water concentrations showed higher values than the surface at all the stations of study [Fig. 9.2(g)]. It may be due to its utilization by phytoplankton. It may not also be the primary factor responsible for lower boron content at surface even though it is an essential micro-nutrient (Narvekar *et al.*, 1981). Some fraction is also removed from seawater during the formation of authigenic silicate and by adsorption onto siliceous oozes (Porrenga, 1967). The order of boron concentrations during study period was summer > pre-monsoon > winter > post-monsoon > monsoon. Highest concentration was observed at Site 1 during summer, whereas lowest at Site 7 during monsoon. During monsoon, the estuary was flooded and almost throughout its course, it was predominant with freshwater. Simultaneously, the concentration of boron was also considerably low during this season and reverse case during summer due to scanty freshwater input to the estuary. According to Rajagopal *et al.* (1981) low values during monsoon are due to high influx of freshwater from the river containing relatively low concentration of boron into the estuary.

The boron showed highly significant positive correlation with other major elements and chlorinity [Tables 9.1(a) and (b)]. It may be due to the seawater origin of major elements like chlorinity. The boron also showed highly significant negative correlation with nutrients for both surface and bottom water [Tables 9.1(a) and (b)]. It may be due to the river water origin of nutrients.

Calcium

In the estuarine waters the calcium concentrations are significant due to land drainage, high rates of biological uptake, precipitation and dissolution processes. The calcium concentration for surface water varied from 0.189–0.249, 0.276–0.322, 0.292–0.343, 0.119–0.211 and 0.146–0.243 g/l during winter, pre-monsoon, summer, monsoon and post-monsoon respectively. The minimum and maximum concentrations for the same were 0.119 and 0.343 g/l (mean 0.226 ± 0.011 g/l) respectively. The coefficient of CV for the same was 0.004. The bottom water calcium concentrations were ranged from 0.195–0.258, 0.278–0.327, 0.292–0.345, 0.137–0.227 and 0.152–0.257 g/l during winter, pre-monsoon, summer, monsoon and post-monsoon respectively. The minimum and maximum concentrations for the same were 0.137 and 0.345 g/l (mean 0.235 ± 0.010 g/l) respectively. The percentage of CV for the same was 0.003.

The calcium concentration showed an increasing trend from (Site 7 to Site 1) riverine, to estuarine end [Fig. 9.2(h)] during all seasons. This variation may be due to gradual reduction in the incursion of seawater into the estuary with increasing distance from mouth (Site 1 to Site 7). Site 1 being mouth of the estuary receives more saline water, which leads more calcium concentration than other stations. Similar reports have been putforth by several workers (Sen Gupta and Naik, 1981; Narvekar *et al.*, 1983; Narvekar and Zingde, 1987; Palanichamy and Balasubramanian, 1989). The bottom water calcium concentrations were more than surface one. Culkin and Cox (1966) opinioned that extraction of calcium by living organisms from surface water leading to lower calcium level in surface water and dissolution of dead animals and precipitated calcium carbonate at bottom region could result increased $CaCO_3$ concentration in bottom water. It was also observed that there was very less to no difference between bottom and surface water concentrations during pre-monsoon and summer. It may be due to the saturation of estuarine water during these two seasons. The calcium concentrations for both surface and bottom water were in the order summer > pre-monsoon > winter > post-monsoon > monsoon. For both surface and bottom water higher calcium concentrations were recorded during summer, whereas lower during monsoon. The higher concentration of calcium during summer may be due to scanty freshwater discharge. Another reason for it may be due to the recovery of calcium for gradual attainment of normal estuarine conditions. This is in agreement with the report of Naik (1978) and Palanichamy and Balasubramanian (1989). The lower calcium concentrations observed during monsoon may be due to high influx of freshwater resulted by the heavy rainfall and land drainage. The more freshwater run-off leads to greater dilution, which accounted lowering of calcium concentration than other seasons. Similar observation was also made for Gopalpur creek by Sahu *et al.* (2002).

The calcium showed highly significant correlations with other major elements, chlorinity and nutrients [Tables 9.1(a) and (b)]. Positive correlations were observed among major elements and chlorinity, whereas negative correlations were observed with nutrients. The reason is same as the reasons mentioned for boron.

Magnesium

In estuarine water magnesium contribute significantly to the biological and geo-chemical processes. High concentrations of this element in the estuary, relative to its bio-geochemical reactions tend to make it conservative. The surface water magnesium concentrations for winter, pre-monsoon, summer, monsoon and post-monsoon varied from 0.427–0.602, 0.706–0.871, 0.801–1.021, 0.261–0.464 and 0.328–0.589 g/l respectively. The minimum and maximum concentrations for the same were 0. 261 and 1.021 g/l (mean.0.553 ± 0.037 g/l) respectively. The percentage of CV for the same was

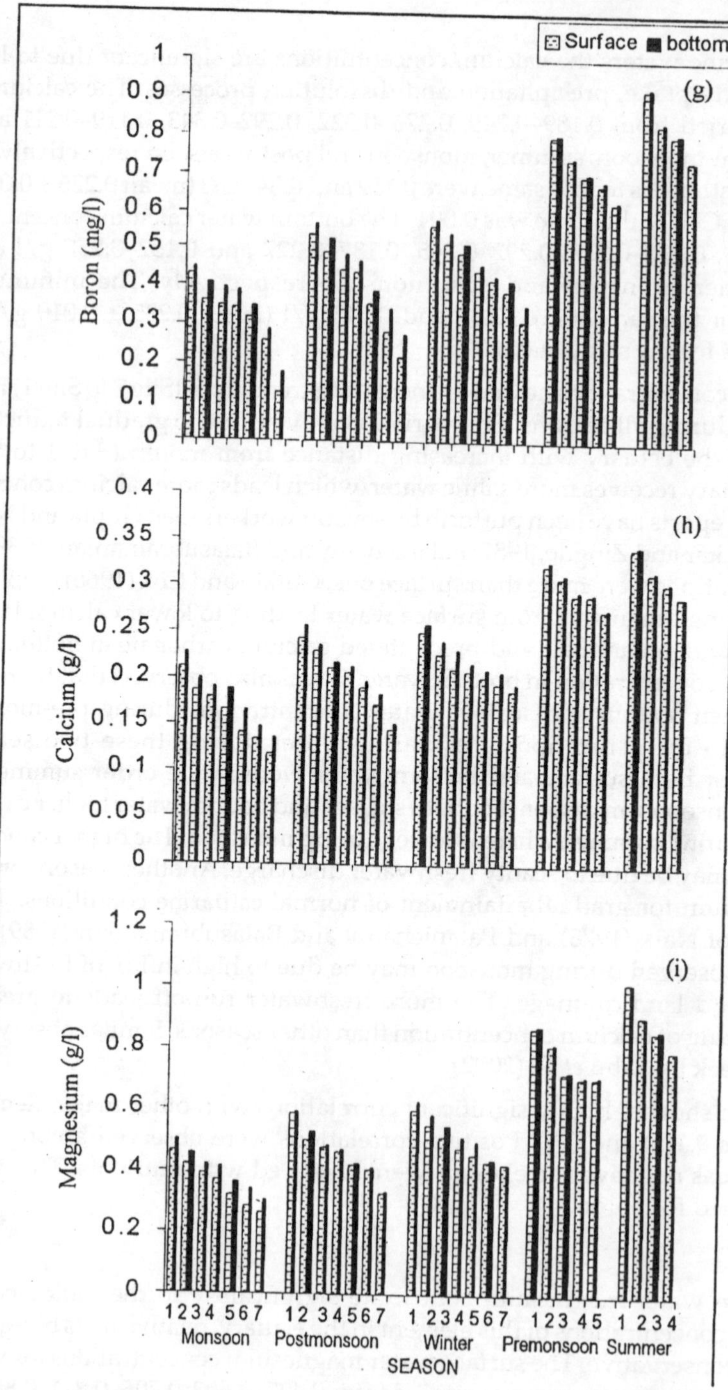

Fig. 9.2(g): Seasonal Variations of Boron in Rushikulya Estuary.
Fig. 9.2(h): Seasonal Variations of Calcium in Rushikulya Estuary.
Fig. 9.2(i): Seasonal Variations of Magnesium in Rushikulya Estuary.

0.042. The bottom water magnesium concentrations for winter, pre-monsoon, summer, monsoon and post-monsoon varied from 0.431–0.631, 0.711–0.871, 0.803–1.025, 0.309–0.498 and 0.334–0.596 g/l respectively. The minimum and maximum concentrations for the same were 0.309 and 1.025 g/l (mean 0.573 ± 0.035 g/l) respectively. The percentage of CV for the same was 0.038.

The distribution pattern of magnesium at seven different stations during five different seasons is given in Fig. 9.2(i). The magnesium concentrations were found to be high at Site 1 (mouth), whereas low at Site 7 (riverine end). This variation may be due to gradual increase in the incursion of freshwater into the estuary with increasing distance from mouth (Site 1 to Site 7). As Site 1 is located at the mouth of the estuary, which is influenced more by neritic waters leading more saline condition, a possible attribute for higher concentration of magnesium at this station. Site 7, which is very much influenced by the freshwater of poor magnesium concentrations, maintains low magnesium level. Other stations are in between, overall there observed an increasing concentration of magnesium from Site 7 to Site 1. Similar reports have been made by several workers elsewhere (Sen Gupta and Naik, 1981; Narvekar *et al.*, 1983; Narvekar and Zingde, 1987; Palanichamy and Balasubramanian, 1989).

The bottom water magnesium concentrations showed higher value as compared to surface one. Bottom to surface water concentration difference was not significant and the minor difference noticed in the present study may be due to some removal mechanisms from surface water. Another reason may be due to the stratification of the estuary by the neritic water at the bottom layer and freshwater dominance at surface layer. Culkin and Cox (1966) are of the same opinion for higher magnesium concentrations in the bottom water.

The magnesium concentrations showed the order of summer > pre-monsoon > winter > post-monsoon > monsoon for both surface and bottom water during the study period. Summer period recorded higher magnesium concentrations, whereas lower during monsoon. The estuary seemed to be influenced mostly by seawater, due to scanty freshwater discharge during most of the periods other than monsoon. This could have resulted in higher concentration of magnesium during summer and pre-monsoon. Much more freshwater had been discharged into the estuary during monsoon period. This could account for the lowering of magnesium concentration during monsoon. The concentration of discussed element was also found to show wider variations during monsoon, whereas more or less stable in nature with higher values observed during summer. During monsoon, the greater dilution would lead to a wide variation, whereas the summer could be recovery period for magnesium along with chlorinity. Palanichamy and Balasubramanian (1989) are of the same opinion for high and low values of magnesium during summer and monsoon respectively.

The magnesium showed highly significant correlations with other major elements, chlorinity and nutrients [Tables 9.1(a) and (b)]. Positive correlations were observed among major elements and chlorinity, whereas negative correlations were observed with nutrients. The reason is same as the reasons mentioned for boron.

Acknowledgement

The author (TRM) is thankful to UGC for the award of JRF and also the Head of the Department for the laboratory facility.

References

Acharya, S.K. (2001). *Studies on water characteristics of Rushikulya estuary (Bay of Bengal)*. M.Phil. Thesis, Berhampur University, Berhampur, Orissa, India.

Anirudhan, T.S. and Nambisan, P.N.K. (1990). Distribution of salinity and silicon in Cochin estuary. *Indian J. Mar. Sci.*, 19: 137-139.

APHA (1975). *Standard methods for the examination of water and wastewater*, 14th edn., Washington, D.C.

Ares, J.O. (1978). Fluoride cycling near a coastal emission source. *J. Air Poll. Cont. Asso.*, 28(4): 344-349.

Bandela, N.N., Vaidya, D.P., Lomte, V.S. and Shivanikar, S.V. (1999). The distribution pattern of phosphate and nitrogen forms and their inter-relationships in Barul Dam water. *Poll. Res.*, 18(4): 411-414.

Barnes, R.S.K. and Green, J. (1970). The estuarine environment. In: *Chemical processes in estuaries*, John Philips (Ed.), Applied Science Pub. Ltd., London.

Bendschneider, K. and Robinson, R.J. (1952). A new spectrophotometric method for the determination of nitrite in seawater. *J. Mar. Res.*, 11: 87-99.

Carpenter, R. (1969). Factors controlling the marine geochemistry of fluoride. *Geochimca et Cosmochima Acta.*, 33: 1153-1167.

Chandran, R. and Ramamorthi, K. (1984). Hydro-biological studies in the gradient zone of the Vellar estuary. I. Nutrients. *Mahasagar-Bull. Natn. Inst. Oceanogr.*, 17: 133-140.

Culkin, F. and Cox, R.A. (1966). Sodium, potassium, magnesium, calcium and strontium in seawater. *Deep-sea Res.*, 13: 789-804.

Das, J., Das, S.N. and Sahoo, R.K. (1997). Semidiurnal variation of some physico-chemical parameters in the Mahanadi estuary, East Coast of India. *Ind. J. Mar. Sci.*, 26: 323-326.

De Sousa, S.N. (1983). Studies on the behaviour of nutrients in the Mandovi estuary during pre-monsoon. *Estuar. Coast. Shelf Sci.*, 16: 299-308.

De Sousa, S.N. and Sen Gupta R. (1988). Distribution and behaviour of fluoride in Mandovi and Zuari estuaries, central west coast of India. *India J. Mar. Sci.*, 17: 195-201.

De Sousa, S.N., Sen Gupta R., Sujata Sanzgiri and Rajagopal, M.D. (1981). Studies on nutrients of Mandovi and Zuari River Systems. *India J. Mar. Sci.*, 10: 314- 321.

Dehardrai, P.V. (1970). Observations on certain environmental features at the Dona Paula Point in Marmagao Bay, Goa. *Proc. Ind. Acad. Sci.*, 52(B): 56.

Dehardrai, P.V. and Bhargava, R.M.S. (1972). Seasonal organic production in relation to environmental features in Mandovi and Zuari estuaries, Goa. *Ind. J. Mar. Sci.*, 1: 52- 56.

Ganapati, P.N. and Subba Rao, D.V. (1958). Qualitative study of plankton off Lawson's, Bay, Waltair. *Proc. of Ind. Asso. of Sci.*, 48(B): 189-209.

Gouda, R. and Panigrahy, R.C. (1992). Seasonal distribution and behaviour of silicate in the Rushikulya estuary, east-coast of India. *Ind. J. Mar. Sci.*, 21: 111-115.

Gouda, R. and Panigrahy, R.C. (1995). Seasonal distribution and behaviour of nitrate and phosphate in Rushikulya estuary east coast of India. *Ind. J. Mar. Sci.*, 24: 233-235.

Grasshoff, K. (1976). *Methods of sea water analysis*. Verlag chemie, Weinhein, 260 pp.

Greenhalgh, R. and Riley, J.P. (1961). Determination of fluoride in natural waters with particular reference to seawater. *Anal. Chim. Acta.*, 25: 179-186.

Harriss, R.C. (1969). Boron regulation in the oceans. *Nature*, 223: 290-291.

Koroleff, F. (1971). Determination of reactive silicate. *ICES paper*. Cm (c) 43 pp.

Legovic, T., Zutic, V., Grzetic, Z., Cauwet, G., Precali, R. and Vilicic, D. (1994). Eutrophication in the Krka estuary. *Marine Chemistry*, 46: 203-215.

Levinson, A.A. and Ludwick, J.C. (1966). Speculation on the incorporation of boron into orgillaceous sediments. *Geochim. Cosmochim. Acta*, 30: 855-861.

Mackey, D.W. and Leatherland, T.M. (1976). Chemical processes in an estuary receiving major inputs of Industrial and domestic wastes. In: *Estuarine chemistry*, J.D. Burton and P.S. Liss (Eds.), Academic Press, London, pp. 185-218.

Mahapatro, T.R. and Padhy, S.N. (2001a). Seasonal fluctuation of physico-chemical parameters of Rushikulya estuary, Bay of Bengal. *Ind. J. Env. and Ecoplan.*, 5(1): 35-40.

Mahapatro, T.R. and Padhy, S.N. (2001b). Behaviour of fluoride in Rushikulya Estuary, Bay of Bengal. *J. of Env. Res.*, 11(2): 141-144.

Mahapatro, T.R., Padhy, S.N. and Patra, S.P. (2001). Assessment of water quality in Rushikulya Estuary. *J. of Indu. Poll. Cont.*, 17(1): 107-112.

Manual of methods in aquatic environment research (1972). FAO Tech. Paper No. 137, FAO, Rome, 145 pp.

Murphy, J. and Riley, J.P. (1962). A modified single solution method for determination of phosphate in natural waters. *Anal. Chim. Acts*, 27: 31-36.

Naik, S. (1978). Calcium and magnesium concentration in the near shore waters of Goa. *Mahasagar-Bull. Natn. Inst. Oceano.*, 11(3&4): 185-189.

Nair, N.B., Abdul Azis, P.K., Dharmaraj, K., Arunchalam, M., Krishna Kumar, K. and Balasubramanian, N.K. (1983). Ecology of Indian estuaries, Part I: Physico-chemical features of water and sediment nutrients of Ashtamudi estuary. *Ind. J. Mar. Sci.*, 12: 143-150.

Nair, N.B., Abdul Azis, P.K., Krishna Kumar, K., Dharmaraj, K. and Arunachalam, M. (1984). Ecology of Indian estuaries Part VI: Physico-chemical conditions in Kadinamkulam backwater, southwest coast of India. *Ind. J. Mar. Sci.*, 13: 69-74.

Narvekar, P.V. and Zingde, M.D. (1987). Behaviour of boron, calcium and magnesium in Puma and Auranga estuaries (Gujarat), west coast of India. *Ind. J. Mar. Sci.*, 16(1): 46- 50.

Narvekar, P.V., Zingde, M.D. and Kamat Dalal, V.N. (1981). Behaviour of boron, calcium and magnesium in Mindola river estuary (Gujarat). *Indian J. Mar. Sci.*, 10: 90- 92.

Narvekar, P.V., Zingde, M.D. and Kamat Dalal, V.N. (1983). Behaviour of boron, calcium and magnesium in a polluted estuary. *Estuar. Coast. Shelf Sci.*, 16: 9-16.

Padmavathi, D. and Satyanarayana, D. (1999). Distribution of nutrients and major elements in riverine, estuarine and adjoining coastal waters of Godavari, Bay of Bengal. *Ind. J. Mar. Sci.*, 28: 345-354.

Palanichamy, S. and Balasubramanian, T. (1989). Distribution of calcium and magnesium in the Vellar estuary. *Mahasagar*, 22(1): 1-11.

Panigrahy, P.K., Das, J., Das, S.N. and Sahoo, R.K. (1999). Evaluation of the influence of various physico-chemical parameters on coastal water quality around Orissa by factor analysis. *Ind. J. Mar. Sci.*, 28: 360-364.

Porrenga, D.H. (1967). Influence of grinding and heating of layer silicates on boron adsorption. *Geochim Coschim Acta.*, 31: 309-312.

Pradhan, S.K. (2000). *Studies on behaviour of nutrients in the Bahuda Estuary (Bay of Bengal)*. M.Phil. Thesis, Berhampur University, Berhampur, Orissa, India.

Rajagopal, M.D., Rajendran, A. and Reddy, C.V.G. (1981). Distribution of dissolved boron in the waters of the Zuari estuary (Goa). *Indian J. Mar. Sci.*, 10: 20-23.

Redfield, A.C. (1934). In: *James Johnstone Memorial volume*, Edited by R.J. Daniel. University Press, Liver Pool, p. 176.

Riley, J.P. and Chester, R. (1971). In: *Introduction to marine chemistry*. Academic Press, London and New York.

Ringbom, A., Pensar, G. and Wanninen, E. (1958). A complexometric titration method for determining calcium in the presence of magnesium. *Anal. Chim. Acta*, 19: 525-531.

Sahu, S.K., Mahapatro, T.R. and Padhy, S.N. (2002). Seasonal variations in calcium and magnesium in the Gopalpur creek, east coast of India. In: *Ecology of polluted waters*, Arvind Kumar (Ed.), APH Publishing Corporation, New Delhi, I: 123-130.

Sankaranarayanan, V.N. and Qasim, S.Z. (1969). Nutrients of the Cochin backwaters in relation to environmental characteristics. *Mar. Biol.*, 2: 236-247.

Sarla Devi, K., Venugopal, P., Remani, K.N., Dominic, Z. and Unnithan, R.V. (1983). Nutrients in some estuaries of Kerala. *Mahasagar-Bull. Natn. Inst. Oceanogr.*, 16: 161-173.

Sarma, V.V., Sudhakar, U. and Varaprasad, S.J.D. (1993). Behaviour of fluoride and dissolved silicon in Gauthami Godavari estuarine environment. *Mahasagar-Bull. Natn. Inst. Oceanogr.*, 26: 105-113.

Satpathy, K.K. (1996). Seasonal distribution of nutrients in the coastal waters of Kalpakkam, east coast of India. *Ind. J. Mar. Sci.*, 25: 221-224.

Sen Gupta, R. and Naik, S. (1981). Studies on calcium, magnesium and sulphate in the Mandovi and Zuari River Systems (Goa). *Ind. J. Mar. Sci.*, 10(1): 24- 34.

Sillen, L.G. (1961). *The physical chemistry of sea water*. Lectures at the International Oceanographic Congress in N.Y., 1959, pp. 549-581.

Solarzano, L. and Ehrlich, B. (1975). Chemical investigations of Loch Etive, Scotland, I. Inorganic nutrients and pigments. *J. Expt. Mar. Bio. and Eco.*, 29: 45-64.

Solorzano, L. (1969). Determination of ammonia in natural waters by phenol-hypochlorite method. *Limnol. Oceanogr.*, 14: 799-801.

Tripathy, S.K., Panigrahy, R.C., Gouda, R. and Panda, D. (1990). Distribution of calcium and magnesium in the Rushikulya estuary (Orissa), east coast of India. *Ind. J. Mar. Sc.*, 19: 212-214.

Uthup, A.T. (1996). *Distribution of nutrients in the sediments of Kayamkulam Estuary*. M.Phil. Thesis, Cochin University of Science and Technology, Cochin, India.

Verlencar, Y.N. (1987). Distribution of nutrients in the coastal and estuarine waters of Goa. *Mahasagar-Bull. Natn. Inst. Oceanogr.*, 20: 205-215.

Wilson, T.R.S. (1975). Salinity and the major elements of sea water. In: *Chemical oceanography*, Vol. I, J.P. Riley and G. Skirrow (Eds.), Academic Press, London, pp. 365-413.

Wright, D.A. and Davison, A.W. (1975). The accumulation of fluoride by marine and intertidal animals. *Env. Poll.*, 8: 1-13.

10

Pesticide and Fish: A Workhouse for the Detection, Evaluation and Abatement of Water Pollution

BIPLAB SARKAR & S. ADHIKARI
Central Institute of Freshwater Aquaculture, Kausalyaganga, Bhubaneswar, Orissa
PARTHA BANDYOPADHYAY & BIDHAN C. PATRA
Aquaculture Research Unit, Department of Zoology, Vidyasagar University, Midnapore, West Bengal
S. AYYAPPAN
Deputy Director General (Fisheries), Indian Council of Agricultural Research Krishi Anusandhan Bhavan, Pusa, New Delhi

ABSTRACT

The aquatic bodies to the earth have been treated as a natural place to discard unwanted waste. In India with occurrence of blue revolution the demand of pesticides were increasing day by day. Now-a-days carbamate and pyrethoid insecticides are getting more preference over traditional organophosphates because of their special characterisation even cypermethrin using to control *Argulas*. Behavioural or morphological changes through pesticide may make the fish more conspicuous in the environment and more susceptible to predation or parasitization, thereby reducing the ability of the population to survive and reproduce. It has been well said that in order to eliminate a species, a pollutant need not be lethal. Though it is the ultimate aim of pollution study, no such antidote or mechanism for controlling pesticide toxicity of fish is developed. But it needs much more studies to come out successfully with some antidotes against these pesticides.

Keywords: *Pesticide toxicity, fish, antidote*

Introduction

Before First World War, toxicity testing was in very infantile stage and sporadic. This testing (One fish-in-a-jar) was done to confirm the causative agents of any fish killings. First major break through came when Power (1917) published his research paper "The gold fish as test animal, in the study of toxicity". Being interested by this, Ellis (1937), Hart *et al.* (1945) started their works and later peoples accepted several species of fishes also, which sometimes proved more sensitive than gold fish. Subsequently an epidemic case of the "gold fish-allergy-syndrome" developed and for therapeutic reasons one began to test everything which will be fitted best into a jar. (Mount, 1977; Macek, 1980).

Thus, since 1950 the acute toxicity testing has become the 'work house' for the detection, evaluation and abatement of water pollution (Buikema *et al.*, 1982). The 96-hour exposure is taken as most common time interval because the effects of test material occur rapidly and are well defined in these time period (Reish and Oshida, 1987). For the assessment of hazard of chemicals in view of legislative measures sequential and tiered testing programme were developed (EPA, 1978; EC, 1979; OECD, 1979) starting with short-term testing and ending with long term test. Later several sophisticated semi-chronic and chronic tests were designed taking into account sub-lethal criteria like growth and reproduction, histology, haematology, toxico kinetics and studying physiological and biochemical effects. (Macek, 1980). Assessment became more complicated by the introduction of lots of hazardous chemicals and their mixtures. Gradually ecological significance were established and the weakness of proposed testing and evaluation methods were realised (Cairns, 1981).

At the last of 20th Century, toxicitogical analysis of different hazardous chemicals and their degraded forms, has been done by new techniques and their residual concentration has been quantified–which gives a new dimension to toxicological research (Murty, 1988).

Acute Toxicity

Acute toxicity denotes relatively short-term lethal or other effect, usually defined as occurring within 4-days for fish and macro-invertebrates and shorter times (2 days) for organisms with shorter life spans. (APHA, 1997). It has two fold application in environmental risk analysis *i.e.* toxicological screening and acute toxicity determination. Acute toxicity is better expressed in the terms of LC_{50} or lethal concentration 50. It is the concentration of a substance that is lethal to 50 per cent of the test organisms in 96 hours. The 96-hour bioassay is advantageous because of minimal laboratory requirements and rapid results (FAO, 1987). A major demerit is that it does not give any information about the site of toxic action. Several factors like environmental parameters, seasonal variation, species multiplicity, age and more over physiological conditions of the organism take great role for the differences of LC_{50} valuations (Singh and Narain, 1982). Especially, in short term experiments, the relatively hydrophobic chemicals, steady state is not achieved and the toxicity may be severely underestimated (Van Hougen, 1985).

Though many reports have been published describing detailed methods and procedures of use in acute toxicity test with aquatic biota (Jones, 1940; Katz, 1961), recommended procedures are provided by the U.S. Environmental Protection Agency (1978), The American Public Health Association (1985), The Organization of European Community Development (OECD, 1981).

A large number of papers had been published on the acute toxicity of pesticides to fishes (Mukhopadhay *et al.*, 1980, Wijngaarden *et al.*, 1993; Fernandez, *et al.*, 1996; Domitrovic 1997; Abdelghani *et al.*, 1997, Farrell *et al.*, 1998; Pathiratne *et al.*, 1998; Dalvi *et al.*, 1998; Rebach, 1999)

Test Animal

According to United States Environmental Protection Agency (USEPA, 1979) the selection of test organisms is to be based on four criteria:

1. The organism must be a representative of an ecologically important group.
2. The organism must occupy a position within a food chain leading to man.
3. The organism must be widely available, amenable to laboratory tests, easily maintained and genetically uniform.
4. There must be adequate background data on the organism (*i.e.* on physiology, genetics, taxonomy etc.)

Rosenberger *et al.* (1978) added other criteria which included economic importance, type of test, sensitivity to the toxicant and consistency in response to the toxicant.

Many researchers has studied several kinds of fish species for some of their merits, Lloyd (1963) worked on the toxicity of *Salmo gairdneri*; Vittozzi and Angelies (1991) on *Lepomis macrochirus*; Gomez *et al.* (1998) on tench (*Tinca tinca* L.); Pathiratne and George (1998) on Nile tilapia (*Oreochromis niloticus*); Dalvi and Davis (1998) on Channel catfish; Soengas *et al.* (1997) on rainbow trout, Tsuda *et al.* (1997) on killifish (*Oryzias latipes*) etc.

Newsome (1980) examined the suitability of several species of fish for use in multi-generation toxicity test and rejected American flag fish (*Jordanella floridae*), gold fish, Zebra danio (*Brachydania rario*) but considered convict cichlid (*Cyclosoma nigro fasciatum*) to be the most suitable.

In India, test species selected by Bhattacharya *et al.* (1988) was *Channa punctatus*, by Mukhopadhay *et al.* (1984) was *Clarias batrachus*, by Dutta (1980) was *Calisa faciata*, by Devi (1984) *Puntius ticto*, by Fernandez, *et al.* (1996) were *Etroplus maculatus*, by Thakur and Bais (1999) on *Heteropneustes fossilis* (Bl.), by Santhakumar and Balaji (2000) were Anabas testudiueus (bl.), etc.

For determining toxicity of a wide varieties of toxic substances, species such as G. giuris, *Clarias batachus*, and *Sacrobranchus fossilis* are not true sensitive fish species. They were included in most cases due to local availability and further easy maintenance in the laboratory (Das, 1996).

Jhingran (1982) has made a literature survey of Indian scientists and published a variable acute toxicity chart of different fish species and analysed their prospective results. Fish size, sex, age, environmental condition, and more over volume of test solution may alter the result.

Chronic Toxicity and Safe Concentration

The ultimate aim of chronic toxicity testing is to determine the significant adverse effects on aquatic ecosystem. The various types of chronic tests are 30 days tests, multi-generation and whole life cycle tests, partial life cycle and embryo-larval tests etc. The life stages of fishes most sensitive to toxicant are generally early life stages (Mckim, 1985; Van leeuwen *et al.*, 1985). Mckim (1985) compared

56 fish life cycle tests involving 34 chemicals and 4 species of fish and showed that the embryo-larval tests offer the same information with comparable dependability as the long-term chronic tests or from full life cycle toxicity test. Fishes survival and activity will be longer when concentration of test chemical decreased. Ten-fold reduction of lethal concentration gives a ten-fold increase in survival time (Jones, 1964).

Safe concentration is the maximum concentration of a toxicant that has no observable harmful effects after long-term exposure over one or more generations. Sprague (1971) emphasized that the term "Safe level" is not an entirely satisfactory one, but in the absence of a better term, it has gained wide acceptance. Lots of scientists and many organizations have given formula and suggestions for the estimation of safe concentration. Hart *et al.* (1948) proposed the formula which is already established and popular. European Fisheries Advisory Commission (EFAC, 1979) suggested that a concentration less than four orders of magnitude lower than the 96h LC_{50} is an acceptable or safe concentration. Another approach to arrive at safe level is to use the one-tenth of the 48h LC_{50} value.

Mount and Stephan (1969) proposed to use Laboratory Fish Production Index (LFPI), which incorporates data generated over at least one generation of the test species, regarding effects on survival, growth, reproduction etc. to calculate the Maximum Acceptable Toxicant Concentration (MATC). It may be recalled that MATC is a range of concentration, the lower end of which does not produce any adverse effect and upper end is the lowest concentration that would cause adverse effect on some life stages. Later Mount and Stephan proposed that the AF (application factor) could be calculated by dividing the MATC by 48 or 96h LC_{50} value. The use of AF for calculating safe levels for other species was explained and discussed in detail by Eaton (1973).

Johnson and Julin (1980) considered that MATC is acceptable more than other methods because it is less time consuming, less expensive to generate and are widely accepted.

Pesticides

Carbamates

Carbamates are moderately toxic to fish, but highly toxic to invertebrates, and introduced into the late 60's for its modified characters over organophosphates. Among them, Carbofuran and Carbaryl are commonly used (Johnson *et al.*, 1980) Singh *et al.* (1990) worked on the pesticides (carbamate) induced changes in circulating thyroid hormones in the freshwater catfish, *Clarias batrachus*. Belisle *et al.* (1988) did specific analysis of carbamate pesticides in sediments using column extraction and gas chromatography. Bhattacharya *et al.* (1992) and Bocquene *et al.* (1993) studied on the effect of carbamate on acetyl cholinesterase system of fish. Hota *et al.* (1993) analysed the metabolic effects of kilex carbaryl on a freshwater teleost, *Channa punctatus* (bloch.).

Few researchers has also concentrated their studies on the impact of carbofuran to several species of fishes. But no or very few of them has emphasised on Indian Major Carps. Verma *et al.* (1982) worked on the determination of LC_{50} value of carbofuran, on fish species like *Mystus vittatus*, *Sacrobranchus fossilis* etc. Gill (1980) concluded from experiments that carbofuran was metabolished by the microsomal-NADPH system in fish, *Trichogaster pectoralis*, oxidative N-hydroxylation was the major route of metabolism of carbofuran. Ram (1988) studied on the carbofuran induced histo-physiological changes in thyroid of teleost fish, *Channa punctatus* (bloch.). Hohreiter *et al.* (1991) worked on the effects of insecticides carbofuran on adenylate parameters (ATP concentration, adenylate energy change etc.) in bluegill sunfish (*Lepomis macrochirus*).

Carbofuran alone or in combination with other carbamate insecticides influences the level of reproductive and metabolic hormones such as thyroxine and corticosterone, and results in impairment of endocrine, immune and behavioural functions (Yoon *et al.*, 2001)

Synthetic Pyrethroid

Among synthetic pyrethroids, important pesticides are Cypermethrin, Deltamethrin, Fenvalerate, Esfenvalerate etc. Although Pyrethroids are extremely toxic under laboratory conditions, in the field their toxicity is significantly reduced, presumably by adsorption of the toxicant to suspended solids (NRCC, 1986). Though previously, very little information was available for pyrethroid research, in last decade of twentieth century, it got a little bit momentum. Symonik *et al.* (1989) studied the effect of fenvalerate on metabolic ion dynamics in fathead minnow (*Pimephales promelas*) and blue gill (*Lepomis macrochirus*). In India Malla-Reddy *et al.* (1992) studied on fenvalerate toxicity to the system of Acctylcholinesterase (ACHE) in *Cyprinus carpio* Ohtsuka (1993) worked on synthesis and insecticidal activity of silicon-containing pyrethroid with low fish toxicity, Barry *et al.* (1995) analysed the effect of age and weight specific respiration rate on toxicity of esfenvalerate to the Australian Crimson spotted rainbow fish (*Melanotacnia fluviatilis*). Bhatnagar (1995) described pyrethroid induced toxicity to Phosphatase in *Clarias batrachus* (Linn.)

Cypermethrin has been developed in England by Elliot *et al.*, 1975. There is very little information available regarding the toxicity of cypermethrin to fish. Only some early life stage tests and detection has been performed. The acute toxicity of cypermethrin to Brown trout, Rainbow trout, Tilapia was studied by Stephenson (1982). Singh *et al.* (1993) worked on the effect of cypermethrin on lactate and succinate dehydrogenase and cytochrome oxidase of fish and Reddy *et al.* (1995) studied on sublethal concentration of cypermethrin to the protein metabolism level in selected tissues of *Cyprinus carpio*. Davies (1993) described on sublethal effects on brown trout, Salmo trouts, caused by cypermethrin spraying on Tasmanian stream and Cagauan (1992) worked on rice-cum-fish culture and the possible use of cypermethrin on it. Very little or almost no works has been done on the cypermethrin toxicity to Indian Major Carps and its different sublethal effects. So, by considering the uprising importance of cypermethrin in aquaculture, it is a new and challenging avenue of research. Ernst *et al.* (2001) worked on the dispersion and toxicity to non-target aquatic organisms of cypermethrin used to treat sea lice on salmon in net pen enclosures and analysed that cypermethrin was having the potential to cause toxic effects over areas of hectares from a single release.

Environmental Parameters

Environmental parameters have a great role in effecting the toxicity of different pesticides. The major parameters are temperature, dissolved oxygen, pH, hardness, alkalinity etc.

Among all the environmental parameters, the effect of temperature on the toxicity of pesticides has been studied a lot and it is very important because it directly influence factors like enzyme activity, metabolic rate, oxygen uptake etc. Generally, toxicity increases with the high temperature. Macek (1969) studied the toxicity of 10 pesticides to the rainbow trout and 11 pesticides to the bluegills at different temperatures and found that the toxicity increased with increasing temperature. Same kind of results has been reported by Singh *et al.* (1982) on endosulfan to *Heteropneustes fossilis*, by Schoettger (1970) on endosulfan to rainbow trout, by Duangsawasdi *et al.* (1979) on fenitrothion and acephate to rainbow trout. But a completely opposite results has been shown by DDT and some pyrethroids where toxicity is less at higher temperature. Ogilvie and Anderson (1965) showed that DDT exposed to Atlantic salmon differs from control. Mauck *et al.* (1976) described that natural pyrethrum and some synthetic pyrethroids (dimethrin, RU-11679, resmethrin) was higher at 12°C than 17°C and dimethrin were more toxic on 17°C than 22°C. Cairns (1975) noted that though vast literature available on effects of temperature on aquatic organisms, yet when one tries to apply this information to specific interactions, often very little evidence is available.

pH has an excellent role in influencing various physio-chemical properties of pesticides like hydrolysis, volatilization and in balancing the dissociated and undissociated forms (Weber, 1972). With few exceptions, the toxicity of organo-chlorine (OC) and organo-phosphate (OP) compounds are not influenced by pH. The toxic effects of 2, 4-D were reduced when the pH was raised by the addition of sodium chloride (Holcombe *et al.*, 1980). Malathion is toxic below pH 7, and loses its toxicity in alkaline pH (Bender, 1969). In the case of carbamate compounds, generally toxicity was reported to increase with pH. Mexacarbate was 38 times more toxic to bluegills at pH 9.5 than at 7.5 (Mauck *et al.*, 1977). The toxicity of picloram to culthroat trout and lake trout increased with a rise in pH from 6.5 to 8.5 (Marking *et al.*, 1981). Generally, the activity of synthetic Pyrethroids was not altered by a change in the pH, the biological activity of natural pyrethrums was influenced by pH, its toxicity being higher at lower pH (Mauck *et al.*, 1976)

Hardness of water does not seem to influence the toxicity of OC and OP compounds (Pickering *et al.*, 1962). The toxicity of three formulations of 2, 4-D, three formulations of endothall, fenoprop, PCP and dichlrobenil was not affected by hardness. (Inglis *et al.*, 1972). Lloyd and Jordan (1964) found that survival time of Rainbow trout in rapidly lethal pH value shortened with decrease of calcium carbonate content of water. Hardness of water had no effect on toxicity of maxacarbate (Mauck *et al.*, 1976) and carbaryl (Johnson *et al.*, 1980). While water hardness had no effect on the toxicity of synthetic Pyrethroids, natural pyrethrums was more toxic in hard water (Mauck *et al.*, 1976).

Not much work seems to have been carried out on the influence of low oxygen on the toxicity of pesticides to fish. Davies (1975) attempted to formulate the criteria for minimum dissolved oxygen requirement of fish. His approaches was on examining the threshold levels of dissolved oxygen that cause changes in some physiological parameters. Oxygen deficiency leads to histological lesions. Channel catfish exposed for 72h to an oxygen content of 1.5 ppm showed anomalies in gill, liver, kidney and spleen (Scott and Rogers, 1980).

Behavioural Study

Behaviour (or activities) of an organism represents the final integrated result of a diversity of biochemical and physiological processes and behavioural patterns are known to be highly sensitive to changes in the steady state of an organism (Warner *et al.*, 1966). Lots of scientist expressed the patterns of behavioural changes which varies according to fish species, concentration of chemical, physico-chemical conditions and on rate of excretion of fish species. Poisons with a resorptive action, the symptoms are manifested primarily through their action on the nervous system, cause a rapid loss of equilibrium in fish (Metlev *et al.*, 1971). Symptoms of Furadon and malathion toxicity were reported by Konar and Ghosh (1982). Within 24 hours of exposure to 0.5 ppm furadon, fish had shown slow opercular movement, lethargy and occasional jumping. After 48 hours, fish showed loss of balance and bending of body while malathion shown loss of balance, lethargy and eratic swimming behaviour. Heath *et al.* (1993) observed that high concentration of carbofuran caused an impairment of swimming performances in Japanese medaka (*Oryzias latipes*). Sagilo *et al.* (1996) worked on the effects of concentration (1, 10, 100 μg/l) and duration (4, 8, 12 h.) of exposure to carbofuran were assessed on the swimming activity, social interactions and behavioural responses of gold fish to a flow of water and observed a rate and concentration dependent increase in behavioural pattern and after 12 hours, it was less apparent.

Shin *et al.* (2001) worked on the effect of diazion on the behaviour of Japanese medaka (*Oryzias latipes*) and gene expression of tyrosine hydroxylase (TH) as a biomarker and concluded that a suppression of TH activity would be one of the causes for the abnormal behaviour of the medaka that could be quantitatively monitored using an image processing system. This study provides molecular and neuro-behavioural bases of a biomonitoring system for toxic chemicals using a model organism such as fish.

Bretaud *et al.* (2002) described about a correlation between biochemical and behavioural effects of carbofuran in gold fish, (*Carassius auratus*). They expressed that behavioural endpoints related to swimming pattern or social interactions had a direct relation to dose and duration of carbofuran exposed biochemical parameters like brain catecholamines, plasma glucose, hepatic glycogen content etc. and relative sensitivities of these different types of responses has been explained in the light of data on neurotoxic effects of carbamate insecticide in fish.

Histopathology

Now-a-days, Histopathological studies of different tissues exposed to toxicants has got enormous attention. Tissues changes due to sublethal doses of toxicants not only giving an idea about the nature of toxicant but also predicting upon its mode of action or mechanism of toxicity, about the probable physiological impairment and assist greatly in the diagnosis of nature of toxicant (Nagler, 1984). Histology of fish and aquatic invertebrates is still a less studied discipline compared to counter parts in mammals (Mayers and Hendricks, 1985).

Liver

The liver is an organ concerned with several vital functions such as absorption of digested food stuff, detoxification, excretion of harmful substances, synthesis of several components of blood plasma, storage of glycogen, release of glucose and control over general metabolism. Hence its study in fish serves as a suitable index for toxicity of polluted water (Sahai and Singh, 1989).

Because of its physiological importance, already some literatures are available on liver. Couch (1975) reviewed the histopathological effects of pesticides and related chemicals on the livers of fish and concluded that many of them were non-specific. Bhattacharya *et al.* (1975) observed liver cord necrosis and cytoplasmic and nuclear disintegration of the hepatopancreas in the teleost *Clarias batrachus* due to endrin. Amminikutty and Rege (1977) reported that chronic exposure to thiodon and agallol caused rapid degeneration and vacuolation of hepatocytes in liver fish. Dubale and Shah (1979) observed coagulation of blood in sinusoids, necrosis and cytoplasmic disintegration in the hepatopancreas of *Channa punctatus* exposed to malathion. Altamirano-Reynoso (1984) analysed that lindane caused hydropic degeneration and the tubular epithelium necrosis were the most frequent microscopic injuries observed in liver of *Tilapia hornorum*. Sahai and Singh (1984) described vacuolation of hepatocytes of *Rasbora dandricus* while exposed to malathion toxicity.

Urdaneta (1989) reported hepatic necrosis due to effect of 12 pesticides in some species of cultivable fishes. Narayan and Singh (1991) observed extensive degeneration of cytoplasm, pyknosis of nuclei and loss of glycogen of *Heteropneustes fossilis* while subjected to acute thiodon. John *et al.* (1993) described the impact of endosulfan on the liver of *Cyprinus Carpio* and expressed that extensive vacuolation of the cells, indistinct cell boundaries, loss of polygonal shape of the cells and degenerative necrosis were the histopathological changes observed in the liver.

Our study on the impact of Carbofuran and Cypermethrin on the hepatocytes of I.M.C. (*Labeo rohita*) is a important and new avenue of research.

Ovary

Some literatures are available on the ovarian damage due to pesticide toxicity but very few or no researches has done on the histopathological changes of ovarian cycle on Indian Major Carps. In India, most of the ovarian research has been concentrated on air-breathing fishes or cat-fishes and in OC and OP pesticides. Kulshrestha *et al.* (1984) worked on the exposure of endosulfan and carbaryl on the ovaries of *Channa straitus* (bloch.) and observed reduction in the number of oocytes, increased number of damaged oocytes, development of inter-follicular spaces, reduction in gonado-somatic index etc. Shukla *et al.* (1984) have noted decreased ovarian activity and atretic oocytes in *Saratherodon mossambica* exposed to Malathion and Ghosh *et al.* (1985) has described on damage in ovary, *viz.*, degeneration of follicular, wall, ooplasm and connective tissues due to malathion toxicity on *Heteropneustes fossilis* and Ansari *et al.* (1987) worked on malathion toxicity on the ovary of zebra fish (*Brachydanio rerio*). Singh *et al.* (1991, 1992, 1993) studied extensively on the impact of gamma-BHC to the steroid hormonal profiles of *Heteropneustes fossilis* and correlated it to ovarian damage, and shown that pesticide suppressed the levels of these hormones, *viz.* testosterone (T), estradiol-17 beta, 17 alpha-OHPg etc. and described that Ovine lutenizing hormone releasing hormone and mystus gonadotropin has modulatory role on it. Pandey (1988) worked on the impact of endosulfan (thiodon) EC 35 on dynamics of oocyte development in the teleostean fish, *Colisa (Trichogaster) fasciatus* and described that ovarian activity was retarded greatly, the diameter of oogonia and stage I oocytes was greatly reduced and ovarian wall become thicken. Hazarika (1998) worked on the toxicological impact of different sublethal doses of BHC on the ovary of the air-breathing catfish *Heteropneustes fossilis* (bloch.) and investigated detailed histopathological changes.

Very few works has been done on the pesticidal effect on ovarian dynamics. Nagler (1984) investigated sublethal effect of penta-chlorophenol (PCP) on ovarian development in rainbow trout (*Salmo gairdneri*). Kling (1986) worked vividly on the toxic effects of sublethal doses of Lebaycid upon the ovaries in *Sarotherodon leucostictus* and observed that exposure caused total atresia, inhibition of vitellogenesis in the gonads and decreased gonado-somatic index. Peczely (1998) described effects of a fungicide (VITAVAX 200 EF) on the reproductive-endocrine functions in mallards and observed ovarian changes and changing pattern of steroid levels.

Reproduction

Reproductive toxicity indicates changes on the pattern of breeding response, on fecundity, on fertilization rate, hatchability of larvae and above all survivability of larvae, fry etc. Though, it is a prime subject of research because it directly relates to productivity of fishes. Very less researches has been conducted in this field. Very less or no research has been done on the effect of carbamate and synthetic pyrethroid on fishes specially to Indian Major Carps.

On Fecundity, Fertilization etc.

At concentration above the threshold toxicity, many organic compounds caused large pregnant female of *Gambussia affinis* to abort (Boyd, 1963). Following the feeding of DDT to brook trout at different doses for 22 weeks, the fishes fed with lower doses showed significantly higher number of ova than those of higher dose (Macek, 1968). In experimental ponds, 2, 4-D butylester at the highest concentration caused a delay in the spawning of blue-gills (Cope *et al.*, 1970). Oviposition of *Oryzias latipes* was suppressed heavily by the 96-hour exposure of chlordecone (Curtis, 1978). Fenitrothion, at sublethal concentration affected egg production, whereas temephos at sublethal concentration affected normal birth (Yasuno *et al.*, 1980). Fecundity and ferility of female sheepshead minnows exposed to different doses of chlordecone shown marked reduction (Goodman *et.al.*, 1982). Several doses of mirex also effected adversely to the spawning and fecundity of fathead minnows. Bhattacharya (1989) worked on histo-pathological changes of ovary due to pesticides and correlated that ovarian damages also interfering with the breeding performances of fishes, Petersen *et al.* (1997) analysed that xenobiotics like pesticide may influence the viability of eggs and larvae. Oern *et al.* (1998) did extensive research of the impact on reproduction of an orally administered mixture of selected PCBs in zebra fish (*Danio rerio*). He treated the fishes with different level of PCB concentration for 13 weeks period and observed that egg production was reduced in all three groups exposed and intermediate and higher groups contained a reduced number of mature oocytes.

On Hatching

Holden (1972) reported that brown trout eggs containing aldrin failed to hatched. Hatching of blue gill eggs exposed to a different doses of a formulation of fenoprop (Kuron) was not affected, but all the fry hatching from eggs exposed to the highest concentration died (Wilbur *et al.*, 1973). On the contrary, hatchability of coho salmon eggs, exposed to PCBs, was much reduced (Halter, 1974). Eggs of DDT exposed adults showed abnormal gastrulation, and 39 per cent of them had vertebral deformities upon hatching. The per cent of deformed larvae was dose-dependent. (Smith *et al.*, 1974). Embryos of Atlantic silverside was treated singly with P. P-DDT, malathion or carbaryl and was observed that they reduced the survival time of the embryoes. These effects were observed even at concentration that may occur temporarily in the environment (Weis *et al.*, 1976). Embryos, sacfry, and larvae of Caranx exposed to lindane had advanced hatching and larval characters. Yolk utilization and sac absorption were also impaired. (Venugopalan *et al.*, 1979). Hatching success of fish eggs exposed to low concentration of mirex increased and hatching of fathead minnows eggs exposed to different concentration of chlordecone was significantly lower than that of the controls. (Buckler *et al.*, 1981). No differences in hatching frequency or median hatching time were recorded in zebra fish (*Danio rerio*) which were orally exposed to a mixture of 20 PCBs in three different dose levels (Oern *et al.*, 1998).

Haematological Parameters

Though physiological effects of toxicant or to diagnose the disease are well established in human and veterinary medicine, yet in fish these clinico-pathological tests are yet to be standardised. In this background haematological studies present variation in blood parameters, as reliable indices of fish health.

These parameters are influenced and modified by many factors such as handling (Houston *et al.*, 1971), metabolism (Booke, 1964), age (Tugarina and Ryzhova, 1970), maturity (Qureshi *et al.*, 1971), sex (Steveke and Atherton, 1965), diet (Smith, 1968), stress (Bouch and Ball, 1966) and pollutants (Spraque, 1971). Some of researches has already been conducted on haematological parameters by several types of pesticides on different fish species by the scientists of India and abroad. Witt *et al.* (1966) found a good correlation between DDT levels in blood and those in adipose tissue. Moss *et al.* (1966) shown that entering the circulatory system, pesticides rapidly bind to the blood proteins. A sublethal concentration of malathion induced decreased erythrocyte and increased leukocyte count and increased serum-free amino acid in a freshwater catfish (Chliomovitch *et al.*, 1977). *Channa punctatus* exposed to sublethal level of quinalphos for 15 days had lowered haemoglobin content (Mebride *et al.*, 1981). *Clarias batrachus* were treated with dichlorvos and shown decreased values of blood parameter like Haemoglobin, red blood cell counts, haematocrit, MCV and MCH values (Rajeswari *et al.*, 1989). Gill *et al.* (1991) worked extensively on the haematological impact created by different pesticides (endosulfan, phosphamidon, aldicarb etc.) to several fish species (*Puntius conchonius, Barbus conchonius* etc.). Their selected parameters were mainly differential count, or biochemical analysis of blood etc. When *Heteropneustes fossilis* were treated with carbomate pesticide, propoxur at different sublethal dose, a decreased result of TLC, TEC, Haemoglobin and increased values of ESR, clotting time was observed. Same sorts of trend was also seen in *Oreochromis mossambicus*, Ekalux (Sampath *et al.*, 1993), in *Channa punctatus* by malathion, in *Cyprinus carpio* by endosulfan (Chandrsekar *et al.*, 1993), in *Heteropneustes fossilis* by lindane (Nath *et al.*, 1996) in *Heteropneustes fossilis* by Thiodan (Ramamjam *et al.*, 1997), in *Cyprinion watsoni* by malathion (Khattak *et al.*, 1996). A opposite data were also presented in *Channa punctatus* with Monocil (Mishra *et al.*, 1994).

Blood glucose and hepatic glycogen can be utilized as a parameter of stress response (Chavin, 1973). Relatively innocuous stimuli such as aquarium transfer (Chavin and Young, 1970), thermal shock (Wede and Meyer, 1973) and putting them in a small shaking cages (Skeda *et al.*, 1975) elicited hyper glycemic response in fish.

Holmberg *et al.* (1972) worked on the impact of Penta-chlorophenol (PCP) to eel *Anguilla anguilla* and analysed that hyperglycemia and increased blood glucose was evident, which persisted for 55 days after transfer to clean water. Blood glucose and hepatic glycogen profile in *Channa punctatus* exposed to non-lethal concentration of single and mixture pollutants respond diversely to two types of treatments (Bhattacharya *et al.*, 1987). According to her, blood glucose profile can be utilized as indicator of pollution stress in fish. Generally blood glucose value were elevated after pesticide exposure to fish. These sorts of result has been observed in *Auguilla anguilla* by the application of endosulfan (Gimeno *et al.*, 1995), in *Cyprinus carpio* by deltamethrin (Balint *et al.*, 1995), and (Chandrasekar *et al.*, 1993), in *Orechronis mossambicus* by quinolphos (Ramakrishnan *et al.*, 1993) etc. But a very exceptional and opposite case was also recorded in *Channa punctatus* by Sastry *et al.* (1982). Gimeno *et al.* (1994) analysed the correlation profile of glucose content between blood and liver of *Anguilla anguilla* and with endosulfan intoxicants commented that liver glycogen content decreased during exposure period and at the same time glucose content increased in blood.

Though lots of works has already been done on haematological profiles in fish in relation to pesticide, no work has been done on the impact of carbofuran and cypermethrin to Indian Major Carps, to especially to *Labeo rohita*.

Antidote Study

To counter the pesticidal effects of water bodies and to increase the survivability level of fishes and other living organisms in water, is the ultimate aim of pollution research and burning work area for contemporary scientists. Almost no references are available in this field except some basic approaches.

Zamfir (1979) worked on the possibilities for the removal of pesticide polluted water in treatment stations and described some methodologies, *i.e.* flocculation and filtration can remove partially-DOT, 2, 4, 5-T, edrin, parathion and lindane, chlorine oxidation can remove parathion, diuron etc., ozone and potassium permanganate appear to exert effects similar to those of chlorimation. Activated charcoal has positive effects in the removal of absorption of most pesticide and U-V ray also can remove a certain amount of pesticides.

Some indirect approaches has also been taken by some scientists and their methods were environmental or nutritional manipulation. Sado *et al.* (1992) reported that increased temperature and optimum level of dissolved oxygen (by aerator) can decrease the pesticidal action and Das (1996) described for the application of lime to increase the pH for counteraction. Ghazahy (1994) and Mukherjee (1996) commented for the efficacy of ascorbic acid (Vitamin-C) for the intoxication of different pollutants including pesticides. But application of different herbal extracts for this purposes are a very purposeful and new avenue of research.

Conclusion

Most of the pesticides acts on the respiratory process and cholinergic nervous system and hampers the cell metabolism. It is necessary to find a suitable antagonist or antidote to check the poisoning caused due to pesticides/pollutant and to establish the life processes. Though it is the ultimate aim of pollution study, no such antidote or mechanism for controlling pesticide toxicity to fish is developed. Considering all the above points some plant extracts and organic matters has been applied as antidote to counteract pesticidal action. But it need much more studies to come out successful sustainable aquaculture with some antidotes against these pesticides for proper environmental impact assessment.

Acknowledgements

The authors are thankful to the Director, Central Institute of Freshwater Aquaculture, Kausalyaganga, Bhubaneswar, Orissa for carrying out this work and the Research Scholars of Soil and Water Chemistry Laboratory, Central Institute of Freshwater Aquaculture, Kausalyaganga, Bhubaneswar, Orissa for their valuable co-operation to prepare this article.

References

Abdelghani, A.A., Tehounwou, P.B., Anderson, A.C., Sujono, H., Heyer, L.R., Monkiedje, A. (1997). Toxicity evaluation of single and chemical mixtures of round up, Garlon-3A, 2, 4-D and syndets surfactant to channel catfish (*Ictalurus punctatus*), blue-gill sunfish (*Lepomis microchirus*) and craw fish (*Procambarus* sp.). *Environ. Toxicol. Water Qual.*, 12(3): 237-244.

Altamirano-Reynoso, R. (1985). Microscopic liver and kidney injuries in *Tilapia hornorum* caused by an acute intoxication with a commercial product that had lindane Thesis or Dissertation-Universidad Nac. Autonoma. dc Mexico (Mexico). Fac. De Med; Vet. Zootec.

APHA (1997). *Standard methods for the examination of water, sewage and industrial waters,* (17th Edn.). Am. Publ. Health. Org. New York, pp. 522.

Balint, T., Szegletes, T., Szegletes, Z., Halasy, K. and Namesok, J. (1995). Biochemical and sub-cellular changes in carp exposed to the organo-phosphorous methidathion and the pyrethroid deltamethrin. *Aquat. Toxicol.*, 33(3-4): 279-295.

Barry M.J., Logan D.C., Van-Dam, R.A., Ahokas, J.T. and Holdway, D.A. (1985). Effect of age and weight specific respiration on toxicity of esfenvalerate pulse- exposure to the Australian Crimson-spotted rainbow fish (*Melanotaenia fluviatilis*). *Aquat. Toxicol.*, 32(2-3): 115-126.

Belisle, A.A. and Swineford, D.M. (1988). Simple, specific analysis of organo-phosphorous and carbamate pesticides in sediments using column extraction and gas chromatography. *Environ. Toxicol. Chem.*, 7(9): 749-752.

Bender, M.E. (1969). The toxicity of the hydrolysis break down products of malathion to the fathead minow (*Pimephales promelas* Raf). *Wat. Res.*, 3: 571-582.

Bhatnagar, M.C., Tyagi, M. and Tamata, S. (1995). Pyrethroid induced toxicity to phosphatase in *Clarias batrachus* (Linn.). *J. Environ. Biol.*, 16(1): 11-14.

Bhattacharya, S., Mukherjee, S. and Bhattacharya, S. (1975). Toxic effects of endrin on hepatopancreas of the teleost fish, *Clarias batrachus* (L). *Ind. J. Experiment Biol.*, 13: 185-186.

Bhattacharya, T., Roy, A.K. and Bhattacharya, S. (1987). Histopathology of the kidney of *Channa punctatus* (Bloch) exposed to sublethal and chronic levels of Industrial effluents. II. *Zool. Jb. Anat.*, 115: 263-272.

Bocquene, G., Galgani, F., Burgeot, T., Le-dean, L. and Truquet, P. (1993). Acetylcholinesterase levels in marine organisms along French Coasts. *Mar. Pollut. Bull.*, 26(2): 101-106.

Booke, H.E. (1964). Blood serum protein and calcium levels in yearling brook trout. *Prog. Fish. Cult.*, 26: 107-110.

Bouck, G.R. and Ball, R.C. (1966). Influence of capture methods on blood character and mortality of rainbow trout (*Salmo gairdeneri*). *Trans. Am. Fish. Soc.*, 95(2). 170-176.

Boyd, C.E. (1963). Insecticides cause mosquito fish to abort. *Prog. Fish. Culturist*, 26: 138.

Bretaud, S., Saglio, P., Saligaut, C. and Auperin, B. (2002). Biochemical and behavioural effects of carbofuran in gold fish (*Carassius auratus*). *Environ. Toxicol. Chem.*, 21(1). 175-181.

Brown, V.M. (1968). The calculation of the acute toxicity of mixtures of poisons to rainbow trout. *Water Res.*, 2: 723.

Brungs, W.A. and Mount, D.I. (1978). Introduction to a discussion of the use of aquatic toxicity tests for the evaluation of the effects of toxic substances. In: *Estimating the hazard of chemical substances to aquatic life*. ASTM STP 657, Cairns, J. Jr., Dickson, K.L. and Maki A.W. (Eds.). American Society for Testing and Materials, Philadelphia, p. 15.

Buckler D.R., Witt, A. Jr., Mayer, F.L. and Huckins, J.N. (1981). Acute and chronic effects of kepone and mirex on the fathead minnow. *Trans. Am. Fish. Soc.*, 110: 270.

Buikema Jr., A.L. Nider-Lehner and Cairns, J. Jr. (1982). Biological monitoring Part IV: Toxicity testing. *Ibid*, 16: 239-262.

Cagauan, A.G. and Arce R.G. (1992). *Overview of pesticide use in rice fish farming in South-east Asia*. Iclarm. Conf. Proc. No. 24.

Cairns, J. Jr. (1981). The effects of the temperature and hardness of water upon the toxicity of zinc to pond water snail *Physa heterosstreplie*. *Notulac Naturae*, 308: 1-11.

Cairns, J. Jr., Heath, A. G. and Parker, B.C. (1975). The effects of temperature upon the toxicity of chemicals to aquatic organisms. *Hydrobiologia*, 47: 135.

Carlson, A.R. (1971). Effect of long-term exposure of carbaryl on survival, growth and reproduction of the fathead minnow *pimephales promelas*. *J. Fish. Res. Bd.* Canada, 29: 583-587.

Carlson, A.R. (1971). Effects of long-term exposure to carbaryl (Sevin) on minnow (*Pimephales promelas*). *J. Fish Res. Board. Can.*, 29: 583.

Chandrasekar, S. and Jayabalan, N. (1993). Haematological responses of the common carp, *Cyprinus carpio* L. exposed to the pesticide endosulfan. *Asian Fish. Sci.*, 6(3): 331-340.

Chavin, W. (1973). *In: Response of fish to in vivo-mental changes*, (Eds.) W.C. Chavin. Springfield, Ilinoisis, p. 199.

Chavin, W. and Young, J.E. (1970). Factors in the determination of normal serum glucose levels of gold fish. *Carassius auratus J. Comp. Biochem. Physiol.*, 33: 629-653.

Chilamovitch, Y.P. and Kuhn, C. (1977). Behavioural, haematological and histological studies on acute toxicity of bis (tri-n-butyltin) oxide on *salmo gairdneri* Richardon and *Tilapia rendalle* Boulenger. *J. Fish. Biol.*, 10: 1575.

Cope, O.B., Wood, E.M. and Watten, G.H. (1970). Some chronic effect of 2, 4D on the blue-gill (*Lepomis macrochirus*). *Trans. Am. Fish. Soc.*, 99: 1.

Couch, J.A. (1975). Histopathological effects of pesticides and related chemicals on the liver of fishes. In: *Pathology of fishes*, Ribelin and Migaki (Eds.). The University of Wisconsin Press, pp. 559-584.

Curtis, L.R. and Beyers, R.J. (1978). Inhibition of oviposition in the telost *Oryzias latipes*, induced by subacute kepone exposure. *Comp. Biochem. Physio.*, 61: 15.

Dalvi, R.R. and Davis, S.W. (1998). Role of beta naphthoflavone in the acute toxicity of chlorpyrifos in channel catfish. *Bull. Environ. Contam. Toxicol.*, 60(2): 335- 339.

Das, B. (1996). Ph.D. Thesis. Orissa University of Agriculture and Technology, Bhubaneswar, Orissa.

Davies, J.S. (1975). Minimal dissolved oxygen requirements of aquatic life with emphasis on Canadian species: A review. *J. Fish. Res. Bd. Can.*, 32(12): 2295- 2332.

Davies, P.E. and Cook, L.S.J. (1993). Catastrophic macro-invertebrate drift and sublethal effects on brown trout, *Salmo trutta*, caused by cypermethrin spraying on a Tasmanian Stream. *Aquat. Toxicol.*, 27(3-4): 201-224.

Days, K.E. (1991). Effects of dissolved organic carbon on accumulation and acute toxicity of fenvalerate, deltamethrin and cyhalothrin to *Daphnia magna* (Straus). *Environ. Toxicol. Chem.*, 10: 91-101.

Decie, S.I.V. and Lewis, S.M. (1991). *Practical haematology* (VII Edn). J. and A Churchill Ltd., Livingston London, Melborne and New York.

Domitrovic, H.A. (1997). The use of autochthonous fishes in toxicity tests. Evaluation of the species Acquidens portalegrensis. *Rev. Ictiol.*, 5(1-2): 37-42.

Donaldson, E.M. (1975). Physiological and physico-chemical factors associated with maturation and spawning. *EIFAC Technical paper*, 25: 53-71.

Duangswawasdi, M. and Klaverkamp, J.F. (1979). Acephate and fenitrothion toxicity in rainbow trout: Effects of temperature stream and investigations on the sites of action. *Aquat. Toxicol.*, 35.

Duke, T.W. and Wilson, A.J. (1971). Chlorinated hydrocarbons in liver of fishes from the northeastern pacific ocean. *Pestic. Monit. J.*, 5: 228.

Eaton, J.G. (1973). Recent developments in the use of laboratory bioassays to determine "SAFE" levels of toxicants for fish, in Bioassay techniques and Environmental Chemistry. *Ann Arbor, Science, Mirch.*, 107.

EC (1979). Sixth amendment directive 67/548. Council of the European Communities, *Off. J. Eur. Comm. L.* 259/24.

Elliot, M., Farham, A.W. Janes, N.F., Netham, P.H. and Pearson, B.C. (1975). Insecticidal activity of the pyrethrins and related compounds. VII. Insecticidal dihalo vinyl analogues of Cis and tras Chrysanthemates. *Pest. Sci.*, 6: 537.

Ellis, M.M. (1937). Detection and Measurement of Stream pollution. *U.S. Bur. Fish Bull.*, 48(22): 365-437.

EPA (1978). *Criteria and rational for decision making in aquatic hazard evaluation.* Am. Inst. Biol. Sc. Arlington, Virginia, p.46.

Ernst W., Jackman, P., Doe, K., Page, F., Julien, G., Mackay, K., Suthenland, T. (2001). Dispersion and toxicity to non-target aquatic organisms of pesticides used to treat sea lice on salmon in net pen enclosures. 42(6): 433-44.

Farrell, A.P., Stockner, E. and Kennedy, C.J. (1998). A study of the lethal and sublethal toxicity of polyphase P-100 an antisapstain fungicide containing 3-iodo-2 propynyl butyl carbamate (IPBC), on fish and aquatic invertebrates. *Arch. Environ. Contam. Toxicol.*, 35(3): 472-478.

Fernandez, T.V., Thomas, G., Vardhanan, Y.S., Kumar, R.M. (1996). An assessment of the toxicity levels of BHC on four aquatic organisms. *J. Environ. Biol.*, 17(1): 21-24.

Ghazaly, K.S. (1994). Efficacy of Ascorbic acid (Vitamin C) on experimental copper intoxication in *Tilapia zilli. Bull. Natl. Inst. Oceanogr. Fish*, Egypt. 20(2): 249-257.

Ghosh, R. and Dutta, J. (1985). Malathion toxicity in air-breathing catfish singhi (*Heteropneustes fossilis*). *Bulletin Ind. J. Fish*, 32: 447-452.

Gill, T.S. Pande, J. and Tewari, H. (1991). Individual and combined toxicity of common pesticides to teleost *P. Conchonius. Ind. J. Expt. Biol.*, 29: 145-148.

Gill, T.S., Pande, J., Tewari, H. (1991). Effects of endosulfan and phosphamidon poisoning on the peripheral blood of fish (*Barbus conchonius*, Hamilton). *J. Environ. Sci. Health*, A26(2): 249-255.

Gills, S.S. (1980). *In vitro* metabolism of carbofuran by liver microsomes of the paddy field fish, *Trichogaster pectoralis. Bull. Environ. Contam. Toxicol.*, 25: 697.

Gimeno, L., Ferrando, M.D., Sanchez, S. and Andreu, E. (1996). Endosulfan effects on liver and blood of the eel, *Anguilla anguilla. Comp. Biochem. Physiol.*, 108C(3): 343-348.

Gomez, L., Masot, J., Martinez, S., Duran, E., Soler, F., Roncero, V. (1998). Acute 24-D poisoning in tench (*Tinca tinca* L.). Lesions in the haematopoictic portion of the kideny. *Arch. Environ. Contam. Toxicol.*, 35(3): 479-483.

Goodman, L.R., Hansen, D.J., Manning, C.S. and Favas, L.F. (1982). Effects of kepone on the sheepshead minnow in an entire life cycle toxicity test. *Arch. Environ. Contam. Toxicol.*, 11: 335.

Grohmann, A. and Sobhani, P. (1979). Beeinflusung der Giflwirkung von stoffen auf Fische durch lonengleichfewiche (Harle des wassers and Toxicitat) Hydrochim. *Hydrogeol.*, 3: 147.

Haldane A.V. (1966). Organo-chlorine insecticide residues in salmonoid fish. *J. Appl. Ecol.*, 3: 45.

Halter, M.T. and Johnson, H.E. (1974). Acute toxicities of a Polychlorinated biphenyl (PCB) and DDT alone and in combination to early life stages of Coho salmon (*Oncorynchus Kisutch*).

Hanke, W., Gluth, G., Buhel, H. and Muller, R. (1983). Physiological changes in carps induced by pollution. *Ecotoxicol. Environ. Saf.*, 7: 229.

Hart, W.B., DoudRoff, P. and Greenback, J. (1945). The evaluation of Industrial wastes, chemicals and other substrial wastes, Chemicals and other substances to freshwater fish. *Wastes Contr. Lab. All. Co.*, 317.

Hart, W.B., Weston, R.F. and Demann, J.G. (1948). An apparatus for oxygenating test solution in which fish are used as test animals for evaluating toxicity. *Trans. Am. Fish. Soc.*, 75: 228.

Hassal, K.A. (1990). *The biochemistry and uses of pesticides, II ed.*, English Language Book Society MacMillan Press Ltd., Hongkong.

Hazarika, R. and Das, M. (1998). Toxicological impact of BHC on the ovary of the air-breathing catfish *Heteropneustes fossilis* (Bloch). *Bull. Environ. Contam. and Toxicol.*, 60(1): 16-21.

Heath, A.G. (1990). *Water pollution and fish physiology*. CRC Press, USA, pp. 1-235.

Heath, A.G., Cech, J.J. Jr., Brink, L., Moberg, P., Zinkl, J.G. (1997). Physiological responses of fathead minnow larvae to rice pesticides. *Ecotoxicol. Environ. Saf.*, 37(3): 280-288.

Heath, A.G., Cech, J.J. Jr., Zinkl, J.G., Steele, M.D. (1993). Sublethal effects of three pesticides on Japanese medaka. *Arch. Environ. Contam. Toxicol.*, 25(4): 485-494.

Hinsen, D.J., Parrish, P.R. Loser, J.I., Wilson, A.J. Jr. and Wilson, P.D. (1971). Chronic toxicity uptake and retention of Aroclor 1254 in two estuarine fishes. *Bull. Environ. Contam. Toxicol.*, 6: 113.

Hohreiter, D.W., Reinert, R.E. and Bush, P.B. (1991). Effects of the insecticides carbofuran and fenvalerate on adenylate parameters in blue-gill sunfish (*Lepomis macrochirus*). *Arch. Environ. Contam. Toxicol.*, 21(3): 325-331.

Holcombe, G.W., Flandt, J.T. and Phipps, G.L. (1980). Effects of pH increases and sodium chloride addition on the acute toxicity of 2-4 dichlorophenol to the fathead minnow. *Water Res.*, 14: 1073.

Holden, A.K. (1972). *The effects of pesticides on life in freshwater*. Proc. R. Soc. London, Ser. B. 180: 383.

Holmberg, B., Jensen, S., Larsson, A., Lewander, K. and Olsson, M. (1972). Metabolic effects of technical penta-chlorophenol (PCP) on the eel *Anguilla anguilla* L. *Comp. Biochem. Physiol.*, 438: 171-183.

Hota, A.K., Mishra D.K. and Tripathy, P.C. (1993). Metabolic effects of kilex carbaryl on a freshwater teleost, *Channa punctatus* (Bloch.) Muzaffarnagar. *India Society of Bioscience*, 335-342.

Inglis, A. and Davis, E.L. (1972). Effects of water hardness on the toxicity of several organic and inorganic herbicides to fish.

James, R., Sampath K. and Alagurathinam, S. (1996). Effects of lead on respiratory enzyme activity, glycogen and blood sugar levels of the teleost, *Oreochromis mossambicus* (Peters) during accumulation and depuration. *Asian Fish. Sci.*, 9: 86-99.

Jhingran, V.G. (1982). *Fish and Fisheries of India*, pp 894.

Johansen, K. (1970). Air-breathing fishes. In: *Fish physiology*, W.S. Hoar and D.T. Randall (Eds.), Academic Press, New York and London, (4): 361-411.

John, K.R., Jayabalan, N. and George, M.R. (1993). Impact of sublethal concentrations of endosulfan on the histology of *Cyprinus carpio* liver and kidney. Proceedings of the national seminar on a Aquaculture-development in India: problems and prospects, 27-29 Nov., 1990, pp. 179-182.

Johnson, W.W. and Finley, M.T. (1980). Handbook of acute toxicity of chemicals to fish and aquatic invertebrates. Resource publ. 137, US Department of Interior Fish and Wildlife Service Washington, DC.

Johnson, W.W. and Julin, A.M. (1980). Acute toxicity of toxaphene to fathead minnows, EPA-600/3-80. 005. U.S. Environmental Protection Agency, Duluth, Minn.

Jones, J.R.E. (1940). The toxicity of the double chlorides of mercury and sodium. I. Experiments with the minow *Phoxinus phoxinus* (L.). *J. Experiment Biol.*, 17: 325-330.

Jones, J.R.E. (1964). Insecticides and Herbicides. In: *Fish and river pollution*, Butterworth and Co. Ltd., pp. 128-140.

Kanazawa, J. (1981). Measurement of the bio-concentration factors of pesticides by freshwater fish and their correlation with physico-chemical properties or acute toxicities. *Pestic. Sec.*, 12: 417.

Kapur, K., Kaur, K. and Toor, H.S. (1978). The effect of fenitrothion on reproduction of a teleost fish, *Cyprinus carpio* (Linn.): A biochemical study. *Bull. Environ. Contam. Toxicol.*, 20: 438-442.

Katz, M. (1961). Acute toxicity of some organic insecticides of three species of salmonids and spike stickleback. *Trans. Amer. Fish. Soc.*, 90: 310.

Khattar, I.U.D. and Hafeez, M.A. (1996). Effect of malathion on blood parameters of the fish *Cyprinion Watsoni. Pak. J. Zool.*, 28(1): 45-49.

Kling, D.B. (1986). Toxic effects of sublethal concentration of Lebaycid upon the ovaries were examined in *sarotherodon leucostictus. Z. Angero. Zool.*, 73(1): 75-92.

Konar, S.K. and Ghosh, T.K. (1982). Effects of Furadon on fish, plankton, worm and water quality. *IAWPC Tech. Annual*, 9: 128-134.

Kulshrestha, S.K. and Arora, N. (1984). Impairments induced by sublethal doses of two pesticides in the ovaries of a freshwater teleost *Channa stratus* Bloch. *Toxicol. Lett.*, 20(1): 93-98.

Lloyd, R. (1960) and Jordan, D.H.M. (1964). Some factors affecting the resistance of rainbow trout (*Salmo gairdneri* Richardson) to acid waters. *Ind. J. Air. Wat. Pollut.*, 8: 393-403.

Lloyd, R. and Jordan, D.H.M. (1963). Predicted and observed toxicities of several sewage effluent to rainbow trout. *J. Proc. Inst. Sewage Purif.*, 2: 16.

Macek, K.J. (1968). Reproduction in brook trout (*Salvelinus fontinalis*) fed sublethal concentration of DDT. *J. Fish Res. Board Can.*, 25: 1787.

Macek, K.J. (1980). Aquatic toxicology: Fact or fiction. *Envir. Health. Persp.*, 34: 159-163.

Macek, K.J. and Sanders, H.O. (1970). Biological variation in the susceptibility of fish and aquatic invertebrate to DDT. *Trans. An. Fish. Soc.*, 99: 89.

Macek, K.J., Hutchinson, C. and Cope, O.B. (1969). The effect of temperature on the susceptibility of blue gills and rainbow trout to selected pesticides. *Bull. Environ. Contam. Toxicol.*, 4: 174.

MacFarlane, R.B. and Livingston, R.J. (1983). Effects of acidified water on the locomotor Behaviour of the gulf killifish, *Funtulus grandis*: A time series approach. *Arch. Environ. Contam. Toxicol.*, 2: 163-168.

Mahajan, C.L. and Juneja, C.J. (1979). Effect of aldrin on peripheral blood of fish, *Channa punctatus* (Bloch.). *Indian J. Environ. Hlth.*, 21: 162-172.

Mahapatra, B.C. and Rengarajan, K. (1995). A manual on Bioassays in the laboratory and their techniques. *CMFRI special publication*, No. 64.

Malhotra, Y.R. and Sharma, K.K. (1986). Effect of malathion on some aspects of early developmental stages of *Puntius conchonius* (Ham.). *Proc. Natl. Symp. Fish and Env.*, 1986, pp. 1-4.

Malla-Reddy, P., Philip, G.H., and Bashamohideeh, M. (1992). Regulation of Ache system of freshwater fish, *Cyprinus carpio*, under fenvalerate toxicity. *Bull. Environ. Contam. Toxicol.*, 48(1): 18-22.

Mani, K. and Saxena, P.K. (1985). Effect of safe concentration of some pesticides on ovarian recrudescence in the freshwater murrel, *Channa punctatus* (BI.). A quantitative study. *Ecotoxicol. Environ. Saf.*, 9: 241-249

Marking, L.L. (1974). Toxicity of the synthetic pyrethroid SBP-1382 to fish. *Prog. Fish Culturist*, 36: 144

Mathur, D.S. (1965). Histological changes in the livers of certain fishes induced by dieldrin. *Sci. Cult.*, 31: 258-259.

Mauck, W.L., Olson, L.E. and. Marking, L.L (1976). Toxicity of natural pyrethrins and five pyrethroids to fish. *Arch. Environ. Contam. Toxicol.*, 4: 18.

Mauck, W.L., Olson, L.E., and Hogan, J.W. (1977). Effects of water quality on deactivation and toxicity of mexacarbate (zectran) on fish. *Arch. Environ. Contam. Toxicol.*, 6: 385.

Mayer, F.L., Mehrle, P.M. and Cruteher, P.L. (1978). Interaction of toxaphone and Vitamin C in Channel catfish. *Trans. Am. Fish. Soc.*, 107: 326.

Mayers, T.R. and Hendricks, J.D. (1985). Histopathology. In: *Fundamentals of aquatic toxicology*, G.M. Rand and S.R. Petrocelle (Eds). Hemisphere Publ. Corp., Washington, pp. 283-331.

Mcbride, J.R., Dye, H.M. and Donaldson, E.M. (1981). Stress response of juvenile sockeye salmon (*oncorhynchus nerker*) to the butoxyethanol ester or 2, 4- dichlorophenoxyacetic acid. *Bull. Environ. Contam. Toxicol.*, 27: 877.

McKim, J.M. (1985). Early life stage toxicity tests. In: *Fundamentals of aquatic toxicology*, G.M. Rand and S.R. Petrocelli (Eds.) Hemisphere Publ. Corp., Washington, pp. 58-95.

Metelev, V.V., Kanaev, A.L. and Diasokhva, N.G. (1971). *Water toxicity*, Amerind Publishing Co. Pvt. Ltd., New Delhi, pp. 3-203.

Metz. J. and Branacin, M. (1975). Biochemical and biophysical aspect of salt excretion by chloride cell in teleost. In: *Excretion* (Wessing, A. Ed.).

Mishra, J. and Srivastava, A.K. (1982). Malathion induced haematological and biochemical changes in the Indian catfish, *Heteropneustes fossilis*. *Environ. Res.*, 30: 393-398.

Moss, J.A. and Hathway, D.E. (1964). Transport of organic compounds in the mammal: partition of dieldrin and telodrin between the cellular components and soluble proteins of blood. *Biochem. J.*, 91: 384.

Mount, D.J. (1977). *An Assessment of Application factors in Aquatic toxicology*. US- EPA-600/3- 77-085. U.S. Environmental Protection Agency. Corvallisore.

Mount, D.J. and Stephan, C.E. (1967). A method for establishing acceptable toxicant limits for fish: Malathion and the butoxyethyl ester of 2, 4-D. Trans. *Amer. Fish. Soc.*, 96: 185-193.

Muir, D.C.G., Hobden, B.R. and Servos, M.R. (1994). Bio-concentration of pyrethroid insecticides and DDT by rainbow trout: uptake, depuration, and effect of dissolved organic carbon. *Aquat. Toxicol.*, 29: 223-240.

Mukherjee, M. (1996). Antidote to combat white spot disease in culture ponds. *Fish. Chimes*, 16(5): 30-31.

Mukherjee, S. and Bhattacharya, S. (1975). Histopathological lesions in the hepatopancreas of fishes exposed industrial pollutants. *Indian J. Exp. Bio.*, 13(6): 571-573.

Mukhopadhyay, P.K. and Dehadrai, P.V. (1980a). Studies on air-breathing catfish, *Clarias batrachus* (Linn) under sublethal malathion exposure. *Ind. J. Experiment Biol.*, 18: 400-404.

Mukhopadhyay, P.K. and Dehadrai, P.V. (1980b). Biochemical changes in the air-breathing catfish *Clarias batrachus* (Linn.) exposed to malathion. *Environ. Pollut.* A22: 149-158.

Mukhopadhyay, P.K. and Dehadrai, P.V. (1980). Biochemical changes in the air-breathing catfish, *Clarias batrachus* exposed to malathion. *Environ. Res.*, 22: 149-158.

Mukhopadhyay, P.K., Mukherjee, A.P. and Dehadrai, P.V. (1984). Phospholipid and fatty acids in liver of *Clarias batrachus* exposed to sublethal malathion. *J. Environ. Biol.* 5: 221-229.

Mulcahy, M.F. (1975). Fish blood changes associated with disease: A haematological study of pike lymphoma and salmon ulcerative dermal necrosis. In: *The pathology of fishes*, W.E. Ribelin and C. Migaki Madison (Eds.), University of Wisconsins, pp. 925-944.

Murty, A.S. (1988): *Toxicity of pesticides to fish*, 1: 19.

Mushthaq, M. and Nazar, A.R. (1994). Effect of pesticide, Nuvan, on *Catla catla*. *J. Aqua. Biol.*, 9 (1 and 2): 80-82.

Nagler, J.J., Aysola, P. and Ruby, S.M. (1984). The seasonal effect of sublethal pentachlorophenol exposure on ovarian development in rainbow trout (*Salmo gairdneri*). 11th annual aquatic toxicity workshop, Richmond, B.C., November, 13-15, 1984 (np.)

Nagler, J.J., Prasad, A. and Ruby, S.M. (1986). Effect of sublethal pentachlorophenol on early oogenesis in maturing female rainbow trout (*Salmo gairdneri*). *Arch. Environ. Contam. Toxicol.*, 15: 549-555.

Narayan, A.S. and Singh, B.B. (1991). Histo-pathological lesions in *H. fossilis* subjects to acute thiodan toxicity. *Acta. Hydrochem. Hydrobiol.*, 19(2): 235-243.

Nath, R. and Jaipuryar, A.R.K. (1996). Effect of sublethal concentration of lindane on peripheral haemogram in the fish *Heteropneuster fossilis* (Bloch). *Environ. Ecol.*, 14(1): 158-162.

Newsome, C.S. (1980). A multi-generation fish toxicity tests as an aid in the hazard evaluation of aquatic pollutants. *Ecotoxicol. Environ. Saf.*, 4: 362.

OECD (1979). *Report on the assessment of potential environmental effects of Chemical*. I. OECD, pp.28.

Oern, S., Anderson, P.L., Foerlin, L., Tysklind, M. and Norrgren, L. (1998). The impact on reproduction of an orally administered mixture of selected PCBs in Zebra fish (*Danio rerio*). *Arch. Environ. Contam. Toxicol.*, 35(1): 52-57.

Ogilvie, D.M., and Anderson, J.M. (1965). Effect of DDT on temperature selection by young Atlantic Salmon, *Salmo Salar*. *J. Fish. Res. Board Can.*, 22: 503.

Pandey, A.C. (1988). Impact of endosulfan (thiodon) EC 35 on behaviour and dynamics of oocyte development in the teleostean fish, colisa (*Trichogastter fasciatus*). *Ecotoxicol. Environ. Saf.*, 15(2): 221-225.

Pathiratne, A. and George, S.G. (1998). Toxicity of malathion to Nile tilapia, *Oreochromis nilotictus* and modulation by other environmental contaminants. *Aquat. Toxicol.*, 43(4): 261-271.

Peczely, P. and Kovacs, K. (1998). Effects of a fungicide (VITAVAX 200 FF) on the reproductive endocrime functions in mallards. *Environ. Toxicol. Water. Qual.*, 13(2): 133-141.

Peterson, R.H., Daye, P.G., Lacroix, G.L. and Garside, E.T. (1982). Reproduction in fish experiencing acid and metal stress. In: *Acid rain/fisheries*, Johnson, R.E. (Ed.), American Fisheries Society, Bethesda, Md., p. 177.

Pickering, A.D., Potliger, T.G. and Christie, P. (1982). Recovery of brown trout, *salmo trutla*, L. from acute handling stress: A time course study. *J. Fish Biol.*, 20: 229.

Pickering, O.R., Henderson, C. and Lemke, A.E. (1962). The toxicity of organic phosphorous insecticides to different species of warm fishes. *Trans. Am. Fish. Soc.*, 91: 175.

Power, E.B. (1917). The gold fish as a test animal in the study of toxicity. *Illinois. Monogr.*, 4: 93-127.

Qureshi, M.D., Heldin, R.V., Vanstone, W.E. and Anastassiadis, P.A. (1971). Levels of constituents of glycoproteins in the sera pacific salmon. *J. Fish. Res. Bd. Canada*, 28(8): 1173-1179.

Rajeswari, K., Reddy, S. J., Reddy, D.C. and Ramamurthi R. (1989). Effect of dichlorvos on certain blood parameters of fish *Clarias batrachus. Environ. Ecol.*, 7(4): 933-934.

Ram, R.N. (1988). Carbofuran induced histo-physiological changes in thyroid of the teleost fish, *Channa punctatus* (Bloch.). *Ecotoxicol. Environ. Saf.*, 16(2): 106- 113.

Ramanujam, S.N. and Mohanty, G. (1997). Thiodan induced sublethal responses on haematological parameters of *H. fossilis. Journal of life sciences-Royal Dublin Society*, 2: 8-13.

Ramkrishnan, R. and Sivakumar, A.A. (1993). Effect of quinolphos on the fish *Oreochromis mossambicus. J. Ecobibl.*, 5(1): 45-50.

Rebach, S. (1999). Acute toxicity of permethrin/piperonyl Butoxide on Hybrid Striped Bass. *Bull. Environ. Contam. Toxicol.*, 62(4): 448-454.

Reddy, P.M. and Bashamohideen, M. (1995). Alterations in protein metabolism in selected tissues of fish, *Cyprinus carpio*, during sublethal concentration of cypermethrin. *Environ. Monit. Assess.*, 36(2): 183-190.

Reish D.L. and Oshida, P.S. (1987). Short-term bioassay. In: *Manual of methods in aquatic environment research*. FAO Fish Tech. Pap. 247(10): 1-62.

Sado, E.K. and Ita, E.O. (1992). The effects of capture methods, temperature and oxygen on the survival of live freshwater clupeids during acclimation and transportation. *J. Aquacult. Trop.*, 7(2): 165-176.

Saglio, P., Trijasse, Azam, S.D. (1996). Behavioural effects of waterborne carbofuran in gold fish. *Arch. Environ. Contam. Toxicol.*, 31(2): 232-8.

Sahai, S. and Singh, S. (1989). Hepatotoxic effects of some pesticides in *Puntius ticto. Environment and Experimental Toxicology*. Jai Research Foundation, Valvada.

Sampath, K., Velammal, S., Kennedy, L.J.J. and James, R. (1993). Haematological changes in *Orieochromis mossambicus* as a function of exposure period and sublethal levels of Ekalux. *Acta Hydrobiol.*, 35: 73-83.

Sastry, K.V., Mallik, P.V. and Singh, S.K. (1982). Alteration in some biochemical and enzymological parameters in the snake head fish *Channa punctatus*, exposed clinically to quinalphos. *Chemosphere*, 11: 1211.

Scott, A.L. and Rogers, W.A. (1980). Histological effects of prolonged sublethal hypoxia of channel cat fish *Ictalurus punctuatus*. *J. Fish. Dis.*, 3: 305.

Shin, S.W., Chung, N.I., Kim, J.S., Chon, T.S., Kwon, O.S., Lee, S.K., Koh, S.C. (2001). Effect of diazinon on behaviour of Japanese medaka (*oryzias latipes*) and gene expression of tyrosine hydroxylase as a biomarker. *J. Environ. Sci. Health*, 36(6): 783-95.

Shukla, L., Shrivastava, A., Merwani, D. and Pandey, A.K. (1984). Effect of sublethal malathion on ovarian histopathology in *Sarothrodon mossambicus*. *Comp. Physiol. Eco.*, 9: 13-17.

Singh, S. and Sahai, S. (1986). Accumulation of malathion in the gills, kidney and liver of *Puntius ticto* as assessed by thin layer chromatography. *J. Environ. Biol.*, 7(2). 107-112.

Singh, S. and Sahai, S. (1987). Pesticide metabolism in a freshwater teleost, *Puntius ticto*. In: *Environment and ecotoxicology*, (Eds. R.C. Dalela, Y.N. Sahai and Saroj Gupta), pp. 211-218.

Singh, A. and Agarwal, R.A. (1993). Effect of cypermethrin on lactate and succinic dehydrogenase and cytochrome oxidase of snail and fish. *Bull. Environ. Contam. Toxicol.*, 51(3): 445-452.

Singh, B.B. and Narain, A.S. (1982). Acute toxicity of thiodan to catfish (*Heteropneustes fossilis*). *Bull. Environ. Contam. Toxicol.*, 28: 122.

Singh, P.B., Kime, D.E. and Singh, T.P. (1993). Modulatory actions of Mystus gonadotropin on gamma-BHC induced histological changes, cholesterol and sex steroid levels in *Heteropneustes fossilis*. *Ecotoxicol. Environ. Saf.*, 25(2): 141-153.

Smith, C.E. (1968). Haematological changes in the coho salmon fed a folic acid deficient diet. *J. Fish. Res. Bd. Can.*, 25: 151-156.

Smith, R.K. and Cole, C.F. (1973). Effects of egg contaminations of DDT and dieldrin on development in winter flounder (*Pseudopleuronectes americanus*). *J. Fish. Res. Board Can.*, 30: 1894.

Soengas, J.L., Strong, E.F., Aldegunde, M. and Andres, M.D. (1997). Effects of an acute exposure to lindane (gamma-hexachlorocyclohexane) on brain and liver carbohydrate metabolism of rainbow trout. *Ecotoxicol. Environ. Saf.*, 38(2): 99-107.

Spraque, J.B. (1971). Measurement of pollutant toxicity to fish. III. Sublethal effects and "safe" concentration. *Water Res.*, 5: 245-266.

Stephenson, C.E. (1982). In analysis of established pyrethroid insecticides. *Residues Review*, 39: 188.

Steveke, E.W. and Atherton, C.R. (1965). Use of micro-haematocrit value to six largemouth. *Bass. Prog. Fish. Cult.*, 27: 87-90.

Thakur, N. and Sahai, S. (1987). Changes in blood serum proteins, and other haematological parameters in the teleost, *Channa punctatus* on exposure to organo-chlorine pesticide, proceedings of the Academy of Environmental Biology. *Environ. Pestic. Toxicol.*, 263-271.

Tsuda, T., Kajima, M., Harada, H., Nakajima, A. and Aoki, S. (1997). Acute toxicity, accumulation and excretion of organo-phosphorous insecticides and their oxidation products in Killifish. *Chemosphere*, 35(5): 939-949.

Tugarina, P.Y. and Ryzhova, L.N. (1970). Age related feature of the blood of the black Baikal graying *Thymallus arcticos* baicalensis. *J. Ichthyol.*, 10: 384-357.

United States Environmental Protection Agency (1979). Guidance for pre-manufacture testing: Discussion of policy issues, alternative approaches and test methods. Federal Register, 44: 16240.

Urdaneta, H. (1989). Organo-chlorine compound in ponds of a fish culture Centre Paez Municipality, Zulia State Venezuela. *Bolcent-Invest. Biol. Maracoitio*, 23: 1-10.

Vaid, S. and Mishra I.M. (1997). Impact.of Macraberin Forte against malathion poisoning in fish brain. *Poll. Res.*, 16(4): 275-276.

Van Hougen and Apperhuizen (1985). In: *APHA*, 1989 New York.

Van Leeuwen, C.J. (1985). Short-term toxicity testing. In: *Manual of aquatic ecotoxocology*, (Eds.) P.K. Ray, P.N. Viswanathan, H.A.M. de Kurijf and D.de. Zwart, Hemisphere Publ. Corp., Washington.

Venugopalan, V.K. and P. Sasibhushana Rao (1979). Pesticide induced impairment in incubation and post-embryonic development of planktonic eggs of esturine fish of Porto NOVO (South India) Waters. *Proc. Symp. Environ. Biol. Academy of Environmental Biology*, Muzzaffarnagar, India, pp. 397.

Verma, S.R., Bansal, S. K., Gupta, A.K., Pal, N., Tyagi, A.K., Bhatnagar, M.C., Kumar, V. and Dalela, R.C. (1982). Bioassay trials with twenty-three pesticides to a freshwater teleost *Sacrobranchus fossilis*. *Wat. Res.*, 16: 525-529.

Vittozzi, L. and. Angelis, G.D. (1991). A critical reiew of comparative acute toxicity data on freshwater fish. *Aquat. Toxicol.*, 19: 167-204.

Warner, R.E., Peterson, K.K. and Borgman, L. (1966). Behaviour pathology in fish: A quantitative study of sublethal pesticide toxication. *J. Appl. Ecol.*, 3: 223.

Weber, J.B. (1972). Interaction of organic pesticides with particulate matter in aquatic and soil systems in fate of organic pesticides in the aquatic environment (*Advances in Chemistry Series III*), Faust, S.D. (Ed.), American Chemical Society, Washington D.C. pp. 55.

Wedemeyer, G. (1973). Some physiological aspects of sublethal heat stress in the juvenile steelhead trout (*Salmo gairdneri*) and coho salmon (*Onchorhynchus kisutch*). *J. Fish. Res. Bd. Can.*, 30: 831.

Weis, J.S. and Weis, P.S. (1976). Optical malformations induced by insecticides in embryoes of the Atlantic silverside, *Menidia menidia*. *Fish. Bull.*, 74: 208.

Wilbur, R.L. and Whitney, E.W. (1973). Toxicity of the herbicide Kuron (Silvex) to blue-gill eggs and fry. *Trans. Am. Fish Soc.*, 102: 630.

Witt, J.M., Brown, W.H., Shaw, G.I., Maynard, L.S., Sullivan, L. M., Whiting F.M. and Stull, J.W. (1966). Rate of transfer of DDT from the blood compartment. *Bull. Environ. Contam. Toxicol.*, 1: 187.

Yoon, J.Y., Oh S.H., Yoo, S.M., Lee, S.J., Lee, H.S., Choi, S.J., Moon, C.K., Lee, B.H. (2001). N-nitrosocarbofuran, but not carbofuran, induces apoptosis and cell cyle arrest in CHL Cells. *Toxicology*, 169(2): 153-161.

Zamfir, Gh (1979). Water pollution by pest-controlling agents and the possibilities for their removal in surface water treatment stations. *Igiena*, 28(2): 97-103.

11

Utilisation of Municipal Wastewater in Aerobic Composting of Solid Organic Waste of Bhubaneswar City

S.P. PANDA & D.K. BEHERA
Orissa Pollution Control Board, Nilakanthanagar, Bhubaneswar
C.S.K. MISHRA
Orissa University of Agriculture and Technology, College of Basic Sciences, Bhubaneswar

ABSTRACT

Approximately 600 m tons of solid organic wastes and a huge bulk of sewage water are generated in Bhubaneswar city daily endangering the urban environment. Solid wastes in windrows with sewage water inoculum decomposed faster compared to cowdung slurry and plain water treated wastes C : N and C : P ratios declined relatively faster in sewage water treated windrows than cowdung slurry and water treated windrows during the experimental period of 45 days. The results indicated that the Bhubaneswar municipal sewage water would be successfully utilised in solid waste composting.

Keywords: Municipal solid wastes (MSW), Sewage water, Aerobic composting, Windrows, C:N ratio.

Introduction

Enormous increase in the quantum and diversity of waste materials generated in urban areas and their detrimental effects on the environment demand urgent need for adoption of feasible scientific methods for their disposal and utilisation (Williams, 1994). In recent years a number of techniques have been developed to help reduce the quantity of wastes, convert the wastes into organic manures and generate substantial decentralised energy (Chalapathi *et al.*, 2000). Among the most popular means for utilisation of organic wastes, aerobic composting has proved to be most efficient and cost effective (Saha *et al.*, 2000; Hajra *et al.*, 2000). Aerobic composting is the biological decomposition process in which organic matter is decomposed to carbon dioxide and water while stabilised products principally humic substances are synthesised (Brunt *et al.*, 1985).

Literatures are available on survey and study of physico-chemical characteristics of municipal solid wastes and its qualitative improvement (Nanda *et al.*, 2000a; Nanda *et al.*, 2000b and Rao and Radhakrishna, 2001). Bhubaneswar, the capital city of Orissa State has witnessed a rapid growth in the last decade and presently has a population of approximately 6.7 lakhs. A huge quantity of solid organic wastes amounting to approximately 600 m tons is produced daily which are disposed of in open dumps. Further, a huge volume of sewage water produced in the city is allowed to accumulate in small lakes which are connected to the rivers Daya and Kuakhai. No systematic study has been done on the municipal solid and liquid wastes of Bhubaneswar city and methods of its utilisation.

Hence the present study was conducted to assess the feasibility of utilising the municipal sewage water in the aerobic composting of solid organic wastes of Bhubaneswar city and its effect on per cent N, P, K, C. C/N ratio and C/P ratio of the decomposing wastes.

Materials and Methods

Municipal solid waste was collected from local urban areas of Bhubâneswar. Sampling points were selected from all the 31 wards of the city. Sampling sites represented all types of localities; such as residential, commercial, slums and markets. The wastes were mixed thoroughly before collection from the sampling sites. Further, the waste samples collected from different wards were mixed thoroughly before the experiment. Per cent C, P, N and K in the waste were measured as per Baruah and Barthakur (1997).

The sewage water sample was collected from the main city drain of Bhubaneswar Municipal Corporation which opens into the Vani Vihar lake. Physical and chemical parameters such as temperature, pH, DO, BOD, COD, NO_3, PO_4, chloride, alkalinity, total hardness, turbidity, SO_4, Na, K, TDS, TSS, TPS, TC, Fe conductivity of the wastewater were measured following standard procedures (APHA, 1985).

The aerobic composting of the MSW was done by windrow method (Brunt *et al.*, 1985). Three windrows of the size (20×3×5 ft) were constructed on a concrete floor. Each windrow contained 3 tons of wastes for experimental purpose. The windrow was designed as follows. There were three layers inside each windrow. In first windrow the first layer contained 1 ton of garbage which was soacked with 80 litres of municipal sewage water. In the second layer 1 ton of garbage was soaked with 80 litres of municipal sewage water and it was spread above the 1st layer. The top or, third layer of 1 ton garbage was also soaked with 80 litres of sewage water and it was spread above the 2nd layer of garbage.

The second windrow (20×3×5 ft) contained 3 tons of garbage, which was soaked with 80 litres of cowdung slurry (2 litres water : 1 kg cowdung) in subsequent layers. The third windrow

(20×3×5 ft) of 1 ton garbage was treated as control which was treated with equal volume of plain water. The moisture level in the windrow was maintained at 50-60 per cent and temperature of the windrow was maintained 44°C–50°C which is the optimal temperature for thermophillic bacteria, which accelerate the composting process. The turning of the organic waste materials were done at regular intervals for forced aeration, since oxygen is required for enhanced activity of the aerobic microbes. The windrows were covered with black polythene sheets to accelerate the thermophillic biodegradation. The per cent of N, P, K of all the windrows were measured at an interval of 15 days for a period of 45 days. The sample size was kept constant at 50 for both control and experimental windrows. The experiment was conducted during the months of May and June of the year 2002.

Results and Discussion

The per cent nitrogen, phosphorous, potassium and carbon of solid waste samples have been depicted in Table 11.1. Table 11.2 represents the physico-chemical characteristics of the municipal wastewater. Figure 11.1 shows a gradual increase in per cent–nitrogen from the day one to 45th day in all the windows. The rate of increase in per cent nitrogen was highest in sewage water treated windrow compared to cowdung treated and negligible increase in control. This corroborates the earlier observations by Horwath *et al.*, 1995 on grass straw decomposition that the per cent nitrogen increased steadily from the first to 30th day in aerobic composting. The increase in per cent nitrogen may indicate an accumulation of microbial by-products with decline in microbial carbon. The minimum per cent of nitrogen was observed in the control.

Table 11.1: Changes in Certain Composting Parameters During Decomposition of Solid Organic Wastes of Bhubaneswar City

Days	N	P	K	C	C:N	C:P
			W_1			
0	0.5 ± 0.3	0.1 ± 0.01	0.4 ± 0.01	12 ± 3.1	24 ± 4.61	120 ± 8.45
15	0.53 ± 0.01	0.2 ± 0.01	0.51 ± 0.11	11.90 ± 2.1	22.45 ± 3.91	59.5 ± 3.36
30	0.56 ± 0.02	0.28 ± 0.03	0.58 ± 0.01	11.65 ± 2.4	20.80 ± 4.12	41.60 ± 2.99
45	0.6 ± 0.02	0.3 ± 0.01	0.9 ± 0.02	11.35 ± 2.7	18.91 ± 3.13	37.8 ± 2.35
			W_2			
0	0.5 ± 0.02	0.1 ± 0.01	0.4 ± 0.01	12 ± 2.1	24 ± 4.51	120 ± 8.4
15	0.52 ± 0.01	0.15 ± 0.01	0.5 ± 0.011	11.97 ± 2.2	23.01 ± 3.5	79.8 ± 3.5
30	0.54 ± 0.02	0.2 ± 0.01	0.51 ± 0.01	11.96 ± 2.2	22.14 ± 4.2	59.8 ± 2.99
45	0.55 ± 0.02	0.28 ± 0.02	0.7 ± 0.021	11.94 ± 2.4	21.70 ± 3.9	42.64 ± 2.34
			W_3			
0	0.5 ± 0.01	0.1 ± 0.01	0.4 ± 0.02	12 ± 2.4	24 ± 4.21	120 ± 8.34
15	0.51 ± 0.02	0.11 ± 0.01	0.4 ± 0.02	12 ± 2.41	23.98 ± 3.9	109 ± 7.89
30	0.51 ± 0.03	0.11 ± 0.01	0.41 ± 0.01	12 ± 2.4	23.98 ± 4.1	109 ± 7.89
45	0.51 ± 0.01	0.11 ± 0.01	0.41 ± 0.01	11.99 ± 2.7	23.98 ± 3.8	109 ± 7.89

W_1 = Windrow with sewage water treatment
W_2 = Windrow with cowdung slurry treatment
W_3 = Windrow with water treatment.

Table 11.2: Certain Physico-chemical Parameters of the Sewage Water

Parameters	*$\bar{X} \pm$ S.D.*
Temp °C	35.4 ± 3.1
pH	7.41 ± 2.1
DO mg/l	4.43 ± 1.1
BOD mg/l	26.6 ± 2.1
COD mg/l	76 ± 4.1
NO_3 mg/l	0.595 ± 0.02
PO_4 mg/l	1.83 ± 0.2
Cl mg/l	74.74 ± 4.1
Alkalinity	76 ± 4.1
Total Hardness	40 ± 3.2
Ca Hardness	32 ± 2.2
Mg Hardness	8 ± 0.4
Turbidity (NTU)	12.6 ± 1.2
Sulphate mg/l	12.85 ± 1.1
Na	21 ± 3.1
K	5.33 ± 0.2
TDS mg/l	266.3 ± 15.9
TSS	53.66 ± 3.1
TFS mg/l	317.33 ± 19.4
TC MPN/100 ml	1,60,000 ± 500
FC MPN/100 ml	70,000 ± 200
Conductivity μmho/cm	507 ± 50.8
Ammonical-nitrogen	17.33 ± 02.2
Total Fe. mg/l	0.508 ± 0.01

The per cent carbon of the windrows declined in all the windrows from the 1st to 45th day (Fig. 11.2). The highest decline was observed in windrow treated with wastewater in comparison to cowdung treated and control windrows. The decrease in per cent carbon in the windrows might be due to utilisation of carbon by aerobic microbes and subsequent oxidation of carbon to carbon dioxide (CO_2). Similar findings have been reported by, Horwath *et al.*, 1995 during aerobic decomposition of grass straw.

The C : N ratio is declined from initial 24 per cent to 23.98 per cent in the control windrow in 45 days (Fig. 11.3). In the windrow with cowdung slurry inoculation the C : N ratio declined from 24 per cent to 21.7 per cent. The highest rate of decline in the C : N ratio was observed in the windrow with municipal sewage water inoculation. The C : N ratio here declined from 24 per cent to 18.91 per cent. This indicates relatively faster decomposition process in the sewage water treated wastes in comparison to cowdung slurry treated and plain water treated wastes. This corroborates earlier reports on decomposition of organic wastes (Reinhart and Trainoi, 1995; Horwath *et al.*, 1995).

There was negligible increase in the per cent phosphorous (Fig. 11.4) and potassium (Fig. 11.5) in the control windrow in 45 days and in the cowdung treated windrow the per cent phosphorous increased from 0.1 to 0.28 and per cent potassium increased from 0.4 to 0.7. Similarly in the wastewater treated windrow the per cent phosphorous increased 0.1 to 0.3 and per cent potassium increased from 0.4 to 0.9.

The highest per cent phosphorous was recorded in the wastewater treated windrow on the 45th day (Fig. 11.4). Brown *et al.* (1998) have reported during composting of MSW that biological activity of microbes is dependent on availability of phosphorous. An increase in the biological activity has been reported with higher percentage of available phosphorous. The C : P ratio in all the windrows declined during 45 days of composting (Fig. 11.6). The maximum decline was observed in wastewater treated windrow and minimum in the control. A relatively low C : P ratio has been reported to be congenial for faster decomposition due to enhanced activity of decomposing microbes (Brown *et al.*, 1998). The present finding supports the above observation.

The above findings clearly indicate that municipality wastewater inoculum accelerates the rate of decomposition processes in municipality solid wastes in comparison to cowdung and water. Hence

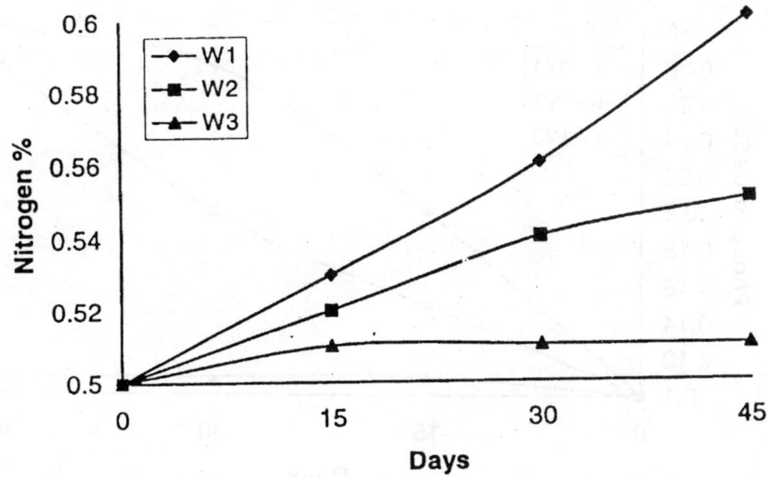

Fig. 11.1: Increase in Per cent Nitrogen During 45 Days of Composting.

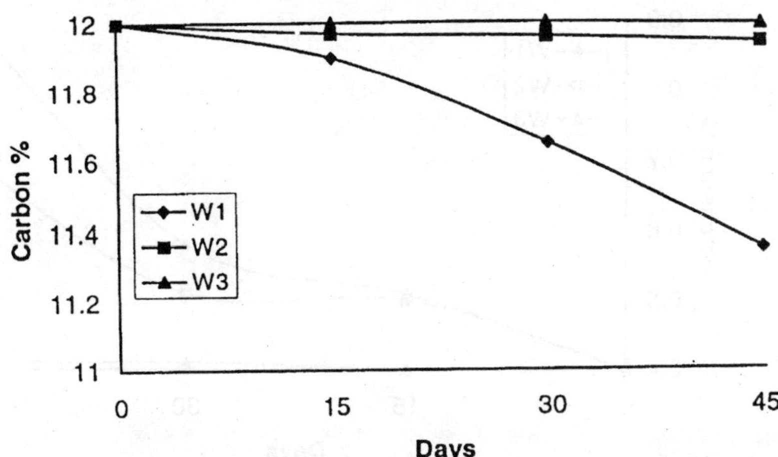

Fig. 11.2: Decrease in Per cent Carbon During 45 Days of Composting.

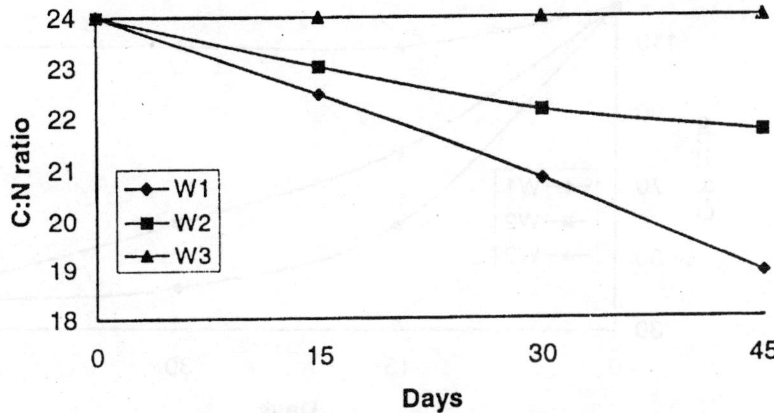

Fig. 11.3: Decrease in C:N Ratio During 45 Days of Composting.

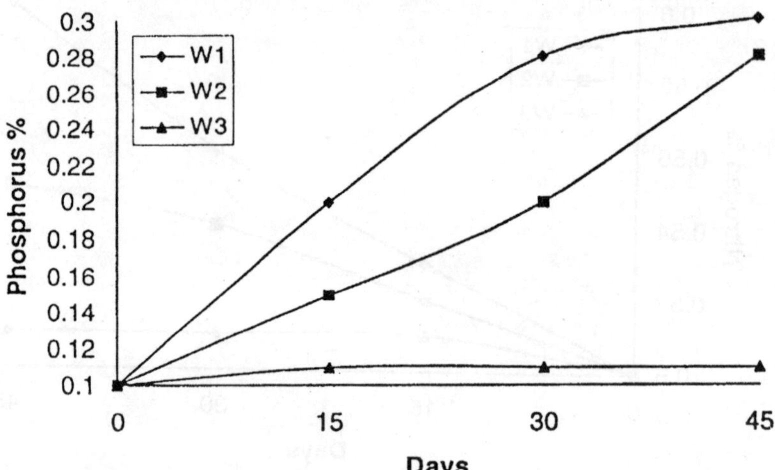

Fig. 11.4: Increase in Per cent Phosphorus During 45 Days of Composting.

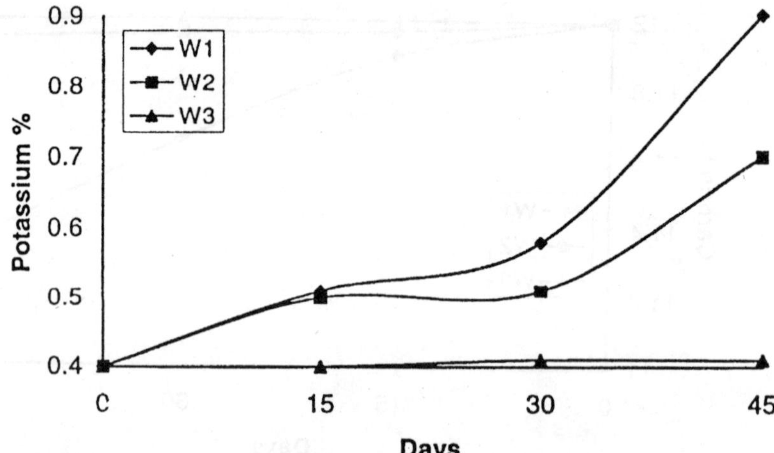

Fig. 11.5: Increase in Per cent Potassium During 45 Days of Composting.

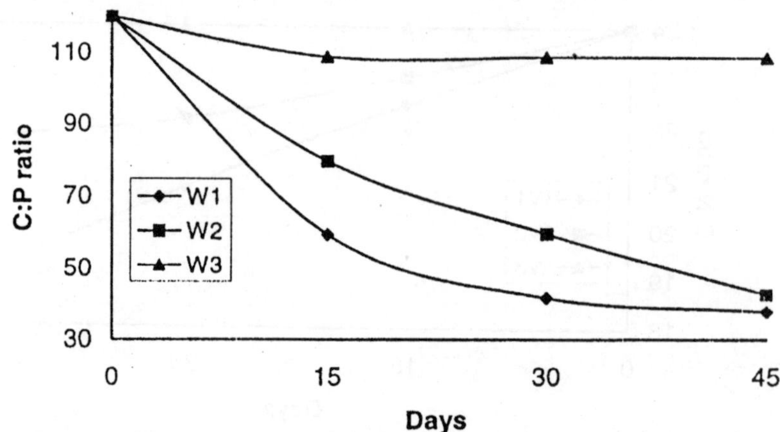

Fig. 11.6: Decrease in C:P Ratio During 45 Days of Composting.

the municipal wastewater from Bhubaneswar City can be utilized for large scale composting. This obviously will minimise water pollution problems and help in urban solid waste management.

References

APHA (1985). *Standard methods for the examination of water and wastewater, 16th Edition.* Washington, D.C.

Baruah, T.C., Barthakur, H.P. (1997). *A textbook of soil analysis*, pp. 1-322.

Brain Kimberly, H., Bouwkamp John, C. and Gouin, Francis, R. (1998). The influence of C : P ratio on the biological degradation of municipal solid waste. *Compost Science and Utilization*, 6(1): 53-58.

Brunt, L.P., Dean, R.B. and Patrick, P.K. (1985). Solid waste management selected topics. *WHO Composting*, pp. 37-77.

Chalapathi, M.V., Mallikarjuna, K., Thimmegowda, S., Devakumar, N., Rao, G. Gangadhar Eswar and Jayaramaiah, R. (2000). Urban garbage recycling: A means to control environment pollution and a way to sustainable agriculture: A Review. *Indian J. Environ. and Ecoplan.* 3(3): 761-770.

Hajra, J.N., Manna, M.C. and Kole, S.C. (2000). Recycling of Calcutta municipal solid wastes through production of enriched compost. *Waste Recycling and Resource Management in the Developing World*, pp. 197-201.

Horwath, R., William Elliott, F. Lloyd and Churchill, B. Donald (1995). Mechanisms regulating composting of high carbon to nitrogen ratio grass straw. *Compost Science and Utilization*, 3(3): 22-30.

Nanda, S.N. and Tiwari, T.N. (2000a). Physico-chemical characteristics of the municipal solid wastes of Sambalpur (India). *Indian J. Env. Prot.* (under publication).

Nanda, S.N. and Tiwari, T.N. (2000b). Characterisation of municipal solid wastes in Sambalpur town (Orissa). *Chemical Weekly* (under publication).

Rao, B., Raghavendra and Radhakrishna, D. (2001). Quality improvement of urban solid waste compost by enrichment with plant growth promoting micro-organisms. Proc. X National Symposium on Environment Assessment Division, BARC, Mumbai, June, 4-6.

Reinhart, R. and Trainor, D. (1995). Windrow composting of municipal bio-solids and yard waste. *Compost Science and Utilization*, 3(2): 38-46.

Saha, N., Bhattacharya, K.K. and Mukhopadhyay, N. (2000). Two-step composting for the management of ligno-cellulosic materials. *Waste Recycling and Resource Management in the Developing World*, pp. 213-218.

Williams Marcia, E. (1994). *Integrated municipal solid waste management*. McGraw-Hill, New York, Ch. 2.

12

Bacteriological Evaluation of Marketed Mineral Water

S. SUMATHY, R. GOWRISANKAR & S. RAMESH

P.G. Department of Microbiology, Sri Paramakalyani College, Azhwarkurichi, Tamil Nadu

ABSTRACT

Commercially available mineral water, both bottled ones and sachets were analysed for the presence of total heterotrophic bacterial population, coliforms and water-borne pathogens. It was observed that all mineral water brands were devoid of coliforms and water-borne pathogens but had considerable heterotrophic bacterial load excluding three bottled brands. It was found that sachets contained a higher bacterial load than the bottled brands. It was inferred that bottled brands should be preferred than sachets and that efforts should be taken by the manufacturers to ensure marketed mineral water free of heterotrophic bacteria.

Keywords: Mineral water, Bottles, Sachets, Heterotrophs, Coliforms.

Introduction

Mineral water is defined as bottled potable water with added minerals obtained from an approved underground source that contains not less than 500 mg/l of the total dissolved solids (TDS). WHO has provided a separate directive on the marketing of natural mineral water (Pike and Ridgeway, 1985). Under this directive, all natural mineral water sources need to be officially recognized and then it becomes a subject to the prescribed conditions for exploitation and treatment. These regulations do control their composition, packaging, labelling and advertising of water. Maximum permitted levels of eight chemical contaminants are listed in the directives, which also states that there must be a satisfactory evidence to show that the water should be free from pollution, as demonstrated by the absence of faecal marker organisms and that it contains no substance in such an amount so as to make it unwholesome. The directive contains specific recommendations on colony counts in water and emphasizes the absence of *Pseudomonas aeruginosa* both at the source and during circulation in trade (Pike and Ridgeway, 1985). Mineral water is generally free of coliforms but the presence of other pathogenic micro-organisms from mineral water has been reported (Warburton *et al.*, 1986). Species of *Aeromonas, Pseudomonas, Xanthomonas, Staphylococcus, Flavobacterium* and *Pasteurella* being isolated from mineral waters in Taiwan focuses their unhygienic status (Tsai *et al.*, 1997). With this as the background, it was planned to carry out a study on the bacterial quality of mineral water, both bottled and sachets marketed to in and around Tirunelveli, which happens to be a semi-urban town in South Tamil Nadu.

Materials and Methods

Mineral water samples that are commercially available in Tirunelveli District were collected. Both bottled ones (11) and those packed in sachets (10) were brought to the laboratory and enumerated for the presence of micro-organisms. Enumeration of heterotrophic bacterial population was carried out by pour plate technique using sterile plate count agar. Morphologically dissimilar colonies were pure cultured and identified on the basis of their microscopic, biochemical and physiological characteristics (Aiso and Sirnudo, 1962; Pallerno, 1992; Slepecky and Hemphill, 1992). Total and faecal coliforms were enumerated using five tubes MPN technique (APHA, 1985). Further, the mineral water samples were also tested for the presence of common water-borne pathogens *viz.*, *Vibrio* sp., *Salmonella* sp. and *Shigella* sp., adopting standard methodologies (APHA, 1985).

Results and Discussion

The mineral water packed both in bottles and sachets were found to contain no coliforms. They were also devoid of the common water-borne pathogens like *Vibrio* sp., *Salmonella* sp. and *Shigella* sp. It was observed that all the mineral water samples tested excluding 3 brands were found to contain a considerable heterotrophic bacterial load. The bacterial load was found to be higher in sachets than in bottled brands (Table 12.1).

Bacteria belonging to the genera *Micrococcus* sp. were found to be predominant (43 per cent) followed by *Pseudomonas* sp. (30 per cent) and *Bacillus* sp. (27 per cent) in the mineral water samples (Fig. 12.1). *Micrococcus* sp. was found in 7 brands of bottle and in 9 sachets with an overall occurrence of 76.19 per cent (16 of 21 samples) while *Bacillus* sp. was found in 4 brands of bottle and sachet each (overall 38.09 per cent). *Pseudomonas* sp. was found in 3 bottled and 7 sachet brands of mineral water samples (overall 47.61 per cent).

In line with the present findings, earlier works too underlines the absence of coliforms in mineral waters. Reports of Warburton *et al.* (1986) on the microbiological quality of bottled water sold in Canada indicated that none of the 114 lots of bottled water contained any faecal coliforms. Tamagnini and Gonzalez (1997) were also able to prove that the sterile water bottled in non-returnable and returnable containers did not harbour any coliforms or *E. coli* in any of the sample.

Table 12.1: Heterotrophic Bacterial Load in Bottled Market Mineral Water

Samples Tested	*Total Heterotrophic Bacterial Load (CFU/ml⁻¹)*
Bottled Water	
Brand–A	6.925 x 10^1
Brand–B	1.190 x 10^2
Brand–C	1.145 x 10^2
Brand–D	Nil
Brand–E	1.167 x 10^2
Brand–F	6.9 x 10^1
Brand–G	3.9 x 10^1
Brand–H	Nil
Brand–I	1.162 x 10^2
Brand–J	1.267 x 10^2
Brand–K	Nil
Sachets	
Brand–a	TLTC
Brand–b	7.6 x 10^1
Brand–c	1.042 x 10^2
Brand–d	1.042 x 10^2
Brand–e	4.45 x 10^1
Brand–f	4.95 x 10^1
Brand–g	5.7 x 10^1
Brand–h	3.975 x 10^1
Brand–I	4.0 x 10^1
Brand–j	6.7 x 10^1

TLTC-Too Low To Count; NIL-No bacterial growth recorded.

On the other hand, the occurrence of heterotrophic bacteria in mineral water has frequently been reported. Tsai *et al.* (1997) while evaluating the uncarbonated bottled mineral water in Taiwan had isolated *Pseudomonas* sp. in 4 of the 48 samples. Apart from *Pseudomonas* they also found that the mineral water was contaminated with *Aeromonas* sp., *Xanthomonas* sp., *Flavobacterium* sp., *Pasteurella* sp. and *Staphylococcus* sp. Gonzalez *et al.* (1987) too have reported the incidence of *Micrococcus* sp. in mineral water samples collected from three Spanish companies. They reiterated that the heterotrophic bacterial population continued to increase from 10^1–10^2 CFU/ml to 10^5–10^6 CFU/ml even when stored at 5°C. Daschner (1997), had reported a heavily contaminated nature of mineral water in a hospital in Germany. The mineral water samples were found to be loaded heavily with Gram-negative, opportunistic pathogens that were mostly non-fermentors and *Aspergillus* sp.

Based on these results, it may be concluded that most of the tested mineral water brands, despite of the absence of coliforms cannot be termed entirely safe bacteriologically as they had significant bacterial load that ranged from 3.9 × 10^1 – 1.042 × 10^3 CFU/ml^{-1} with a variety of heterotrophic bacteria *viz.*, *Micrococcus* sp., *Pseudomonas* sp., and *Micrococcus* sp. While the heterotrophic bacteria were recovered from both bottle and sachet mineral water samples, the percentage incidence of the bacteria was much higher in sachets (90 per cent; 9/10) than the bottles (73 per cent; 8/11). Nevertheless, on a comparative basis, mineral water in disposable bottles are rather bacteriologically safer compared to those that are marketed in sachets. Further, it is recommended to store these marketed mineral water bottles in refrigerated conditions till use in order to control the proliferation of heterotrophic bacteria during storage (as they are mostly mesophiles). Moreover, analytical authorities should include heterotrophic bacterial load as a part of standards slammed on mineral water to ensure the true safety of mineral water.

Acknowledgements

The authors wish to thank the Tamil Nadu State Council for Science Technology, Chennai for financial support.

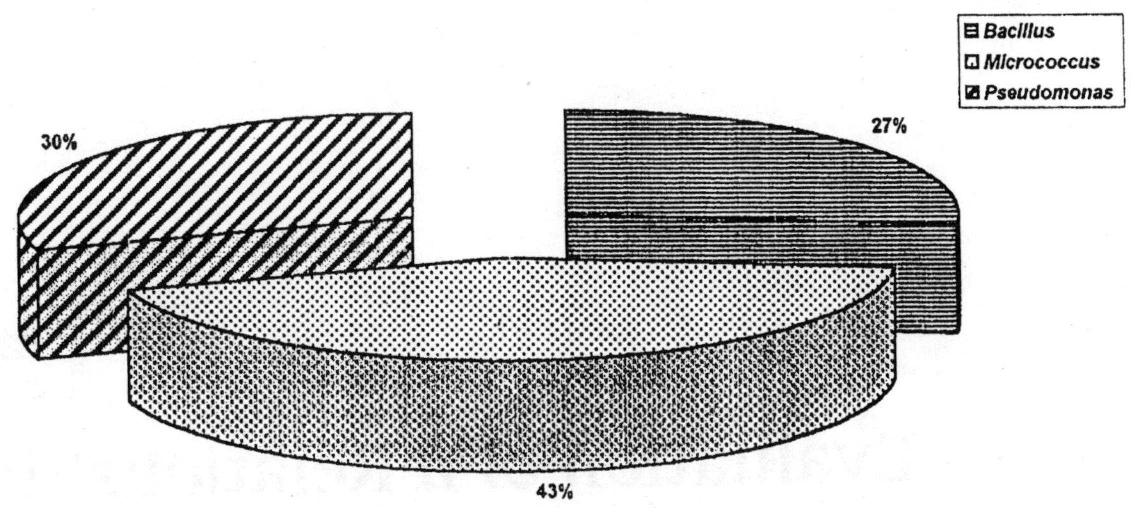

Fig. 12.1: Generic Distribution of Heterotrophic Bacteria in Mineral Water Samples.

References

Aiso, K and Simudo, V. (1962). Occurrence and distribution of heterotrophic bacteria in the sea water from the sea Kamogawa Bay. *Bull. Jap. Sci. Fish.*, 28: 1133.

APHA, AWWA, WPCF (1989). American Public Health Association, Washington, D.C.

Daschner, F.D. (1997). Holy water, tap water, mineral water or water filters. *J. Hospital. Infection*, 35(1): 71-72.

Gonzalez, C., Gutierrez, C. and Grande, T. (1987). Bacterial flora in bottled uncarbonated mineral drinking water. *Can. J. Microbiol.*, 33: 1120-1125.

Palleroni, N.J. *The Prokaryotes* (Eds. Balows, A., Truper, H.G., Dworkin, M., Harder, W., and Schleifer, K.H.), Springer-Verlag, New York, (1): 3071-3103.

Pike, E.B., and Ridgway, J.W. (1985). The microbiological implications in European Community Directives related to water cycle. *J. Applied Bacteriol. Symposium Suppl.*, 1058 -1268.

Slepecky, R.A., and Hemphill, H.E. *The Prokaryotes* (Eds. Balows, A., Truper, H.G., Dworkin, M., Harder, W., and Schleifer, K.H.), Springer-Verlag, New York, (1): 1663 -1696.

Tamagnini, L.M, and Gonzalez, R. (1997). Bacteriological stability and growth kinetics of *Pseudomonas aeruginosa* in bottled water. *J. Applied Microbiol.*, 83(1): 91 -94.

Tsai, Guo-Jane, and Yu-Shou-Chin (1997). Microbiological evaluation of bottled uncarbonated mineral water in Taiwan. *International J. Food Microbiol.*, 37(2-3): 137- 143.

Warburton, W. Donald, Pearl I. Peterkin, Karl F. Weiss, and Murray A. Johnston. (1986). Microbiological quality of bottled water sold in Canada. *Can. J. Microbiol.*, 32(11): 891 -893.

13

Evaluation of a Relationship Between BOD and COD for River Nagavali and River Kolab in Koraput District, Orissa

SASWAT KUMAR MOHANTY
Scientific Division, Orissa State Pollution Control Board, Bhubaneswar
DIPIKA PATNAIK & SWOYAM P. ROUT
Environmental Laboratory, Department of Chemistry, Utkal University, Vani Vihar, Bhubaneswar

ABSTRACT

A systematic relationship of correlation coefficient has been found out between BOD and COD of river Nagavali and river Kolab flowing near Jaykaypur and Jeypore in Koraput district. The results shows a positive correlation between BOD and COD of both the rivers. The values r_1 and r_2 are 0.989 and 0.991 for river Nagavali and river Kolab respectively. The well known least square method was applied to the experimental data to find out the calculated values by using equation Y=Ax+B. The calculated values are in good agreement with experimental values.

Introduction

Water is the essential ingredient of life. In fact every aspect of civilization directly or indirectly depend on the water resources. The non-judicious disposal of industrial and domestic waste contaminated our water bodies with toxic elements. In Koraput district two rivers namely river Nagavali flows near T.K. Paper Mill, Jaykaypur and river Kolab is flowed near Ballarpur Paper Industries, Jeypore. The two paper mills discharge their wastewater to the two rivers. As a result the BOD and COD level of both the rivers increase remarkably. To determine both the parameters quickly it is necessary to find out the relationship existing between two parameters. This is achieved with the help of correlation study.

Materials and Methods

Composite sampling method was adopted for collection of river water sample. The samples were collected in clean plastic bottles thoroughly cleaned with 1:1 HNO_3 and rinsed several times. For BOD and COD parameters suitable preservative was added. The samples were collected during summer, monsoon and winter. The samples were collected at five different locations of both the rivers, and the description of the sampling points are reflected in Table 13.1 and Table 13.2. The samples were analysed following standard procedure (APHA, AWWA, WPCF, 1995). The analysis result of BOD and COD of both the rivers are shown in Table 13.3.

Table 13.1: River Nagavali

Sl.No.	*Location of the Sampling Points*	*Code*
1.	Upstream to discharge point of industrial effluent	NSW_1
2.	Industrial effluent before discharged to river Nagavali	NSW_2
3.	Downstream of river Nagavali, 1 km away from the discharge point of industrial effluent	NSW_3
4.	Downstream of river Nagavali, 4 km. away from the discharge point of industrial effluent	NSW_4
5.	Near liatipathar, 6 km away from the discharge point of industrial effluent	NSW_5

Table 13.2: River Kolab

Sl.No.	*Location of the Sampling Points*	*Code*
1.	Upstream to discharge point of industrial effluent	KSW_1
2.	Industrial effluent before discharged to river Kolab	KSW_2
3.	Downstream of river Kolab, 1 km away from the discharged point of industrial effluent	KSW_3
4.	Downstream of river Kolab 4 km away from the discharge point of industrial effluent	KSW_4
5.	Further downstream, 6 km away from the discharge point of industrial effluent	KSW_5

Statistics

Correlation coefficient was calculated by using the following equation

$$R = \frac{N\Sigma x_i y_i - \Sigma x_i y_i}{[\{-\Sigma x_i^2 - (\Sigma x_i)^2\}\{N\Sigma y_i^2 - (\Sigma y_i)^2\}]^{1/2}} \quad \text{.............................. eqn. (i)}$$

Where

x and y are two variables

N is the No. of observations.

While A and B of the parameters are calculated by using eqn. (ii) and eqn. (iii)

$$A = \frac{N\Sigma x_i y_i - \Sigma x_i y_i}{N\Sigma x^2 - (\Sigma x_i)^2} \quad \text{..................} \quad \text{eqn. (ii)}$$

$$B = 1/N\,[(\Sigma y_i - A(\Sigma x_i)]^2 \quad \text{..................} \quad \text{eqn. (iii)}$$

Y values can be calculated by using A and B values in eqn. (iv)

$$Y = \quad AX + B \quad \text{..................} \quad \text{eqn. (iv)}$$

The values of A, B and r between BOD and COD of both the rivers have been found out and by putting these values two new equations are obtained. The equations are:

$$BOD_1 = A \times COD_1 + B_1 \quad \text{..................} \quad \text{for river Nagavali}$$

$$BOD_2 = A_1 \times COD_2 + B_2 \quad \text{..................} \quad \text{for river Kolab}$$

The values of A = 0.1582 and B_1 = 1.238 for river Nagavali.

The values of A_1 = 0.1406 and B_2 = 0.659 for river Kolab.

Hence the above equation becomes

$$BOD_1 = 0.1582 \times COD_1 + 1.238 \quad \text{...........} \quad \text{for river Nagavali.}$$

$$BOD_2 = 0.1406 \times COD_2 + 0.659 \quad \text{..................} \quad \text{for river Kolab.}$$

By putting COD values in the above equation, BOD values for both the rivers can be calculated. It was observed that the calculated BOD values are in good agreement with the experimental values.

Table 13.3: BOD and COD Values of Nagavali and Kolab for the Year 1997 and 1998

Code No.	*Year 1997*		*Year 1998*	
	BOD (mg/l)	*COD (mg/l)*	*BOD (mg/l)*	*COD (mg/l)*
NSW_1	02	09	02	11
NSW_2	52	322	56	348
NSW_3	39	248	44	276
NSW_4	13	80	16	108
NSW_5	06	42	08	54
KSW_1	02	10	03	14
KSW_2	32	228	39	278
KSW_3	22	162	27	192
KSW_4	14	106	16	114
KSW_5	04	32	07	46

Results and Discussion

The relationship between BOD and COD of river Nagavali and river Kolab reveals that the parameters are highly correlated. The correlation coefficient r_1 for river Nagavali is 0.989 and r_2 for river Kolab is 0.991 indicating high correlation between BOD and COD. With the help of correlation equations BOD values can be easily determined after knowing the COD values. The calculated and experimental results of BOD values for both the rivers are shown in Table 13.4 and Table 13.5 which are in good agreement.

Table 13.4: Predicted Values of BOD for River Nagavali for the Year 1997 and 1998

Sampling Code	Parameter	Year 1997		Year 1998	
		Calculated Value	Experimental Value	Calculated Value	Experimental Value
NSW_1	BOD_1	2.66	2.0	2.98	2.0
NSW_2	BOD_1	52.18	52.0	56.29	56.0
NSW_3	BOD_1	40.47	39.0	44.9	44.0
NSW_4	BOD_1	13.89	13.0	18.32	16
NSW_5	BOD_1	7.88	6.0	9.78	8.0

Table 13.5: Predicted Values of BOD for River Kolab for the Year 1997 and 1998

Sampling Code	Parameter	Year 1997		Year 1998	
		Calculated Value	Experimental Value	Calculated Value	Experimental Value
KSW_1	BOD_2	2.1	2.0	2.63	3.0
KSW_2	BOD_2	32.7	32.0	39.7	39.0
KSW_3	BOD_2	23.4	22.0	27.6	27.0
KSW_4	BOD_2	15.6	14.0	16.03	16.0
KSW_5	BOD_2	5.2	4.0	7.1	7.0

The correlation study is a very good tool for developing equation by applying $y=AX+B$ between two parameters. Here BOD of both the rivers are calculated within few minutes by knowing the respective COD values of both the rivers instead of waiting for five days to get the BOD result. By the help of correlation studies number of parameters can be calculated within very short period which in return saves time, energy and money.

References

APHA, AWWA, WPCF (1995). *Standard methods for examination of water and wastewater, 19th Edition.*

Kumar, S. (1993). Correlation among water quality parameters for groundwater in Barmer district. *Indian J. Env. Prot.*, 13(7): 487.

Mohanty, S.K. (2001). *Studies on strategies to control environmental pollution in major industries of Koraput area.* Ph.D. Thesis, Utkal University.

Rao, Somasekhar K. (1993). Correlation among water quality parameters of Nuzvid Mandalam. *Indian J. Env. Prot.*, 13(4): 261.

Tiwari, T.N. and Ali, Manzoor (1998). Correlation among water quality parameters of industrial waste, sugar industry. *Ind. J. Env. Prot.*, 8(1): 93-97.

14

Seasonal Variations in the Water Quality Index for Vani Vihar Lake in Bhubaneswar, Orissa

HRUSHIKESH BEHERA & SWOYAM P. ROUT

Environmental Analytical Laboratory, Department of Chemistry, Utkal University, Vani Vihar, Bhubaneswar, Orissa

LAXMIDHARA PAL

Orissa State Pollution Control Board, Bhubaneswar

ABSTRACT

This paper deals with the water quality of Vani Vihar lake and a simple method has been suggested to calculate the water quality index taking physico-chemical and biological parameters. From water quality index it is observed that water quality of the lake is not suitable for designated use, seasonal variation indicates that the quality of water gets highly polluted during summer and least in winter season. Suitable remedial measures have been suggested to mitigate the problem.

Keywords: Vani Vihar lake, Water quality index, Seasonal variation.

Introduction

Water is one of the valuable natural resources and the quality of water is of vital concern for the mankind since it is directly linked with human welfare. Growing population, industrial proliferation, urbanisation, increasing living standards and wide sphere of human activities exert heavy pressure on our vast but limited water resources. Since water pollution of a growing city is always in the rising trend, its effect on the environment must be studied thoroughly.

The new capital Bhubaneswar though has limited water resources, but its rivers, lakes, ponds etc. provides water for various purposes. A recent study reveals that the groundwater in Bhubaneswar has also been depleted to a large extent. Thus surface water is the major source to cater to the needs of the people in the city.

Considering the above factors an attempt has been made to assess the water quality of Vani Vihar lake. In this paper we have suggested a simple method to evaluate the water quality index from the physico-chemical and microbiological characteristics of water by comparing with inland surface water classified as Class-B (BIS 2296-1982). This paper also deals with various mitigative suggestions including remedial measures and public awareness.

Study Area

Bhubaneswar the capital city of Orissa is located at latitude of 20°15′ N and longitude of 85°52′ E at an altitude of 45m above the M.S.L. An important city of the temple triangle of Orissa, Bhubaneswar has direct air, road, and rail connection with the other major cities of India. The population of Bhubaneswar was 16,512 in 1951 and at present the population is believed to be more than 7 lakhs.

Description of Study Site: Vani Vihar Lake

Vani Vihar lake is enclosed within the premises of Utkal University campus in Bhubaneswar. It is a shallow freshwater body situated at about 65 km inland from the coastal line.

It is an attractive wetland ecosystem located by the side of National High Way No. 5 (NH-5) has an area of about 27,000 sq.m. and has a depth of about 3.5 meters at the centre. The total catchment area in all the streams is 3,400 sq.m. The prominent streams supplying water to this lake are Naharathali nallah, Guali nallah, Kashia nallah, Bauli nallah, Chandaka nallah, and Baramunda nallah. All these are united before Jayadev Vihar and flow into the lake as a single stream which passes through extensive areas of crop fields. The outlet of Vani Vihar lake meets the river Kuakhai passing through the industrial estate. To the south of Vani Vihar lake, there is a vast stretch of low lying marshy land of about 80 sq.km. separated from the lake by NH-5 and a number of residential locations situated nearby it. Pressure due to human activities has been increasing since 1962, the year in which university campus was established. People of adjoining regions of the university *viz.*, Sahid Nagar, Acharya Vihar, Rasulgarh, VSS Nagar, and Unit-IX, use the lake in various ways. The lake is chiefly used for bathing, recreation, washing clothes, waste disposal, cleaning of motor vehicles etc. Near the catchment area of Vani Vihar lake a number of villages are located and the Chandaka Industrial Complex comprising of small scale industries like plastic and polythene goods, pharmaceutical and chemicals, dyeing are also located.

With all these considerations in view of deterioration of water quality, an attempt has been made to study the biological characteristics and effect of some physio-chemical parameters of Vani Vihar lake.

The samples were collected from three different sites of Vani Vihar lake *viz.*, Site-I, Site-II and Site-III. Site-I represents the inlet of a waste channel, Site-II is a washermen point and Site-III is a point where motor vehicles are cleaned and bathing point. There is an outlet of the lake located very

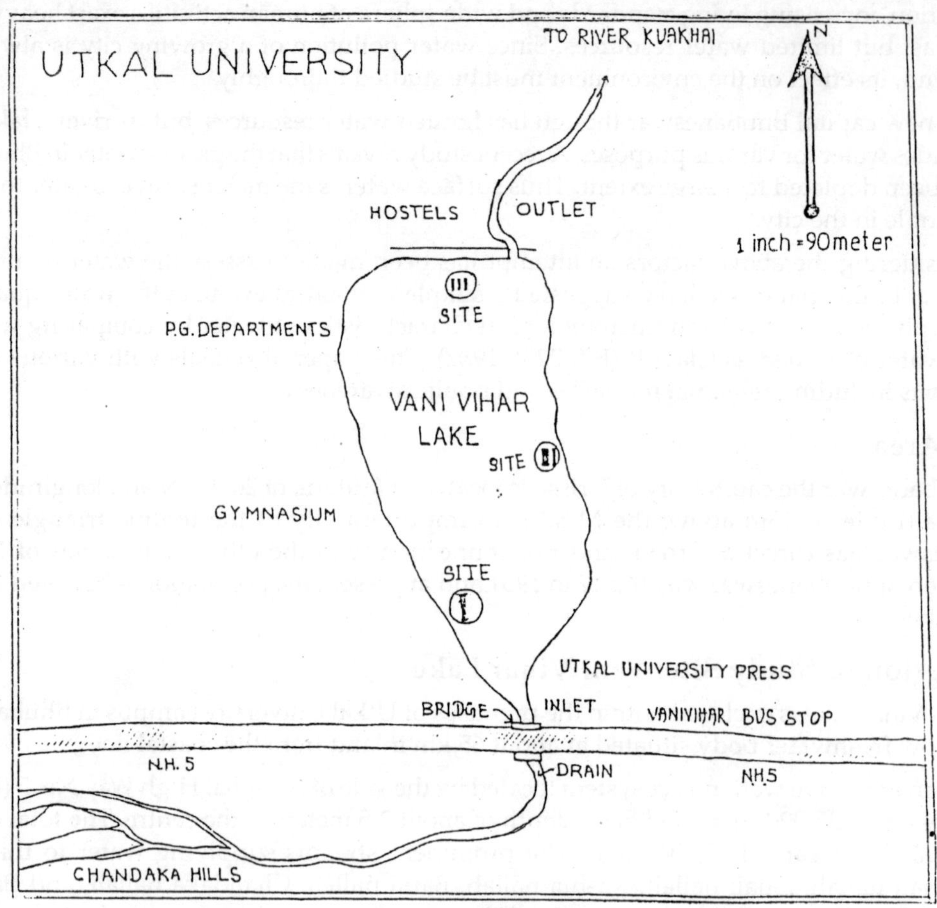

Map of Vani Vihar Lake

near to Site-III. Site-I is present near the entrance gate of the university campus. It is deeper in comparison to other sampling stations. Here the lake is connected to a sewage channel. Most of the time the channel is covered with hydrophytes. Site-II is located near the University press. The growth of the hydrophytes is less prominent. Site-III is present near Students Hostel and is used by the workers of the hostels for bathing and washing. This site is also used for washing of utensils of the nearby hotels of Govinda Sahu Market complex. Trucks, buses and other motor vehicles are also cleaned at this site. The water from the lake flows into a nearby canal connected to river Kuakhai. The growth of hydrophytes is not very prominent at this site.

Methodology

Water samples were collected during February 2000 to December 2000 at monthly intervals from different sites of Vani Vihar lake. Determination of pH, specific conductance and dissolved oxygen were done at the sampling sites for analysing the parameters. The samples were collected in one litre polythene bottles, sterilised bottles were taken for examining faecal coliform as well as total coliform. The parameters such as alkalinity, TSS, TDS, total hardness, calcium hardness, magnesium hardness, nitrate, phosphate, sulphate, chloride, total iron, chromium (Cr^{+6}), COD, BOD, total coliform (TC) and faecal coliform (FC) were analysed within 24 hours according to standard methods.

Results and Discussion

The value of pH, DO, BOD, Total Coliform (TC) and Faecal Coliform (FC) for three different sites of Vani Vihar lake have been reported in Table 14.1. Table 14.2 represents the variation of mean values of above parameters for Class B (Inland Surface Water). Table 14.3 depicts inland surface water standards.

Table 14.1: Variation of Parameters at Different Sites in 2000

Sl.No.	*Parameters*	*Months*	*Sample I*	*Sample II*	*Sample III*
1.	pH	Feb	6.9	7.7	7.6
		April	7.94	8.32	8.30
		June	8.30	8.00	7.90
		Aug	7.42	7.18	7.43
		Oct	7.1	7.6	7.5
		Dec	6.9	7.3	7.2
2.	DO (mg/l)	Feb	4.5	6.7	6.8
		April	1.0	2.1	2.5
		June	1.8	11	13.6
		Aug	4.4	9.4	11.8
		Oct	4.5	5.8	5.4
		Dec	4.8	6.9	6.2
3.	BOD (mg/l)	Feb	17.2	13.0	13.3
		April	28.2	12.1	16.1
		June	29	23	26
		Aug	27	20	24
		Oct	17	12	14
		Dec	18	12	15
4.	Total Coliform (MPN/100ml)	Feb	28000	25000	25000
		April	16000	12000	11000
		June	20000	16000	16000
		Aug	18000	13000	10000
		Oct	2100	1800	1900
		Dec	97000	90000	93000
5.	Fecal Coliform (MPN/100 ml)	Feb	13000	10000	10000
		April	2400	1800	1600
		June	3500	3000	2400
		Aug	3000	2400	2000
		Oct	1800	1550	1650
		Dec	96000	91000	93000

Table 14.2: Variation of Mean Values of Parameters in Different Months of 2000

Sl.No.	Sampling Month	pH	Range mg/l	DO mg/l	Range mg/l	BOD_5 at 20°C mg/l	Range mg/l	TC MPN/ 100 ml	Range MPN/ 100 ml	FC MPN/ 100 ml	Range MPN/ 100 ml
1.	Feb	7.4	7.2–	6.0	1.87–	14.5	14.0–	26000	1900–	11000	1650–
2.	April	8.19	8.19	1.87	8.8	18.8	26.0	13000	93000	1930	93000
3.	June	8.07		8.8		26.0		17300		2970	
4.	Aug	7.34		8.53		23.7		13700		2470	
5.	Oct	7.5		5.4		14.0		1900		1650	
6.	Dec	7.2		6.2		15.0		93000		93000	

Table 14.3: Inland Surface Water Classified as Class-B (BIS: 2296–1982)

Sl.No.	Characteristic	Tolerance Limits
1.	pH	6.5–8.5
2.	DO mg/l	5 (min)
3.	BOD mg/l	3 (min)
4.	TC MPN/100ml	500 (max)
5.	FC MPN/100ml	100 (max)

The Water Quality Index of different months are tabulated in Table 14.4. Table 14.5 represents the yearly average value of Water Quality Index. Variation of mean values of different parameters during various months is depicted in Figure 14.1. Figure 14.2 shows WQI variation.

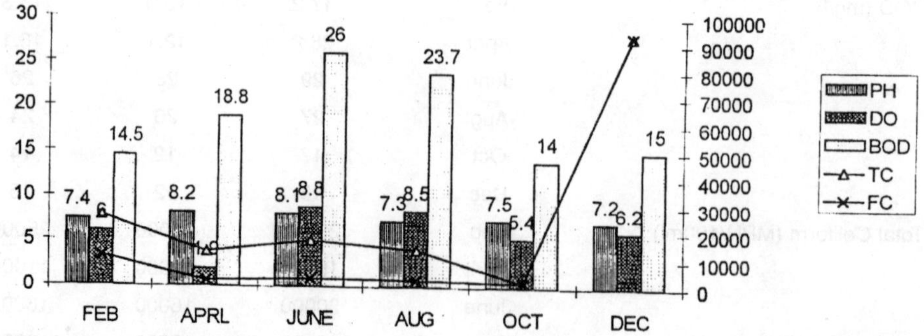

Fig. 14.1: Variation of Mean Values of Different Parameters During Various Months

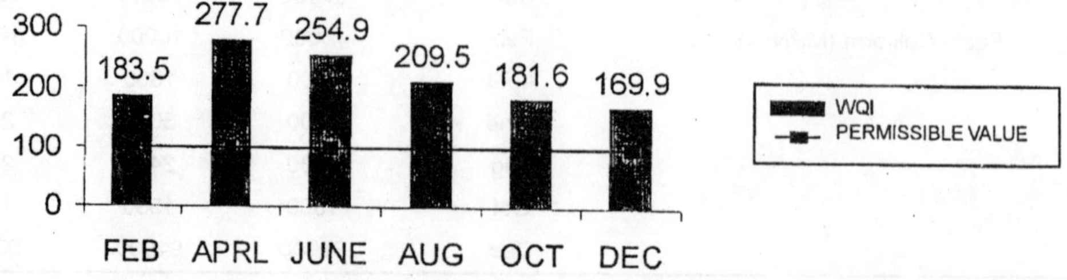

Fig. 14.2: WQI Variation at Vani Vihar Lake

Table 14.4: WQI for Different Months in 2000

Sl.No.	Parameters	Observed Value (V_n)	Standard Value(V_s)	W_n=K/S_n	Quality Rating(q_n)	$log_{10}q_n$	$W_n log_{10}q_n$
1.	pH	7.400000	8.500000	0.177452	26.666674	1.425969	0.253041
2.	DO	6.000000	5.000000	0.301668	89.583336	1.952227	0.588925
3.	BOD	14.500000	3.000000	0.502780	483.333344	2.684247	1.349586
4.	TC	26000.000	500.0000	0.003017	5200.00000	3.716003	0.011210
5.	FC	11000.000	100.0000	0.015083	11000.0000'	4.041393	0.060958

$\Sigma W_n log_{10} q_n$ = 2.263719

$(WQI)_{February}$ = 183.535078

Sl.No.	Parameters	Observed Value (V_n)	Standard Value(V_s)	W_n=K/S_n	Quality Rating(q_n)	$log_{10}q_n$	$W_n log_{10}q_n$
1.	pH	8.1900000	8.500000	0.177452	79.333305	1.899456	0.337062
2.	DO	1.8700000	5.000000	0.301668	132.604172	2.122557	0.640308
3.	BOD	18.799999	3.000000	0.502780	626.666626	2.797037	1.406294
4.	TC	13000.000	500.0000	0.003017	2600.00000	3.414973	0.010302
5.	FC	1930.0000	100.0000	0.015083	1930.00000	3.285557	0.049557

$\Sigma W_n log_{10} q_n$ = 2.443523

$(WQI)_{April}$ = 277.666145

Sl.No.	Parameters	Observed Value (V_n)	Standard Value(V_s)	W_n=K/S_n	Quality Rating(q_n)	$log_{10}q_n$	$W_n log_{10}q_n$
1.	pH	8.0700	8.500000	0.177452	71.333313	1.853292	0.328870
2.	DO	8.8000	5.000000	0.301668	60.416668	1.781157	0.537318
3.	BOD	26.000	3.000000	0.502780	866.666687	2.937852	1.477094
4.	TC	17300.000	500.0000	0.003017	3460.00000	3.539076	0.010676
5.	FC	2970.0000	100.0000	0.015083	2970.00000	3.472756	0.052381

$\Sigma W_n log_{10} q_n$ = 2.406339

$(WQI)_{June}$ = 254.881861

Sl.No.	Parameters	Observed Value (V_n)	Standard Value(V_s)	W_n=K/S_n	Quality Rating(q_n)	$log_{10}q_n$	$W_n log_{10}q_n$
1.	pH	7.3400000	8.500000	0.177452	22.666677	1.355388	0.240516
2.	DO	8.5300000	5.000000	0.301668	63.229172	1.800918	0.543279
3.	BOD	23.700000	3.000000	0.502780	790.00000	2.897627	1.456869
4.	TC	13700.0000	500.0000	0.003017	2740.0000	3.437751	0.010371
5.	FC	2470.00000	100.0000	0.015083	2470.0000	3.392697	0.051173

$\Sigma W_n log_{10} q_n$ = 2.302208

$(WQI)_{August}$ = 200.543423

Sl.No.	Parameters	Observed Value (V_n)	Standard Value(V_s)	W_n=K/S_n	Quality Rating(q_n)	$log_{10}q_n$	$W_n log_{10}q_n$
1.	pH	7.500000	8.500000	0.177452	33.333332	1.522879	0.270238
2.	DO	5.400000	5.000000	0.301668	95.833336	1.981517	0.597760
3.	BOD	14.00000	3.000000	0.502780	466.666656	2.669007	1.341923
4.	TC	1900.000	500.0000	0.003017	380.000000	2.579784	0.007782
5.	FC	1650.000	100.0000	0.015083	1650.00000	3.217484	0.048531

$\Sigma W_n log_{10}q_n$ = 2.266234

$(WQI)_{October}$ = 181.601148

Sl.No.	Parameters	Observed Value (V_n)	Standard Value(V_s)	W_n=K/S_n	Quality Rating(q_n)	$log_{10}q_n$	$W_n log_{10}q_n$
1.	pH	7.200000	8.500000	0.177452	13.333321	1.124938	0.199622
2.	DO	6.200000	5.000000	0.301668	87.500000	1.942008	0.585842
3.	BOD	15.00000	3.000000	0.502780	500.00000	2.698970	1.356988
4.	TC	93000.00	500.0000	0.003017	18600.000	4.269513	0.012880
5.	FC	930000.0	100.0000	0.015083	93000.000	4.968483	0.074942

$\Sigma W_n log_{10}q_n$ = 2.230274

$(WQI)_{December}$ = 169.931528

Table 14.5: Yearly Average Value of WQI

Sl.No.	Month	WQI	Range of WQI	Yearly Average Value of WQI
1.	February	183.5		
2.	April	277.7		
3.	June	254.9	169.9–277.7	211.35
4.	August	200.5		
5.	October	181.6		
6.	December	169.9		

The results obtained are described below.

pH

It is clear from Table 14.1 that variation in pH is negligible for various water samples for different sites. However, from Table 14.2, pH value ranges from 8.19 to 7.2 from the month of April to December. The pH value of the lake water is alkaline in nature. Slight fluctuation in pH value in some months were noticed because of temperature difference of water body and inflow of drain water. A comparison of pH values of different sampling sites with the standard values showed that observed pH value was well within the limits.

DO

The DO value in water sample was found to be in the range of 1.0 mg/l to 13.6 mg/l at all the sampling sites. It is interesting to note that DO value at sampling Site-I was lowest when compared with Site-II and Site-III. This may be due the fact that Site-I is mainly due to sewage having high organic matter leading to low DO value. The increase in DO value at Site-II and Site-III may be due to natural purification, low organic content and aeration of the water body.

Dissolved oxygen content of water sample in the month of April was low (1.87 mg/l) which may be due to high temperature leading to evaporation of water, with the advent of eutrophication, there is an overall decrease in oxygen content. DO content of water sample in other months were appreciably better which was mainly due to low temperature, low organic content, high algal activity and self-purification capacities of water body during that period. A comparison of DO content with its standard limit (5 mg/l) showed that it was much less in April (1.87 mg/l) but was appreciably better in the rest of the months.

BOD

It is apparent from Table 14.1 that values of biochemical oxygen demand of different months is low at Site-II and Site-III and is high at Site-I as indicated in Fig. 14.1. The highest BOD value at Site-I may be due to the decomposition of algal cells and sewage disposal. The presence of high BOD load at sampling Site I indicates that the lake is receiving a high quantity of organic wastes at this point. The comparatively lower BOD values of Site-II and Site-III were due to self-purification of water body at these sites. It is important to note that a comparison of BOD values of different sampling sites with its standard limits show that Vani Vihar lake water is not at all suitable for bathing purpose. BOD value was found to be more in June and August which may be due to high microbial activity. Regular increase in BOD values from April to August may be due to decrease in temperature. Again decrease and constancy of BOD value for the month of October to February may be due to high human activities.

Coliform Bacteria (TC, FC)

The coliform density (both TC and FC) at Site-I is maximum and Site-III is minimum. The maximum density of coliform at Site-I may be due to fact that raw domestic sewage is discharged into the lake through this site. Fluctuation in coliform density may be due to irregular human activities, presence of bacteria from the rain water. A comparison of inland surface water tolerance limits with data for coliform group of bacteria indicates that water of Vani Vihar lake is unsuitable for human purpose. The presence of coliform organisms in high numbers was indicative to the fact that pathogenic organisms were also present in Vani Vihar lake.

The mean value of parameter like electrical conductivity, alkalinity, TSS, TDS, total hardness, calcium hardness, magnesium hardness, chloride, nitrate phosphate, sulfate, total iron, chromium (hexavalent) and COD are represented in Table 14.6.

Table 14.6: Variation of Mean Values of Parameters in Different Months

Sl. No.	*Sampling Month*	*Mean Value of Parameters*													
		E.C.	*Alkalinity*	*TSS*	*TDS*	*Total Hardness*	*Ca Hardness*	*Mg Hardness*	*Cl^-*	*NO_3^{1-}*	*PO_4^{3-}*	*SO_4^{2-}*	*Total Iron*	*Cr^{+6}*	*COD*
1.	Feb.'00	295	80.2	20	195	67	40	27	23	0.424	1.62	115	2.12	0.085	30.9
2.	April'00	351	103	29	350	72	41	30	30	0.603	2.45	126	2.92	0.481	48.7
3.	June'00	339	90.7	254	220	65	45	24	26	0.741	1.76	145	199	1.380	56.0
4.	Aug.'00	493	77.3	227	22.7	70	59	12	0.6	0.679	1.51	118	3.07	0684	59.3
5.	Oct.'00	278	74.0	10	190	68	40	28	20	0.412	0.12	102	2.81	0.021	28.0
6.	Dec.'00	258	98.0	14	200	72	42	30	27	0.511	3.61	110	1.92	0.092	33.1

Water Quality Index

Chemical analysis of any sample of water will give us a complete picture of its physical and chemical constituents. This will give us only certain numerical value but for estimating exact quality of water an indexing system has been developed known as Water Quality Index (WQI) which gives us the idea of quality of water of whole system.

Definition

Water Quality Index may be defined as a rating, reflecting the composite influence on overall quality characteristics of water of individual quality parameters, which is being regarded as one of the most effective way to communicate water quality.

Statistical Analysis

For the calculation of water quality index the following equation have been used in different steps.

Step 1

$$\text{Quality Rating } q_n = 100 \times \frac{(Vn - Vi)}{(Vs - Vi)}$$

Where,

Vn = Actual amount present in polluted water on n^{th} parameter.

Vi = The ideal value of parameter

= 0 for all parameters except pH and DO which is 7.0 mg/l and 14.6 mg/l respectively.

Vs = Its standard.

Step 2

Unit weight (Wn) for various parameters is inversely proportional to the recommended standard (Sn) for the Corresponding parameter.

$$Wn_1 = K/Sn_1$$

Where,

$$K = \frac{1}{1/Vs_1 + 1/Vs_2 + \ldots. + 1/Vs_n}$$

$$\sum_{n=1}^{n} W \text{ is then determined}$$

Step 3

$$WQI = \text{Anti} \log_{10} \left(\sum_{n=1}^{n} Wn \log_{10} q_n\right)$$

For the purpose of present investigation, we have selected five physical, chemical and bacteriological parameters of polluted water. These parameters are pH, DO, BOD, TC and FC as per BIS specification for Class B inland surface water.

K is calculated to be 1.508340 and calculation of WQI is depicted in Table 14.4.

Water Quality Index is the most effective way to communicate water quality. WQI=0 means a complete absence of pollutants, when 0 < WQI < 100 indicates that the water is under consideration and is fit for human use and WQI > 100 reflects its unsuitability for use.

Water Quality Index of different months has been indicated in Table 14.4. Water Quality Index was found to be greater than 100 in all sampling months. It was highest in the month of April and lowest in the month of August. This indicates that the lake water was most polluted in April and least polluted in December. The yearly average value of water quality index was found to be 211.35 which was more than the permissible limits. It was found that the water of Vani Vihar lake was unfit for human use.

Conclusion

From the above observation it is concluded that the lake water is not fit for the designated use as per class B surface water quality standard. From the water quality index it is apparent that quality of lake water has been deteriorated *i.e.* permissible range is 100 and the observed range is 169.9 to 277.7.

The yearly average value of Water Quality Index was found to be 211.35 which was more than the permissible limit. From this it was found that the water of Vani Vihar lake was unfit for human activities.

However due to rapid population migration to Bhubaneswar in coming years this would result in deterioration of water qualities of Vani Vihar lake unless adequate steps are taken by the University authorities in Bhubaneswar to protect the lake.

Suggestions

1. Complete check of entry of municipal sewage to the lake.
2. Complete drainage of existing water and disilting and cleaning of bottom debris and filling it with freshwater.
3. Arrangement of regular flushing of freshwater with provision of outlet.
4. Arrangement should be made to provide lavoratories near the lake to serve the local inhabitants, thereby restricting pollution of the lake.
5. Attempt to be made to clean the algal bloom preferable by mechanical means by suitable algicide and by introducing suitable fish species which are capable of scavenging both biological, organic and inorganic load of the lake.
6. Public awareness to be created about the potential hazard of using polluted water and citizen's role in maintaining the water bodies so that the desired uses can be sustained.

Acknowledgement

We are thankful to Chairman, Orissa State Pollution Control Board, Bhubaneswar for providing us research facilities.

References

APHA, AWWA, AWEF (1995). *Standard methods for the examination of water and wastewater*, 19th Edition.

Central Pollution Control Board, *Pollution Statistics Bhubaneswar*, Calcutta.

Liarkins, R.D. (1974). An objective water quality index. *J. Water Pollution Control Federation*, 46: 589

Lohani, B.N. (1981). Water quality indices. In: *Water pollution and management review*, (Ed. C.K. Varshney), South Asian Publishers Pvt. Ltd., New Delhi, pp. 53-69.

Singh, J.P., Yadava, P.K. and Singh, L. (1988). Pollution status on Sangam and its adjoining rivers before the Kumbh Mela at Allahabad. *Indian J. Environmental Protection*, 8(11): 839-942.

Singh, J.P., Yadava, P.K. and Singh, L. (1989). Mass bathing effect on water quality of Sangam during Mahakumbh Mela at Allahabad. *Indian J. Environmental Protection*, 9(3): 189-193.

15

Groundwater Quality of Ghataprabha Command Area, Karnataka

C.K. JAIN, C.P. KUMAR & M.K. SHARMA

National Institute of Hydrology, Roorkee, Uttaranchal

ABSTRACT

The groundwater quality of Ghataprabha Command Area (Karnataka) has been assessed to see the suitability of groundwater for domestic applications. Sixty groundwater samples from shallow and deep wells were collected each during pre- and post-monsoon seasons in the month of May and November 2001 respectively. Various water quality constituents, *viz.*, pH, conductance, total dissolved solids, alkalinity, hardness, sodium, potassium, calcium, magnesium, chloride, sulphate, nitrate, phosphate and fluoride have been determined. The data was analysed with reference to BIS and WHO standards, ionic relationships were studied and hydro-chemical facies were determined. Distribution of various constituents indicates that about 10-20 per cent samples of the study area crosses the maximum permissible limit for TDS, hardness, calcium, magnesium, sulphate and fluoride while 20-30 per cent samples crosses the limit for nitrate. The groundwater of the command area has also been classified to study various hydro-chemical processes.

Keywords: Groundwater quality, Hydro-chemical facies, Chadha's classification

Introduction

The quality of groundwater is the resultant of all the processes and reactions that act on the water from the moment it condensed in the atmosphere to the time it is discharged by a well or spring and varies from place to place and with the depth of the water table. Groundwater has unique features, which render it particularly suitable for public water supply. It has excellent natural quality, usually free from pathogens, colour and turbidity and can be consumed directly without treatment. Groundwater is widely distributed and can be frequently developed incrementally at points near the water demand, thus avoiding the need for large-scale storage, treatment and distribution system. Groundwater is particularly important as it accounts for about 88 per cent safe drinking water in rural areas, where population is widely dispersed and the infrastructure needed for treatment and transportation of surface water does not exist. Unfortunately, the availability of groundwater is not unlimited nor it is protected from deterioration. In most of the instances, the extraction of excessive quantities of groundwater has resulted in drying up of wells, damaged ecosystems, land subsidence, salt-water intrusion and depletion of the resource. The problem of, groundwater pollution in several parts of the country has become so acute that unless urgent steps for detailed identification and abatement are taken, extensive groundwater resources may be damaged.

There are many sources that contribute contaminants to the groundwater, *e.g.*, land disposal of solid wastes, sewage disposal on land, agricultural activities, urban run off and polluted surface water (Jain *et al.*, 1995). Earlier we have carried out groundwater quality studies in different regions of the country (Jain and his coworkers, 1995, 1996a, b, 1997a, b, 1999a, b, 2001, 2002; Jain and Ram, 1996; Jain, 1998). In continuation of our earlier studies, an attempt has been made to examine the suitability of groundwater of Ghataprabha Command Area in the State of Karnataka. The suitability of groundwater has been examined with reference to BIS (1991) and WHO (1996) standards. An attempt has also been made to classify the groundwater on the basis of different classification schemes.

Study Area

The study area lies between 16°00′ and 16°40′ N latitude and 74°15′ and 76°00′ E longitude in the State of Karnataka (Fig. 15.1). The area is bounded by the Krishna river in the north, Maharashtra State to the west, the confluence of Krishna river and Malprabha river in the east and the water divide or the basin boundary between Ghataprabha and Malaprabha rivers in the south. The total area taken up for the study is 10,370 km^2. The general topographic elevation varies from 520 to 890 m above mean sea level. The general slope of the ground is toward the Ghataprabha River in the southern part and toward Krishna River in the north.

Geologically, the area is underlain by rock types of Archaean crystalline to recent alluvium. The average annual rainfall in the area is about 650 mm, out of which 65 per cent occur during the southwest monsoon (June to September). The contribution from northeast monsoon is nearly 20 per cent and the rest is received during the pre-monsoon period. The climate is generally tropical with hot summers and cold winters. The relative humidity varies with seasonal changes with values less than 30 per cent during non-monsoon to about 85 per cent during monsoon months.

Methodology

Sixty groundwater samples were collected from the study area each during pre- and post-monsoon seasons (Fig. 15.1). The samples were collected from hand pumps, open wells and bore wells, which are being extensively used for drinking and other domestic purposes. Some parameters

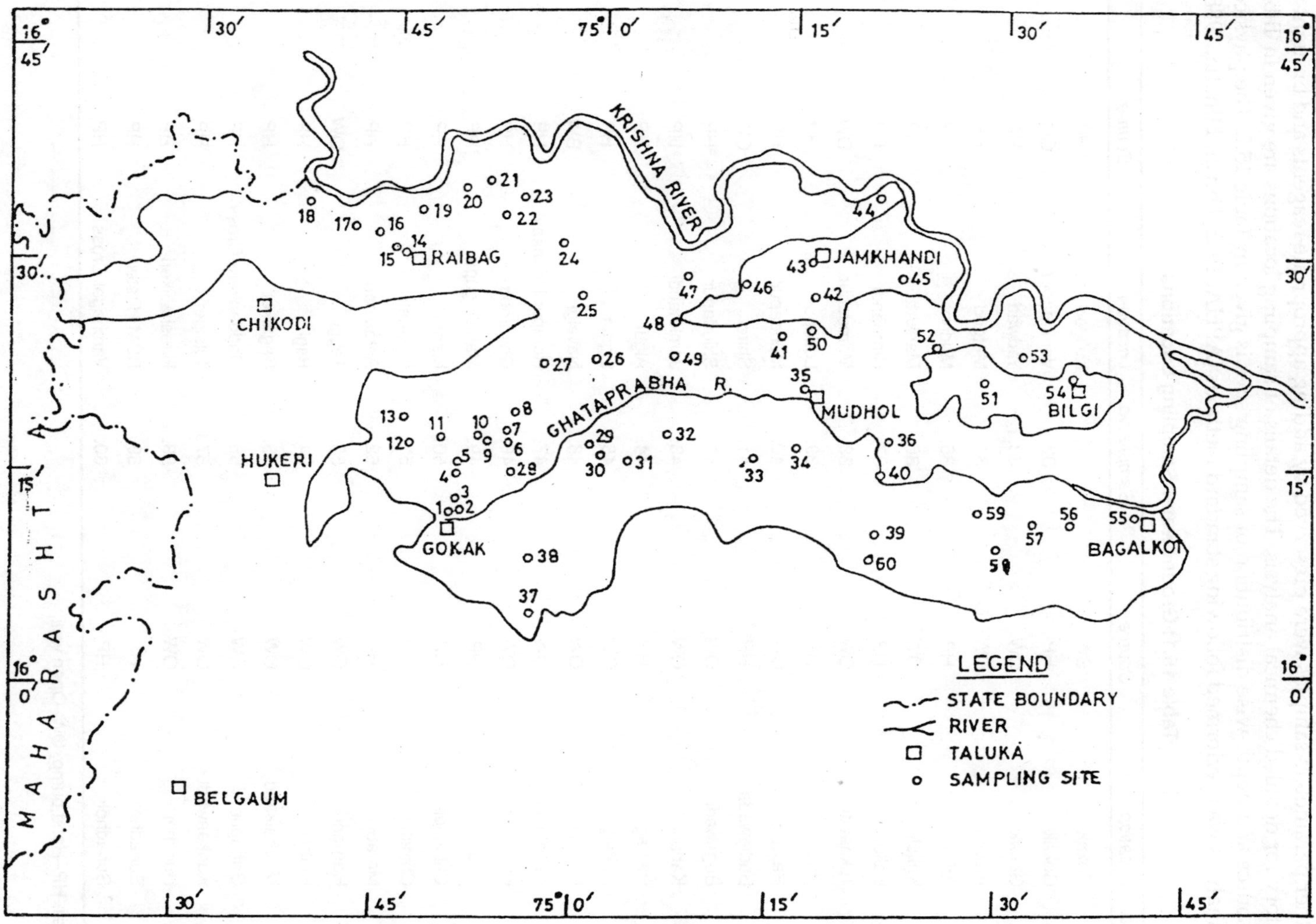

Fig.15.1: Study Area Showing Location of Sampling Points

like pH and conductance were measured in the field at the time of sample collection using portable kits. For other parameters samples were preserved by adding appropriate reagents and brought to the laboratory for detailed chemical analysis. The details of sampling locations are given in Table 15.1. The source and depth-wise distribution of sampling sites is given in Table 15.2. The physico-chemical analysis was performed following standard methods (APHA, 1985; Jain and Bhatia, 1988).

Table 15.1: Groundwater Sampling Locations

Sample No.	*Location*	*Source*	*Sample No.*	*Location*	*Source*
1.	Gokak	BW	31.	Kuligod	HP
2.	Gokak	HP	32.	Venkatapur	OW
3.	Gokak	OW	33.	Yadwad	HP
4.	Arbhavi	OW	34.	Metgud	HP
5.	Arbhavi	HP	35.	Mudhol	HP
6.	Kalloli	OW	36.	Marikatti	HP
7.	Kalloli	HP	37.	Hirenandi	HP
8.	Tuknautti	OW	38.	Mamdapur	OW
9.	Rajapar	HP	39.	Lokapur	HP
10.	Rajapar	OW	40.	Muddapur	HP
11.	Badigwada	HP	41.	Shirol	OW
12.	Bogewadi	OW	42.	Shiddapur	HP
13.	Kabbur	OW	43.	Jamkhandi	HP
14.	Raibag	HP	44.	Algur	HP
15.	Raibag	OW	45.	Lignur	HP
16.	Raibag	OW	46.	Navalgi	BW
17.	Nasalapur	OW	47.	Rampur Rabakavi	HP
18.	Kallol	OW	48.	Chimmod	HP
19.	Birdi	HP	49.	Mahalingapur	HP
20.	Chincholi	HP	50.	Mantur	HP
21.	Chincholi	HP	51.	Halagali	HP
22.	Morab	HP	52.	Amalzhari	HP
23.	Harugeri	OW	53.	Teggi	OW
24.	Hidkal	OW	54.	Bilgi	HP
25.	Mughalkhod	OW	55.	Bagalkot	HP
26.	Gurlapur	OW	56.	Gaddankere Cross	HP
27.	Kankanwadi	OW	57.	Tulsigeri	HP
28.	Hunshal	OW	58.	Hireshallikeri	HP
29.	Sunadholi	HP	59.	BCM Hostel	HP
30.	Sunadholi	HP	60.	Varchagal Cross	HP

BW–Bore well, HP–Hand pump, OW–Open well.

Table 15.2: Source and Depth-wise Distribution of Sampling Sites

Source Structure	*Depth Range*			*Total Number*
	< 20 m	*20-40 m*	*> 40 m*	
Hand Pumps	–	55,59,60	2,5,7,9,11,14,19, 20,21,22,29,30, 31,33,34,35,36, 37,39,40,42,43, 44,45,47,48,49, 50,51,52,54,56, 57,58	37
Bore Well	–	1	46	2
Open Wells	3,4,6,8,10,12,13, 15,16,17,18,23, 24,25,26,27,28, 32,38,41,53	–	–	21
Total	21	4	35	60

Results and Discussion

The hydro-chemical data for the two sets of samples collected during pre- and post-monsoon seasons are presented in Table 15.3. The pH values in the study area are mostly confined within the range 6.6 to 8.5 during both pre- and post-monsoon seasons indicating alkaline nature of the groundwater and are well within the limits prescribed for various uses of water including drinking water supplies (BIS, 1991; WHO, 1996). The electrical conductivity and dissolved salts are primary indicators of the total mineral contents of water and may be related to problems such as excessive hardness, corrosive characteristics or other mineral contamination. The conductivity values vary from 356 to 15726 μS/cm during pre-monsoon season and 246 to 12710 μS/cm during post-monsoon season with more than 50 per cent samples having conductivity value above 1000 μS/cm during both pre- and post-monsoon seasons. The maximum conductivity value of 15726 and 12710 μS/cm was observed at Algur during pre- and post-monsoon season respectively. Conductivity measurements are directly related to the concentration of ionized substance in water and commonly used to determine the purity of de-mineralized water and total dissolved solids in boiler, cooling tower water, irrigation and domestic supply. Solutions of most inorganic acids, bases and salts are relatively good conductors.

In natural waters, dissolved solids consists mainly of inorganic salts such as carbonates, bicarbonates, chlorides, sulphates, phosphates and nitrates of calcium, magnesium, sodium, potassium, iron etc. and small amount of organic matter and dissolved gases. In the present study the TDS values varies from 228 to 10065 mg/l during pre-monsoon season and 157 to 8134 mg/l during post-monsoon season with more than 85 and 60 per cent samples having TDS values above the desirable limit of 500 mg/l during pre- and post-monsoon seasons respectively. The spatial distribution map of TDS in pre- and post-monsoon season is shown in Figs. 15.2 and 15.3 respectively. Water containing more than 500 mg/l of TDS is not considered desirable for drinking water supplies, though more highly mineralized water is also used where water with low TDS value is not available. For this reason, 500 mg/l as the desirable limit and 2000 mg/l as the maximum permissible limit has been suggested for drinking water (BIS, 1991). Water containing TDS more than 500 mg/l causes

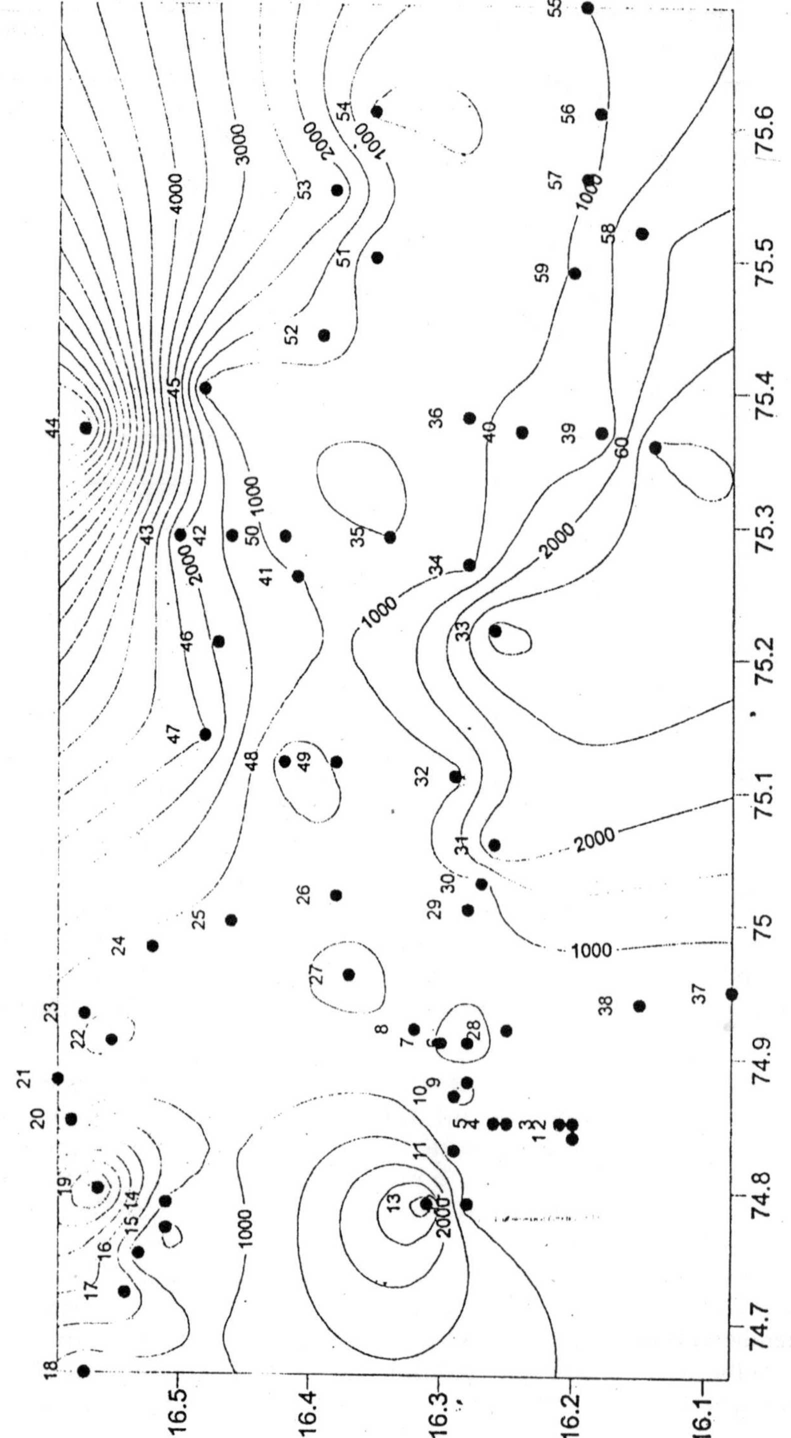

Fig.15.2: Distribution of TDS in Ghataprabha Command Area (Pre-monsoon 2001)

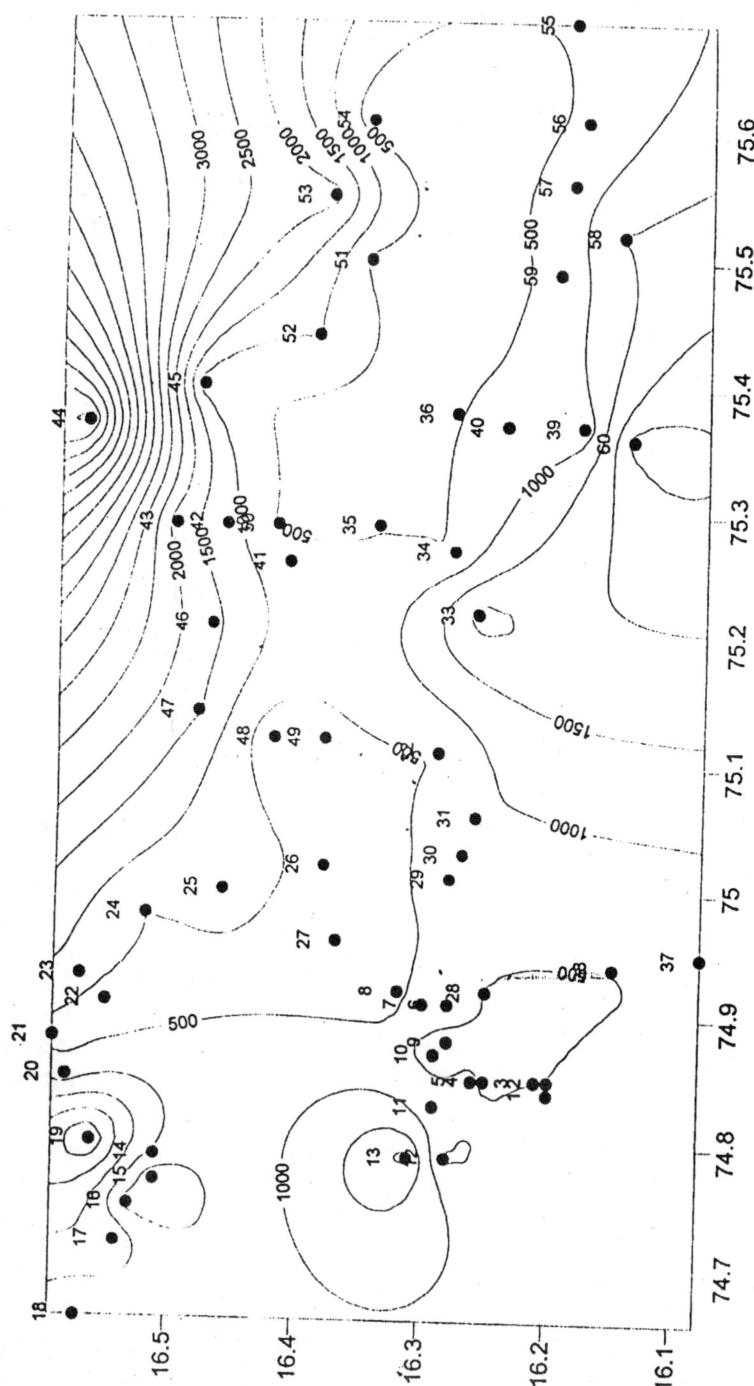

Fig.15.3: Distribution of TDS in Ghataprabha Command Area (Post-monsoon 2001)

gastrointestinal irritation (BIS, 1991). In the study area, 15 per cent sample even crosses the maximum permissible limit of 2000 mg/l during pre-monsoon season and 10 per cent samples during post-monsoon season. In general TDS content decreases during post-monsoon season due to dilution effect of rain water.

Table 15.3: Hydro-chemical Data of Groundwater Samples of Ghataprabha Command Area

Parameter	*Min*		*Max*		*Mean*	
pH	6.8	(6.6)	8.8	(8.5)	7.35	(7.44)
Conductivity, μS/cm	356	(246)	15726	(12710)	1996	(1456)
TDS, mg/l	228	(157)	10065	(8134)	1277	(932)
Alkalinity, mg/l	116	(65)	744	(307)	335	(168)
Hardness, mg/l	64	(80)	4382	(3361)	551	(409)
Chloride	2.5	(5.0)	2500	(2102)	191	(133)
Sulphate, mg/l	2.5	(3.5)	1950	(1976)	172	(184)
Nitrate, mg/l	ND	(ND)	2222	(1452)	111	(115)
Phosphate, mg/l	ND	(ND)	1.20	(0.71)	0.12	(0.03)
Fluoride, mg/l	ND	(ND)	6.13	(5.40)	1.21	(0.87)
Sodium, mg/l	12	(8.1)	1563	(1342)	154	(117)
Potassium, mg/l	1.8	(ND)	468	(406)	34	(27)
Calcium, mg/l	11	(12)	1275	(876)	144	(95)
Magnesium, mg/l	3.9	(7.0)	292	(286)	47	(41)

N = 60; Values given in parenthesis indicate post-monsoon data.

The presence of carbonates, bicarbonates and hydroxides are the main cause of alkalinity in natural waters. Bicarbonates represent the major form since they are formed in considerable amount from the action of carbonates upon the basic materials in the soil. The alkalinity values in the study area vary from 116 to 744 mg/l during pre-monsoon season with majority of samples (>90 per cent) having alkalinity values above the desirable limit of 200 mg/l but within the maximum permissible limit of 600 mg/l. Only one sample of Kalloli (OW) crosses the maximum permissible limit during pre-monsoon season. The high alkalinity values in the study area may be attributed to the action of carbonates upon the basic materials in the soil. Such water gives unpleasant taste. During post-monsoon season, the alkalinity values vary from 65 to 307 mg/l with more than 80 per cent samples having alkalinity values within the desirable limit.

Calcium and magnesium along with their carbonates, sulphates and chlorides makes the water hard, both temporary and permanent. A limit of 300 mg/l has been recommended as a desirable limit and 600 mg/l as the maximum permissible limit for potable water (BIS, 1991). The total hardness values in the study area ranges from 64 to 4382 mg/l during pre-monsoon season and 80 to 3361 mg/l during post-monsoon season. Distribution of hardness values clearly indicates that during pre-monsoon season about 43 per cent samples lies within the desirable limit and 38 per cent samples within the maximum permissible limit of drinking water while during post-monsoon season about 66 per cent samples falls within the desirable limit and about 16 per cent samples within maximum permissible limit. About 15-20 per cent samples even crosses the maximum permissible limit of drinking water in both pre- and post-monsoon seasons.

The desirable limit for calcium and magnesium for drinking water are 75 and 30 mg/l respectively (BIS, 1991). In groundwater of the study area, the values of calcium and magnesium ranges from 11 to 1275 and 3.9 to 292 mg/l respectively during pre-monsoon season. The same was found to vary from 12 to 876 and 7.0 to 286 mg/l during post-monsoon season. The low concentration of calcium and magnesium in the post-monsoon season may be attributed to the dilution effect of rain water. In groundwater, the calcium content generally exceeds the magnesium content in accordance with their relative abundance in rocks. The increase of magnesium is quite proportionate with calcium in both the seasons. Distribution of calcium and magnesium clearly indicates that about 10-20 per cent samples crosses the maximum permissible limit of drinking water in both the seasons.

The concentration of sodium in the study area varies from 12 to 1563 mg/l during pre-monsoon season and 8.1 to 1342 mg/l during post-monsoon season. The sodium concentration more than 50 mg/l make the water unsuitable for domestic use. The sodium concentration was found higher at most of the sites. The concentration of potassium in groundwater varies from 1.8 to 468 mg/l during pre-monsoon season and nil to 406 mg/l during post-monsoon season.

The chloride content ranged from 2.5 to 2500 mg/l during pre-monsoon season and 5.0 to 2102 mg/l during post-monsoon season. The limits of chloride have been laid down primarily from taste considerations. A limit of 250 mg/l chloride has been recommended as desirable limit and 100 mg/l as maximum permissible limit for drinking water (BIS, 1991; WHO, 1996). However, no adverse health effects on humans have been reported from intake of waters containing even higher content of chloride. Distribution of chloride indicates that more than 80 per cent of the samples lies well within the desirable limits of drinking water during both pre- and post-monsoon seasons.

The concentration of sulphate in the command area varies from 2.6 to 1950 mg/l during pre-monsoon season and 3.5 to 1976 mg/l during post-monsoon season. Sulphate generally occurs as soluble salts of calcium, magnesium and sodium. The sulphate content of water may change significantly with time during infiltration of rainfall and groundwater recharge, which mostly takes place from stagnant pools, puddles and surface run off water collected in low lying areas. It is evident from the distribution of sulphate that about 80 per cent samples lies well within the desirable limit of drinking water and about 10 per cent sample crosses the maximum permissible limit of drinking water.

The nitrate content was found to vary from nil to 2222 mg/l during pre-monsoon season and 3.5 to 1976 mg/l during post-monsoon season with wide range of variations. Increased level of nitrate at various locations may be attributed due the surface disposal of sewage and agricultural wastes. In general concentration of nitrate was found higher during post-monsoon season at most of the locations. Nitrate is effective plant nutrient and moderately toxic and is considered important for its adverse health effects. A limit of 45 mg/l has been prescribed by BIS (1991) and WHO (1996) for drinking water supplies. Its concentration above 45 mg/l may prove detriment to human health. In higher concentrations, nitrate may produce a disease known as methaemoglobinaemia, which generally affects bottle-fed infants. Repeated heavy doses of nitrates may also cause carcinogenic diseases. The spatial distribution of nitrate in pre- and post-monsoon seasons is shown in Figs. 15.4 and 15.5 respectively. Distribution of nitrate indicates that about 61 per cent samples has depicted nitrate content less than 45 mg/l, whereas in about 20 per cent samples the nitrate exceeded the permissible limit of 100 mg/l during pre-monsoon season. The higher nitrate concentration may be attributed due to combined effect of contamination from domestic sewage and runoff from fertilized fields.

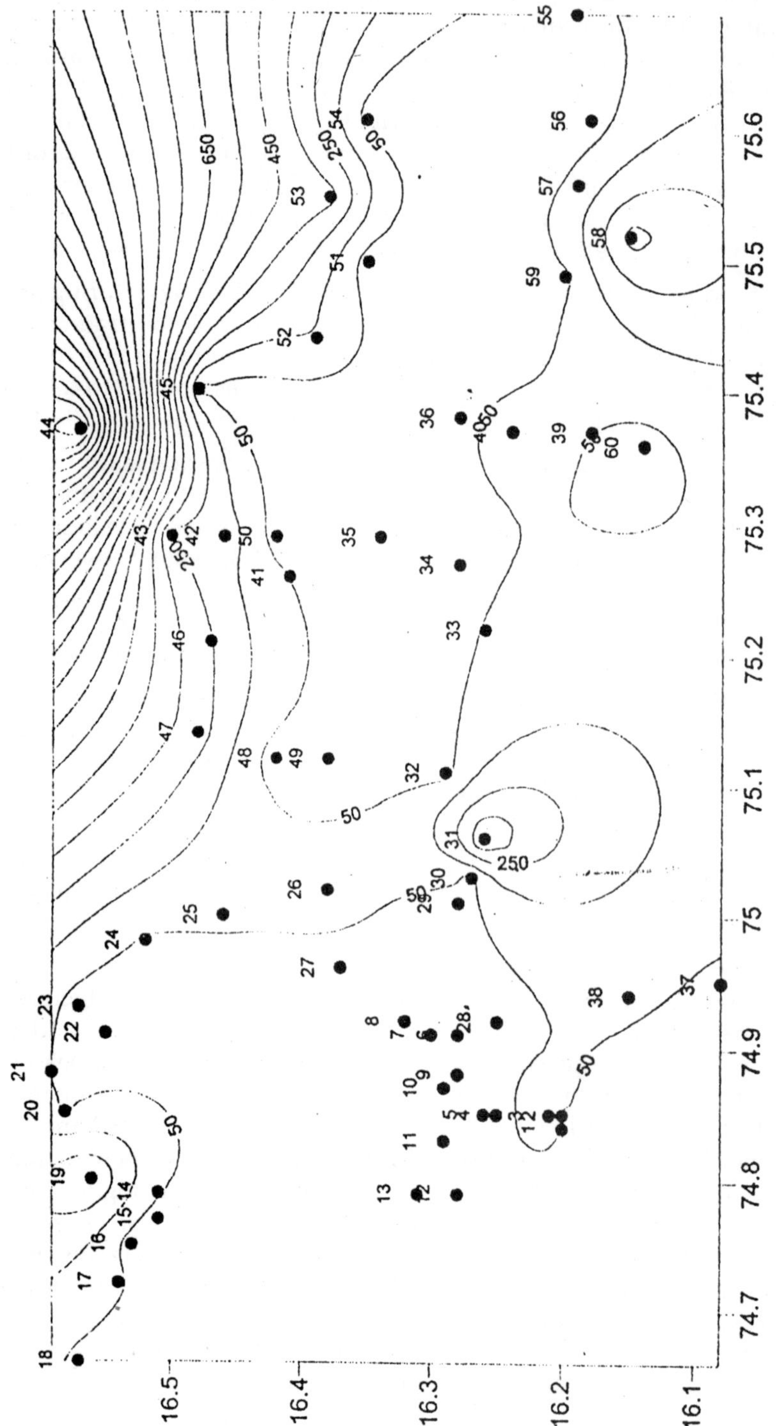

Fig.15.4: Distribution of Nitrate in Ghataprabha Command Area (Pre-monsoon 2001)

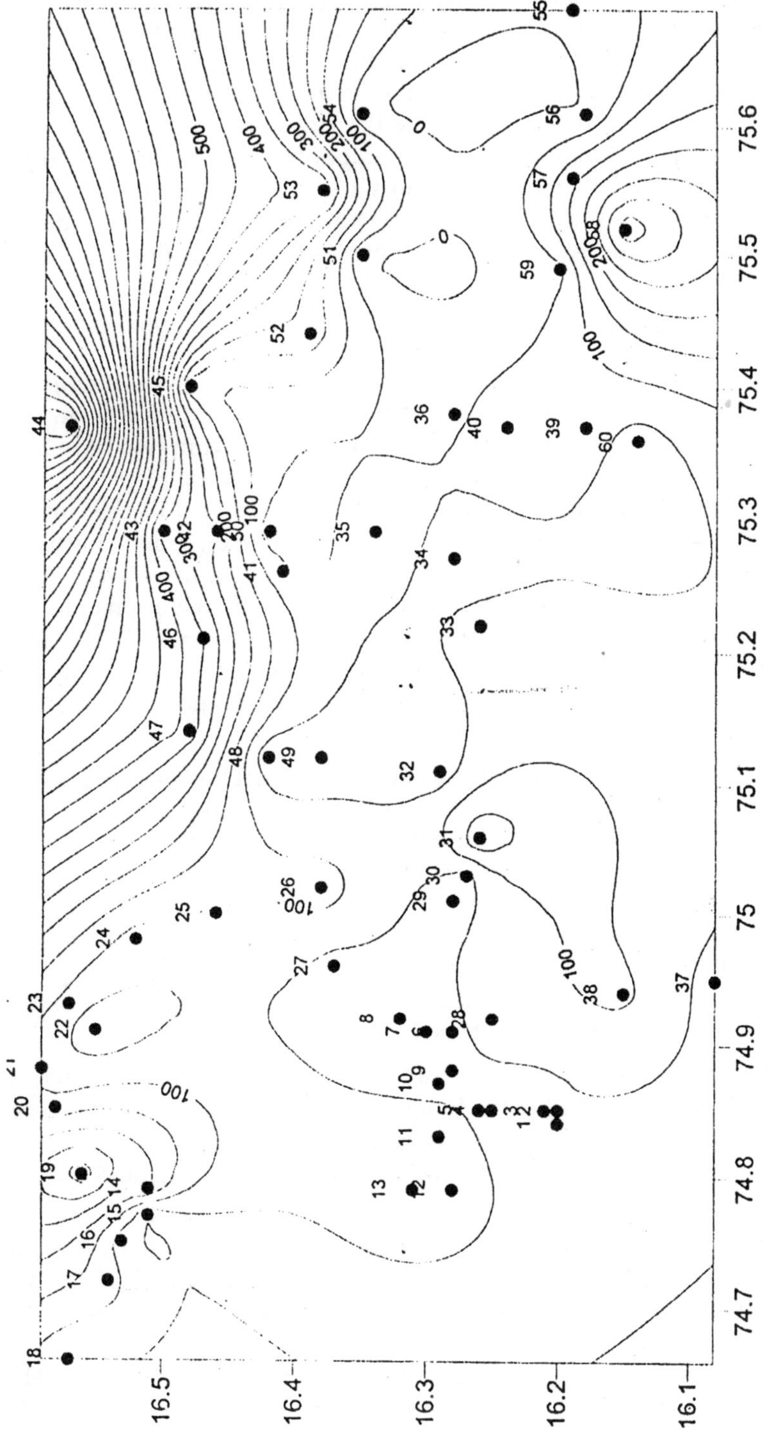

Fig.15.5: Distribution of Nitrate in Ghataprabha Command Area (Post-monsoon 2001)

The fluoride concentration varies from nil to 6.13 mg/l during pre-monsoon season and nil to 5.4 during post-monsoon season with about 85 per cent samples having fluoride concentration within permissible limit of 1.5 mg/l. The fluoride concentration beyond permissible limit of 1.5 mg/l was observed in about 15 per cent samples. The spatial distribution of fluoride in pre- and post-monsoon season is shown in Figs. 15.6 and 15.7 respectively. The fluorides are present in soil strata due to natural geological formations in the form of fluorspar, fluorapatite, ampheboles such as hornblinde, trimolite and mica. Weathering of alkali, silicate, igneous and sedimentary rocks specially shales contributes a major portion of fluorides to natural waters. The accumulation of fluoride in soil strata eventually results in its leaching due to percolating water pressure resulting increase fluoride concentration in groundwater,

Classification of Groundwater

The groundwater of the command area has been classified as per Chadha's diagram (1999). The diagram is a somewhat modified version of the piper trilinear diagram. In the piper diagram the milli-equivalent percentages of the major cations and anions are plotted in two base triangles and the type of water is determined on the basis of position of the data in the respective cationic and anionic triangular fields. The plottings from triangular fields are projected further into the central diamond field, which represents the overall character of the water. Piper diagram allows comparisons to be made among numerous analyses, but this type of diagram has a drawback, as all trilinear diagrams do, in that it does not portray actual ion concentration. The distribution of ions within the main field is unsystematic in hydrochemical process terms, so the diagram lacks certain logic. The method is not very convenient when plotting a large volume of data. Nevertheless, this shortcoming does not lessen the usefulness of the Piper diagram in the representation of some geochemical processes.

In contrast, in Chadha's diagram, the difference in milli-equivalent percentage between alkaline earths (calcium plus magnesium) and alkali metals (sodium plus potassium), expressed as percentage reacting values, is plotted on the X-axis and the difference in milli-equivalent percentage between weak acidic anions (carbonate plus bicarbonate) and strong acidic anions (chloride plus sulphate plus nitrate) is plotted on the Y-axis. The resulting field of study is a square or rectangle depending upon the size of the scales chosen for X and Y co-ordinates. The milli-equivalent percentage differences between alkaline earth and alkali metals and between weak acidic anions and strong acidic anions would plot in one of the four possible sub-fields of the diagram. The main advantage of this diagram is that it can be made simply on most spreadsheet software packages.

The square or rectangular field describes the overall character of the water. The diagram has all the advantages of the diamond-shaped field of the Piper trilinear diagram and can be used to study various hydro-chemical processes, such as base cation exchange, cement pollution, mixing of natural waters, sulphate reduction, saline water (end-product water) and other related hydro-chemical problems (Chadha, 1999).

The chemical analysis data of all the samples have been plotted on Chadha's diagram for two sets of samples (Figs. 15.8 and 15.9) and results have been summarized in Table 15.4. From the figures it is evident that during pre-monsoon season 55.0 per cent samples fall in sub-field 5 (Ca-Mg-HCO_3 type), 18.3 per cent samples in sub-field 6 (Ca-Mg-Cl-NO_3-SO_4 type), 8.3 per cent samples in sub- field 7 (Na-K-Cl-NO_3-SO_4 type) and 18.4 per cent samples in sub-field 8 (Na-K-HCO_3 type). However, during post-monsoon season 31.7 per cent samples fall in sub-field 5 (Ca-Mg-HCO_3 type), 40.0 per cent samples in sub-field 6 (Ca-Mg-Cl-NO_3-SO_4 type), 16.7 per cent samples in sub-field 7

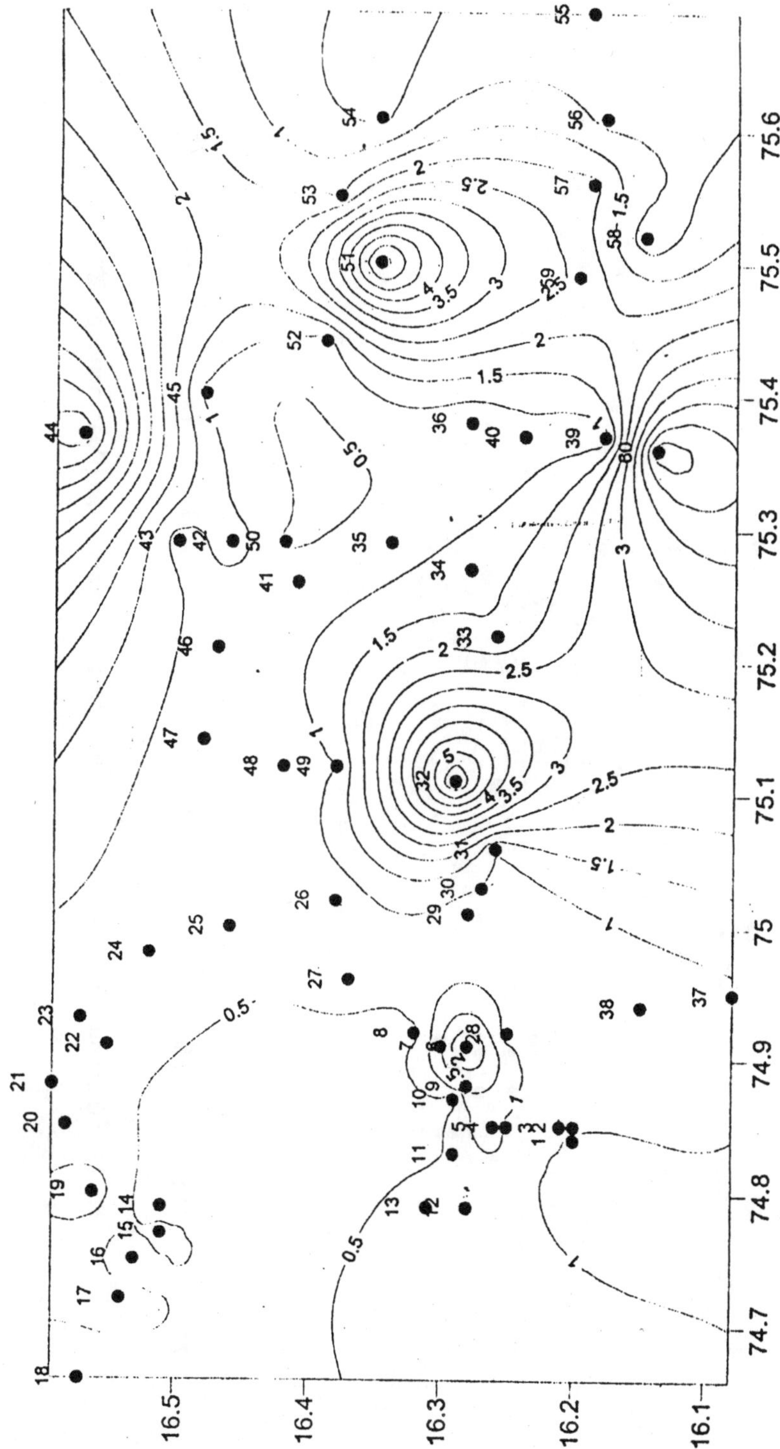

Fig.15.6: Distribution of Fluoride in Ghataprabha Command Area (Pre-monsoon 2001)

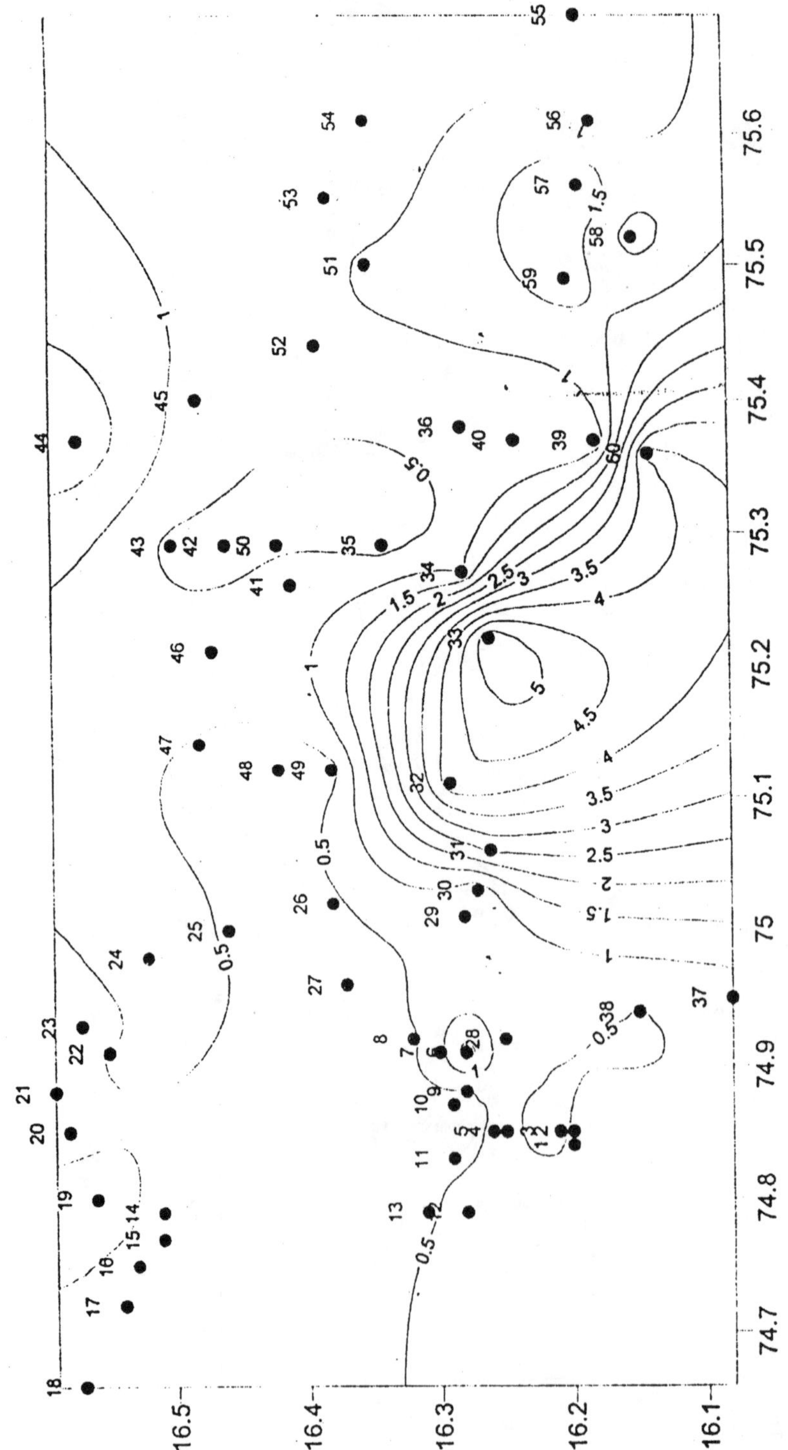

Fig.15.7: Distribution of Fluoride in Ghataprabha Command Area (Post-monsoon 2001)

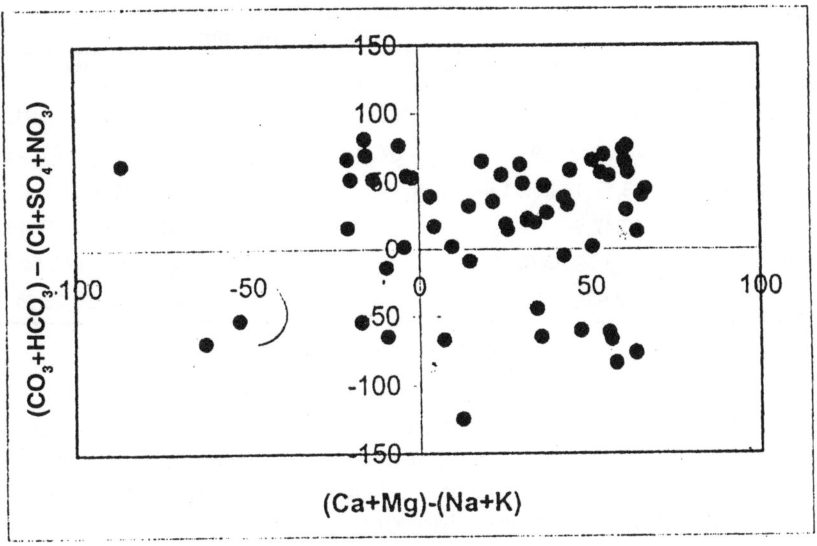

Fig.15.8: Chadha's Diagram (Pre-monsoon 2001)

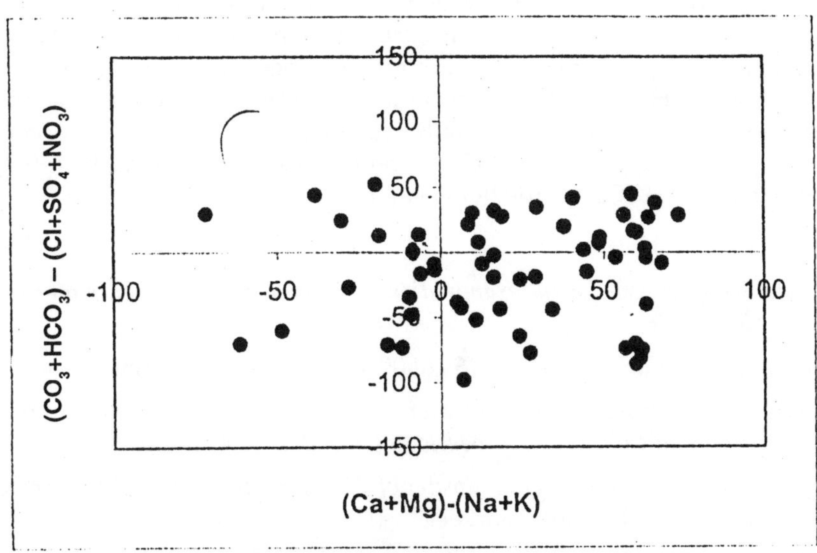

Fig.15.9: Chadha' s Diagram (Post-monsoon 2001)

(Na-K-Cl-NO_3-SO_4 type) and 11.6 per cent samples in sub-field 8 (Na-K-HCO_3 type). The modified diagram has all the advantages of the diamond-shaped field of the Piper trilinear diagram and can be conveniently used to study various hydrochemical processes.

Table 15.4: Summarized Results of Water Classification as per Chadha's Diagram

Classification/Type	*Sample Numbers*	
	Pre-monsoon	*Post-monsoon*
Ca-Mg-HCO_3	3,4,8-11,12,14,15,18,21,22, 25-30,34-39,41,48-50,54-57,59	3,4,8-10,15,16,21,22,26,27 29,30,34,35,37,50,54,56
Ca-Mg-Cl-NO_3-SO_4	13,17,19,20,31,42,43,44, 46,47,58	1,12,13,14,17-20,23,25,28,31,38, 39,41-44,46,47,48,49,55,58
Na-K-Cl-NO_3-SO_4	7,33,52,53,60	7,11,33,40,45,52,53,57,59,60
Na-K-HCO_3	1,2,5,6,16,23,24,32,40,45,51	2,5,6,24,32,36,51

Conclusion

The groundwater quality of command area varies from place to place with the depth of water table. It also shows significant variation from one season to another. The survey of the command area has indicated higher concentration of total dissolved salts, electrical conductivity, hardness, calcium, magnesium, sulphates, nitrates, fluorides, etc. About 90 per cent samples have TDS values above the desirable limit of 500 mg/l during pre-monsoon season and about 65 per cent samples have TDS values above the desirable limit during post-monsoon season. About 10-15 per cent samples even crosses the maximum permissible limit of 2000 mg/l during both pre- and post-monsoon seasons. From the hardness point of view, about 15-20 per cent samples crosses the maximum permissible limit. The concentration of nitrate exceeds the maximum permissible limit in about 20-30 per cent samples. The grouping of samples according to their hydrochemical facies clearly indicates predominance of Ca-Mg-HCO_3 hydro-chemical facies during pre-monsoon season. While during post-monsoon season 31.7 per cent samples falls in Ca-Mg-HCO_3 facies, 40.0 per cent samples in Ca-Mg-Cl-NO_3-SO_4 hydro-chemical facies, 16.7 per cent samples in Na-K-Cl-NO_3-SO_4 facies and 11.6 per cent samples in Na-K-HCO_3 hydro-chemical facies.

References

APHA (1985). *Standard methods for the examination of water and wastewater.* American Public Health Association, New York.

BIS (1991). *Specifications for drinking water, IS:10500:1991.* Bureau of Indian Standards, New Delhi.

Chadha, D.K. (1999). A proposed new diagram for geochemical classification of natural waters and interpretation of chemical data. *Hydrogeology Journal*, 7(5): 431-439.

Jain, C.K. (1998). *Arsenic pollution in groundwater of West Bengal.* Technical Report, CS(AR) 6/97-98, National Institute of Hydrology, Roorkee.

Jain, C.K. and Bhatia, K.K.S. (1988). *Physico-chemical analysis of water and wastewater UM-26.* National Institute of Hydrology, Roorkee, 1987-88.

Jain, C.K. and Rakesh Kumar (2002). *Groundwater quality in Patna Town (Bihar).* Technical Report, 2001-2002, National Institute of Hydrology, Roorkee.

Jain, C.K. and Ram, D. (1996). *Impact of sewage waste disposal on soil strata and groundwater quality.* Technical Report, TR(BR) 141, National Institute of Hydrology, Roorkee, 1995-96.

Jain, C.K., Bhatia, K.K.S. and Sharma, M.K. (1997a). *Groundwater quality evaluation in Doon valley, Dehradun.* Technical Report, CS(AR) 10/96-97, National Institute of Hydrology, Roorkee.

Jain, C.K., Bhatia, K.K.S. and Vijay Kumar (1997b). *Groundwater quality monitoring and evaluation in Sagar district, M.P.* Technical Report, CS(AR) 11/96-97, National Institute of Hydrology, Roorkee.

Jain, C.K., Bhatia, K.K.S. and Vijay, T. (1995). *Groundwater quality monitoring and evaluation in and around Kakinada, Andhra Pradesh.* Technical Report, CS(AR) 172/94-95, National Institute of Hydrology, Roorkee.

Jain, C.K., Bhatia, K.K.S. and Kumar, S.R. (1999a). *Groundwater quality in Greater Guwahati, Assam with reference to trace elements.* Technical Report, CS(AR) 8/98-99, National Institute of Hydrology, Roorkee.

Jain, C.K., Ali, I. and Sharma, M.K. (1999b). Fluoride contamination in groundwater: Indian scenario. *Indian J. Environ. Prot.*, 19(4): 260-266.

Jain, C.K., Bhatia, K.K.S., Kumar, C.P. and Purandara, B.P. (2001). *Groundwater quality in Malprabha River Basin, Karnataka.* Technical Report, 2000-2001, National Institute of Hydrology, Roorkee.

Jain, C.K., Omkar and Sharma, M.K, (1996a). *Groundwater quality monitoring and evaluation in district Jammu (J&K).* Technical Report, CS(AR) 196, National Institute of Hydrology, Roorkee, 1995-96.

Jain, C.K., Ram, D. and Bhatia, K.K.S. (1996b). *Groundwater quality monitoring and evaluation in district Hardwar, U.P.* Technical Report, CS(AR) 198, National Institute of Hydrology, Roorkee, 1995-96.

WHO (1996). *Guidelines for drinking water, Vol. 2: Recommendations.* World Health Organization, Geneva.

16

Study on Water Quality of Subansiri River in Assam: An EIA Approach for a Proposed Hydroelectric Power Project

B.K. BARUAH
Department of Zoology, Cotton College, Guwahati, Assam
D. BARUAH
Assam Pollution Control Board, Guwahati, Assam

ABSTRACT

Water quality of Subansiri river, a major tributary of mighty river Brahmaputra at a proposed hydel project site was studied for 12 months. Physico-chemical and bacteriological analysis revealed that the river water is suitable for drinking, irrigation and industry. The data would be useful for EIA study in terms of water quality for the proposed hydel project.

Keywords: Water quality, Subansiri river, EIA

Introduction

Water quality study is an integral part of evaluation of health of any wetland ecosystem. In general, the riverine ecosystem, while traversing through hills, before reaching plains, maintain

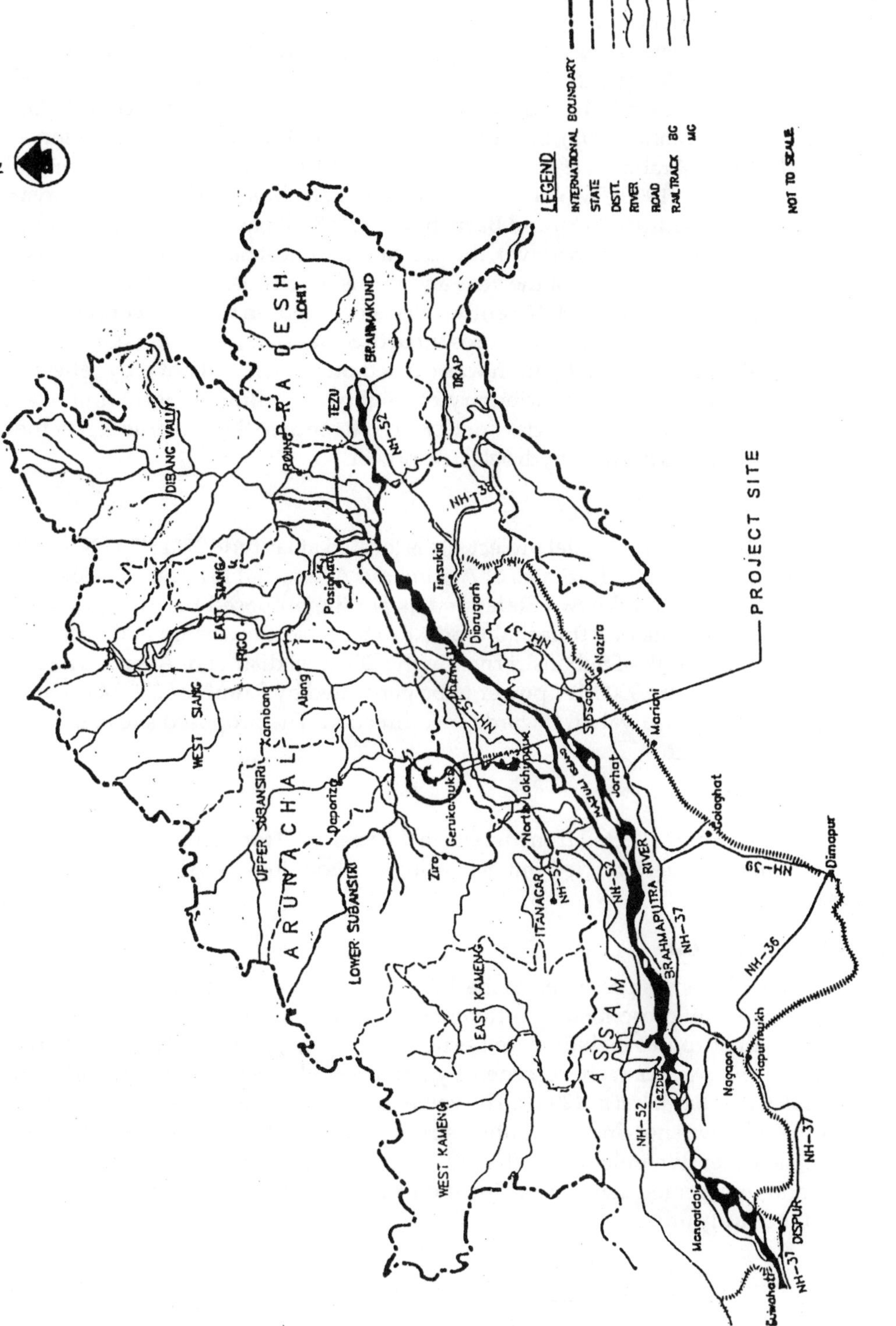

Project Site

unpolluted water quality due to less human activities and can be regarded as good and potable water and suitable for sustenance of aquatic life. All over the world, the rivers in high altitude are subjected to exploitation to generate hydel energy resulting disturbance in the riverine ecosystem including deterioration of water quality and affecting associated abiotic and biotic components. The study of water quality of rivers in India are well documented by Sangu and Sharma (1985), Ghatak and Konar (1992), Das and Sinha (1993), Sahu *et al.* (1994), Chandra *et al.* (1996), Jameson and Rana (1996), Abbasi *et al.* (1996), Arabinda *et al.* (1998), Sharma and Agarwal (1999), Kumar and Sarma (1999), Dhanapakiam *et al.* (1999), Mohanta and Patra (2000), Koshi and Vasudevnayar (2000), Gyananath *et al.* (2000), Musaddie (2000) and Baruah *et al.* (2003). It has been established that for any water resource development project on a river, the first and foremost task is to assess the availability and quality of water at different times of the year and this is the primary criteria for any successful completion of any multi-purpose project. Recently, a decision has been taken by concerned authority to build a hydel project by exploiting the water resource of Subansiri river in the state of Assam, and keeping this in view, the present investigation was carried out which will not only reflect the water quality of Subansiri river, an important tributary of mighty river Brahmaputra, but also provide baseline data for environmental impact assessment with reference to the construction of the dam for the proposed hydel project to harness hydroelectric power.

Location

The proposed Subansiri lower hydel project site is located at latitude 27°15′30″ N and longitude 94°15′30″ E, in the border of both the States of Assam and Arunachal Pradesh. The dam abutment will be equally divided between the two States and as per the proposed plan the powerhouse and switchboard will be in the State of Arunachal Pradesh. The project township and headquarter will be at Gerukamukh, on the bank of Subansiri river in the Dhemaji district of Assam. The project will have the installing capacity of 2000 MW power to be generated by 8 units of 250 MW each. The site is about 480 kms away from Guwahati towards east direction. The proposed site area has a rainfall of about 2350 mm per annum.

Materials and Methods

The water samples were collected from the site in the year 1999 for 12 months following standard methodology and analysis were carried out according methodology of APHA (1985) and Trivedy and Goel (1986).

Results and Discussion

The result of the analysis is presented in the Table 16.1. The values of different parameters were within the permissible limit prescribed by different standard institutions. The river water is potable and can be used for all purposes including domestic, municipal, agricultural and industrial. Similar nature of upstream river water were recorded by Abbasi *et al.* (1996), Jameson and Rana (1996), Arabinda *et al.* (1998) and Baruah *et al.* (2003). The values of different parameters in present experiment suggested that the river was not under pollution stress due to less human activities in the upstream course. Besides, the water dynamics along the hilly slopes in the upstream reduced the organic and inorganic load and augmented the self-purification process leading to maintenance of standard water quality in the study area.

Table 16.1: Result of Physcio-chemical Analysis of Water of Subansiri River (Year 2000)

Parameters	*Months*												
	Jan.	*Feb.*	*Mar.*	*Apr.*	*May*	*June*	*July*	*Aug.*	*Sep.*	*Oct.*	*Nov.*	*Dec.*	*Mean*
pH	7.5	8	8	7.6	7.5	7.6	7.5	7.5	7.5	7	7	7	7.47
Temperature °C	23	23	23.5	26	28	27	28	29	28	27	27	22	25.95
Turbidity (NTU)	9	6	9	15	20	48	90	95	42	22	28	2	32.16
Conductance (mmho/cm)	180	200	180	130	70	120	107	100	120	155	170	150	140
DO (mg/l)	7.5	8	8	7.5	7.8	8	7	8	8.5	7	8	9	7.85
COD (mg/l)	2.5	3	8	5	3.5	9	6	9	7.5	7	5	8	6.12
BOD (mg/l)	1	1	4	1	1.5	1	1.5	1	0.5	1	1	1.5	1.33
Alkalinity (mg/l)	80	70	70	50	40	50	50	55	50	60	60	80	59.58
Hardness (mg/l)	90	100	95	50	35	45	50	55	55	75	70	70	65.83
Calcium (mg/l)	65	60	65	50	20	30	30	30	35	50	50	50	44.58
Magnesium (mg/l)	30	30	20	15	10	20	20	20	15	20	10	10	18.33
Chloride (mg/l)	5	6	6	5	6	6	7	5	5	7	7	5	5.83
Sulphate (mg/l)	20	15	18	19	11	15	22	25	21	3	17	11	16.41
Nitrate-N (mg/l)	0.1	0.1	0.05	0.05	0.1	0.2	0.2	0.2	0.1	0.2	0.1	0.1	0.12
Phosphate-P (mg/l)	1.5	1	1.5	1	1	1	0.1	1.5	3	1.5	0.1	1	1.18
Ammonia-N (mg/l)	Nil	Nil	Nil	Nil	Nil	Nil	Nil	Nil	0.1	Nil	Nil	0.1	0.1
Kjheldel-N (mg/l)	1	0.5	0.1	1	0.3	0.6	0.9	0.9	1.5	2	1.5	1	0.94
Sodium (mg/l)	1	1	1	3.5	1.5	1.5	5	3	3	5	6	4	2.95
Potassium (mg/l)	8	1	1	3	1	1	2	1	1	2	3	1	2.08
Total Dissolved Solids (mg/l)	130	120	120	100	50	90	80	85	90	110	120	100	99.58
Fixed Dissolved Solid (mg/l)	30	30	30	30	20	30	25	20	30	35	40	30	26.66
Total Coliform (MPN/100ml)	Nil	Nil	Nil	300	300	Nil	Nil	2000	2000	Nil	300	800	950
Feacal Coliform (MPN/100ml)	Nil	Nil	Nil	Nil	300	Nil	Nil	300	Nil	Nil	Nil	Nil	300

A proposed hydro-electric power project in the study area and consequent construction of a dam on Subansiri river for storage of water will have severe impact on the riverine ecology including change in the water quality, aquatic life and overall functioning of river ecosystem. The creation of dam will inundate new areas affecting settlement of people and lead to ecological change as well as ecosystem dynamics resulting deforestation and loss of biodiversity. The dam will disrupt river function, change in physico-chemical characteristics of water, cause flood in new areas, dry out flood plain and result elimination of flood plain dependent organisms, destroy riverine ecology, prevent exchange of micro-nutrients and consequent loss of micro-communities such as diatom, encourage growth of harmful algae and death of aquatic life. These are the immediate outcome of construction of dam for harnessing hydel energy, therefore a complete environmental impact assessment has to be done to make the project optimum eco-friendly.

The present study has clearly provided information regarding the water quality of Subansiri river and possible environmental impact due to construction of dam for the proposed hydel project. The success of any multi-purpose water resource development project totally depends on the accuracy of the estimation of the availability of water and its quality in the different periods of the year. Besides, the routine technical and economic feasibility of a mega project should not be considered alone without consideration of environmental impact after operation of the project. Therefore, further studies are required to complete the total environmental impact assessment and follow up actions for achieving a complete balance between development and environment in the fragile ecosystem of North-East India.

References

Abbasi, S.A., Arya, D.S., Hameed, A.S. and Abbasi, N. (1996). Water quality of a typical river of Kerala, Punnurpuzha. *Poll. Res.*, 15(2): 163-166.

APHA (1985). *Standard methods for examination of water and wastewater, 16th Edition*. APHA, AWWA, WPCA.

Arabinda, H.B., Manjappa, S. and Puttiak, E.T. (1998). Correlation coefficient of some physico-chemical parameters of river Tungabhadra, Karnataka. *Poll. Res.*, 17(4): 371-375.

Baruah, B.K., Baruah, D., Sengupta, S. and Das, A.C. (2003). Study of wetlands of Guwahati: Water quality of rivers and drains. *Poll. Res.*, (In Press).

Chandra, R., Babadur, Y. and Sharma, B.K. (1996). Monitoring the quality of river Ramganga water at Bareilly. *Poll. Res.*, 15(1): 31-33.

Das, N.K. and Sinha, R.K. (1993). Assessment of water quality of Ganga river, Patna, India. *Environment and Ecology*, 11(4): 829-832.

Dhanapakiam, P., Sampurani, V. and Kavitha, R. (1999). Assessment of water quality of the river Cauvery. *Poll. Res.*, 20(4): 347-352.

Ghatak, D.B. and Konar, S.K. (1992). Impact of various industrial effluents on the Hooghly river ecosystem at Naihati, West Bengal. *Environment and Ecology*, 10(1): 144-150.

Gyananath, G., Shewdikar, S.V. and Samiuddin, S. (2000). Water quality analysis of river Godabari during Holimela at Nandad. *Poll. Res.*, 19(4): 673-674.

Jameson, J. and Rana, B.C. (1996). Pollution status of the river complex Sabarmati at Kheda region of Gujarat. I. Physico-chemical characteristics. *Poll. Res.*, 15(1): 53-55.

Koshy, M. and Vasudevnayar, T. (2000). Water quality of river Pamba at Kozhencherry. *Poll. Res.*, 19(4): 665-668.

Kumar, A. and Sharma, R.K. (1999). Physico-chemical characteristics of river Sanger in district Etawah (UP). *Ind. J. Env. Sci.*, 3(2): 201-203.

Mahanta, B.K. and Patra, A.K. (2000). Studies on the water quality index of river Sanamachhakandana at Keonjhar Garh, Orissa, India. *Poll. Res.*, 19(3): 371-385.

Musaddie, M. (2000). Surface water quality of Moma river at Akola. *Poll. Res.*, 19(4): 685-691.

Sahu, B.K., Rao, R.J. and Behera, S.K. (1995). Studies of some physico-chemical characteristics of the Ganga river water (Rishikesh-Kanpur) within 24 hours during winter. *Ecol. Env. and Cons.*, 1(1-4): 35-38.

Sangu, R.P.S. and Sharma, K.D. (1985). Studies of water pollution of Yamuna river at Agra. *Ind. J. Environ. Health,* 27(3): 257-261.

Sharma, B.S. and Agarwal, A. (1999). Assessment of water quality of river Yamuna at Agra. *Poll. Res.,* 18(1) 109-110.

Trivedy, R.K. and Goel, P.K. (1986). *Chemical and biological methods for water pollution studies.* Environmental Publications, Karad, India.

17

Bacteriological Assessment of Boiling Water and Point of Use Aqua Purifying Systems

S. VANAJA INDHUMATHY, R. GOWRISANKAR & S. RAMESH
P.G. Department of Microbiology, Sri Paramakalyani College, Azwarkurichi, Tamil Nadu

ABSTRACT

Water purifying equipments along with an indigenous traditional household processing technique *viz.*, boiling were assessed on the basis of their ability to remove the bacterial pathogens from the water sample. Among these, the UV based system was found to be extremely potential in eliminating 7 of the 8 tested bacterial pathogens except *Pseudomonas aeruginosa*. On the other hand, the system employing iodophores was effective in removing *P. aeruginosa*, but was not effective against many other pathogens tested. The water purifying tap attachment that incorporates absorbent cotton wool was ineffective in removing most pathogens. Boiling method warranted a minimum holding time of 15-20 minutes to achieve inactivation of most of the tested bacterial pathogens.

Keywords: Water purification, boiling, systems–UV; iodophores.

Introduction

The bacterial quality of drinking water in India is indeed a cause of great concern. Irrespective of source, the drinking water used by the public has often been reported to be bacteriologically unacceptable. Reports of Gowrisankar *et al.* (1998) have indicated high bacteriological loads in the drinking water samples screened from in and around Azhwarkurichi. Incidence of water-borne epidemics has been on the rise despite the advancement of science. Clark *et al.* (1996) have reported that a water-borne outbreak due to *Salmonella typhimurium* occurred in Missouri, USA. Angulo *et al.* (1997) too have reported a community water-borne outbreak of *Salmonellosis* in USA. Houang *et al.* (1998) remarked the risk of epidemic outbreak coinciding with the occurrence of *Shigella* sp. in Hongkong. These reports reinforce the necessity for comprehensive water treatment systems. The enormous volume of water handled, depletion of filter materials, online contamination during supply to various areas have made the municipal water less reliable on the point of end user. This is true especially in countries like India wherein protected water supply is still restricted to few metro cities.

Realizing the inherent drawbacks in municipal water treatment systems, many household water treatment procedures are still practised from olden days. Owing to its simplicity, boiling of water is still the most popular among the communities Angulo *et al.* (1997). With an increase in awareness on the necessity of potable and safe drinking water, a wide range of water purifying systems are now marketed with each claiming its superiority. Among these water purifying systems, UV disinfection of drinking water is fast emerging as a credible alternative. At a given flow rate, the intensity of the UV radiation in the reactor is tuned to guarantee safe disinfections. Obviously, changes in water quality have an influence on the disinfecting efficiency of UV, because a decrease in transmission could occur in water due to dissolved or suspended matter that might reduce the radiation flux. Other negative effects caused are the deposits on the surface of the UV source and the simultaneous weakening of the UV arch with age (Hoyer *et al.*, 1992). Despite the tall claims made by respective marketing companies, any water purifying system needs to be monitored to obtain an unbiased profile of its efficacy. Wallis *et al.* (1974) were able to note an increase in bacterial densities in activated carbon units following an overnight period of non-use. Similar opinion on the demerits of water filter systems was reported by Daschner *et al.* (1996). With this background, an investigation was carried out to understand the bacteriological merits and demerits of boiling and 3 other points of use water purifying systems.

Materials and Methods

In this study, eight bacterial pathogens comprising *Escherichia coli, Enterobacter aerogenes, Salmonella paratyphi, Staphylococcus aureus, Citrobacter* sp., *Vibrio parahalmolyticus, Pseudomonas aeruginosa* and *Serratia marcescens* were used. These organisms were the laboratory stocks isolated from various clinical and environmental samples. Bacterial strains were pure cultured and their log growth population was correlated with their OD_{600}. For all experiments, bacterial turbidity was adjusted to OD_{600}=0.2 (approx. 10^6 $CFU.ml^{-1}$) and replicates and suitable controls were respectively maintained.

250 ml of tap water was collected in a clean conical flask, autoclaved and after cooling, overnight broth culture of the bacterial pathogen was incorporated individually at 10 per cent (v/v) concentration aseptically. The first set of conical flask with pathogen inoculated sterile water was maintained in a boiling water bath (100°C) for 5 minutes. It was later cooled; serial diluted with sterile distilled water and the bacterial load was determined by streaking on air-dried nutrient agar plates. The inoculated plates were incubated at 37°C for 24 hours. Similarly, the other sets of conical flasks incorporated with the test pathogens were held in the boiling water bath for 5 minutes and evaluated as aforesaid. The effect of the different holding times *viz.*, 10,15 and 20 minutes was also tested.

Sterile water inoculated with bacterial pathogen was made to pass through the UV based water purification equipment with the help of its inbuilt suction unit. The outlet water was collected in a sterile conical flask taking care not to expose it to the external environment and the water sample was enumerated for the residual load of the respective pathogen adopting the technique cited above.

In the case of iodophores incorporated point of use tap attachment, the inner seal of the equipment was removed aseptically. Later, pathogen incorporated sterile water was injected into the tap attachment outfit (inlet) with the help of a sterile 50 ml glass syringe. The outlet water was collected in a sterile conical flask serial diluted, and enumerated using nutrient agar plates.

The commonly used point of use tap filter that employs absorbent cotton wool was tested in this experiment. For each bacterial pathogen, fresh sterile absorbent cotton wool was replaced in the equipment. With the help of sterile 50 ml glass syringe, pathogen incorporated sterile water was injected into this tap outfit (inlet) and the filtered water was collected in a sterile conical flask, which was then serially diluted, and enumerated on nutrient agar plates.

Results and Discussion

It is practically not possible for the Government to satisfy the need of the vast population for a protected and safe drinking water supply. However, realizing the importance of safe drinking water, several water purifying techniques are employed at municipal, industrial and domestic levels. Many water purifying systems have entered the market staking claims for total bactericidal potentials. The most popular and traditional method of domestic water purification has been the boiling of water that was tested for its potential. Analyzing the data presented in Table 17.1, it is evident that to obtain a complete destruction of bacterial pathogens it requires a minimum of 20 minutes holding time at 100°C. Despite of this, *E. aerogenes* was found to have retained a certain amount of viability (0.001 per cent), which probably warrants extended holding time beyond 20 min. Nevertheless, owing to its simplicity and economical nature, boiling water technique is still popular among the semi-urban and rural communities. Realizing these advantages, in many instances, the government has recommended the consumption of well boiled and cooled water to avoid major epidemic outbreaks. Angulo *et al.* (1997) have attributed an outbreak of *Salmonellosis* in a community to the failure of adhering to the government regulations advising them to consume boiled water in the absence of the supply of chlorinated water.

Table 17.1: Comparative Analysis of Different Water Purifying Procedures on the Survival of the Bacterial Pathogens

	Purifying Mode						
	Boiling-Time (minutes)				*UV Based System*	*Iodophore Based System*	*Cotton Wool Filter*
Pathogens Used	*5'*	*10'*	*15'*	*20'*			
P. aeruginosa	0.1	0.1	0.1	0	0.001	0	0.1*
E. coli	0.0001	0.0001	0	0	0	0	0
E. aerogenes	0.001	0.001	0.001	0.001	0	0.1	0.1
Vibrio sp.	0.1	0	0	0	0	0.01	0.1
S. aureus	0.1	0	0	0	0	0.01	100
S. paratyphi	0.1	0	0	0	0	0.1	0.1
Citrobacter sp.	10	0	0	0	0	1	0.1
S. marcescens	100	0	0	0	0	0.1	0

* Numbers indicate percentage of survival.

This study absolutely reiterates the water treatment potentials of UV based purification systems, wherein 7 of the eight bacterial pathogens tested were completely eliminated. Nevertheless, persistence of *P. aeruginosa* even at a low recovery rate (0.001%) is indeed a cause of concern. Hoyer *et al.* (1992) have remarked such deficiencies in the disinfection efficiency of UV radiation. In fact, they have recommended that the UV sensor need to be calibrated according to the bacteriological parameters of local water. Grabow *et al.* (1997) remarked that employing UV irradiation was effective in the removal or inactivation of viruses, fungi, bacteria and even *Cryptosporidium* sp. Slade *et al.* (1986), Karanis *et al.* (1992) and Gardner (1995) have put forth similar reports on the efficiency of UV based water purifying systems.

Water purifying system incorporating iodine compounds are equally popular among the low economic groups as it is relatively cheaper. Further, it does not demand large equipment or sophistication. In fact, these are typical point of use water purifying systems that can be fitted to the delivery point or water tap in the household. Despite the popularity, this investigation indicates mild to moderate bactericidal potential of this system (Table 17.1). While the most persistent *P. aeruginosa* was effectively removed, the common enteric pathogen such as *S. paratyphi* and *Vibrio* sp. were incompletely removed. The sensitivity of individual pathogens to iodine compounds seems to be one of the major deciding factors in the success of this water purifying system. Further, gradual dissipation of iodine could lower the efficiency of this system over a period of time, warranting a frequent change of the cartridge or recharging of the same. Grabow *et al.* (1997) have remarked the limitations of the presence of residual iodine in the potable water treated with iodophoric point in microbiological purifier systems. Their analysis, too have indicated limited bactericidal potentials of iodophores.

Conventional filters employing activated charcoal, asbestos and other compounds are frequently employed in low cost water treatment technology. In tune with this, a point of use water tap attachment system employing absorbent cotton wool was also evaluated for its potential to disinfect/ remove challenged bacteria and the results are described in Table 17.l. Analysis of this data clearly indicates its poor efficiency in removing most of the introduced pathogens. It has been widely reported that conventional water purifying systems employing indigenous adsorbent filter materials are less efficacious in the complete treatment of water. Works of Oie *et al.* (1998) clearly portray the contaminated nature of the so-called sterile water purified with conventional filtration. Similarly, Daschner *et al.* have also strongly reported the unsafe nature of such household filter systems.

In short, water purifying system employing UV disinfection was found to be the most effective for domestic use compared to the other systems. However, the persistence of *P. aeruginosa* needs a further investigation. It will be premature to declare that this system is the best among the existing point of use water purifying systems as there are other equipments that utilize reverse osmosis, membrane filter, ozone that need to be compared and evaluated.

Acknowledgement

The authors wish to thank the Management, Secretary and Principal, Sri Paramakalyani College, Azhwarkurichi for providing the necessary facilities to carry out this research work.

References

Angulo, F.J., Tippen, S., Sharp, D.J., Payne, B.J., Collier, C., Hill, J.E., Barrett, T.J., Clark, R.M., Geldreich, E.E., Donnell, H.D., Swerdlow, J.R. and Swerdlow, D.L. (1997). A community water-borne outbreak of salmonellosis and the effectiveness of a boil water order. *American J. Public Hlth.*, 87: 580-584.

Clark, R.M., Geldreich, E.E., Fox, K.R., Rice, E.W., Johnson, C.H., Goodrich, Barnick, J.A., Abdesaken, F., Hill, J.E. and Angulo, F.J. (1996). A water-borne *Salmonella typhimurium* out break in Gideon, Missouri. *Int. J. Environ. Health. Res.*, 6: 187-193.

Daschner, F.D., Rueden, H., Simon, R. and Clotten, J. (1996). Microbial contamination of drinking water in commercial household water filter systems. *Eur. J. Clin. Microbiol.*, 15: 233-237.

Gardner, T. (1995). Ultraviolet disinfection of water. In: *New world water*, pp. 71-73.

Gowrisankar, R., Palaniappan, R. and Sankar, S. (1998). Bacterial evaluation of rural drinking water systems. *J. Ecotoxicol. Environ. Monit.*, 7(4): 255-258.

Grabow, W.O.K., Jagals, P. and Williams, E. (1997). The effects of supplied water quality on human health in an urban development with basic subsistence facilities. *Water Institute of Southern Africa*, 23: 373- 378.

Houang, E.T.S., Chu, Y.W., Ng, T.K. and Cheng, A.F.B. (1998). Study of the relatedness of isolates of *Shigella flexnerl* and *Shigella sonnel* obtained in 1986 and 1987 and in 1994 and 1995 from Hongkong. *J. Clinical. Microbiol.*, 36: 2404-2407.

Hoyer, O., Kryschi, R., Piecha, I., Mark, G., Schuchmann, M.N., Schuchmann, H.P. and Vonsonntag, C. (1992). UV and fluence rate determination of the low-pressure mercury arc in the UV disinfection of drinking water. *J. Water SRT-Aqua.*, 41: 75-81.

Karanis, P., Maier, W.A., Seilz, H.M. and Schoenen, D. (1992). UV sensitivity of protozoan parasites. *J. Water SRT-Aqua.*, 41: 95-100.

Oie, S., Oomaki, M., Yonoka, K., Tatsumi, T., Amasaki, M., Jukuda, T., Hakuno, H., Nargano, K., Matsuda, M., Hirata, N., Miyano, N.A. and Kamlya, A. (1998). Microbial contamination of 'sterile water' used in Japanese hospitals. *Journal of Hospital Infection*, 38: 61-65.

Slade, J.S., Harris, N.R. and Chisholm, R.G. (1986). Disinfection of chlorine resistant enterovirus in groundwater by ultra violet irradiation. *Water Sci. Tech.*, 18: 115-123.

Wallis, C., Stagg, C.H. and Melnick, J.L. (1974). The hazards of incorporating charcol filters into domestic water systems. *Water Res.*, 8: 111-113.

18

Physico-chemical Analysis of the Water Samples in the Freshwater Ponds of Canchipur, Manipur

L. GEETABALI DEVI & B. MANIHAR SHARMA

Ecology Research Laboratory, Department of Life Sciences, Manipur University, Canchipur, Imphal, Manipur

ABSTRACT

The present paper deals with the seasonal variations in some important physico-chemical characteristics of the freshwater ponds of Manipur University, Canchipur. The overall concentration of the DO_2 level appeared to be low (0.6 to 5.74 mg l^{-1}) while the concentration of free CO_2 significantly varied from 4.2 to 19.5 mg l^{-1} in Site-I and 14.2 to 31.4 mg l^{-1} in Site-II. The nature of the water is found to be acidic to slightly alkaline (pH 5.6 to 7.2). The maximum depths of the water as well as maximum turbidity index values were observed during rainy season.

Keywords: Freshwater ponds, Canchipur, Seasonal variations, Biosphere, Eutrophic.

Introduction

Water is not only the most important essential constituent of all animals, plants and other organisms but also it is pivotal for the survivality of the mankind in the Biosphere (Sharma, 2000). But most of our natural water bodies are gradually becoming degraded to a great extent due to rapid progress of industrialization and population explosion. There is no doubt that the physical and chemical characteristics of water can adversely affect the aquatic environment and ecological balance in many different ways. In order to safeguard such water bodies, there is need to examine the physico-chemical characteristics of the water bodies. Considerable investigations have been done on the physico-chemical characteristics of different types of water bodies in India. But little information is available on this line from Manipur. Thus the present study was aimed to study some important physico-chemical characteristics of the two freshwater ponds of Canchipur, Manipur.

Study Site

The present sites are located within the Manipur University Campus, Canchipur situated at about 6 km due south from the heart of the Imphal town on the western side of the Indo-Burma road. The topography of the area is plain and low lying. Canchipur is a famous historical place in Imphal district located at the intersection of 25°68′ N latitude and 94°78′ E longitude at an altitude of 782m above sea level. The region is characterised by the subtropical climatic condition with distinct summer, rainy and winter seasons. The ponds located in front of the Humanity Block, Manipur University have been categorized into two sites *i.e.*, the southern pond as Site-I while Site-II represents the northern pond. These two ponds are medium sized eutrophic ponds. The morphometric as well as the bathymetric characters of these ponds are given in Table 18.1.

Table 18.1: Morphometry and Bathymetry of the Study Sites

Parameters	*Site-I*	*Site-II*
Maximum length (km)	0.09	0.10
Maximum breadth (km)	0.05	0.06
Maximum depth (m)	1.01	0.95
Minimum depth (m)	0.28	0.23
Mean depth (m)	0.69	0.67
Mean depth/Maximum depth ratio	0.68	0.71
Shore line (km)	0.73	0.75
Shore line development ratio	0.92	0.86
Volume (km^3)	0.0003	0.0005
Surface area (km^2)	0.05	0.06
Surface area/Volume ratio	166.66	120.00
Development of Volume	2.05	2.12

Materials and Methods

The water samples were collected at monthly intervals from different spots by using the procedure employed after Michael (1984). Depth of water was measured with a straight and pointed bamboo pole while turbidity was measured by Secchi Disc. pH was determined with the help of electronic pH meter. Other chemical parameters were done following standard methods given by Misra, 1968; Misra *et al.*, 1970 and Michael, 1984.

Results and Discussion

Oxygen is required for many biological processes. Therefore, the quantity of DO_2 in the water body is of prime importance. During the present investigation, the concentration of DO_2 ranged between 2.2 to 5.74 mg l^{-1} in Site-I and 0.6 to 3.8 mg l^{-1} in Site-II which has been shown in Table 18.2. In the present study the overall DO_2 level appeared to be low (0.6 to 5.74 mg l^{-1}). The values in the present study are found to be higher than the values reported by Singh and Yadav, 1985 (1.0 to 4.0 ppm) in the Nambul River, Manipur. But the present ranges are found lower than those reported by a number of workers in different regions of India *i.e.* Shardendu and Ambasht, 1988 (6.6 to 15.7 mg l^{-1}); Rao *et al.*, 1996 (2.3 to 14.8 mg l^{-1}); De, 1999 (7.0 to 12.0 mg l^{-1}) and Devi, Ch. U., 2002 (0.18 to 14.54 mg l^{-1}) in the freshwater ecosystems of Canchipur, Manipur. But the range in the present study are found comparable with the range reported by Devi, O.I., 1993 (2.2 to 7.4 ppm) in the Waithou Lake, Manipur.

Table 18.2: Seasonal Variations in the Physico-chemical Characteristics of Water in the Study Sites

Parameters	*Site-I*						*Site-II*					
	Summer		*Rainy*		*Winter*		*Summer*		*Rainy*		*Winter*	
	Min.	*Max.*	*Min.*	*Max.*	*Min.*	*Max.*	*Min.*	*Max.*	*Min.*	*Max.*	*Min.*	*Max.*
Dissolved Oxygen (mg l^{-1})	2.20	5.74	2.18	3.40	3.50	4.80	0.60	3.80	0.64	2.64	2.98	3.70
Free Carbon dioxide (mg l^{-1})	4.20	9.00	11.60	19.50	4.40	5.20	14.2	28.6	29.0	31.4	14.4	30.0
Hydrogen ion concentration (pH)	6.6	7.2	6.2	7.1	6.7	7.2	5.7	6.9	5.6	6.7	5.8	6.4
Depth (m)	0.61	0.93	0.84	1.00	0.72	0.77	0.70	0.96	0.75	1.01	0.70	0.74
Turbidity Index (cm)	21.00	35.00	30.00	42.00	24.00	27.00	31.00	56.00	47.00	56.00	30.00	32.00
Total Solute Contents (g %)	0.003	0.020	0.009	0.014	0.003	0.004	0.020	0.034	0.003	0.020	0.004	0.017

The present values showed low concentrations of DO_2 which may be due to indiscriminate use of these ponds by the local people. It may also be due to decomposition of organic matter as well as due to bacterial decomposition (Das and Verma, 1993). During the present study, the concentration of DO_2 was higher during summer season (March-April). De, (1999) also observed such higher concentration of DO_2 during summer season (March-April) *i.e.* 12.0 and 10.0 mg l^{-1} in the two freshwater ponds of Kolkata. However in some freshwater ecosystems in India maximum concentration of DO_2 during winter months have also been reported by several authors (Shardendu and Ambasht, 1988; Misra *et al.*, 1999). Such varied observations are resulted mainly due to difference in climatic and edaphic characteristics of the physiographic factors in the different regions of India.

The concentration of free CO_2 significantly varied from 4.2 to 19.5 mg l^{-1} in Site-I and 14.2 to 31.4 mg l^{-1} in Site-II. The overall present range (4.2 to 31.4 mg l^{-1}) are found higher than the range obtained by Roy, 2000 (1.2 to 17.8 mg l^{-1}); Devi Ch. U., 2000 (0.17 to 8.26 mg l^{-1}) and Pandey *et al.*, 2002 (8.8 to 13.4 mg l^{-1}). However, the present ranges are found lower than the findings of Devi, O.I., 1993 (10 to 35 ppm) and Kumar *et al.*, 1996 (3.10 to 42.25 mg l^{-1}). The lowest concentrations (4.2 and 14.2 mg l^{-1}) of free CO_2 during April in both the study sites may be due to high photosynthetic activity of the aquatic macrophytes (Billore and Vyas, 1982). The maximum concentrations (19.5mg l^{-1} in Site-I and

31.4 mg l^{-1} in Site-II) were recorded during rainy season which may be due to diffusion of atmospheric CO_2 into the rainwater producing carbonic acid (Chakrabarty *et al.*, 1995). Misra, *et al.* (1999) also observed higher values of free CO_2 during rainy season.

Hydrogen ion concentration (pH) also showed seasonal variations. During the present study, the pH value ranged from 5.6 to 7.2 which is comparable with the values reported by Devi, K.I. (1998) in the Utrapat Lake, Manipur (6.28 to 8.88). Highest values occurred during summer months (March and May) and the lowest values during rainy season (July). The highest values during summer months may be due to removal of sufficient amount of CO_2 by photosynthetic process of the aquatic macrophytes (Shardendu and Ambasht, 1988). Billore and Vyas (1982) also reported maximum pH value during hot summer season. Sah *et al.* (2000) also reported highest pH value (9.4) during summer season at the Narayani River, Nepal. The occurrence of lowest pH value during rainy season (July) in Site-II may be attributed to the active decomposition of organic matter during the study period (Sharma *et al.*, 2001). During the present study, pH value in Site-II was found higher than that in Site-I. This may be due to the insignificant consumption of free CO_2 by the aquatic macrophytes in Site-II since, the site is rich in macrophytic population as compared to Site-I.

The depths of the two ponds varied in different seasons which has been shown in Table 18.2. The maximum depths (1.00 m in Site-I and 1.01 m in Site-II) were observed during rainy season and the minimum depths (0.61 m in Site-I and 0.70 m in Site-II) were found during summer in both the study sites. The maximum depths which were observed during rainy season (July) may be attributed to the supply of rain water while the minimum depths (0.61 m and 0.70 m) during April may have resulted mainly due to very less rainfall during this period (summer) substantiated by the excessive loss of water due to high evapo-transpiration rate. Maximum depths during rainy season (July) has also been reported by Misra *et al.* (1999).

The present range of the values of turbidity index in the present investigation varied from 21.00 to 42.00 cm in Site-I and 31.00 to 56.00 cm in Site-II. The overall range of the turbidity index value (21.00 cm to 56.00 cm) are found lower than the findings of Devi, K.I., 1998 (36.23 to 107.93 cm) and Avasn *et al.*, 2000 (51.5 to 279.2 cm) at the polluted and unpolluted sites of the River Sarada (Andhra Pradesh) and Sharma *et al.*, 2001 (7.0 to 270.0 cm). But the present ranges are higher than the values reported by Devi, O.I., 1993 (15.0 to 45.0 cm) in the Waithou Lake, Manipur. The peak value (0.034%) of the total solute content during rainy season may be due to the presence of high organic matter as well as soil particles during the study period. The overall present range (0.003 to 0.034%) was found lower than those reported by Devi, N.B., 1993 (47 to 592.6) in the Loktak Lake, Manipur and Avasn *et al.*, 2000 (47.0 to 508.0%).

Conclusion

From the above investigations, it may be inferred that the physico-chemical characteristics of the present study sites varied considerably and showed characteristic trends in relation to the seasonal changes during the study period. The values obtained during the present study are found comparable to the other freshwater bodies in the state and elsewhere. The values of surface area/volume ratio, development of volume and the physico-chemical characteristics of water reveal the eutrophic status of the ponds under the present study.

However, all the physico-chemical parameters in general appeared to be within manageable levels of ponds were properly maintained. It is observed that ponds of the state (Manipur) were fast disappearing due to excessive exploitation by the expanding population. Therefore, steps must be taken to check the misuse of the ponds otherwise it will further deteriorate the quality of the water to a great extent.

References

Avasn Maruthi, Y., M.V. Subba Rao and S. Raffia Krishna Rao (2000). Pollution status of River Sarada at Anakapalli, Andhra Pradesh. *Indian J. Environ. and Ecoplan.*, 3(1): 45-48.

Billore, D.K. and L.N. Vyas (1982). Distribution and production of Macrophytes in Pinchhola Lake, Udaipur (India). In: B. Gopal, R.E. Turner, R.G. Wetzel and D.F. Whingham (Eds.), *Wetlands ecology and management*. National Institute of Ecology and International Scientific Publications, India, pp. 45-54.

Chakraborty, R.D., P. Roy and S.P. Singh (1995). The quantitative study of the plankton and the physico-chemical conditions of the River Yamuna at Allahabad. *Ind. J. Fish*, 6: 186-203.

Das, H.N. and Verma, B.N. (1993). Limnology of a Pond and Lake at Raj Nagar (Madhubani) Bihar. *Bio. Journal*. 5(1 and 2): 41-46.

De, U.K. (1999). Seasonal variation of some abiotic factors of two freshwater ponds. *Geobios New Reports*, 18: 119-121.

Devi, K.I. (1998). *Ecological studies of the freshwater Macrophytes in Utrapat Lake, Manipur*. Ph.D. Thesis, Manipur University, Imphal, India.

Devi, Ch. U. (2000) *Phyto-sociology and primary production of the macrophytes in the freshwater ecosystems of Canchipur*. Ph.D. Thesis, Manipur University, Imphal, India.

Devi, N.B. (1993). *Phyto-sociology, primary production and nutrient status of macrophytes of Lokiak Lake, Manipur*. Ph.D. Thesis, Manipur University, Imphal, India.

Devi, O.I. (1993). *Distribution, primary production and nutrient status of the macrophytic communities in Waithou Lake, Manipur*. Ph. D. Thesis, Manipur University, Imphal, India.

Kumar, A., H.P. Gupta and D.K. Singh (1996). Impact of sewage pollution on chemistry and primary productivity of two freshwater bodies in Santal Pargana (Bihar). *Indian J. Ecol.*, 23(2): 86- 92.

Michael, P. (1984). *Ecological methods for field and laboratory investigations*. Tata McGraw-Hill. Publs. Company Limited, New Delhi.

Misra, A.P., B.K. Borah and M. Sarma (1999). Limnological investigation of the Freshwater Tributary. *J. Freshwater Biol.*, 11(1-2): 1-5.

Misra, K.C., D.N. Rao, R.S. Dwivedi, K.L. Mukherjee and P.V. Ananthekrishnan, (1970). *Ecology: Study of ecosystem*. A.H. Wheeler and Co. (P) Ltd., Allahabad.

Misra, R. (1968). *Ecology work book*. Oxford and IBH Publs. Co., New Delhi.

Pandey, A.K., S. Dutta and K.C. Sharma (2002). Physico-chemical characteristics of Hamor pond of Kishangarh, Ajmer (Rajasthan). *Indian J. Environ. and Ecoplan.*, 6(1): 167-169.

Rao, A.M., V. Narendra Rao and S.K. Mahmood (1996). Assessment of water quality and pollution of Narsingi Pond. *Ecol. Env. and Cons.*, 2(1-2): 45-49.

Roy, P.N. (2000). Studies on hydrological status of a stream in Santhal Parganas (South Bihar) with special reference to pollution. *Indian J. Environ and Ecoplan.*, 3(1): 131- 135.

Sah, J.P., S.K. Sah, P. Acharya, D. Pant, V.A. Lance (2000). Assessment of water pollution in the Narayani River, Nepal. *International Jour. of Eco. and Environ. Sc.*, 26: 235-252.

Shardendu and R.S. Ambasht (1988). Limnological studies of a rural and an urban tropical aquatic ecosystem: Oxygen forms and ionic strength. *Trop. Ecol.*, 92(2): 98- 109.

Sharma, B.M. (2000) *Environmental studies*. Manipur University, Imphal.

Sharma, K.C., S. Ojha and A. Dutta (2001). Impact of industrial pollution on two water bodies of Bhilwara, Rajasthan. *Bulletin of the National Institute of Ecology*, 11: 49-54.

Singh, L. Gyaneshwar and P.S. Yadava (1985). *Primary production and limits of tolerance of river Nambul and Imphal town*. M.Sc. Dissertation, Manipur University.

19

Groundwater Quality Index Near Industrial Area

DEEPALI A. SOHANI, G.R. CHAUDHARY & V.S. SHRIVASTAVA
Centre for P.G. Research in Chemistry, G.T.P. College, Nandurbar, Maharashtra

ABSTRACT

Water is most essential and precious thing on earth for every life. The freshwater has always demanding for any biological activities, due to increasing population. These demands have some limitations. WQI gives quick assessment of any water system. The present work focuses on the water quality of groundwater and near industrial area of Jalgaon and Dhule (Maharashtra). The physico-chemical characteristics were determined (before and after rainy season) for the parameters such as pH, hardness, Ca hardness, Mg hardness, alkalinity, dissolved oxygen, total solids, chlorides, sulphates, nitrates, iron etc. of seven sampling sites, WQI studies suggest that, water is not suitable for direct consumption, they categorized in slightly polluted.

Keywords: Water quality index.

Introduction

Water quality index is contributing factor for water quality of any water system. It is one of the effective, helpful parameter and provides informative data, which is important to citizens, government and public health policies for improvement of water quality programmes (Singh and Ghosh, 1999).

The sample site near industrial area of Jalgaon and Dhule (Maharashtra) which was rapidly growing industrial area. The present work was done seasonally (before and after rainy season) with seven sample sites. The domestic waste and industrial effluent was run-off through drainage. By this system, it may changes or disturbs the groundwater quality due to percolation of water contains high attendance of some parameters like hardness, biological oxygen demand, dissolved oxygen, chlorides, nitrates and some heavy metals generated from human and industrial activities (Yazdandoost and Katdare, 2000). Common ions and nitrates of sodium, potassium, magnesium and calcium etc. keep accumulating in the stream once added (Geldrich, 1972; Elengo, 1992; Jain *et al.*, 1998). In this, the water quality index was based on Dhamija and lain (1995).

Some literatures are studied on water quality index (Sharma and Pande,1999; Adak and Purohit, 2001; Musaddiq, 2000; Patil and Tijare, 2001).

Materials and Methods

The groundwater samples were collected in plastic cans from bore wells located adjoining to the MIDC Jalgaon and Dhule industrial area. The samples were analysed as per standard procedures (APHA, 1989 and Trivedy and Goel, 1986). The statistical calculations made by following literature method (Dhamija and Jain, 1995).

In this method the rating scale was apply to the parameters according to its relative importance in the overall water quality. The maximum weight is 4 given to the parameters like pH, dissolved oxygen, total solids etc. and the minimum weight is 2 given to the parameters like hardness, chlorides, sulphates etc. due to its importance in assessment of water quality which is shown in Table 19.1.

Table 19.1: Water Quality Parameter (WHO Standards)

Parameters	*Weight*	*Unit weight*
pH	4	0.16
Hardness	2	0.08
Ca hardness	2	0.08
Mg hardness	2	0.08
Alkalinity	3	0.12
Dissolve oxygen	4	0.16
Total solids	4	0.16
Chloride	2	0.08
Sulphate	2	0.08
Nitrate	3	0.12
Iron	3	0.12

The unit weight (Wi) is calculated from the formula,

$$Wi = \sum_{I=1}^{Wi/11} Wi \qquad \text{as} \qquad \sum_{I=1}^{q} Wi = 1$$

The rating scale from eleven physico-chemical parameters is given in Table 19.3. Parameters divided into four stages (permissible, slight, moderate, severe) and quality rating (qi) ranges from 100 to 0.

For calculating WQI, the sub index (SI) I is first found out for each parameter which is,

$$(SI)\ i = qiWi$$

$$WQI = \frac{\sum_{I=1}^{11} (SI)\ i}{\sum^{11} Wi}$$

$$WQI = \sum_{I=1}^{11} qiwi \qquad \text{as} \qquad Wi = 1$$

Results and Discussion

The consideration of permissible limits of standards (Table 19.1) for various beneficial uses, values of some relevant variables for each class of various uses are given in Tables 19.2. The five classes adopted are giving on next page.

Table 19.2: Seasonal Physico-chemical Changes of Groundwater Samples of Before Rainy Season

Parameter	*No. of Samples*													
	X_1	X_2	X_3	X_4	X_5	X_6	X_7	X_8	X_9	X_{10}	X_{11}	X_{12}	X_{13}	X_{14}
pH	7.2	7.1	7.0	7.9	8.5	6.8	7.9	9.2	6.8	6.4	7.2	7.2	7.4	7.0
	(100)	(100)	(100)	(100)	(100)	(80)	(100)	(50)	(80)	(0)	(100)	(100)	(100)	(100)
Hardness	472	692	600	332	236	596	316	404	724	472	328	264	356	412
	(50)	(0)	(0)	(50)	(80)	(0)	(50)	(50)	(0)	(50)	(50)	(80)	(50)	(50)
Ca hardness	150.70	176.35	176.35	48.10	84.97	184.37	96.19	88.176	131.46	134.7	44.89	24.05	56.11	59.32
	(50)	(50)	(50)	(100)	(80)	(50)	(80)	(80)	(80)	(80)	(100)	(100)	(100)	(100)
Mg hardness	23.4	61.43	39.0	51.68	5.85	42.9	18.53	44.83	96.49	33.14	52.63	49.71	52.63	64.32
	(100)	(80)	(80)	(80)	(100)	(80)	(100)	(80)	(50)	(80)	(80)	(80)	(80)	(80)
Alkalinity	840	380	–	560	100	–	300	300	–	–	220	200	140	340
	(0)	(0)	(100)	(0)	(50)	(100)	(0)	(0)	(100)	(100)	(0)	(0)	(50)	(0)
D.O.	11.90	12.15	15.19	10.12	10.25	11.14	12.03	10.26	11.90	8.48	9.87	10.13	8.86	6.33
	(100)	(100)	(100)	(100)	(100)	(100)	(100)	(100)	(100)	(100)	(100)	(100)	(100)	(100)
T.S.	480	1120	760	720	960	1280	480	440	2200	680	1320	160	520	760
	(100)	(50)	(80)	(80)	(80)	(50)	(100)	(100)	(0)	(80)	(50)	(100)	(80)	(80)
Chloride	170.4	198.8	187.4	153.3	386.2	170.4	96.56	389.1	116.6	156.2	156.2	68.2	161.88	142.0
	(100)	(100)	(100)	(100)	(80)	(100)	(100)	(80)	(100)	(100)	(100)	(100)	(100)	(100)
Sulphate	3752.88	2584.2	2897.9	2699.4	3144.8	2667.5	2650.1	2485.5	1547.3	1283.8	3802.3	1382.6	4625.3	2765.3
	(0)	(0)	(0)	(0)	(0)	(0)	(0)	(0)	(0)	(0)	(0)	(0)	(0)	(0)
Nitrate	35	60	50.0	30.2	100	42.0	48.0	30.1	110	51.7	90.0	48.0	18.0	40.0
	(50)	(0)	(0)	(80)	(0)	(50)	(50)	(80)	(0)	(0)	(0)	(0)	(100)	(50)
Iron	2.20	2.45	–	3.0	4.50	1.65	–	1.05	1.60	1.75	1.60	1.05	3.65	1.65
	(0)	(0)	(100)	(0)	(0)	(0)	(100)	(0)	(0)	(0)	(0)	(0)	(0)	(0)
WQI	77.99	58.4	75.2	92.8	78.0	69.2	92.4	72.8	59.2	65.6	66.4	76.8	92.4	80.4

Class 1 (WQI 90 and above) – Excellent

Class 2 (WQI between 65 and 89) – Good

Class 3 (WQI between 35 and 64) – Satisfactory

Class 4 (WQI between 10 and 34) – Poor

Class 5 (WQI 10 and below) – Unaccepted

The pH of all the samples in before rainy season was in accepted range except, sample No. X_6, which was in the range 7.0–8.5. In after rainy season pH was 6.4–9.2 *i.e.* the water of this site can be used for irrigation purposes.

The hardness was in the range 238 to 692 mg/l in before rainy season and in after rainy season it was 264 to 724 mg/l. Both are in the moderate range. The Mg hardness and Ca hardness are under slightly permissible limit.

The dissolved oxygen value of before and after rainy season was 10.12 to 15.19 mg/l and in after rainy season, it was 6.33–11.90 mg/l. These values are in permissible limit.

Chlorides were found to be in the range of 480 to 1280 and 160 to 2200 mg/l in before and after rainy season. The limit of chloride in both seasons is slightly permissible. But the Sulphate present in these sample was very high. In before rainy season, it was 2584 to 3752.88 mg/l and in after rainy season, it exists in the range 1283.8 to 4625.3 mg/l. The Nitrates was found to be high in some samples. It ranges from 30.2 to 100 mg/l and 18.0 to 110 mg/l in before and after rainy season.

The Iron content was also high *i.e.* 0–4.50 mg/l and 1.05–3.65 mg/l in before and after rainy season.

Therefore, the results indicate that some parameters like dissolved oxygen, chlorides, hardness etc, were under the permissible limits. But sulphate, nitrate, iron were found beyond the permissible limit.

Table 19.3: Scale Rating for Water Quality Parameters (qi)

Parameters	*Standards*	*Permissible (100)*	*Slight (80)*	*Moderate (50)*	*Severe (0)*
pH	7.0–8.5	7.0–8.5	8.6–8.8	8.9–9.2	> 9.2
			6.8–7.0	6.5–6.7	< 6.5
Hardness	100–500	< 100	101–300	310–500	> 500
Ca hardness	75–200	< 75	76–137	138–200	> 200
Mg hardness	30–150	< 30	31–90	91–150	> 150
Alkalinity	< 120	50	51–85	86–120	> 120
Dissolve oxygen	> 6	> 6	4.4–4.9	3–4.5	< 3
Total solids	500–1500	< 500	500–1000	1000–1500	> 1500
Chlorides	200–500	200	201–400	401–500	> 500
Sulphate	200–400	200	201–300	301–400	> 400
Nitrate	> 45	20	21.0–32.5	33.0–45	> 45
Iron	0.1–1.0	0.1	0.2–0.5	0.6–1.0	> 1.0

A Water Quality Index (WQI) denoting the integrated effect of the various parameters that are relevant and significant to a particular use that is proposed to express the water quality for different uses. From Table 19.2, it can be clear that, from these sites show good WQI values. The average of WQI in before and after rainy season, the value was high in before rainy season *i.e.* 77.71, it may be due to settlement or accumulation of suspended dissolved matter. In after rainy season, the average value was 73.37 may be due to percolation of water with dissolved matter. These values indicate the slightly pollution in all the samples.

In view of the above it has been observed that, the samples of all the sites are not suitable for drinking and can be used for irrigation purposes. After the treatment like osmosis, electrodialysis, ion exchange and solar distillation etc. can be used for drinking purposes. So our investigations are useful for the Government authorities, for taking care and alteration against the groundwater pollution. Therefore, WQI is most useful for the water quality assessment.

Acknowledgement

The authors are thankful to college authorities for providing laboratory facilities.

References

Adak, M.D. and Purohit, K.M. (2001). Status of surface and groundwater quality of Mandiakudar Part-II: Agricultural utilities. *Poll. Res.*, 20: 219-225.

APHA (1989). *Standard methods for examination of water and wastewater, 17th edition*. AWWA and WPCF, Washington.

Dhamija, S.K. and Jain, Y. (1995). Studies on the water quality index of a lentic water body at Jabalpur, M.P. *Poll. Res.*, 14: 341-346.

Elangp, L.S., Ramchandran and Choudhary, Y.S.N. (1992). Groundwater quality in costal region of South Madras. *IJEH*, 34: 318-325.

Geldreich, E.C. (1972). Buffalo take recreation water quality: A case study of bacteriological data interpretation. *Water Resource*, 6: 913-916.

Jain, C.K., Bhatia, K.K. and Seth, S.M. (1998). Effect of waste disposal on the water quality of river Kali. *IJEH*, 40: 372-375.

Musaddiq, M. (2000). Surface water quality of Moma river at Akola. *Poll. Res.*, 19: 685-691.

Patil, D.B. and Tijare, R.V. (2001). Studies on water quality of Gadchiroli lake. *Poll. Res.*, 20: 257-259.

Sharma, S.D. and Pande, K.S. (1999). Use of water quality index for Ramganga river classification and pollution control strategy. *Poll. Res.*, 18: 335-338.

Singh, A.P. and Ghosh, S.K. (1999). Water quality index for river Yamuna. *Poll. Res.*, 18: 435-439.

Trivedy, R.K. and Goel, P.K. (1986). *Chemical and biological methods for water pollution studies*. Environmental Publications, Karad.

Yazdandoost, M.Y. and Katdare, M.S. (2000). Water quality index of major rivers in Pune. *J. Industrial Pollution Control*, 16: 231-238.

20

Analysis of Heavy Metals in Groundwater from Coal Mining Area in Jamtara District, Jamtara, Jharkhand

K.K. PRASAD

Pollution Research Laboratory, Jamtara College, Jamtara, Jharkhand

ABSTRACT

Heavy metals *viz.* Mn, Cu, Fe, Zn, Pb and Cd in natural spring water and dugwell water from Lote Industrial Area and nearby villages have been monthly analysed from April 2001 to March 2002. This is the initial and preliminary effort to study the heavy metal pollution of groundwater resources in a newly developed Lote industrial area of Jamtara District. The results on trace metals indicated varying degrees of contamination in groundwater which may cause a serious health problems to domestic animals and human beings.

Introduction

Jamtara district in Jharkhand shows slow industrial development. Due to the major water resources has developed in industrial zone at Lote in Jamtara district. Total area of Lote district is about 575 hect. including 270 industries with 79 unit manufacturing pesticides, chemicals, paints, dyes, pharmaceuticals, detergents etc. and discharging gaseous, liquid and solid effluents in air, water and soil and thus polluting local and nearby environment. Most of the industries do not have effluent plants. Liquid effluents of various industries are carried through the drainage pipe and discharged into Dabhol crick. Due to frequent leakage problems in drainage pipe, effluents spill into the nearby streams and open land. Some industries release untreated effluents directly into the streams through some bypass arrangement. In addition to this, solid waste disposal in and around the industries has blocked the free flow of water making it more contaminated leading to frequent kills of domestic animals and field crops.

Materials and Methods

Lote M.I.D.C. is established at the top of Parshuram Ghat. Flow of water is always towards the low-lying areas from the top and hence seven spots (4 natural springs and 3 dugwells) around the industrial zone from low-lying areas were selected for sampling. Spots A, B, D and F were natural springs from Jamtara district, Dhangarwadi, Awashi and Songaon respectively while spots C, E and G were dugwell from Chalkewadi, Ghanekhunt and Patawardhan Lote respectively. Water samples were collected in 500 ml polythene bottles and analysed on 15th day of each month during April 2001 to March 2002. The collected samples were acidified with concentrated nitric acid on the spot to bring the pH of the samples blow 2 in order to kill all the biota utilizing trace metals present in the water.

The pre-treatment for heavy metals determination in water was done by using standard methods for analysis of water and wastewater (APHA, 1989). Pre-concentration of the samples was done by evaporation method. Glass distilled water was used for all preparations. The analysis of heavy metals was done with the Perkin Elmer Atomic Absorption Spectrophotometer (Model 3030, USA). The results of the analysis are reported in mg/l and compared with WHO (1984) and Indian Standards (1983) for drinking water.

Results and Discussion

The results of heavy metal contents in all samples have been represented in Tables 20.1 to 20.8). The findings are compared either with WHO (1984) or Indian Standards (Prasad and Iyer, 1983) for drinking water.

Table 20.1: Concentration (mg/l) of Calcium in Groundwater from Lote Industrial Area

Month/Spot	*A*	*B*	*C*	*D*	*E*	*F*	*G*
Apr. 2001	158.10	1.00	8.00	208.00	3.40	0.48	219.90
May 2001	136.8	1.31	1.92	10.99	27.70	2.88	187.50
Jun. 2001	12.37	1.32	1.32	4.24	19.27	1.74	18.25
Jul. 2001	12.01	9.04	6.51	9.50	10.12	0.91	7.74
Aug. 2001	12.14	0.38	0.52	0.92	1.67	0.24	3.10
Sep. 2001	13.21	0.86	0.40	5.90	0.64	0.43	11.10
Oct. 2001	14.00	0.79	0.55	0.57	0.87	0.55	13.29
Nov. 2001	15.37	1.65	0.97	1.14	1.27	0.64	14.27
Dec. 2001	16.00	1.98	0.97	0.81	1.53	0.59	14.70
Jan. 2002	46.38	4.76	5.02	2.66	1.80	0.76	29.00
Feb. 2002	83.25	2.41	1.58	1.11	3.75	1.64	36.83
Mar. 2002	115.68	2.10	3.18	14.27	3.56	0.93	48.36

Calcium remains present in rocks and leaches from there to water and is quite abundant in water. Sewage and industrial wastes are also important sources of calcium (Mishra and Saxena, 1989). Calcium is responsible for hardness of water. Although hardness has no known adverse on health, there is evidence that hard water plays a role in heart disease (Sastry and Rathee, 1998). During the present work Ca was detected in all the samples and ranged from 0.246 to 219.9 mg/l. The Indian Standards for Ca is 200 mg/l. While taking into consideration these limits, the values of calcium content at all spots except D and G during April 2001 (Table 20.1) were much below the limits given by IS and WHO.

Manganese (Mn) is one of the important trace elements essential for organisms. It has low acute oral toxicity, however, the Chronic Manganese poisoning leads to progressive deterioration of the central nervous system and symptoms resembling to those of Parkinson's disease are observed. Mn stains laundry and other fixture and gets deposited on food during cooking (Sayani *et al.*, 1988). In the present investigation Mn ranged from below detectable level (BDL) to 66.5 mg/l. In general, the values of Mn content of water samples from all die spots were much below the limits given by WHO. The WHO standards for Mn is 0.05 mg/l. This limit was exceeded at spots A, B and G during May, Dec. and Sep.-Oct. 2001 respectively (Table 20.2).

Table 20.2: Concentration (mg/l) of Manganese in Groundwater from Lote Industrial Area

Month/Spot	*A*	*B*	*C*	*D*	*E*	*F*	*G*
Apr. 2001	38.40	BDL	BDL	BDL	BDL	BDL	0.006
May 2001	66.50	0.023	0.014	BDL	BDL	BDL	BDL
Jun. 2001	5.34	0.018	0.01	BDL	BDL	BDL	0.006
Jul. 2001	0.01	0.01	BDL	BDL	BDL	BDL	BDL
Aug. 2001	9.79	BDL	BDL	BDL	BDL	BDL	0.006
Sep. 2001	6.35	0.023	0.007	BDL	BDL	BDL	0.183
Oct. 2001	9.861	0.008	BDL	BDL	BDL	0.013	0.596
Nov. 2001	10.312	0.024	BDL	0.009	0.011	0.008	BDL
Dec. 2001	11.124	1.091	BDL	BDL	BDL	BDL	0.02
Jan. 2002	31.347	0.022	0.011	0.009	0.01	0.01	0.022
Feb. 2002	32.298	BDL	BDL	BDL	BDL	BDL	0.015
Mar. 2002	35.357	BDL	BDL	BDL	BDL	BDL	0.014

BDL = Below Detectable Level.

Excessive ingestion of Copper (Cu) is responsible for specific disease of the bone (Verma *et al.*, 1995). The range of copper observed in various water samples was from BDL to 0.454 mg/l (Table 20.3).

Iron (Fe) is abundantly present in the Earth's crust. There may be 1 to 10 mg/l concentration of Fe in groundwater (Hem, 1970). Although it is abundant in the earth's crust, it is absorbed in different forms at different rates and iron deficiency is quite common among people throughout the world (NRC, 1980). In the present study, the iron contents vary between BDL to 39.5 mg/l (Table 20.4). The recommended Indian Standards (1983) and WHO (1984) standard of iron in drinking water is 0.3 mg/l. At spot 'A' the iron content of water was much above the WHO and IS limits throughout the year, while at spot 'E', permissible limit was exceeded during April and June to September 2001 and March 2002 (Table 20.4).

Table 20.3: Concentration (mg/l) of Copper in Groundwater from Lote Industrial Area

Month/Spot	*A*	*B*	*C*	*D*	*E*	*F*	*G*
Apr. 2001	0.01	BDL	BDL	BDL	0.065	BDL	BDL
May 2001	0.012	BDL	BDL	BDL	BDL	BDL	BDL
Jun. 2001	0.009	0.01	0.018	0.011	0.039	BDL	0.01
Jul. 2001	0.007	0.011	0.021	0.019	0.041	0.008	0.011
Aug. 2001	0.008	BDL	0.008	BDL	0.011	0.008	0.012
Sep. 2001	0.079	0.021	0.026	0.019	0.012	0.015	0.014
Oct. 2001	0.011	0.014	0.01	0.014	0.011	0.013	0.013
Nov. 2001	0.454	BDL	BDL	0.011	BDL	BDL	BDL
Dec. 2001	0.015	BDL	BDL	0.007	0.014	0.015	0.018
Jan. 2002	0.032	0.015	0.011	0.011	0.009	0.015	0.026
Feb. 2002	0.024	0.006	0.007	0.008	0.01	0.012	0.011
Mar. 2002	0.013	BDL	BDL	BDL	0.019	0.01	0.01

BDL = Below Detectable Level.

Table 20.4: Concentration (mg/l) of Iron in Groundwater from Lote Industrial Area

Month/Spot	*A*	*B*	*C*	*D*	*E*	*F*	*G*
Apr. 2001	31.70	0.027	0.066	0.009	1.303	0.013	0.016
May 2001	38.90	0.019	0.052	0.01	0.015	0.011	0.012
Jun. 2001	39.50	0.012	0.05	BDL	0.728	0.01	0.01
Jul. 2001	37.85	0.01	0.074	BDL	0.728	0.01	0.01
Aug. 2001	35.78	0.01	0.07	BDL	0.427	0.009	0.012
Sep. 2001	30.00	0.009	0.065	BDL	0.568	0.008	0.011
Oct. 2001	375.0	0.008	0.05	BDL	0.931	0.01	0.014
Nov. 2001	2.049	BDL	0.008	BDL	0.014	0.04	0.019
Dec. 2001	1.554	0.021	BDL	BDL	0.016	0.007	0.013
Jan. 2002	16.289	0.01	BDL	BDL	BDL	BDL	0.02
Feb. 2002	6.801	BDL	BDL	BDL	BDL	BDL	0.015
Mar. 2002	5.123	BDL	0.032	BDL	0.639	BDL	0.007

BDL = Below Detectable Level.

Zinc (Zn), when present above permissible limit of 0.5 mg/l (WHO, 1984), imparts upleasant taste to drinking water. Symptoms of Zn toxicity include irritability, muscular stiffness, loss of appetite and nausea (NRC, 1977). Zinc content of all the water samples except spot G ranged between BDL and 0.407 mg/l which is well within the permissible limits. From October 2001 to February 2002 at spot G, concentration of Zn was exceeded (Table 20.5). Such increased concentration of Zn is dangerous to the human beings.

Table 20.5: Concentration (mg/l) of Zinc in Groundwater from Lote Industrial Area

Month/Spot	*A*	*B*	*C*	*D*	*E*	*F*	*G*
Apr. 2001	0.134	BDL	BDL	BDL	0.011	BDL	BDL
May 2001	0.068	0.025	0.229	1.274	0.035	0.335	0.21
Jun. 2001	0.05	0.022	0.248	0.083	0.029	0.057	0.092
Jul. 2001	0.03	0.029	0.035	0.041	0.029	0.046	0.076
Aug. 2001	0.076	BDL	0.02	0.078	0.029	0.054	0.563
Sep. 2001	0.117	0.042	0.049	0.036	0.021	0.017	1.251
Oct. 2001	0.272	0.035	0.019	0.038	0.032	0.056	2.036
Nov. 2001	0.246	0.014	0.007	BDL	BDL	BDL	1.313
Dec. 2001	0.347	0.043	BDL	0.024	0.008	0.011	1.794
Jan. 2002	0.407	0.22	BDL	BDL	BDL	0.011	1.326
Feb. 2002	0.249	BDL	BDL	0.01	BDL	BDL	0.648
Mar. 2002	0.187	BDL	BDL	BDL	BDL	BDL	0.277

BDL = Below Detectable Level.

Table 20.6: Concentration (mg/l) of Cadmium in Groundwater from Lote Industrial Area

Month/Spot	*A*	*B*	*C*	*D*	*E*	*F*	*G*
Apr. 2001	0.017	BDL	BDL	BDL	0.008	0.007	BDL
May 2001	0.007	BDL	BDL	BDL	BDL	BDL	BDL
Jun. 2001	0.006	BDL	BDL	BDL	0.01	0.007	0.008
Jul. 2001	0.007	BDL	BDL	BDL	0.013	0.009	0.019
Aug. 2001	BDL	BDL	BDL	BDL	BDL	BDL	BDL
Sep. 2001	BDL	BDL	BDL	BDL	BDL	BDL	BDL
Oct. 2001	0.011	BDL	BDL	BDL	BDL	0.782	0.032
Nov. 2001	0.024	0.007	BDL	0.008	BDL	0.008	0.01
Dec. 2001	0.008	BDL	BDL	0.007	BDL	0.007	0.009
Jan. 2002	0.01	BDL	BDL	BDL	BDL	BDL	0.008
Feb. 2002	0.009	BDL	BDL	BDL	BDL	BDL	0.007
Mar. 2002	0.01	BDL	BDL	BDL	BDL	BDL	0.007

BDL = Below Detectable Level.

Cadmium (Cd), even in low concentration is quite toxic to human health (Verma *et al.*, 1995). The cadmium content in groundwater have been observed below detectable limits (Sayani *et al.*, 1988). Present findings revealed trace amount of cadmium ranging from BDL to 0.01 mg/l, which was well within the standard limit of 0.01 mg/l laid by WHO (1984). Spot 'A', in April and November 2001, Spot 'E', in July 2001, Spot 'F' in October 2001 and Spot 'G' in July and October 2001 exhibited exceeded level of cadmium concentration (Table 20.6).

Table 20.7: Concentration (mg/l) of Lead in Groundwater from Lote Industrial Area

Month/Spot	*A*	*B*	*C*	*D*	*E*	*F*	*G*
Apr. 2001	0.046	0.018	0.026	0.025	0.013	0.022	0.035
May 2001	0.065	0.022	0.043	0.032	0.04	0.037	0.055
Jun. 2001	0.057	0.034	0.067	0.052	0.044	0.02	0.007
Jul. 2001	0.056	0.078	0.08	0.074	0.036	0.018	BDL
Aug. 2001	0.019	0.008	0.015	BDL	0.01	BDL	BDL
Sep. 2001	0.01	BDL	BDL	BDL	BDL	BDL	BDL
Oct. 2001	0.13	BDL	BDL	BDL	BDL	0.156	0.058
Nov. 2001	0.183	BDL	BDL	BDL	BDL	BDL	BDL
Dec. 2001	0.031	BDL	BDL	BDL	BDL	BDL	BDL
Jan. 2002	0.062	BDL	BDL	BDL	0.056	0.058	0.049
Feb. 2002	0.086	BDL	BDL	BDL	0.037	0.025	0.032
Mar. 2002	0.092	BDL	BDL	BDL	0.019	0.02	0.034

BDL = Below Detectable Level.

Main sources of lead (Pb) in groundwater includes run off from gas stations and effluent from battery recycling operations (Rajgopal, 1984). Lead accumulates in bone and tissues. The accumulation of Pb may lead to Pb poisoning or plumbism, which disturbs hemosyndiesis and causes kidney damage and encephalopathy. It has been associated with miscarriages and with reproductive and sperm abnormalities. The safe concentration of lead in drinking water given by WHO (1984) is 0.06 mg/l. In the present investigation, all spots except 'E' and 'G' exhibited exceeded level of lead during certain months of die study period, otherwise, Pb concentration was within the permissible limit (Table 20.7). During October 2001 at spot 'A' and 'F', the concentration of lead was 0.130 and 0.156 mg/l, respectively, which gives warning of heavy pollution of groundwater by lead.

The results of the present study indicated that the groundwater at spot' A' is severely polluted and waters from other spots are also at the verge of severe pollution if not cared properly. Some incidences of air and water borne diseases like nausea, headache, vomiting etc. in human beings and deaths in domestic animals and crop failure due to air, water and land pollution have been recorded in Lote industrial area which might be due to heavy concentration of some metal ions in discharges. The major sources of pollution of groundwater are the surface discharge of effluents, disposal of solid wastes and percolation of toxic effluents in the industrial area, which needs monitoring, and management of waste disposal.

References

APHA (1989). *Standard methods for the examination of water and wastewater.* American Public Health Association, Washington, D.C.

Hem, J.D. (1970). *Study and interpretation of the chemical characteristics of natural water,* Second Edn. USGS Water Supply Paper 1473, U.S. Govt. Printing Office, Washington, D.C.

Mishra, S.R. and Saxena, D.N. (1989). Industrial pollution at Birla Nagar, Gwalior. *Poll. Res.,* 8:77-86.

NRC (National Research Council) (1977). *Drinking water and public health,* Volume 1. Safe Drinking Water Committee. National Academy Press, Washington, D.C.

NRC (National Research Council) (1980). *Drinking water and public health,* Volumes 2 and 3. Safe Drinking Water Committee. National Academy Press, Washington, D.C.

Prasad, V.S. and Iyer, S.K. (1983). Drinking water quality aspects. *Science Reporter.*

Rajgopal (1984). *Groundwater quality assessment for public policy in India.* First Annual Report, Department of Geography, IOWA University, IOWA.

Sastry, K.V. and Rathee, Pratima (1998). Physico-chemical and microbiological characteristics of water of village Kanheli, (Distt. Rohtak) Haryana. *Proc. Acad. Environ. Biol.* 7: 103-108.

Sayani, S., Srinivas, K., Muley, E.V., Kodarkar, M.S. and Vasantrao (1988). Groundwater characteristics and their significance with specific reference to public health in Sanathager-Balanagar Industrial Area, Hyderabad, India.

Verma, N.K., Jai, O.P. and Shrivastava, P.K. (1995). Preliminary studies on heavy metals in groundwater of Mandideep by atomic absorption spectroscopy. *Proc. Acad. Environ. Biol.,* 4: 123-126.

WHO (1984). *World Health Organisation Technical Report, 1973.*

21
Luni River: A Case Study

N.K. BOHRA

Introduction

The people of arid and semi-arid zones live on the margins of scarcity. A crisis is not impending rather crises have been recurrent. They have been experienced before and they will be experienced again. In Western Rajasthan a part of Indian arid zone, there is acute shortage of good quality water for drinking and irrigation. The rivers of Rajasthan except for the Chambal are ephemeral and flow only during the rainy season. The river catchments areas, which are largely influenced by local topography and amount of precipitation, very largely showing an indelible impact upon the drainage system. In the absence of any formidable mountain barrier in the State and the low quantum of annual rainfall, the river catchments areas are mostly small. The flat, rugged and undulating rolling lands of the low hill and plateaus, which are the origins of the rivers and streams, are highly dissected and devoid of deep channels to add stream density and soil the catchments are of low stream density.

Luni is the only major stream on the west of the Aravalli ranges, which meet the sea in the Rann of Kutch. It originates in Nag Pahar in Ajmer and flows southwest through Jodhpur, Barmer and Jalore over a distance of about 320 kms. covering a total catchment areas of 37,363 kms. Its catchment's area has spread over some parts of Nagaur district also. It flows only during the rainy season and even at this time, the river is choked with advancing sands at many places. Even in the monsoon months when the water flow rate is high and at its peak, the flowing water is not able to cut these sand deposits on the river bed as a result of which the run off spreads to the adjoining fields on both the banks of the river braiding it. This results in sand deposition on arable and non-arable lands, thus adversely affecting their productivity.

History of River Basin

It is believed that the river basin planning may be originated in the United States. According to Weber and Huff Schmidt (1962) it has two major elements:

1. The multiple use of individual units of river system especially the multiple use of reservoir storage and
2. The treatment of river basin as a hydrologic and physiographic unit.

In India the concept of river basins had attracted the attention of renowned geographers in the mid fifties. According to Sivcar (1954) the river basin concept require the pre-requisite of organization. The civilization of India began with organized water control of river basins. Even today in India emphasis is given for the optional development of land and water resources, taking river basin as a unit. To appreciate and visualize river basin as a geographical region. We quote Kirk (1950) "The catchments basin is the vital natural region, not so much on the basis of similarity, which is a generalized myth in any case, but the process. It can be considered as a unit in which according to the second law of thermodynamics the interplay of focus strives towards an equilibrium in which the stress and energy are at the minimum and into which come human group to disturb or sustain the natural processes.

Extending these premises into the arid western Rajasthan the Luni is the only well defined drainage system with a total length of 445 kms. and an average slope of 1:500, depth of 2.4 m, width of 1159 m and an approximate basin area of 45,800 sq.km. The geological history of the upper Luni basin is principally a history of geological activities in the pre-Cambrian era. Before the deposition of the so-called Vindhyan Sediments (now known as the Marwars) the basic format of the solid geology as we see today was already built through the formation and erosion of the rocks of the Aravalli and the Delhi systems on the gneissic basement, especially in the Aravalli mountainous tract and through widespread igneous intrusions and extrusions. After the Vindhyan formation there has been little deposition to solidify into rocks. Presumably the region experienced a long period of sub-aerial erosion. The only deposits over the pre-tertiary formations are the semi-compact and loose sediments of Pleistocene and Holocene periods.

Luni Basin

Luni River basin is located in south-western Rajasthan between latitude 23°41' and 27°5' and longitudes 71°4' and 74°42'. It is bounded by the arid western districts (the outside basins) in the west by Banas basin in the east, Shekhwati Basin in the north and Sukli and west Banas basins in the south. The basin extends over parts of Ajmer, Barmer, Jalore, Jodhpur, Nagaur, Pali, Rajsamand, Sirohi and Udaipur districts.

Luni river basin lies to the west of the Aravalli hills and forms part of the mid-west alluvial plain. Geographically, the eastern part of the basin is marked by hilly terraces belonging to the Aravalli chain, west of the hills lies a narrow alluvial plain that gently slopes westwards. Luni river originates in Ajmer and flows to the Rann of Kutch in Gujarat State. Its total cultivable area is 1.42 million hac. and total culturable command area is 2,48,706 ha. By the use of Luni river basin water several projects in nearly 164617 ha. area exists and proposal for another 8415 hac. is under consideration.

Luni river originates at an elevation of 550 metre near Ajmer and after flowing nearly 495 kms. in south-west Rajasthan, it disappears in marshy land of Runn of Kutch. The water of river Luni is sweet upto Balotra and becomes more and more saline further downstream. The main tributaries of Luni on the left are Sukri, Mithri, Bandi, Khari, Jawai, Guhiya and Sagi, whereas only Jojri river joins it on the right side. The mean annual rainfall over the Luni basin is normally 320 mm of which nearly 95 per cent falls in between July-Sept. By the Luni river water 4 irrigation projects are under construction and 25 more projects are proposed.

Soil and Land Capability

In the Luni basin nearly 30 series of soils, based on soil taxonomy are present. The soil type range from fine textured soils (soil with clay, silt-clay or heavy clay loam in sub-soil). Moderately fine textured soils (soil having clay loam, silty clay loam, sandy clay loam in sub-soil), Medium textured soil (soil having sandy loam, loam, silty loam texture in sub soil) and coarse textured soils (soil having sand, loamy sand and gravel in the sub-soil. Inadequate moisture in the soil is the major limiting factor for crop production in the basin. In major parts of the basin, the groundwater is of poor quality. The probability of occurrence of drought is also substantially high. Due to these constraints the land of Luni river is not of the best quality and therefore not much productive. However, there are certain areas along river banks where reasonably good quality water is available at comparatively shallow, depth where double cropping is practised. The majority of land is suitable for rain fed cropping. In the fine textured soil jowar, maize and gram are the major crops. In moderately fine textured soil wheat, rayada, rijika, barley etc. are taken. In medium textured soil bajra, til, Moong, Guar etc. are the major crops whereas in course textured soil bajra, guar, moth, moong etc. are the major crops grown.

Plant Resources

In the Luni river basin the major tree species found are:

Anogeissus pendula, Boswellia serrata, Acacia leucophloea, Acacia Senegal, Acacia catechu, Acacia nilotica, Salvadora persica, Zizyphus nummularia, Prosopis cineraria, Maytenus emerginata, etc. Beside these several shrub species also exist namely *Grewia tenax, Capparis deciduas, Rhus mysorensis, Euphorbia caducifolia* etc. as well as *Prosopis juliflora* in shrub form. These plants have not only fodder value but also used for fuel wood, fruit and medicinal value too.

Water Level

Water requirement for crop is normally met from groundwater storage or from medium to nahar storage reservoir in this region. There are 2 major (Sardar Samand and Jawai dam) 99 medium and 344 minor irrigation projects in the Luni river basin as well as some small irrigation systems (covering less than 20 ha.) constructed and operated by Panchayat Samities. Besides these 2514 ha. will be irrigated by completion of 4 irrigation projects and an additional area of 8399 is plan by proposed 25 irrigation projects. Most of water requirement for irrigation, drinking and industrial purposes within the basin is met from groundwater. As a result, water level depletion is recorded in major parts of the area. The groundwater in Luni catchments is generally fresh, but in Jojri and part of Guhiya and Bandi catchment it is saline. The bicarbonate type of characteristics is noticed in low salinity groundwater in Luni catchments. They are mostly associated with alkaline earths. The associated cations are either sodium or alkaline earth.

A study carried out by CAZRI between 1984-92 indicates that there have been significant changes in agriculture, industrial development and urban expansion. All these factors have considerably influenced the groundwater regime especially in Pali district. Reports indicate that the groundwater resources in most of the zones of Luni are under stress due to over exploitation. The stage of groundwater development in crystalline is low but scope for construction of additional wells is very limited due to poor storage capacity of aquifers. The part of Pali and Jodhpur districts are affected by chemical pollution due to effluent from textiles and other industries and leaching from chemical fertilizers. The overdraft conditions are deteriorated in chemical quality of groundwater needs attention. The conservation of water resources and augmentation through recharge by artificial methods utilizing rainwater is needed. Small rain water harvesting structures such as surface barrier, percolation tanks etc. constructed at suitable sites would and augmentation of recharge and in turn groundwater potential. The adoption of modern methods of agriculture such as dry farming, mixed cropping, agro forestry etc. with drip irrigation and other techniques would promote water conservation. The parts of Pali and Jodhpur districts affected by chemical population require immediate corrective measures.

Water Level Fluctuation

Water level fluctuates in response to local recharge/draft condition. A fraction of rainfall infiltrates and joins the groundwater, resulting in rise of water levels. Variation in infiltration rate in the vadose zone and specific field of aquifer cause wide range of water level fluctuations. Most of water requirements for irrigation, drinking and industrial purposes within the basin is met from groundwater. As a result, water level depletion is recorded in major parts of the area. The water level two decades ago before industries entered in Luni basin the water table was 15-40 metres below ground level (mbgl), which was up to 3-5 mbgl in 1997, which raises somewhat due to good rainy season. In fact it is still below what it was two decades ago but it is fluctuating depending on the area and climatic conditions of the river catchments area. The scarcity of water in summer may lead to water crisis and the high rate of water tanker (some times Rs.200-250 for 4000-5000 litres water tank) gives people in a real problematic condition for their livehood. The last survey report on Luni basin indicates that the Luni and its tributaries have affected a large sum of area and that affects reduction in crop production too.

Land Use

Among the Luni basin a detailed study on the upper Luni basin which include Pali in south, Borunda in the north, Jodhpur and Samdari in the west. It indicates that cultivated land cover nearly 66.88 per cent (out of which 90 per cent are rainfed). Wastelands 15.27 per cent, pasture 6.61 per cent and forest 6.41 per cent. Due to rainfed land the crop production varies depending on the rain. The decrease in forest area and pastureland. Cultivable land is significant in the past. The land use in Luni basin is subjected to many limitations and susceptible to natural hazards and as such each parcel of land should be properly and cautiously utilized with suitable management practices.

Agriculture and Agricultural Production

In the Luni basin nearly 66.89 per cent of the total area is used as agricultural land and 73.5 per cent of the total populations subsist on agriculture. Out of the total cultivated land 84 per cent is put to rainfed agriculture. The annual cropped area, cropping pattern and yield are governed by changing pattern of soil texture and depth, slope surface and groundwater potential, severity of natural hazards, rainfall and other climatic parameters as well as socio-economic conditions of the people. The main crops of the basin are cereals, oilseeds, pulses, and other crops. Cereals are nearly 49 per cent and include bajra (29 per cent) Jawar (11 per cent), maize (2.5 per cent), wheat (6 per cent) and barley (0.64 per cent) in different regions and their yield vary considerably. Oilseeds which is nearly (25 per cent) include sesamum (13.6 per cent) and mustard (10.11 per cent) along with other crops like groundnut, taramira, castor and linseed with different yield in different parts. Pulses constitute about 10 per cent of the total area and include moong, moth, cowpea and gram. Besides these several other cash crop like cotton, chilli, methi, cumin also produced in nearly 1.3 per cent of the total area.

Morphdynamic Process and Associated Hazard

A land reform of Luni basin is formed by both fluvial and aeolian process that at many places beneficial but at several places creating problem. It has been calculated that 55.12 per cent of the total basin area is affected by fluvial erosional hazards. Among the other hazards aeolian depositional hazards cover only 1.33 area, salinity alkalinity covers 9.21 per cent area and different combinations of all these three principal hazard types cover 3.86 per cent area. Thus 69.60 per cent area of the basin is affected by different hazards leaving only 30.4 per cent area free from them. In the Luni river basin area there were drought year in 1958, 61, 63, 64, 67, 68, 69, 71, 72, 74, 77, 78 and 79 etc. whereas flood years were 1980, 86, 87, 88 and 93 etc.

Surface Water Development

That although numerous streams collecting surface water normally be stored in human and live stock consumption, the actual situation is far from satisfactory. Nature is not only or alone to be blamed for the vagaries, however human follies like construction of some reservoirs at places where it is favouring salinity has also added to the problem like Kharda lake on Jalore granite terrain. Therefore for any rational planning of the water resources of the basin both natural and man induced factors are to be kept in mind. In order to conserve the surface water there is need to renovate the village nadis, construction of water storage structures, construction of small undergroundwater reservoirs (tanks) etc. Besides these there is also need for better management of this scarce commodity in its existing set up. Since interaction with other environmental parameters play important roles in the mutual benefit or otherwise to the interacting processes, the management of the surface water of the basin should also entail those environmental factors.

Soil Conservation

In order to control the process of soil loss several sites have been selected in the Luni river basin and its catchments area for soil conservation.

Groundwater Development

Due to low groundwater recharge capacity the basin area have the groundwater level very low. In its catchments the groundwater have EC value ranging from 2000 to 8000. Owing to these conditions it is recommended that too much tapping of the groundwater should be stopped and there should be efficient distribution and management of groundwater resources. Although groundwater with 8000 EC may be used for irrigation in sandy soil but it is suggested that such water with gypsum should be used to nullify any adverse effect.

Pali and Jodhpur Industries and their Effluents

There are large number of small-scale textile processing units in Pali districts, where as nearly 250 units of different types are working in Jodhpur. Rainwater harvesting is the only source of raw water in river Bandi and other water bodies. River Bandi remains dry throughout the year. The groundwater is being used for industrial purposes heavily. The treated effluent from CETP along with city sewage is discharged in river Bandi. River Bandi join river Jojri at upstream Jodhpur and join Luni.

River Jojri and Kaylana Lake are the two important surface water bodies in Jodhpur. The freshwater in River Jojri flows only in upstream reach at Kurigaon, as a result of rainwater harvesting during monsoon. Downstream of this location, the river remains dry throughout the year and no freshwater except the domestic and industrial wastewater of Jodhpur flows in the downstream. Mostly textile and pickling industrial wastewater is discharged from various industrial areas such as Jodhpur industrial area, Basni industrial areas (Phase I and II), heavy industrial area and light industrial area behind Power House.

The river Jojri joins Luni at downstream during monsoon. The main environmental problem for restoring the water quality of river Bandi and Jojri (accessories of Luni) are:

1. Maintenance of minimum flow of freshwater in the downstream stretches of rivers.
2. Discharge of industrial and domestic wastewaters in the rivers.

Pali

This city has mainly textile processing units located in various industrial complexes. There are more than 767 small-scale industrial units, and M/s Maharaja Shree Ummaid Mills Ltd. (the largest textile industry manufacturing cotton, synthetic yarn and cotton and textile fabrics). The total water requirement of Pali industrial area is about 42 MLD that is completed by either undergroundwater and from Municipal Corporation supplies water (which is mainly rain water or canal water). The entire untreated polluted wastewater generated from textile processing units from various industrial areas discharge into the river Bandi through various drains. Thus, the river Bandi takes up the entire pollution load of the city. The river flows along the side of the city. The river Bandi is dry except few days in monsoon period and only industrial wastelands is flowing mostly throughout the year. Its pollution impact is observed up to 40 kms downstream of the river. The CETP, Pali report indicates that nearly 30 MLD wastewater is discharged into drains generated from industrial areas besides 5 MLD wastewater from Mandia Road Industrial Area, Phase III presently being treated at existing CETP.

Bleaching Ghats

In the process of cotton textile manufacturing after kiering the cloth, it is subjected to the bleaching operation to remove the colouring matter from cloth so as to tender the fabric for the subsequent dyeing operations. For this the cloth is steeped in a cold solution of sodium hypochlorite of chlorine squeezed between the rollers and exposed to the air. The operation requires large quantity of water and wastewater from bleaching contains spent bleach liquor, alkalinity, solids and some organic matters. After the bleaching, mercerising or causticizing is done and then dyeing and printing. In printing of Khaddi and Prussian type, again large amount of water is used.

Process Diagram

Raw cotton → Desizing (gives neutral pH wastewater) → Scouring (wastewater with high temperature, high alkalinity and high BOD) → Bleaching (wastewater with high total dissolved solids) → Mercerizing (wastewater with Alkaline pH) → Finishing and Packing.

During this whole process, a large No. of chemicals like sodium bicarbonate (causting), Bleaching Powder (Bleaching/Kiering), Azodyes etc. (Printing/dyeing) and sodium silicate (finishing) are used and their wastewater is gone to the river catchment's area.

There is a classification of wastewater which indicates category A (Biodegradable without Pre-treatment), category B (Biodegradable after Pre-treatment), category C (Non-biodegradable) and D (Toxic to biological system). They are further sub-categorized.

Treatment Plant

With the help of the NEERI and Pali, Water Pollution Control Board, a treatment plant of 1 MGD is working and two more are in progress with 1.5 MGD (unit 2) and 2.0 MGD (Unit 3). By this, the 4.5 MGD polluted water can be treated and the unit 1 is increased by 0.5 MGD and therefore total 5 MGD is the maximum power of these treatment plants which is very less in comparison to the industrial wastewater discharged.

Cost of Treatment

There is general support throughout the world for the "Polluter Pays Principle" *i.e.* industry must pay the cost of the disposing waste in an environmentally acceptable manner. It there follows that the cost of operating a CETP must be borne by the industry, which the CETP serves. Even if sludge's and possible the final effluent could be utilized it would result in the reduction of total expenses.

Use of Industrial Effluent

In Pali, more than 700 small and large units are discharging nearly 405 million gallons/day of industrial effluents (in the form of wastewater discharge) into nearby agricultural lands and drinking water resources including seasonal river basin areas (Anonymous, 1991). The chemical composition of textile effluent shows the presence of coloured dyes, high pH (7.7–10.8) EC 4.4–16.3 dsm^{-1} having predominance of sodium content 2736 ppm and near absence of calcium and magnesium. Manganese, copper, iron and zinc are present in agreeable amount with traces of cobalt and cadmium. Such industrial effluent can thus possibly be utilized for establishment of certain economic tree species of this region and soil deterioration can be reduced by some soil amendments. A study by CAZRI (R.S. Singh *et al.*) shows that industrial effluent is of high saline quality and it can also be utilized for afforestation with some soil treatments in arid zone. The *Colophospermum mopane* (fuel and fodder tree), *Eucalyptus camaldulensis* (timber and fuel wood tree) and *Acacia nilotica* (fodder, fuel and timber tree) was used in study, which performed satisfactory.

As per reports 9-17 tonnes of gypsum per ha can be applied to reduced solidicity. Beside this salt tolerant crop and vegetation could reclaim degraded land (Singh, S. and Balakram, 1997).

Jodhpur Case Study

In Jodhpur, textile and chemical industry are nearly 300-400 in different industrial areas. Their effluent has high pH, electrical conductivity (EC), sodium absorption ratio (SAR), and residual sodium carbonate (RSC), where as the bivalent cations were in traces. The study carried out by Arid Forest Research Institute, Jodhpur (G. Singh *et al.*, 2001) indicates that wood ash treated with soil irrigated with effluent could be used to attain good plant biomass. These studies suggest that the industrial effluent in Jodhpur, which is also giving out in river basin catchment areas could be used for plantation purpose. The Rajasthan Pollution Control Board neither has staff nor more care to control the pollution. Regarding health hazard, they say that it is the work of Health Department and our concern is only the pollutants of effluent for which we take random sampling. In fact, in reality it is quite a documentary work and not practicable. Everyday tons of wastewater containing harmful chemicals are flowing into the river catchment's area there by polluting them.

Health Impact of Water Pollution

It is well known fact that clean water is absolutely essential for healthy living. Adequate supply of fresh and clean drinking water is a basic need for all human beings on the earth, yet it has been observed that millions of people worldwide are deprived of this.

Freshwater resources all over the world are threatened not only by over exploitation and poor management but also by ecological degradation. The main source of freshwater pollution can be attributed to discharge of untreated waste, dumping of industrial effluent and run off from agricultural fields. Industrial growth, urbanization and the increasing use of synthetic organic substances have serious and adverse affect on health of freshwater bodies. It is generally accepted fact that the developed and developing countries suffer from problem of chemical discharge into the water resources mainly groundwater resources. Chemicals mixed polluted water causes problem to health and leads to water-borne diseases that can be prevented by taking appropriate measures.

Groundwater and its Contamination

Many areas of groundwater and surface water are now contaminated with heavy metals, POP (persistent organic pollutants) and nutrients that have an adverse effect on health. Water-borne diseases and water-caused health problems are mostly due to inadequate and incompetent management of water resources.

Chemicals in Drinking Water

Chemicals in water can be either naturally occurring or introduced by human interference and can have serious health effects. Synthetic organics of many types are in use today and they are found in aquatic environment and accumulate in food chain. Pops or persistent organic pollutants represent the most harmful element for the ecosystem and for human health too. These chemicals can accumulate in fish and cause serious damage to human health. Acidification of surface water mainly lakes. Untreated sewage, industrial effluent and agricultural waste are often discharged into the water bodies such as the lakes, coastal areas and rivers endangering their use for recreational purposes such as swimming and canoeing.

Exposure to polluted water can cause diarrhoea, skin irritation, respiratory problems and other diseases, depending on the pollutant that is in the water body. Stagnant water and other untreated water provide a habitat for the mosquito and a host of other parasites and infect that cause a large number of diseases.

Minamata-Environmental Contamination with Merthyl Mercury

In Minamata, Japan, inorganic mercury was used in the industrial production of acetaldehyde. It was discharged into the nearby bay as wastewater and was ingested by organisms in the bottom sediments. Fish and other creature in the sea were soon contaminated and eventually residents of the area who consumed fish suffered from MeHg (methyl mercury) intoxication, later known as the MINAMATA disease. The disease was first detected in 1956 but the mercury emissions continued until 1968. But even after the emission of mercury was stopped the bottom sediments of the polluted water contained high levels of mercury. The MINAMATA disease was proved a turning point towards progress in environmental protection measure. In India several reports of river basin pollution induced diseases and their effect have already been reported.

Preventive Measures

Water borne epidemics and health hazards in the aquatic environment are mainly due to improper management of water resources. Proper management of water resources has become the need of the hour as this would ultimately lead to the cleaner and healthy environment.

Conclusion

Rivers are considered sacred, goddess and purifier of all sins. However it is unfortunate that our holy rivers are still holy but not pure. In India nearly 113 rivers and thousands of their tributaries carry annually about 1645 cubic km of water into the seas. Restricted flow during monsoon periods, soil erosion leading to siltation and unchecked inflow of untreated municipal sewage and industrial effluent are the major problem faced by our rivers.

The national river action plan (NRAP) trying best to combat this situation *i.e.* flow of river pollution but it seems that their efforts are very limited. In the major rivers of India approximately 17,600 million litres of wastewater is thrown away where as everyday only 4000-5000 million litres can be treated through various plants and rest of all is drawn away into these rivers (as per central pollution control reports). In these rivers BOS (bio-oxygen demand) is above 3 mg per litre which indicate that it is not fit for drinking purpose. In Yamuna river it is 35-40 mg per litre, which is dangerous for health.

In Rajasthan the analysis of industrial effluent of Pali and Jodhpur clearly indicate that the wastewater flowing out from industries contain the hazardous substances beyond the limit fixed by Central Pollution Control Board and their high pH value and BOD reveal that it is bound to produce ill effect to both the ecology of river basin as well as human health (Tables 21.1, 21.2 and 21.3).

Table 21.1: Chemical Composition of Textile Effluent at Jodhpur

Effluent Parameters	*Range*	*Mean*
Optical density (OD)	0.08–0.40	0.28
pH	8.82–11.8	9.7
Ec (dSm^{-1})	3.82–10.7	5
Total solids	3.42–19.3	7.5
TDS	3.12–18.8	6.5
Residual sodium carbonate	12–125	41
Sodium adsorption ratio (SAR)	30–153	128
Phenolphthalein alkalinity (mg $CaCO_3^{-1}$)	49–5200	1025
Total alkalinity (mg $CaCO_3^{-1}$)	560–6200	1535
Chemical oxygen demand	160–800	377
Ionic Composition		
Sodium	600–5500	2069
Potassium	25–30	38
Calcium	Traces–30	10.2
Magnesium	Traces–8	4
Copper	Traces	Traces
Zinc	Traces	Traces
Cobalt	Traces	Traces
Iron	Traces–0.67	0.35
Carbonate	29–278	86
Bicarbonate	280–1136	592
Chloride	410–806	680

Source: Singh *et al.*, 2001.

Table 21.2: Characteristics of Textile Industrial Effluent, Pali (Rajasthan)

Characters	*Range*
pH	7.7 to 10.8
EC (dSm^{-1})	4.4 to 16.3
Colour	Light to dark green; reddish and purple
Activity of Industries	Desizing, bleaching, mercerizing, dyeing and printing etc.
No. of Textile Industries	700 (small to big units)
Discharge of Textile Effluent (Industrial wastewater)	4 million gallons/day
Effluent contains	Chemicals, dyes, caustic, oils, detergents, etc.
Effect of industrial effluent	Water resources, agricultural land, seasonal river, health, fauna and flora of the region

Source: Singh *et al.*, 2001.

Table 21.3: Soil Characteristics of Industrial Wastewater Irrigated Experimental Field at Jodhpur

Tree Species	*Soil Sample*	*pH 1:2*		*EC 1:2 (dSm^{-1})*		*SAR*	
		0–15 cm	*15–30 cm*	*0–15 cm*	*15–30 cm*	*0–15 cm*	*15–30 cm*
Colophospermum mospane	I	8.4	7.6	0.19	1.13	1.17	3.70
	II	9.1	7.8	3.70	4.11	8.75	17.30
Eucalyptus camaldulensis	I	7.8	7.8	0.88	0.94	2.83	4.17
	II	–	8.5	–	0.70	–	3.27
Acacia nilotica	I	7.4	8.2	1.13	0.29	4.00	2.50
	II	8.3	8.3	4.11	0.90	6.79	2.83

I = initial, II = four months after effluent irrigation of freshwater in river Jojri
Source: Singh *et al.*, 2001.

The limited capacity of treatment plant for these wastewater and not proper control by pollution control board leads to increase the severity of the problem. Neither people nor industrialists are serious about the consequences and the Luni river basin catchments getting worst and worst. The pollution river wasting millions of area every year. Who will be responsible for all this? Who will take initiatives? The question is still unanswered and we are waiting for the Indian Minamata-Environmental contamination with Merthyl Mercury like situation.

References

Balak Ram (1998). *Wastewater management in textile and other industries, paper impact assessment of industrial effluents on land use and cropping pattern in arid regions.* CAZRI, Jodhpur, p. 42. 101 J4 39681 WAS.

Gupta, I.C. (1995). *Assessment and management of water and soil pollution due to industrial effluents in western Rajasthan.* Central Arid Zone Research Institute, Jodhpur, p. 2. 104 Al, SAT 39657.

Gupta, I.C. (1995). *Assessment and management of water and soil pollution due to industrial effluents in western Rajasthan.* Central Arid Zone Research Institute, Jodhpur, pp. 8-9. 104 Al, 39682, SAT.

Gupta, Pragya (1996). *Evaluation of water pollution hazards in western Rajasthan and their remedies.* Thesis, Department of Zoology, Jai Narayan Vyas University, Jodhpur, p. 5.

Gupta, Pragya (1996). *Evaluation of water pollution hazards in western Rajasthan and their remedies.* Thesis, Department of Zoology, Jai Narayan Vyas University, Jodhpur, pp. 71-72.

Singh, *et al.* (2001). Effect of textile industrial effluent on tree plantation and soil chemistry. *Journal of Environmental Biology,* 22(1): 59-66.

Singh, R.S. *et al.* (1995) Effect of textile industrial effluents on establishment of certain tree species. *Journal of Tropical Forestry,* July-Sept, 19(3).

Singh, Surendra and Balak Ram (Eds.) (1997). *Impact assessment of industrial effluent on natural resources along the Jojri, Bandi and Luni rivers.* Central Arid Zone Research Institute, Jodhpur, pp. 133-137.

Singh, Surendra and Balak Ram (Eds.) (1997). *Impact assessment of industrial effluent on natural resources along the Jojri, Bandi and Luni rivers.* Central Arid Zone Research Institute, Jodhpur, pp. 78.

Singh, Surendra and Balak Ram (Eds.) (1997). *Impact assessment of industrial effluent on natural resources along the Jojri, Bandi and Luni rivers.* Central Arid Zone Research Institute, Jodhpur, pp. 118.

Singh, Surendra and Balak Ram (Eds.) (1997). *Impact assessment of industrial effluent on natural resources along the Jojri, Bandi and Luni rivers.* Central Arid Zone Research Institute, Jodhpur, pp. 8. 101 J4, IMP 39682.

The Hindu (2001). Folio 6th July.

22

Influence of Freshwater Influx on Calcium and Magnesium Concentrations in the Rushikulya Estuary

TAPAN RANI MAHAPATRO
Department of Marine Sciences, Berhampur University, Berhampur, Orissa

ABSTRACT

Calcium and magnesium concentrations during five different seasons at seven different stations covering the estuarine water of Rushikulya was studied for a period of one year (December 1998 to November 1999). The overall calcium concentrations for surface and bottom water were ranged from 0.086–0.363 g/l and 0.114–0.366 g/l respectively. Similarly, the overall magnesium concentrations for surface and bottom water were 0.058–1.153 g/l and 0.084–1.156 g/l respectively. Both calcium and magnesium concentrations increased from riverine to estuarine end, coinciding with high chlorinity condition. Higher values were observed during summer season than other four seasons, whereas lower values observed during post-monsoon season. The lower values during post-monsoon may be due to high efflux of freshwater by the super cyclonic activity. The Ca/Cl and Mg/Cl ratios were found to be high during post-monsoon and low during summer season. The surface water Ca/Cl and Mg/Cl ratios were more than the bottom one. There observed a decreasing trend in the ratio of the said elements to chlorinity towards estuary. This indicated higher freshwater inflow from river to the estuary.

Keywords: Calcium, Magnesium, Increasing, Decreasing, Influx.

Introduction

Calcium and magnesium are two of the major constituents of seawater contributing significantly to the biological and geo-chemical processes in the sea. High concentrations of these elements in the sea, relative to their biogeochemical reactions tend to make them conservative especially in the open sea. However, in the estuarine water the concentration variations specially for calcium are quite significant due to land drainage, high rates of biological uptake, precipitation and dissolution processes characteristic of shallow systems (Naik, 1978). Seasonal variability of different characteristics in a tide-dominated river system demands an understanding of the freshwater discharge into the system and out of it to the sea associated with the compensatory flow of saline water (Sen Gupta and Naik, 1981). In estuaries due to mixing of freshwater from river and saline water from sea, the calcium and magnesium concentrations vary at different chlorinity. The freshwater from river alters the concentration of the said elements and their ratios to chlorinity. Tripathy *et al.* (1990) reported the distribution and behaviour of Ca and Mg in Rushikulya estuary. Das and Sahoo (1996) reported influence of freshwater influx on the ratio of Ca, Mg and F with respect to Cl in the coastal water of Orissa. But there has not been any report on the influence of freshwater influx on calcium and magnesium concentrations in this estuary. Here is an attempt to study the calcium and magnesium concentrations and these elements to chlorinity ratio by the influence of freshwater and land drainage.

Materials and Methods

As Rushikulya estuary is in tropical belt, like other estuaries of this belt, it experiences change in water quality by monsoon cycles and the tidal changes. This area also experiences a delayed onset and an early withdrawal of monsoon. Monsoon usually lasts from early July to early September. The climate of this region is semi-arid type. Basing on the meteorological condition of the area five different seasons nameiy monsoon (July-Sep), post-monsoon (Oct-Nov), winter (Dec-Jan), pre-monsoon (Feb-Apr) and summer (May-June) are experienced. The tides of this estuary are mixed and semi-diurnal type.

The sampling was undertaken at selected seven stations (Fig. 22.1), which cover the entire estuary, during the seasons mentioned above. The surface water samples were collected using polythene bucket, while bottom water samples were collected 1.5 m below the surface using shallow water sampler. Before filling the acid cleaned polythene bottles were rinsed 2/3 times with the respective station water. These filled polythene bottles with leak-proof stoppers were preserved in a refrigerator till analyses, which was completed within a month from the date of collection. Generally, samples preserved in refrigerator gives no change in result even after 1 month of storage.

Chlorinity determinations were made using argentometric titration procedure (Grasshoff, 1976). Calcium and magnesium were estimated by a complexometric titration. In this method (Culkin and Cox, 1966) total concentration of calcium and magnesium was first estimated in an aliquot of sample by titrating with standard EDTA solution at pH 10 using NH_4OH/NH_4Cl buffer and Eriochrome Black-T (EBT) as indicator. At the end point the colour changes from wine red to purple blue. Calcium was then estimated by titrating another (same) aliquot of sample with standard EGTA solution at pH 12 using a buffer (sodium tetra borate, NH_4Cl and NaOH) in presence of Zincon as indicator. At the end point the blue colour of the solution changes to orange-red (Ringbom *et al.*, 1958). The interference of strontium and magnesium in the titration of calcium using EGTA is negligible in riverine and estuarine waters. The amount of magnesium present in the aliquot of sample was computed from the difference in the two titres. The mean, standard error (SE) and percentage of coefficient of variation (CV) were calculated by MS Office-2000, Excel Package.

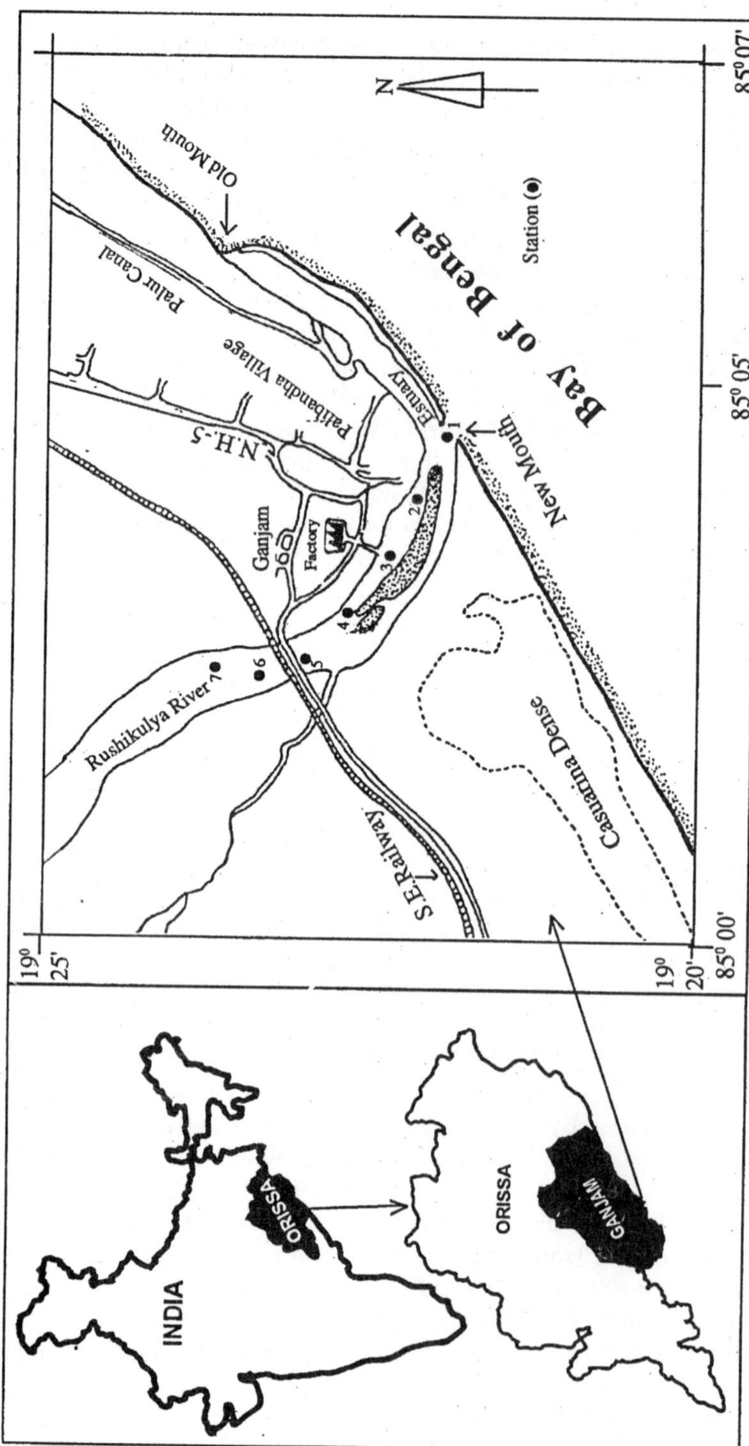

Fig. 22.1: Location Map of Rushikulya Estuary Showing Sampling Stations

Results and Discussion

Chlorinity

The chlorinity concentrations for surface water ranged from 5.403–11.896 per cent, 9.931–14.547 per cent, 11.674–16.401 per cent, 1.129–7.163 per cent and 0.537–5.214 per cent during winter, pre-monsoon, summer, monsoon and post-monsoon respectively. The minimum, maximum and mean surface chlorinity values were 0.537 per cent, 16.40 per cent and 7.508 ± 0.845 per cent (SE) respectively. The calculated percentage of CV was 21.443 for surface water chlorinity. The bottom water chlorinity concentrations ranged from 6.609–13.213 per cent, 11.475–14.923 per cent, 11.752– 16.562 per cent, 1.212–7.561 per cent and 0.697–6.061 per cent during winter, pre-monsoon, summer, monsoon and post-monsoon respectively. The minimum, maximum and mean bottom chlorinity values were 0.697 per cent, 16.562 per cent and 8.076 ± 0.855 per cent (SE) respectively. The calculated percentage of CV for bottom water chlorinity was 21.195.

Both surface and bottom chlorinity concentrations showed marked variation during all the seasons of study. The bottom water chlorinity concentration showed higher value than surface one. But during pre-monsoon and summer there observed very less difference from bottom water to surface water chlorinity concentration. Maximum and minimum concentrations were noticed during summer and post-monsoon respectively. The minimum concentration during post-monsoon may be due to the effect of super cyclone, which effluxed a large volume of freshwater and land runoff. The chlorinity concentrations during this season showed a fairly marked deviation from the normal distribution in the estuarine environment. Maximum concentration during summer may be due to minimum influx of freshwater to the estuary and maximum evaporation. This estuary (Rushikulya) is entirely dominated by semi-diurnal tides with seawater penetrating deep upstream. It may also be another reason for maximum chlorinity at St-1. There observed an increasing chlorinity concentration from (St-7 to St-1) riverine to estuarine end (Fig. 22.2a) during all seasons. St-7 being upper reaches of the estuary receives plenty of freshwater and St-1 being mouth of the estuary receives plenty of sea water. Though sea water penetrates upto the upstream of the estuary, but is diluted by the fresh river water. So, there observed increasing trend of chlorinity concentration towards mouth of the estuary. Similar observations were reported elsewhere (Naik, 1978; Sen Gupta and Naik, 1981; Narvekar *et al.*, 1983; Narvekar and Zing de, 1987; Palanichamy and Balasubramanian, 1989; Das *et al.*, 1997 and Padmavathi and Satyanarayanan, 1999).

Calcium

The surface water calcium concentrations during winter, pre-monsoon, summer, and post-monsoon were ranged from 0.203–0.316 g/l, 0.316–0.361 g/l, 0.328–0.363 g/l, 0.115–0.256 g/l and 0.086–0.195 g/l respectively. The minimum and maximum and mean values for surface water concentration were 0.086 g/l, 0.363 g/l and 0.238 ± 0.017 g/l (SE). The calculated percentage of CV was 0.008. The bottom water calcium concentrations during winter, pre-monsoon, summer, monsoon and post-monsoon were ranged from 0.275–0.346 g/l, 0.331–0.363 g/l, 0.328–0.366 g/l, 0.121–0.273 g/l and 0.114–0.224 g/l respectively. The minimum, maximum and mean values were 0.114 g/l, 0.366 g/l and 0.259–0.015 g/l (SE) respectively. The calculated percentage of CV was 0.007.

The calcium concentration showed an increasing trend from (St-7 to St-1) riverine to estuarine end (Fig. 22.2b) during all seasons. This variation may be due to gradual reduction in the incursion of seawater into the estuary with increasing distance from mouth (St-1 to St-7). St-1 being mouth of the estuary receives more saline water, which leads more calcium concentration than other stations.

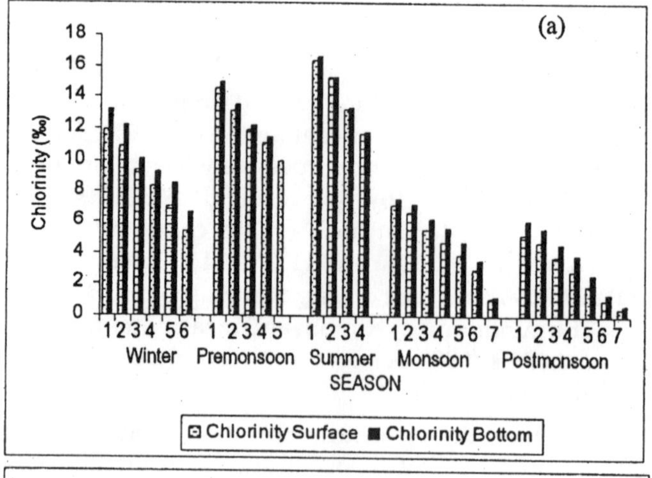

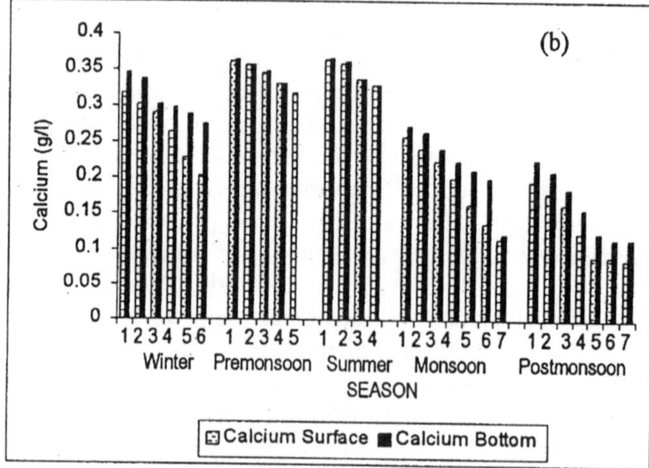

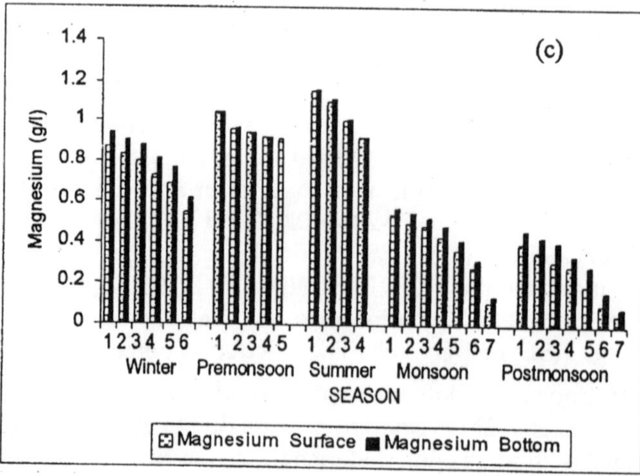

Fig. 22.2: Seasonal Variations of Chlorinity, Calcium and Magnesium in Rushikulya Estuary

Similar reports have been putforth by several workers (Sen Gupta and Naik, 1981; Narvekar *et al.*, 1983; Narvekar and Zingde, 1987; Palanichamy and Balasubramanian, 1989). The bottom water calcium concentrations were more than surface one. Culkin and Cox (1966) opined that extraction of calcium by living organisms from surface water leading to lower calcium level in surface water and dissolution of dead animals and precipitated calcium carbonate at bottom region could result increased $CaCO_3$ concentration in bottom water. It was also observed that there was very less to no difference between bottom and surface water concentrations during pre-monsoon and summer. It may be due to the saturation of estuarine water during these two seasons. For both surface and bottom water higher calcium concentrations were recorded during summer, whereas lower during post-monsoon. The higher concentration of calcium during summer may be due to scanty freshwater discharge. Another reason for it may be due to the recovery of calcium for gradual attainment of normal estuarine conditions. This is in agreement with the report of Naik (1978) and Palanichamy and Balasubramanian (1989). The lower calcium concentrations observed during post-monsoon may be due to high influx of freshwater resulted by the super cyclonic effect. The more freshwater run-off lead to greater dilution, which accounted lowering of calcium concentration than other seasons.

The said element to chlorinity ratio revealed that St-7 recorded higher values than other stations. In other words there appeared a decreasing trend from (St-7 to St-1) riverine to estuarine end (Table 22.1). It may be due to the higher value of Ca/Cl ratio for rivers than seawater, which indicates that the very low chlorinity of river water not only brings down the value at estuarine points but also increases the Ca/Cl ratio. This was in agreement with the report of Noronha *et al.* (1981). Surface water sample showed higher Ca/Cl ratio when compared to bottom water sample. The variation of Ca/Cl ratio from surface to bottom might be due to non-conservative nature of calcium in view of its greater involvement in the biological process as described by Naik (1978). Seasonal variation also indicated the involvement of chlorinity in changing the ratio as high values were noticed during post-monsoon and monsoon due to lower chlorinity and vice versa during summer season. Palanichamy and Balasubramanian (1989) observed the same result during monsoon and summer season. As St-1 is mouth of the estuary, the Ca/Cl value of St-1 was closer to the mean value (0.2229) reported for the Bay of Bengal (Sen Gupta *et al.*, 1978).

Magnesium

The magnesium showed well marked seasonal variations during all the seasons. The surface water magnesium concentrations varied from 0.546–0.867 g/l, 0.907–1.038 g/l, 0.912–1.153 g/l, 0.109–0.538 g/l and 0.058–0.396 during winter, pre-monsoon, summer, monsoon and post-monsoon respectively. The minimum, maximum and mean surface magnesium concentrations were 0.058 g/l, 1.153 g/l and 0.607 ± 0.060 g/l (SE) respectively. The calculated percentage of CV of it was 0.107. The bottom water magnesium concentrations varied from 0.619–0.935 g/l, 0.915–1.042 g/l, 0.913–1.156 g/l, 0.138–0.569 g/l and 0.084–0.461 g/l during winter, pre-monsoon, summer, monsoon and post-monsoon respectively. The minimum, maximum and mean bottom water magnesium concentrations were 0.084 g/l, 1.156 g/l and 0.645 ± 0.058 g/l (SE) respectively. The percentage of CV was 0.098.

The distribution pattern of magnesium at seven different stations during five different seasons is given in Fig. 22.2c. The magnesium concentrations were found to be high at St-1 (mouth), whereas low at St-7 (riverine end). This variation may be due to gradual increase in the incursion of freshwater into the estuary with increasing distance from mouth (St-1 to St-7). As St-1 is located at the mouth of the estuary, which is influenced more by neritic waters leading more saline condition, a possible attribute for higher concentration of magnesium at this station. St-7, which is very much influenced by the freshwater of poor magnesium concentrations, maintains low magnesium level. Other stations are in between, overall there observed an increasing concentration of magnesium from St-7 to St-1. Similar reports have been made by several workers elsewhere (Sen Gupta and Naik, 1981; Narvekar *et al.*, 1983; Narvekar and Zingde, 1987; Palanichamy and Balasubramanian, 1989).

Table 22.1: The Ratio of Chlorinity with Calcium and Magnesium in the Estuarine Water of Rushikulya

Para-meter	Season	*Stations* 1		2		3		4		5		6		7	
		S	*B*	*S*	*B*	*S*	*B*	*S*	*B*	*S*	*B*	*S*	*B*	*S*	*B*
Ca/Cl	Winter	0.027	0.026	0.028	0.028	0.031	0.030	0.032	0.032	0.033	0.034	0.038	0.042		
	Pre-monsoon	0.025	0.024	0.027	0.027	0.029	0.029	0.030	0.029	0.032					
	Summer	0.022	0.022	0.024	0.024	0.026	0.025	0.028	0.028						
	Monsoon	0.036	0.036	0.037	0.037	0.041	0.039	0.043	0.040	0.043	0.046	0.047	0.057	0.102	0.100
	Post-monsoon	0.037	0.037	0.039	0.038	0.043	0.040	0.043	0.041	0.047	0.047	0.091	0.082	0.160	0.163
	Winter	0.073	0.071	0.076	0.074	0.086	0.087	0.087	0.088	0.098	0.090	0.101	0.094		
Mg/Cl	Pre-monsoon	0.071	0.070	0.073	0.071	0.079	0.077	0.082	0.080	0.091					
	Summer	0.070	0.070	0.072	0.073	0.076	0.076	0.078	0.078						
	Monsoon	0.075	0.075	0.076	0.076	0.089	0.084	0.093	0.087	0.094	0.089	0.095	0.090	0.097	0.114
	Post-monsoon	0.076	0.076	0.078	0.077	0.081	0.089	0.101	0.090	0.102	0.110	0.105	0.116	0.108	0.120

S = Surface, B = Bottom.

The bottom water magnesium concentrations showed higher value as compared to surface one. Bottom to surface water concentration difference was not significant and the minor difference noticed in the present study may be due to some removal machanisms from surface water. Another reason may be due to the stratification of the estuary by the neritic water at the bottom layer and freshwater dominance at surface layer. Culkin and Cox (1966) are of the same opinion for higher magnesium concentrations in the bottom water.

Summer period recorded higher magnesium concentrations, whereas lower during monsoon and post-monsoon. The estuary seemed to be influenced mostly by seawater, due to scanty freshwater discharge during most of the periods other than monsoon and post-monsoon. This could have resulted in higher concentration of magnesium during summer and pre-monsoon. Due to super cyclone in October 1999, much more freshwater had been discharged into the estuary than monsoon period. This could account for the lowering of magnesium concentration during post-monsoon. The concentration of discussed element was also found to show wider variations during post-monsoon, whereas more or less stable in nature with higher values observed during summer. During monsoon and post-monsoon, the greater dilution would lead to a wide variation, whereas the summer could be recovery period for magnesium along with chlorinity. Palanichamy and Balasubramanian (1989) are of the same opinion for high and low values of magnesium during summer and monsoon respectively.

As regards to conservative behaviour of magnesium in seawater, the ratio remains absolutely constant throughout ocean. The change in the ratio occurs mostly at near shore regions under the influence of freshwater (Noronha *et al.*, 1981). There is annual discharge of 1800 million M^3 freshwater to the sea through Rushikulya estuary (Panigrahy *et al.*, 1999), which alters the Mg and Chlorinity concentrations in the estuary. During the study period Mg/Cl ratio increased with decrease in chlorinity value. Freshwater of river is responsible for the higher ratio of Mg/Cl at riverine end than estuarine end (Narvekar *et al.*, 1987). From this it may be concluded that highest values at St-7 (which is the riverine end), whereas lowest values at St-1 (which is the estuary mouth). Surface water sample showed higher Mg/Cl ratio when compared to bottom water sample (Table 22.1). It may be due to the less chlorinity value of surface water as compared to bottom one. Highest value of Mg/Cl was noticed during post-monsoon, whereas lowest during summer. The highest value during post-monsoon may be due to drastic drop in chlorinity associated with heavy cyclonic floods. Lowest value during summer may be due to the dominance of sea water.

Acknowledgement

The author (TRM) is thankful for the award of S R F and also the Head of the Department for the laboratory facility.

References

Culkin, F. and Cox, R.A. (1966). Sodium, potassium, magnesium, calcium and strontium in seawater. *Deep Sea Research*, 13: 780-804.

Das, J. and Sahoo, R.K. (1996). Influence of freshwater influx on the ratio of Ca, Mg and F with respect to Cl in the coastal water of Orissa, east coast of India. *Ind. J. Mar. Sci.*, 25(1): 74-77.

Das, J., Das, S.N. and Sahoo, R.K. (1997). Semidiurnal variation of some physico- chemical parameters in the Mahanadi estuary, east coast of India. *Ind. J. Mar. Sci.*, 26: 323-326.

Grasshoff, K. (1976). *Methods of sea water analysis*. Verlag Chemie, Weinhein, p. 260.

Naik, S. (1978). Calcium and magnesium concentration in the nearshore waters of Goa. *Mahasagar-Bull. Nat. Inst. Oceano.*, 11(3 and 4): 185-189.

Narvekar, P.V., Zingde, M.D. and Kamat Dalal, V.N. (1983). Behaviour of Boron, Calcium and Magnesium in a polluted estuary. *Estuar. Coast. Shelf Sci.*, 16: 9-16.

Nervekar, P.V. and Zingde, M.D. (1987). Behaviour of Boron, Calcium and Magnesium in Purna and Auranga estuaries (Gujarat), west coast of India. *Ind. J. Mar. Sci.*, 16(1): 46-50.

Noronha, R.J., Caroline, M. and Sen Gupta, R. (1981). Calcium, magnesium and fluoride concentrations in the Andaman Sea. *Ind. J. Mar. Sci.*, 10: 234- 237.

Padmavathi, D. and Satyanarayana, D. (1999). Distribution of nutrients and major elements in riverine, estuarine and adjoining coastal waters of Godavari, Bay of Bengal. *Ind. J. Mar. Sci.*, 28: 345-354.

Palanichamy, S. and Balasubramanian, T. (1989). Distribution of calcium and magnesium in the Vellar estuary. *Mahasagar*, 22(1): 1-11.

Panigrahy, P.K., Das, J., Das, S.N. and Sahoo, R.K. (1999). Evaluation of the influence of various physico-chemical parameters on coastal water quality around Orissa by factor analysis. *Ind. J. Mar. Sci.*, 28: 360-364.

Ringbom, A., Pensar, G. and Wanninen, E. (1958). A complexometric titration method for determining calcium in the presence of magnesium. *Anal. Chim. Acta.*, 19: 525-531.

Sen Gupta, R. and Naik, S. (1981). Studies on calcium, magnesium and sulphate in the Mandovi and Zuari River Systems (Goa). *Ind. J. Mar. Sci.*, 10(1): 24-34.

Sen Gupta, R., Naik, S. and Singhal, S.Y.S. (1978). A study of fluoride, calcium and magnesium in the northern Indian Ocean. *Mar. Chem.*, 6: 125-141.

Tripathy, S.K., Panigrahy, R.C., Gouda, R. and Panda, D. (1990). Distribution of calcium and magnesium in the Rushikulya estuary (Orissa), east coast of India. *Ind. J. Mar. Sci.*, 19(3): 212-214.

23

Ecological Study of the Macrophytes of Ikop Lake, Manipur: Morphometry and Qualitative Analysis

CH. NIVANONEE & B. MANIHAR SHARMA

Ecology Research Laboratory, Department of Life Sciences, Manipur University, Canchipur, Imphal, Manipur

ABSTRACT

The Lake Ikop is a freshwater, saucer-shaped natural lake. It is situated in the south-eastern part of Imphal at a distance of about 40 km with an area of 13.5 sq.km. in the Thoubal District, Manipur. It lies at an intersection of 93°52′ to 94°05′ E longitude and 24°31′ to 24°40′ N latitude and it is situated at an altitude of 772 m above the mean sea level. A total of twenty nine (29) macrophytic species were recorded during the study period which were categorised in submerged (2 species), Rooted submerged (5 species), Rooted with floating leaves (5 species), Free floating (5 species) and emergent (12 species). Five life form categories were recorded *viz.* Chamaephytes 1-species (35 per cent), Hemicryptophytes 4 species (13.8 per cent), Therophytes 7 species (24.1 per cent), Errant vascular hydrophytes 8 species (27.6 per cent) and Geophytes 9 species (31.0 per cent). The biological spectrum of Ikop lake revealed a "Geo-Errant Vascular hydrophytes" type of phytoclimate. Sixteen growth form categories of macrophytic species were recorded during the tenure of study.

Keywords: Ikop Lake, Eutrophic, Macrophytes, Life form, Growth form, Morphometry.

Introduction

The structure and dynamics of the vegetation provide an indication of the nature of the whole environment biological and physical. The description of the vegetation leads any person to draw an overall picture of an area and then ultimately classify the different units of vegetation. Macrophytic communities playa significant role in regulating the structure of a lake ecosystem. The macrophytes are also found indispensable for the regulation of mineral cycling in the water bodies and they can serve as the indicators for the possible degree of damage in the ecosystem. (Pieczynska and Ozimek 1976) Pearsall (1921) studied the dynamic relationship between the macrophytic plant communities and their environment in the lake. The structural attributes through which a plant community is studied are conveniently viewed as analytic and synthetic characters. The analytical characters may be both qualitative and quantitative. The qualitative or the subjective analysis comprises the Floristic composition, Life-form analysis and Growth-form, classification. Floristic list of a particular area gives a reliable background information about species diversity in a community as each plant species has its own specific ecological amplitude and the same indicates the ecological nature of the habitat. It is one of the major anatomical characters of the plant community. The significant pioneering works on the subjective approach were made by Zurich Montpeliar school led by Braun-Blanquet (1932). In India floristic studies of macrophytic vegetation were started quite earlier by Biswas and Calder (1936) and Kachroo (1959). In the North-eastern region of India which is one of the 25 "*Biodiversity Hotspots*" of the world (Myers *et al.*, 2000), the bioresource are found to be much degraded. The situation is more acute in Manipur as most of the few existing lakes including Loktak, (a Ramsar site) Waithou, Ikop and Utrapat lakes, have seen seriously degraded due to eutrophicatian (Sharma 1999). There is very little information available pertaining to the freshwater ecosystem, specially the ecology of macrophytes till date. In the present study an attempt has been made to investigate the qualitative aspects of the macrophytes in Ikop Lake, Manipur.

Study Sites

Ikop Lake is situated in the south eastern part of Imphal at a distance of about 40 km. The lake having an area of about 13.5 sq km., is located in Thoubal District, Manipur at an intersection of 93°52′ to 94°05′ E longitude and 24°31′ to 24°40′ N latitude. It lies at an altitude of 772 m above the mean sea level. The climate of the lake is sub-tropical monsoonic type. The maximum temperature varied from 21.6°C (January 1999) to 30.6°C (May 1999) while the minimum temperature varied from 4.3°C (January 1999) to 11.5°C (March 1999). The maximum monthly rainfall was recorded to be 310.7 mm (June 1999) and the minimum rainfall was 1.5 mm (December 1998). The depth of the lake varied from a minimum of 0.27 m to a maximum of 1.59 m during the different seasons of the year with an estimated average depth of 0.93 m. The morphometry and the bathymetric characters of Ikop Lake are presented in Table 23.1.

Materials and Methods

Regular periodical survey and sampling of the vegetation was done during October 1998 to September 2000 at four different study sites, *viz.* Site-I – Tarapan, Site-II – Tarahumdoipan, Site-III – Ganjapan, Site-IV – Patchao. The morphometric and bathymetric characters of the lake were determined as described by Russel-Hunter (1970), Sharma and Pant (1979) and Moss (1989).

Table 23.1: Morphometry of Ikop Lake, Manipur

Altitude (m)	772 msl
Latitude	24°31' to 24°40' N
Longitude	93°52' to 94°05' E
Maximum length (km)	7.5
Maximum breadth (km)	1.8
Surface area (sq. km)	13.5
Maximum depth (m)	1.59
Minimum depth (m)	0.27
Mean depth (m)	0.93
Mean depth/maximum depth ratio	0.58
Basin shape	Saucer
Basin slope	Gentle
Bottom texture	Silted
Shore-line length (km)	27.5
Shore-line development ratio	2.12
Volume of lake (km^3)	0.013
Surface area/volume ratio	1038.5
Development of volume	1.75
Maximum temperature (°C)	24.9–30.6
Minimum temperature (°C)	4.3–17.8
Annual rainfall (mm)	1476.8

Floristic Composition

Floristic studies of the macrophytic vegetation is a pre-requisite to understand the structure of a lake ecosystem. The life forms and growth forms were determined after detailed floristic analysis. Raunkiaer's life form system (1934) and its further extension by Braun-Blanquet (1951) is widely accepted which is generally based on the position of the perrenating buds or organs from which new shoots or foliage develops. Life form analysis was revised by Ellenberg and Mueller-Dombois (1967) and Mueller-Dombois and Ellenberg (1974) taking into consideration various criteria like, the structure and seasonalty of crown foliage and shoot system. The new system separates the five life forms of Raunkiaer which are expanded into 23 major life forms. In the present study the plants were classified into different life forms following the revised life form classification Mueller-Dombois and Ellenberg (1974).

The growth form classification was adopted after the method given by Hogeweg and Brenkert (1969) who attempted to apply the growth form classification of Hartog and Segal (1964) in the Indian aquatic vegetation with due modification.

Results and Discussions

Lake Morphometry

Morphometric and bathymetric characters of the lake (Table 23.1) revealed saucer shaped basin with gentle slope and silted bottom. The shore line of the lake measured 27.5 km with shore-line development ratio of 2.12 which according to Moss (1989) is indicative of the productive nature of the lake. A number of limnologists (Russel-Hunter, 1970; Fee, 1979; Vollenweider and Petalas, 1980; Moss, 1989), have earlier observed the existence of a linear co-relationship among the morphometric and bathymetric characters and richness of species as well as productive status (tropic nature) of the lakes. According to this observations, the mean depth and maximum depth ratio (0.58) as well as surface area/volume ratio (1038.5) are indicative of the luxuriant growth of the macrophytes with possibilities of high magnitude of production in the Ikop Lake.

Floristic Composition

A total of 29 macrophytic species were recorded from the different study sites, 24 species each in site I and IV, and 21 species each site II and III. Floristic distribution is incorporated in the Table 23.2. The plants were categorised into:

1. Submerged with 2 species
2. Rooted submerged with 5 species

3. Free floating species with 5 species
4. Rooted with floating leaves also with 5 species and
5. Emergents with 12 species.

Table 23.2: Floristic Composition of Ikop Lake at Different Sites (+ and – signs indicate presence and absence of species)

Name of Species	*Sites*			
	I	*II*	*III*	*IV*
Alternanthera philoxeroides Griseb.	+	+	+	+
Azolla pinnata R. Br.	+	+	+	+
Carex cruciata Wahlenb.	+	–	–	+
Ceratophyllum demersum Linn.	+	+	+	+
Ceratopteris thallictroides (L) Brogn.	+	–	–	–
Chara zeylanica Wild.	+	–	+	+
Echinochloa colonum (Linn) Link.	+	+	+	+
Eichhornia crassipes (Mart) Solms.	+	+	+	+
Enhydra fluctuans Lour.	+	+	+	+
Hydrilla verticillata (Linn f) Royle	+	+	+	+
Hydrocharis cellulosa Buch-Ham	+	+	–	+
Hygrorhyza aristata (Retz) Nees.	+	+	+	+
Ipomoea aquatica Forsk	+	–	–	+
Jussiaea repens Linn.	+	+	+	+
Limnophila heterophylla Benth.	–	+	–	–
Marsilia quadrifoliata Linn.	–	–	–	+
Nympheae stellata Willd.	+	+	+	+
Nymphoides cristatum (Roxb) Kuntz.	+	–	–	–
Nymphoides indicum (Linn) Kuntz.	+	+	+	+
Potamogeton crispus Linn.	+	+	+	+
Potamogeton natans Linn.	–	–	+	–
Pseudoraphis spinescence R. Br.	–	+	–	–
Salvinia cucullata Roxb.	+	+	+	+
Salvinia natans Hoffim.	+	+	+	+
Scripus supinus Linn.	+	+	+	+
Trapa natans Linn.	+	+	+	+
Utricularia flexuosa Vahl.	+	+	+	+
Vallisneria spiralis Linn.	+	+	+	+
Zilznia latifolia (Griseb) Stapf.	–	–	+	+

The group-wise distribution of the 29 macrophytes species have been set in Table 23.3. The present findings are in conformity with the findings of Shah and Abbas (1979) who reported 28 macrophytic species in Ganga river, Bhagalpur out of which 22 species were emergent, 4 submerged species and 2 floating species. Devi K.I. (1998) reported 26 macrophytes consisting 6 species in the submerged groups, 3 rooted floating leaved species, 3 free floating species and 11 emergent species in Utrapat Lake, Manipur.

Table 23.3: Group-wise Distribution of Macrophytes in Four Study Site

Name of Group	*Number of Species*				
	Site-I	*Site-II*	*Site-III*	*Site-IV*	*Whole Lake*
Submerged	2	2	2	2	2
Rooted with floating leaves	4	3	4	3	5
Free floating	5	5	4	5	5
Emergent	9	7	4	10	12
Rooted submerged	4	4	7	4	5
Total	24	21	21	24	29

A comparatively higher number of macrophytic species were reported by Devi N.B. (1993) with 86 species in the Loktak Lake Manipur out of which 73 species were emergents from the phumdi floatingmat zone. While 13 species were reported from the non-phumdi zone, of which 6 species were submerged, 4 species in the free floating and 3 species in the rooted with floating leaved species; Devi O.I. (1993) reported 35 species from the Waithou Lake Manipur of which 18 species were emergents, 5 free floating, 6 with rooted floating leaved species.

Handoo and Kaul (1982) reported 58 species in the wetlands of Kashmir out of which 23 species were in the emergent category, 21 in the ground layer species, 7 species in floating leaved and 7 species in the submerged category. Billore and Vyas (1982) reported lower number of macrophytic species with only 14 species of which 3 were floating leaved, 7 submerged and 4 in the emergent category. Swindale and Curtis (1957) and Schmid (1965) reported heterogenous compositions in the submerged vegetations of U.S.A. Crowder *et al.* (1977) also reported heterogenous composition of submerged vegetation in lake Opinicon, Canada, Seshavatharam *et al.* (1982) also reported intermixed mats of macrophytic species of emergent, submerged and free floating groups in the Kolleru Lake. Similarly in the present study also a clear cut zonation of floating, emergent and submerged species could not be observed and hence all such communities were found in intermixed mats. The nature of occurrence as well as the variations in the distribution pattern of the macrophytic species in different groups in the present study might have been influenced by the morphometric and bathymetric characters of the lake, physiographic and environmental factors, mainly the physico-chemical characters of water.

Life Form

In the present study, the twenty nine macrophytic species were categorised into five life-form categories *viz.* 1 species in Chamaephytes (3.5 per cent), 4 species of Hemicryptophytes (13.8 per cent), 7 species in Therophytes (24.1 per cent), 8 species in errant vascular Hydrophytes (27.6 per cent) and 9 species in Geopyhtes (31.0 per cent). The different life-forms along with the percentage composition of the various macrophytes of Ikop Lake have been furnished in Table 23.4.

The percentage composition of the plants in the different life-form classes are termed as Life-form Spectrum or the Biological Spectrum. When the life form spectra of a particular plant community is compared to the normal life-form spectrum (Raunkiaer, 1934), the adaptational capabilities of the plants to the adverse climatic conditions are revealed. The phytoclimate of a region reflects the dominant life form which comprises the largest number of plant species occurring in the community. Therefore based on the various life form classes of the Ikop Lake, this region may be assigned to the "Geo-Errant vascular hydrophytic" type of phytoclimate as the Geophytes and Errant vascular hydrophytes represent the dominant life form comprising the higher number of plant species. The biological spectrum of Ikop Lake is incorporated in Table 23.5. Similar type of phytoclimate was reported by Devi K.I. (1998) in Lake Utrapat (Manipur) in which 8 species were Therophytes, 6 species in Geophytes, 4 species in Hemicrytophytes and 8 species in Errant vascular hydrophytes. In Loktak Lake, Manipur Devi N.B. (1993) reported 34 species of Therophytes, 8 species of Geophytes, 15 species in Hemicryptophytes, 7 Chamaephytes, 11 Phanerophytes and 8 species of Lianas.

Varying values have also been reported by Devi O.I. (1993) from the Waithou Lake, Manipur, with a record of 20 species under Geophytes, 9 species in Therophytes, 5 species in Errant vascular hydrophytes and 1 species in Thallo-therophytes. Bebika and Sharma (2002) reported 10 Therophytes, 8 Geophytes, 6 hydrophytes and 10 Errant vascular hydrophytes with a "Thero-Errant Vascular Hydrophytic" type of phytoclimate from the Senapat Lake Manipur. Okram *et al.* (1999) observed "Geo-Therophytic" phytoclimate in the Waithou Lake Manipur while Devi Ch.U. (2000) reported Thero-Geophytic type of phytoclimate in the freshwater bodies (4 ponds and a Stream) in Canchipur, Manipur.

Growth Forms

In the present study an attempt has been made to differentiate the various growth forms of plants based on the scheme of Hogeweg and Brenkert (1969). The twenty-nine macrophytic species reported in the present study belong to sixteen growth forms out of the twenty three growth forms recognised by the Hogeweg and Brenkert. The growth form categories of the macrophytes of Ikop Lake have been set in Table 23.6. The Helophytes, Nymphaids and Rhizopleustohelophytes represented the dominant growth forms over the other forms. The helophytes comprising of 5 species (*Carex cruciata, Echinochloa colonum, Pseudorophis spinescence, Scirpus supinus, Zizania latifolia*) represent the rooted and emergent plants. Rhizopleustohelophytes comprised 4 species (*Nymphaea stellata, Nymphoides* sp., *Potamogeton* sp. etc.) also exhibit rooted plants with floating stems and aerial shoots and roots. The Nymphaids consisted of 4 species which are rooted and submerged with rosette floating leaves. The Parvopotamids had 2 species which are rooted and submerged. The Magnolemnids comprising of *Salvinia cucullata* and *Salvinia natans* represented the free-floating forms with rosette or stoloniferous stems and emergent sessile leaves. The remaining growths forms comprising one species each exhibited different habits in relation to soil, aquatic and aerial environments. Lower number of growth form categories were reported by Devi O.I. (1993) from Waithou Lake Manipur with only 13 growth forms whereas 12 growth forms have been reported by Devi K.I. 1998 from Utrapat Lake Manipur and Devi N.B. (1993) reported 9 growth forms from Loktak Lake Manipur.

Table 23.4: Life Form Classification of the Macrophytic Species in Ikop Lake (After Ellenberg and Mueller-Dombois 1967 and Mueller-Dombois and Ellenberg 1974)

Life-form	*Name of Species*	*Number of Species*	*Percentage Composition*
Chamaephytes	*Ceratopteris thallictrodes*	1	3.5
Errant Vascular Hydrophytes	*Azolla pinnata*	8	27.6
	Chara zeylanica		
	Ceratophyllum demersum		
	Eichhornia crassipes		
	Marsilia quadrifoliata		
	Salvinia cucullata		
	Salvinia natans		
	Utricularia flexuosa		
Geophytes	*Hydrocharis cellulosa*	9	31.0
	Nympheae stellata		
	Nymphoides cristatum		
	Nymphoides indicum		
	Potamogeton crispus		
	Potamogeton natans		
	Trapa natans		
	Vallisneria spiralis		
	Zizania latifolia		
Hemicryptophytes	*Enhydra fluctuans*	4	13.8
	Hydrilla verticitalla		
	Ipomoea aquatica		
	Hygrorhyza aristata		
Therophytes	*Alternanthera philoxeroides*	7	24.1
	Carex cruciata		
	Echinochloa colonum		
	Jussiaea repens		
	Limnophila heterophylla		
	Pseudoraphis spinescence		
	Scirpus supinus		

Table 23.5: Biological Spectrum of the Flora of Ikop Lake

Parameters	*Life-form*						
	Th	*G*	*H*	*Ch*	*Ph*	*EVH*	*Total*
Total number of species	7	9	4	1	–	8	29
Life-form percentage	24.1	31.0	13.8	3.5	–	27.6	100
Raunkiaer's normal spectrum and composition (per cent)	13.00	6.00	26.00	9.00	46.00	–	100

Abbreviations: Th = Therophytes, G = Geophytes, H = Hemicryptophytes, Ch = Chamaephytes, Ph = Phanerophytes, EVH = Errant Vascular Hydrophytes.

Table 23.6: Growth Forms of the Macrophytes of Ikop Lake

Sl. No.	*Growth Forms*	*Names of Species*	*Family*	*Relation to (Soil) Substratum*	*Relation to Aquatic and Aerial Envionment*	*Habit Form*
1.	Ceratophyllids	*Ceratophyllum demersum*	Ceratophyllaceae	Free floating	Completely submerged	Whorls of finely dissected foliage submerged flowering
2.	Charids	*Chara zeylanica*	Characeae	Attached	Completely submerged	Branched (verticillate) thallus
3.	Eichhorniids	*Eichhornia crassipes*	Pontederiaceae	Free floating, rooting in soft partly organic bottom sediment	Leaves emergent (at least the greater part of the leaf and petiole)	Stoloniferous rosettes with floating to emergent petiolate leaves
4.	Hydrocharids	*Hydrocharis cellulosa*	Hydrocharitaceae	Free floating	Floating leaves when plants well developed	Stoloniferous rosettes with emergent floating petiolate leaves
5.	Helophytes	*Carex cruciata* *Echinochloa colonum* *Pseudoraphis spinescence* *Scirpus supinus* *Zizania latifolia*	Cyperraceae Poaceae Poaceae Cyperaceae Poaceae	Rooted Rooted Rooted Rooted Rooted	Emergent Emergent Emergent Emergent Emergent	When full grown, vegetative parts and generative parts (almost) completely emerged as a rule
6.	Lemnids	*Azolla pinnata*	Salvinaceae	Free floating	Floating on the surface	Small, not differentiated into leaves and stems
7.	Marselids	*Marsilia quadrifoliata*	Marsiliaceae	Rooted	Floating leaf and blade	Rhizomatous, stemless sometimes
8.	Magnolemnids	*Salvinia cucullata* *Salvinia natans*	Salvinaceae	Free floating Free floating	Floating on the surface	Stoloniferous or branching rosettes or stem like structure with floating or emergent sessile leaves
9.	Myriophyllids	*Limnophila heterophylla*	Scrophulariaceae	Rooting at least a part of the life cycle	Submerged sometimes emergent	Developing long sterns with finely dissected (submerged) foliage
10.	Nymphaeids	*Nymphee steltata* *Nymphoides cristatum* *Nymphoides indicum* *Potamogeton natans*	Nymphaceae Nymphaceae Nymphaceae Pontederiaceae	Rooted Rooted Rooted Rooted	Submerged but the large leaf blades floating on the surface or emerging	Leaves in rosette or from short rhizome on long petioles

Contd...

Table 23.6–Contd...

Sl. No.	*Growth Forms*	*Names of Species*	*Family*	*Relation to (Soil) Substratum*	*Relation to Aquatic and Aerial Enviornment*	*Habit Form*
11.	Parvopotamids	*Potamogeton crispus* *Hydrilla verticillata*	Pontederiaceae Hydrocharitaceae	Rooted during at least a considerable part of life cycle. Rooting during a considerable part of the life cycle	Submerged	Usually developing long stems with undissected narrow leaves
12.	Pseudro hydrophytes	*Ceratopteris thallictroides*	Polypodiaceae	Rooted	Submerged	When full grown helophytic but when younger with differentiation in submerged foliage adapted to an aquatic habitat and emerged aerial organs
13.	Rhizopleusto helophytes	*Alternanthera philoxeroides* *Hygrorhyza aristata* *Ipomoea aquatica* *Jussiaea repens* *Enhydra fluctuans*	Amaranthaceae Poaceae Convolvulaceae Onagraceae Asteraceae	Rooted Rooted Rooted Rooted Rooted	Emergent Emergent Emergent Emergent Emergent	Stems floating or just below the surface producing aerial leave and shoots. Stems floating or just below the surface producing aerial shoots. Stems floating or just below the surface bearing shoots vegetative and generative parts emerged
14.	Trapids	*Trapa natans*	Trapaceae	Rooted	Floating leaves	Profusely branched long stems with tufts of leaves, Chlorenchymatic adventitious roots
15.	Utriculareids	*Utricularia flexuosa*	Lentibulariaceae	Free floating never rooting	Submerged	Finely dissected leaves emerged inflorescence
16.	Vallisnerids	*Vallisneria spiralis*	Hydrocharitaceae	Rooted	Submerged	Long linear flabby leaves from rosette or in bundle

Conclusion

The Ikop Lake under the present investigations exhibited high values of morphometric and bathymetric characters which vividly indicates the eutrophic nature of the lake in the present study. Human activities and eutrophication in the lake have been found to affect the population distribution of the submerged macrophytes in the present study.

Acknowledgement

The authors are grateful to the State Department of Science and Technology for providing financial assistance to carry out the investigations.

References

Bebika, Ch. and Sharma, B.M. (2002). Floristic distribution and Life form analysis of Macrophytes of Sanapat lake, Manipur. *Indian J. Environ. and Ecoplan.*, 6(1).

Billore, D.K. and L.N. Vyas (1982). Distribution and production of Macrophytes in Pichhola lake, Udaipur (India). In: Wetlands ecology and management, B. Gopal, R.E. Turner, R.G. Wetzel and D.F. Whigham (Eds). National Institute of Ecology and International Scientific publications, India, pp. 45-54.

Biswas, K. and Calder, C.C. (1936). *Handbook of common water and marsh plants of India and Burma.* Government of India Publication, Calcutta.

Braun Blanquet, J. (1932). *Plant sociology* (translated by G.D. Fuller and H.S. Conrad). McGraw-Hill Pub. Co., New York.

Braun Blanquet, J. (1951). *Pflanzensoziologie.* 3 Aufl., Springer Wien.

Crowder, A.A., J.M. Bristow, M.R. King and S. Vanderkloet (1977). Distribution, seasonality and biomass of aquatic macrophytes in Lake Opinicon. *Nautraliste Can.*, 104: 441-456.

Devi, Ch. Umabati and B.M. Sharma (2000). *Phyto-sociology and primary production of the macrophytes in the freshwater ecosystems of Canchipur, Manipur.* Ph.D. Thesis, Manipur University, Manipur.

Devi, K. Indira (1998). *Ecological studies of freshwater macrophytes in Utrapat lake, Manipur.* Ph.D. Thesis, Manipur University, Manipur.

Devi, N. Beenakumari (1993). *Phyto-sociology, primary production and nutrient status of macrophytes of Loktak lake, Manipur.* Ph.D. Thesis, Manipur University, Manipur.

Ellenberg, H. and D. Mueller-Dombios (1967). *A key to Raunkiaer plant life-forms with revised sub-divisons.* Ber. Goebot. Inst. ETH. Stiftg Rubel, Zurich, 37: 56-73.

Fee, F.J. (1979). A relation between lake morphometry and primary production in its use in interpreting whole lake eutrophication experiments. *Limnol. Oceanogr.*, 24: 401-416.

Handoo, J.K. and V. Kaul (1982). Phyto-sosiological and standing crop studies in wetlands of Kashmir. In: *Wetlands ecology and management*, B. Gopal R.E. Turner, R.G. Wetzel and D.F. Wigham (Eds.). National Institute of Ecology and International Scientific Publications, India, pp. 187-195.

Hartog, den and S, Segal (1964). A new classification of water plant communities. *Acta. Bot. Neerl.* 13: 367-393.

Hogeweg, P. and A.L. Brenkert (1969). Structure of aquatic vegetation: A comparison of aquatic vegetation in India, Netherlands and Czechoslovakia. *Trop. Ecol.*, 10(1): 139-162.

Kachroo, P. (1959). Aquatic vegetation of Damodar valley. I. Phanerogamic flora of freshwater ponds and marshy lands with particular reference to Damodar-Eden Canal area of West Bengal. *J. Asiatic Soc.*, 1: 271-297.

Moss, B. (1977). Conservation problems in the norfork Broads and rivers of East Anglia, England-Phytoplankton, boats and the causes of turbidity. *Biological Conservation*, 12: 95-114.

Moss, B. (1989). *Ecology of freshwaters*. Man and Medium. Blackwell Scientific Publication, Oxford.

Mueller-Dombois, D. and H. Ellenberg (1974). *Aims and methods of vegetation ecology*. John Wiley and Sons, New York.

Myers Norman, Russell A. Mittermeler, Cristina, G. Mittermeler, Gustavo, A.B. da fonseca and Jenifer Kent (2000). Biodiversity hotspots for conservation priorities. *Nature*, (3): 853-858.

Okram, I.D., Sharma, B.M. and Singh, E.J. (1999). Ecological exploration of the macrophytic vegetation of the Waithou Lake, Manipur, North-east India: Phyto-sociological study of aquatic macrophtyes. *Indian J. Environ and Ecoplan.*, 2(2): 191-191.

Pearsall, W.H. (1921). A suggestion as to factors influencing the distribution of free floating vegetation. *J. Ecol.*, 9: 224-253.

Petalas, K. (1980). Common on "A relation between lake morphometry and primary productivity and its use in interpeting whole lake eutrophication experiments". (F.J. Free, 1979). *Limnol. Oceanogr.* 25: 1147-1149.

Phillips, G.L., Eminson, D. and Moss, B. (1978). A mechanism to account for macrophyte decline in progressively eutrophicated freshwaters. *Aquat. Bot.*, 4: 105-126.

Pieczynska, E. and T. Ozimek (1976). Ecological significance of macrophytes. *Int. J. Ecol. Environ. Sci.* 2: 115-128.

Raunkiaer, C. (1934). *The life-forms of plants and statistical plants geography.* Clarendon Press, Oxford.

Russel-Hunter, W.D. (1970). *Aquatic productivity: An introduction to some basis aspects of biological oceanography and limnology.* Macmillan Publishing Co. Inc., New York. p. 310.

Schmid, W.D. (1965). Distribution of aquatic vegetation as measured by line intercept with SCUBA. *Ecology*, 46(6): 816-823.

Seshavatharam, V., E.S.M. Dutta and P. Venu (1982). An ecological study of the vegetation of the Kolleru lake. *Bull. Bot. Surv. India*, 24(1-4): 70-75.

Shah, J.D. and S.G. Abbas (1979). Seasonal variation in frequency, density, biomass and rate of production of some aquatic macrophytes of the river Ganges at Bhagalpur (Bihar). *Trop. Ecol.*, 20(2): 127-134.

Sharma, A.P. and M.C. Pant (1979). Certain physio-chemical features, chlorophyll concentration and phytoplankton population in high altitude lake. *Trop. Ecol.*, 20(1): 101-113.

Sharma, B.M. (1999). Wetland ecosystems of Manipur: An ecological appraisal. In: *Wetland of Manipur*, R.K.R. Singh and H.P. Sharma (Eds.). Manipur Assn. for Sciences and Society (MASS), Imphal, (1): 12-19.

Swindale, D.N. and T.T. Curtis (1957). Phyto-sociology of the larger submerged plants in Winsconsin Lake. *Ecology*, 38: 397-407.

24

Physico-chemical Analysis of the Bhavani River Water Collected from the Kalingarayan Dam, Tamil Nadu

B. REGINAA
Department of Zoology, Alamelu Angappan College for Women, Komarapalayam
B. NABI
Department of Biological Sciences, Pondicherry University, Pondicherry

ABSTRACT

The present study was undertaken to find out the effect of industrial pollution on the biological qualities of the Bhavani river water collected at Kalingarayan dam. The water samples collected at monthly intervals for a year was used for physico-chemical parameters. The effect of pollution was recorded in the different parameters analysed.

Keywords: Bhavani river, Primary productivity, COD, BOD.

Introduction

Water, the universal solvent, takes the first priority of life on our planet Earth which serves as an unavoidable medium for the livelihood of the human beings, plants and animals. Nowadays due to reckless exploitation and misuse, river water gets polluted. Of the total amount of global water, only 2.4 per cent is distributed on the main land. The available freshwater to man is hardly 0.3 to 0.5 per cent of the total water supply on earth and therefore, its judicious use is imperative. The increase in suspended and dissolved material in the river water affects its physical, chemical, biological and aesthetic value beyond certain desirable limits. The increase demand for water, as a consequence of population growth, agricultural and industrial development, has usurped environmentalists to determine the chemical, physical and biological characteristic of natural water resources. The hydrochemistry of an aquatic body is influenced by a set of inter-dependent physical factors such as morphometry, ratio of sediment area to volume, nature of catchment area.

The present investigation was carried out to assess the effect of industrial pollution on the hydro-biological quality and the primary production of the river Bhavani.

Materials and Methods

Kalingarayan dam a major reservoir of Tamil Nadu constructed over the Bhavani river was selected as the sampling station. The water quality was analyzed for one year at regular monthly intervals. The samples were collected from just below the surface in the morning hours (8 A.M. to 12 noon).

Table 24.1: Physico-chemical Parameters of the River Bhavani at Kalingarayan Dam During September 1997 to August 1998

Sl.No.	*Parameters*	*Month*											
		Sep. 97	*Oct. 97*	*Nov. 97*	*Dec. 97*	*Jan. 98*	*Feb. 98*	*Mar. 98*	*Apr. 98*	*May 98*	*Jun. 98*	*Jul. 98*	*Aug. 98*
1.	Temp.	30	30	27	30	30	27	29	32	32	31	30	30
2.	pH	7.3	7.3	7.3	7.5	7.3	8.0	7.3	8.1	7.7	7.2	7.4	7.8
3.	TSS	16	48	14	10	18	12	20	34	22	20	12	10
4.	TDS	110	226	218	142	172	232	186	216	294	295	216	220
5.	Cl	30	36	30	20	30	30	26	9.0	10	18	9.0	14
6.	SO_4	9.0	26	22	25	26	42	20	46	52	20	16	12
7.	DO	5.3	6.45	6.56	6.56	7.01	6.17	7.01	6.45	7.01	6.1	6.45	6.1
8.	BOD	<1.0	2.5	<1.0	1.0	1.2	1.1	1.6	1.2	1.1	1.9	1.2	1.0
9.	COD	8.0	46	8.0	18	20	110	30	18	56	46	30	10
10.	Hardness	52	110	110	120	90	120	120	220	170	190	110	90
11.	Na	16	26	35	22	20	26	16	26	20	49	36	74
12.	Ca	4.0	16	20	10	10	16	9.0	20	40	46	20	13
13.	Mg	6.0	14	16	10	12	10	14	32	22	16	4.0	10
14.	Alkalinity	35	110	70	60	80	106	106	110	160	110	90	50
15.	PO_4	0.01	0.3	0.06	0.03	0.02	BDL	0.03	BDL	0.01	0.9	0.3	0.3

All the parameters are expressed in mg/l except temperature and pH.
BDL – Below the Detectable Limit.

The water samples were analyzed immediately on the spot for temperature, pH, dissolved oxygen (Winkler method), alkalinity (using Phenolphthalein and methyl orange indicators), hardness of water (EDTA titrimetric method) and chloride by Mohr's argentometric method (APHA, 1989). The bio-chemical oxygen demand, chemical oxygen demand, sulphate content (using flame emission photometry at 589 nm), magnesium, calcium, potassium content and primary productivity were estimated within six hours after collection using standard procedures (APHA, 1989).

Table 24.2: Monthly Variation in Phytoplankton Productivity (gC/m³/day) of Bhavani River at Kalingarayan Dam During September 1997 to August 1998

Months	*Initial DO mg/l*	*Light Bottle DO mg/l*	*Dark Bottle DO mg/l*	*NPP (gC/m³/day)*	*GPP (gC/m³/day)*
Sep'97	5.3	5.6	5.05	0.45	0.82
Oct'97	6.45	6.73	5.89	0.42	1.26
Nov'97	6.56	7.29	6.45	1.09	1.26
Dec'97	6.56	7.01	6.45	0.67	0.84
Jan'98	7.01	7.29	6.56	0.42	1.09
Feb'98	6.17	6.73	5.89	0.84	1.26
Mar'98	7.01	7.29	6.56	0.42	1.09
Apr'98	6.45	6.56	5.89	0.16	1.00
May'98	7.01	7.29	6.56	0.84	1.51
Jun'98	6.17	7.01	5.89	1.26	1.68
Jul'98	6.45	7.01	6.17	0.84	1.26
Aug'98	6.17	6.56	5.89	0.58	1.00

Results and Discussion

The temperature of the sample water collected at regular monthly interval varied from 27°C–32°C. The minimum temperature was recorded in the months of November 1997 and February 1998 respectively. The maximum was recorded in the month of April and May 1998. The subsequent decrease and increase in temperature of the samples analyzed might be due to the low water level, low velocity, atmospheric condition and greater solar radiation. This was similar to the finding of Vyas (1968). This was mainly due to the presence of thermal power station near the investigated area.

Minimum pH value of the sample analyzed was recorded as 7.2 in the month of June 1998 and that of maximum 8.1 in the month of April 1998. The difference in the pH value of the sample analyzed might be due to the fact that the accumulation of industrial discharges along with domestic sewage. Similar finding was also made by Patil (1982) and Bhatnagar and Sharma (1973).

Maximum total suspended solid content 48 mg/l was found in the water sample collected during the month of October 1997. The minimum value of 10 mg/l was recorded in the month of December 1977 and August 1998. Generally the disposal of sewage and industrial pollutants, contribute suspended matter to rivers and streams. The significant amount of suspended solids was found in the rainy season, as the water flow is very high and it tends to carry all the natural wastes.

The values of total dissolved solids in water varied from a minimum of 110 mg/l in the month of September 1997 and to a maximum of 295 mg/l in the month of June 1998. The highest yearly average value of total dissolved solids was due to the accumulation of all the related chemicals which are present in the form of industrial pollutants.

During the period of study the values of chloride ranged from a minimum 9 mg/l in the month of April 1998 and July 1998 to a maximum of 36 mg/l in the month of October 1997. The high chloride content in the summer may be due to input of highly soluble chloride salts through run-off from the catchments and high evaporation coupled with the low rate of the river water as observed by Zafer (1964) and Qayyum (1976). Minimum chloride content in the rainy season was due to its dilution in the large amount water.

The values of sulphate varied between a minimum 9.0 mg/l in the month of September 1997 and to a maximum of 52 mg/l in the month of May 1998.

The dissolved oxygen concentration of river sample water varied from a minimum of 5 mg/l in the month of September 1997 and a maximum of 7 mg/l in the month of January 1998, March 1998 and May 1998. The change in the dissolved oxygen content was mainly due to the solubility of dissolved oxygen increased with the decrease in water temperature. The change in the content was mainly due to the utilization these molecules by the aerobic bacteria present in the sample and release free CO_2 as a result of decomposition. This was in accordance to the findings of Moss (1972) and similar findings by Morrissette and Mavinic (1978).

Maximum BOD value 2.5 mg/l was recorded in the sample collected during the month of October 1997. The lowest value was recorded as less than one in the sample collected during the month of September 1997 and November 1997. The highest BOD was due to the highest biological activity at higher temperature and vice versa. Although the water level was reduced in the summer, but the level of pollutants from the near industries added was the same throughout the year. This view was in agreement with the works of Rai (1978).

The lowest value of COD in the present study observed was 8.0 mg/l in the month of September 1997 and November 1997 respectively and the highest value of 110 mg/l in the month of February 1998. The higher level of COD was mainly due to the higher concentration of industrial pollutants, hence the biologically resistant organic substances and level of toxicity was greater in most of the monthly samples analyzed.

Total hardness values varied from 52 mg/l to 220 mg/l .The lowest value 52 mg/l was noticed in the sample collected during the month of September 1997 and that of the highest 220 mg/l in the month of April 1998. The discharges of industrial pollutants increase the hardness during the summer and their dilution decrease during the winter months.

The monthly values of the samples recorded a minimum value 16 mg/l in the months of September 1997 and March 1998. The maximum sodium content 74 mg/l was noticed in the month of August 1998. Significant value 46 mg/l in calcium hardness was noticed from the sample collected during the month of June'1998 and the minimum values recorded 1.5 mg/l in the month of September 1997 .The values of magnesium hardness were ranging from a minimum 4.0 mg/l in the month of July 1998 to a maximum of 32 mg/l in the month of April 1998. The values of alkalinity in water sample varied from a minimum of 35 mg/l in the month of September 1997 and to a maximum of 160 mg/l in the month of May 1998. The significant increase in calcium, sodium and magnesium alkalinity content might be due to the fact that the accidental mixing of enormous amount of industrial substances in low water quantity and in high evaporation rate. The change in alkalinity might be due to the increased rate of decomposition. Similar observation was also reported by Hegde and Bharathi (1985).

The phosphate content was not significant in all the samples analyzed. The lowest value was 0.01 mg/l in the month of September 1997 and May 1998 respectively and the highest value observed was 0.9 mg/l in the month of June 1997. The non-significant phosphate content might be that the industrial pollutants which enters the sample zone is merely free from phosphate content. This was supported by the work of Ray *et al.* (1966) and Stephenson (1949).

The biological parameters measured were Gross Primary Productivity (GPP) and Net Primary Productivity (NPP) from September 1997 to August 1998. All values are represented in $gC/m^3/day$. GPP was maximum (1.68 $gC/m^3/day$) in the sample collected during the month of June 1998 and minimum (0.82 and 0.84 $gC/m^3/day$) was recorded from those collected in the month of September, December 1997. The highest NPP (1.26 $gC/m^3/day$) was recorded in the sample collected from the month of June 1998 and the least NPP (0.42 $gC/m^3/day$) was noticed in the sample collected during October 1997, January 1998 and March 1998. The low level of productivity was attributed to the hardness of its nature. The high value was due to the presence of essential minerals distributed in the water sample without much variation. This study was supported by the findings of Kamat (2000).

Acknowledgements

The author wishes to thank Mr. S. Dhanabal, Scientist, Tamil Nadu Pollution Control Board, Hosur for valuable suggestions during this work.

References

APHA (1989). *Standard methods for the examination of water and wastewater,* 17th edn. Amm. Publ. Engg. Centre, Cincinnati, Ohio, USA.

Bhatnagar, G.P. and Sharma, G.P. (1973). Physico-chemical features of sewage polluted lower lake, Bhopal. "Environ-agents and their biological effect". *Proc. Intel. Symp. Int.*, 4-5: 214-223.

Hegde, C.R. and Bharathi, S.G. (1985). Comparative phytoplankton ecology of freshwater ponds and lakes of Dharwad, Karnataka State, India. Proc. Nat. Sym. Pure and Appl. Limnology (Ed.) Adoni A.D. *Bull. Bot. Soc. Sagar,* 32: 24-29.

Kamat, S.V. (2000). Hydro-biological studies of two temple ponds in Ponda taluka, Goa. *Ecol. Env. Con.*, 6: 361-362.

Morrissette, D.G. and Mavinic, D.S. (1978). BOD test variables. *J. Environ. Engg. Division*, EP 6: 1213-1222.

Moss, B. (1972). Studies on Gull lake, Michigan II. Eutrophication evidence and prognosis freshwater boil. 2: 309-320.

Patil, P. (1982). *An ecological study of the algal flora of the lakes of Bhopal.* Ph.D. Thesis, Bhopal Univ., Bhopal.

Prasad, D.Y. and Qayyam, M.A. (1976). Pollution aspects of blue green algal in river Gomathi. *Ind. J. Environ. Hlth.*, 22: 151-168.

Rai, L.C. (1978). Ecological studies of algal communities of the Ganges river at Varanasi India. *J. Ecol.*, 5: 1-6.

Ray, P., Singh, S.B. and Sehgal, K.L. (1966). A study of some aspects of ecology of the river Ganga and Jamuna at Allahabad (U.P.) in 1958-1959, *Proc. Nat. Acad. Sci. India*, Sec-B, 3: 235-272.

Stephenson, W. (1949).Certain effects of agitation upon the release of phosphate from mud. *J. Mar: Biol. Assoc.*, V.K.N.S. 28: 371-380.

Vyas, L.N. (1968). Studies in phytoplankton ecology of Picchola lake, Udaipur, Proc. *Symp. Recent Adv. Trop. Ecol.*, pp. 334-337.

Zafar, A.R. (1964). On the ecology of algal in certain ponds of Hyderabad, India. Physico-chemical complex. *Hydrobiologia*, 23: 179-195.

25

Water Supply of Kollam Municipality of Kerala: Problems and Solutions

M.K.P. ROYEE

Department of Zoology, Site Gregorios College, Kottarakkara, Kerala

V.R. PRAKASAM

Department of Environmental Sciences, University of Kerala, Trivandrum

ABSTRACT

Physico-chemical characteristics of municipal drinking water supply of Kollam showed that all parameters were within the desirable limit of BIS. However, water was contaminated with coliforms and fecal coliforms the presence of which was directly influenced by the amount of residual chlorine in water. Through a survey, the problems of consumers including the causes of water borne and water related diseases were identified and solutions are suggested for better management.

Keywords: Water supply, Municipal, Water home disease, Kollam.

Introduction

Municipalities of India face drinking water crisis due to shortage of groundwater and lack of safe water supply. It is a consequence of environmental degradation by human action, inadequate sanitation and poor management. This situation has led to several water borne and water related diseases among the consumers. In Kollam municipality of Kerala which has a population of 1.39 lakh, water borne outbreak of diseases have been common (Aravindan, 1989; Various newspapers report), casting doubts on the purity of piped water supply. No detailed studies on the quality of municipal water supply of Kollam and also on the causes of contamination and measures to check contamination are available. Therefore, the present study was undertaken aiming at qualitatively analyzing the water and surveying the various water related problems, and suggesting remedies.

Materials and Methods

The study consisted of two phases. The first phase included a detailed monthly analysis of physico-chemical and bacteriological characteristics of municipal pipe water for a period of one year from June 1997 to May 1998.

Sampling Area

Kollam municipality is located between North latitude 9°00′–9°05′ and East longitude 76°45′–76°50′. It consists of 18.48 km² with 36 municipal wards.

Sampling Station

Sl.No. 1 is located in ward no. 20-Pallithottam; Sl.No. 2 is located in ward No. 21-Vaddy; Sl.No. 3 is located in ward No. 23-Iron bridge; Sl.No.4 is located in ward no.16-State bus stand; Sl.No. 5 is located in ward No. 5-Kollam railway station; Sl.No. 6 is located in ward No. 8-FMN College. In Station 5, water is obtained from a large dug up pond. In station 6, occasionally water is also pumped from a nearby dug well.

Characteristics Analyzed

The physical, chemical and bacteriological parameters tested consisted of the temperature, colour, pH, turbidity, conductivity, total alkalinity, acidity, total dissolved solids (TDS), chloride, total hardness, Ca and Mg hardness, fluoride, iron, sodium, potassium, nitrate and coliforms (APHA, 1985). Residual chlorine in S1 to 4 was analyzed by Orthotolidine test. The data obtained are presented as yearly mean ± standard deviation. Season-wise analysis (Rainy season–June to November and non rainy season–December to May) of data was also carried out.

Survey

The second phase consisted of a household survey conducted during June to August 1998. It was through a questionnaire to find out the various water sources available in the municipal area, the amount of water consumption, the scarcity experienced, the opinion of consumers on the purity of water, the difficulties faced, consumers satisfaction on municipal services and their suggestions. Out of 42,016 households of Kollam municipality, a total of 800 households from 19 wards were randomly selected for the survey.

Results and Discussion

The water analysis revealed that all the physico-chemical characteristics of drinking water (Table 25.1) were within the permissible limits set by Bureau of Indian Standards (BIS, 1991). One observation of the study was that the amount of the fluoride was less than 0.6 mg/l in all water sampling stations. It is likely that this low concentration would have affected the dental health of consumers, as evidenced by a large number of reports on dental caries due to low concentration of

Table 25.1: Water Characteristics of Kollam Municipality
Values are Yearly Mean ± S.D of Each Station

Sl.No.	*Characteristics*	*Station No.*					
		S1	*S2*	*S3*	*S4*	*S5*	*S6*
1.	Temperature (°C)	32.13 ± 1.93	31.46 ± 1.31	31.96 ± 1.61	31.17 ± 1.42	30.75 ± 2.17	32.34 ± 2.49
2.	Colour(pt/l)	18.34 ± 4.93	20.0 ± 6.39	18.75 ± 4.83	18.34 ± 4.44	21.67 ± 2.46	12.92 ± 7.53
3.	pH	7.10 ± 0.41	6.99 ± 0.41	7.19 ± 0.44	7.08 ± 0.38	6.67 ± 0.15	6.66 ± 0.33
4.	Turbidity (NTU)	2.27 ± 1.16	2.40 ± 0.91	1.88 ± 0.85	1.86 ± 0.66	1.85 ± 0.82	1.15 ± 1.01
5.	Conductivity (μmhos)	12.03 ± 1.37	11.71 ± 1.33	11.90 ± 1.36	12.64 ± 4.35	43.39 ± 13.93	54.43 ± 7.76
6.	Total alkalinity (mg/l)	20.0 ± 3.68	18.75 ± 3.11	19.17 ± 2.88	20.84 ± 5.15	32.92 ± 3.96	62.92 ± 8.11
7.	Acidity (mg/l)	3.84 ± 1.34	3.96 ± 1.91	3.92 ± 1.56	3.94 ± 1.34	7.82 ± 3.93	11.96 ± 7.04
8.	TDS (mg/l)	116.67 ± 46.58	126.67 ± 95.09	110.0 ± 44.72	121.67 ± 62.94	286.67 ± 94.33	335.0 ± 126.82
9.	Chloride (mg/l)	17.03 ± 4.13	16.79 ± 4.20	17.29 ± 4.56	17.31 ± 4.08	39.08 ± 12.41	38.27 ± 4.16
10.	Total hardness (mg/l)	18.17 ± 2.33	17.84 ± 2.33	18.67 ± 2.74	19.17 ± 5.99	57.67 ± 11.91	105.84 ± 12.95
11.	Ca hardness (mg/l)	6.52 ± 3.26	5.98 ± 3.06	6.57 ± 2.85	7.03 ± 3.24	19.62 ± 4.33	34.32 ± 3.81
12.	Mg hardness (mg/l)	2.85 ± 0.53	2.85 ± 0.53	2.95 ± 0.74	2.94 ± 1.04	9.05 ± 1.79	17.47 ± 2.29
13.	Fluoride (mg/l)	0.39 ± 0.14	0.39 ± 0.14	0.39 ± 0.15	0.37 ± 0.14	0.47 ± 0.14	0.47 ± 0.15
14.	Iron (mg/l)	0.22 ± 0.08	0.19 ± 0.07	0.26 ± 0.13	0.24 ± 0.07	0.26 ± 0.06	0.16 ± 0.09
15.	Sodium (mg/l)	2.25 ± 2.04	2.17 ± 1.84	2.13 ± 1.87	2.32 ± 1.98	6.64 ± 2.06	11.57 ± 6.62
16.	Potassium (mg/l)	1.17 ± 1.03	1.34 ± 0.98	1.34 ± 0.98	1.34 ± 0.28	2.25 ± 0.87	5.92 ± 2.19
17.	Nitrate (mg/l)	0.42 ± 0.22	0.34 ± 0.21	0.34 ± 0.21	0.41 ± 0.22	0.64 ± 0.04	0.17 ± 0.09
18.	Total Coliforms (MPN/100ml)	100.0 ± 128.37	257.67 ± 673.28	1181.34 ± 183.34	240.0 ± 480.0	3533.34 ± 2589.91	2370.84 ± 1661.81
19.	Fecal Coliforms (MPN/100ml)	34.17 ± 44.32	20.34 ± 48.83	23.34 ± 38.27	37.67 ± 72.12	845.0 ± 1313.27	270.0 ± 245.52
20.	Residual Chloride (mg/l)	1.17 ± 1.56	2.08 ± 1.63	1.34 ± 1.33	1.26 ± 1.39	–	–

fluoride in food and drinking water (Naujoks, 1978; Musaiger and Khunji, 1990; Selwitz *et al.*, 1998; Grobleri *et al.*, 2001). As per Indian standard (BIS, 1983), 0.6 to 1.2 mg/l is the desirable limit of fluoride in drinking water and low fluoride level, below 0.6 mg/l, may cause dental caries and above 1.5 fluorosis. However, investigation on dental health is required for confirmation.

The survey showed that the average consumption of water for drinking and domestic purposes by municipal households was 192 l/c/d. Whereas 63 per cent of households depended on piped water sources for drinking and cooking the rest made use of dug and tube wells.

That municipal water supply was intermittent was the response of 89 per cent of households. Similarly 85 per cent of public standard point users replied that they had to stand in long queue for water collection both during morning and evening hours. About 37 per cent of households experienced scarcity of drinking water which was mainly due to drying up of surface and groundwater, and discontinuous supply of piped water. Besides, majority of households noticed organisms including worms and larvae in municipal water supply and complained that the water was contaminated. Whereas 21 per cent of consumers were of opinion that the municipal water contained excess amount of chlorine, a small percentage believed that it is the leakage of pipes and subsequent contamination that was responsible for the poor quality of water. It is evident that the negative pressure in the distribution system due to intermittent supply might pave way for contamination. In many places, water pipes ran parallel to drainage canals. In the survey indicated that 47 per cent of the consumers believed that the water they drank was not safe and astonishingly 96 per cent of consumers were not satisfied with the services of the municipality in maintaining adequate and proper supply of uncontaminated water. It was in 1963 that the present water supply of Kollam municipality was started with co-operation of Indo- Norwegian Foundation (Menon, 1964). Water from Sasthamcotta lake which is located 29 km north-east of Kollam caters to the needs of municipal dwellers and the existing distribution system is more than 38 years old. Consequently the pipelines that were too old and worn out got broken occasionally.

Thus outbreak of water borne diseases occurred in Kollam municipal areas due to contamination of water supplies through leaks in pipelines. A recent one was in November 1997 in Mundakkal ward of Kollam where it started with breakage of the septic tank of post-metric hostel and subsequent contamination of leaky municipal pipeline. More than 200 people, mostly children were severely affected through consumption of pipe water.

The present survey data on water borne diseases showed that 33 per cent of diseases were diarrhoea, 30 per cent gastroenteritis, 16 per cent ascariosis, 8 per cent typhoid, 7 per cent enterobiasis, 2 per cent cholera, 2 per cent polio and 1 per cent amoebic dysentery. Similarly, 36 per cent of water-related diseases registered were conjunctivitis, 25 per cent allergy, 24 per cent malaria and 15 per cent skin ulcers.

In municipal water supply, the annual mean values of total coliforms ranged from 100 to 1181 MPN/100 ml, in spite of disinfection steps carried out by municipality. In an earlier study, it was reported that raw Sasthamcotta lake water was contaminated with fecal matter and that the counts of coliforms and fecal coliforms were 1983 and 1533 MPN/100 ml respectively. Further coliform values were still higher in S-5 and S-6 both of which were not under direct control of the municipality.

Contamination of water with coliforms makes water unsatisfactory for drinking (Powar and Daginawala, 1988). Fecal coliforms by themselves are not disease causing while some strains are known to cause diarrhoea (SEU, 1991) Avila *et al.* (1989) documented high incidence of diarrhoeal diseases due to consumption of drinking water contaminated with fecal coliforms in Mexico. Welch *et al.* (2000) reported microbial contamination of both treated and untreated water in rural Trinidad revealing thereby that such contaminated drinking water was grossly unfit for human consumption. Strauss *et al.* (2001) reported high prevalence of bacterial contamination in private wells but observed no association between indicator bacteria and gastrointestinal illness. However, in the present study 30 per cent of water diseases were of gastroenteritis type. Payment *et al.* (1991) observed that 35 per cent of the reported gastrointestinal illness were water related due to the consumption of contaminated tap water.

Records of District Medical Officer, Kollam also indicated that in the district there were 32,423 cases of diarrhoea, 206 cases of typhoid, 580 cases of malaria besides other contagious diseases during 1998.

With reference to residual chlorine it was noticed that its amount was higher during rainy season than non-rainy season. When the residual chlorine was 2.22 mg/l during rainy season, the number of fecal coliforms got reduced in piped water supply whereas the number registered an increase during non rainy season when the residual chlorine was 0.67 mg/l. It implied that municipal authorities should take more care through regular monitoring of chlorine level.

The suggestions of consumers on improvement of drinking water supplies were: (1) Proper treatment of drinking water; (2) Replacement of the old pipes by new ones; (3) Proper drainage and sanitation; (4) Construction of more bore wells; (5) Regular monitoring of water quality and scientific application of chlorine; (6) Conducting awareness programmes on water disinfection; and (7) Laying of additional pipelines. In addition to these suggestions, it is recommended that: (1) Efforts should be made to protect Sasthamcotta lake water which is the major source of drinking water of Kollam residents; (2) Water from small water supplies should be consumed after boiling or addition of bleaching powder or chlorine tablets; (3) Detailed studies on dental caries and fluorosis among the consumers are to be made.

The pumping rate of water from Sasthamcotta lake is 9 million litres per day (mld) for supply to Kollam municipality. In addition 2.4 mld are obtained from tube wells of municipal area. By merging certain adjoining panchayats, the area of Kollam municipality was expanded and upgraded to a municipal corporation in 2000, and as per 2001 Census Report, the total population is 3.61 lakhs. In view of this, the present water supply system is insufficient and proposals to cover the demand upto 2004 are now underway (Information from Kerala Water Authority, Kollam).

References

APHA (1985). *Standard methods for the examination of water and wastewater.* American Public Health Association, Washington, DC.

Aravindan, K.P. (1989). A study of epidemic diarrhoeal deaths in Kerala. *Proc. Ker. Sci. Con.*, pp. 361-366.

Avila, H.G., Winkler, S.B and Carmona, H.B. (1989). *Quality of potable water and incidence of gastroenteritis in two cities in the State of Sonora, Mexico.* Salud Publica Mex., 31(3): 299-304.

BIS (1983). *Drinking water specification. IS 10500: 1983.* Bureau of Indian Standards, New Delhi.

BIS (1991). *Drinking water specification. IS 10500: 1991.* Bureau of Indian Standards, New Delhi.

Grobleri, S.R, Louw A.J, and Kotze, T.J. (2001). Dental fluorosis and caries experience in relation to three drinking water fluoride levels in South Africa. *Int. J. Paediatr. Dent.*, 11(5): 3721-3729.

Musaiger, A:D. and Khunji, Z.A. (1990). Chemical quality of drinking water in Baharain. *J. R. Soc. Health*, 110(3): 104-105.

Naujoks, R. (1978). Prevention of the caries through the application of fluoride. *MMW Munch Med Wochenschr*, 120(49): 1633-1636.

Payment, P. (1999). Poor efficiency of residual chlorine disinfectant in drinking water to inactivate water borne pathogens in distribution systems. *Can. J. Microbiol.*, 45(8): 709-715.

Powar, C.B. and Daginawala, H.F. (1998). *General Microbiology*, Vol. II. Himalaya Publishing House, Bombay.

Selwitz, R.H, Nowjack-Raymer, R.E., Kingman, A. and Driscoll, W.S. (1998). Dental caries and dental fluorosis among school children who were life long residents of communities having either low or optional levels of fluoride in drinking water. *J Public Health Dent.*, 58(1): 28-35.

SEU (1991). The bacteriological quantity of water in selected wells in Kerala. Integrated water supply and sanitation programme. Socio-economic Units, Kerala Water Authority, Trivandrum, pp. 1-23.

Strauss, B., King, W., Ley, A. and Hocy, J.R. (2001). A prospective study of rural drinking water quality and acute gastrointestinal illness. *BMC Public Health*, 1(1): 8-12.

Welch, P., David, J., Clarke, W., Trindale, A., Penner, D., Bernstein, S., McDougall, L. and Adesiyum, A.A. (2000). *Microbial quality of water in rural communities of Trinidad.* Rev Panam. Salud. Publica. Mex. 8(3): 172-80.

26

Removal of Dyes by Adsorption Technique: A Review

SATISH N. VAISHNAV & V.S. SHRIVASTAVA
Centre for P.G. Research in Chemistry G.T.P. College, Nandurbar

ABSTRACT

In present communication, available information in literature for adsorption of dyes from its aqueous solution on low cost adsorbents such as tree barks, leaves, rice husk ash, charcoal, clays, fly ash, chitin, waste biomass (Aspergillus niger) and mixed adsorbents are summarised. The principal attention is given to examination of the principles involved in the adsorption process studied by various authors. The process parameters such as dye solution pH, contact time, initial dye concentration, temperature, adsorbent dose and its characteristics influencing the adsorption of dyes are also considered.

Introduction

The effluents coming out from industries like textile dyeing, printing, carpet, woollen, pulp and paper are highly coloured. For industrial effluents, colour is the first indication of water pollution. In textile industries wet processing operations are the major cause for the water pollution because they consume substantial volume of water and chemicals for their wet processing (Sharma *et al.*, 2001). The chemical used for desizing, scouring bleaching, dyeing, printing and finishing include organic and inorganic compounds. Synthetic dyes are major contaminant in textile wastewater. These dyes include several structural varieties such as acidic, disperse reactive, basic, azo, diazo, anthraquinone based and metallic complex dye. The metal complex dye mainly containing chromium

(Tyagi and Yadav, 1990) and even in trace amount chromium is considered as carcinogenic (www. atsdr.cdc.gov/toxfaq.html). In addition some of the dyes might be toxic to some organisms. These dyes though present in only small amounts are visually detectable and thereby are capable of causing serious problem of an aesthetic nature in receiving water bodies. Unless properly treated, these dyes can significantly affect photosynthetic activity due to reduced sunlight penetration and may also toxic to certain aquatic life (Kouba and Zhaung, 1884). The dyes and dye intermediates are stable to light and heat and resistant to biodegradation because of their complex aromatic molecular structure, intentionally constructed to resist fading on exposure to light, soap and water. (Sheshadri *et al.*, 1994). Some of the basic dyes were inhibitory to the growth of micro-organisms (Ogawa *et al.*, 1989). Complex aromatic molecular structure and non-biodegradable nature of dyes makes conventional wastewater treatment methods unsuitable for treatment of coloured effluents. In view of pollution potential, toxicity and carcinogenicity of dyes its removal from effluents has been the target of great attention in last few years.

Several methods are reported in literature for the treatment of coloured effluents. The chemical coagulation (Niranjan and Karthikeyan, 1992) using alum, ferric chloride ammonium ferrous sulphate effectively removes basic and disperse dyes from wastewater. The disadvantage of this method is that it requires high chemical dosages and large volume of sludge production take place. The chemical treatment using stronger oxidizing agents such as chlorine, ozone, hydrogen peroxide (Perkins *et al.*, 1980; Kao, 1992; Antaony and Perkins, 1995; Charchley, 1994) has been reported. Formation of toxic by product after chemical treatment and high capital cost are main disadvantage of this method. Reverse osmosis can remove all types of dyes from wastewater. The filtration process leaves a concentrated dye solution which must be disposed by some other method. The disadvantage of reverse osmosis is a high capital cost and the possibility of clogging the membrane with dyes (Diaper *et al.*, 1996; Watson and Hornburg, 1989). The photocatalytic oxidation is one of the method for degradation of dye in aqueous media. Some of the important textile dyes can effectively degraded in presence of photocatalyst (TiO_2, ZnO, CdS, ZnS etc) (Munevver *et al.*, 2001; Slvakumar and Shanthi, 2001; Sharma *et al.*, 2001; Hustert and Zepp, 1992; Sharma *et al.*, 2001). Among the various available methods, adsorption process have received considerable attention for colour removal due to its low cost, simple design, sludge free operation and superior colour removal. Activated carbon (Mittal and Venkobachar, 1996) is the most widely used adsorbent for wastewater treatment because of its high adsorption capacity. But its use is limited due to its high manufacturing cost and necessitates regeneration. Hence low cost adsorbent as an alternative to activated carbon are being investigated. They include wood (Mckay *et al.*, 1980), china clay (Gupta *et al.*, 1992), fly ash (Khare *et al.*, 1988) forest and animal byproducts, sludge, chitin and chitosan (Annaduri *et al.*, 1997).

In this communication, removal of some dyes from aquous solution by adsorption onto various adsorbent are reviewed. This article is based on the recent available information in literature.

Earlier Work

Recently, Sanghi and Bhattacharya used different tree barks (Eucalyptus, Guava, Papaya) for the adsorption of malachite green from aquous solutions. The dye adsorption experiment were carried out at original pH–7.2. The adsorbent dose was optimised using 150 ml 20 mg/l malachite green solution. Authors reported that significant removed was recorded in dye solution with 0.05 g adsorbent dose; about 97, 93 and 96 per cent removed was achieved with eucalyptus, gauva and papaya tree bark respectively. When initial concentration increase from 5 to 30 mg/l the percentage of dye adsorbed increases from 42 to 85 for eucalyptus, 46-84 for guava and 40 to 87 for papaya tree bark. The tree barks attained equilibrium at maximum adsorption capacity after 20 mg/l dye concentration. Equilibrium was attained within 50 min rapid removal was observed for first 10 min. which slowed with further increase in time. Sorption data had been analysed using Langmuir and Freundlich isotherm. Authors shown the non-linear Langmuir plots at constant temperature suggested that the monolayer theory may not be applicable for present system. However, freundlich plots gives straight line showing favourable adsorption.

Kannan and Meenakshi investigate the adsorption of Acid violate, Congo Red, Methylene Blue and Rhodamine B onto commercial activated carbon. Authors reported the experimental pH 7.4 for Methylene Blue and Rhodamine B and for Acid violate and Congo Red experimental pH was 6.6 and 7.1 respectively. They found the decrease in percentage removal of dyes with increase in initial dye concentration. When initial concentration of Acid violate, Congo Red, Methylene Blue and Rhodamine B increases from 250 to 450, 200 to 450, 600 to 900 and 300 to 550 mg/l percentage removal decreases from 86.4 to 77.8, 87.5 to 76.7, 99.6 to 99 and 94.9 to 84.5 respectively. The equilibrium was established within 35 min of contact time. Authors analysed the batch adsorption data by Freundlich and Langmuir model and stated that system obeys both the adsorption model with correlation coefficient value ranging from 0.913 to 0.998. Statistical analysis of adsorption data for Natarajan and Khalal equation Langergren equation, Bhattacharya and Venkobachar equation and intraparticle diffusion model was also reported. Authors concluded that Methylene Blue and Rhodamine B adsorbed rapidly on commercial activated carbon than Acid violet and Congo Red.

Markorska *et al.* (2001) investigated the solid diffusion control model for adsorption of Maxilon Goldgel B GLEC 400 (MG-400) and Maxilon Schwarz FBL 01 300 per cent (ms-300) onto granular activated carbon and natural zeolite in a fixed bed column. The column were prepared from glass tubes having internal diameter 1.5 cm and height of 15 cm. 40 gm natural zeolite and 23 g Granular activated carbon (GAC) were randomly introduced into the column. The average temperature during the experiments was maintained at 20°C. Batch adsorption experiments were carried out using 0.5 gm GAC and 2 g zeolite with 100 ml dye solution. Equilibrium was established after 4 to 10 days. The adsorption data for all the systems were fitted by Langmuir, Freundlich, Langmuir-Freundlich, Redlich-Peterson equation. A computer program has been developed using SPEEDUP software capable of solving ordinary and partial differential equation for solid diffusion control model. Authors studied the effect of initial dye concentration (50, 100, 200 mg/l) on the breakthrough curve. The slope of breakthrough curve increases with increasing dye concentration reducing the volume treated before adsorbent regeneration. This may be caused by saturation of the active sited on adsorbent more quickly with high dye concentration.

Ho and Chiang studied the sorption of Acid blue-9 onto mixed adsorbent prepared from activated clay and activated carbon. In their communication authors reviewed the work done by Gupta and Shukla (1996) for the treatment of carpet effluents containing mainly Metomega Chrom Orange GL, the popular chrome dye. The various process parameters such as initial dye concentration, adsorbent dose, temperature, pH, contact time, influencing the adsorption of dye had been studied. The equilibrium isotherm for the sorption of Acid blue-9 on mixed adsorbent was determined using 6.5 gm of mixed adsorbent (6 g clay and 0.5 g activated carbon) with 800 ml dye solution of various concentration for 2.5 hrs at pH-3. The sorption data was analysed by Langmuir model they found that present system very well obey Langmuir isotherm having correlation coefficient value 0.996. Authors reported that sorption of Acid Blue 9 is highly dependent on pH of the dye solution, the sorption capacity for Acid blue 9 on mixed adsorbent decreased 41.1 to 29.7 mg/g when pH increased from 2-9. The removal of dye increases rapidly in the beginning and more slowly upto the equilibrium.

This was caused by strong attractive forces between dye molecule and mixed adsorbent and fast surface and pore diffusion to attain rapid equilibrium similar results were reported by Gupta *et al.* (1990) for removal of chrome dye by fly ash and coal. The sorption capacity for Acid Blue 9 increased from 7.28 to 52.9 mg/g when initial dye concentration increased from 59.7 to 475 mg/l. However rate constants decreases from 1.34 to 2.79 × 10^{-2} g/mg/min. Increase in temperature from 20-40°C adsorption decreases from 41.0 to 40.1 mg/g. Ahmed and Rao (1992) reported similar observation for adsorption of Basic Blue 3 on silica. The adsorption process system follows pseudo second order rate equation. Kannan and Ganesh (2001) used mixed adsorbent prepared from fly ash (FA) and commercial activated carbon (CAC) for removal of Malachite Green from aqueous solution. Authors studied the effect of various parameter such as composition of FA-CAC blend, initial pH of dye solution, contact time for adsorption of Malachite Green. They carried out sorption experiment at optimised pH 4.5 and 30 min of contact time. The optimum composition of FA-CAC blend was reported to 86 and 14 per cent by weight respectively. The FA and CAC gives 2.67 and 100 per cent removal of malachite green with initial concentration 600 mg/l indicate that adsorption capacity of fly ash is poor. The per cent removal of Malachite Green exponentially decreases with increase in initial dye concentration and increases with adsorbent dose at fixed initial dye concentration. Adsorption data were modelled with Freundlich and Langmuir adsorption model both the models gives straight line indication monolayer formation and favourable adsorption. Authors also used Langergran equation for modelling of adsorption data they found that removal of malachite green by FA-CAC blend follow fist order kinetics the values of q_{max} and K_{ads} was found to be 8.13 mg/g and 7.36×10^{-2}/min respectively.

Anjaneyulu and Hima Bindu reported the application of waste biomass (Aspergillus niger) for removal of basic blue, malachite green, congo red and fast red. Authors reported the optimum pH 2-3 for removal of congo red and fast red and pH 7-8 for basic blue and malachite green. The equilibrium was established within 15 and 60 minute of contact time for acidic (congo red and fast red) and basic (basic blue and malachite green) dyes respectively. The adsorption of dyes increases with decrease in particle size. The adsorption process obeys Langmuir isotherm model. They conducted batch absorption studies at various temperature (30, 40, 50 and 60°C) for all four dyes, the concentration of dye adsorbed decreases with increase in temperature. Author explained this observation on the basis of effect of temperature on solubility and chemical potential of adsorbate. As the temperature increases, solubility of adsorbate increases, chemical potential decreases causing decrease in adsorption. Authors reported the thermodynamic parameters like change in free energy (ΔG), entropy (ΔS), enthalpy (ΔH) and mass transfer coefficient for adsorption process. The positive ΔH value indicating endothermic nature of adsorption process and positive ΔS value shows increase in randomness at solid solution interface during adsorption process. Negative ΔH value indicating feasible adsorption. Adsorption rate constant was calculated using Lagergran plot for basic blue and results indicate that with increase in initial concentration from 50 to 1200 mg/l, K_{ad} (min^{-1}) decreases from 0.70 to 0.30 while K_{id} (mg/gm $min^{-½}$) increases from 0.0407- to 0.0687. Authors concluded that waste biomass is economically reliable as compared to activated carbon for treatment of textile, pulp and paper mill effluents.

Samanjit and Prasad (2001) used rice husk ash (RHA) as an adsorbent for removal of chemical oxygen demand of five industrially important dyes namely acid violet 54, acid violet 17, acid blue 15, acid violet 49 and acid red 119 from their aqueous solutions. Batch adsorption studies were carried out using 100 ml dye solution for various concentration (1.65-16.80×10^{-4} M) and 1 gm of RHA for optimum contact time 100 min. The adsorbent from solutions was removed by filtration and COD values of filtrate was determined. Adsorption data was fitted to Freundlich and Langmuir isotherm by least square method. The correlation coefficient value (0.96 to 0.99) confirm that, the adsorption of dyes on RHA obeys Freundlich equation and these results were in agreement with the results reported by Converti *et al.* (1992) for COD removal from organic effluents. The Langmuir model was also applicable for present system and linearity of plot indicate the formation of monolayer coverage of dyes on RHA. Author studied the effect of pH on Freundlich and Langmuir isotherm. For this study RHA was treated with 6N HCI and 6 N NaOH which results the RHA having pH 2.5 and 11.4 respectively. Authors found the appreciable increase in adsorption capacity of base treated RHA which may be due to the adsorbent surface modification.

Anjaneyulu and Hima Bindu (2001) also studied the adsorption of Basic Blue and Basic Red onto mixed adsorbent prepared from oxygenated coconut shell activated carbon, (OCSAC), fly ash and china clay. OCSAC was prepared by carbonisation of coconut shell at high temperature (400-600°C) then carbonised product was treated with CO_2 and heated at 800-1000°C for widening of existing pores, then carbon sample was treated with 0.1 N HCl for 24 hrs, washed, dried at 120°C, surface modification was done by wet oxidation process using ammonium persulphate. The ratio OCSAC, fly ash, china clay in mixed adsorbent was 1:0.2:0.2 respectively. Author reported optimum pH 7, contact time 60 min for maximum dye adsorbent. The batch adsorption data was examined by Langmuir model. The value of Langmuir constant Q^0 (mgg^{-1}) and b ($L\ mg^{-1}$) were 10.55 and 9.98 and 0.025 and 0.029 for Basic Blue and Basic Red respectively. Authors studied the rate constant for adsorption of dyes on OCSAC pellets using Langergran equation when concentration Basic Blue and Basic Red increases from 100 to 400 mg/l K_{ads} min^{-1} decreases from 0.1396 to 0.1137 and 0.1628-0.1515 respectively. The applicability of mixed adsorbent for treatment of highly coloured effluent from pulp and paper and textile mill has also been examined. They design parameters for batch and column reactor and also gives economic evaluation for treatment of effluent. Authors concluded that OCSAC-Fly ash-china clay pellet has good adsorption capacity and pellets can be used as column material at higher flow rates than that of pure OCSAC for treatment of industrial waste.

Khattri and Singh (2000) reviewed the work done by Yeh and Thomas (1995) for the treatment of effluents from Taiwanees textile industry using powdered activated carbon (PAC). They observed the ultimate colour removal ΔE_{*ab} in the range 11-55 for different initial dye concentration of 25-200 mg/l. Further investigation was done on effect of particle size, they found that best results for colour removal was obtained using reduced particle size of PAC. Khattri and Singh reported the application of Neem saw dust for removal of Crystal Violate, Methylene Blue, Malachite Green, Rhodamine B decreased from 91.56 to 78.94, 88.36 to 75.47, 84.93 to 71.21, 62.02 to 49.08 respectively. When initial concentration increases from 6 to 12 mg/l. The decrease in particle size of the adsorbent from 50 to 100 mesh, per cent removal of crystal violate was increased from 70.09 to 82.54. Authors reported that present adsorption process follows first order kinetics. The per cent removal of Crystal Violet and Malachite Green decreases from 91.21 to 73.56 and 83.57 to 65.83 respectively as temperature increased from 25-45°C. Authors reported the spontaneous and exothermic nature of present adsorption process on the basis of negative ΔG and ΔH values. The negative values of ΔS suggest the probability of favourable adsorption. The equilibrium data was fitted in linearised form of Langmuir model at various temperature.

Singh and Srivastava (1999) used chemically treated Psidium Guyava leaves for the removal of Methylene Blue, Malachite Green Safranine and Crystal Violet from their aqueous solution. Adsorbent was prepared by treating tree leaves with 39 per cent formaldehyde and sulphuric acid at 50 ± 10°C. Optimum pH range was 8-10 for complete removal of 100 mg/l Methylene Blue with 2 gm adsorbent dose. When pH increases from 2 to 9.5 the sorption of Methylene Blue increases from 53-98 mg g^{-1}. Increased sorption at higher pH seems may be due to greater interaction of dye molecule with negatively charged adsorbent. The effect of initial concentration of Methylene Blue was studied in the range 80-100 mg/l. Authors observed the sorption rate is very rapid during initial period of 30 min contact time. However equilibrium is attained within 180 min. They carried out column studies for Methylene Blue and Malachite Green. Breakthrough and exhaustive capacity of sorbent for Methylene Blue and Malachite Green was 105.127 and 180.204 mg/g respectively. The sorption data was analysed by Freundlich and Langmuir models. The values of Langmuir constant R lies between 0 and 1 and Freundlich constant R lies between 0.2 to 0.4 represent the favourable sorption of dyes and good sorption potential of the adsorbent. The sorption of Methylene Blue and Malachite Green increases with temperature indicating the endothermic nature of process and this observation was confirmed by positive ΔH value 7.35 for Methylene Blue and 400 Kj/mol for Malachite Green. Negative ΔG value indicating spontaneous nature of process. The sorption process of Methylene Blue and Malachite Green follows first order rate kinetics.

Ramkrishna and Vivarghavan (1997) reported the use of waste material, steel plant slag for the removal of Basic Blue 9, Acid Blue 29, Acid Red 91 and Disperse Red 1 from aqueous solution. The optimum pH 11 was reported for removal of Basic Blue 9 and pH 2 for Acid Blue 29 and Disperse Red 1 while pH studies on Acid Red 91 could not be conducted, since the dye behave as an pH indicator and changed colour with change in pH. The equilibrium time for removal of Basic Blue 9, Acid Red 91, Acid Blue 24, Disperse Red 1 was 4, 18, 6, 7 hrs and per cent removal was 4, 74, 46, 94 respectively. Authors compared dye removal capacity of slag with granular activated carbon (GAC). Adsorptive capacity of slag is much higher than GAC for Disperse Red 1. Disperse Red 1 satisfied the Freundlich and BET isotherm model exhibiting the higher correlation while other dyes confirmed only Freundlich model. Authors also conducted column studies for 50 mg/l Disperse Red 1 and find out sorption capacity of slag is 1.3 mg/g for Disperse Red 1. Authors concluded that application of slag as an adsorbent for dye removal may not be environmentally safe especially at acid pH range because of tendency of heavy metals present in slag to go into the solution.

Longhinotti *et al.* (1998) used biopolymer chitin derived from exoskeleton of shrimp with 1 M NaOH for the removal of Orange G, Orange IV and Xylenol Orange. Authors stated that adsorption of anionic dyes on chitin is pH dependent, in acid pH the polymer amino groups are protonated and the polymer chain is positively charged with predominance of adsorption through ion exchange. The maximum amount of dye was adsorbed in pH range 3-4. Authors reported the partition coefficient of Nernst equation. The value of K_L corresponds to the solubility of the dye in water and in solid phase. For xylenol Orange K_L value was 0.726 L mg^{-1}. In their investigation authors found that increase in temperature reduce the adsorption capacity by chitin, due to the enhancement of the desorption step in the mechanism. Increase in temperature from 25 to 60°C adsorption of Orange G decrease from 16.4 to 12.8 mg g^{-1} and for Orange IV adsorption decreases from 1.64 to 1.50 mg g^{-1} when temperature increases from 25 to 53°C. The ΔH value for Orange G and Orange IV was –10.9 and –28.9 KJ mol g^{-1} respectively. Authors stated that the ΔH values are in good agreement with those reported by Mckay *et al.* (1982) for adsorption of Acid Blue 25. Acid Blue 158 and Mordant Yellow on chitin. The negative ΔH values indicate the exothermic nature of adsorption process and small ΔH values confirm the physical nature of adsorption.

Youssef (1993) studied the adsorption of Acid Blue 13 onto chemically modified cellulose. The chemical modification was carried out using N-methyl derivative tri and bis (2-carbomoyl ethyl) ethyl amine. The adsorption capacity of bis derivative was reported to 0.013 mol/kg. The equilibrium was established within 30 min. The adsorption of dye decreases with increase in pH. The desorption of dye can be achieved with NaOH.

Potgieter (1991) used commercial activated carbon for adsorption of Methylene Blue from 25 mg/l of its aqueous solution. Adsorption isotherm were obtained by shaking 1 to 100 mg of adsorbent with 100 ml 25 mg/l Methylene Blue solution for about 72 hrs. Analysis of supernet was carried out spectrophotometrically at 630 nm. All the experiment were conducted at 25°C. Author found that increase in adsorption of Methylene Blue from 0.51 to 2.48 mg when dose of adsorbent increases from 1 to 100 mg with 2.5 mg Methylene Blue solution. He found that present adsorption process obeys Freundlich, Langmuir and Frumkin isotherm with correlation coefficient value 0.987, 0.998, 0.940 respectively.

References

Ahmed, M.K. and Rao, R.K. (1992). Removal of basic dyes from wastewater using silica as adsorbent. *Environmental Pollution*, 77: 79-86.

Anjaneyulu, Y. and Hima Bindu, V. (2000). Application of waste biomass for the removal of coloured organics from industrial effluent. *Chem. Env. Res.*, 9 (1 and 2): 105-119.

Anjaneyulu, Y. and Hima Bindu, V. (2001). Application of mixed adsorbent (Oxygenated coconut shell activated carbon OCSAC-Flay Ash-China clay) for removal of basic dyes from industrial effluents. *Jr. of Env. and Poll.*, 8(1): 117-127.

Annadurai, G. and Krishna, M.R.V. (1997). Adsorption of acid dye from aqueous solutions by Chitin: Equilirium studies. *Ind. J. Chem. Tech.*, 4: 217-222.

Annadurai, G., Chellapandian, M. and Krishna, M.R.V. (1997). Adsorption of basic dye from aqueous solution by chitosan: Equilibrium studies. *Ind. Jr. of Env. Prot.*, 17(2): 95-98.

Anthony, F. and Perkins, W.S. (1995). Decoloursation of continuous dyeing wastewater by ozonisation. *Textile Chemist and Colourist*, 27(5): 11-15.

Bumpus, J.A. and Borck, B.J. (1988). Biodegradation of crystal violet by the white rot fungus phanerochaete chryososporium. *Applied and Environmental Microbiology*, 54: 1143-1150.

Churchley, J.H. (1994). Removal of dye wastewater colour from sewage effluent: The use of full scale ozone plant Water. *Sci. Tech.*, 30: 275-284.

Converti, A., Borghi, M.D., Rovatti, M., Zilli, M. and Ferrailo, G. (1992). *Chem. Biochem Eng.*, (a) 6: 195.

Diaper *et al.* (1996). The use of membranes for recycling of water and chemicals from dye house effluents of economic assessment. *J.S.D.C.*, 112: 277-280.

Gupta, G.S. and Shukla, S.P. (1996). An inexpensive adosrption technique for the treatment of carpet effluents by low cost materials, *Adsorp. Sci. Technol.*, 13: 15-26.

Gupta, G.S., Prasad, G. and Singh, V.K. (1990). Removal of Chrome dye from aqueous solution by mixed adsorbent fly ash and coal. *Water Research*, 24: 45-50.

Gupta, G.S., Shukla, G.P. and Singh, V.K. (1992). China clay as an adsorbent for dye house wastewater. *Env. Technol.*, 13: 925-936.

Ho, Y.S. and Chiang, C.C. (2001). Sorption studies of acid dye by mixed adsorbents. *Adsorption*, 7: 139-147.

http:/www.atsdr.cdc.gov/toxfaq.html.

Hustert, K. and Zepp, T.G. (1992). Photocatalytic degradation of selected azo dyes. *Chemisphere*, 24: 335-342.

Kannan, N. and Jeya Ganesh, R.V. (2001). Studies on removal of malachite green from aqueous solution by mixed adsorbents. *Ind. Jr. of Env. Prot.*, 20(2): 127-133.

Kannan, N. and Meenakshi Sundaram (2002). Kinetics of adsorption of dyes on activated carbon: A comparative study. *Ind. Jr. of Env. Prot.*, 22(1): 9-16.

Khare, S.K., Pandey, K.K., Shrivastava, R.M. and Singh, V.N. (1988). Removal of Victoria blue from aqueous solution by fly ash. *J. Chem. Tech. Biotech.*, 38: 281-290.

Khattri, S.D. and Singh, M.K. (2000). Colour removal from synthetic dye wastewater using a biosorbent. *Water, Air and Soil Pollution*, 120: 283-294.

Kouba, J.F. and Zhaung, P. (1994). Colour removal for textile dyeing wastewater. *Fluid Particle Separation Journal*, 7(3): 87.

Kuo, W.G. (1992). Decolourising dye wastewater with Fentons reagent. *Water Research*, 26: 881-886.

Longhinotti, E., Pozza, F., Furlan, L., Mario, N.M.S., Klug, M., Mauro, C.M.L. and Fdvere, V.T., (1998). Adsorption of anionic dyes on the biopolymer chitin. *J. of Serb. Chem. Soc.*, 9(5): 435-440.

Markovsk, L., Meshko, V., Noveski and Marinkovski, M. (2001). Solid diffusion control of the adosrption of basic dyes onto granular activated carbon and natural zeolite in fixed bed columns. *J. Serb. Chem. Soc.*, 66(7): 463-475.

Mckay, G. and Poots, V.J.P. (1980). Kinetics and diffusion processes in colour removal from effluents using wood as an adsorbent *J. Chem Tech. Biotech.*, 30: 279-292.

Mckay, G., Blair, H.S. and Gardner, J.R. (1982). *J. Appl. Polym. Sci.*, 27: 3043.

Mittal, A.K. and Venkobachar, C. (1996). Uptake of cationic dyes by sulphonated coal: Sorption methanism. *Ind. Engg. Chem. Res.*, 35(4): 1472-1474.

Munevver, Sokmen, David, W. Allen, Fatih Akkas Nuket Karta and Filiz Acar (2001). Photodegradation of some azo dyes using Ag Loaded titamium dioxide. *Water, Air and Soil Pollution*, 132: 153-163.

Niranjan, P. and Karthikeyan, J. (1992). Colour removal from dye effluents by chemical coagulation-II Basic and Disperse dyes. *Ind. J. Env. Prot.*, 12(8): 599-603.

Ogawa, T., Fuji, I., Kawal, K., Yatome, C. and Idaka E. (1989). Growth inhibition of Bacillus subtilis upon interaction between basic dyes and DNA. *Bulletin of Environmental Contamination and Toxicology*, 42: 402-408.

Perkins, W.S., Judkins, J.F. and Perry, W.D. (1980). Renovation of dye bath water by chlorination or ozonization. *Textile Chemistry and Colour*, 12: 182-187.

Potgieter, J.H. (1991). Adsorption of Methylene Blue on activated carbon. *Chem. Edu.*, 68(4): 349-350.

Ramkrishna, K.R. and Viraraghavan, T. (1997). Use of slag for dye removal. *Waste Management*, 17(8): 483-488.

Sanghi, Rashmi and Ban Bhattacharya (2002). Decoulorisation of malachite green dye solution using tree barks as low cost adsorbents. *J. of Ind. Ass. Env. Mgyt.*, 29: 129-135.

Sharma, Ajay, Sharma, Bhoopendra, Mathur, R.P. and Ameta Suresh C. (2001). Use of zinc oxide particulate system as a photocatalyst: Photobleaching of brilliant green, *Poll. Res.*, 20(3): 419-423.

Sharma, Deepti, Sharma, Poonam, Tiwari, Swati and Ritu Kesarwani (2001). Use of Solar energy for Decolorisation of dye bearing effluent by photocatalytic oxidation. *Eco. Env. and Conservation*, 7(4): 415-419.

Sharma, J.K. and Arora, M.K. (2001). Decolourization of textile effluent. *Poll. Res.*, 20(3): 453-457.

Sheshadri, S., Bishop, P.L. and Agha, A.M. (1994). Anaerobic/aerobic treatment of selected azo dyes in wastewater. *Waste Management*, 15: 127-137.

Singh, D.K. and Srivastava, B. (1999). Removal of basic dyes from aqueous solutions by chemically treated Psidium Guyava Leaves. *Ind. Jr. of Env. Hlth.*, 41(1): 333-345.

Sivakumar, T. and Shanthi, K. (2001). Photocatalytic studies on some textile reactive dyes using TiO_2. *Ind. Jr. of Env. Prot.*, 21(2): 101-104.

Sumanjit and Prasad, N. (2001). Adsorption of dyes on rice husk ash. *Ind. J. of Chem.*, 40(A): 388-391.

Tyagi, O.D. and Yadav, M. (1990). *A text book of synthetic dyes* 1st edition. Anmol Publications, New Delhi.

Watson, D. and Hornburg, C.D. (1989). *Low energy membrance nanofiltration for removal of colours organics and hardness, desalination*, 72: 11.

Yeh, R.Y. and Thomas, A.J. (1995). *Chem. Biotechnol.*, 63: 48.

Youssef, B.M. (1993). *Am. Dyestuff Rep.* 82: 30.

27

Geochemistry and Environmental Evaluation of the Bharalu River Sediments

P.K. DAS & RANJAN BORAH

Department of Geological Sciences, Gauhati University, Guwahati, Assam

ABSTRACT

The chemical characteristics and their probable impact on the environment have been studied in the Bharalu river sediments. The chemical forms of some major oxides and trace elements have been determined from the bed sediments. From the study, minor spatial variation has been observed from upstream to downstream. This is due to anthropogenic incorporation of elements of the basin. This is due to increased loading of industrial effluents and domestic wastes. So, sediment characteristics played a significant role in defining the chemical forms of the elements present in the sediments.

Keywords: Geochemistry, Environment, Bharalu river sediments.

Introduction

Sediments play an important role in the chemical characteristics of river water. Rivers are the major means of transport of water and sediments from the continents to the seas and oceans. The water quality of a river depends on the terrain through which it flows and the sediment it carries along. To know the contamination of the external elements, metal content of the sediments is essential. Moreover the chemical characterization of the sediments in a river helps to measure the anthropogenic contamination and to assess the probable impact on the environment.

The Bharalu river is an important river of the Guwahati city. It flows through the heart of the city. It is one of the South bank tributary of the Brahmaputra and meets the Brahmaputra at Bharalumukh (Fig. 27.1). The main natural drainage channel of the city, the river Bharalu, known as Bahini in its upper reaches, is a small tributary of the river Brahmaputra. Emanating from the Khasi hills, the river enters Guwahati through the south-eastern corner of the city near Basistha and passing through thickly inhabited areas across the city joins the river Brahmaputra at Bharalumukh. The river has a stretch of about 10.2 km from Basistha to Bharalumukh and the total catchments area is about 120 sq.km. The average depth of flow is 3 m and the bed width is from 5 to 9 m.

Table 27.1: Percentage of Major Oxides in the Sediments of the Bharalu River Around Guwahati

Sample No.	*SiO_2*	*Al_2O_3*	*Fe_2O_3*	*CaO*	*MgO*	*Na_2O*	*K_2O*	*MnO*	*TiO_2*	*P_2O_5*
1	64.15	15.57	7.21	2.10	5.12	2.35	1.10	0.08	1.26	0.52
2	65.10	18.11	4.19	2.04	3.79	2.23	2.42	0.07	1.54	0.28
3	66.51	20.78	2.24	1.72	2.20	1.77	2.67	0.07	1.62	0.20
4	63.37	20.23	2.82	3.62	2.40	1.84	2.58	0.10	1.67	1.47
5	60.31	24.10	3.98	1.94	3.55	1.54	2.84	0.04	1.18	0.31
6	59.75	24.94	3.47	2.23	2.96	1.99	2.21	0.05	1.76	0.33
7	58.83	26.32	3.71	1.93	3.28	1.57	2.15	0.08	1.44	0.41

It is known that environmental distribution in a river system depends upon various factors such as basin geology, physiography, chemical reactivity, vegetation, land-use pattern, grain size, organic content, biological productivity etc.

Metal pollution in various river systems of India have been studied from time to time by different scientists like Chakrapani and Subramanian (1990), Jha *et al.* (1990), Sarin *et al.* (1984, 1992), Subramanian (1987), Saikia *et al.* (1988), Subramanian *et al.* (1985), Borah *et al.* (2001), Baruah *et al.* (1998). The present study was designed to focus element contamination in the bed sediments of the Bharalu river, Guwahati, Assam.

Methods of Study

Considering the accessibility and representativeness in sampling, seven sites were selected for sediment samples covering 10 km. The bed sediment samples were collected during post-monsoon and pre-monsoon periods for upstream *i.e.* Basistha to downstream *i.e.* Bharalumukh (Fig. 27.1).

Degens (1965) is of the opinion that chemical study of sediments rather than the entire bulk sample. Therefore, for chemical analysis the bulk sample was not considered.

In the present study, sediments finer than 0.0625 mm size that represents the matrix and clays in the sediments were taken for analysis. Major oxides such as SiO_2, Al_2O_3, CaO, MgO, Na_2O, K_2O, MnO, TiO_2, R_2O_5, and trace elements like Ba, Co, Ni, Cu, Zn, Ga, Rb, Sr were determined quantitatively with the help of XRF using parameters 8–VS GS, Kr-50 and mA–40 in the laboratory.

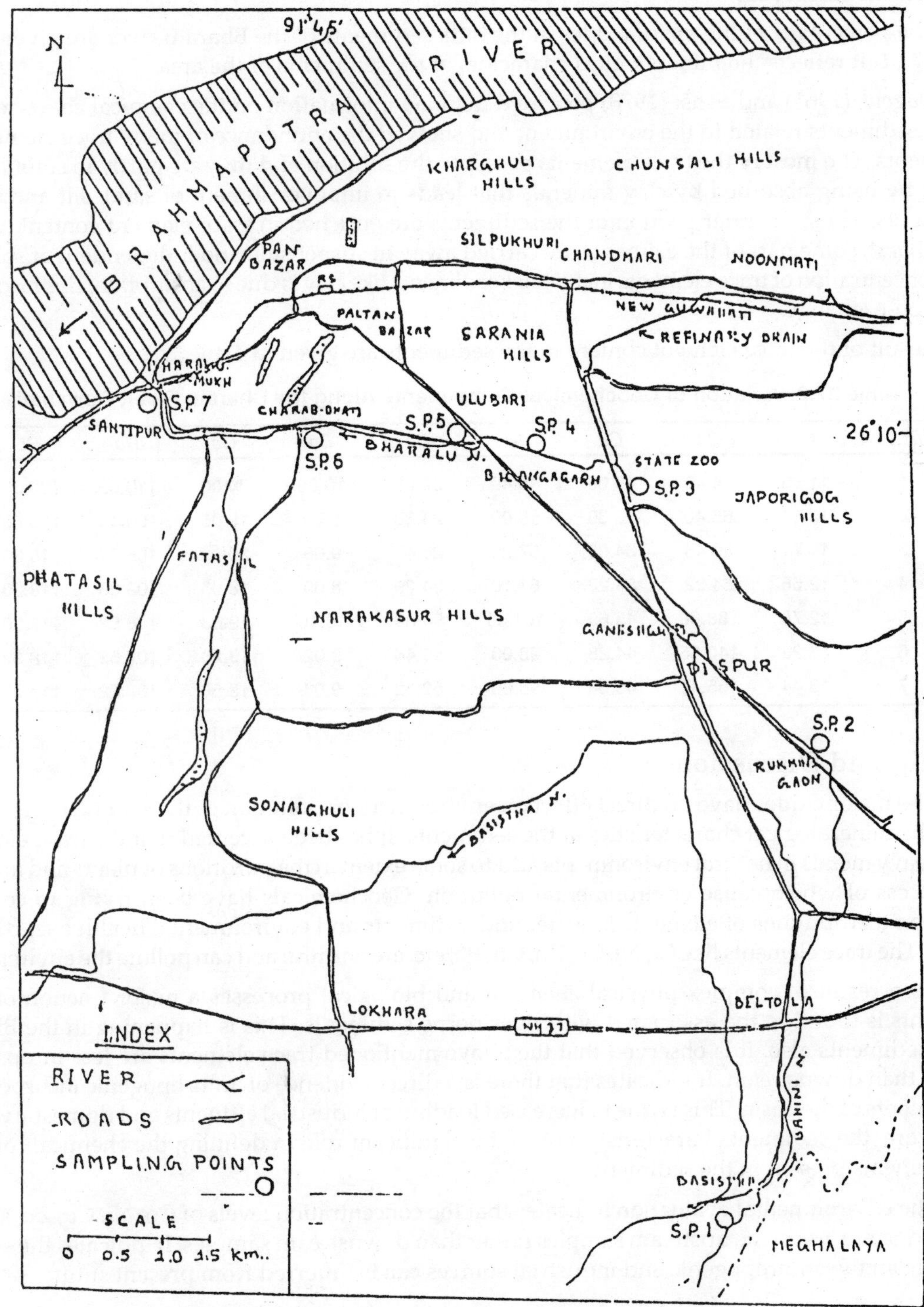

Fig. 27.1: Locations of Sampling Points Along Bharalu River

Chemical Analysis

The percentage of the major oxides in the bulk sediments of the Bharalu river are given in the Table 27.1. It reflects the mineralogical characters of the sediments of the area.

Degens (1965) and Ernst (1970) provided detailed information on the chemical characteristics of the sediments related to the environment and showed the abundance of many trace elements in sediments. The most of the trace elements pass into the solution and migrate partly in solution and partly by being absorbed by clay minerals that leads to impoverishment of sand–silt and clayey sediments. Thus, the finer portion of the sediments are enriched in trace element content. On the other hand, some part of the elements are carried away in suspension and independent of solution. The concentration of trace elements in the finer sediments like clay is due to absorption, chemisorption and iron-exchange.

Result of the trace element content in the sediments are given in Table 27.2.

Table 27.2: Variation of Geochemical Components Along the Bharalu River, Guwahati

Sample No.	*Pb*	*Zn*	*Cu*	*Ni*	*Co*	*Ba*	*Ga*	*Rb*	*Sr*
1	11.63	34.40	48.00	48.40	22.09	10.00	20.00	110.02	120.30
2.	12.14	66.40	33.30	55.00	24.33	8.00	19.05	109.15	118.50
3	11.71	90.45	34.05	57.20	25.47	9.00	20.00	108.04	119.30
4	12.66	64.22	34.22	61.10	34.29	8.00	18.05	109.03	118.20
5	12.78	188.08	45.69	101.50	54.33	10.10	19.03	108.52	117.50
6	13.23	140.66	44.28	98.00	50.46	9.02	19.20	107.55	118.32
7	13.44	155.40	45.68	98.80	52.12	9.03	18.50	109.02	117.03

Results and Discussions

The major oxides have no direct effect on environment or pollution of the sediments. It reflects only the mineralogical characteristics of the sediments. It has been accepted that the trace elements and heavy metals in natural environments add to some extent in the nutritions of plants and animals, the excess of which cause environmental pollution. Geochemicals have been trying to correlate between distributions of elements in water and sediments and environmental health hazards they cause. The trace elements like Cr, Ni, Co, Zn, Cu, Pb, etc. are harmful and can pollute the environment.

As a result of complex physical chemical and biological processes a major fraction of trace elements is found to the associated with river borne sediments. This is happening in the Bharalu river sediments also. It is observed that the above mentioned trace elements are less in upstream rather than downstream. It indicates that there is a direct evidence of anthropogenic incorporation of elements of the basin. This is due to increased loading of industrial effluents and domestic wastes. Therefore, the sediment characteristics played a significant role in defining the chemical forms of the elements present in the sediment.

The environmental evaluation indicates that the concentration levels of Co, Ni, Cu, Zn, Ga, Ba, Rb, and Sr are less in the upstream samples rather than downstream samples. It indicates the specific input from the anthropogenic and industrial sources can be inferred from present study.

References

Baruah, J., Baruah, N.K., Kotoky, P. and Sarma, J.N. (1988). Geochemistry and environmental evolution of the Jhanji River sediments, Assam. *Journal Indian Association of Sedimentologists,* 17(2): 271-285.

Borah, Ranjan, Das, P.K. and Bhattacharyya, K.G. (2001). Studies on interaction between surface and groundwaters at Guwahati, Assam (India). *Journal of Environment and Pollution,* 8(4): 361-369.

Chakrapani, G.J. and Subramanian, V. (1990). Preliminary studies on the geochemistry of the Mahanadi River Basin. *India Chemical Geology,* 81: 241-253.

Degens, E.T. (1965). *Geochemistry of Sediments.* Prentice Hall Inc., Englewood Cliff, New Jersey, p. 342.

Ernst, W. (1970). *Geochemical Facies Analysis.* Elsevier, Amsterdam, p. 152.

Jha, P.K.V., Subramanian, R., Sitasawad, R. and Van Grieken (1990). Heavy Metals in sediments of the Yamuna river (Attributary of the Ganges) India. *The Science of the Total Environment,* 95: 7-27.

Saikia, D.K., Mathur, R.P. and Srivastava, S.K. (1988). Heavy metal in water and sediments of upper Ganga. *Ind. Jour. Env. Hlth.,* 31(1): 11-17.

Sarin, M.M. and Krishnaswami, S. (1984). Major ion chemistry of the Ganga Brahmaputra river system, India. *Nature,* 312(5994): 538-541.

Sarin, M.M., Krishnaswami, S. Trivedi, J.R. and Sarma, K.K. (1992). Major ion chemistry of the Ganga source water. Weathering in the high attitude Himalaya. *Proc. Ind. Aca. Sci. (Earth and Plan. Sci),* 101(1): 88-98.

Subramanian, V., Vant Dack, L. and Van Greiken, R. (1985). Chemical Composition of the river sediments from the Indian subcontinent. *Chemical Geology,* 48: 271-279.

28

Potability of Dug Wells of Mayyanad Panchayat, Kerala

S. RESHMA & V.R. PRAKASAM

Department of Environmental Sciences, University of Kerala, Kariavattom, Thiruvananthapuram, Kerala

ABSTRACT

Water quality analyses of six wells in Mayyanad panchayat of Kollam district (Kerala) were carried out for a period of one year (July 2002–June 2003) based on the essential characteristics specified by Bureau of Indian Standards (BIS). The study revealed that (1) faecal coliforms were invariably present in large numbers in all the wells (2) iron content was higher than desirable limit in all wells (3) the amount of fluoride was lower than the desirable limit in all the wells (4) in two wells (No. 4 and 5) pH was acidic and it was below the desirable limit set by BIS. Groundwater contamination was found to be so widespread that extensive efforts at community level with people's participation are suggested for eliminating the menace.

Keywords: Dug wells, water quality, Mayyanad, Kerala, groundwater.

Introduction

The dug wells are the major groundwater extraction structures in Kerala. The average density of dug wells in its coastal belt is about 200 per sq.km. The average density of dug wells of midland region is about 150 per sq.km and that of the high land is about 70 per sq.km (Anon, 1991). All India

Census 2001 reported that 76 per cent of people of Kerala are depending on well water for drinking use (Mathrubhumi, 2003).

During the last two decades a large number of investigations on the groundwater quality of Kerala have been carried out, and it has been reviewed by Reshma (2003). Investigations were focused on areas such as groundwater utilization pattern, recharge through rainfall, geophysical survey, salt water intrusion, physico-chemical and bacterial quality, influence of industrial and domestic wastes on groundwater, the growth pattern of wells and sustainability approach to protect quality and quantity of shallow aquifers. It was understood from the review that only a few studies have been made from Kollam district (Sajikrishnan *et al.*, 1994; Nazimudeen *et al.*, 1995; Narasimha Prasad, 1998; Nair, 2000 and Royee, 2001). Therefore, in the present study, one of the panchayats of the district was chosen for the detailed investigation.

The main objectives of the study were to make an extensive study on groundwater quality of wells that are spread over the different parts of Mayyanad panchayat and to find out the impact of environmental factors on the well water quality, especially season, location of latrines and solid waste dumping sites.

Materials and Methods

The study consisted of analysis of water quality of six wells located in Kottiyam (well No. 1), Umayanalloor (well No. 2), Mayyanad (well No. 3), Kulangara (well No. 4), Kuttikkad (well No. 5) and Mukkam (well No. 6) of Mayyanad panchayat of Kollam district. It is located near to the west coast of Kerala.

Monthly collections were made from the six wells after making measurements of water level (depth of water column). Collections were made during the first week of every month in the afternoon (2 to 4 pm) using polyethylene bottles. Temperature was noted at the site itself. The samples were then stored in a refrigerator and later analyzed for various parameters such as colour, odour, taste, turbidity, pH, total hardness, iron, chloride and fluoride following APHA (1985). The analyses were mostly confined to the essential characteristics of drinking water, as specified by BIS. Microbiological analysis was carried out following Mackie and McCartney (1996).

For analysis of soil texture, soil from a depth of 2 feet was collected from the surroundings of each well and the percentages of sand, silt and clay were estimated (Galehouse, 1971; Ingram, 1971).

Results

The annual means of physico-chemical and bacteriological parameters are given in Table 28.1. The results showed that there were fluctuations in water level during the one year period of study. There was an increase in water level during rainy season in all the wells. All the water samples collected from the six dug wells appeared to be colourless. These showed desirable taste and unobjectionable odour.

The temperature of water ranged from 28 to 30°C in well No. 1 whereas in well No. 2 it varied from 29 to 32°C and in well Nos. 3, 4, 5 and 6 it varied from 29 to 32°C. The temperature was higher during rainy season than non-rainy season. But the increase was only by a margin of 0.5°C.

Table 28.1: Annual Mean of Various Physico-chemical and Bacteriological Parameters

of Six Wells in Mayyanad Panchayat of Kollam District

Parameters	*Well No. 1*	*Well No. 2*	*Well No. 3*	*Well No. 4*	*Well No. 5*	*Well No. 6*
Depth of water column (cm)	110.20 ± 149.76	163.75 ± 86.03	150.83 ± 69.76	205.83 ± 112.21	99.58 ± 38.59	96.46 ± 33.96
Temperature	29 ± 0.71	29.58 ± 1.04	30.42 ± 0.86	30.33 ± 0.94	30.33 ± 0.94	30.33 ± 0.94
Turbidity (NTU)	8.10 ± 4.42	4.05 ± 2.56	1.43 ± 1.76	3.61 ± 4.59	2.45 ± 2.87	18.22 ± 13.26
pH value	6.04 ± 0.43	6.51 ± 0.43	6.49 ± 0.41	5.61 ± 0.58	5.98 ± 0.78	6.62 ± 0.35
Total hardness (mg/l)	37.58 ± 39.94	93.33 ± 34.88	140.50 ± 30.12	40.50 ± 29.45	36.25 ± 13.24	116.33 ± 42.20
Iron (mg/l)	0.59 ± 0.08	0.57 ± 0.07	0.59 ± 0.07	0.56 ± 0.10	0.56 ± 0.09	0.59 ± 0.08
Chloride (mg/l)	69.87 ± 86.63	53.31 ± 13.42	74.91 ± 27.52	45.91 ± 4.93	30.41 ± 16.20	57.14 ± 8.55
Fluoride (mg/l)	0.07 ± 0.02	0.06 ± 0.02	0.06 ± 0.02	0.06 ± 0.02	0.07 ± 0.02	0.06 ± 0.02
Coliforms (MPN/100ml)	1211 ± 507	1637 ± 353	1750 ± 150	943 ± 422	1750 ± 150	1750 ± 150
Faecal coliforms (MPN/100ml)	78 ± 22	149 ± 36	169 ± 28	95 ± 13	153 ± 34	104 ± 20

The turbidity of the six wells varied from 0.1 to 45.4 NTU. In well No. 1 it varied from 1.8 to 10.2 NTU while in well No. 2 it showed a range of 0.4 to 9.2 NTU. The turbidity noted in well No. 3 ranged between 0.1 and 2.5 NTU and in well No. 4 it ranged between 0.2 and 13.5 NTU. The turbidity of well No. 5 varied between 0.35 and 10.6 NTU. Well No. 6 showed a variation of 3 to 45.4 NTU. It was noted that in all the wells except 1 and 5 an increase in turbidity was noticed during non-rainy season. In well No. 1 and 5 there was an increase in turbidity during rainy season. The desirable limit of turbidity for drinking water, according to BIS and WHO, is 5 NTU and the permissible limit of turbidity is 10 NTU. Accordingly Well No. 1 was above desirable limit but within permissible limit of BIS, while the well No. 2 was below desirable limit except slight increase during non-rainy season. Well No. 3, 4 and 5 were below desirable limit in both seasons and well No. 6 showed a turbidity value much above the desirable and permissible limits during both the seasons. It may be noted that this well is located nearer to the coast and it is most shallow. The water is constantly stirred due to the frequent use of bucket for water collection.

The pH of the 6 wells ranged between 5.43 and 7.69. In well No. 1 it ranged from 5.43 to 6.95. Well No. 2 and 3 showed a range of 6.09 to 7.69 and 5.96 to 7.65 respectively. In well No. 4 it varied between 4.91 and 6.93 and in well No. 5 it ranged from 5.08 to 8.09. Well No. 6 ranged from 6.08 to 7.12. The desirable limit of pH according to BIS is 6.5 to 8.5. Accordingly well No. 2, 3 and 6 were within the desirable limit and thus suitable for drinking. It was quite noteworthy that well No. 4 and 5 recorded much decrease in pH from the desirable limit. In all wells a lower pH was noticed during non-rainy season than rainy season.

The total hardness of well No. 1 ranged from 11 to 160 mg/l and that of well No. 2 varied from

64 to 192 mg/l. Total hardness range of 90 to 188 mg/l and 16 to 132 mg/l were noticed in well No. 3 and 4 respectively. In well No. 5 it was between 20 and 62 mg/l, and in well No. 6 it ranged from 32 to 228 mg/l. It was generally noted that (1) there was wide variation in hardness in the same well from time to time (2) there was variation between wells (3) total hardness was higher during non-rainy season in well Nos. 1, 2, 3 and 5 whereas it was lower in well Nos. 4 and 6 than rainy season. The desirable and permissible limits of total hardness for drinking water, according to BIS, are 300 mg/l and 600 mg/l respectively. The data indicated that all the water samples collected from the 6 wells were below these limits and hence good for drinking purpose.

The study showed that in the well No. 1 there was a range of iron from 0.4 to 0.68 mg/l and in well No. 2 from 0.42 to 0.66 mg/l. Well No. 3 showed a range from 0.48 to 0.68 mg/l and well No. 4 between 0.39 and 0.7 mg/l. A range of 0.42 to 0.68 mg/l was noticed in well No. 5 and 0.48 to 0.7 mg/l in well No. 6. In general (1) there was variation in iron content within and between wells (2) there was increase in iron content during non-rainy season and such an increase was noticed in all wells except No. 2. In well No. 2, however, the difference was negligible. The desirable limit for iron according to BIS and guidelines of WHO is 0.3. The data obtained from the 6 wells in the present study showed that iron content was above permissible limit and thus not suitable for drinking.

The chloride of well No. 1 varied between 25.56 and 353.58 mg/l and well No. 2 ranged between 35.50 and 78.1 mg/l. The chloride estimated in well No. 3 was between 48.28 and 133.48 mg/l. The chloride in well No. 4 ranged from 39.76 to 56.80 mg/l. In well No. 5 the chloride content was in the range from 48.28 to 79.52 mg/l, while in well No. 6 it was between 14.2 and 65.32 mg/l. It was observed that *(i)* chloride content in all the collections was below 100 mg/l and *(ii)* in three wells (No.1, 3, 4) it was lower in rainy season and in three wells it was higher in rainy season (No.2, 5, 6). According to the BIS the desirable limit of chloride in drinking water is 250 mg/l and the permissible limit is 1000 mg/l. The limit of chloride, according to WHO, is also 250 mg/l. Chloride in all the samples collected from the six wells during both the seasons except well No. 1 was below desirable limit. In well No. 1 a high in chloride content (353.58 mg/l) was noted in March 2003. It was understood that this was due to addition of salt and ash into the well by the owner of the well.

The fluoride content in the well No. 1 ranged from 0.04 to 0.09 mg/l and in well Nos. 2, 4, 5 and 6 it ranged between 0.02 and 0.09 mg/l. The fluoride content in well No. 3 ranged from 0.03 to 0.09 mg/l. It was generally noted that there was only narrow fluctuation in fluoride in the water samples of the wells studied, however, it was slightly lower during rainy season in all wells except No. 4. In the latter no variation between seasons was noticed. According to the desirable limit of BIS, and the guideline of WHO, the fluoride standard for drinking water is 1.5 mg/l. The present data indicated that all the samples collected from the six wells were very below the standards of fluoride.

The MPN of coliforms obtained from 6 wells varied between 540-1900 MPN/100ml. In well No. 1, 2, and 4 it ranged from 540 to 1900 MPN/100ml. and in well No. 3, 5 and 6 it ranged from 1600 to 1900 MPN/100ml. It was noted that in 3 wells (No. 3, 5 and 6) the MPN/100ml of coliforms was same for both seasons; in 2 wells (No. 1 and 2) it was higher during rainy season and in 1 well (No.4) it was higher during non-rainy season.

It was further observed that in 3 wells (No. 2, 5, 6) the MPN of faecal coliforms was greater during rainy season; in 2 wells (No. 1 and 4) it was almost same and in 1 well (No.3) it was slightly lower during rainy season. According to BIS and WHO no faecal coliforms should be present in drinking water. Here in the present study all the wells were found to be highly contaminated with the presence of faecal coliforms which indicated that the water was not suitable for drinking.

The soil texture analysis showed that sample 1 consisted of 88.99 per cent of sand, 0.46 per cent silt and 10.55 per cent clay. In sample 2, 97.55 per cent of sand, 0.51 per cent salt and 1.94 per cent clay were present. In sample 3, it contained 95.17 per cent sand, 0.06 per cent silt and 4.77 per cent clay. Sample 4 consisted of 93.5 per cent sand, 0.23 per cent silt and 6.27 per cent clay. In sample 5, 94.23 per cent sand, 0.16 per cent silt and 5.61 per cent clay were found. The composition of sample 6 was 97.49 per cent sand, 0.25 per cent of silt and 2.26 per cent clay. The results indicated that soil around all wells was sandy.

A comparative study showed that there was similar fluctuation in coliform count and water level (depth of water column) during most of the periods of observation (Fig. 28.1).

Results and Discussion

One of the observations made in the study was the low chloride levels in the wells irrespective of their location whether it was near to the shore or away from the shore. That chloride level was below desirable limit of drinking water standard was indicative of the fact that salt water intrusion from the sea had not taken place although there is a previous report of occurrence of saline groundwater in Mayyanad panchayat (Narasimha Prasad, 1998). In well numbers 2, 5 and 6 there were increases in chloride during rainy season. This also supported the fact that the source of chloride in the wells was not the nearby sea. Similarly water did not show any salty taste which indicated that the cation involved was not sodium.

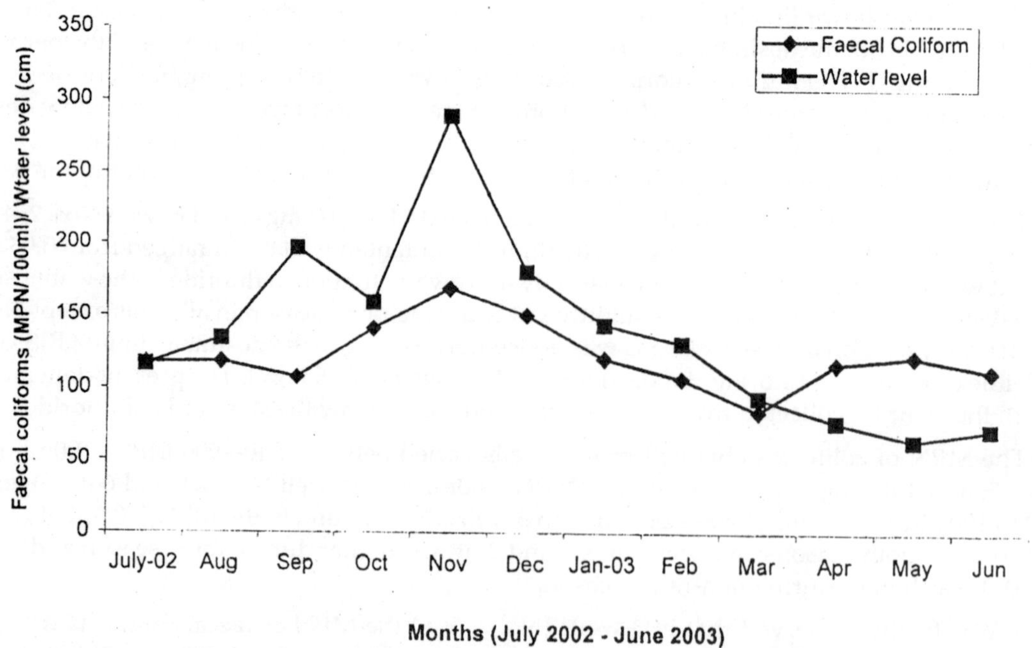

Fig. 28.1: Graph Showing Relationship Between Water Level (cm) and Faecal Coliforms (MPN/100ml) in the Wells of Mayyand Panchayat During July 2002-June 2003
(Scale: 1cm = 100 Faecal coliforms/100cm water level)

Another observation regarding water quality was the below desirable limit of fluoride in the wells monitored. This was in contrast to the reports of higher fluoride content in certain other parts of Kerala (Kokkal and Varadan, 1995; Harikumar and Madhavan, 1996; Remani *et al.*, 1997)

The desirable limit of fluoride in drinking water is 1.0 mg/l as per BIS. Fluoride concentration of 1 mg/l in drinking water helps in preventing dental caries in children and promotes the formation of hard, strong and decay resistant teeth (Abbasi, 1998). In a recent study on surface and groundwater of Kollam municipality Royee (2001) found that most samples contained less than 0.6 mg/l of fluoride and he opined that this low concentration might have affected the dental health of the consumers. A similar possibility is likely in Mayyanad panchayat also since the fluoride value was below 1.00 mg/l in all the wells. Naujoks (1978) has suggested administration of fluoride tablets or fluoridation of salt as preventive measure against dental caries. Based on a survey on dental health of children in the panchayat, similar curative measures could be attempted.

Out of six wells in the Panchayat four exhibited pH values within the desirable limit whereas two (Nos. 4 and 5) showed acidic pH below the limit. pH also registered a decrease during non-rainy season in all wells. This low pH of water might be due to the geological peculiarities of the surroundings as suggested by Anon (1992). A few previous studies in Kerala have indicated that well water was mostly acidic (Suvarnakumari *et al.*, 1997; Harikumar *et al.*, 1999). It may be recollected that low acidity can cause health hazards, corrosion to metals and detrimental effects to algae and fish (Jacob, 1991; Murali Krishna, 1998). It is recalled in this context that the acidity of water could be removed by using calcium oxide (Harikumar, 2002).

Iron is known to occur in groundwater, the quantity of which is dependent on several factors. Karanth (1996) stated that higher concentration of iron was attained in water with low pH. The present results indicated that pH of well water was always low and that it decreased during non-rainy season. It was noteworthy that a corresponding increase in iron was noticed in all wells during non-rainy season when there was decrease in pH also. Buy there was no such similar increase in iron in well Nos. 4 and 5 which registered acidic pH. With regard to iron, another fact invariably noticed was its high level beyond desirable limit in all the wells. Previous studies carried out in other parts of Kerala state had also shown the presence of high iron content (Saji Krishnan *et al.*, 1994; Suvarna Kumari *et al.*, 1997; Remani *et al.*, 1997; Harikumar, 2002; Namboodiri, 2003).

The present study revealed that the coliform and faecal coliform counts in the well waters were much above the desirable limit. Similar observations have been made in a number of studies carried out in different parts of Kerala (Royee, 2001).

Another finding of the study was that in 50 per cent of wells there was increase in faecal coliforms during rainy season and in other wells there were no differences between seasons. In addition, there were increase in faecal coliforms with peaks in rainfall and also with increase in water level (depth of water column) of the wells. Ward *et al.* (1985) noticed that during rainy season the coliform count was very high in groundwater. Pillai and Ouseph (2000) found alarmingly high faecal coliform count during rainy season which was an indication of seepage through unplastered wells. They also observed that unhygienic use of wells by children, external contamination by birds were also reasons for this. Another study by Prajapati and Mathur (2002) also noted that the contamination of drinking water sources was highest in rainy season compared with summer season.

It was derived from the study that use of septic latrines instead of open defecation did not eliminate the problem of faecal contamination in the wells. As for example, both well No. 2 where there was open defecation in the surroundings and in well No. 1 which was 70 feet away from the septic latrine were contaminated with faecal coliforms. Similarly wells were found to be contaminated with faecal coliforms irrespective of their location near or away from septic latrines. For example, well No. 1 was located 70 feet away from the septic tank whereas well No. 5 was stationed just about 7 feet away from the septic tank.

Previous bacteriological studies undertaken in the different parts of Kerala indicated that faecal coliform contamination was quite widespread in the groundwater of Kerala and this situation has resulted from various factors such as open defecation and careless disposal of solid and liquid wastes into the surroundings. It was also evident from the present study that even possession of septic latrines and distant location of them do not save their wells from faecal contamination. The sandy soil of the panchayat seemed to be much favourable for travel of coliform bacteria from the source to groundwater. Controlled experiments had shown that bacterial pollutants can travel more than 200 feet in groundwater. In actual outbreaks of disease, the pathogens of dysentery and typhoid have been shown to spread in groundwater laterally to as far as 700 feet (Aester *et al.*, 1995). It is inferred that better sanitation of the total environment is imperative for maintaining the groundwater clean. A massive drive at community level is quite necessary to eradicate bacterial pollution. It is however, suggested that water consumed for drinking should be boiled before use or disinfected using chlorine tablets so that water borne outbreaks of diseases could be checked.

References

Abbasi, S.A. (1998). *Water Quality Sampling and Analysis*. Discovery Publishing House, New Delhi.

Aester, E.W., Roberts, E.C. and Aester, M.T. (1995). *Microbiology: A Human Perspective.* Wm.C. Brown publishers, United States of America.

Anon. (1991). *Water Atlas of Kerala*. CWRDM. San Jose Process, Thiruvananthapuram.

Anon. (1992). *Development of Low Cost Sanitation in India*. Jalasandesh, SEU, 5 (3 and 4): 2-9.

APHA. (1985). *Standard Methods for the Examination of Water and Wastewater.* American Public Health Association, Washington.

Galehouse, J.S. (1971). Sedimentation analysis In: *Procedures in Sedimentary Petrology*. Wiley Interscience Publishers, America.

Harikumar, P.S. (2002). *Drinking Water and Water Quality*. Kudivella Paripalanam, Seminar material (Malayalam), CWRDM, Kozhikode.

Harikumar, P.S. and Madhavan, K. (1996). *Water Quality Problems Related to Excess Fluoride in Alappuzha and Cherthala Regions.* Proceedings of the Eighth Kerala Science Congress, pp. 61-62.

Harikumar, P.S., Gopakumar, R., Madhavan, K., Manju, V.S., Vasu, K. and Basak, P., (1999). *Effect of Industrial Effluents on Groundwater Quality*. CWRDM, Kozhikode.

Ingram, R.L. (1971). Sieve analysis. In: *Procedures in Sedimentary Petrology*. Wiley Interscience Publishers, America.

Jacob, V.C. (1991). *Groundwater Pollution in Kerala*. Seminar on groundwater development and management in irrigation and other water sectors, Thiruvananthapuram, CWRDM, Kozhikode.

Karanth, K.R. (1996). *Hydrogeology*. Tata McGraw Hill Publishing Company Limited, New Delhi.

Kokkal, K. and Varadan, K.M. (1995). *Water Quality Studies in Tripallur Kudallur Area of Palakkad District, Kerala*. Proceedings of Seventh Kerala Science Congress, p. 119.

Mackie and McCartney (1996)). *Practical Microbiology*. Longman Singapore Publishers (Private) Ltd., Singapore.

Mathrubhumi (2003). Thamasa saukaryangal mechappettu; Ernakulam Munnil (Malayalam), Census 2001 report. *Mathrubhumi daily*, April 24.

Muralikrishna, K.V.S.G. (1998). *A Treatise on Rural Municipal and Industrial Water Management*. Environmental protection society, Kakinada.

Nair, N.V. (2000). Environmental impact assessment with particular emphasis to the groundwater quality problems and its management in parts of the coastal areas of Kollam district, Kerala. Kerala Research programme on local level development, CDS, Thiruvananthapuram.

Namboodiri, N.T.V. (2003). *Sustainability Approaches to Protect Quality and Quantity of Shallow Aquifer in a Restricted Land Environment*. CWRDM, Kozhikode.

Narasimha Prasad, N.B. (1998). Groundwater status in Kerala. In: *Water quality Problems with Special Reference to Drinking Water in Kerala*. Training programme on water quality assessment with special reference to drinking water, CWRDM, Thiruvananthapuram.

Naujoks, R. (1978). Prevention of the caries through the application of fluoride. *MMW Mech. Med.*, Wochenscher, 120(49): 1633-1636.

Nazimuddin, M., Basak, P. and Narasimha Prasad, N.B. (1995). *Groundwater Availability and Utilization in the Unconfined Aquifer: A Case Study from Quilon District in Kerala State*. CWRDM, Kozhikode.

Pillai, M.G. and Ouseph, P.P. (2000). *Water Quality Management in Wells: A Case Study*. Proceedings of the Twelfth Kerala Science Congress, pp. 17-20.

Prajapati, R. and Mathur, R. (2002). Groundwater quality in Sheopurkalan town. *Indian Journal of Environmental Protection*, 22(2): 197-200.

Remani, K.N., Harikumar, P.S., Madhavan, K., Joseph, J. and Basak, P. (1997). *Pollution of Groundwater Due to Domestic Sewage*. CWRDM, Kozhikode.

Reshma, S. (2003). *Water Characteristics of Dug Wells of Mayyanad Panchayat*. M.Phil dissertation, University of Kerala, Thiruvananthapuram.

Royee, M.K.P. (2001). *Environment Related Water Quality of Kollam Municipality, Kerala*. PhD. Thesis, University of Kerala, Thiruvananthapuram.

Saji Krishnan, K., Ouseph, P.P. and Ajay Varma, R. (1994). *Locally Adaptable Measures for Water Quality Improvement and Management in a Typical Coastal Panchayat-Clappana, Quilon District*. Proceedings of Sixth Kerala Science Congress, Thiruvananthapuram, pp. 71- 73.

Suvarnakumari, N., Upendran, N., Nazimuddin, M. and Rajagopalan Nair, S. (1997). *Water Quality Status of Groundwater in Chittar Sub Basin: A Case Study in Southern Kerala*. CWRDM, Thiruvananthapuram.

Ward, C.H., Giger, W. and McCarty, P.L. (1985). *Groundwater Quality*. Wiley Interscience Publication, USA.

29

Environmental Impact of Limestone Mining on River Yamuna, Giri and Tons in Sirmour in H.P. with Special Reference to Biological Water Quality Monitoring

T.B. SINGH

H.P. State Environment Protection and Pollution Control Board, Paonta Sahib, Distt. Sirmour, H.P.

DEVENDERA SINGH

I.R.I. Roorkee (Presently in Department of Environment, U.P. Govt.)

ABSTRACT

In order to assess the impact of mining on river water quality a study on aquatic benthic macroinvertebrates, physical and chemical characteristics has been carried out for all rivers *viz.* Giri, Tons and Yamuna during the year 1999-2000. This paper illustrates that limestone mining activities have little affect on aquatic life. In downstream of mining area, results show increase in hardness, turbidity, TDS, TSS, calcium and pH. The high mineral content in the river might be due to the dissolution of lime and minerals in to running water during mineral exploitation. However, study indicates that the COD, BOD, TDS, TSS, Turbidity are affecting factors to the saprobity (oxygen availability) of river water for presence or absence of species which establish a co-relation between chemical and physical with biological factors.

Introduction

Himachal Pradesh is an ecological fragile area and limestone mining has become a contentious issue for years. The main Boundary Thrust is a deep seated tectonic dislocation extending all along the lower Himalaya from Kashmir in the west to Arunachal in the east. It has a proven reserve to the tune of 550 million tonnes. Sirmour mining area is located in the southern part of Himachal Pradesh in foothills of the Himalayas. Physiographically most of the area is very steep and rugged terrain. The elevation varies from plains to high hills and ranges between 1500 feet and 12000 feet above sea level. The district lies in outer Himalayan ranges commonly called as Shivalik between 70°49′ East longitude and 30°22′ and 31°01′ North latitude. The district is predominantly mountainous with deep valleys lying between ranges of varying elevations. The river Yamuna flows at south-eastern boarder and Tons river forms north-eastern boundary of the district. The Giri river, a tributary of Yamuna is the biggest river in the district, which flows down in the south-east direction divides the district into two parts *viz.* Trans-Giri and Cis-Giri. The river Yamuna, Tons and Giri control the drainage system of the district. Debris produced, roughly five to six times the mineral lifted was strewn around and left to be washed down by rains into rivers and other water bodies. In order to assess the impact of mining on river water quality a study on aquatic macroinvertebrates, physical and chemical characteristics has been carried out for all rivers *viz.* Giri, Tons and Yamuna during the year 1999-2000. During the last two decades, concentration of mining activities in Sirmour (HP) has led to disequilibrium in the ecological balance and significant deterioration in the quality of the Environment. It is recognized now that continuing unscientific mining cannot be sustained on a long term basis because of adverse environmental impacts. The impact of various environmental pollutants cannot be assessed comprehensively by determination of limited physico-chemicals parameters, where as assessment through biological system can integrated the impact of all environmental variables occurring over a large period of time and can be easily evaluated and quantified. The bio-monitoring of environmental components of such areas highlight the state of ecological environment of surface water bodies in the vicinity of mining area.

Biomonitoring methodology established the fact that among all the biological components of the aquatic ecosystem, benthic macroinvertebrates are the best suitable for evaluation of water quality. Most of these benthic macroinvertebrates particularly insect larvae, which are aquatic, share their biological life in freshwater, while their adults flyover for miles together in search of suitable environment for reproduction, breeding and laying eggs for establishment of biological communities in suitable habitats.

The benthic biota provide a method for assessment of pollution; Biological examinations have been stressed to provide a more accurate picture than the physico-chemical examinations (Hirsch, 1958; Hynes, 1965; Singh, 1988 and Sinha *et al.*, 1989).

In aquatic system, the macroinvertebrates community is most often investigated due to absence of mobility and sensitivity towards physico-chemical stress. Several workers have pointed out that benthic organisms provide a valuable indicator of past and present water quality conditions and prove to be the most useful in the assessment of pollution (Hynes, 1965; Rama Rao, 1978 and Sinha *et al.*, 1989) because of their life-cycle length, central position in food chain and ease of sorting and preservation. The ecological sustainability of the recipient freshwater resources of some of these area's are constantly threatened despite of various anticipative and preventive policies.

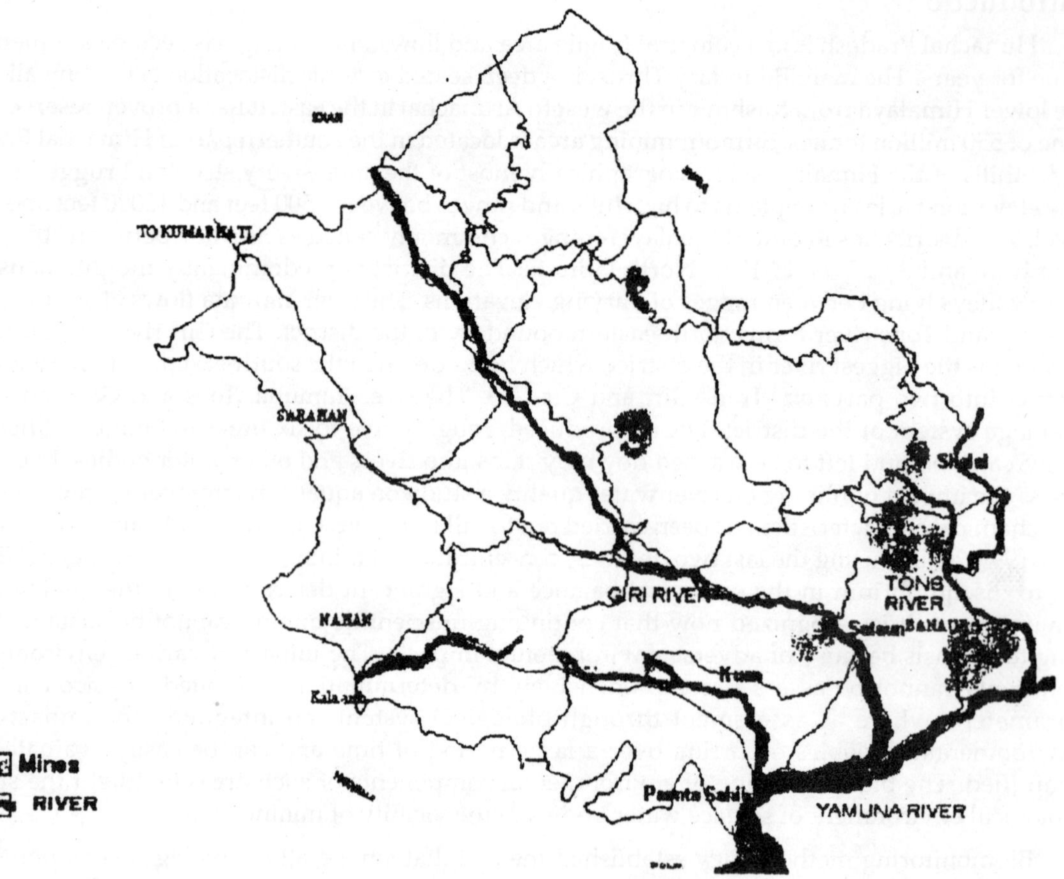

Fig. 29.1: Sirmour Mine Area Map

The macroinvertebrates are limited to specific ranges of physical environments by their physical tolerances (Fiance, 1978; Vannote and Sweeney, 1980). These organisms depend on the type of catchment's drainage/discharge who restrict the distribution and abundance of individuals. In addition, they may be restricted to particular ranges of current velocities, substrata types or food availability.

Material and Methods

The approach of this study is to describe the impacts of limestone mine waste on aquatic macroinvertebrate fauna in selected locations a concise normal ecology of the aquatic fauna from the upstream of rivers under similar environmental conditions has been dealt with to provide a clear pollution impact.

Seven sampling locations were identified and selected on river Giri, Tons and Yamuna for biomonitoring of aquatic macroinvertebrates and physico-chemical determination. Samples of macroinvertebrate were collected through handnet (mesh size 0.5 mm) alongwith river water samples from the ecologically similar locations to increase biological and chemical comparability on quarterly basis *i.e.* March, June, September and December. BMWP (British Biological Monitoring Working Party) evaluation method was applied for biological classification of river water quality. Physico-chemical analysis was carried out by following (Greenberg *et al.*, 1995) and Trivedy and Goel (1984).

Results and Discussion

The study carried out on upstream of river Tons, Giri and Yamuna for one year reveals that macroinvertebrate community is composed of some twenty one families, the maximum being from insects and minimum being oligochaets. The presence of different population of the community are well expressed by data presented in Table 29.1.

Biological Evaluation

The macrozoobenthic taxa of Giri river revealed decrease in sensitive families and increase in number of tolerant families from 7 to 12 in downstream (d/s). The biological water quality class of river indicates slight pollution in upstream (u/s) of mine, moderate in d/s of mine and further too moderate in d/s of Sataun town stockyard of limestone (Table 29.1). The macrozoobenthic taxa of Tons river revealed decrease in sensitive families and increase in number of tolerant families in u/s to d/s of river. The biological water quality class of river indicates slight pollution in d/s of mining area (Table 29.1).

The macrozoobenthic taxa of Yamuna river revealed decrease in sensitive families/species and increase in number of tolerant species in u/s to downstream. The biological water quality class of river indicates slight pollution in d/s of Paonta (Tables 29.1 and 29.2).

The majority of benthic macroinvertebrates occur from a pH of 7.0 to 8.4. Dissolved Oxygen is a important factor for aquatic life. However, some species have been reported from very low DO content, where they survive by virtue of lowering their oxygen consumption (Berg and Ockelman, 1959).

Chemical Evaluation

The water quality of river Giri, Tons and Yamuna are shown in Table 29.3. The surface water of mining area shows high buffering capacity indicating the pH values varying 7.79 to 8.88. The water quality of all three rivers are showing high mineral content in terms of hardness (220-330 mg/l), TDS (436-678 mg/l), TSS (3.0-16 mg/l), Calcium (56-88 mg/l). The water quality of rivers in u/s locations indicates no impact of mining due to negligible mining activities in the u/s regions. In downstream of mining area, results show increase in hardness, turbidity, TDS, TSS, calcium and pH. The high mineral content in the river might be due to the dissolution of lime and minerals into running water during mineral exploitation. The COD, BOD, TDS, TSS, Turbidity are affecting factors to the saprobity (oxygen availability) of river water for presence or absence of species which establish a co-relation between chemical and physical with biological factors (Tables 29.2 and 29.3). This study indicates limestone mining activities have very little affect on water quality of river Giri, Tons and Yamuna as far as aquatic life is concern.

Table 29.1: Biological Classification of River Giri, River Tons and River Yamuna on the Basis of Macrozoobenthic Taxa

River Giri			*River Tons*		*River Yamuna*	
U/S Manal Mine	*D/S Manal Mine*	*D/S Sataun*	*Before confluence of Nehra River*	*Before confluence of Yamuna River*	*U/S Paonta Sahib*	*D/S Paonta Sahib*
Ephemeroptera *Ephemerellidae*	Ephemeroptera *Ephemerellidae*	Ephemeroptera	Ephemeroptera *Ephemerellidae*	Ephemeroptera *Ephemerellidae*	Ephemeroptera *Ephemerellidae* *Leptophlebiidae*	Ephemeroptera *Ephemerellidae*
Ephemeridae *Heptageniidae* *Caenidae*	*Heptageniidae* *Caenidae* *Baetidae*	*Heptageniidae* *Caenidae* *Baetidae*	*Ephemeridae* *Heptageniidae* *Caenidae*	*Heptageniidae* *Caenidae* *Baetidae*	*Ephemeridae* *Heptageniidae* *Caenidae* *Baetidae*	*Heptageniidae* *Caenidae* *Baetidae*
Plecoptera *Nemouridae*	Plecoptera *Nemouridae*	Plecoptera *Nemouridae*	Plecoptera *Nemouridae*	Plecoptera *Nemouridae*	Plecoptera *Nemouridae*	Plecoptera
Hemiptera *Aphelocheiridae*	Hemiptera	Hemiptera	Hemiptera *Aphelocheiridae*	Hemiptera	Hemiptera *Aphelocheiridae*	Hemiptera
Odonata *Gomphidae*	Odonata *Gomphidae*	Odonata *Gomphidae*	Odonata *Gomphidae*	Odonata *Gomphidae*	Odonata	Odonata *Gomphidae*
Trichoptera	Trichoptera *Hydropsychidae*	Trichoptera *Hydropsychidae*	Trichoptera *Hydropsychidae* *Rhyacophilidae*	Trichoptera *Hydropsychidae*	Trichoptera *Hydropsychidae*	Trichoptera *Hydropsychidae*
Diptera	Diptera *Chironomidae* *Ceratopogonidae* *Tipulidae*	Diptera *Chironomidae* *Ceratopogonidae* *Tipulidae*	Diptera *Ceratopogonidae*	Diptera *Chironomidae* *Ceratopogonidae*	Diptera	Diptera *Chironomidae*
			Hemiptera	*Hemiptera*	*Hemiptera* *Simulidae*	*Hemiptera* *Corixidae*
Megaloptera	Megaloptera *Sialidae*	Megaloptera *Sialidae*	Megaloptera	Megaloptera *Sialidae*	Megaloptera	Megaloptera
Coleoptera	Coleoptera *Elminthidae*	Coleoptera *Elminthidae*	Coleoptera	Coleoptera	Coleoptera	Coleoptera
			Dryopoidae *Hydrophilidae*	*Dryopoidae*		*Dytiscidae*
Water Quality Slight Pollution Class-B	Water Quality Moderate Pollution Class-C	Water Quality Moderate Pollution Class-C	Water Quality Clean Class-A	Water Quality Slight Pollution Class-B	Water Quality Clean Class-A	Water Quality Slight Pollution Class-B

Table 29.2: Biological Water Quality Evaluation (on the basis of one year biomonitoring results)

Sampling Location	*Saprobic Score*	*Diversity Score*	*Water Quality Characteristics*	*Water Quality Class*
River Giri				
1. U/S Manal Mine	6.2	0.53	Slight Pollution	B
2. D/S Manal Mine	5.80	0.35	Moderate Pollution	C
3. Sataun stockyard	5.86	0.38	Moderate Pollution	C
River Tons				
1. Before confluence of Nehra river	7.0	0.156	Clean	A
2. Before confluence of Yamuna river	6.4	0.277	Slight Pollution	B
River Yamuna				
1. U/S Paonta Sahib	7.8	0.26	Clean	A
2. D/S Paonta Sahib	6.45	0.29	Slight Pollution	B

Table 29.3: Physico-chemical Assessment

Sampling Location	*pH*	*DO*	*COD*	*BOD*	*Total Hardness*	*Ca*	*Mg*	*TSS*	*TDS*
River Giri									
1. U/S Manal Mine	7.79	8.45	19.0	2.67	290	76	24	8.0	436.0
2. D/S Manal Mine	8.10	8.20	20	2.70	330	88	27	10	590.0
3. Sataun Stockyard	8.38	8.15	33.62	3.40	300	80	24	16	678.0
River Tons									
1. Before confluence of Nehra river	8.64	8.0	12.0	2.0	220.0	56.0	19.44	3.0	454.0
2. Before confluence of Yamuna river	8.88	7.7	16.0	3.5	310.0	80.0	26.73	8.0	602.0
River Yamuna									
1. U/S Paonta Sahib	8.39	9.13	12.93	1.25	290	68	29	4.0	530.0
2. D/S Paonta Sahib	8.43	8.46	48.0	4.12	300	72	29	6.0	580.0

References

Fiance, S.B. (1978). Effect of pH on the biological and distribution of Ephemerotera. *Oikos*, 31: 332-339.

Greenberg, A.E., Clesseeri, L.S. and Eaton, A.D. (1995). *Standard Methods for Examination of Water and Wastewater*, 19th edn. APHA, New York.

Hirsch, A. (1958). Biological evaluation of organic pollution in New Zealand stream. *N. Z. J. Sci.*, 1: 500-553.

Hynes, H.B.N. (1965). The significance of macroinvertebrates in the study of mild river pollution. In: *Biological Problems in Water Pollution.* 3rd Seminar U.S. Deptt. Health, Education and Welfare, Cincinnati.

Rama Rao, S.V., Singh, V.P., and Mall, L.P. (1978). Biological methods for monitoring water pollution levels: Studies of Ujjain. *Water Reasearch*, 341-348.

Rohm, M.C. (1987). Evaluation of an aquatic ecoregion classification of stream in Arkansas. *Journal of Freshwater Ecology*, 4(1): 127-140.

Shina, M.P. and Trivedy, R.K. (1990). *Impact of Mining on Environment*. Ashish Pub., New Delhi.

Sinha, M.P., Pandy, P.N. and Mehrotra, P.N. (1989). Some aspects of biological studies of organically polluted urban stream in Ranchi II. Marco benthic fauna. *Indian Zoologist*.

Trivedy, R.K. and Goel, P.K. (1984). *Chemical and Biological Methods for Water Pollution Studies.*

30

Bio-Ecology of Potable Water

N.K. BOHRA, S. MUTHA & P.K. AGGARWAL

Arid Forest Research Institute, Jodhpur, Rajasthan

Introduction

Water is a precious natural resource it is used an essential for living beings. We cannot imagine the existence of nature without water. It is true that, according to the Hindu mythology earth is made up of Punch Mahabhut namely Agni, Water, Air, Akash and Earth. Nearly 75 per cent of the earth comprises of water still the world is facing the problem of drinking water. Now it is assumed that in future the water is the main reason for war among the countries. In the year 2025 the world population will be six hundred million. If this is kept in mind, we need more water. The requirement of water will meet only if its management and conservation will be done judiciously. So that our Prime Minister told if we would not take proper attention towards its proper utilization and judicious conservation, the water will not be available to the next generation.

Urbanization, industrialization and increasing population have raised pressure on drinking water. Drinking water and clear environment is the main concern for the world. Everybody needs water. People can live for two months without food, but will die in less than a week without water. Our own bodies are two-thirds water, and our brains are at least 85 per cent water. While access to drinking water in India has increased over the past decade, the tremendous adverse impact of unsafe water on health continues. The World Bank estimates 21 per cent of communicable diseases in India are water related. It is presumed that within the next eight seconds, a child will die from a water-related disease. An estimated one billion people around the world still lack safe drinking water. Statistics are staggering. Over 480 million people in India don't have access to safe drinking water. Similarly in America Scientists estimate that each year up to seven million Americans become sick from contaminated tap water. Keeping in view the present situation, it is advised for proper management, judicious use of water and to adopt effective conservation measures.

3/4th of the earth has the water but this water is not for the use of drinking water because only 2.7 per cent water is available for freshwater and most of the water is trapped as ice cap in polar region. If represented by a half gallon bottle, the quantity of freshwater would be about ½ teaspoon; a single droplet would represent the surface flowing water, the rest being groundwater (UNWWC, 1977). The water that enters the hydrological cycle constitutes just 0.03 per cent of the world's total water that comes down as rain and fills the river, lakes, ponds and streams. So water is the prime need of the human for his existence but safe drinking water is also essential for better health. In ancient literature, there are million of examples how to make the water for drinking purpose. In ancient time, all the civilization was developed along the water resources. The water we use, on it way down, dissolves gases like carbon dioxide and dissolves some salts, become slightly acidic. The quality of the drinking water, particularly groundwater depends upon the quality and quantity of salt that prevail in the soil. The potable water is usually clear, tastes sweet, free from unpleasant flavour and odour and food well and quickly. In contrast water, which is not fit for drinking, is unclear, has a tinge of colour, salty, smell, metallic taste and take a long time for working.

Keeping the view in mind that water is essential or basic source of life, health and food production the Government of India declare year 2003 is the year of safe and pure drinking water. This article with special emphasis to their quality, their impacts on human health, feasible purification method and the measure taken by the GOI to improve the quality of drinking water.

Drinking Water

Drinking water affects health. An Excite search using the phrase "drinking water" came up with 57890 documents. A rather recent book *Chemistry of Water Treatment* by S.D. Faust and O.M. Aly, 2nd Ed. (1998) [TD433 F38 1998] addresses the standards for drinking water. Safe drinking water is a suitable combination of minerals and electrolytes. Potable water is usually clear, tastes sweet, free from unpleasant odour. Usually, one should not drink soft water or distilled water. Hard water with calcium and magnesium ions is good for drinking, washing and ablution. Water quality may be broadly classified under three categories *i.e.* physical, chemical and microbiological.

The quality of water is an extremely important factor for drinking water. The WHO states that about 21 per cent of diseases in developing countries are belonging to the water. Therefore, the quality of water is an important aspect because it is a life sustaining prerequisite to all living organisms. Water that is absolutely pure is never found in nature. Hence, provision of good quality of drinking water is an important as it provided for drinking, cooking, bathing.

Contaminant	*Standard*	*Contaminant*	*Standard*
Aluminum	0.05 to 0.2 mg/l	Iron	0.3 mg/l
Chloride	250 mg/l	Manganese	0.05 mg/l
Colour	15 (color units)	Odour	3 threshold odour number
Copper	1.0 mg/l	pH	6.5-8.5
Corrosivity	Non-corrosive	Silver	0.10 mg/l
Fluoride	2.0 mg/l	Sulphate	250 mg/l
Foaming Agents	0.5 mg/l	Total Dissolved Solids	500 mg/l
		Zinc	5 mg/l

Water Quality

The quality of water is an extremely important factor of drinking water. The WHO states that about 21 per cent of diseases in developing country are belonging to the water. Therefore, the quality of water is an important aspect because it is a life sustaining prerequisite to all living organism. Water that is absolutely pure is never found in nature. Hence, provision of good quality of drinking water is an important as it provided for drinking, cooking, bathing, washing utensils and ablution. So every effort should be made to maintain water quality as high as possible. Water quality may be broadly classified under three categories *viz.* physical, chemical and microbiological.

Physical Quality

The drinking water needs to be fit from the aesthetic point of view also *i.e.* its colour, turbidity, taste and odour. If these are not satisfactory, it may discourage consumers from using. The colouring of water is caused by natural minerals, vegetative colouring material, and suspended impurities and also due to discharge from agriculture, industrial and mining activities (WHO, 1993). Permissible colour limit is measured as True Colour Unit (TCU). Removal of excess colour through chemical coagulation, filtrate and slow sand filters. Disagreeable taste and odour are due to microscopic pollutants like algae, bacteria, fungi etc. Chlorination reduces the unpleasant odour and taste to a great extent. Water is a life sustaining prerequisite to all organisms. Water that is absolutely pure is never found in nature. Even rainwater collects dust bacteria and gases while its way from clouds to earth. Drinking water needs to be fit from aesthetic point of view *i.e.* colour, turbidity, taste and odour.

Colour

Colouring of water is caused by natural mineral, vegetable colouring matter and suspended impurities and also due to discharge from agricultural, industrial activities (WHO, 1993). Permissible colour unit is measured as True colour unit (TCU). It is measured by the spectrophotometer. For drinking water TCU should not exceed 3, (threshold number). Removal of excess colour or unwanted colour is through chemical coagulation, filtration and slow sand filters.

Taste

Disagreeable taste and odour in water is due to microscopic pollutants like algae, bacteria, fungi and sewage and industrial wastes. Chlorination of water reduces the unpleasant odour and taste of water to great extent. pH *i.e.* hydrogen ion concentration affects the taste of water. A pH value of 6.5-8.5 is recommended for drinking water.

Turbidity

Turbidity is due to suspended and colloidal matters that disturbs the clarity of water and interferes with the passage of light through it. Turbidity below 5 nephelo metric units is acceptable. High turbidity is normally associated with sewage pollution and industrial effluents. Filtration and chemical coagulation will reduce the turbidity.

Presence of Minerals, Chemicals, Metals and Salts

Presence of mineral acids, salts of strong acids and weak bases will make the water acidic. Excessive acidity is neutralized by addition of lime or soda ash. Similarly presence of salts of weak acids, carbonates, bi-carbonates, hydroxides, silicates, etc contribute to the alkalinity of water. Chemical coagulation, softening, corrosion control process will correct alkalinity of water.

Most of the toxic chemicals found in drinking water are from pharmaceutical, paints and pigment industries. The chemicals like lead, mercury, arsenic, cadmium, strontium are generally found in the discharge of these industries. Fluoride is another important chemical in water. The permissible limit is 1.5 mg/litre. Excess fluoride causes flurosis of teeth and bones with neurological complication. Inadequate fluoride may cause weak and fragile bones. Presence of nitrates greater than 50 mg/litre shows sewage contamination of drinking water and usage of nitrogenous fertilizers in that area apart from natural resources. Homogenate hydrocarbons like dichloromethane, tribromomethane, hexachlorocylohexane etc. present in drinking water may not pose any immediate effect on the users, but on long-term usage they are harmful (Packham, 1995).

Chemical Quality

Assessment of water quality by its chemistry includes measures of many elements and molecules dissolved or suspended in the water. Chemical measures can be used to directly detect pollutants such as lead or mercury. Chemical measures can also be used to detect imbalances within the ecosystem. Such imbalances may indicate the presence of certain pollutants. For example, elevated acidity levels may indicate the presence of acid mine drainage. Commonly measured chemical parameters include pH, alkalinity, harness of water, and presence of nitrates, fluorides, ammonia and ortho and total phosphates and dissolved oxygen and biochemical oxygen demand. The presence of bacteria is also determined using a chemical test. In addition, some "chemical" measurements actually indicate the physical presence of pollutants in water. These include measurements such as conductivity and density.

The chemical quality is the most important factor of safe drinking water. Following are the chemical features of the water quality. The pH affects the taste of water. High pH favours corrosion control while low pH helps in effective colourizations. A pH value of 6.5 to 8.5 is recommended for drinking water. Presence of salts of weak acids, carbonate, bicarbonate, hydroxide etc. contributes to the alkanity of water. Highly alkaline water is not potable. Alkanality is corrected by chemical coagulation, softening, corrosion control process etc. Presence of heavy metals in drinking water like Hg, As, Cd, Strontium etc. is also found because these chemical sink in water. But the chemical constituents may vary according to geomorphologic condition of the area and the qualitative and quantitative variation in the use of chemicals. Fluoride also an is important chemical in water. Fluoride contents in water are determined by the geo-morphological nature of the area, which may have rocks containing fluoride to a large extent (Gupta *et al.*, 1993). Its permissible limit is 1.5 mg/l. Excess fluoride causes flurosis. Concentration of nitrates greater than 50 mg/lit shows contamination of drinking water. Nitrate possess various hazards to health. Vijaya Kumar *et al.* (1993) found to cause methemglobimemia in infant leading to a disease in the oxygen carrying capacity of blood and causes blue baby disease. Some homogenates hydrocarbons like dichlomethane, tribromo methane, hexachlorocyclo hexane etc. present in drinking water. They may not pose immediate effect (Pae Khan, 1995) but long term usage they are causing hepatotoxic and neurotoxic.

Microbiological Quality

Microbial quality of water is also important because it is an indicator of health determinant. The presence of coliform organism in water is an indication of the preserve of pathogens microbes. For each physical chemical and microbiological constituents WHO, ICMR, ISI, USPHS and CPHEEO have framed guideline value which denotes the permissible amount.

Microbial contamination is the most important health determinant in the supply area. This occurs due to lack of sanitation, improper waste disposal, faulty construction of well and lack of source protection (WHO, 1993). Water borne diseases caused by bacteria, virus etc and other organisms are transmitted by contaminated water. Contaminated water causes epidemics by cholera, dysentery, typhoid etc.

Deterioration of Water Quality

Environmentally, quality of water is an most important aspects of drinking water. In India drinking water gets deteriorated due to many reasons, stating from a rapid growth of population, industrialization, urbanization and excess use of chemical etc. In rural areas improper management and maintenance of hand pumps sites and open air defecation cause havoc on water quality. There is also an unresolved controversy over the quality of surface vis-à-vis groundwater. One version is that groundwater is safer than the surface water, as the latter is more susceptible for contamination. The counter argument is the irreversibility of groundwater contamination. Since once polluted its difficult to restore its quality. There are five major factors causing deterioration of water quality. They include geological formation, release of industrial effluents, sea water intrusion, improper use of chemical fertilizer and pesticides and excess of fluoride. Electrical conductivity (EC) is used to determine the chemical quality of water. There is a positive correlation between EC value and the chloride concentration. Contamination water is no doubt, a serious health hazards. Natural water contain inorganic chemicals like Fe, Ca, Mg, Mn, Si, F, NO_3, PO_4, SO_4, Cl etc. When the quantity of these chemicals are excluded permissible level than they affect the body system: For instance arsenic may cause cancer and cadmium affect kidney. Similarly fluoride damages teeth and causes fluorosis. Nitrates in soil is enhanced due to decaying organic matter, use of nitrogen, rich fertilizer and the water discharge from animals and poultry that ultimately contaminate the drinking water.

Water Quality and Health

In India, it is fact that there is a large area of public ignorance about health is affected by consuming unsafe drinking water. In the absence of factual knowledge, misconceptions and vague guesses are used to explain the relationship between quality of drinking water and health. Therefore, awareness among the people is very much needed in the present context.

Table: Water Related Diseases in India (in millions of DALYs)

Diseases	*Female*	*Male*
Diarrhoeal diseases	14.39	13.64
Intestinal helminthes	1.00	1.06
Trachoma	0.07	0.04
Hepatitis	0.17	0.14
Total water related diseases	15.73	14.88

Source: World Development Report, 1993, pp. 216-219.

The major sources and the routes of transmission of the water related infection are:

1. Water borne bacterial and viral disease like cholera, gastro-enteritis, diarrhoea, dysentery, vital hepatitis, etc. are borne by contaminated water , food or dirty hand

2. Water washed parasitic diseases like, scabies, lice, trachoma, conjunctivities etc. are caused due to inadequate water for washing and lack of personal hygiene in bathing and laundering and warm infections like and chistosomiassr and guinea warm infection that depend on aquatic intermediate hosts to complete their life cycle.
3. Diseases through insect vector route like malaria, yellow fever, dengue, encephalitis etc. that are spread by insects that breed in water.

Some water borne diseases causes by pathogenic organisms are listed below:

Pathogenic Organism	*Disease*
Salmonella typhi	Typhoid
Salmonella paratyphi	Paratyphoid
Salmonella sp.	Food poisoning
Vibro cholerae	Cholera
Shigella sp.	Dysentry
Entamoeba histolytica	Dysentry
Schistosoma sp.	Bilharzia
Taenia sp.	Tapeworm
Leptospira sp.	Weil's disease
Enteric virus	
i) Polio virus	Poliomyelitis
ii) Coxsackie virus	Myocarditis, Bornholm disease
iii) Echo virus	Diarrhoea, respiratory infections

Although it is a formidable task to give an exhaustive account of deterious effect of water related diseases, a few instances can sutfice to understand the magnitude of the problem in our country. According to a study conducted by the Centre for Science and Environment, the drinking water in Delhi was found to be carrying industrial toxins and pesticides residue causes gastro intestinal disorder besides cancer. Nitrate contamination of drinking water in excess level has causes respiratory infections among 40-80 per cent of the children in Rajasthan village as reported by Down to Earth (December 31, 2000, p. 48). According to survey conducted by MORD to assess the health problem caused by drinking water, about 1,42,336 habitation were found affected. Their results are summarized below.

Purification Method

For whatever purpose it is to be used, water quality is important, but of paramount concern is the provision of a safe water supply. The provision of a safe water supply involves processing a raw material of variable quality to a final product, which must be the highest quality. Therefore, purification of water involves a number of operations to improve the physical, chemical and microbiological properties of a source to an acceptable level of hygiene. Water treatment consists of following methods (WHO, 1993):

Number of Rural Habitations Affected in Various States (Due to Impaired Drinking Water Supply)

State	*Fluoride*	*Iron*	*Salinity*	*Arsenic*
Andhra Pradesh	4858	441	3977	–
Bihar	12	669	–	–
Gujarat	2513	–	1072	–
Haryana	738	450	106	–
Himachal Pradesh	860	274	769	–
Karnataka	237	502	26	–
Kerala	201	2215	87	–
Madhya Pradesh	–	157	–	–
Manipur	–	52	–	–
Mizoram	1138	42835	304	–
Orissa	1113	4861	5349	–
Punjab	14643	240	39623	–
Rajasthan	527	604	NA	–
Tamil Nadu	1072	3720	4426	–
Uttar Pradesh	NA	NA	NA	–
West Bengal	–	–	–	427

Source: Report of survey on the status of drinking water supply in rural habitation, Rajiv Gandhi Drinking Water Mission, Ministry of Rural Development, India (1991-1994).

Chemical Contaminants of Water

Chemical	*Diseases*	*Maximum Permissible Limit (mg/l)*
Aluminium	Alzheimer diseases	0.2
Antimony	Carcinogenic	0.005
Arsenic	Carcinogenic	0.17
Barium	Cardio vascular	0.7
Boron	Gastro intestinal disorder	0.3
Cadmium	Nephrotoxic	0.003
Chromium	Carcinogenic	0.5
Copper	Liver cirrhosis	2.00
Cyanide	Thyro toxic, neurotoxic	0.07
Fluoride	Flurosis	1.5
Lead	Neuro toxic, Metabolic disease	0.01
Manganese	Neuro toxic	0.5
Mercury	Nephro toxic, neuro toxic	0.001
Nickel	Carcinogenic	0.02
Nitrate	Increase risk of cancer	50.00
Nitrite	Metheaemoglobiraemia	3.00
Selenium	Deposit in hair and nails	0.01

Source: World Development Report, 1993.

1. Reservoir storage or pre-treatment
2. Coagulation, flocculation and sedimentation
3. Filtration
4. Disinfection

Pre-treatment involves impounding surface waters in the lakes or reservoirs where the lethal effect of ultraviolet rays from sun reduces the microbial content of the water to a great extent. When water is not stored, pre-disinfections with chlorine are carried out. Coagulation by the addition of chemicals like aluminium or iron salts, facilitates agglomeration of suspended impurities. The formed floccules precipitate at the bottom, which can be removed at regular intervals. Flocculation, co-precipitate mineral particles protozoan, bacteria, viruses and improved the colour and reduces turbidity of water.

Rapid filter or slow sand filtration does filtration. Efficiency of rapid filters is variable since frequent backwashing is needed. On the other hand slow sand filters (sand velocity 0.03 mm/s approximately) are effective and simple to operate but demand an extensive land area. A properly filter system can remove 98-99.5 per cent of the pathogens.

Terminal disinfections are of paramount importance in drinking water supply. For this chloride chemicals or treatment with ozone and ultra violet radiation may be resorted. Under normal conditions, free residual chlorine of 0.5 mg/litre with at least 30 minutes contact can reduce 99 per cent of pathogenic bacteria, viruses, parasites, etc.

The final and important stage in purification of water is the point of use treatment followed in individual household by filtration, boiling etc. Now a day, two types of water purification systems are adopted: (1) On line (2) Offline. The offline system is cheaper, required no electricity continuous water supply and easy to carry out. However the safest and simplest way is to boil water for everyday use at 90°C for a minimum of 20 minute to make it safe to drink. Use of candle filters and ion exchange filters like zero B iodine are not 100 per cent fit. The online system is more efficient and employs the UV or ozone treatment that needs electricity for operation with continuous water supply.

This multiple barrier concept of water treatment it can be adopted to both urban and rural water supply. Alternate or any additional process can be interposed in the sequence of this treatment to meet the local needs. Apart from the above four stages of purification, periodic check-up of the pipe lines for leakage, taps and storage tank, corrosion in the pipe, and the utensil used in household to collect and store water also play a significant role in maintaining the quality of water. Education of the consumers to make them aware of the quality of water and modes of transmission of water borne disease is of paramount importance to enhance the water quality.

Government Programme

Government introduced the Accelerated Rural Water Supply Prograrne (ARWSP) in 1972-73 to assist the states and union territories to accelerate the pace of drinking water supply and to ensure maximum inflow of scientific and technical input for improved performance, cost effectiveness and safety of drinking water (GOI, 2000). In 1986 a National Drinking Water Mission (NDWM) was launched and renamed in 1991 as Rajiv Gandhi National Drinking Water Mission (RGNDWM) are also progress in various states. In addition to that, there are various supportive activities or programmes are also launched to eradicate guinea worm, control of brackishness, removal excess of iron etc. For water quality testing, testing laboratories have also been established in various states where different testing facilities are available. All this indicate a growing need to provide sustainable drinking water with great importance to its quality.

Environmentally, quality of water is an most important aspects of drinking water particularly in rural areas where improper management and maintenance of land purposes sites and open air defecation cause havoc on water quality. Contamination water is no doubt, a serious health hazards. Natural water contain mega chemical like Fe, Ca, Mg, Mn, Si, F, NO_3, PO_4, SO_4, Cl etc. When the quantities of these chemicals are exceeded permissible level than they affect the body system: For instance arsenic may cause cancer and cadmium affect kidney. Similarly fluoride damages teeth and causes flurosis. Concentration of nitrates greater than 50 mg/litre in drinking water possess various hazards to health. Vijaya Kumar *et al.* (1993) found to cause methemglobimemia in infants leading to a disease in the oxygen carrying capacity of blood and causes blue baby disease.

Conclusion

Water is a life sustaining prerequisite to all living organism. Water that is absolutely pure is never found in nature because ultimately all the source of drinking water is rain water. The rain water collects dust, bacteria and gases while travelling from the clouds to the earth through the atmosphere. The water issue is by no means an issue only for experts. It constitutes nothing less than a major question of human survival and is closely related to both eradication of poverty and the need to secure health as well as adequate food supports (Falkinmark Malin, 2000). Therefore, knowledge about the ecological aspects of drinking water should reach to the people at grass root level since it is they who suffer most due to water quality problems. Provision of good quality drinking water is an important as providing the required quantity of water for drinking, cooking, bathing and ablution. Therefore, every effort should be made to attain water quality as high as possible.

References

Down to Earth (1999). Disasters in the making, August 31.

Falkenmark, Malin (2000). Water for people, food and ecosystems. Opening session remarks. The jubilee Stockholm Water Symposium, Water Security for the 21st Century: Innovative approaches. *Stockholm Water front: Forum for Global Water Issues.* No. 3, October.

Government of India (2000). Rural water supply program, Annual report, 1999-2000, Ministry of Rural Development.

Gupta, S.C., Rathore, G.S. and C.S. Doshi (1993). Flouride Distribution in groundwaters of southeastern Rajasthan. *Indian Journal of Environment and Health,* 35(2): 97-109.

Packham, R.F. (1995). Water Quality and Health Water Research Centre, Med menham Laboratory, Marlow Bucks, Sl. 7 2HD, U.K.

Vijaykumar, V., Sai, C.S.T., Swami, M.S.R. and Rao, P.L.K.M. (1993). Nitrates in groundwater sources in medchal, Andhra Pradesh. *Indian Journal of Environment and Health,* 35(1): 40-46.

WHO (1993). *Guidelines for Drinking Water Quality.*

31

Pollution Impact on the Hydro-biology of River Nakatia at Bareilly

NEELIMA GUPTA & V.K. VERMA

Department of Animal Science, M.J.P. Rohilkhand University, Bareilly, UP

D.K. GUPTA

Department of Zoology, Bareilly College, Bareilly, UP

ABSTRACT

River Nakatia was chosen as a site for monitoring hydrobiological parameters and it was found to provide an excellent model for observing pollution impact on aquatic biota. Water samples from 3 sites: Site A (Non-polluted–Behind, M.J.P. Rohilkhand University, Bareilly); Site B_1 (Polluted–Bareilly Shahjahanpur Nakatia bridge, Bareilly) and Site B_2 (Polluted–Village Bahuapur, Bareilly) were assayed for hydro-biological parameters: temperature, pH, turbidity, DO, BOD chlorides, carbonates and oil and grease. The water temperature, chloride, carbonate, oil and grease and BOD increased to varying levels whereas pH, turbidity and DO decreased at the polluted sites (Sites B_1 and B_2. At Site B_2, city sewage and cantonment sewage merges whereas at Site B_2, some more sewage, effluents and slaughter house wastes are added. The studies indicate that pollution caused by the above mentioned factors directly affect the water body at Nakatia and is also directly proportional to the pollution load of water.

Keywords: Nakatia, Temperature, pH, Turbidity, DO, BOD, Chloride, Carbonate, Oil and Grease.

Introduction

Nakatia river flows in the northern part of district Bareilly and approximately ten drainage systems fall in this tributary. This river covers 8 kms distance in Bareilly district and merges in river Ramganga at village Raipurhans.

The rapid industrialization, disposal of domestic wastes, human and animal excreta and city debris and sewage without any pre-treatment into this water body not only changes the physico-chemical parameters of water but also affect the aquatic fauna dwelling therein.

Hydr-obiological monitoring at river Ganga (Pandey, 1985), Cauvery (Somashekhar, 1985), Alaknanda (Tiwari *et al.*, 1991), Godawari (Somashekhar *et al.*, 1993), Deorania (Mazhar and Kapoor, 1992), Ramganga (Chandra *et al.*, 1996) and Shankha (Habib, 1996) have been conducted during the past laying stress on physico-chemical analyses.

The present study has been conducted with an objective to assess the physical (temperature, pH and turbidity) and chemical (DO, BOD, chloride, carbonate and oil and grease) parameters of river Nakatia indicating some interesting results.

Materials and Methods

Sampling Stations

The sampling stations were selected along the water course of river Nakatia (Fig 31.1).

Site A (Unpolluted)

Village Doralalpur, about 2 kms southwest to M.J.P. Rohilkhand University campus.

Site B_1 (Polluted)

Bareilly Shahjahanpur bridge, Nakatia village, Bareilly. It contains polluted water of dhobighat, cantt. sewage and city sewage.

Site B_2 (Polluted)

Faridpur Bukhara road, Village Bahuapur. At this site, in addition to pollutants of B_1, the slaughter house wastes are also discharged.

Procedure

The water samples were collected from three different sites of river Nakatia in company made clean polythene containers. Hydro-biological analyses were conducted according to different specialized procedures. Temperature was measured on the site by a thermometer at a depth of approximately 20 cm from the surface of water, pH by electrometric pH-meter and turbidity was estimated by Nephlometer. Chlorides and carbonates were assayed by simple titration and sulphate by spectrophotometer (APHA, 1995). For the estimation of DO and BOD, the samples were collected in 300 ml BOD bottles at the sites and were estimated by Winkler's method according to Trivedi and Goel (1986), oil and grease were estimated according to Saxena (1994).

Results and Discussion

Physical Parameters

Temperature

The temperature increased at both the pollution sites B_1 and B_2. The air temperature increased from 31°C ± 0.27 (Site A) to 32°C ± 0.28 (Site B_2), but the water temperature increased to a greater extent from 26.5°C ± 0.32 (Site A) to 28.2°C ± 0.22 (Site B_2). Temperature is one of the most important ecological factor, which plays a significant role in the existence of biota. It is well known that organisms have evolved various methods to meet the extremes of temperature during the course of evolution.

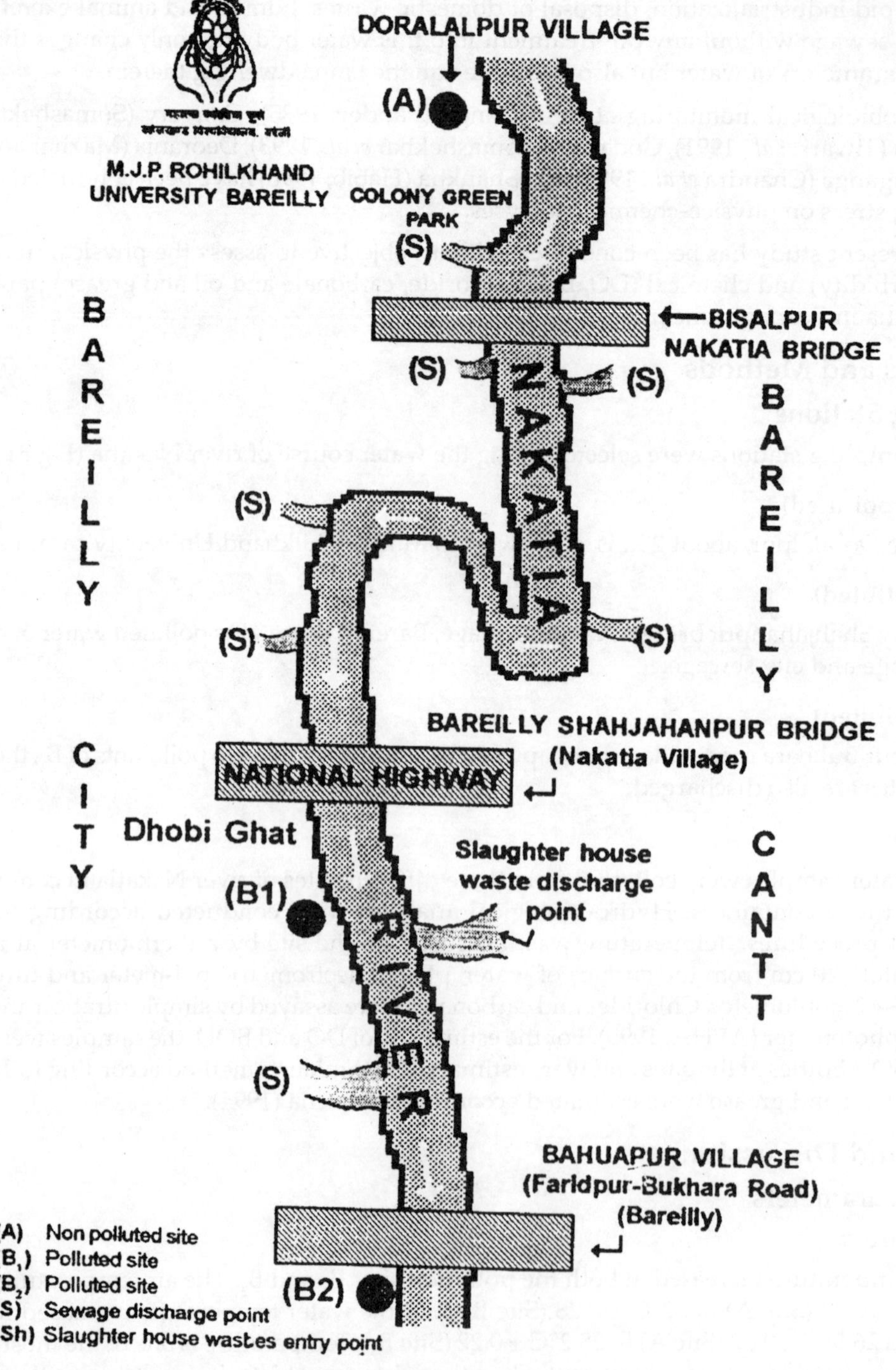

Fig. 31.1: Sampling Stations Along the Water Course of River Nakatia

Temperature is one of the chief factor influencing other biotic and abiotic parameters of aquatic the solubility of oxygen and free carbon dioxide. A rise in temperature of the waters leads to the acceleration of chemical reaction in water, reduces the solubility of gases and amplifies the taste and odour.

The present studies indicate that the temperature of water body is affected by effluents. At both the pollution sites, an increase in water temperature was observed and moreover, at Site B_2 which contained additional slaughter house wastes, the temperature was the highest. Studies on water temperature have been conducted by various authors. Ruttner (1953) stated that temperature affects the quantity of oxygen level. According to Lakshminarayana (1965), the temperature variation of river Ganga are controlled through changes in density of water. Variation in water temperature during different seasons has also been recorded by Bohra (1976) in Padamsagar and Ranisagar; Mishra *et al.* (1978) in Balsamand and Vyas and Nama (1991) in Jodhpur. Kapoor (1993) observed that the water temperature varied from 16°C to 32°C in a comparative study of four rivers, Alakhnath, Deorania, Nakatia and Ramganga. Similar observations were also recorded by Kumar (2002).

Rawat and Jakher (2002a) observed a correlation between variation in DO, free CO_2 and pH in Takhat Sagar lake at Jodhpur. However, there is dearth of literature on impact of effluents and wastes on the temperature of different water bodies. The present work thus indicates the correlation between the effluents, sewage and slaughter house wastes on temperature of river Nakatia.

pH

pH is an index of acidity or alkalinity and measures the concentration of hydrogen ions presents in water. Determination of pH is one of the important objective for the treatment of wastes. Significant changes in pH occur due to the disposal of industrial wastes and acid mine drainage (Trivedi and Goel, 1986). During the present studies, similar observations were recorded. In normal unpolluted water (Site A) the pH was slightly alkaline (7.50 ± 0.43) which increased in alkalinity to 7.89 ± 0.26 (Site B_1) and 8.52 ± 0.26 (Site B_2). Apparently, the waste discharge into the river are responsible for inducing alkalinity in the water.

Blum (1956) observed that the majority of flowing waters are neutral to alkaline in nature. Attempts to study the seasonal variation in pH have been made by various authors. Adoni (1975) observed changes in pH during different seasons. Rawat and Jakher (2002b) studied the influences of seasonal temperature on the pH values. The present studies are indicative of the fact that the wastes released in river Nakatia tend to make the water more alkaline.

Turbidity

Turbidity of water is correlated with the penetration of light and is caused by the substances which are not present in the form of true solution. It is the actual expression of optical property (Tyndall effect) in which the light is scattered by particles present in water. Clay, silt, organic matter, phytoplankton and other microscopic organisms are responsible for causing turbidity in natural waters.

Turbid water is unfit for domestic purposes, food and beverage industries and many other industrial uses. Turbidity in natural water restricts light penetration for photosynthesis. Bhatt and Negi (1985), Joshi and Bisht (1993) and Habib (1996) observed changes in turbidity during different months. Sinha *et al.* (1986) reported higher values of turbidity after confluence of industrial wastes and bathing site of river Ganga, Dalamau. In our studies, also the water was more turbid in the polluted site (Site B_2=22.00 ± 0.97 NTU) and it was less turbid at Site A (11.57 ± 0.52 NTU). The observations are thus in agreement with Sinha *et al.* (1986) indicating higher turbidity in polluted waters.

Chemical Parameters

Dissolved Oxygen

Dissolved oxygen (DO) content has been identified as one of the important parameter for the assessment of water quality and reflects the physical and biological conditions which are prevalent in the water. It is determined by the oxygen balance of the system and is essential for maintaining higher forms of biota in the water. Good quality non-polluted surface waters are saturated with DO but polluted water tend to lower the DO content which can even kill fish and other organisms dwelling in the water.

The present study reveals that the DO concentration rapidly declines in polluted water. The normal value of 9.54 ± 0.33 mg/l (Site A) reduced to 4.39 ± 0.61 mg/l at Site B_1 and further depleted to 2.91 ± 0.44 mg/l at site B_2. The result indicates that the percentage fall at site B_1 was 53.98 per cent whereas that at B_2 was 69.49 per cent. Venkateswarlu (1970) stated that low values of DO are generally associated with high organic matter. The DO content has also been estimated by various other workers. Raju and Padmawathi (1993) studied the pollution level of the river Krishna in a stretch of 90 km at four sampling stations. Haniffa *et al.* (1994) investigated the DO concentration in the channel of river Tambaraparani. Rawat and Jakher (2002b) investigated limnological status of few water bodies at Jodhpur and observed the highest DO concentration of 6.85 ± 1.89 mg/l at Gulabsagar. However, the impact of pollution sewage and waste disposal on water quality has been investigated to a lesser extent. Siddiqi *et al.* (1994) observed changes in the biochemical parameters of Satluj river laying special emphasis on the effect of domestic water quality on the river water. The municipal wastes of Muzaffarnagar city, composite industrial wastes of industrial complex, Mansurpur Sugar mill and distillery wastes were characterised physico-chemically in order to assess the impact of the effluent on water quality of river Kali (Jain *et al.*, 1997). The present investigation also indicates variation in DO concentration at the different sampling stations and was found to decline in the water which contained the effluents. The findings are in concurrence to Habib (1996).

BOD

Measurement of BOD is crucial for the assessment of organic pollution load on a natural water course and is the amount of oxygen utilised by micro-organisms in stabilizing the organic matter. BOD values approximate the amount of oxidisable organic matter present in solution and can be used as a measure of waste strength. During the present course of studies, the BOD values increased significantly at the population sites. The value of 2.43 ± 0.14 mg/l at site A increased to 3.95 ± 0.37 mg/l at site B_1 with a further enhancement at site B_2 (5.23 ± 0.34 mg/l).

A perusal of literature indicates that the normal value of BOD also changes during different seasons as have been recorded by various workers. Joshi and Bisht (1993) observed the BOD concentration in western Ganga canal. Patel and Patel (1993) assessed the water quality of river Sankha in western Orissa. Sharma and Habib (1994) made investigations on the effluent channel of the industry, Synthetics and Chemicals Ltd., Bareilly and noted highest BOD values at the confluence point of the effluent channel with river Shankha during February. The impact of effluents and polluted water on BOD values have attracted attention of a few workers. Notable contributions are those of Habib (1993), Ruparelia *et al.* (1993) and Jain *et al.* (1997) who observed the role of wastes, effluents and sewage on the BOD of different waters. It has been observed that the BOD concentration rises in polluted water. Ruparelia *et al.* (1993) postulated that increases in BOD values were due to the influx of the washing ghats. It was further opined by some authors that the rise at the sampling stations could be due to the entry of effluents from the dying, printing industries as well as domestic sewage of the town. The present study also infers that the BOD values at sites B_1 and B_2 which receive the wastes show a rise in the BOD values and can be used as an indicator for investigating the pollution load of water.

Chloride

Chloride practically occurs in all waters and wastewaters and is usually a major inorganic constituent. Discharge of agricultural, industrial and domestic wastewater can contribute to the presence of chloride in natural waters. Chloride concentration is found to serve as an indicator of pollution by sewage as the most important source of chloride in water is the discharge of domestic sewage. During the present course of study, it was found that the chloride contents increased by 65.03 per cent in polluted water and the findings are indicative of the fact that the discharge of domestic sewage and pollutants increase the chloride concentration. Thus the findings can be used as an indicator of pollution. Kofoid (1903) pointed out the significance of chlorides and stated that its principle source is animal matter, sewage or drainage from refusal of animal matter. Taylor (1949) suggested that chlorides indicate the presence of organic matter, presumably of animal origin and in most cases this may bring about an increase in the amount of ammonical nitrogen and organic matter. The importance of chloride as an important anion has in a lotic ecosystem been pointed out by Mitra (1982). Krishna Muthy and Bharati (1997) studied the concentration of chloride of river Kali in Karnataka at four different stations and observed low values of chloride during the winter months. Sahota and Gill (1997) noticed higher level of chloride in a study of pollution level of different rivers. The chloride concentration of water is thus directly correlated with its pollution level as reported by earlier workers and confirmed by the present studies.

Carbonates

Carbonate is one of the bases which contributes to the alkalinity of water. The alkalinity of water is its capacity to neutralise a strong acid and is characterised by the presence of hydroxyl (OH^-) ions capable of combining with hydrogen (H^+) ions. A close relationship between pH and carbonates is known to exist which is also observed in the present findings whereby the values of both the parameters rise in polluted water and is directly proportional to each other. The normal values of carbonates of 23.00 ± 0.62 mg/l at site A increased to 28.71 ± 1.07 mg/l at site B_1 and 42.57 ± 1.53 mg/l at site B_2 which was most polluted. Habib (1996) observed that waters rich in carbonate are more alkaline. Suvarna and Somashekhar (1997) observed minimum carbonate alkalinity in winter which was higher during summer. The present findings correlate the relationship between pH and carbonate at the different sampling stations.

Oil and Grease

Oil and grease enter the water bodies as a result of pollution. Being insoluble in water and low in density, it floats and forms a film on the surface which is disastrous for the surface dwellers. The source of oil and grease in the water bodies has been postulated to be industrial effluents, oil refineries and storage tanks, automatic waste oil and petrochemical plants. Khan *et al.* (1992) observed the oil and grease contents of rubber factory effluents and studied the effect of such waters on the growth of leguminous crops. The present studies indicate an increase in the oil and grease content from 20.71 ± 1.32 mg/l (site A) to 87.50 ± 5.14 mg/l (site B_1) (Table 31.1) and show 322.50 per cent increase in polluted water. Thus the high concentration of oil and grease at sites B_1 and B_2 may be correlated with the heavy pollution load of the sampled water.

Table 31.1: Physico-chemical Characteristics of River Nakatia

No.	*Investigated Parameters*	*Selection of Sites*		
		A (*Non-polluted*)	*B_1* (*Polluted*)	*B_2* (*Polluted*)
1.	Temperature (°C)			
	(i) Air –	31.00 ± 0.27	31.50 ± 0.35	32.00 ± 0.28
	(ii) Water –	26.50 ± 0.32	27.50 ± 0.29	28.20 ± 0.22
2.	pH	7.50 ± 0.04	7.89 ± 0.02	8.52 ± 0.02
3.	Turbidity (NTU)	11.57 ± 0.52	14.80 ± 0.95	22.00 ± 0.97
4.	Dissolved Oxygen (mg/l)	9.54 ± 0.33	4.39 ± 0.61	2.91 ± 0.44
5.	BOD (mg/l)	2.43 ± 0.14	3.95 ± 0.37	5.23 ± 0.34
6.	Chloride (mg/l)	26.08 ± 0.85	33.35 ± 0.54	43.04 ± 2.20
7.	Carbonate (mg/l)	23.00 ± 0.62	28.71 ± 1.07	42.57 ± l.53
8.	Oil and Grease (mg/l)	20.71 ± 1.32	65.16 ± 4.72	87.50 ± 5.14

Acknowledgements

The authors express their thanks to the Ministry of Environment and Forest for granting a major research project entitled *"Monitoring of hydro-biological parameters in polluted water and their impact on the haematology, haematopathogenesis and growth index of catfishes"* vide their sanction order No.19-1/99-RE.

References

Adoni, A.D. (1975). *Studies on Microbiology of Sagar Lake*. Ph.D. thesis, University of Sagar, Sagar.

APHA, AWWA and WPCF (1995). *Standard Methods for the Examination of Water and Wastewater*, 19th edn. American Public Health Association Inc., NewYork, U.S.A., pp. 10-157.

Bhatt, S.D. and Negi, U. (1985). Hydrology and phytoplankton population in river Kosi of western Himalaya (U.P.). *Indian J. Ecol.*, 122(1): 141-146.

Blum, J.L. (1956). The ecology of river algae. *Bct. Rev.*, 22: 292.

Bohra, O.P. (1976). *Some Aspects of Limnology of Padam Sagar and Ranisagar, Jodhpur*. Ph.D. thesis, University of Jodhpur, Jodhpur.

Chandra, R., Bahadur, Y. and Sharma, B.K. (1996). Monitoring the quality of river Ramganga waters at Bareilly. *Poll. Res.*, 15(1): 31-33.

Habib, I. (1993). Hydro-biological studies on river Shankha at Bareilly with references to algal indicators of water pollution. *Nat. Environ.*, 10(11): 55-57.

Habib, I. (1996). Hydro-biological studies on river Shankha at Bareilly. *Recent. Adv. Biotch.*, 162-186.

Haniffa, M.A., Martin, P. and Jeevaraj, J. (1994). Hydro-biological studies on the channels of river Tambaraparani for the assessment of water quality. *Indian J. Environ. Poll.*, 14(11): 821-828.

Jain, C.K., Bhatia, K.K.S. and Seth, S.M. (1997). Characterization of waste disposals and their impact on the water quality of river Kali. *Indian J. Environ. Poll.*, 17(4): 287-295.

Joshi, B.D. and Bisht, R.C.S. (1993). Some aspects of physico-chemical characterstics of western Ganga canal near Jwalapur at Hardwar. *Him. J. Environ. Zool.*, 7: 76-82.

Kapoor, C.P. (1993). Physico-chemical and biological study of four rivers at Bareilly (U.P.). *Poll. Res.*, 12(4): 267-270.

Khan, M.G., Gahlot, A.K.S. and Varshney, K.A. (1992). Effect of rubber factory effluents on seed germination and early seedling growth of some leguminous crops. *Bangladesh. J.*, 21(2): 261-263.

Kofoid, C.A. (1903). The plankton of the Illinois river 1894-1899 with introductory notes upon the hydrography of the Illinois river and its basin, Part I: Quantitative investigations and general results. *Bull. III. Lab. Nat. Hist.*, 6: 95-629.

Krishna Murthy, S.R. and Bharati, S.G. (1997). A study on concentration of chloride of the river Kali near Dandeli, Karnataka (India). *J. Environ. Poll.*, 4(1): 9-15.

Kumar, A. (2002). Ecology of wetlands in Jharkhand, India. In: *Ecology and Conservation of Lakes, Reservoirs and Rivers* (Ed. A. Kumar). Ashish Publ. House, New Delhi, pp. 1-32.

Lakshminarayana, J.S.S. (1965). Studies in the phytoplankton of river Ganges, Varanasi, India, Pt. I: Physico-chemical characterstics of river Ganges, Pt. II: The seasonal growth and succession of the plankton algae in the river Ganges. *Hydrobiologia*, 25: 119.

Mazhar, M.D. and Kapoor, C.P. (1992). Limnological studies of Deorania river at Bareilly (U.P.). *J. Freshwater Biol.*, 4(2): 155-158.

Mishra, S.D., Bhargava, S.C., Jakher, G.R. and Dey, T. (1978). Hydrobiology and productivity of some freshwater reservoirs and lakes of semi-arid zone, Jodhpur (Rajasthan). UGC, Project Report, University of Jodhpur (India).

Mitra, A.K. (1982). Chemical characterstics of surface water at selected Gauging stations in the Godavari, Krishna and Tungabhadra. *Indian J. Environ. Hlth.*, 24(2): 165-179.

Pandey, N.C. (1985). Pollution of river Ganga in U.P. with specific reference to Varanasi. *Civic Affairs*, 32(11): 52-59.

Patel, M.K. and Patel, T.K. (1993). Assessment of water quality in the rivers of western Orissa, Part 1: River-Sankh. *Indian J. Environ. Poll.*, 13(12): 909-916.

Raju, M.V.S. and Padmavathi, V. (1993). Pollution studies of the river Krishna near Vijayawada. *Indian J. Environ. Poll.*, 13(6): 431-435.

Rawat, M. and Jakher, G.R. (2002a). Influence of seasonal temperature variations on the level of dissolved oxygen, free carbon dioxide and pH in Takhat Sagar lake, Jodhpur (Rajasthan). *J. Curr. Sci.*, 2(2): 169-172.

Rawat, M. and Jakher, G.R. (2002b). Limno-microbiological studies of few water reservoirs of Jodhpur, Rajasthan. *J. Environ. Ecoplan.*, 6(2): 355-358.

Ruparelia, S.G., Verma, V. and Hargan, M.C. (1993). A short-term study on the pollution status of river Bhadar with special reference to BOD and COD. *Indian J. Environ. Poll.*, (10): 742-744.

Ruttner, F. (1953). *Fundamentals of Limnology*. University of Toronto Press.

Sahota, H.S. and Gill, S.K. (1997). Effect of Mukerian paper mill effluent on Beas river water. *Indian J. Environ. Poll.*, 17(10): 749-751.

Saxena, M.M. (1994). *Environmental Analysis of Water, Soil and Air*. Agro-Botanical Publishers (India), pp. 184.

Sharma, B.K. and Habib, I. (1994). Ecology of effluent channel flowing from Synthetics and Chemicals Ltd. Bareilly. *Environ. Appld. Biol.*, 179-202.

Siddiqi, Z.M., Panesar, R.S. and Rani, S. (1994). Biochemical effect of few sewerage disposals on the water quality of Sutlej river. *Indian J. Environ. Poll.*, 14(10): 740- 743.

Sinha, A.K., Shrivastava, D.K., Pandey, D.P., Singh, R.P., Srivastava, S., Pandey, D.C., Santhibhushan, K., Shrivastava, K.N., Kumar, A. and Tayal, S. (1986). Water quality assessment of Ganga water at Dalmau (Rae Bareli) India on Kartika Purnima, 1985. *Bull. Environ. Sci.*, 3: 3-6.

Somashekhar, R.K. (1985). Studies on water pollution of river Cauveri physico-chemical characteristics. *Int. J. Environ. Studies*, 23: 115-124.

Somashekhar, R.K., Padmarthy, D. and Babu, R. (1993). Monitoring the quality of Godavari waters during and after the 1991 Pushkaram at Rajahmundry. *Poll. Res.*, 12(13): 191-195.

Suvarna, A.C. and Somashekhar, R.K. (1997). Ecological study on the riverine ecosystem of Karnataka. I: Physico-chemical characterstics of river Cauvery. *J. Environ. Poll.*, 4 (1): 57-63.

Taylor, F.V. (1949). *The Examination of Water and Water Supplies* (Thresh, Beas and Sucking). 6th Edn. J. and Churchill. Ltd., London, pp. 819.

Tiwari, N.C., Sagar, G., Tiwari, D. and Singh, H.R. (1991). Monitoring the water pollution in snow feed rivet, Alaknanda Rudraprayag at Chamoli. *Environ. Ecol.*, 9(1): 202-266.

Trivedi, R.K. and Goel, P.K. (1986). *Chemical and Biological Methods for Water Pollution Studies*. Environmental Publications, Karad, (India), pp. 217.

Venkateswarlu, V. (1970). An ecological study of the algae of the river Moosi, Hyderabad (India) with special reference to water pollution. *Hydrobiologia*, 35: 45.

Vyas, N. and Nama, H.S. (1991). Pollution ecology of freshwater reservoir at Jodhpur with special references to micro-organisms. *Geobios*, 18: 33-37.

32

Status of Freshwater in 21st Century: A Review

ANIL KUMAR*, SEEMA TRIPATHI, P. GHOSH

**Asstt. Professor, Deptt. of Zoology Govt. V.Y.T.P.G.A College (Science College) Durg, (C.G.), Bhilai, M.P.*

ABSTRACT

In this paper, an attempt has been made to study the present status of freshwater, its sources, man's water requirement, the future scenario, the causes of wastage of freshwater resources and its conservation in India as well as foreign country. Freshwater is a natural resource of fundamental importance. Due to wide applicability of water, it becomes necessary to study the global distribution of freshwater and the nature of water. Now a days, water is contaminated by a variety of ways. For convenience the sources of contamination of natural waters are classified as natural, agricultural, mining, municipal and industrial and their impacts on global environments will be very serious in future.

Introduction

The history of human civilization reveals that water supply and civilization are almost synonymous. Several cities and civilization have disappeared due to water shortage originating from climatic changes. Millions of people all over the world, particularly in the developing countries, are losing their lives every year from water-borne diseases. An understanding of water chemistry is the basic of knowledge of the multi-dimensional aspects of aquatic environmental chemistry which involve the sources, composition, reactions and transport of water.

Water Quality Parameters and Standards

The parameters for freshwater quality characterisation are listed in Table 32.1. The permissible limits, as laid by the United States Public Health Drinking Water Standard (USPH) and Indian Standard Institution (ISI) are listed for comparison. These refer to domestic water supplies for drinking water (De, 1996).

Table 32.1: Parameters for Water Quality Characterisation and Standard (Domestic Water Supplies)

Parameters	*USPH Standard*	*ISI Standard (IS: 2296-1963)*
Colour,	Colourless	–
odour,	Odourless,	
taste	Tasteless	
Inorganic Chemicals		
pH	6.0–8.5	6.0–9.0
Specific conductance	300 mmho ch^{-1}	–
Dissolved Oxygen (DO)	4.0–6.0 (ppm)	3.0
Total dissolved solids	500	–
Suspended solid	5.0	–
Chloride	250	600
Sulphate	250	1000
Cynide	0.05	0.01
Nitrate + nitrite	<10	–
Fluoride	1.5	3.0
Phosphate	0.1	–
Sulphide	0.1 mg L^{-1} (ppb)	–
Ammonia	0.5	–
Boron	1.0	–
Calcium	100	–
Magnesium	30	–
Arsenic	0.05	0.2
Barium	1.0	–
Cadmium	0.01	–
Chromium (VI)	0.05	0.05
Copper	1.0	–
Iron (filterable)	<0.3	--
Lead	<0.05	–
Manganese (filterable)	<0.05	0.01
Mercury	0.001	–
Selenium	0.01	0.05
Silver	0.05	–
Uranium	5.0	–
Zinc	5.5	–

Contd....

Table 32.1–Contd...

Parameters	*USPH Standard*	*ISI Standard (IS: 2296-1963)*
Organics		
COD	4.0	–
Carbon $CHCL_2$ extract (CCE)	0.15	–
Methylene blue active substances	0.5	–
Phenols	0.001	0.005
Pesticides (total)	0.005	–
Polycyclic aromatic Hydrocarbons (PAH)	0.2 pb (0.002 ppm)	
Surfactants	200	–
Radioactivity		
Gross beta	1000 pc/L	–
Radium – 226	3 pc/L	–
Strontium – 90	10 pc/l	–
Bacteriological Parameters		
Coliform cells/1000 ml.	100	<5000
Total bacteria count/100 ml	1×10^6	–

1. The official journal of the European Committees L 229, p.16-21 (1980) quotes figures close to those of USPH. The II values are considerably on the high side, which remain open to question and these do not cover all the parameters.
2. All the units except otherwise mentioned and pH, specific conductance and radioactivity are in ppm, *i.e.* parts per million or mg/l. The radioactivity units are in picacuries/l, *i.e.* 10^{-12} curies/l (2.2 disintegrations min^{-1} l^{-1}).

Global Distribution of Freshwater

We have a huge quantity of freshwater on our planet, almost 1500 million cubic kms; but most of it is useless for us as it contains too much of salt. Total amount of available freshwater on our planet is only about 84.4 million cubic kms and is given in Table 32.2. Any withdrawal may be tends to disturb the natural system and the resource base in future. (Vohra, 1990)

Table 32.2: Global Distribution of Freshwater

Sl.No.	*Sources*	*Water as Cubic kms*
1.	Water in snow-caps, Ice sheets, glaciers etc.	24,000,000
2.	Surface ponds, lakes and reservoirs	280,000
3.	Water in streams and rivers	1,200
4.	Water present as soil moisture	85,000
5.	Groundwater	60,000,000
	Total amount of freshwater on our planet	84,366,200

Sources of Freshwater for Terrestrial Life

An important source of freshwater for terrestrial life is precipitation. Solar radiations vaporise large quantities of water both from land and oceans. The moisture condenses in the atmosphere and forms rains, dew or snow which are brought back to earth's surface. Table 32.3 provides information about the quantities of freshwater which evaporates from earth's surface, the total amount of precipitation and annual surface run offs etc. It is concluded from the Table 32.3, that oceans contribute about 4,52,600 cubic kms of water annually to atmosphere but receive only 4,11,600 cubic kms of surface and sub-surface run off when they receives. On the other hand land surface contributes about 72,500 cubic kms of water to the atmosphere while it receives 1,13,500 cubic kms of water as precipitation. The excess amount is drained off as surface and groundwater run off to oceans. So the amount of water present in sea, on land surface, under groundwater, water present in atmosphere as vapours etc. are in a state of dynamic equilibrium. The excess water received by land surface, about 41,000 cubic kms has to flow back to sea–it can not be retained on earth's crust ordinarily.

Table 32.3: Annual Water Budget of Our Planet

	Water in Cubic kms
Evaporation from sea surface	452,600
Evaporation from land surface	72,500
Precipitation on sea surface	411,600
Precipitation on land surface	113,500
Surface and groundwater run off	41,000
Total evaporation from land sea surface	525,100
Total precipitation on land and sea surface	525,100

Freshwater Resources of India

India receives about 4,000 cubic kms of freshwater as precipitation every year. About 700 cubic kms of water thus received evaporate immediately and are lost to the atmosphere. About 2,150 cubic kms go the soil moisture while about 500 cubic kms are permeate through the soil surface to undergroundwater deposits. Only 1,150 cubic kms of freshwater received annually are retained on land surface. Water resources in our country can be classified as follows (Vohra, 1994):

Surface Water

The total surface flow per year is about 1,800 cubic kms which are distributed among a number of river basins as listed in Table 32.4.

Table 32.4: Distribution of Surface Flow in Rivers of India

Sl.No.	*Rivers*	*Total flow*	*Utilizable flow*	*1994–Use*
1.	Indus basin	77.0	46.0	37.0
2.	Ganga basin	510.0	250.0	85.0
3.	Brahamputra basin + barak	540.0	24.0	5.0
4.	Mahanadi + other east flowing rivers	123.0	90.0	28.0
5.	Godavari, Krishna + other east flowing rivers	225.0	190.0	73.0
6.	West flowing river south of Tapti	218.0	30.0	10.0
7.	Narmada and Tapti	62.0	49.0	7.0
8.	West flowing rivers north of Narmada	25.0	20.0	5.0
	Total Flow	1780.0	700.0	250.0

All quantities in cubic kms.

Groundwater

The total of about 670 cubic kms of freshwater enters the groundwater annually. It is upto this amount that we can withdraw freshwater from our sub-surface deposits. Any withdrawal above this limit shall be detrimental to the surface base.

The Future Scenario Requirement of Freshwater in India

The water is needed in almost every sphere of human activity. It is required for direct consumption or indirectly for washing, cleaning, cooling, transportation. Freshwater requirement in India for important sectors of human activity are summarised in Table 32.5. The approximate distribution of total annual freshwater resources in India is summarised in Table 32.4. From Table 32.5 we come to the conclusion that by 2000 A.D., the total water requirement is expected to double itself about 2025 A.D. we shall be requiring almost thrice as much water as we did in 1974 A.D. The extraordinary rise of demand for freshwater shall not be uniform in various sectors of our country. A glance at Table 32.4 reveals that the utilizable surface flow in India amounts to about 700 cubic kms only and our today's requirement has already exceeded beyond 700 cubic kms (Asthana *et al.*, 1997).

Table 32.5: Estimates of Freshwater Requirement in India

Sl.No.	*Water needed for*	*1974*	*2000*	*2025*
1.	Irrigation	350.0	630.0	770.0
2.	Thermal Power generation	11.0	60.0	160.0
3.	Industries	5.5	30.0	120.0
4.	Domestic requirement	8.8	26.6	39.0
5.	Live stock Management	4.7	7.4	11.0
	Total	380.0	754.0	1100.0

All quantities in cubic kms.

Sources of Water Pollution

The sources of contamination of natural waters are classified as natural, agricultural, mining, municipal and industrial (Pandey *et al.*, 1998).

Natural Pollution

This type of pollution is either accidental or occasional. Aerial contaminants like gases and dusts get transformed to a body of water in the form of rainfall, soil silts stripped from the land by heavy rainfall or melting snow or as soluble salts, leached from the earth by surface streams. Lactates from animal excreta, decaying bodies of organic matter as in swamps or deep ponds also introduces appreciable quantities of micro-organisms.

Agricultural Pollution

Use of fertilizers, insecticides, herbicides and killers in agricultural practices, poses serious pollution problems.

Mining Pollution

Fines or tailing from ore washing are disposed of in water suspensions which are ultimately transferred to the natural stream. Today one of our major unsolved problems is the acid drainage from coal mines.

Municipal Pollution

Municipal wastes, mainly domestic sewage includes the discharges from toilets, wash rooms, kitchens and similar areas in dwellings, institutions and commercial and industrial buildings are also serious problems.

Industrial Pollution

Industries wastes which are considered the principal sources of water pollution may be classified as apparel, food, drugs, chemicals, material and energy.

Action Plan for Conservation of Freshwater Resources in India

Water is the basic need of a living organism. No one can live without water. The demand for freshwater is likely to exceed its supply by the first or second decade of the 21st century. For this the conservation of freshwater is therefore an absolute necessity of today. The suggested steps for conservation of freshwater resources in India are as follows:

1. Water economy, reuse and recycling
2. Development of an efficient distribution system
3. Reduction of pollution and recycling of water
4. Enhancement of surface storage capacity
5. Improvement of underground storage capacity
6. Desalination of sea water
7. Artificial rain making
8. Population growth rate must be reduced by at least 30 per cent in the next five years.

References

Asthana, D.K. and Asthana, Meera (1997). *Environment Problems and Solutions*. S. Chand & Company Ltd., New Delhi, 63-73.

De, A.K. (1996). *Environmental Chemistry*. New Age International (P) Ltd, New Delhi, 226-227.

Pandey, G.N. and Corney, G.C. (1998). *Environment Engineering*. Tata McGraw-Hill Publishing Company Ltd., New Delhi, 104-108.

Vohra, B.B. (1990). *Managing India's Water Resources*. INTACH Environmental Series 11, New Delhi.

Vohra, B.B. (1994). Agenda for water: Myth and reality. In: *The Hindu Survey of Environment*, Madras.

33

Assessment of Water Quality of Mosam River of Baglan of Maharashtra by Saprobity System

S.N. NANDAN

P.G. Department of Botany, S.S.V.P.S's L.K.Dr.P.R.Ghogrey Science College, Dhule, Maharashtra

N.H. AHER

Arts, Commerce and Science College, Sakri, Dhule, Maharashtra

ABSTRACT

Earlier workers of taxonomists had attempted to arrange saprobiotic algae which are capable of tolerating various degrees of pollution. The study area was selected the Mosam river which runs west east from Haranbaree Malegaon. It joins the Girna river about 3 km below Malegaon city. In present study 3 stations of Mosam river were selected for collection of water and algal samples. The saprobity indices for 3 stations of river were calculated according to Pantle and Buck (1955). Kolkwitz and Marsson (1908) investigated the saprobity index for assessment of biologically polluted waters of different habitats. In present study the system of Kolkwitz and Marsson is followed to find out saprobity index of Mosam river. All stations of Mosam river were found to be β-mesosaprobic in nature.

Keywords: *Saprobity index, algae, Mosam river.*

Introduction

The Indian rivers with relation to algae are not extensively studied. Recently workers have paid attention on hydro-biological work of algae. Some major contributions from Maharashtra are those by Gunale and Balakrishnan (1981), Pingale (1981), Goel *et al.* (1985), More (1997), Thorat and Marsarrat and Nandan and Mahajan (2002).

The Mosam river flows from west-east direction in Baglan. It runs from Haranbaree → Mulher → Antapur → Taharabad → Jaykheda and Nampur. Further the Mosam river joins the Girna river about 3 km below Malegaon city.

Materials and Methods

Water samples and algal samples were collected from 3 stations of river at monthly intervals during July 1998 to June 2000. The chemical analysis of water samples were carried out following standard methods of APHA (1995). The phytoplanktons, epiphytic and floating forms of algae were collected in acid washed collection bottles and preserved in 4 per cent of formalin for identification of algae.

The saprobity indices for 3 stations of Mosam river were calculated according to Pantle and Buck (1956).

$$M = \{S\} = \Sigma\, sh / \Sigma\, h$$

Where

'S' is saprobity index.

's' is the degree of saprobity .

'h' is the frequency which the single species occurs.

For calculations the following numerical degrees are used.

Oligo-saprobic indicator organism	S = 1
β-mesosaprobic indicator organism	S = 2
α-mesosaprobic indicator organism	S = 3
Poly-mesosaprobic indicator organism	S = 4
Species found per chance	h = 1
Species found frequently	h = 3
Species occurring in abundance	h = 5

The following numerical values for the individual zones have been followed:

1. 1–1.5	denotes oligo (O) saprobic zone
2. 1.5–2.5	denotes oligo β-mesosaprobic zone
3. 2.5–3.5	denotes oligo α-mesosaprobic zone
4. 3.5–4.0	denotes polysaprobic zone

In present study the system of Kolkwitz and Marsson (1908) was used for assessing the water quality of Mosam river.

Results and Discussion

The saprobity indices for 3 stations of Mosam river were calculated according to Pantle and Buck (1955) as shown in Table 33.1. The system of Kolkwitz and Marsson (1908) comprised 4 saprobic levels *viz.*

1. Very heavily contaminated – Polysaprobic zone.
2. Heavily contaminated – α-mesosaprobic zone (α)
3. Moderately contaminated – β-mesosaprobic zone (β)
4. Scarcely contaminated – Oligosaprobic zone (O)

Table 33.1: Saprobity Index of 3 stations of Mosam River

Sl.No.	*Stations*	*Saprobity Index*	*Degree of Pollution*
1.	MRH_1	1.56	β-Mesosaprobic zone
2.	MRM_2	1.89	β-Mesosaprobic zone
3.	MRA_3	2.14	β-Mesosaprobic zone

MRH_1 – Mosam river Haranbaree station I
MRM_2 – Mosam river Mulher station II
MRA_3 – Mosam river Antapur station III.

The following algal taxa were observed at 3 stations of Mosam river and classified in above saprobic levels according to Kolkwitz and Marsson (1908) as under:

1. Polysaprobic zone : Nil
2. β-mesosaprobic zone : *Oscillatoria princeps* Voucher
   Stigeoclonium tenue Kuetz.
3. α-Mesosaprobian : *Euglena acus* Ehrenb.
   Pediastrum sp.
   Trachelomonas sp.
4. Oligo saprobian : *Cyclotella meneghiniana* Kuetz.
   Ulothrix zonata Kuetz.

The presence of species of *Spirogyra, Closterium* and *Cosmarium* indicated β-mesosaprobic zones as agreed with the observations of Thomas (1944). The presence of *Spirogyra* and *Oscillatoria* indicates the organic pollution as agreed with views of Patrick (1965) and Nandan and Patel (1984). Species of *Scenedesmus, Pediastrum, Chlorella* and *Pandorina* are capable of different degrees of myxotrophic and heterotrophic growth (Jeejibai and Rajendra, 1980). All these forms were recorded in present study.

As per Kolkwitz and Marsson (1908) most of the forms in present study were placed into β-mesosaprobic category suggesting to β-mesosaprobic nature of water of Mosam river. Saprobity index of all 3 stations showed β-mesosaprobic zone. Thus the saprobity indices and system of Kolkwitz and Marsson (1908) hold good for assessing the water quality of Mosam river as agreed with Nandan and Mahajan (2002).

Acknowledgement

The authors are grateful to principal B.M. Patil, S.S.V.P.S's L.K.Dr.P.R. Ghogrey Science College, Dhule for providing facilities for this work.

References

APHA (1995). *Standard Methods of the Examination of Water and Wastewater,* 16th Edn.. American Publ. Hlth. Asso., New York.

Goel, P.K., Trivedy, R.K. and Bhave, S.V. (1985). Studies on the limnology of a few freshwater bodies in south-western Maharashtra. *Indian J. Envir. Prot.,* 5(1): 19-25.

Gunale, V.R. and Balkrishnan, M.S. (1981). Biomonitoring of eutrophication in the Pavana, Mula and Mutha rivers flowing through Poona. *Indian J. Environ. Hlth.,* 23(4): 316-322.

Kolkwitz, R. and Marsson, M. (1908). Olologie de Flanzlichen saprobein, Deutsch. *Bot. Ges.* 25: 505-519.

More, Y.S. (1997). *Limnological Study of Panzara Dam and River.* Ph.D. Thesis, North Maharashtra University, Jalgoan.

Nandan, S.N. and Mahajan, S.R. (2002). Saprobity system for the assessment of water quality of Hartala lake of Jalgaon (Maharashtra). *Oikoassay,* 15(1 and 2): 35-37.

Nandan, S.N. and Patel, R.J. (1984). Saprobity system for the assessment of river water quality by algal community. *Indian Botanical Reporter,* 3(2): 129-132.

Pantle, R. and Buck, H. (1955). Die biologischa uberwachung der Gewassawer and die Brtellunder Ergebnisse. *Gas and Wassorfach,* 96: 604.

Patrick, R. (1965). Algal as indicator of pollution. In: *Biological Problem in Water Pollution.* 3rd Seminar Bot. A. Tuft. Sanitary Eng. Centre Cincinati, Ohio, pp. 223-232.

Pingale, S.D. (1981). *Studies on the Algae of Impoundments and Streams in Maharashtra,* Ph.D. Thesis, Poona University, Pune.

Thomas, E.A. (1944). Versuchuber die selbstreiugung Flie Bendon, Wassers, Mitt. Gebiet Lebonsmittluntor such. *U. Hyg.,* 35: 199.

Thorat, S.R. and Masarrat Sultana (2000). Pollution status of Sam Ali lake Aurangabad (M.S.). *Poll. Res.,* 19(2): 307-309.

34

Assessment of Irrigation Water Qualities

A. SOLAIMALAI & R. SARAVANAKUMAR

Research Associate, Krishi Vigyan Kendra, Tamil Nadu Agricultural University, Vriddhachalam, Tamil Nadu

M. BASKAR

Research Associate, A.D. Agricultural College and Research Institute, Tamil Nadu Agricultural University, Trichy, Tamil Nadu

K. SANKARANARAYANAN

Senior Scientist, Central Institute for Cotton Research, Indian Council of Agricultural Reseach, Coimbatore, Tamil Nadu

With the gross irrigated area of about 64 million hectares and net irrigated area of about 44.8 million hectares. India has the highest irrigated area in the world followed very closely by China. India has planned to enhance its irrigation potential further by about 2.5 million hectares per year. About 51.2 per cent of this area is irrigated by well. Hence, groundwater is an important source of irrigation. Majority of the groundwater in the arid, semi-arid and coastal region are of poor quality. The problem and the management of poor quality water is different in coastal region than in the arid and semi-arid region. Due to tidal effects and ingress of the seawater the majority of the soils of the coastal region becomes saline. Such salinity problem becomes more hazardous particularly in the post-rainy season which hampers the pace of production and further poses a serious management problem due to the non-availability of good quality of water for leaching and irrigation purpose. In majority of the places, poor quality groundwater is the only source of irrigation. Therefore, some of the important physical and chemical properties of irrigation water are necessary to be known to assess its suitability for irrigation. By knowing the salt content in meq/litre, its value in parts per million (ppm) can be estimated. By multiplying meq/litre to equivalent weight, the estimated value in ppm can be obtained, which is fairly accurate for irrigation water.

Collection of Water Sample

For reliable assessment, besides accurate analysis, collection of representative samples are important. The sampling technique will depend on the use to which the information is to be put, the chemical characteristics of the water, and the nature of the supply to be sampled whether a well, a stream or a reservoir. Maximum care should be taken to collect a representative sample. Collect at different times and mix the portions collected. Rinse the container 3 or 4 times with the water sample before the sample is drawn. If samples from wells fitted with electric motors are to be collected. Collect after the pump has been running for at least 15 minutes. If the well has only a mhote wheel, run the mhote wheel for one hour and then draw the samples. Samples from streams should be collected from running water avoiding floating impurities. Glass bottles made of borosilicate glass may release boron and polythene bottles may absorb boron. So, for boron estimation glass bottle made of boron free glass are to be used. To avoid changes from chemical and biological activities minimum time lag should be given between the time of collection and laboratory evaluation.

Sample should be collected at a gauging point so that the analytical data can be related to the discharge or to the run off. In case of running stream, sample should be collected from the flowing water and few cm below the surface to avoid floating oil. In case of large and deep reservoirs/well samples should be collected from several depths and at a number of location in the reservoir with the help of some sampler. Water in small reservoir is usually homogenous and sampling from the outlet are representative.

Estimation of pH

The pH is defined as the negative logarithm of hydrogen ion concentration or simply the log of the reciprocal of the hydrogen ion concentration.

Materials Required

pH meter, 100 ml beakers, glass rods and buffer solution (pH 4.0, 7.0 and 9.2).

Standardisation of pH Meter

Switch on the instrument and allow it to warm for 10 minutes. Keep the pH selector switch on zero position. Set the temperature compensation control to the solution temperature. Adjust the zero adjustment knob so that the pointer in the meter reads exactly zero, when the electrodes are immersed in distilled water. Lift the electrodes from distilled water and wipe with filter paper and dip them in standard buffer solution of known pH (4.0, 7.0 and 9.2). Change the function switch to particular pH range (0-7 or 7-14) and adjust the standardisation knob till the pointer reads the correct pH value of the buffer solution. Do not disturb the zero knob adjustment.

pH Measurement

Wash the electrodes carefully with a jet of distilled water and wipe with a piece of filter paper. Immerse the electrodes into the beaker containing water sample and change the function switch to the particular pH range. Record the meter reading.

Estimation of Electrical Conductivity

Electrical conductivity is a very simple and quick measurement to assess the quality of irrigation water. In irrigation water, electrical conductivity is reported in μ mhos/cm at 25°C. At the time of measurement of electrical conductivity, if temperature is more than 25°C then for each increase of 1°C the conductivity will increase by 2 per cent and thereby correction is required. Therefore, 2 per cent of electrical conductivity values for each °C should be deleted from the observed value for temperature higher than 25°C. Similarly 2 per cent for each °C should be added in the observed value for temperature lower than 25°C.

Table 34.1: Temperature Factors (Ft) for Correcting EC Values to the Standard Temperature of 25°C

°C	Ft	°C	Ft	°C	Ft	°C	Ft
3.0	1.709	20.2	1.107	25.4	0.992	30.6	0.897
4.0	1.660	20.4	1.102	25.6	0.988	30.8	0.894
5.0	1.613	20.6	1.097	25.8	0.983	31.0	0.890
6.0	1.560.	20.8	1.092	26.0	0.979	31.2	0.887
7.0	1.528	21.0	1.087	26.2	0.975	31.4	0.884
8.0	1.488	21.2	1.082	26.4	0.971	31.6	0.880
9.0	1.448	21.4	1.078	26.6	0.967	31.8	0.877
10.0	1.411	21.6	1.073	26.8	0.964	32.0	0.873
11.0	1.375	21.8	1.068	27.0	0.960	32.2	0.870
12.0	1.341	22.0	1.064	27.2	0.956	32.4	0.867
13.0	1.309	22.2	1.060	27.4	0.953	32.6	0.864
14.0	1.277	22.4	1.055	27.6	0.950	32.8	0.861
15.0	1.247	22.6	1.051	27.8	0.947	33.0	0.858
16.0	1.218	22.8	1.047	28.0	0.943	34.0	0.843
17.0	1.189	23.0	1.043	28.2	0.940	35.0	0.829
18.0	1.163	23.2	1.038	28.4	0.936	36.0	0.815
18.2	1.157	23.4	1.034	28.6	0.932	37.0	0.801
18.4	1.152	23.6	1.029	28.8	0.929	38.0	0.788
18.6	1.147	23.8	1.025	29.0	0.925	39.0	0.775
18.8	1.242	24.0	1.020	29.2	0.921	40.0	0.763
19.0	1.136	24.2	1.016	29..4	0.918	41.0	0.750
19.2	1.131	24.4	1.012	29.6	0.914	42.0	0.739
19.4	1.127	24.6	1.008	29.8	0.911	43.0	0.727
19.6	1.122	24.8	1.004	30.0	0.907	44.0	0.716
19.8	1.117	25.0	1.000	30.2	0.904	45.0	0.705
20.0	1.112	25.2	0.996	30.4	0.901	46.0	0.694

Materials Required

1. Wheatstone resistance bridge,
2. Conductivity cell with platinised platinum electrodes
3. Potassium chloride solution 0.01 normal (0.7456 g of dry KCl dissolved in one litre of water at 25°C).

Procedure

Clear the new cells with chromic sulphuric acid. Fill the conductivity cell with the reagent. Read the resistance R_{25} at 25°C and calculate the cell constant K from the relation.

$$K = EC_{25} \times R_{25}$$

Rinse the cell with the solution of which EC is to be found out. Fill it again with the unknown solution and record the resistance of the cell (Rt) and temperature (t) of the solution. Calculate the (ECt) by the relation,

$$Ect = K/Rt$$

Where

$K = EC_{25} \times R_{25}$

Adjust the EC value to 25°C with a temperature conversion factor (Ft), obtained from table, Thus,

$$EC_{25} = Ect \times Ft = KFt/Rt$$

Estimation of Total Soluble Salts (TSS)

A known volume of water is evaporated to dryness and the amount of the soluble salts present in the water is estimated gravimetrically.

Materials Required

(i) Potash basin *(ii)* 100 ml pipette *(iii)* Balance *(iv)* Water basth and *(v)* Desiccator

Procedure

Pipette out 100 ml of water into a clean already weighed dry potash basin. Evaporate the water in the basin to dryness over a water bath. Wipe outside of the basin, dry in an air oven at 105°C for an hour to remove moisture, cool and weigh. The difference in the weights is the weight of total soluble salts. Express the results as ppm.

Calculation

Volume of water taken	=	100 ml
Weight of empty basin (g)	=	A
Weight of basin + total soluble salts (g)	=	B
Weight of total soluble salts alone	=	B – A
Total soluble salts content of water (ppm)	=	$\frac{(B-A)}{100} \times 10^6$

Estimation of Carbonate and Bicarbonate

Carbonate and bicarbonate in the water are determined by titrating a known volume of water with standard sulphuric acid using phenolphthalein and orange as indicators. When a drop of phenolphthalein is added to the water sample, if pink colour does not develop, it shows the absence of carbonate in water. When the water sample containing carbonate and bicarbonate are titrated against standard sulphuric acid, phenolphthalein loose its pink colour when half of the carbonate is converted into bicarbonate. So twice this value is a measure of carbonates present in water. To the colourless solution, a few drops of methyl orange is added and titrated against sulphuric acid till the colour changes to pinkish red. This value gives the amount of acid required to neutralise the bicarbonates originally present and that formed from the carbonates. By subtracting the first titre value from the second one, acid required to neutralise the bicarbonate originally present in the water is obtained.

Materials Required

(i) 250 ml conical flask *(ii)* 50 ml pipette *(iii)* 50 ml burette *(iv)* 0.1 NH_2SO_4 *(v)* Phenolphthalein indicator and *(vi)* Methyl orange

Procedure

Pipette out 50 ml of the water into a 250 ml conical flask. Add few phenolphthalein (if a pink colour develops, carbonate is present). Titrate this against 0.1 NH_2SO_4 till the pink colour disappears (The titre value gives the value for half the amount of carbonate present). To the same solution in the conical flask, add one or two drops of methyl orange and continue the titration till the colour changes to pinkish red. The titre value gives the value for half the carbonate plus all the bicarbonate present in the water.

Calculation

Volume of water taken = 50 ml

Volume of 0.1 NH_2SO_4 used with phenolphthalein (1/2 CO_3 to HCO_3) = A ml

Volume of 0.1 NH_2SO_4 used with methyl orange = B ml

Volume of 0.1 NH_2SO_4 required to neutralise carbonate alone = (2 × A) ml

Volume of 0.1 NH_2SO_4 required to neutralise bicarbonate alone = (B – A) ml

Carbonate

1 ml of 0.1 NH_2SO_4 = 0.003 g of CO_3

2 × A ml of 0.1 NH_2SO_4 = 0.003 × (2 × A) g

This is present in 50 ml of water

In 10^6 ml $= \dfrac{0.03 \times (2 \times A)}{50} \times 10^6$ ppm

In terms of me/litre $= \dfrac{\text{ppm}}{30}$

Bicarbonate

1 ml of 0.1 NH_2SO_4 = 0.0061 g of HCO_3

(B – A) ml of 0.1 NH_2SO_4 = (0.0061) × (B – A) g of HCO_3

This is present in 50 ml of water
In 10^6 ml $= \dfrac{(0.0061) \times (B - A)}{50} \times 10^6$ ppm

In terms of meq/litre $= \dfrac{\text{ppm}}{61}$

Estimation of Chloride

The chloride present in the water is precipitated as silver chloride by titration with standard silver nitrate in the presence of potassium chromate. After all the chloride is precipitated, further drop of silver nitrate added combines with potassium chromate forming flesh red coloured precipitate of silver chromate which serves as the end point.

Materials Required

(i) Porcelain basin *(ii)* Glass rod *(iii)* 50 ml pipette *(iv)* 50ml burette *(v)* 0.1 N $AgNO_3$ and *(vi)* Potassium chromate (5 per cent aqueous solution)

Procedure

Pipette out 50 ml of the water into a porcelain basin. Add a few drops of potassium chromate indicator and titrate against 0.1 N$AgNO_3$ till a flesh red precipitate of silver chromate appears. From the amount of 0.1 N$AgNO_3$ consumed, calculate the chloride content.

Calculation

Volume of water taken = 50 ml

Volume of 0.1 N$AgNO_3$ used = A ml

1 ml of 0.1 N$AgNO_3$ = 0.00355 g of Cl

'A' ml of 0.1 N$AgNO_3$ = 0.00355 × Ag of Cl

This is present in 50 ml of water
Therefore in ppm $= \dfrac{(0.00355 \times A)}{50} \times 10^6$

In terms of meq/litre $= \dfrac{\text{ppm}}{35.5}$

Estimation of Sulphate

The sulphate in water is precipitated as barium sulphate by the addition of barium chloride in hydrochloric acid medium. The precipitate is filtered, washed free of chloride, ignited and weighed as barium sulphate.

Materials Required

(i) 50 ml pipette *(ii)* 250 ml beaker *(iii)* Silica crucible *(iv)* balance *(v)* Water bath *(vi)* desiccator *(vii)* dilute hydrochloric acid *(viii)* 10 per cent barium chloride solution *(ix)* silver nitrate solution *(x)* solid ammonium chloride (which aids in the granulation of the precipitate).

Procedure

Pipette out 50 ml of irrigation water into a clean 250 ml beaker. Add 10 ml of diluted hydrochloric acid and 1 g of solid ammonium chloride. Heat to boiling and add 10 ml of 10 per cent barium chloride solution drop by drop with constant stirring. Continue boiling for another 2 to 3 minutes. Allow the precipitate to settle and test for completion of precipitation by adding a small amount of barium chloride solution through the sides of the beaker. If any turbidity is noticed, add sufficient quality of barium chloride to precipitate all the sulphate. Transfer the contents to sand bath and digest for half an hour to promote granulation of the precipitate. Filter through whatman No.42 filter paper and wash with boiling water till the filtrate runs free of chloride (test with silver nitrate solution). Transfer the filter paper along with the precipitate to a weighed silica crucible and dry it in an air oven. Ignite over a low flame initially, taking care to ash the filter paper completely, then ignite strongly over a rose head flame to constant weight. From the weight of barium sulphate obtained calculate the sulphate content of the irrigation water.

Calculation

Volume of water sample taken = 50 ml

Weight of empty silica crucible = A g

Weight of crucible + precipitate = B g

Weight of precipitate alone = (B – A) g

233 g of $BaSO_4$ contains 96 g of SO_4

(B – A) g of $BaSO_4$ will contain = $96/233 \times (B - A)$ g of SO_4

This is present in 50 ml of irrigation water

In 10^6 ml = $\frac{96}{233} \times (B - A) \times \frac{10^6}{50}$ ppm

In terms of meq/litre = $\frac{\text{ppm}}{48}$

Estimation of Nitrate

Materials Required

(A) Phenol disulphonic acid [dissolve 25 g of phenol in 150 ml of concentrated sulphuric acid, add 75 ml of fuming sulphuric acid (13 to 15 per cent SO_4) and heat at 100°C for 2 hours]. (B) Potassium nitrate 0.01 N. (C) Silver sulphate 0.02 N. (D) Ammonium hydroxide 7 N. (E) Calcium oxide.

Procedure

First determine the concentration of chloride in an aliquot. Pipet another aliquot containing 0.004 to 0.04 meq of nitrate into a 25 ml volumetric flask. Add an amount of reagent C equivalent to the amount of chloride present. Dilute to volume and mix. Transfer most of the suspension to a 50 ml centrifuge tube and separate the precipitate by centrifuging. After transferring the solution to another centrifuge tube, flocculate any suspended organic matter by adding about 0.1 g of E and clear by again centrifuging. Pipet a 10 ml aliquot representing two-fifth of the sample into a 8 cm, evaporating dish. Evaporate the aliquot to dryness, cool and dissolve the residue in 2 ml of phenoldisulphonic acid. After 10 minute, add 10 ml of water and transfer to a 100 ml volumetric flask. Make alkaline by addition of ammonium hydroxide 7 N, dilute to volume and mix. Measure the light transmission through a 460 ml filter in an optical cell against that of water in a similar cell. Prepare a calibration curve by pipeting 0, 0.2, 0.4, 0.8, 1.2 and 1.6 ml of potassium nitrate 0.01 N into evaporating dishes and treating as above omitting the additions of silver sulphate 0.02 N and calcium oxide and the clarifying procedure.

Calculations

NO_3 (meq/litre) = (Meq of NO_3 in aliquot as found by interpolation on NO_3 curve) × 1000/(ml in aliquot).

Estimation of Calcium and Magnesium (Versenate Method)

Ethylene diamine tetra acetic acid (EDTA) forms stable complexes with various polyvalent cations at different pH levels. Calcium and magnesium are complexed by EDTA in the order calcium first and magnesium afterwards at pH levels of 12 and 10 respectively. Hence, calcium is first estimated by titration with EDTA using murexide indicator in the presence of sodium hydroxide at pH 12, then magnesium is estimated together with calcium using eriochrome black-T indicator at pH 10 in the presence of ammonium chloride-ammonium hydroxide buffer solution. Sn, Cu, Zn, Fe and Mn may interfere in the determination of calcium and magnesium. The interference of Sn, Cu, Zn, Fe and Mn is prevented by the use of 2 per cent NaOH solution. Usually in the irrigation waters the quantities of interfering of ions are negligible and can be neglected.

Materials Required

(i) Dissolve 67.5 g of ammonium chloride in 570 ml of concentrated ammonium hydroxide and make to one litre. *(ii)* Sodium hydroxide, approximately 4 N. *(iii)* Standard calcium chloride solution 0.01 N, *(iv)* Eriochrome black T-indicator, *(v)* Ammonium purpurate indicator, *(vi)* Ethylene–diamine–tetra–acetate (versenate) solution approximately 0.01 N (dissolve 2.0 g of disodium dihydrogen ethylenediamine–tetra acetate and 0.05 g of magnesium chloride hexhydrate in water and dilute to a volume of one litre. Standardise the solution against calcium chloride solution 0.01 N using the titration procedures given below. The solution is standardized, using each of the eriochrome black T-indicator and ammonium purpurate indicator, as the normality with ammonium purpurate indicator is 3 to 5 per cent higher than that with eriochrome black T-indicator.

Procedure

Pipet out 5 to 25 ml aliquot containing not more than 0.1 meq of calcium plus magnesium into a 125 ml Erlenmeyer flask. Dilute to a volume of approximately 25 ml. Add 0.5 ml (10 drops) of ammonium chloride-ammonium hydroxide buffer solution and three or four drops of ammonium purpurate indicator. Titrate with versenate solution, using a 10 ml microburette. The colour change is from wine-red to blue or green. No tinge of the wine-red colour should remain at the end point.

Calculations

Ca + Mg (meq/litre) = Versenate solution (ml) × normality of versenate solution × 1000

Estimation of Sodium

Sodium is estimated by flame photometry. It emits a bright yellow colour (589 millimicrons) when excited in the flame. The intensity of emission is proportional to the concentration of sodium in the sample.

Materials Required

(i) Flame photometer *(ii)* volumetric flasks 100 and 1000 ml *(iii)* 100 ml beaker *(vi)* A.R grade sodium chloride.

Procedure

Standardise the flame photometer before feeding the water sample. Set galvanometer to zero using 0 ppm Na. Using 100 ppm Na solution adjust the meter reading to 100. Feed the water sample in the flame photometer and note the meter reading. From the standard curve for Na, find out the concentration of Na (ppm) in the water sample.

Calculation

Concentration of Na in the water sample as read from the standard curve = A ppm

In terms of meq/litre = A/23

Preparation of Standard Curve

About 0.254 g of AR grade sodium chloride dissolved in one litre of water gives 100 ppm Na solution. From this prepare a series of Na solution of concentrations ranging from 10-100 ppm. Record the flame photometer reading for the above Na solutions and plot the readings against the corresponding concentrations of Na. Using this as a standard curve find out the concentration of Na in water sample.

Estimation of Potassium

Potassium emits an yellow colour (404 millimicrons) flame when excited in the flame. The intensity of emission is proportional to the concentration of potassium in the sample.

Materials Required

(i) Flame photometer *(ii)* Voiumetric flasks 100 and 1000 ml *(iii)* 100 ml beaker *(iv)* A.R. grade potassium chloride salt.

Procedure

Standardise the flame photometer as done in the previous estimation. Adjust the meter reading to 100 by feeding 100 ppm K solution. Now feed the water sample and note the meter reading. From the standard curve for K as prepared already, find out the concentration of K (ppm) in the water sample, from the meter reading.

Calculation

Concentration of K in the water sample as read from the standard curve } = A ppm

In terms of meq/lire = A/39

Estimation of Boron

Boron reacts with the red dye carmine in the presence of concentrated sulphuric acid to form a blue compound, the concentration of which may be measured colorimetrically at 585 nm. The reaction proceeds slowly and about one hour is needed for full colour development. Interference of nitrate ion is removed by adding hydrochloric acid.

Materials Required

(i) 50 ml Erlenmeyer flasks *(ii)* Spectrophotometer *(iii)* Conc. HCl *(iv)* H_2SO_4 *(v)* 0.025 per cent carmine solution (by weight in sulphuric acid) *(vi)* Boric acid (A.R. grade).

Procedure

Transfer 2 ml of water sample to a 50 ml Erlenmeyer flask. Similarly transfer 2 ml of pure water to another 50 ml of Erlenmeyer flask. To each flask, add one drop of concentrated hydrochloric acid followed by 10 ml of concentrated sulphuric acid stopper the flasks and allow to cool. Add 10 ml carmine solution, mix well, stopper and leave for one hour. Determine the absorbance or transmittance of light at 585 nm wave length. Use the red liquid obtained with 2 ml pure water for reference. From standard curve prepared, calculate the boron content of the irrigation water.

Preparation of Standard Curve

About 2.860 g AR grade boric acid dissolved in one litre of water gives 500 ppm boron. From this prepare standards containing 1, 2, 3, 4 and 5 ppm boron. In 2 ml portions of each of these standards determine absorbance or transmittance of light (585 nm) by using the red liquid obtained with 2 ml pure water for reference as done in the case of water sample. Plot the values of absorbance or transmittance against concentration of boron and use it as a standard curve for finding out the concentration of boron in the water sample.

Calculation

Concentration of boron in the water sample as read from the standard curve = A ppm

In terms of meq/litre = A/11

Interpretation of Water Analysis Results

Irrigation water contains dissolved salts. These salts are sodium, calcium and magnesium as cations and bicarbonate, carbonate, sulphate and chloride as anions. The quality of irrigation water is judged based on the quantity and proportion of these salts. Various workers have suggested different methods of expression of results of the analysis of irrigation water.

Puri's Salt Index

Puri (1949) proposed an index for evaluating the suitability of irrigation water. The salt index will vary from – 24.5 to 0 for all good waters and from 0 to any positive value for all bad waters. An irrigation water with a positive index is considered unsuitable for irrigation purposes and the one with a negative index is considered suitable.

$$\text{Salt Index} = (\text{Total Na}^+ - 24.5) - (\text{Total Ca}^2 + -\text{Ca}^2 + \text{in CaCO}_3) \times 4.85$$

Sodium Adsorption Ratio (SAR)

The alkali hazard produced by the irrigation water is determined as sodium adsorption ratio (SAR). Sodium adsorption ratio is defined by the following equation:

$$SAR = \frac{Na^+}{\sqrt{Ca^{2+} + Mg^{2+}/2}}$$

Na^+, Ca^{2+} and Mg^{2+} concentration in milliequivalents per litre

Adjusted SAR

Adjusted SAR is calculated by using the formula

$$\text{Adjusted SAR} = \frac{Na^+}{\sqrt{Ca^{2+} + Mg^{2+}/2}} \times 1 + (8.4 - PHC)$$

Where,

Na^+, Ca^{2+} and Mg^{2+} concentration in milliequivalents per litre

PHC is a theoretical value, calculated from pH of the irrigation water in contact with time and in equilibrium with the soil CO_2.

$$PHC = (Pk^1_2 - Pk^1C) + P\,(Ca + Mg) + P\,(Alk)$$

PK^1C = Negative log of solubility product of $CaCO_3$

P (Ca + Mg) = Negative log of concentration of sum total of Ca and Mg

P (Alk) = Negative log of concentration of sum total of CO_3 and HCO_3

Soluble Sodium Percentage (SSP)

Yet another way of expressing the sodium hazard of irrigation water is soluble sodium percentage (SSP). This is calculated by the formula:

$$SSP = \frac{Na^+}{\text{Sum of all cations}} \times 100$$

Na^+ in milliequivalents per litre.

Potential Salinity

Salinity is mainly due to presence of chloride and sulphate ions in the irrigation water. The salinity hazard is known from the potential salinity as given below:

$$\text{Potential Salinity} = Cl + \tfrac{1}{2} SO_4^{2-}$$

Cl and SO_4^{2-} in milliequivalents per litre.

Residual Sodium Carbonate (RSC)

The carbonate and bicarbonate ions over and above calcium and magnesium is called residual sodium carbonate (RSC):

$$RSC = (CO_3 + HCO_3) - (Ca^{2+} + Mg^{2+})$$

Where

CO_3^{2-}, HCO_3^-, Ca^{2+} and Mg^{2+} in milliequivalents per litre.

Permeability Index (PI)

$$PI = \frac{Na + HCO_3}{Ca + Mg + Na} \times 100s$$

Higher values indicate less permeability

Exchangeable Sodium Percentage (ESP)

The SAR is related with ESP and the relationship between SAR and ESP is given as:

$$ESP = \frac{100\,(-0.0126 + 0.01475 \times SAR)}{1 + (-0.0126 + 0.01475 \times SAR)}$$

Classification of Irrigation Water

The suitability of irrigation water depends upon amount and nature of salts present, nature of soils to be irrigated, climatic conditions and nature of crops grown.

Total alkalinity (At) = Ab + Ac + Ah

Where

Ab = Alkalinity due to bicarbonate,

Ac = Alkalinity due to carbonate and

Ah = Alkalinity due to hydroxide.

Gypsum Requirement of Irrigation Water (GR)

GR (lb/acre foot of water) = a + b + c × 2.24

A = Na × 0.43 – (Ca + Mg)

b = $(CO_3 + HCO_3)$ 0.7

c = 0.30

USDA System (1954)

In this system, the irrigation waters are classified based on their electrical conductivity (micromhos/cm) and SAR. The horizontal axis represents conductivity in μ mhos per cm and the vertical axis represents SAR. The curves running from left to right are given a negative slope to take into account the dependence of the sodium hazard on the total concentration.

Salinity Classification

Low Salinity Water (C1)

It can be used for irrigation with most crops on most soils with little likelihood that soil salinity will develop. Some leaching is required, but this occurs under normal irrigation practices except in soils of extremely low permeability.

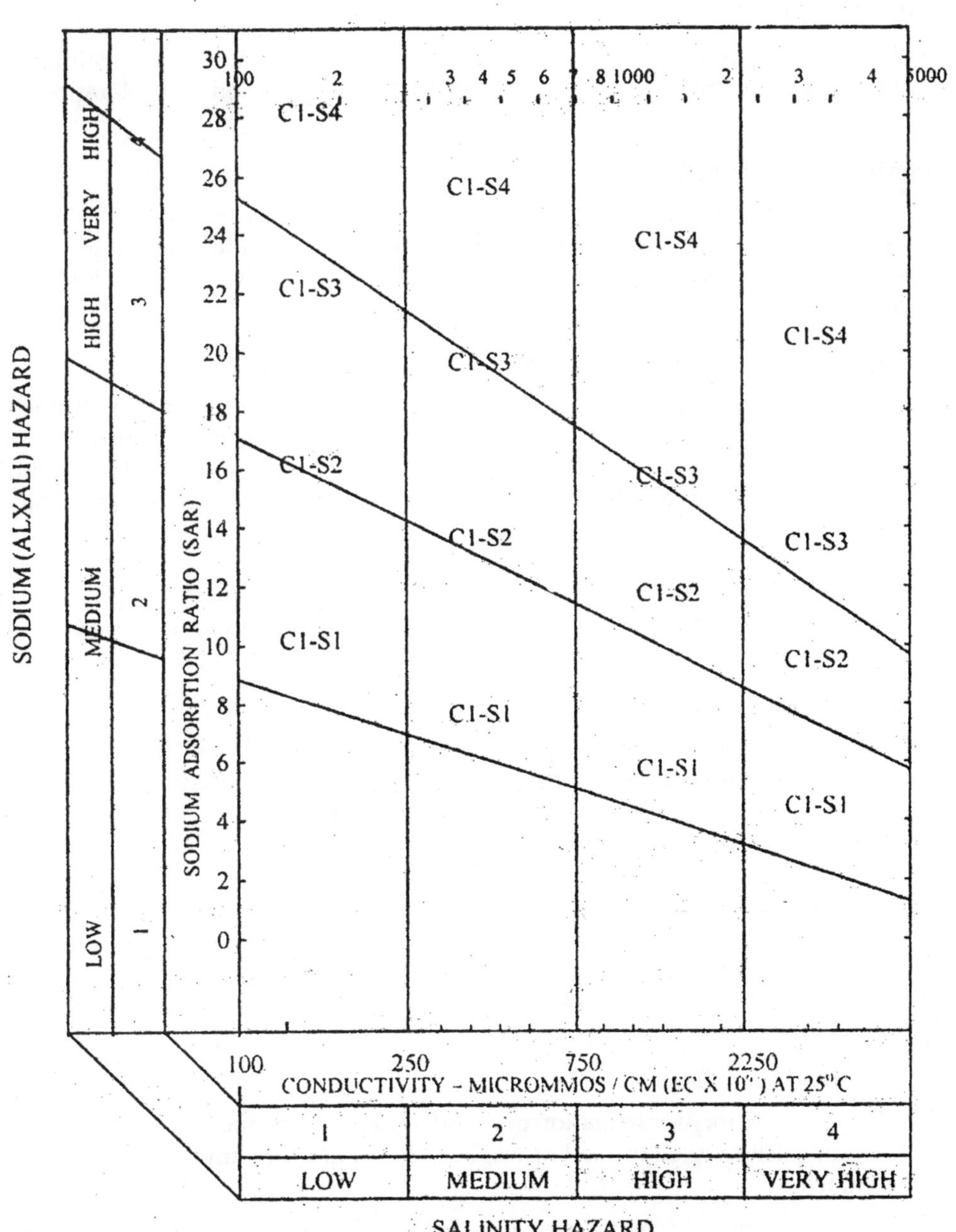

Medium Salinity Water (C2)

It can be used if a moderate amount of leaching occurs. Plants with moderate salt tolerance can be grown in most cases without special practices for salinity control.

High Salinity Water (C3)

It can not be used soils with restricted drainage. Even with adequate drainage, special management for salinity control may be required and plants with good salt tolerance should be selected.

Very High Salinity Water (C4)

It is not suitable for irrigation under ordinary conditions, but may be used occasionally under very special circumstances. The soils must be permeable, drainage must be adequate, irrigation water must be applied in excess to provide considerable leaching and very high salt tolerant crops should be selected.

Sodium Classification

The classification of irrigation waters with respect to SAR is based primarily on the effect of exchangeable sodium on the physical condition of the soil. However, sodium sensitive plants may suffer injury as a result of sodium accumulation in plant tissues when exchangeable sodium values are lower than those effective in causing deterioration of the physical condition of the soil.

Low Sodium Water (S1)

It can be used for irrigation of almost all soils with little danger of the development of harmful levels of exchangeable sodium. However, sodium sensitive crops may accumulate injurious concentrations of sodium.

Medium Sodium Water (S2)

It will present an appreciable sodium hazard in fine textured soils having high cation exchange capacity, especially under low-leaching conditions, unless gypsum is present in the soil. This water may be used on coarse-textured or organic soils with good permeability.

High Sodium Water (S3)

It may produce harmful levels of exchangeable sodium in most soils and will require special soil management, good drainage, high leaching and organic matter additions. Gypsiferous soils may not develop harmful levels of exchangeable sodium from such waters. Chemical amendments may be required for replacement of exchangeable sodium, except that amendments may not be feasible with waters of very high salinity.

Very High Sodium Water (S4)

It is generally unsatisfactory for irrigation purposes except at low and perhaps medium salinity, where the solution of calcium from the soil or use of gypsum or other amendments may make the use of these waters feasible.

Sometimes the irrigation water may dissolve sufficient calcium from calcareous soils to decrease the sodium hazard appreciably and this should be taken into account in the use of C1-S3 and C1-S4 waters. For calcareous soils with high pH values or for non-calcareous soils, the sodium status of waters in classes C1-S3, C1-S4 may be improved by the addition of gypsum to the water. Similarly, it may be beneficial to add gypsum to the soil periodically when C2-S3 and C3-S2 waters are used.

Computation of Water Soluble Salts

When the determination of various constituents in irrigation water is completed, it is customary to combine the ions to represent the probable saline constituents of water. The number of salts found in ordinary water is very few and generally it is presented in the following order:

1.	$CaCO_3$	2.	$Ca(HCO_3)_2$	3.	$CaSO_4$
4.	$CaCl_2$	5.	$MgCO_3$	6.	$Mg(HCO_3)_2$
7.	$MgSO_4$	8.	$MgCl_2$	9.	Na–CO_3
10.	$NaHCO_3$	11.	Na_2SO_4	12.	NaCl

This sort of computation gives abetter picture of the composition of water, for better evaluation besides serving as a means to check the correctness of the analysis.

Factors for Computation of Salts

Generally Ca and Mg ions are computed first followed by slats of Na. The acid ions are apportion to Ca and Mg in the order of HCO_3, SO_4 and Cl and remaining acid ions are apportioned to Na.

To Convert	*Into*	*Factor*	*To Convert*	*Into*	*Factor*
CO_3	$CaCO_3$		Ca^{++}	$CaCO_3$	2.500
	$MgCO_3$			$Ca(HCO_3)^2$	4.050
	Na_2CO_3			$CaSO_4$	3.398
	K_2CO_3			$CaCl_2$	2.771
				$Ca(NO_3)^2$	4.094
HCO_3	$Ca(HCO_3)^2$		Mg^{++}	$MgCO_3$	3.469
	$Mg(HCO_3)^2$			$Mg(HCO_3)^2$	6.017
	$NaHCO_3$			$MgSO_4$	4.945
	$KHCO_3$			$MgCl_2$	3.915
				$Mg(NO_3)^2$	6.098
Cl	$CaCl_2$		Na^+	Na_2CO_3	2.305
	$MgCl_2$			$NaHCO_3$	3.652
	NaCl			Na_2SO_4	3.080
	KCl			NaCl	2.542
				$NaNO_3$	3.696
SO_4	$CaSO_4$		KT	K_2CO_3	1.767
	$MgSO_4$			$KHCO_3$	2.566
	Na_2SO_4			K_2SO_4	2.228
	K_2SO_4			KCl	1.907
				KNO_3	2.586

Boron Concentration of Quality

The occurrence of boron in toxic concentration in irrigation waters makes it necessary to consider this element in assessing the water quality.

Permissible Limits of Boron for Several Classes of Irrigation Water

Boron Class	*Sensitive Crops (ppm)*	*Semitolerant Crops (ppm)*	*Tolerant Crops (ppm)*
1.	<0.33	<0.67	<1.0
2.	0.33 to 0.67	0.67 to 1.33	1.0 to 2.0
3.	0.67 to 1.00	1.33 to 2.00	2.0 to 3.0
4.	1.00 to 1.25	2.00 to 2.50>	3.0 to 3.75
5.	>1.25	2.50	>3.75

Classification of Irrigation Waters

In the soil testing labs, the irrigation waters are analysed for EC and pH and classified as follows:

Ratings on Soil Reaction (pH)

Status	*pH*
Acidic	Below 6.0
Normal	6.0 to 8.5
Tending to alkaline	8.5 to 9.0
Alkaline	9.0 and above

But when the given irrigation water sample contains more of soluble salts and sodium as indicated by electrical conductivity and pH, then the detailed analysis for various cations and anions is done and rating on the quality of the water determined.

Ratings on Electrical Conductivity

Status	*EC (dSm^{-1})*
Normal	<1.0
Critical	1.0–3.0
Injurious	>3.0

Classification of Irrigation Water Under Detailed Analysis

Sl.No.	*Quality of water*	*EC (dSm^{-1})*	*pH*	*Na (%)*	*Cl (Me/litre)*	*SAR*	*RSC*
1.	Excellent	<0.5	6.5–73.5	<30	<2.5	<1	<1.0
2.	Good	0.5–1.5	7.5–8.0	30–60	2.5–5.0	1–2	1.0–1.25
3.	Fair	1.5–3.0	8.0–8.5	60–75	5.0–7.5	2–4	1.25–2.0
4.	Poor	3.0–5.0	8.5–9.0	75–80	7.5–10.0	4–8	2.0–2.5
5.	Very poor	5.0–6.0	9.0–10	80–90	10.0–12.5	8–15	2.5–3.0
6.	Unsuitable	>6	>10	>90	>12.5	>15	>30

References

AOAC (1960). *Association of agricultural chemists, methods of analysis*. Washington.

Gupta, I.C. (1990). *Use of saline water on agriculture*. Oxford & IBH Publishing Company Pvt. Ltd., New Delhi.

Gupta, R.K. *et al.* (1994). *Water quality for irrigation in India*. Tech. Bull. No. 19, CSSRI, Karnal, p. 6.

Hatcher, J.T. and Wilcox, L.V. (1950). *Analytical Chemistry*, 22: 567-569.

Minhas, P.S. and Gupta, Raj, K. (1992). *Quality and irrigation water assessment and management*. CSSRI, Karnal, p. 102.

Muthuvel, P. and Udayasoorian, C. (1999). *Soil, plant, water and agrochemical analysis*. Tamil Nadu Agricultural University, Coimbatore, p. 298.

Richards, L.A. (1954). *Diagnosis and improvement of saline and alkali soils*. Oxford & IBH Publishing Company Pvt. Ltd., New Delhi.

Venkateswarlu, P. (1962). Ph.D. thesis, University of Minnesota, Minneapolis, USA.

35

Irrigation with Poor Quality Water on Soil and Crop

A. SOLAIMALAI & R. SARAVANAKUMAR

Research Associates, Krishi Vigyan Kendra, Tamil Nadu Agricultural University, Vridhachalam, Tamil Nadu

M. BASKAR

Research Associate (Soil Science and Agricultural Chemistry), A.D. Agricultural College and Research Institute, Trichy, Tamil Nadu

K. SANKARANARAYANAN

Senior Scientist (Agronomy), Central Institute for Cotton Research (ICAR), Coimbatore, Tamil Nadu

In India, about 7 million hectares of land is salt affected (Abrol and Bhumbla, 1971). Salinity problem is ever increasing at least in those arid and semi-arid regions where there is no choice but to irrigate the land with poor quality water. In old Mesopotamia, an area of fertile soil between the Tigris and Euphrates river produced large quantities of grain and other crops sufficient to feed large populations for a prolonged period. In modern times, this area has been transformed into a bare desert. It is also well known that in ancient China, in the Indus Valley and in South America, vast territories have turned into deserts because of salinity problems resulting from irrigation by ancient societies. The other source may be surface water that are used for irrigation purposes. In major rivers, the water is of good quality has a low salt concentration but in other places, particularly in small tributaries, lakes, lagoons and swamps, saline water often exists. Sooner or later, generally, when saline water is utilized for irrigation, secondary salinization and alkalization results. On sea shores and in nearby areas the salt content of sea water has a direct or indirect effect on soil salinity.

Considerable differences are found in the chemical composition of river waters on the different continents. Even larger differences can be observed in the concentration of some cations and anions in river waters.

Since all irrigation waters contain certain quantity of soluble salts, some changes will always occur in the physical, chemical and microbiological properties of soils upon the introduction of irrigation, resulting in either an improvement or deterioration of the soil and groundwater properties. Some times the increased level of salt in the soil solution results in coagulation of soil particles, better structure and better permeability, especially in heavy soils, than soil irrigated with good water of low salinity. On the other hand, presence of high proportion of Na in the exchange complex affects the swelling and dispersion of clay particles and the extent of swelling and dispersion depend on the total electrolyte concentration of the ambient solution. The swelling and dispersion of soil clay minerals result in a structural breakdown and eventually in reduction of soil permeability which, in turn, results in difficulties in irrigation and drainage. Ancient irrigation systems in North Africa, West Asia and Rajasthan (India) have shown that highly saline waters from surface-dug wells have been successfully used over hundreds of years on well drained soils with excellent permeability. A state of permanent leaching under the prevalent conditions of soil and climate ensures removal of accumulated salts and replacement by the new irrigation waters. But if natural drainage is bad or permeability restricted, even slightly saline waters can cause strong soil salinization. Under no drainage conditions, the salt content of soils can increase through repeated waterings with saline waters of different salt concentrations. Unless adequate steps are taken to safeguard against this hazard, saline irrigation waters should never be used in practice.

Excess salinity within the plant root zone has a general deleterious effect on plant growth, which is manifested as nearly equivalent reductions in the transpiration and growth rates (including cell enlargement and the synthesis of metabolites and structural compounds). This effect is primarily related to total electrolyte concentration and is largely independent of specific solute composition. Excessive salinity reduces plant growth primarily because it increases the energy that must be expended to acquire water from the soil of the root zone and to make the biochemical adjustments necessary to survive under stress. This energy is diverted from the processes, which lead, to growth and yield. Growth suppression is typically initiated at some threshold value of salinity which varies with crop tolerance and external environmental factors which influence the need of the plant for water especially the evaporative demand of the atmosphere (temperature, RH, wind speed etc.) and the water supplying potential of the root zone and increases as salinity increases until the plant dies.

Saline Water Effects

Irrigating the soil with high saline water (high EC waters), enhances the soil EC and the soils tend to become saline. The physical properties of soil are however, not affected because of high electrolyte concentrations particularly due to neutral salts (chlorides and sulphates of sodium, calcium and magnesium). The principal quality of saline water injurious to plants is the high osmotic pressure of the soil solution, which reduces the availability of the water for absorption leading to physiological drought.

Sodic Water Effects

Irrigation with water containing high sodium and plenty of carbonate and bicarbonate ions relative to calcium leads to the development of high exchangeable sodium percentage and pH in soil. These together adversely affect the soil physical properties. It disperses the soil particles leading to the destruction of soil structure and aggregation. It also affects water infiltration and soil aeration. These adverse effect of high sodicity water is more evident in the shrinking swelling types of soil (Montmorillinate clay) rather than in non-swelling types (soils dominant with illite, vermiculite and kaolinitic clay minerals). The increased soil pH results in soil nutritional imbalances, the phosphorous and zinc being complexed and nitrogen being expressed.

Soil Physical Properties

Due to continuous use of irrigation water having high SAR, the soil exchange complex is enriched with sodium as a result of cation exchange. The water holding capacity of medium black soils treated with high concentration of sodium carbonate increased proportionately. The dispersion and expansion of volume led to reduced permeability. The per cent pore space in soil irrigated with high sodium water was observed to be less as compared to unirrigated soil. This is because of the structural deterioration and formation of compact sub-horizons due to translocation of silt and clay.

The possibility of salinity increase is much more in heavy soils as compared with lighter ones, if saline waters are used for irrigation purposes. Increased salinization with increase in the clay content of soils. An increase in salt content of the calcareous silty to clayey soils in Gujarat with the use of saline irrigation waters of two quality groups, one with a value of 2,000–3,000 µmhos/cm and the other of EC 5,000–7,000 µmhos/cm after a period of 3–10 years. Irrigation with saline water (EC 6,500 µmhos/cm) for seven years, created 0.18–0.36 per cent total salt content in heavy black soil, where cotton is being cultivated with 1–3 irrigations. In general, salinity build up with the use of saline water irrigation was more in the surface layer than in the lower depths and that the accumulation of salts progressed with years. The salinity level of the soil, however, tended to decrease in the rainy season due to the leaching effect. Highly saline waters are of help in improving water infiltration rates in soils of poor permeability. When sea waters (8.37 to 56.85 µmhos/cm) were used on sandy and sandy clay, the permeability of soils increased with increase in salinity of water and the electrical conductivity of saturation extracts varied from one-third to one-half that of sea water (Narayana *et at.*, 1966). Much lower EC values of the soil solutions of sea-shore sand with the use of sea water, containing 40 to 60 µmhos/cm of salts under the prevailing climate.

Sodium Hazard

High salt water (TCC ≥14.4 mg/l) containing divalent cations reduced ESP of alkali soil and increased its permeability (Badiger *et al.*, 1969). Highly saline water, in conjunction with calcium, seemed to increase hydraulic conductivity of saline sodic soil (Singh and Jaiswal, 1973). Higher Na concentration in the irrigation water replaces Ca ions from the exchange complex of clay colloids, thereby the structural make up, aggregate stability and water stable aggregates gets collapsed due to dispersion of soil colloid. The swelling and dispersion of clay particles in the pores was the main mechanism responsible for reducing the hydraulic conductivity. Irrigation water containing high proportions of sodium bicarbonate increase the exchangeable sodium in the field which result in deflocculation of clay which decreases soil permeability and aeration and ultimately the crop growth. Texture of the soil modifies the effect to large extent as it influences moisture retention and transmission properties of soil.

The bulk density of soil is increased considerably as the ESP increases due to dispersion. Higher bulk densities of these soils result in low infiltration and hydraulic conductivity, as a consequence of decreased bulk volume, total pore space and impeded drainage (Gupta and Verma, 1982). The cumulative infiltration in a highly alkaline soil is increased with increasing EC and decreased with increasing SAR of water used. Irrigation water containing high sodium increase the exchangeable sodium in the field which result in deflocculation of soil, decreased soil permeability and aeration which ultimately affect crop growth. The infiltration rate was higher when sodic soils were irrigated with water dominated by Na and Ca than that dominated by Na and Mg (Saha and Palkhiwala, 1979). A significant reduction in saturated hydraulic conductivity with increasing levels of RSC (Raghav *et al.*, 1991).

Table 35.1: RSC Levels on Saturated Hydraulic Conductivity of Soil

RSC Levels (meq/l)	*Saturated Hydraulic Conductivity (mm/hr)*
2.8	0.77
12.0	0.20
16.0	0.002

(Raghav *et al.*, 1991).

Porosity

Due to use of Na rich waters and pore spaces of the irrigated soil is reduced mainly because of the structural and the composition of the sub-surface soil due to translocation of the clay. The increase in bulk density due to use of poor quality waters, particularly sodic waters, is also responsible for bringing down the pore space as bulk density and pore space have inverse relationship.

Crust Formation

Irrigation with sodic waters creates alkalinity in the soil leading to low stability of the soil structure and dispersion of clay particles. This crust reduces infiltration rate causing over saturation and aeration problems and creates a strong barrier to the emergence of seedings. The factors responsible for the development of hard crust on soil surface are high exchangeable Na, low organic matter, puddling and wetting the soil to zero tension, which occurs in the field. The crust strength increases with increasing ESP and dispersion percentage of silt and clay (Goyal and Jain, 1982). Increasing crust thickness and crust strength was recorded with increasing salinity of irrigation water.

Table 35.2: Saline Water Irrigation on Crust Properties and Crack Width

Salinity of Irrigation Water (dSm^{-1})	*Crust Thickness (mm)*	*Crust Strength (kg/cm^2)*	*Crack Width (mm)*
2.7	1.03	4.55	4.00
6.0	1.56	20.20	4.66
9.0	1.90	26.41	6.30
12.0	2.26	19.85	6.36
Unirrigated	0.94	–	3.12

(Goyal and Jain, 1982).

The formation of crust in saline water irrigated soils could be attributed to the stagnation of water containing high sediments which accumulate in surface and sub-surface layers. The increased strength of soil crust was found to be associated with higher amount of silica deposition (Goyal, 1985). A very high degree of correlation was observed between the concentration of Si in irrigation water and soil crust force because the dissolved Si is absorbed instantaneously on both fine and coarse textured soils and adsorption had a highly significant relationship with its concentration of water.

Table 35.3: Salinity of Irrigation Water on Physico-Chemical Properties of the Soil

Salinity of Irrigation Water (mmhos cm^{-1})	*EC (dSm^{-1})*	*ESP*	*Silica Adsorption (μg/100 g soil)*	*Crust Strength (kg/cm^2)*
3	0.57	14.4	3.54	4.41
6	0.61	20.5	3.27	11.19
9	0.68	22.2	4.57	15.00
12	0.66	32.2	4.28	12.33

(Goyal, 1985).

Hydraulic Conductivity and Permeability

Both hydraulic conductivity and diffusivity decreased with an increase in SAR and reduction in the electrolyte concentration of percolating water. The decrease was more at SAR value of 10 with electrolyte concentration of 10-40 meq/litre as compared to that of 80 meq/litre (Ponnia and Pal, 1979). The Saline water having low proportion of Na either increase the hydraulic conductivity or may not have any significant effect but Na dominant water results in a definite reduction in hydraulic conductivity. The hydraulic conductivity decreased with an increase in SAR of irrigation water while it increased with increasing salt concentration (Sharma *et al.*, 1981; Khandelwal, 1986).

Table 35.4: EC and SAR of Irrigation Water on Hydraulic Conductivity of Soil (cm hr^{-1})

Levels of SAR	*Levels of EC (dSm^{-1})*	
	4.0	8.0
16	6.43	6.46
20	6.14	6.25

(Khandelwal, 1986).

Crack Formation

Swelling of clay in soils-irrigated with sodic water results in cracking, which poses problem of water management in heavy textured soils. A part of rain or irrigation water flows through these cracks directly to the groundwater. These infiltration losses diminish water storage in the soil, decrease the rain and irrigation water efficiency and results in a rise of the water table which leads to the secondary salinization and alkalization of soil. These soils not only dry quickly during dry periods but to a greater depth as a consequence of shrinkage which results in the formation of deep, wide cracks that accentuates evaporation losses. The tendency of a clay soil to swell when wet and to shrink and form wide deep cracks is greater with Na than with Ca in the soil solution and on the exchange complex of the soil.

Infiltration

The infiltration rate of soils irrigated with sodic water is known to decrease with increasing Na in the exchange complex. Under field conditions, cumulative infiltration in a highly alkali soil increased with increasing electrolyte concentration (0-40 meq/litre) and decreased with increasing SAR (5-20) of the water used. The infiltration rates of a soil with irrigation waters of the same salinity and SAR varied to a great deal depending on the kind of divalent cations (Ca and Mg) present in the waters in addition to Na. The steady state infiltration rate increased from 2.45 to 4.2 cm/day during salinization but was reduced to 0.75 cm/day when the soil was leached with good quality water (Khosla *et al.*, 1982).

Chemical Properties

Irrigation with saline water increased the EC of the soil by three to seven times as compared to before irrigation. EC of the soil after irrigation was less than EC of irrigation water used. The lower values of EC as compared to EC of irrigation water prevailed because of the light and medium texture of soils with good drainage. There were no significant changes in pH of soils. Considering root zone depth (0-20 cm), ESP very much pronounced. The differential increase in ESP seems to be the difference in amount of clay content in soil (Gupta and Abichandani, 1970). The ECe, SAR and boron content of soil increased while pH decreased with increase in the salt concentration of water. An increase in SAR of irrigation water resulted in higher ECe, SAR, pH, hydraulic conductivity and boron content of soil while hydraulic conductivity decreased.

The HCO_3, HCO_3 + Cl and HCO_3 types of groundwaters occupy the major aquifers of India. However, the waters with serious alkalinity hazards (RSC > 4. 0 meq/litre) are restricted to the localized pockets. Low RSC waters (< 4.0 meq/litre) generally have EC_{iw} of 2.5 dSm^{-1}. When alkali waters low in calcium are used for irrigation, sodication/alkalinization of the soils increase with RSC levels in the waters. Under the transient field conditions, sodication of surface soil is increased at higher EC_{iw} and also at higher RSC concentration for the similar levels of SAR/adj. SAR of waters, pH and ESP of the soil increased with the rise in RSC of irrigation water (Manchanda *et al.*, 1982).

Table 35.5: Different RSC Levels on pH and ESP of Soil

RSC Levels (meq/litre)	*pH*	*ESP*
2.8	8.86	29
11.0	9.50	50
16.0	9.73	60

(Manchanda *et al.*, 1982).

Irrigation of an established white clover grass pasture with saline water (2.4 dSm^{-1}) diminished the annual yield of biologically fixed N by 27 to 47 per cent compared to irrigation with channel water (0.07 dSm^{-1}) (Smith *et al.*, 1993).

Table 35.6: Changes in Soil ESP after Seven Year Irrigation with Sodic Water

Soil Depth (cm)	*Na in Irrigation Water (meq/litre)*							
	5.7				*18.4*			
	After		*Before*		*After*		*Before*	
	Na	*ESP*	*Na*	*ESP*	*Na*	*ESP*	*Na*	*ESP*
0–30	4.2	16.0	3.7	14.2	7.2	22.0	3.4	12.9
30–60	5.0	18.9	4.1	14.5	7.6	28.9	4.4	14.6
60–90	7.9	26.1	5.0	16.8	8.9	29.0	5.1	16.0

(Manchanda *et al.*, 1982).

Soil Salinization

The excessive accumulation of soluble salts in the soil is the most important single factor limiting plant growth. The nature and extent of salt accumulation in the soil depends upon soil texture, climate (mainly rainfall) and groundwater table. Accumulation of salts increases with the increase in clay content. Irrigation with water of EC higher than 3 dSm^{-1} or even 2 dSm^{-1} on heavy soil may inhibit the growth of any crop (Talati, 1969). The adverse effect of saline irrigation water on soil salinization was less on light textured soils than on heavier textured soils. The degree of adverse effect increased with the increasing salinity of irrigation water. The adverse effect of saline water irrigation increased with the clay content of the soil and number of irrigations and the increasing salt content of the waters. The relative amount of rains, their seasonal timings, sizes and intensities of individual storms and the frequency of rain storms affect the salinization of soil following irrigation with saline water. Irrigation with highly saline water (EC 15-19 dSm^{-1}, SAR 12-16) resulted in soil salinization (EC 15-55 dSm^{-1}) during the rabi season. Rainfall is helpful in the replacement of Na with Ca on the exchanged complex.

Soil Alkalization

The nature and amount of cations as well as anions in an irrigation water governs the adsorption of cations on the basis of their chemical affinity, activity and the solubility product in the soil solution. Soil alkalization is more pronounced when the water used for irrigation contains excess of (CO_3 + HCO_3) over (Ca^{3+} and Mg^{2+}) which is referred to as RSC. Irrigation with higher RSC water accentuates the formation of Na_2CO_3 in soil solution, which leads to excessive adsorption of Na on the exchange complex of the soil. As the soil solution gets concentrated due to evaporation and transpiration, the CO_3 and HCO_3 ions tend to precipitate Ca and Mg as their carbonates which have low solubility as compared to Na_2CO_3 increasing the relative proportion of Na in the soil solution and ultimately on the exchange complex of soil irrigated by such water. Of the two ions CO_3 and HO_3, soil alkalization is higher with CO_3 at the same level of salt concentration (Gupta, 1983).

Soil Microbial Properties

Physical and chemical properties of a soil determine the nature of the environment. This environmental characteristics affect soil microflora both qualitatively and quantitatively. The salts, which lower the production of crops also, affect the activities of desirable of soil micro-organisms. The total number of bacteria, actionmycetes and fungi are usually higher in productive normal soils followed by saline and saline-alkali soils.

In salt affected soils, the counts of nitrifying bacteria range from 1500-70000/g of soil depending upon the nature and extent of salt accumulation (Gupta and Bajpai, 1974). Salinity of soil affects the activity of nitrifiers. The harmful effect of different salts on nitrification was found to be in the order $Na_2CO_3 > NaCl > Na_2SO_4$ (Lipman and Sharp, 1912). Like ammonifiers, nitrifiers also tolerate higher concentrations of salts in the presence of $CaCO_3$. Many of the soils irrigated with saline water are calcareous and thus, $CaCO_3$ may provide a stimulatory effect with salt concentration which would be toxic to nitrifiers.

Saline and alkali soils constitute an unfavourable habitat for the growth and multiplication of rhizobia. An almost complete failure of nodule development of soybean, when the EC of the soil around the root exceeds 7 dSm^{-1}. The tolerance to salts varied considerably for strains of each rhizobia. Some strains could not grow in the absence of salts, whereas some other could easily tolerate high concentration of salts. The strains of rhizobia specific to cowpea and grams were sensitive to $NaHCO_3$ but were resistant to NaCl and Na_2SO_4 (Yadav and Vyas, 1971). Fungi being acidophilic, their population in general is very low in soils irrigated with saline water. Some fungi are capable of tolerating high salinity.

High magnesium saline waters exert marked influence in releasing calcium from Kankar, usually present in alkali soils (Mondal, 1972, 1973; Mondal and Bhumbla, 1972). An increase in availability of nitrogen in soils of Pali, Bhilwara and Jodhpur districts of Rajasthan with application of saline waters rich in nitrate contents. Favourable effects were observed with use of saline water having high content of nitrate and potassium on light textured soils of Gurgaon district of Haryana.

Boron Hazard

Greater adsorption of boron occurred in heavy soil than in light soil, with irrigation water containing high concentration of boron (4-5 ppm) whereas boron was retained equally by both the soils at low concentration (0-5 ppm B). Accumulation of boron in normal soil increased with an increase in EC and SAR of irrigation water and fineness of soil texture. An increase in water soluble boron of soil with an increase in SAR of irrigation water (Jain and Saxena, 1971). An increase in total and available boron contents of irrigated soils was recorded under irrigation with high boron waters. Positive correlation in boron content of the irrigated soil and that in the irrigation water was observed in the arid and semi-arid districts of Uttar Pradesh. There was sharp fall in the redox potential of soil within one month after submergence. The redox potential of soil did not decrease much afterwards. Along with the fall in redox potential of soil, there was an enormous increase in the exchangeable + water soluble Fe content of soil. With the fall in redox potential and continuance of submergence of soil there was decrease in the soil salinity.

Table 35.7: Irrigation Water Quality on Redox Potential (mv) of Soil

H_2S in Water Applied (ppm)	*Redox Potential (mv)*					
	ECe 3.1 dSm^{-1}			*ECe 5.9 dSm^{-1}*		
	0 DAT	*30 DAT*	*65 DAT*	*0 DAT*	*30 DAT*	*65 DAT*
0	205	85	80	210	90	85
25	200	80	85	205	90	85
50	200	65	65	210	75	60
100	200	58	50	210	60	45

(Bandyopadhyay *et al.*, 2001).

The salinity of soil also decreased slightly due to application of higher concentration of H_2S. Compared to other treatments, the pH of soil was slightly higher at treatment with 100 ppm H_2S water.

Table 35.8: Irrigation Water Quality on Exchangeable plus Water Soluble Fe (ppm)

H_2S in Water Applied (ppm)	*Exchangeable + Water Soluble Fe (ppm)*					
	ECe 3.1 dSm^{-1}			*ECe 5.9 dSm^{-1}*		
	0 DAT	*30 DAT*	*65 DAT*	*0 DAT*	*30 DAT*	*65 DAT*
0	18.3	60.4	58.5	21.3	61.2	60.0
25	18.3	58.7	52.9	21.3	57.2	53.5
50	18.3	53.2	47.6	21.3	52.0	46.3
100	18.3	45.6	39.0	21.3	44.4	40.0

Table 35.9: Irrigation Water Quality on Soil Salinity (ECe dSm^{-1})

H_2S in Water Applied (ppm)	Soil Salinity (ECe dSm^{-1})					
	ECe 3.1 dSm^{-1}			ECe 5.9 dSm^{-1}		
	0 DAT	30 DAT	65 DAT	0 DAT	30 DAT	65 DAT
0	3.1	2.4	1.7	5.9	4.8	2.9
25	3.1	2.4	1.7	5.9	4.8	2.8
50	3.1	2.2	1.6	5.9	4.6	2.5
100	3.1	1.9	1.3	5.9	4.3	2.1

Table 35.10: Salinity Status of Soils after Harvest of the Crop

Irrigation Water Salinity (meq/lit)	Salinity Status (mmhos/cm)			
	0-30 cm		30-60 cm	
	1973	1974	1973	1974
20	3.16	3.40	1.36	1.36
40	4.76	5.01	1.70	1.70
80	6.46	7.14	2.04	2.04
160	10.62	10.88	2.04	2.04

(Paliwal and Yadav, 1978).

Most of the salt was deposited in upper surface layers (0-30 cm) through the soil capillaries and root zone at lower depth was comparatively less saline (Paliwal and Yadav, 1978). ESP of soil is positively correlated with RSC and SSP of irrigation water with low coefficient value with SAR_{iw} (Paliwal and Maliwal, 1968; Lal *et al.*, 1988). Use of saline well waters for irrigation increased elctrolyte concentration and ESP which might deteriorate chemical and physical properties of soils (Maliwal and Timbadia, 2000). The chemical and physical properties of light textured soils will be much less adversely affected than heavy soils indicating safe utilization of poor quality irrigation waters (Gandhi, 1973).

Poor Quality Water on Crop Growth

Irrigation with poor quality groundwaters may cause salinity, specific ion toxicity or infiltration problem in soils. Such effects of irrigation may adversely affect crop production. Indiscriminate use of poor quality water affects the chemical properties of soil solution (increase in pH and EC) which has a definite negative influence on germination per centage, root growth, absorption of water and nutrients. The increase in the osmotic pressure of soil solution has a direct bearing on the growth behaviour of crops and their yield response pattern. The harmful effects of saline water irrigation are mainly associated with accumulation of salts in the soil and are manifested through reduced availability of water to plants, poor to delayed germination and slow growth rate. Excessive salts in the soil can induce early wilting and the effects are almost similar to those of drought. The visual symptoms are stunted growth, smaller but thicker leaves which have often dark green colour as compared to plants growing in a salt free soil irrigated with good quality water.

Fine textured soils remain dispersed and puddled when wet and hard when dry due to sodic water irrigation and do not attain proper soil moisture condition for cultivation. A thin crust is often formed at the surface, which acts as a barrier to penetration of water and seedling emergence. Increase in soil pH reduces availability of a number of plant nutrients like N, Zn, Fe, Ca and Mg. Excessive availability leads to toxicity of elements like B, Mo, F, Li and Se. Thus, the plants suffer from poor soil physical conditions, reduced nutrients availability and toxicity some of ions, when sodic waters are used for irrigation. The effect of saline water irrigation on crop growth varies with the composition of water, kind, variety and growth stages of the crop, planting method, irrigation technique, soil and climatic conditions and other related agronomic factors.

Rice

A clay loam soil irrigated with waters containing up to 1 per cent NaCl showed greater reduction in yield of rice in summer than in rainy season (Rao *et al.*, 1969). A 50 per cent reduction in height of seedlings took place at salt concentration of 60 mg/l in all varieties except China 4 (Paliwal and Mehta, 1973). Germination was delayed and per cent emergence decreased with increase in SAR (5 to 100) and salinity (4 to 64 meq/l) or both. Of the anions, bicarbonates and carbonates are more toxic, sulphates least toxic and chlorides and nitrates intermediate. Irrigation with saline water is more detrimental after transplanting and during tillering to flowering stages of crop growth (Yadav, 1975). The seedling emergence of rice cultivars was found to be delayed and affected to a greater extent with increasing levels of water salinity.

Table 35.11: Seedling Emergence of Rice Cultivars at Different Levels of Water Salinity

Salinity	*Rice Cultivars Seedling Emergence (%)*				
(dSm^{-1})	*BR 10*	*BR 11*	*BR 3*	*Patnai 3*	*DA 29*
0.2 (control)	95	96	83	89	81
4.0	95	95	79	89	79
8.0	93	93	76	88	73
12.0	91	91	74	75	65
16.0	86	86	25	73	27
20.0	21	21	01	24	06

(Ahmed *et al.*, 1990).

All the rice cultivars were found to be susceptible to emergence of seedling particularly at 20.00 dSm^{-1} salinity level. The seedling emergence was significantly reduced by 30 per cent and 74 per cent as the salinity of water increased from 0.2 (control) to 16.0 and 20.0 dSm^{-1} respectively, regardless of rice cultivars. Water salinity beyond 12 dSm^{-1} was found to be the critical level for seedling emergence of rice cultivars (65-91 per cent) (Ahmed *et al.*, 1990).

Table 35.12: Irrigation Water Quality on Dry Matter Yield of Rice (Cv Pankaj)

H_2S in Water Applied	*Dry Matter Yield (g/pot)*	
(ppm)	*ECe 3.1 dSm^{-1}*	*ECe 5.9 dSm^{-1}*
0	39.9	30.2
25	39.6	30.8
50	35.9	28.5
100	22.1	19.6

There was a significant decrease in dry matter yield of rice at both the soil salinity levels when H_2S concentration in irrigation water was 100 ppm (Bandyopadhyay *et al.*, 2001). The concentration of Fe, Mn, Zn, Cu and P in plant also reduced drastically when the concentration of H_2S in irrigation water was 100 ppm.

Table 35.13: Irrigation Water Quality on Nutrients Content on Rice Plant (ppm)

H_2S in Water Applied (ppm)	*ECe 3.1 dSm^{-1}*					*ECe 5.9 dSm^{-1}*				
	Fe	*Mn*	*Zn*	*Cu*	*P*	*Fe*	*Mn*	*Zn*	*Cu*	*P*
0	580	230	75	13	3050	620	245	78	15	2830
25	550	205	68	10	2909	568	220	70	12	2740
50	505	200	60	8	2530	530	205	64	9	2400
100	295	120	20	4	2050	410	135	22	5	1870

Plants show symptoms of scorching and leaf burning. High amounts of carbonate and bicarbonate ions injure the plants directly as has been observed with paddy seedlings following their transplantation on soils irrigated with sodic waters.

Wheat

A decrease in the yields of maize and wheat by irrigation with waters of increasing salinity (7.5, 22.5, 50.0 and 100.0 mg/l) and SAR (4, 8, 16 and 26). In areas where rainfall is 500 mm or higher and the soils are sandy, highly saline waters can be used for irrigating salt tolerant crops (wheat). The yield of wheat grain and maize fodder showed a 50 per cent reduction at EC of 20,00 μmhos/cm, but was not affected at EC of 5,600 μmhos/cm of the irrigation water. Use of saline tubewell waters (EC 2,700 μmhos/cm, SAR 9.1 and RSC 3.1 meq/l), for three years, did not affect the growth of wheat variety Pb 591 in the Agra tract of Uttar Pradesh (Tripathi *et al.*, 1971). An excess concentration of Na was shown to induce K deficiency leading to lower K/Na ratio in wheat irrigated saline water which resulted in poor yield (Finck, 1977).

Germination of six varieties of wheat was delayed and emergence decreased with increase in salinity (4, 8, 16, 32 and 64 mg/l) and SAR (5, 10, 25, 50 and 100), the effect being more marked with salinity than with alkalinity. The varieties NP 830 and Kharchia were found to be more tolerant (Maliwal and Paliwal, 1967). A reduction of 24.2 and 25.9 per cent in the yield of wheat at EC 4,800 μmhos/cm and SAR of 32 respectively. The uptake of N, P, K, Ca and Mg decreased with an increase in EC and SAR of irrigation waters (Lal and Singh, 1972). Good growth of wheat under irrigation with saline waters was reported up to RSC of 7.52 mg/l (Singh and Sharma, 1970).

Control plots produced higher grain and dry matter yield of wheat than those getting saline irrigation. Increase in salt concentration of irrigation water reduced grain yield. Saline irrigation at tillering stage was least harmful (Verma and Bains, 1974). The adverse effect of RSC was observed only on the third crop. The increase in RSC decreased the calcium content and increased the sodium content of the plants. RSC from CO_3 source was more harmful than HCO_3 source. Increasing SAR of irrigation water decreased wheat growth. An increase of sodium in proportion to calcium, magnesium or potash decreased the wheat yield in the order K > Mg > Ca (Ram Deo and Kanwar, 1968). Saline water irrigation at the pre-sowing stage is very harmful and thereafter at 50 per cent flowering stage also appears to be sensitive. Varieties Kharchia 65 and HD 1553 were found to be comparatively more tolerant.

Table 35.14: Wheat as Influenced by Saline Irrigation

Treatments	*Total Dry Matter Yield (g/60 × 60 cm plot)*	*Grain Yield (g/60 × 60 cm plot)*
Salt content of irrigation water (ppm)		
9000	281.0	111.2
18000	231.6	85.8
Control	396.0	153.1
Growth stages		
Tilleri mg	328.0	130.9
Jointing to ear emergence	253.7	109.3
Ear emergence to harvest	266.5	100.5
Tillering to harvest	177.0	53.7

(Verma and Bains, 1974).

Salinity build up increased with an increase in the salinity of irrigation water. After two seasons, the average EC_{iw} values of 8.39 were observed under field irrigated with 12 dSm^{-1} saline water. Seed yield showed reduction of 46.2 per cent at 12 dSm^{-1} over control. Reductions in yield components at 12 dSm^{-1} over control was 25.2, 44.7, 12.3 and 21.8 per cent for plant height, spikes/m row length, length of spike and 100 seed weight (Chopra and Chopra, 1997).

Table 35.15: Salinity Water on Yield Attributes and Yield of Wheat

Salinity of Water (dS/m)	*Plant Height (cm)*	*Spikes/m Row Length*	*Length of Spike (m)*	*100 Grains Weight (g)*	*Grain Yield (q/ha)*
4.0	64.7	82.1	6.3	4.63	23.68
6.0	55.3	66.8	6.1	4.26	19.12
9.0	49.9	52.1	5.3	4.22	13.19
12.0	48.3	45.4	5.5	3.62	12.74
LSD (5 %)	7.3	17.0	0.48	0.76	3.41

(Chopra and Chopra, 1997).

Groundwater often poor in quality is the major source of irrigation on 0.29 million ha land in western Rajasthan (Dhir, 1977). In Pali district alone 0.15 million ha land is irrigated by poor quality groundwater which has been classified as highly saline sodic in nature and therefore unsuitable for irrigation (lain, 1979). Wheat is the main cereal crop of the salt affected areas and rated semi-tolerant to salinity (Francois *et al.*, 1986). N, P and K concentration of wheat plants were not significantly influenced by irrigation waters of different salt concentration (EC 2, 4, 6, 8, 12 and 16 µmhos/cm). However, there was a decrease in N, P and K concentration of wheat plants where salty irrigation waters were applied (Lunin and Gallatin, 1965; Gauch and Wadleigh, 1945). Higher straw yield and grain yield of wheat and lower soil EC and pH were obtained under control (2.1 µmhos/cm irrigation water) (Rai, 1977).

Table 35.16: N, P and K Content of Wheat under Different Saline Irrigation Water

EC of water (mmhos/cm)	*Nitrogen (per cent)*			*Phosphorous (per cent)*			*Potassium (per cent)*		
	Tillering	*Jointing*	*Flowering*	*Tillering*	*Jointing*	*Flowering*	*Tillering*	*Jointing*	*Flowering*
0.6	1.66	1.27	0.82	0.32	0.24	0.12	3.32	2.63	2.40
2.0	1.65	1.25	0.79	0.32	024	0.13	3.34	2.67	2.41
4.0	1.63	1.25	0.78	0.30	0.25	0.12	3.33	2.69	2.40
6.0	1.63	1.21	0.78	0.31	0.23	0.11	3.36	2.60	2.37
8.0	1.56	1.23	0.78	0.31	0.23	0.11	3.32	2.64	2.39
12.0	1.56	1.21	0.76	0.31	0.22	0.13	3.33	2.62	2.29
16.0	1.56	1.21	0 78	0.29	0.23	0.11	3.30	2.62	2.29
CD (5 %)	NS	NS	NS	NS	NS	NS	NS	NS	NS

(Singh *et al.*, 1974).

Reduction in germination differ with different varieties of wheat (Maliwal and Paliwal, 1967; Sarin and Narayana, 1968; Singh and Dastane, 1970) and also in barley (Bhumbla *et al.*, 1968). Wheat and barley irrigated with saline waters having EC 6.0 to 6.5 μmhos/cm increase growths by 15-45 per cent. This increase could be due to increased availability of nutrients and their effective utilization in various physiological functions in the plant system (Paliwal, 1972; Sarin, *et al.*, 1972). Spread of salinity in the irrigated lands is to be so serious when long-term irrigation with poor quality water (Casey, 1972).

Table 35.17: Use of Saline Water for Irrigation on Germination and Yield of Wheat

EC of Water (mmhos/cm)	*Salinity Build in the Soil at Crop Harvest*		*Germination (%)*	*Straw Yield/Running meter (g)*	*Grain Yield/Running meter (g)*
	ECs (mmhos/cm)	*pH (1:2)*			
2.1	4.1	9.1	70.9	106.3	88.4
4.0	9.0	8.8	73.8	106.0	82.7
6.0	10.2	9.2	59.8	94.1	79.3
8.0	13.2	9.2	43.5	77.5	66.0
10.0	16.7	9.0	36.2	70.8	61.4
12.0	18.0	8.7	30.0	62.1	56.0
CD (5 %)	–	–	23.5	23.3	17.7

(Rai, 1977).

In general, wheat yield decreased as the salinity of the irrigation water increased. At all the salinity levels, the reduction in grain yield was always more than the dry matter yield. At the higher salinity level (160 meq/litre), the reduction in grain yield was 13.6 and 18.5 per cent in the two years respectively. It is primarily due to the fact that in light textured soils, salt accumulation does not occur much in the root zone if frequent irrigations are provided (Paliwal and Yadav, 1978). Wheat can successfully be grown with saline water having EC value of 8 μmhos/cm with a maximum reduction of 5 per cent (Paliwal and Yadav, 1976).

Table 35.18: Different Quality Waters on Relative Yield of Wheat

Salt Content (meq/lit)	*Grain Yield (q/ha)*		*Dry Matter Yield (q/ha)*	
0	1973	1974	1973	1974
3	29.5	29.0	78.5	79.2
20	30.1	28.5	79.6	79.0
40	29.1	28.4	77.7	79.2
80	28.2	27.2	76.5	79.5
160	25.5	23.6	72.5	68.5
CD (5%)	2.1	1.8	3.8	2.5

(Paliwal and Yadav, 1978).

Barley

Germination of barely was delayed and emergence decreased with increase in salinity and/or alkalinity. But in general, barely appeared to be more tolerant than wheat (Maliwal and Paliwal, 1967). Application of fertilizers and manures improved the soil properties and increased availability of nitrogen and phosphorous to barley irrigated with waters of varying quality (Paliwal and Maliwal, 1971). Varieties Amber, Vijay and Ratna have been found to be more tolerant at Agra (Yadav, 1975).

Table 35.19: Use of Saline Water for Irrigation on Germination and Yield of Barley

EC of water (mmhos/cm)	*Salinity build in the soil at Crop Harvest*		*Germination (%)*	*Straw Yield/Running meter (g)*	*Grain Yield/Running meter (g)*
	ECs (mmhos/cm)	*pH (1:2)*			
2.1	4.1	9.0	73.3	81.0	94.4
4.0	7.2	9.1	63.2	75.3	86.5
6.0	7.8	9.3	64.9	64.5	76.2
8.0	10.2	9.2	54.5	62.3	73.3
10.0	12.9	9.1	46.8	61.8	73.0
12.0	15.6	9.0	34.7	64.2	72.7
CD (5%)	–	–	13.0	13.3	NS

(Rai, 1977).

Irrigation with 2.1 μmhos/cm of EC registered higher germination (73.3 per cent), grain (81.0 g) and straw (94.4 g) yields of barley with lower soil salinity build up than other qualities of irrigation water (Rai, 1977). Decline in yield was recorded only at ECe 8 and 6 μmhos/cm in barley and wheat respectively. Yield declines 5 per cent in barley and 7 per cent in wheat per unit increase in salinity beyond threshold level (Rana, 1977). Salinity build up in the soil following saline water irrigation depends on soil type, amount of precipitation, drainage condition, crop rotation, crop management, wind velocity, sun shine intensity and duration, relative humidity etc.

Sorghum

In general, bicarbonates and carbonates were found to be most toxic, sulphates least and chlorides and nitrates intermediate (Paliwal and Gandhi, 1968). Sorghum has shown tolerance to saline water irrigation (Yadav, 1975). Of the potential yield (dry matter produced where good water was used throughout) 82, 67 and 32 per cent were obtained with EC_{iw} of 2.5, 4.7 and 6.4 dSm^{-1} respectively when higher salinity waters were used after pre-sowing with good water, yields were improved to about 85 per cent of the potential level whereas straw mulch did not show much benefits.

Pearl Millet

Normal yields were obtained with irrigation water having EC up to 28 mg/l and SAR of 25 on clay loam soil (Maliwal and Paliwal, 1971). Direct irrigation with sea water (24,000–35,000 ppm salt content) on sea shore sand for three years brought about reduction in grain yield of pearl millet. There was more accumulation of salts in leaves than in grains or stems.

Maize

Satisfactory growth of maize was obtained on clay loam soil in green house on irrigation with water of salinity up to 6,500 µmhos/cm and SAR 32.

Green Gram

Significant reduction in grain yield was not observed with water having EC up to 4.3 dSm^{-1} but reduced at higher levels of water salinity in all soils. Nitrogen content of grain was decreased with increasing salinity in water beyond EC of 4.3 dSm^{-1}. No significant reduction in N content of straw upto EC of 12.3 dSm^{-1} and after that it reduced to 18 dSm^{-1} EC. This is due to lack of partitioning of N because no grain formation occurred with higher salinity. Uptake of P was reduced after 8.9 dSm^{-1} EC water and K uptake was not affected with salinity level in grain but in straw it followed the pattern of N content in straw.

Oilseed Crops

The cultivation of groundnut may not be profitable when irrigated with saline water having EC 4.0 µmhos/cm and above. At EC 16.0 µmhos/cm, the groundnut plants did not survive. Reduction in yield of groundnut due to saline water irrigation was observed to be more during the period of pegging to pod formation stages (Yadav, 1975). Grain yield of safflower was better when irrigated with control than with water of higher EC and SAR. The yield and oil content were decreased significantly with an increase in EC, SAR and boron content of irrigation water. Application of brackish water decreased the plant height and seed yield of Indian mustard in both the soils. The seed yield reduction was more in Vetric Ustochrept than in Fluventic Eutrochrept. Plant height was more under Vertic Ustochrept than under Fluventic Eutrochrept. Significant reduction in yield was observed with water of 8 and 4 dSm^{-1} EC respectively in Fluentic Eutrochrept and Vertic Ustochrept soils. Application of saline water increased the soil salinity upto a depth of 120 cm, but the increase was more within the top 60 cm soil. Less saline water can be used on fine textured soil whereas more saline water can be used on coarse textured soil (Ravender Singh and Bhargava, 1997). There was a significant reduction in seed yield and plant height of sunflower when waters of higher salinity (EC_{iw} 9 and 12 dSm^{-1}) were used throughout for irrigation as compared to the substitution of pre-sowing irrigation with good quality water. The crop was more sensitive to salinity during later growth period of the crop. Pre-sowing irrigation with good quality water can help in the utilization of much higher salinity subsequently.

Tobacco

Irrigation waters containing up to 4000 μmhos/cm of dissolved salts did not have much adverse effects on tobacco in the well drained soils of low clay content (Ramakrishnan Kurup and Sastry, 1962). A decrease in the growth, as the salinity of sea water used for irrigation increased. On the other hand, irrigation with artificial brackish water increased the yield and improved the quality of tobacco. The presence of potassium nitrate in some well waters proved beneficial to tobacco crop (Talati, 1969). A substantial quantity of nitrogen and potassium is added to the soil annually through nitrate rich brackish irrigation water and this helped in producing higher yields of tobacco.

Cotton

Satisfactory yields of seed cotton was recorded in black clay soil when irrigated with saline water (EC 6500 μmhos/cm SSP 50 and SAR 8.25). Pre-sowing irrigation with saline water proved very harmful and resulted in delayed and reduced germination (Yadav, 1975). Saline, water irrigation at early seedling and boll development stages was also detrimental.

Horticultural Crops

Differential performance of fruit crops was observed to salinity (Minhas and Dubey, 1998). Salinity of irrigation water reduced growth of robusta banana crop. Decrease in girth was more pronounced at 5.0 dSm^{-1} level. There was significant reduction in the number of leaves due to salinity at the time of harvest. Bunch yield varied significantly. Salinity of water had significant effect on finger weight of banana. Likewise there was reductions in the pulp and peel weight as well as pulp/peel ratio of banana fruits and no effect on TSS content of fruit and uptake of nutrients reflected the negative effect of salinity (Palaniappan, 1996).

Table 35.20: Salt Tolerance of Fruit Species

Sensitive (EC dSm <4)	*Moderately Tolerant (EC 4-6 dSm^{-1})*	*Tolerant (EC 6-8 dSm^{-1})*	*Highly Tolerant (EC 8-16 dSm^{-1})*
Lemon	Peach	Sapota	Ber
Plum	Papaya	Pomegranate	Date palm
Grape fruit	Banana	Fig	
Orange	Grape	Olive	
Apple	Mango	Guava	
Pear	Custard Apple	Aonla	
Lime	Phalsa	Bael	
Avocado			
Cheery			
Strawberry			

(Minhas and Dubey, 1998).

The number of tillers/plant, height of the plants and total dry matter yield of citronella both cuttings were decreased with increasing levels of RSC and even the lowest RSC level had an adverse effect on yield and yield attributing characters as compared to control. The increasing levels of RSC decreased the oil content and yield significantly as compared to control.

Table 35.21: Saline Water Irrigation on Growth and Bunch Yield of *Robusta banana*

Salinity Level (dSm^{-1})	*Harvest Stage*			*Bunch Yield (kg/plant)*	*TSS brix (%)*
	Height (cm)	*Girth (cm)*	*No. of leaves/plant*		
0.3	158.7	58.7	8.0	18.18	24.5
2.5	148.5	58.5	7.0	15.52	23.5
5.0	123.5	47.5	6.2	10.25	23.5
CD (p = 0.05)	NS	8.8	0.7	1.71	NS

(Palaniappan, 1996).

Table 35.22: RSC Levels of Irrigation Water on *Citronella jawa* Plant

RSC Levels (mmol/litre)	*Tillers/Plant*	*Plant Height (cm)*	*Total Yield (g/plot)*	*Dry Yield (g/plot)*	*Oil Yield (ml/pot)*
0.0	8.50	54.80	12.30	1.740	0.630
7.5	7.80	50.80	12.10	1.690	0.600
15.0	7.10	48.20	11.70	1.650	0.570
22.5	6.0	46.30	10.90	1.620	0.550
CD (p = 0.05)	0.76	0.53	0.24	0.008	0.011

(Ramhod Singh *et al.*, 1994).

Soil pH was decreased from 5.7 to 3.8 with the addition of acidic water. The organic carbon content was slightly increased with decreasing pH due to lesser decomposition of organic matter.

Table 35.23: Acid Water Irrigation on Soil Properties and Fenugreek Plant

Treatment	*Plant Weight (g/dish)*	*Shoot Length (cm)*	*Post Harvest Soil Fertility (kg/ha)*				
			pH (1:2)	*Organic Carbon (%)*	*Available P*	*Available K*	*Available Na*
Distilled water (control) (pH – 7.0)	1.27	5.88	5.7	0.78	6.96	56.9	84.6
Groundwater (pH – 6.4)	1.29	6.19	5.6	0.79	7.11	54.4	85.6
Treated groundwater (pH – 5.5)	1.23	5.69	5.3	0.81	6.37	58.4	89.2
Treated groundwater (pH – 3.0)	1.07	5.29	4.9	0.83	5.94	58.0	89.0
Treated groundwater (pH – 2.2)	0.78	4.88	3.8	0.84	5.31	58.1	88.9

(Sanniigrahi, 1999).

Available phosphorous was also decreased but no change in K and Na content. When only groundwater was applied, both shoot length and plant weight of fenugreek were found to be higher which were decreased further with the acidity of irrigation water. The yields of radish were reduced when irrigation water of increasing salinity was used at early growth stage. Salinity tolerance, however, increased at later growth stages.

Coconut

Growth and yields of coconuts were not affected by irrigation with waters having EC in the range of 548–2181 μmhos/cm. But there was a reduction in yield and size of the nuts when irrigated with water of EC 2900 μmhos/cm containing 5.66 mg/l sulphates (Kulkarni *et al.*, 1973).

Poor Quality Water on Crop Quality

Besides the yield of crops, quality of the produce is also very much affected in green gram (Tripathi and Praveen Kumar, 1999). Oil content of crops like sunflower, mustard and safflower is significantly affected by irrigation with saline and sodic water (Ashok Kumar *et al.*, 1997). In general soil salinity caused by saline irrigation water my result in reduction in size of produce, change in colour and appearance and change in the composition of produce. A reduction of seed size in groundnuts beginning at soil salinity levels (ECe) of 3 dSm^{-1}. However, there is an increase in seed oil content with increasing salinity up to a point (Shalhevet *et al.*, 1969). In case of tomatoes for every increase in 1.5 dSm^{-1} in mean ECe beyond 2 dSm^{-1}, there was a 10 per cent reduction in yield. The yield reduction was due to reduction in fruit size and weight and not to reduction in fruit number. However, there was a marked increase in soluble solids in the extract, which may be an important criterion for tomato juice production. Fruits from higher salinity levels were less liable to damage and the number of spoiled fruits was less (Shalhevet and Yaron, 1973). Increased salinity reduced fruit size in muskmelons. However, ripening was accelerated by salinity (Meiri *et al.*, 1981). Grapefruits yield decreased with increase in chloride ion concentration. The yield reduction was caused more by reduction in fruit size and weight. Salinity effects on fruit quality were similar to those caused by water stress. Comparing the low and high salinity levels, there is an increase in soluble solids and titratable acidity in the juice. There were no differences in juice content (Bielorai *et al.*, 1978). Increase in the quality of wheat, melons and alfalfa from use of saline drainage water for irrigation (Rhoddes *et al.*, 1989).

References

Abrol, I.P. and Bhumbla, D.K. (1971). *Saline and Alkali Soils in India: Their Occurrence and Management.* FAO Report, FAO, Rome, 41: 42-51.

Ashok Kumar, K.L. *et al.* (1997). *J. Indian Soc. Soil Sci.*, 45(2): 409-410.

Bandyopadhyay, B.K. *et al.* (2001). *J. Indian Soc. Coastal Agric. Res.*, 19(1 and 2): 112-119.

Bernstein, L. (1962). In: *Proc. Symp. on Problems of Arid Zone*, Paris, pp. 139-171.

Bhumbla, D.R. and Abrol, I.P. (1972). *Indian Fmg.*, 21(4): 15-16.

Bhumbla, D.R. *et al.* (1968). *Indian J. Agron.*, 13: 181-185.

Casey, H.E. (1972). Salinity problems in arid land irrigation. *Arid Land Resource Information Papers*, 1: 28-36.

Dhir, R.P. (1977). In: *Desertification and Its Control.* ICAR, New Delhi, pp. 130-148.

Francois, L.E. *et al.* (1986). *Agron. J.*, 78: 1053-1058.

Gandhi, A.P. (1973). Ph.D. thesis, Udaipur University, Udaipur.

Gauch, H.G. and Wadleigh, C.H. (1945). *Soil Sci.*, 59: 139-153.

Goyal, R.S. (1985). *J. Indian Soc. Soil Sci.*, 33: 235-245.

Goyal, R.S. and Jain, B.L. (1982). *J. Indian Soc. Soil Sci.*, 30: 447-454.

Gupta, B.R. and Bajpai, P.D. (1974). *J. Indian Soc. Soil Sci.*, 22: 176-180.

Gupta, I.C. (1979). *Use of Saline Water in Agriculture.* Oxford and IBH Publishing Company Pvt. Ltd., New Delhi, pp. 210.

Gupta, I.C. (1983). *Curr. Agric.*, 7: 97-113.

Honura, J.F. *et al.* (2000). In: *Recent Advances in Irrigation Management for Field Crops.* Tamil Nadu Agricultural University, Coimbatore, pp. 252-257.

Jain, B.L. (1979). *Annals of Arid Zone*, 18: 135-139.

Khandelwal, R.B. (1986). Ph.D. thesis, Rajasthan Agricultural University, Bikaner.

Khosla, B.K. *et al.* (1982). In: *47th Ann. Conven. Indian Soc. Soil Sci.*, Nagpur, Abt. No. 12.

Lal, P. *et al.* (1988). *J. Indian Soc. Soil Sci.*, 46: 119-123.

Lipman, C.B. and Sharp, L.D. (1912). *Soil Sci.*, 1: 291-297.

Lunin, J. and Gallatin, M.H. (1965). *Agron. J.*, 57: 339-342

Maliwal, G.L. and Paliwal, K.V. (1967). *Indian J. Pl. Physiol.*, 10: 26-35.

Maliwal, G.L. and Paliwal, K.V. (1972). *Agro Chimica*, 16: 450-468.

Maliwal, G.L. and Timbadia, N.K. (2000). *J. Indian Soc. Soil Sic.*, 18 (1): 30-35.

Manchanda, H.R. and Bhandari, D.K. (1976). *J. Indian Soc. Soil Sci.*, 24: 432-435.

Paliwal, K.V. (1972). *Indian Fmg.*, 22: 70-71.

Paliwal, K.V. and Maliwal, G.L. (1968). *Annals of Arid Zone*, 7: 127-131.

Paliwal, K.V. and Yadav, B.R. (1976). Irrigation water quality and crop management in the union territory Delhi. *Tech. Bull.*, IARI, New Delhi, 9: 166.

Paliwal, K.V. and Yadav, B.R. (1978). *Indian J. Agron.*, 23(4): 334-336.

Paliwal, K.V. *et al.* (1975). *Annual Report*. Udaipur University, Udaipur.

Poonia, S.R. and Pal, R. (1979). *Agric. Water Mgmt.*, 2: 163-175.

Rai, M. (1977). *Indian J. Agron.*, 22(4): 206-211.

Rana, R.S. (1977). In: Proc. Indo-Hungarian Seminar on Management of Salt Affected Soils, February 7-12, CSSRI, Karnal, pp. 29-31.

Rhoddes, J.D. *et al.* (1992). *The Use of Saline Waters for Crop Production*. Scientific Publishers, Jodhpur, pp. 23-40.

Sarin, M.N. and Narayanan, A. (1968). *Physiologia, Pl.*, 21: 1201-1209.

Sarin, M.N. *et al.* (1972). In: *Proc. Symp. on New Development in the Field of Salt Affected Soils.* ISSS, Cairo, pp. 108-109.

Shah, R.K. and Palkhiwala, A.B. (1978). *Curr. Agric.* 3: 53-58.

Sharma, O.P. *et al.* (1981). *Curr. Agric.*, 5: 48-52.

Singh, D.V. *et al.* (1974). *Indian J. Agron.*, 19(4): 281-282.

Singh, N.P. and Dastane, N.G. (1970). *Indian J. Agron.*, 15: 84-85.

Talati, R.P. (1969). In: *Proc. Symp. on Soil Water Management*. Hisar, pp. 300-301.

Tripathi, K.P. and Praveen Kumar (1999). *J. Indian Soc. Soil Sci.*, 47(2): 383-385.

Verma, U.K. and Bains, S.S. (1974). *Indian J. Agron.*, 19(1): 80.

Yadav, J.S.P. (1978). In: *Proc. Indian Nat. Sci. Acad.*, 44B: 205-212.

Yadav, N.K. and Vyas, S.R. (1971). *Indian J. Agri. Sci.*, 41: 875-881.

Yadav, R.K. (1986). *Indian Hort.*, 30: 2-5.

36

Sustainability of Paddy Cultivation in a Tannery Effluent Polluted Agricultural Environment

R. VENKATTAKUMAR
Scientist, Extension NRCC, Puttur, D.K., Karnataka

ABSTRACT

The study aimed to measure the sustainability of paddy cultivation in a tannery effluent polluted agricultural environment. The study was conducted in randomly selected 6 moderately affected and 6 severely affected villages from 2 affected Taluks of Vellore District with 60 moderately and 60 severely affected respondents. To measure sustainability of paddy cultivation, indicators were identified and standardized for using at the field level, through a well-structured and pre-tested interview schedule. Principal component analysis and ANOVA were used to derive further interpretation.

There were significant differences existed among the moderately affected and severely affected respondents with respect to sustainability of paddy cultivation, while both the categories had low to medium level of sustainability of paddy cultivation. Useful implications were drawn from the study and pointers for future research also were suggested.

Keywords: *Sustainability of paddy cultivation, Tannery effluent pollution, Indicators measuring Sustainability.*

Introduction

With over 700 tonnes of leather being processed daily, Vellore district of Tamil Nadu, accounts for 40 per cent of the country's and 80 per cent of Tamil Nadu's exports, earning over Rs.2000 crores in foreign exchange annually. (Krishnakumar, 1995). So there is no doubt the tannery industries in Vellore district are contributing a considerable amount for the Indian exchequer.

But at the same time Singaram and Pothiraj (1989) found that the location of 75 per cent of tanneries in Vellore district of Tamil Nadu had caused damage to soil and irrigation water quality. According to the report of Department of Industrial Development (1990) the soil fertility was affected due to:

1. Salinity of wastewater
2. High sodium absorption ratio causing concentration of suspended solids clogging soils because of leather pollution.

Tannery effluent pollution in Vellore district led to a shift in cropping pattern in affected farms from paddy, sugarcane, groundnut and vegetables to crops such as sorghum, maize, finger millet, coconut, fodder crops and fruit trees. The intensity of shift in cropping pattern was greatly influenced by the pollution intensity. The increase in pollution intensity resulted in increased fallow lands and also increased the number of years of fallow lands (Amarnath, 1997). Murthi (1998) reported that the agricultural lands were becoming unfit for cultivation as could be seen from the stunted growth of crops in the tannery effluent affected areas. According to Ambur Farmers Agricultural Association vice-president C.A., Vadivelu, paddy yield declined from 6 tonnes/ha, in the 1970's to 3.7 tonnes/ha in the 1980's, it is less than 3.5 tonnes/ha now. (Krishnakumar, 1999)

The perusal of the above background leads to the idea that due to the continuous pollution of agricultural lands by the tannery effluent pollution, the sustainability of agriculture in the district, especially the sustainability of paddy cultivation becomes a big question mark. Since this condition is not existing over the recent but being a tragedy over a period of more than twenty years of time, it raises a stage strong basis for "measuring the sustainability of paddy cultivation in this area". So the study was conducted with the following objective.

"To measure the sustainability, of paddy cultivation in a tannery effluent polluted agricultural environment."

Concept Clarification

Alteri (1987) defined sustainability as the ability of an agro-ecosystem to maintain production through time in the face of long-term ecological constraints and socio-economic pressure.

Based on this concept which is relevant to the present study, that test the agricultural environment polluted continuously by tannery effluent, the sustainability of the agriculture in this condition could be measured. Since the study area along the Palar river was cultivated with paddy, the sustainability of paddy cultivation was taken for the study.

Research Methodology

Sampling Procedure

Vellore district of Tamil Nadu was selected as the locale for this study due to the relatively highest number of tanneries in the district and also due to the acute pollution problem reported in the district.

A multi-stage sampling technique, was used to select the study area from the selected locale two taluks were considered; two taluks affected by tannery effluents. The affected taluks were selected based on the concentration of maximum number of tanneries per unit agricultural area.

From the affected two taluks 6 severely affected and 6 moderately affected villages were selected randomly after getting their list from the Assistant Director of Agricultural Offices (ADA) and the concerned taluks.

Measuring Indicators for Sustainability

In the present study the thrust was to identify the components that truly reflect the form level sustainability, amenable to operationalisation, accessible to farmer's information domain, in the given tannery effluent agricultural environment. So sustainability of paddy cultivation in a tannery effluent polluted environment could be operationalised as the process by which a farmer manages to maintain the productivity of paddy cultivation to meet out the family needs *viz.*, food and economic requirement for both present and future generations. In this way, all the indicators that may measure the sustainability of paddy cultivation in the given condition were collected based on the review of literature and by discussion with the experts from environmental sciences, soil science and agronomy which were 14 in number initially. These selected indicators were discussed with the resource persons who had knowledge and experience in pollution studies and then scrutinized for their suitability, for operationalisation of measurement and possibility of eliciting data from farmers. After this process eight indicators were retained as essential for measuring the sustainability.

The selected indicators were, subjected to relevancy rating by 33 judges to obtain the respective relevancy coefficients. The indicators with relevancy coefficients more than 0.6 were selected, thus 7 indicators were selected. Then the scale values of each indicator were computed based on their relative importance to measure the sustainability, as given by the judges.

The sustainability of paddy cultivation was measured based on their relevancy coefficient and scale values, after measuring each of the selected indicators based on the operationalisation (given in Annexure) using the following formula. (Nagabhusanam, 1997):

$$\text{Sustainability index} = \frac{Uij \cdot S_j}{\text{Total Scale value} \times 100}$$

Where

Uij = Unit value of the ith respondent on j^{th} component.

Sj = Scale value of the jth component.

Collection of Data

The data were collected using a well structured and pre-tested interview schedule. The data were collected between September to November 1999 by personally interviewing the respondents.

Statistical Tools Used

The collected data were subjected to statistical analysis with tools *viz.*, mean, standard deviation, percentage analysis, analysis of variance (ANOVA) and principal component analysis, for further meaningful interpretation.

Results and Discussion

Sustainability Levels for Overall Respondents

A glance at the Table 36.1, revealed that slightly more than one-tenth (13.33 per cent) of the overall respondents belonged to low sustainability group, whereas 75.00 per cent and 11.67 per cent of the overall respondents belonged to medium and high groups respectively with respect to the sustainability of paddy cultivation in a tannery effluent polluted agricultural environment.

Table 36.1: Distribution of Overall, Moderately and Severely Affected Respondents According to the Sustainability of Paddy Cultivation

Sl.No.	Categories	Moderately Affected Farmers (n=60)		Severely Affected Farmers (n=60)		Overall Affected Farmers (n=120)	
		No.	Per cent	No.	Per cent	No.	Per cent
1.	Low	9	15.00	10	16.67	16	13.33
2.	Medium	43	71.67	43	71.67	90	75.00
3.	High	8	13.33	7	11.66	14	11.67
	Total	60	100.00	60	100.00	120	100.00

It could be concluded that about ninety per cent (88.33 per cent) of the affected respondents belonged to low to medium level category with respect to the sustainability of paddy cultivation in the given environment.

So from the above results it could be inferred that in general, due to the continuous exposure of the agricultural ecosystem to the tannery effluent pollution, the sustainable paddy production became limited due to the stress developed by the pollution. So the sustainability of paddy cultivation among the overall affected people was observed from low to medium level.

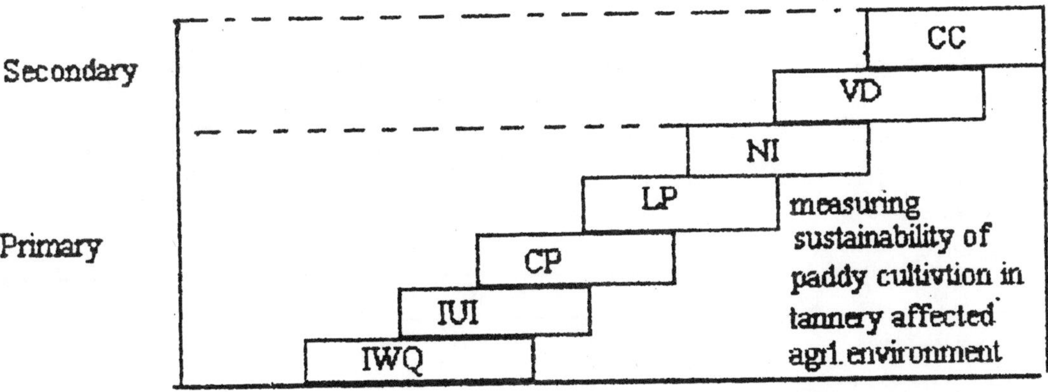

Fig. 36.1: Primary and Secondary Indicators to Measure Sustainability of Paddy Cultivation in a Tannery Pollutant Agricultural Environment

Category-wise Sustainability of Paddy Cultivation

Sustainability Level of Moderately Affected Respondents

A perusal of Table 36.1, reveals that 15.00 per cent of the moderately affected respondents belonged to low category, whereas 71.67 per cent and 13.33 per cent of them were belonging to medium and high category respectively.

In general, it could be concluded that more than four-fifth of the moderately affected respondents (86.67 per cent) belonged to low to medium level of sustainability with respect to paddy cultivation.

It could be further interpreted that due to the exposure of agricultural lands in the moderately affected areas to the tannery effluent pollution, the paddy cultivation was forced to be less sustainability even though the crop was not affected seriously when compared to severely affected areas. But there was reduction in the net production per unit area and so, profit also, which might have caused the above result.

Sustainability Level of Severely Affected Respondents

It could be observed from the Table 36.1, that 16.67 per cent of the severely affected respondents belonged to low level of sustainability whereas 71.67 per cent and 11.66 per cent of them belonged to medium and high levels of sustainability with respect to paddy cultivation.

It could be concluded in general that slightly less than nine-tenth of the severely affected respondents (88.34 per cent) had come under low to medium level category with respect to sustainability of paddy cultivation.

It could be further interpreted that in the severely affected areas due to the severe problem, the paddy cultivation was resulted in unsustainability and hence the results obtained.

Analysis of Variance for Sustainability of Paddy cultivation among Moderately and Severely Affected Farmers

It is evident from the Table 36.2, that there was existence of significant difference among the moderately and severely affected respondents with respect to the sustainability of paddy cultivation in the tannery effluent polluted agricultural environment. This was statistically confirmed as the 'F' value was 11.40, which found significant at 0.01 level of probability.

Table 36.2: Analysis of Variance for Sustainability of Paddy Cultivation Among Moderately and Severely Affected Respondents

Sources of Variation	*Degrees of Freedom*	*Sum of Squares*	*Mean Squares*	*'F' value*
Treatment	1	1247.844	1247.844	11.39909**
Error	118	12917.310	109.4688	1.0
Total	119	14165.160		

** Significant at 0.01 level.

Treatment Mean Values:

Moderately Affected	Severely Affected
55.5175	49.0683

The treatment mean values of the groups which were arranged in descending order and compared with the calculated critical difference showed that, the mean value difference was more than the critical difference, so it was proved significant. The mean value representation in the Table 36.2 showed that sustainability was more for moderately affected farmers and less for severely affected farmers.

These results observed might be possible due to the fact that the problem of tannery effluent pollution was more in the severely affected areas than in the moderately affected areas, even though both areas were affected. So there was the possibility for less sustainability in severely affected areas than in the moderately affected areas.

Indicators Influencing the Sustainability of Paddy Cultivation in a Tannery Effluent Polluted Agricultural Environment

The principal component analysis (PCA) was used to identify the indicators or group of indicators influencing the sustainability of paddy cultivation in a tannery effluent polluted agricultural environment.

Table 36.3 gives the Eigen values and cumulative variation percentage of the given components. The results of the PCA emphasized that the first component has exerted 43.014 per cent variation, Out of the 7 components, the component upto 5 accounted for more than 90.00 per cent of the variation. The first two components had contributed more than fifty per cent of variation and thereafter, the variation had steadily increased.

Table 36.3: Eigen Values from Principles Component Analysis of Indicators for Sustainability

Component Number	*Eigen Values*	*Cumulative Variation per cent*
I	3.017	43.014
II	1.178	59.936
III	0.961	73.734
IV	0.854	85.874
V	0.455	92.379
VI	0.301	96.683
VII	0.232	100.00

Further by using this data the per cent variation was calculated in components, This would give clear picture about the components, which had the larger contribution. It is possible to obtain the zero order correlation between variables and components from the coefficient of Eigen vector. This correlation in the square of the coefficient associated with the components in the vector, which was used to calculate the per cent variation. These statistics are presented under Table 36.4.

It could be seen from the Table 36.4, that only two components *viz.*, III and I had contributed the maximum variation for the sustainable paddy cultivation in the descending order. The component III expressed the variation to the extent, of 38.81 per cent followed by component number I with the variation level of 27.04 per cent. All the significant and highest valued indicators in each of these two components were identified and presented in Table 36.5.

Table 36.4: Variation in Sustainability Explained by the Indicators

Component Number	*Eigen Values*	*Zero Order Correlation with Sustainability*	*Percentage of Variation in Component*
I	1.737	0.520	27.04
II	1.085	0.261	6.81
III	1.980	0.623	38.81
IV	0.924	0.345	11.90
V	0.675	0.366	13.40
VI	0.549	0.066	0.44
VII	0.482	0.241	5.81

Table 36.5: Principle Components Significantly Associated in the Components where Higher Percentage of Variation Established

Sl.No.	*Indicators*	*Components*		*Rank*
		III	*I*	
1.	Crop productivity	–	0.88	III
2.	Irrigation Water Quality	-0.902	0.292	I
3.	Land Productivity	–	0.823	IV
4.	Net Income	–	0.802	V
5.	Input Use Index	–	0.908	II
6.	Varietal Diversity	0.274	–	VI
7.	Carrying Capacity	–	–	VII

The findings in Table 36.5, reveals that the indicator irrigation water quality was observed in both the components with highest value of correlation, whereas the indicators *viz.*, input use index, crop productivity, land productivity, net income and varietal diversity were observed in one component with highest value of correlation. The negative sign of the coefficient indicated an invades association with the component.

From foregoing analysis according to the level of significance and number of times explained in two different components, these indicators could be grouped into two as prime and secondary indicators for explaining sustainability.

Prime Indicators	*Secondary Indicators*
Irrigation Water Quality (IWQ)	Variety Diversity (VD)
Input Use Index (IUI)	Carrying Capacity (CC)
Crop Productivity (CP)	
Land Productivity (LP)	
Net Income (NI)	

The indicators IWQ, IUI, CP, LP and NI were found as prime indicators whereas VD and CC were found to be secondary indicators. Primary indicators could be visualized that these could be of more existence in the entire polluted environment where as these secondary indicators could be visualized depending upon the availability of resistant varieties and extent to which the paddy cultivation is becoming unsustainable.

Conclusion

More than four-fifth of the affected farmers in the study area had only low to medium level of sustainability with respect to paddy cultivation. Significant difference was existed among the moderately and severely affected respondents with respect to sustainability of paddy cultivation. The indicators IWQ, IUI, CP, LP and NI were found as prime indicators and the VD and CC were found to be secondary indicators influencing the paddy cultivation sustainability.

Implications

The researchers may identify, standardise and popularize the sustainable paddy cultivation practices for this tannery effluent polluted agricultural environment. The effective alternative land use pattern should be enticed and popularized, if the, paddy cultivation will remain unsustainable in the study area.

Pointers for Future Research

1. Sustainability of various crops either system-wise (or) crop-wise can be studied in the polluted areas.
2. Sustainability of socio-economic conditions and even culture of the farming community can be studied as an impact of pollution.

References

Alteri, M.A. (1986). *Agro-ecology: The Scientific Booms of Alternative Agriculture.* Div. of Biological Control, Univ. of California, Berkeley.

Amarnath, J.S. (1997). *Externalities Analysis and Economic Valuation of Tannery Pollution in Vellore District.* Ph.D. thesis, TNAU, Coimbatore.

Department of Industrial Development Report (1990). Report of the working group on leather and leather goods industries for the Eighth Five Year Plan (1990-95), Department of Industrial Development, Ministry of Industry, Government of India, New Delhi.

Krishnakumar, Asha (1995). Streams of poison: Tannery effluents; their land and people. *Frontline,* August 11.

Krishnakumar, Asha (1999). Along the Polluted Palar. *Frontline,* October 8.

Murthi, P.V.V. (1998). CEPT's fail to convince farmers. *The Hindu,* September 10.

Nagabhusanam, M. (1997). *Sustainability of Agriculture in Watershed Environment: An Analysis.* Ph.D. thesis, TNAU, Coimbatore.

Singaram, P and Pothiraj, P. (1989). *Soil Health Care in Tannery Polluted Areas of North Arcot District.* Paper presented in the national symposium, on Environment and its Conservation, Abdul Hakim College Melvisharam, Tamil Nadu.

Annexure

Selected indicators with their operational definition:

1. *Crop productivity*–is operationalised as the ratio between the yield of paddy crop obtained with that of the expected yield level.
2. *Irrigation water quality*–is operationalised as the perceived quality of the irrigation water obtained from both well and tank irrigation sources as perceived by the farmers.
3. *Land productivity*–is operationalised as yield per unit area, expressed in terms of tons per hectare.
4. *Net income*–is operationalised as the sum of rupees as profit obtained from paddy cultivation for the past three years.
5. *Input use index*–is operationalised as the level of adoption of the production inputs for paddy crop grown by farmers as against the level to sustain the crop production.
6. *Cropping pattern*–is operationalised on the pattern of utilization of available area.
7. *Varietal diversity*–is operationalised as the genetic diversification of the paddy crop as adopted by the farmers in a tannery effluent polluted agricultural environment in order to deriving maximum benefits and causing minimum damage to the crop by not allowing a single variety too continuous expense.
8. *Carrying capacity*–is operationalised as the level of pressure exerted on the limit of land by human and cattle population to protect the ecological degradation.

37

Nutrient Uptake and Yield of Sorghum as Influenced by Irrigation Methods, Levels of Coir Waste Incorporation, Placement of Hydrophillic Weirs and Sectors Under Saline Water Irrigation

A. SOLAIMALAI

Research Associates, Krishi Vigyan Kendra, Tamil Nadu Agricultural University, Vriddhachalam, Tamil Nadu

K. SANKARANARAYANAN

Senior Scientist (Agronomy), Central Institute for Cotton Research (ICAR), Coimbatore, Tamil Nadu

M. BHASKAR

Research Associate (Soil Science and Agricultural Chemistry), A.D. Agricultural College and Research Institute, Trichy, Tamil Nadu

ABSTRACT

Field experiments were conducted during kharif 2000, summer 2001 and kharif 2002 at Coimbatore to study the influence of irrigation methods, levels of coir waste incorporation, placement of hydrophillic weirs and sectors on nutrient uptake and yield of sorghum under saline water irrigation. Irrigation methods had no significant influence on nutrient uptake and yield of sorghum. Higher nutrient uptake and yield of sorghum were observed with coir waste applied @ 20 t/ha. Placement of hydrophillic weir improved nutrient uptake and yield in sector 3. Lower nutrient uptake and yield were recorded in sector 3 (50-75 m) while sector 1 gave higher nutrient uptake and yield.

Introduction

In India, sorghum is grown in an area of 11.57 million ha with a production productivity and area under irrigation are 11.09 million tonnes, 958 kg/ha and 6.7 per cent respectively (Anonymous, 1998). This less productivity of sorghum may be due to crop grown under moisture stress, use of saline water for irrigation and poor utilization of applied nutrients. Total uptake of N and K in sorghum was usually higher with irrigation (Bennett *et al.*, 1964). The transport of nutrients in the soil, absorption by the roots, translocations within the plant and their metabolism were influenced by soil moisture availability (Singh, 1972). Surge irrigation saves irrigation water and increases irrigation efficiency. But a penultimate depression (50-75 m) of crop growth and yield is observed in tropical condition. Coir pith application improves nutrient uptake, grain yield and soil physico-chemical properties than no coir waste application (Krishnan, 1986). Hence, attempts were made through the present investigation to study the effect of irrigation methods, levels coir waste incorporation, placement of hydrophillic weirs and sectors on nutrient uptake and yield of sorghum under saline water irrigation.

Materials and Methods

The field experiment was carried out at the Tamil Nadu Agricultural University, Coimbatore during kharif 2000, summer 2001 and kharif 2002. The soil of the experimental site was sandy clay loam with pH 8.2, low in available N (186 kg/ha), medium in available P (11 kg/ha) and high in available K (413 kg/ha). The treatments consisted of combination of two irrigation methods (I_1 – continuous irrigation, I_2 – surge irrigation), two coir pith levels (10 t/ha, 20 t/ha), placement of hydrophillic weirs (H_0 – No hydraulic weir, H_1 – Hydrophillic weir) and four sectors (S_1: 0-25 m, S_2: 25-50 m, S_3: 50-75 m, S_4: 75-100 m) was compared with common control (I_3 – Basin furrow method) in factorial randomized block design replicated thrice. Hundred metre long furrows with 0.5 per cent gradient towards one end were formed at an interval of 90 cm both for surge and continuous irrigations. Inlet pipes (6 cm diameter with 50 cm length) were placed at the head of each furrow. Sorghum variety CO 26 seeds were sown on both sides of ridges by adopting a spacing of 90/2 × 15 cm (45 × 15 cm). Hydrophillic weirs were designed with help of coconut epicarb and bamboo peg. Coconut epicarb of 5.0 and 7.5 cm height was fixed at the middle of the bamboo peg of 15 cm length with the help of nail. Two hydrophillic weirs one (5.0 cm height) at 60 m distance and another (7.5 cm height) at 75 m distance were placed across the furrow from 30 DAS onwards. For surge irrigation, irrigation water was given at 10 minutes 'on' and 'off' fashion. Irrigation was scheduled at 0.6 IW/CPE ratio. The irrigation water was saline in nature (pH 8.5, EC 1.27 dSm^{-1}, SAR 4.86 and RSC 2.24). Plant samples were collected and were analyzed for N, P and K contents (Tandon, 1994) and their uptake values were computed at vegetative, flowering and maturity stages. Yield was recorded at harvest.

Results and Discussion

Nutrient Uptake

The results showed that the uptake of N, P and K were not influenced by irrigation methods during all the three seasons (Tables 37.1, 37.2 and 37.3). Between the coir pith levels, higher uptake of N, P and K by sorghum were registered with soil incorporation of coir waste @ 20 t/ha than coir waste applied @ 10 t/ha. Increased availability of nutrients due to contribution from coir waste and adequate supply of soil moisture might have favoured better mineralization and solubilization of the nutrients resulting in higher availability and uptake by crop. These results corroborate with the findings of Flaig (1982). Among the sectors, sectors 1, 2 and 4 were comparable in N, P and K uptake at all the growth stages. Sector 3 recorded significantly lower nutrient uptake as compared to other sectors during all the three seasons and growth stages of crop. This might be due to lesser availability of soil moisture and lesser infiltration opportunity time in sector 3 (50-75 m distance). Placement of two hydrophillic weirs (5.0 cm height at 60 m distance and another 7.5 cm height at 75 m distance) registered higher N, P and K uptake as compared to the no hydrophillic weir. This might be due to obstruction of irrigation water in sector 3 and stagnated and then allowed the water front to over flow which in turn increased soil moisture content in sector 3 (50-75 m) under placement of hydrophillic weirs. The interaction effect between hydrophillic weirs and sectors was significant but all the other interactions were not significant.

Table 37.1: Influence of Irrigation Methods, Coir Waste Levels, Barrier and Sectors on Nutrient Uptake and Yield of Sorghum During Kharif 2000

Treatments	*Nitrogen Uptake (kg/ha)*			*Phosphorous Uptake (kg/ha)*			*Potassium Uptake (kg/ha)*			*Yield (kg/ha)*	
	30 DAS	*60 DAS*	*Harvest*	*30 DAS*	*60 DAS*	*Harvest*	*30 DAS*	*60 DAS*	*Harvest*	*Grain*	*Stover*
Irrigation Methods											
Continuous	27.33	60.14	120.2	7.41	12.63	30.25	33.08	76.04	130.8	6126	13501
Surge	27.25	60.00	119.4	7.35	12.54	30.17	32.97	75.78	129.5	6116	13483
Basin furrow	27.47	60.27	120.6	7.47	12.71	30.34	33.14	76.40	131.6	6135	13521
CD (P = 0.05)	NS	NS	NS	NS	NS	NS	NS	NS	NS	NS	NS
Coir Waste (t/ha)											
10	26.20	56.89	114.5	7.09	12.01	28.95	31.57	72.89	124.5	5828	12838
20	28.38	63.26	124.9	7.67	13.71	31.47	34.49	78.92	135.9	6415	14146
CD (P = 0.05)	0.21	0.44	0.98	0.17	0.20	0.28	0.37	0.55	1.1	52	87
Barrier											
No hydrophillic weir	27.23	59.20	117.9	7.30	12.34	29.68	33.00	75.18	128.3	6036	13296
Hydrophillic weir	27.36	60.94	121.5	7.46	12.83	30.74	33.03	76.64	132.1	6207	13688
CD (P = 0.05)	NS	0.44	0.98	NS	0.20	0.28	NS	0.55	1.1	52	87
Sectors											
0-25 m	28.24	61.74	123.5	7.62	12.96	31.01	33.93	78.27	134.2	6285	13853
25-50 m	28.15	61.56	122.4	7.57	12.89	30.93	33.83	78.07	133.1	6270	13832
50-75 m	24.77	55.53	110.9	6.81	11.66	28.06	30.90	69.38	121.2	5672	12480
75-100 m	28.01	61.45	122.0	7.51	12.83	30.84	33.45	77.91	132.2	6259	13803
CD (P = 0.05)	0.29	0.62	1.3	0.19	0.29	0.36	0.48	0.71	1.4	76	128

Table 37.2: Influence of Irrigation Methods, Coir Waste Levels, Barrier and Sectors on Nutrient Uptake and Yield of Sorghum During Summer 2001

Treatments	*Nitrogen Uptake (kg/ha)*			*Phosphorous Uptake (kg/ha)*			*Potassium Uptake (kg/ha)*			*Yield (kg/ha)*	
	30 DAS	*60 DAS*	*Harvest*	*30 DAS*	*60 DAS*	*Harvest*	*30 DAS*	*60 DAS*	*Harvest*	*Grain*	*Stover*
Irrigation Methods											
Continuous	24.87	55.93	109.7	6.51	12.38	28.41	29.41	68.24	115.5	5422	12018
Surge	24.78	55.88	109.2	6.45	12.25	28.26	29.11	68.20	115.0	5413	12003
Basin furrow	24.93	55.98	109.8	6.68	12.45	28.57	29.24	68.31	116.3	5431	12029
CD (P = 0.05)	NS	NS	NS	NS	NS	NS	NS	NS	NS	NS	NS
Coir Waste (t/ha)											
10	24.07	53.42	104.5	6.36	11.76	27.13	27.45	64.11	108.5	5089	11290
20	25.59	58.38	114.5	6.60	12.87	29.54	30.80	72.34	122.0	5725	12731
CD (P = 0.05)	0.18	0.36	0.8	0.14	0.18	0.26	0.32	0.51	0.8	47	80
Barrier											
No hydrophillic weir	24.85	54.41	106.5	6.44	12.01	27.61	29.10	65.56	111.1	5202	11574
Hydrophillic weir	24.81	57.40	112.3	6.51	12.63	29.06	29.15	70.89	119.3	5612	12447
CD (P = 0.05)	NS	0.36	0.8	NS	0.18	0.26	NS	0.51	0.8	47	80
Sectors											
0-25 m	25.65	56.78	112.4	6.75	12.57	28.84	30.42	69.70	118.3	5546	12274
25-50 m	25.60	56.72	111.4	6.70	12.51	28.74	30.37	69.65	117.6	5531	12254
50-75 m	22.56	53.46	103.7	5.83	11.72	27.04	25.39	63.95	108.0	5035	11277
75-100 m	25.50	56.65	110.3	6.63	12.46	28.66	30.32	69.59	117.0	5517	12237
CD (P = 0.05)	0.27	0.56	1.2	0.19	0.26	0.34	0.44	0.67	1.1	67	113

Yield of Sorghum

Higher grain yield of sorghum was recorded in kharif seasons than in summer season. Irrigation methods did not influence the grain and stover yields of sorghum significantly. With reference to coir pith levels, application coir pith. @ 20 t/ha increased grain (6415, 5725 and 6851 kg/ha) and stover (14146, 12731 and 14471 kg/ha) yields significantly. A yield increase of 10.07, 11.11 and 7.65 per cent was recorded under coir waste incorporation @ 20 t/ha over coir pith applied @ 10 t/ha in the three seasons respectively. This might be due to higher uptake of nutrients by crop and favourable soil condition under higher level of coir waste incorporation. Among the sectors, higher grain yield (6285, 5546 and 6697 kg/ha in the three seasons respectively) and stover (13853, 12274 and 14323 kg/ha) yields were registered in sector 1 (0-25 m distance) which was on par with sector 2 and sector 4. Sector 3 produced significantly lesser grain (5672, 5035 and 6303 kg/ha in the three seasons respectively) and stover (12480, 11277 and 13493 kg/ha) yields. The yield reduction was 9.75, 9.21 and 15.88 per cent under sector 3 as compared to sector 1. This might be due to lower uptake of nutrients and lower availability of soil nutrients. Hydrophillic weir placed across the furrow resulted in higher grain (6267, 5612 and 6769 kg/ha) and stover (13688, 12447 and 14650 kg/ha) yields. The yield improvement was 7.31, 7.58 and 5.63 per cent under placement of hydrophillic weir than no hydrophillic weir treatment. Placement of hydrophillic weirs might have increased soil moisture content and uptake of nutrients, which in turn improved yield of crop. The interaction effect between hydrophillic weir and sectors was significant. All other interactions were not significant. It can be concluded that surge irrigation with application of coir waste @ 20 t/ha and placement of hydrophillic weirs increases the nutrient uptake and productivity of sorghum under saline water irrigation.

Table 37.3: Influence of Irrigation Methods, Coir Waste Levels, Barrier and Sectors on Nutrient Uptake and Yield of Sorghum During Kharif 2002

Treatments	*Nitrogen Uptake (kg/ha)*			*Phosphorous Uptake (kg/ha)*			*Potassium Uptake (kg/ha)*			*Yield (kg/ha)*	
	30 DAS	*60 DAS*	*Harvest*	*30 DAS*	*60 DAS*	*Harvest*	*30 DAS*	*60 DAS*	*Harvest*	*Grain*	*Stover*
Irrigation Methods											
Continuous	30.18	64.82	126.8	7.92	14.49	32.18	35.32	81.17	138.7	6594	14106
Surge	30.16	64.78	126.3	7.89	14.45	32.13	35.27	81.14	138.1	6583	14091
Basin furrow	30.27	64.94	127.2	7.97	14.53	32.22	35.44	81.22	139.9	6614	14120
CD (P = 0.05)	NS	NS	NS	NS	NS	NS	NS	NS	NS	NS	NS
Coir Waste (t/ha)											
10	28.97	62.34	121.8	7.59	13.92	30.84	33.91	78.24	132.7	6327	13726
20	31.37	67.26	131.8	8.22	15.02	33.46	36.68	84.08	144.1	6851	14471
CD (P = 0.05)	0.25	0.48	1.0	0.20	0.26	0.33	0.40	0.58	1.1	66	94
Barrier											
No hydrophillic weir	30.10	62.99	123.6	7.91	14.08	31.29	35.24	78.91	134.8	6408	13547
Hydrophillic weir	30.21	66.60	130.0	7.89	14.86	33.02	35.35	83.40	142.0	6769	14650
CD (P = 0.05)	NS	0.48	1.0	NS	0.26	0.33	NS	0.58	1.1	66	94
Sectors											
0-25 m	31.08	65.72	129.2	8.16	14.74	32.68	36.36	82.39	141.1	6697	14323
25-50 m	31.03	65.66	128.5	8.13	14.68	32.62	36.30	82.34	140.5	6685	14303
50-75 m	27.59	62.22	121.4	7.26	.13.83	30.75	32.27	77.62	132.1	6303	13493
75-100 m	30.97	65.50	128.0	8.08	14.63	32.56	36.25	82.27	139.8	6670	14275
CD (P = 0.05)	0.33	0.67	1.4	0.25	0.34	0.40	0.49	0.76	1.3	94	133

References

Anonymous (1998). *Agricultural Statistics at a Glance.* Ministry of Agriculture, New Delhi.

Bennett, O.L., Doss, B.D., Ashby, D.A. and Richardson, E.C. (1964). Effects of soil moisture regime on yield, nutrient content and ET for three annual forage species. *Agron. J.*, 56: 195-198.

Flaig, W. (1982). *Dynamics of Organic Matter Decomposition in Soils.* In: 12th International Congress of Soil Science, New Delhi, pp. 115-124.

Krishnan, P.K. (1986). *Integrated Use of Organic Wastes and Fertilizer Nitrogen.* M.Sc (Ag) Thesis, Tamil Nadu Agricultural University, Coimbatore.

Singh, N.P. (1972). Irrigation and fertilizer use in crop production. *Indian Fmg.*, 22: 65-68.

Tandon, B.L.S. (1994). *Soil, Water and Fertilizer Analysis.* FEDCO, New Delhi.

38

Irrigation Water Quality of Ghataprabha Command Area, Karnataka

C.K. JAIN, C.P. KUMAR & M.K. SHARMA
National Institute of Hydrology, Roorkee, Uttaranchal

ABSTRACT

The groundwater quality of Ghataprabha Command Area (Karnataka) has been assessed to see the suitability of groundwater for irrigation applications. Sixty groundwater samples from shallow and deep wells were collected during pre- and post-monsoon seasons in the month of May and November 2001 respectively. The values of sodium absorption ratio (SAR) ranged from 0.48 to 22.0 during pre-monsoon season and 0.32 to 10.8 during post-monsoon season. The values of SAR indicate that majority of samples of the area falls under the category of low sodium hazard indicating no risk of sodification. According to U.S. Salinity Laboratory Classification of irrigation water, more than 50 per cent samples fall under water type C3-S1 during both pre- and post-monsoon seasons. Such type of water cannot be used on soils with restricted drainage. Even with adequate drainage special management for salinity control may be required and plants with good tolerance should be selected.

Keywords: *Groundwater quality, TDS, SAR, U.S. Salinity Laboratory Classification.*

Introduction

Groundwater plays an important role in agriculture, for both watering of crops and for irrigation of dry season crops. It is estimated that about 45 per cent of irrigation water requirement is met from groundwater sources. Indiscriminate use of fertilizers and pesticides in agricultural fields has become common practice in India and has already resulted in very high concentrations of these constituents in groundwater. It has been estimated that once pollution has entered the sub-surface environment, it may remain concealed for many years, becoming dispersed over wide areas of groundwater aquifer and rendering groundwater supplies unsuitable for consumption and other uses.

Assessment of groundwater quality in command areas assume great importance in view of its role in irrigated agriculture. Many problems originate during inefficient management of water for agriculture use especially, when it carries high salts. The problem of groundwater contamination in several parts of the country has become so acute that unless urgent steps for detailed identification and abatement are taken, extensive groundwater resources may be damaged. Tamta (1994) reported high concentrations of fluoride in groundwater, in Karnataka. Bhadale *et al.* (1996) and Suresh (1996) has carried out groundwater quality studies in Belgaum city and reported contamination in old Belgaum, Khade Bazar and Maruti Galli. In certain areas of the city nitrate content is above the permissible limits. In the present study, groundwater quality of Ghataprabha Command Area has been examined with reference to irrigation purposes.

Study Area

The study area lies between 16°00′ and 16°40′ N latitude and 74°15′ and 76°00′ E longitude in the state of Karnataka (Fig. 38.1). The area is bounded by river Krishna in the north, Maharashtra State to the west, the confluence of Krishna river and Malprabha river in the east and the water divide or the basin boundary between Ghataprabha and Malaprabha rivers in the south. The total area taken up for the study is 10,370 km^2. The command area and its environs essentially form part of the Krishna river basin of northern Karnataka and is drained by Krishna river and its tributary, Ghataprabha. A small portion towards the eastern part of the basin is drained by the Malaprabha river, which also joins the Krishna river near Sangam in Bagalkot taluk.

The average annual rainfall in the area is about 650 mm, out of which 65 per cent occur during the southwest monsoon (June to September). The contribution from northeast monsoon is nearly 20 per cent and the rest is received during the pre-monsoon period. The climate is generally tropical with hot summers and cold winters. The major land use in the study area is agriculture and about 60 per cent of population is dependent on it. The major crops grown in the area are Paddy, Sugarcane, Ragi, Bajra, Jowar, Maize, Pulses, Groundnut, Cotton and Oilseeds.

Methodology

Sixty groundwater samples were collected from the study area each during pre-monsoon (May 2001) and post-monsoon (November 2001) seasons. Some parameters like pH and conductance were measured in the field at the time of sample collection using portable kits. For other parameters samples were preserved by adding appropriate reagents and brought to the laboratory for detailed chemical analysis. The details of sampling locations are given in Table 38.1. The source and depth-wise distribution of sampling sites is given in Table 38.2. The physico-chemical analysis was performed following standard methods (APHA, 1985; Jain and Bhatia, 1988).

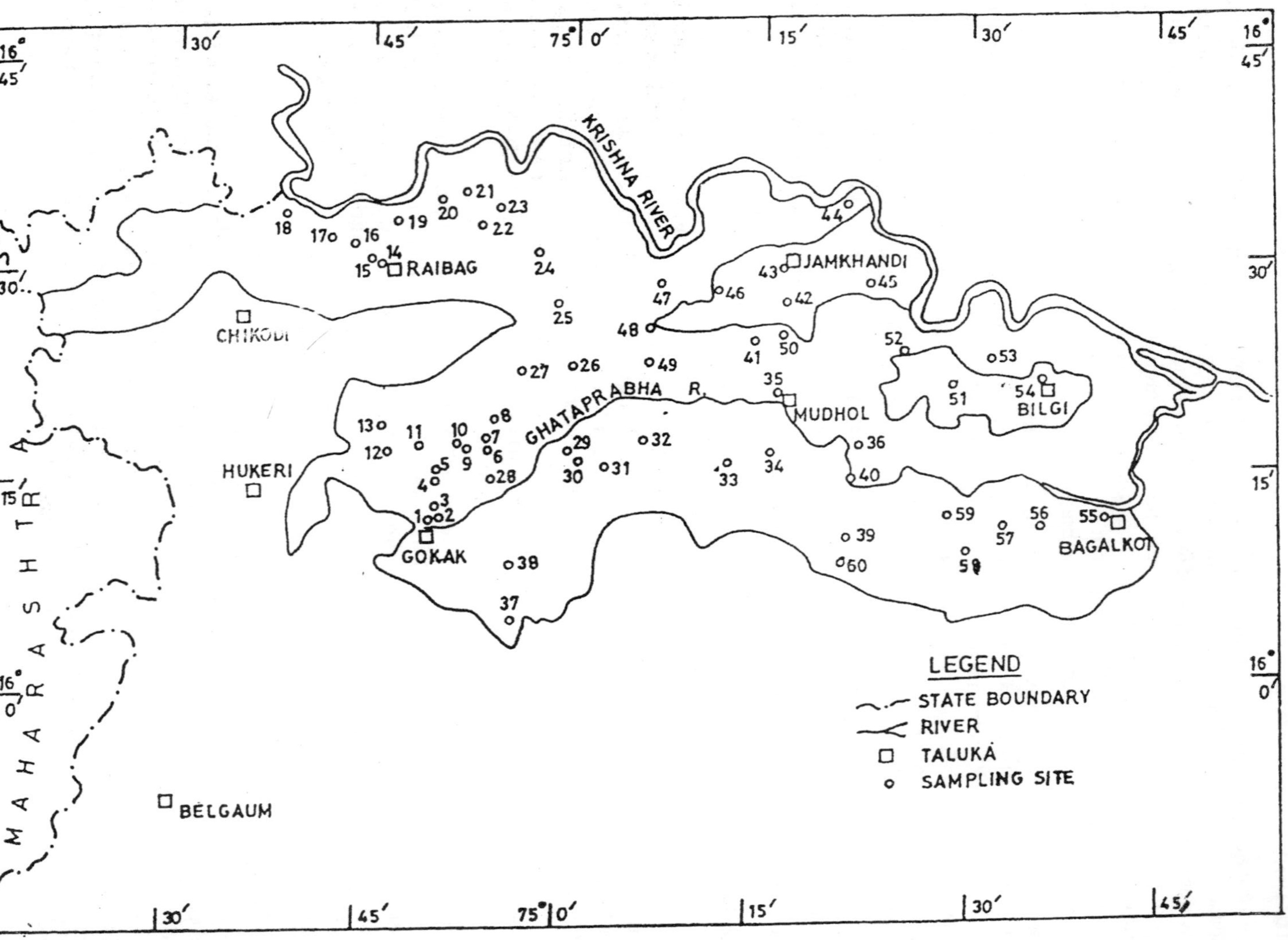

Fig. 38.1: Study Area Showing Location of Sampling Points

Table 38.1: Description of Groundwater Sampling Locations

Sample No.	Location	Source	Sample No.	Location	Source
1.	Gokak	BW	31.	Kuligod	HP
2.	Gokak	HP	32.	Venkatapur	OW
3.	Gokak	OW	33.	Yadwad	HP
4.	Arbhavi	OW	34.	Metgud	HP
5.	Arbhavi	HP	35.	Mudhol	HP
6.	Kalloli	OW	36.	Marikatti	HP
7.	Kalloli	HP	37.	Hirenandi	HP
8.	Tuknautti	OW	38.	Mamdapur	OW
9.	Rajapar	HP	39.	Lokapur	HP
10.	Rajapar	OW	40.	Muddapur	HP
11.	Badigwada	HP	41.	Shirol	OW
12.	Bogewadi	OW	42.	Shiddapur	HP
13.	Kabbur	OW	43.	Jamkhandi	HP
14.	Raibag	HP	44.	Algur	HP
15.	Raibag	OW	45.	Lignur	HP
16.	Raibag	OW	46.	Navalgi	BW
17.	Nasalapur	OW	47.	Rampur Rabakavi	HP
18.	Kallol	OW	48.	Chimmod	HP
19.	Birdi	HP	49.	Mahalingapur	HP
20.	Chincholi	HP	50.	Mantur	HP
21.	Chincholi	HP	51.	Halagali	HP
22.	Morab	HP	52.	Amalzhari	HP
23.	Harugeri	OW	53.	Teggi	OW
24.	Hidkal	OW	54.	Bilgi	HP
25.	Mughalkhod	OW	55.	Bagalkot	HP
26.	Gurlapur	OW	56.	Gaddankere Cross	HP
27.	Kankanwadi	OW	57.	Tulsigeri	HP
28.	Hunshal	OW	58.	Hireshallikeri	HP
29.	Sunadholi	HP	59.	BCM Hostel	HP
30.	Sunadholi	HP	60.	Varchagal Cross	HP

BW–Bore well, HP–Hand pump, OW–Open well.

Results and Discussion

The hydro-chemical data for the two sets of samples collected during pre- and post-monsoon seasons are presented in Table 38.3 and the results have been discussed in an earlier report (Jain *et al.*, 2002). The quality of water plays an important role in irrigated agriculture. Many problems originate during inefficient management of water for agriculture use especially, when it carries high salts. Irrigation water quality refers to its suitability for agricultural use. The concentration and composition of dissolved constituents in water determine its quality for irrigation use. Quality of water is an important consideration in any appraisal of salinity or alkali conditions in an irrigated area. Good quality water has the potential to cause maximum yield under good soil and water management practices. The most important characteristics of water which determine suitability of groundwater for irrigation purpose are: *(i)* Total concentration of soluble salts, *(ii)* Relative proportion of sodium to other cations, *(iii)* Relative concentration of bicarbonate to calcium and magnesium and *(iv)* Concentration of boron.

Table 38.2: Source and Depth-wise Distribution of Sampling Sites

Source Structure	*Depth Range*			*Total Number*
	< 20 m	*20-40 m*	*> 40 m*	
Hand Pumps	–	55,59,60	2,5,7,9,11,14,19, 20,21,22,29,30, 31,33,34,35,36, 37,39,40,42,43, 44,45,47,48,49, 50,51,52,54,56, 57,58	37
Bore Well	–	1	46	2
Open Wells	3,4,6,8,10,12,13, 15,16,17,18,23, 24,25,26,27,28, 32,38,41,53	–	–	21
Total	21	4	35	60

Table 38.3: Hydro-chemical Data of Groundwater Samples of Ghataprabha Command Area

Parameter	*Min*		*Max*		*Mean*	
pH	6.8	(6.6)	8.8	(8.5)	7.35	(7.44)
Conductivity (μS/cm)	356	(246)	15726	(12710)	1996	(1456)
TDS (mg/l)	228	(157)	10065	(8134)	1277	(932)
Alkalinity (mg/l)	116	(65)	744	(307)	335	(168)
Hardness (mg/l)	64	(80)	4382	(3361)	551	(409)
Chloride	2.5	(5.0)	2500	(2102)	191	(133)
Sulphate (mg/l)	2.5	(3.5)	1950	(1976)	172	(184)
Nitrate (mg/l)	ND	(ND)	2222	(1452)	111	(115)
Phosphate (mg/l)	ND	(ND)	1.20	(0.71)	0.12	(0.03)
Fluoride (mg/l)	ND	(ND)	6.13	(5.40)	1.21	(0.87)
Sodium (mg/l)	12	(8.1)	1563	(1342)	154	(117)
Potassium (mg/l)	1.8	(ND)	468	(406)	34	(27)
Calcium (mg/l)	11	(12)	1275	(876)	144	(95)
Magnesium (mg/l)	3.9	(7.0)	292	(286)	47	(41)

N = 60; Values given in parenthesis indicate post-monsoon data.

Total Concentration of Soluble Salts

In natural waters, dissolved solids consists mainly of inorganic salts such as carbonates, bicarbonates, chlorides, sulphates, phosphates and nitrates of calcium, magnesium, sodium, potassium, iron etc. and small amount of organic matter and dissolved gases. In the present study the TDS values varies from 228 to 10065 mg/l during pre-monsoon season and 157 to 8134 mg/l during post-monsoon season with more than 85 and 60 per cent samples having TDS values above the desirable limit of 500 mg/l during pre- and post-monsoon seasons respectively (Table 38.4). In the study area, 15 per cent sample even crosses the maximum permissible limit of 2000 mg/l during pre-monsoon season and 10 per cent samples during post-monsoon season. The spatial distribution map of TDS in pre- and post-monsoon season is shown in Figs. 38.2 and 38.3 respectively. In general TDS content decreases during post-monsoon season due to dilution effect of rainwater.

Table 38.4: TDS Distribution in Groundwater of Ghataprabha Command Area

Sl.No.	*TDS Range, mg/l*	*Depth Range, m*	*Sample Numbers*		*Areal Distribution %*	
			Pre-monsoon	*Post-monsoon*	*Pre-monsoon*	*Post-monsoon*
1.	0-500	< 20	10,15,27	3,4,8,10,12,15,16, 18,26,27,28,38		
		20-40	–	–	11.7	36.7
		> 40	22,35,48,54	21,22,35,36,37, 48,49,50,51,54		
2.	500-2000	< 20	3,4,6,8,12,16,18, 23,24,25,26,28, 32,38,41	1,55,59		
		20-40	1,55,59	2,5,7,9,11,14, 20,29,30,31,34,39, 40,42,43,45,46,47, 52,56,57,58	73.3	53.3
		>40	2,5,7,9,11,14,20,21, 29,30,34,36,37,39, 40,42,43,45,46,49, 50,51,52,56,57,58	13,53		
3.	> 2000	< 20	13,17,53	–		
		20-40	–	–	15.0	10.0
		> 40	19,31,33,44,47,60	19,33,44,60		
Total number of samples		60	60	100	100	100

The quality of irrigation water depends primarily on its silt and salt content. Water used for irrigation always contains measurable quantities of dissolved substances, which, as a general collective term, are called salts. They include relatively small but important amounts of dissolved solids originating from dissolution or weathering of the rocks and soils and dissolving of lime, gypsum and other salt sources as water passes over or percolates through them. The salts present in the water, besides affecting the growth of the plants directly, also affect the soil structure, permeability

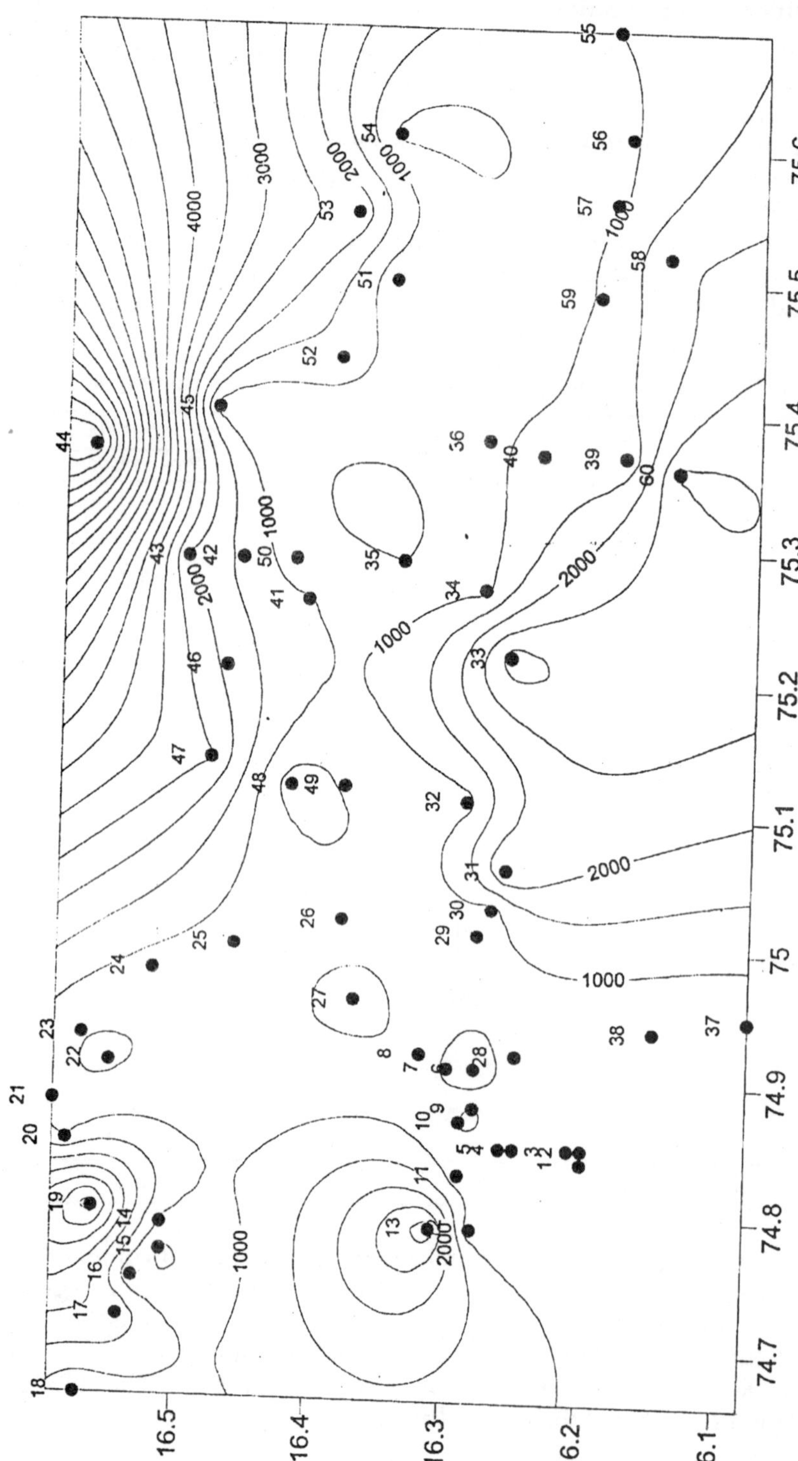

Fig. 38.2: Distribution of TDS in Ghataprabha Command Area (Pre-monsoon 2001)

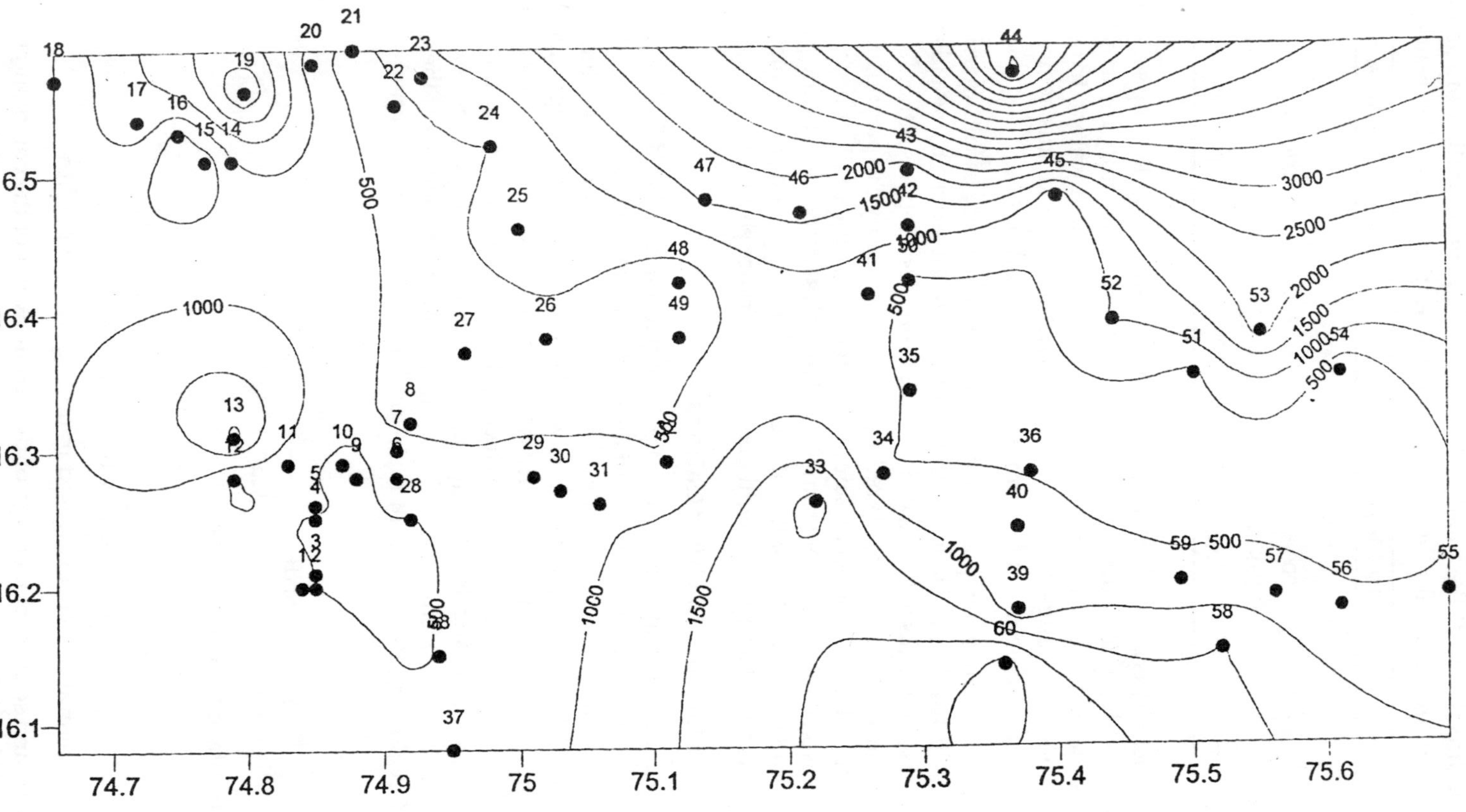

Fig. 38.3: Distribution of TDS in Ghataprabha Command Area (Post-monsoon 2001)

and aeration, which indirectly affect the plant growth. The total concentration of soluble salts in irrigation water can thus be expressed for the purpose of classification of irrigation water as follows:

Zone	*TDS (mg/l)*	*Conductance (μS/cm)*
Low Salinity Zone	< 200	< 250
Medium Salinity Zone	200-500	250-750
High Salinity Zone	500-1500	750-2250
Very High Salinity Zone	1500-3000	2250-5000

The electrical conductivity is a useful parameter of water quality for indicating salinity hazards. In general waters with conductivity values below 750 μS/cm are satisfactory for irrigation except for salt sensitive crops that may be adversely affected by the use of such waters. Irrigation waters having conductivity values in the range 250-750 μS/cm are widely used and satisfactory crop growth is obtained under good management and favourable drainage conditions.

A salinity problem related to water quality occurs if the total quantity of salts in irrigation water is high enough for the salts to accumulate in the crop root zone to the extent that yields are affected. If excessive quantities of soluble salts accumulate in the root zone, the crop has difficulty in extracting enough water from the salty soil solution. This reduced water uptake by the plant usually results in slow or reduced growth. About 60 per cent samples of the study area fall in high salinity zone during pre-monsoon season while 35 per cent samples fall in high salinity zone during post-monsoon season. Such water should not be used on soils with restricted drainage and special management for salinity control may be required and plants with good salt tolerance should be selected.

Relative Proportion of Sodium to Other Cations

A high salt concentration in water leads to formation of a saline soil and high sodium leads to development of an alkali soil. The sodium or alkali hazard in the use of water for irrigation is determined by the absolute and relative concentration of cations and is expressed in terms of Sodium Adsorption Ratio (SAR). If the proportion of sodium is high, the alkali hazard is high; and conversely, if calcium and magnesium predominate, the hazard is less. There is a significant relationship between SAR values of irrigation water and the extent to which sodium is absorbed by the soil. If water used for irrigation is high in sodium and low in calcium, the cation-exchange complex may become saturated with sodium. This can destroy the soil structure owing to dispersion of the clay particles. A simple method of evaluating the danger of high-sodium water is the sodium-absorption ratio, SAR (Richards, 1954):

$$SAR = \frac{Na^+}{\sqrt{(Ca^{2+} + Mg^{2+})/2}}$$

The sodium percentage is calculated as:

$$Na\ \% = \frac{Na^+ + K^+}{\sqrt{Ca^{2+} + Mg^{2+} + Na^+ + K^+}} \times 100$$

Where all ionic concentrations are expressed in milliequivalent per litre.

The values of SAR and sodium percentage of the groundwater of the command area are given in Table 38.5. Calculation of SAR for given water provides a useful index of the sodium hazard of that water for soils and crops. A low SAR (2 to 10) indicates little danger from sodium; medium hazards are between 7 and 18, high hazards between 11 and 26, and very high hazards above that. The lower the ionic strength of the solution, the greater the sodium hazards for a given SAR (Richards, 1954).

Table 38.5: Values of Sodium Adsorption Ratio (SAR), Sodium Percentage (Na%) and Residual Sodium Carbonate (RSC)

Sl.No.	*Sample Location*	*Type*	*Pre-monsoon*			*Post-monsoon*		
			SAR	*Na%*	*RSC*	*SAR*	*Na%*	*RSC*
1.	Gokak	BW	4.31	59.9	3.69	3.02	41.9	-3.63
2.	Gokak	HP	4.15	57.7	3.59	3.67	59.4	0.46
3.	Gokak	OW	3.15	48.3	1.62	1.43	35.2	-0.40
4.	Arbhavi	OW	1.45	35.1	1.18	1.68	44.2	-0.76
5.	Arbhavi	HP	4.00	60.4	3.86	2.49	54.3	-0.47
6.	Kalloli	OW	22.0	93.3	13.6	10.8	85.9	4.56
7.	Kalloli	HP	11.6	81.2	-1.23	7.66	74.4	-1.23
8.	Tuknautti	OW	0.77	19.6	0.32	0.61	18.0	-1.47
9.	Rajapar	HP	0.70	19.7	-0.11	0.63	19.9	-1.38
10.	Rajapar	OW	0.72	19.9	-0.04	0.71	20.4	-1.71
11.	Badigwada	HP	1.14	27.9	0.47	1.82	51.0	-1.31
12.	Bogewadi	OW	1.90	37.9	1.25	1.21	17.6	-2.80
13.	Kabbur	OW	5.97	46.4	-22.3	3.45	36.3	-21.1
14.	Raibag	HP	1.98	42.7	0.88	1.49	32.9	-6.54
15.	Raibag	OW	0.92	23.4	-0.34	0.32	13.2	-0.89
16.	Raibag	OW	2.87	51.0	1.23	0.78	21.8	-1.27
17.	Nasalapur	OW	2.43	26.5	-16.8	1.56	19.7	-17.7
18.	Kallol	OW	1.63	34.1	-0.72	1.85	37.8	-2.34
19.	Birdi	HP	2.86	21.3	-48.9	2.18	20.0	-34.8
20.	Chincholi	HP	1.84	28.9	-3.73	0.98	18.4	-6.60
21.	Chincholi	HP	0.70	16.8	-1.09	0.63	18.5	-1.79
22.	Morab	HP	0.83	23.0	0.33	0.64	20.6	-0.77
23.	Harugeri	OW	2.91	56.6	2.12	2.13	40.9	-4.94
24.	Hidkal	OW	3.98	57.9	4.50	4.15	65.2	1.51
25.	Mughalkhod	OW	0.95	18.1	-3.86	0.83	15.9	-5.32
26.	Gurlapur	OW	1.32	28.3	-0.73	1.61	28.0	-1.58
27.	Kankanwadi	OW	0.67	20.1	0.17	0.46	16.9	-0.60
28.	Hunshal	OW	1.01	22.2	-0.16	1.01	23.2	-2.68
29.	Sunadholi	HP	2.09	37.3	-0.48	2.23	41.7	-0.04
30.	Sunadholi	HP	1.90	34.2	-0.69	2.38	45.6	-0.23
31.	Kuligod	HP	1.92	32.3	-16.7	3.01	47.5	-2.86

Contd...

Table 38.5–Contd...

Sl.No.	Sample Location	Type	Pre-monsoon			Post-monsoon		
			SAR	Na%	RSC	SAR	Na%	RSC
32.	Venkatapur	OW	3.45	52.9	3.86	4.53	69.2	2.45
33.	Yadwad	HP	8.29	54.7	-15.6	7.04	56.0	-11.4
34.	Metgud	HP	1.37	24.8	-3.02	1.16	25.6	-2.07
35.	Mudhol	HP	0.91	24.7	0.38	0.96	25.8	-1.46
36.	Marikatti	HP	1.86	39.2	0.51	2.45	53.3	0.42
37.	Hirenandi	HP	1.24	40.8	1.32	1.37	45.0	-0.05
38.	Mamdapur	OW	1.33	37.0	-1.55	1.58	46.9	-1.93
39.	Lokapur	HP	1.94	31.4	-0.87	2.38	41.8	-1.73
40.	Muddapur	HP	3.14	52.2	0.43	2.33	50.9	-1.42
41.	Shirol	OW	2.02	33.1	-1.08	2.01	35.4	-2.79
42.	Shiddapur	HP	2.90	42.6	-3.71	2.80	44.6	-5.57
43.	Jamkhandi	HP	1.86	22.0	-17.3	1.76	21.7	-17.3
44.	Algur	HP	10.3	43.8	-83.8	10.1	46.5	-65.9
45.	Lignur	HP	1.43	60.4	1.37	1.33	64.1	-0.04
46.	Navalgi	BW	1.74	22.3	-16.6	1.42	20.2	-15.9
47.	Rampur Rabakavi	HP	1.84	18.4	-26.7	1.43	19.0	-16.2
48.	Chimmod	HP	0.48	19.4	-0.08	0.44	19.4	-0.21
49.	Mahalingapur	HP	0.77	19.7	-1.29	0.70	18.4	-2.02
50.	Mantur	HP	1.43	28.9	-0.28	1.74	40.5	0.14
51.	Halagali	HP	2.81	51.8	2.24	3.22	59.9	1.37
52.	Amalzhari	HP	1.85	54.8	-0.23	1.64	54.6	-2.68
53.	Teggi	OW	2.59	58.5	-4.43	2.60	58.2	-9.94
54.	Bilgi	HP	0.60	17.5	-1.50	0.72	29.7	-0.14
55.	Bagalkot	HP	2.13	34.7	0.29	2.22	43.5	-0.92
56.	Gaddankere Cross	HP	1.86	31.7	0.58	1.46	31.0	-0.79
57.	Tulsigeri	HP	2.09	47.8	-0.02	2.11	53.1	-0.53
58.	Hireshallikeri	HP	2.00	32.8	-9.23	2.23	37.9	-9.32
59.	BCM Hostel	HP	3.47	45.2	-2.18	3.51	54.8	-1.27
60.	Varchagal Cross	HP	14.9	76.3	-1.48	16.0	80.8	-1.80

HP–Hand pump, OW–Open wel1, BW–Bore well.

The values of SAR in the groundwater of the study area ranged from 0.48 to 22.0 during pre-monsoon season and 0.32 to 10.8 during post-monsoon season. As evident from the SAR values, all the samples of the study area except Kalloli and Varchagal Cross falls under the category of low sodium hazard, which reveals that groundwater of the study area is free from any sodium hazard.

As per the BIS standard, a maximum limit of 60 per cent is recommended for irrigation water. The sodium percentage in the study area was found to vary from 16.8 to 93.3 during pre-monsoon season and 13.2 to 85.9 during post-monsoon season. About 8 per cent of the samples in the study area exceed the recommended value of 60 per cent during pre- as well as post-monsoon season.

Residual Sodium Carbonate (RSC)

In addition to total dissolved solids, the relative abundance of sodium with respect to alkaline earths and boron, and the quantity of bicarbonate and carbonate in excess of alkaline earths also influence the suitability of water for irrigation purposes. This excess is denoted by Residual Sodium Carbonate (RSC) and is determined by the following formula:

$$RSC = (HCO_3^- + CO_3^{--}) - (Ca^{++} + Mg^{++})$$

Where all ionic concentrations are expressed in epm. If the RSC exceeds 2.5 epm, the water is generally unsuitable for irrigation. If the value is between 1.25 and 2.5, the water is of marginal quality, while values less than 1.25 epm indicate that the water is safe. The RSC values in the study area vary from -83.8 to 13.6 during pre-monsoon season and -65.9 to 2.45 epm during post-monsoon season with more than 10 per cent samples having RSC values greater than 2.5 epm during pre-monsoon season. During post-monsoon season only one sample crosses the limit of 2.5 epm.

Boron

Boron is essential to the normal growth of all plants, but the concentration required is very small and if exceeded may cause injury. Sometimes it is added to fertilizers in small amount to make up for the boron deficiency particularly in humid regions. Plant species vary both in boron requirement and in tolerance to excess boron, so that concentrations necessary for the growth of plants having high boron requirement may be toxic for plants sensitive to boron. Though boron is an essential nutrient for plant growth, generally it becomes toxic beyond 2 ppm in irrigation water for most of the field crops. It does not affect the physical and chemical properties of the soil, but at high concentrations it affects the metabolic activities of the plant. However, during the course of present investigations boron could not be analysed.

U.S. Salinity Laboratory Classification

U.S. Salinity Laboratory Classification (Wilcox, 1955) is used to study the suitability of groundwater for irrigation purposes. In classification of irrigation waters, it is assumed that the water will be used under average conditions with respect to soil texture, infiltration rate, drainage, quantity of water used, climate and salt tolerance of crop.

Sodium concentration is an important criterion in irrigation water classification because sodium reacts with the soil to create sodium hazards by replacing other cations. The extent of this replacement is estimated by Sodium Adsorption Ratio (SAR). A diagram for use in studying the suitability of groundwater for irrigation purposes is based on the sodium adsorption ratio (SAR) and electrical conductivity of water expressed in $\mu S/cm$.

The chemical analysis data has been processed as per U.S. Salinity Laboratory Classification for the two sets of data (Figs. 38.4 and 38.5) and the results have been summarized in Table 38.6. It is evident from the results that during pre-monsoon season majority of samples (65 per cent) fall under water type C3-S1 (high salinity and low SAR) followed by 11.7 per cent samples under water type C2-S1 (medium salinity and low SAR) and 11.7 per cent samples under water type C4-S1 (very high salinity and low SAR). Few samples were also found under water type C3-S3, C4-S1, C4-S2, C4-S3 and C4-S4 category. C3-S1 type of water cannot be used on soils with restricted drainage. Even with adequate drainage special management for salinity control may be required and plants with good tolerance should be selected. C2-S1 type of water can be used if a moderate amount of leaching

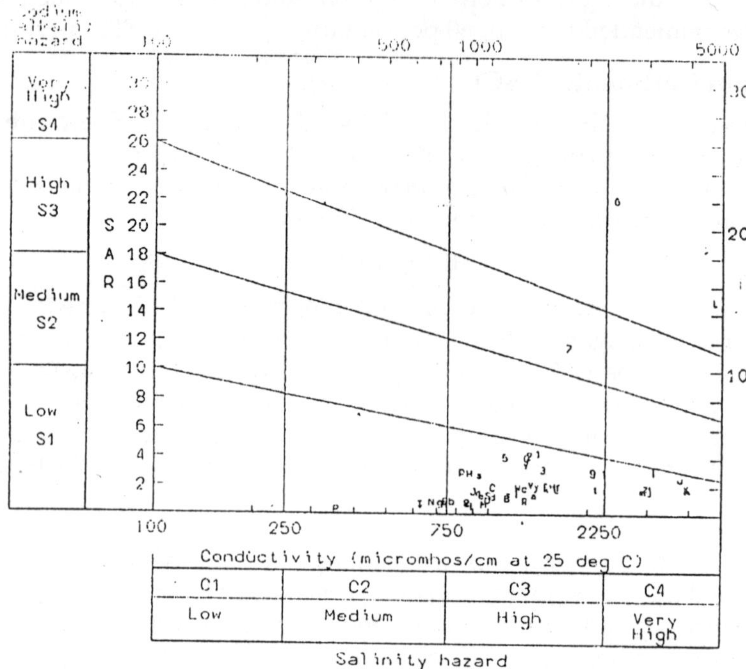

Fig. 38.4: U.S. Salinity Laboratory Classification (Pre-monsoon 2001)

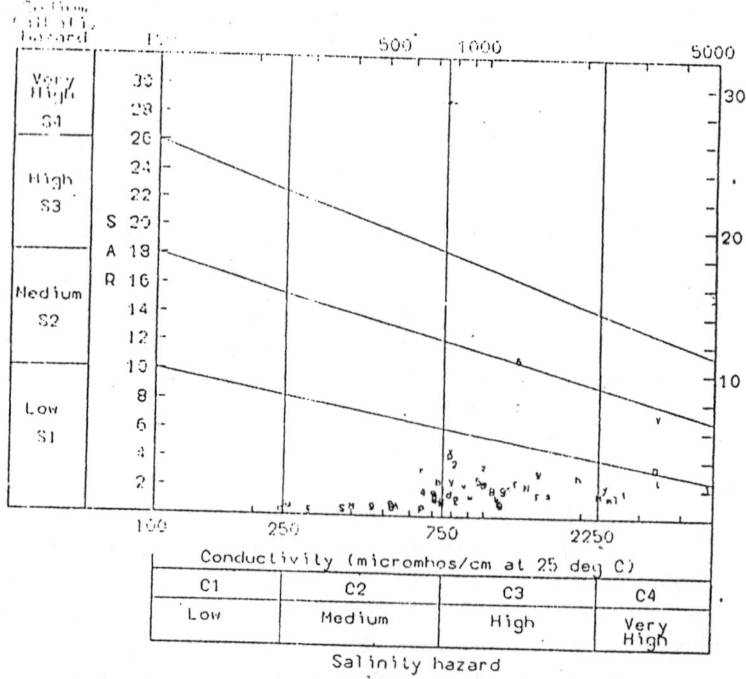

Fig. 38.5: U.S. Salinity Laboratory Classification (Post-monsoon 2001)

occurs and plants with moderate salt tolerance can be grown in most cases without special practices for salinity control. C4-S1 type of water cannot be used on any type of soils.

Table 38.6: Summarised Results of U.S. Salinity Laboratory Classification

Classification/Type	*Sample Numbers*	
	Pre-monsoon	*Post-monsoon*
C2-S1	9,10,15,22,27,48,54	3,4,8-10,12,15,16,21 22,26,27,35,36,49-51,54
C3-S1	1-5,8,11,12,14,16,18,20 21,23-26,28-30,32,34-42 45,49-52,55-57,59	1,2,5,11,14,17,18,20 23-25,28-32,34,37-42 45,52,55-57,59
C3-S2	–	7
C3-S3	7	6
C4-S1	17,31,43,46,47,53,58	19,43,46-48,53,58
C4-S2	13,19	13, 33
C4-S3	33,44	44
C4-S4	6,60	60

Conclusion

The values of sodium absorption ratio in the command area indicate no risk of sodification. About 60 per cent samples of the study area fall in high salinity zone during pre-monsoon season while 35 per cent samples fall in high salinity zone during post-monsoon season. The U.S. Salinity Laboratory classification for irrigation water indicates that more than 50 per cent samples fall under water type C3-S1 (very high salinity and low SAR). Such water cannot be used on soils with restricted drainage. Even with adequate drainage, special management for salinity control may be required and plants with good tolerance should be selected.

References

APHA (1985). *Standard Methods for the Examination of Water and Wastewater*. American Public Health Association, New York.

Bhadale, S., Sahu, D., Velani, B., Singh, A. and Samant, K. (1996). *Groundwater Quality: Studies of Deep Aquifers of Belgaum*. B.E. Project Report, GIT, Belgaum.

Jain, C.K. and Bhatia, K.K.S. (1988). *Physico-chemical Analysis of Water and Waastewater UM-26*. National Institute of Hydrology, Roorkee, 1987-88.

Jain, C.K., Kumar, C.P. and Sharma, M.K. (2002). *Groundwater Quality in Ghataprabha Command Area, Karnataka*. Technical Report, National Institute of Hydrology, Roorkee, 2001-2002.

Richards, L.A. (ed.) (1954). *Diagnosis and Improvement of Saline and Alkali Soils*. Agricultural Handbook 60, U.S. Deptt. Agric., Washington, D.C., 160 p.

Suresh (1996). *Conjunctive Use Studies in Belgaum Taluk*. M.Sc. thesis, University of Goa, Goa.

Tamta, S.R. (1994). Possible mechanism for contamination of fluoride in groundwater. *Bhujal News*, 9(2).

Wilcox, L.V. (1955). *Classification and Use of Irrigation Water*. U.S. Dept. of Agr. Circular, 969, Washington, D.C., 19 p.

39

Management of Poor Quality Water for Irrigation

A. SOLAIMALAI & R. SARAVANAKUMAR

Research Associates, Krishi Vigyan Kendra,
Tamil Nadu Agricultural University, Vridhachalam, Tamil Nadu

K. SANKARANARAYANAN

Senior Scientist (Agronomy), Central Institute for Cotton Research (ICAR),
Coimbatore, Tamil Nadu

M. BASKAR

Research Associate (Soil Science and Agricultural Chemistry),
A.D. Agricultural College and Research Institute, Trichy, Tamil Nadu

In irrigated agriculture, the hazard of salt water is a constant threat. Poor quality irrigation water is generally more concerning as the climate changes from humid to arid conditions. Salinity is not normally a threat where precipitation is a major source of salt free water for crop production. Water entering the soil, which is not stored or consumed by evapotranspiration moves through the crop root zone ultimately reaching the water table. This percolating process leaches soluble salts. Lesser rainfall means smaller amount of precipitation available to leach salts. If the amount of water leaching through the soil is too low to remove salts, the soil salt content increases and crop yields may decrease.

Saline soils can be brought under production by utilizing the stored rain water in the soil profile, shallow water table and also by using multi quality waters (Minhas, 1996). Use of saline groundwater for crop production must aim at preventing the build up of salinity and toxic ion accumulation in the root zone to levels that limit the productivity of soils, control the salt balances is soils water systems and minimize the damaging effects of salinity on crop growth. The basic ways to use saline water are to blend with good quality water and apply for irrigation and to apply good quantity water at sensitive growth stages and saline water at comparatively tolerant stages (Rhodes *et al.*, 1992).

Selection of Crops

The choice of a crop for a given salinity water depends upon water potential to build up salts in the root zone. Plants are known to exert the absorptive forces throughout the rooting depth and absorb water from zones where it is most readily available. Surface soil horizons, where roots are most prevalent are depleted most rapidly of water to meet the evapotranspirational demands of plants. This zone also is replenished first by irrigation and rainwater. Between the irrigation cycles as the surface soil is depleted of moisture, in order to meet further demands, roots increasingly probe the lower depths of the soil profile. In general, oilseed crops requiring less water can tolerate higher levels of EC_{iw} whereas, most of the pulses are very sensitive to salts. Late sown crops (wheat) can tolerate only lower levels of EC_{iw} than timely seeded crops.

Same crop when grown during summer shows less tolerance to EC_{iw} sorghum than monsoon grown crop. Also the salinity tolerance of the rabi season crops are generally higher. Climatic parameters are important in reducing the negative impacts of salinity on crop performance. During the monsoon season, ET demands are reduced and saline solutions are diluted and leached due to rainwater received during the growth of kharif crops. Depending upon the percentage of the total annual rainfall received during kharif, the crops grown during this season may at best need only 1 or 2 irrigations with saline waters or in a good rainfall year, saline water may not be used at all for irrigations. Hence, the tolerance of the kharif season crops is based upon the salinity of the irrigation water used during the previous rabi season crop. Salt build up during the previous rabi season crop does affect the yield of these crops. Similarly, plants also vary in their tolerance to soil sodicity/ alkalinity. The range of ESP tolerance limits should serve as general guidelines for selection of crops. The sodicity tolerance limits of crops may differ in case of soils undergoing sodications process due to irrigation with alkali waters.

Table 39.1: Relative Tolerance of Crops to Alkalinity of Soils

ESP	*Crops*
10-15	Safflower, blackgram, peas, lentil, pigeonpea
16-20	Chickpea, soybean
20-25	Groundnut, cowpea, onion, pearlmillet
25-30	Linseed, garlic, clusterbean
30-50	Raya, wheat, sunflower
50-60	Barley, sesbania
60-70	Rice

(Gupta and Abrol, 1990).

Use of alkali waters increases sodicity and soil pH, reduce the water infiltration into the soils as result of added salts through irrigation concentrate in the upper soil layers. The salt build up as a result of use of alkali waters may also influence the sodicity tolerance of crops. In addition to ESP tolerance, irrigation water requirement should also be considered for selection of crops in the management of alkali waters. Crops having high irrigation water requirements hasten the pace of sodication of soils. For instance, rice is rated high in sodicity tolerance but its water requirement is also high. Consequently, rice should be avoided when irrigation is intended with waters having high alkali hazards. A common practice in many regions particularly those receiving <400 mm rainfall is to grow rabi crops and keep the fields fallow during the kharif to take advantage of rainfall. Rainwater leaches the accumulated salts and dissolves a portion of the precipitated or *in situ* $CaCO_3$ to furnish Ca^{2+} ions to soil solution.

Selection of Varieties

There is also a wide variation in the inherent salt tolerance of the crop cultivars. Cultivars such as Damodar in rice and Kharchia in wheat have high relative tolerance to salinity stress. Vigorously growing cultivars of rice (Jaya) and wheat (HD 2304) having high yield potential show considerable yield reductions owing to salinity but may still out yield their respective counterparts having high relative tolerance. Cultivars which are stable in saline environment should be preferred for commercial cultivation (Gupta *et al.*, 1989).

Table 39.2: Cultivars of Crops for Saline Environment

Crop	*Varieties*
Rice	CSR 6, MTU 4870, NLR 3098, Phalguna
Wheat	DWR 137, K 65, MP 157, Raj 2325, Raj 2560, Raj 2936, Raj 2917, Raj 3077, WH 157
Barley	BL 2, C 164, CS 157, CS 19, DL 348, DL 356, RD 103, RD 137, Ratna
Maize	Deccan 103, Vijaya
Sorghum	CSH 11, SPH 296, SPH 488
Pearlmillet	MH 269, MH 324, MH 333
Finger millet	HR 911, HR 10181, HR 10541, Indaf 9
Triticale	DTS 972, DTS 976, Maharaja Local
Bengal gram	C 7, KB 8
Red gram	BDNH 7, PBNA 53
Green gram	LGG 127 .
Safflower	A 1, Bheema, C 83
Mustard	DIRA 342, Kranti, Laha 101, Pusa bold
Sunflower	Ramson Record
Cotton	DB-3-12, GDH 9, MESR 16
Chillies	X 235, CA 960

Wheat variety kharchia 65 yield maximum whereas minimum yield was obtained under WH 416. Variety kharchia 65 had lower yield reduction per unit increase in salinity beyond threshold and higher EC_{iw} values for 50 per cent yield decline followed by HD 2189 (Chopra and Chopra, 1997). Variety with lower regression slope and higher values of EC_{iw} for 50 per cent yield decline is characterized as fertility tolerant variety (Mass and Hoffman, 1977).

Table 39.3: Performance of Wheat Varieties as Influenced by Saline Water Irrigation

Varieties	*Plant height (cm)*	*Spike/m row length*	*Length of spike (cm)*	*100 grain weight (g)*	*Grain yield (q/ha)*
HD 2189	65.7	74.8	6.2	3.78	22.05
WH 157	56.5	47.6	6.4	4.90	18.34
WH 147	47.5	57.2	6.0	4.22	13.56
WH 283	42.7	52.1	5.4	3.90	11.81
Kharchia 65	71.1	85.6	6.2	4.05	22.10
LSD (5%)	5.4	13.1	0.61	0.46	3.10

(Chopra and Chopra, 1997).

Table 39.4: Relative Varietal Tolerance of Wheat to the Application of Saline Water

Varieties	*Germination (per cent)*		*Grain Yield/Running Meter (g)*		*Straw Yield/Running Meter (g)*	
	1973	*1974*	*1973*	*1974*	*1973*	*1974*
K 68	75.5	63.0	76.7	81.7	121.1	112.2
K 816	73.7	51.2	77.6	52.5	96.6	67.9
HD 1553	74.6	55.3	102.0	92.1	137.6	105.9
HD 1925	77.5	60.9	117.0	95.6	131.7	120.0
HD 1941	76.0	59.3	72.0	63.2	80.5	69.5
HD 1949	74.9	48.2	75.1	71.1	94.5	68.1
HD 1955	76.2	53.3	64.4	44.0	98.3	59.0
HD 2009	73.9	52.3	74.8	75.4	92.4	74.7
HD 2016	78.9	59.9	103.4	104.0	139.1	120.4
UP 301	75.0	50.0	68.7	42.9	98.3	53.9
S 307	77.4	62.3	80.7	93.9	130.8	114.2
S 331	73.6	52.9	72.3	76.2	103.5	99.8
WG 377	70.8	54.1	98.5	75.4	139.1	94.7
CD (5%)	NS	NS	21.2	16.2	25.8	18.8

(Rai, 1977).

Table 39.5: Cultivars of Crops Suitable for Alkali Soils

Rice	CSR 4, CSR 5, IR 8, IR 20, IR 24, Jaya, Jhona 349, Kalarata, Sita, Usar 1
Wheat	Kalyansona, KRL 1-4, Raj 3077, WH 157
Barley	BHS 12, DL 4, 106, DL 120, DL 157
Sorghum	MSH 71, CSV 10, SPV 475, 61-1-1, SPV 235
Pearl millet	MH 269, MH 280
Linseed	EPS 233, Himalini, Neelam
Mustard	DIRA 343, NRL 8604, Pusa Bold, Varuna
Safflower	IC 11750, JSF 144
Cotton	IH 452, Khandwa 2, Maljari, MESR 16, Vikram 70

High grain and straw yields of wheat were produced under variety HD 1925 whereas HD 1955 gave lower yields under saline water irrigation (Rai, 1977).

Table 39.6: Relative Varietal Tolerance of Barley to Saline Water Irrigation

Varieties	*Germination (per cent)*		*Grain Yield/Running Meter (g)*		*Straw Yield/Running Meter (g)*	
	1973	*1974*	*1973*	*1974*	*1973*	*1974*
Amber	85.1	59.8	85.7	78.7	141.5	91.7
Vijay	88.5	56.8	92.7	74.4	121.9	82.0
Jyoti	85.1	58.7	95.5	69.5	115.3	86.9
Ratna	87.2	58.0	88.2	75.9	195.7	85.7
K 12	88.9	59.7	113.7	67.4	117.9	76.5
K 18	86.5	59.6	91.5	71.0	142.3	78.7
DL 69	82.2	55.1	63.5	39.7	96.8	47.3
RS 6	84.1	52.5	79.7	63.3	106.3	75.7
RD 102	87.9	57.9	92.0	76.5	148.7	86.7
C 24	85.3	58.7	82.7	76.3	126.8	92.0
C 50	88.3	57.6	93.6	59.9	154.8	68.3
C 251	87.5	60.3	104.9	78.3	164.5	90.1
CD (5%)	NS	NS	15.2	14.7	40.4	14.9

(Rai, 1977).

Varieties Amber, Jyoti, K 12 and DL 69 were recorded higher grain yield of barley than other varieties under saline water irrigation (Rai, 1977).

Susceptible Growth Stages of Crop

All crops do not tolerate salinity equally well at different stages of their growth. Difference in salt sensitivity of crops at various growth stages should help in planning appropriate irrigation management measures especially where both saline and non-saline waters are available. The use of saline waters can be avoided at some of the sensitive stages to minimize the salinity damage.

Table 39.7: Relative Sensitivity of the Crop at Their Growth Stages

Crop	*Relative Sensitivity*
Mustard	Presowing > Flower initiation > Secondary branching
Wheat	Presowing > Flowering > Milking > Crown root initiation > Jointing
Barley	Crown root initiation > Presowing > Flowering/booting > Jointing
Safflower	Presowing > Rosette > Flower initiation > Main head opening
Paddy	Transplanting > 50% Flowering > End of tillering
Maize	Silking > Tasselling > Presowing > Knee height
Pigonpea	Presowing > Flowering to pod development > Flowering > Seedling to flowering
Sunflower	Seedling > Flower bud initiation > Grain filling > Germination to flowering
Onion	Transplanting to primordia formation > Primordia formation to flowering > flowering to maturity
Italian millet	Transplanting to primordia formation > Primordia formation to flowering > flowering to maturity
Groundnut	Germination to pegging > Pegging to pod formation > Pod formation to maturity
Black gram	Flowering to maturity > Seedling > Seedling to flowering

Irrigation Frequency

When using saline water, irrigation should meet both the water requirement of crops and the leaching requirement to maintain a favourable salt balance in the root zone. During the intervening periods in between the irrigation cycles, evapotranspiration by crops reduces the soil water content which in turn decreases the matric and solute potentials. Consequently, the water deficits between the irrigation cycles. However, a consequence of small irrigation intervals will be the increased salt load of the soil. An increase in depth of saline water irrigation though slightly enhanced the yield of crops but there was no interactive effect with salinity of irrigation water.

Table 39.8: Wheat Yield as Affected by Leaching Fractions at Different Salinities of Waters

IW/CPE ratio	Yield (q/ha) at EC_{iw} (dS/m)							
	BAW	4	8	12	BAW	4	8	12
	Pearl millet – Wheat				Sorghum – Wheat			
1.0	43.3	42.7	40.7	36.2	32.7	27.3	27.3	24.2
1.25	44.0	46.1	44.5	39.3	35.1	31.5	30.0	28.9
1.5	46.6	46.6	45.8	40.4	33.5	29.3	26.6	24.7

During the second year of barley crop, yields obtained with 2 irrigations of good quality canal water were equal to 4 irrigations with saline waters containing 10,000-30,000 ppm salts.

Methods of Irrigation

Surface irrigation methods including the border strips, check basins and furrows are most commonly practised in most part of India. These irrigation methods generally result in excessive irrigation and non-uniformity in water application. Consequently, the on-farm irrigation efficiency is low (60-70 per cent). However, properly designed and operated surface irrigation methods can maintain the salt balance and minimize salinity hazards of saline waters. To meet these objectives, land needs to be properly levelled to ensure even distribution of water.

Application of highly saline (EC_{iw} = 12 dS/m) water through sprinkler to pearl millet and cotton is detrimental whereas it can be safely used for wheat and barley. Although water use efficiency decreased with salinity of water, it was higher when the water was applied by using the sprinkler than by surface method to winter crops. For saline water use, sprinklers should better operated in the evening or night time when evaporation rates are lower. The accumulation of salts in soils was lower under sprinkler than surface method of irrigation. Higher cane yield (107.4 t/ha) and sugar content (12.3 per cent) were recorded with sodic water (RSC 12.5 me/litre) irrigation through drip at 80 per cent pan evaporation and gypsum application @ 50 per cent GR (Nagarajan *et al.*, 2001).

Leaf burns in citrus orchards even with the use of waters containing as low as 3 meq/litre of Cl^- or Na^+ whereas, irrigation with such waters using surface methods did not pose any problem. Drip system of irrigating crops has been observed to enhance the threshold limits of their salt tolerance by modifying the patterns of salt distribution and maintenance of constantly higher matric potentials. Owing to enhanced leaching and accumulation of the salts at the wetting front and the soil between the drip laterals, the salt accumulation below the drippers remain very low whereas the water contents are maintained at higher levels at the latter sites. As the crop roots are known to follow the path of least resistance, most roots are found below the surface drippers. Hence, the drip system seems to be the best system of saline water application as it avoids leaf injury to plants as with sprinklers and maintains optimum condition for the uptake of water by the plant roots. Application of saline water of EC_{iw}=3 dS/m through drip did not limit the yield of potato and tomato but the yields obtained under furrow irrigation were significantly reduced. At EC_{iw}=10 dS/m, the yield of potato and tomato were reduced by 91 and 35 per cent respectively (Singh *et al.*, 1978). With drip lines installed at 90 and 150 cm drawback of irrigation with drippers is high salt concentration that develops at the wetting front.

Table 39.9: Yields of Crops Under Different Salinities of Water and Irrigation Methods

EC_{iw} (dS/m)	*Yield (q/ha)*			
	Wheat	*Barley*	*Cotton*	*Pearl millet*
		Sprinkler		
Canal	36.9	34.8	22.8	25.4
6-7	37.1	29.1	16.3	21.6
12	35.4	25.9	13.4	15.0
		Surface		
Canal	40.0	35.1	23.0	23.8
6-7	35.7	28.0	18.2	19.1
12	36.2	23.2	17.1	20.7

(Aggarwal and Khanna, 1983).

Table 39.10: Yield and Water Use Efficiency of Raddish for Different Irrigation Methods

Irrigation Method	*Yield (t/ha)*	*WUE (kg/ha cm)*
Drip Method – Surface	15.7	17.5
Drip Method – Subsurface	23.6	26.2
Furrows	9.9	8.7

(Aggarwal and Khanna, 1983); (EC_{iw} = 6.5 dS/m).

Accumulated salts cause difficulties in the planting of subsequent crops because effective leaching of salts will require the use of flood or sprinkler irrigation (Jain, 1984). Another problem is the clogging of drippers due to precipitation of salts. Some of the alternatives to drips like the earthen pitchers (Modal *et al.*, 1987) and the specially designed earthen cups (Yadav, 1986) have been tried for irrigation of vegetables crops.

Conjunctive Use of Waters

The canal irrigation has been extended to nearly 30 million ha out of the total cultivated area of 165 million ha. In many cases, the canal water suppliers are either unassured or in short supply such that farmers are often forced to pump groundwaters of varying quality for crop production. This calls for using the limited quantities of non-saline (canal) waters most judiciously in combination with poor quality waters. Two options available for making conjunctive use of saline and canal waters are either to apply after mixing or to apply canal water at sensitive growth stages and make use of saline water at comparatively tolerant stages (Minhan and Gupta, 1990; Rhoades *et al.*, 1989).

The ratio of mixing can be worked out depending upon the salinities of water and the desired salinity by using the equation:

$$[EC_{cw} \times F_{cw}] + [EC_{tw} \times F_{tw}] = EC_{mw}$$

Where,

EC_{cw} = EC of canal water,

EC_{tw} = EC of tube well and

EC_{mw} = EC of mixed water

F_{cw} = Fractions of canal and

F_{tw} = Fractions of tube well water.

Table 39.11: Conjunctive Use of Saline Water with Best Available Water on Yield of Crops (q/ha)

Treatments	*Mustard*	*Safflower*	*Dill*
BAW at all the stages	7.35	8.25	8.32
Saline water at branching stage (BR) + BAW at flower initiation stage (FI) and grain filling stage (GF)	6.65	7.45	7.93
Saline water at FI + BAW at BR and GF	6.29	7.65	7.84
Saline water at GF + BAW at BR and FI	6.45	7.75	7.68
Saline water at BR and FI + BAW at GF	5.59	6.35	7.52
Saline water at BR and GF + BAW at FI	5.19	6.85	7.44
Saline water FI and GF + BAW at BR	5.35	6.88	7.36
Saline water at BR, FI and GF stages	3.26	4.95	6.82
CD (5%)	0.75	0.21	0.06

(Gururaja Rao *et al.*, 2001).

Seed yield of Indian mustard decreased with increase in the number of saline water irrigations. Application of three irrigations with water having salinity of 4 dS/m resulted in 3.26 q/ha seed yield as compared to yield of 7.35 q/ha with three irrigations with BAW. However, significant reduction in the seed yield over control was observed when same saline water was applied at flowering initiation and grain filling stage. Application of two saline water irrigations in conjunction with one BAW in rotation caused significant increase in yield over three saline water irrigations (3.26 q/ha) which indicated that by compromising the yield loss of about 2 q/ha, 66 per cent of irrigation with BAW can be saved (Gururaja Rao *et al.*, 2001).

Mixing does not reduce the total salt load but may allow for more area to be planted. Mixing can either be done via network dilution or the soil dilution (Shalhevet, 1984). In network dilution, water suppliers are blended in the irrigation conveyance system. In soil dilution, saline and the non-saline waters are applied alternately and soil acts as a medium for dilution. Farmers alternating canal and saline waters for irrigation were getting higher production of the kharif crops (cotton and millets) than those using saline water after mixing it with canal water (Bouwmans *et al.*, 1988).

Alternate Use of Waters

In most of the crops, the germination and seedling establishment has been identified as the most sensitive stage to saline water use. A failure at this stage will lead to poor stands and considerable reduction in yield. Moong bean, sorghum and Indian mustard tolerated the salinity build up due to use of higher salinity waters, once the non-saline water was substituted for pre-sowing irrigation to leach out salts of the seeding zone (Minhas *et al.*, 1989). Substitution of non-saline water for pre-sowing irrigation markedly enhanced germination, crop growth and also resulted in better utilization of soil-water even from the lower soil depths. Thus, it is better to use non-saline waters in early growth stages and switch over to poor quality waters later when the crops can tolerate higher salinity.

Crop yields were higher when cropping intensities were low but the crop responses to various modes of saline and canal water usage in general did not vary. Yields could be maintained close to those obtained under BAW by delayed substitution of the saline water (after two initial irrigations with BAW). The next best is to alternate irrigations with BAW and saline water. Irrigation with saline water should not match with the critical growth stages, especially it should be avoided at sowing time as high salinity hinders germination and seedling establishment. Higher salinity waters could be used for moong bean if non-saline water is substituted at pre-sowing stage to leach out the salts at the seeding zone for its better germination (Minhas *et al.*, 1989).

Table 39.12: Mixed Use of Saline Water with Best Available Water on Yield of Crops (q/ha)

Treatments	*Mustard*		*Safflower*		*Dill*	
	Two irrigation	*Three irrigation*	*Two irrigation*	*Three irrigation*	*Two irrigation*	*Three irrigation*
BAW	2.54	6.52	7.63	8.44	8.34	9.14
4 dS/m	2.08	2.71	7.00	7.50	8.15	9.06
8 dS/m	1.04	1.35	5.69	6.15	4.17	5.67
12 dS/m	0.70	1.16	7.99	5.12	2.92	3.67

(Gururaja Rao *et al.*, 2001).

With the application of three saline irrigations, the reduction of seed yield of mustard was about 58, 79 and 82 per cent respectively at 4, 8 and 12 dS/m water salinity as compared to BAW. Safflower also showed a significant yield reduction when three saline water irrigations (4 dS/m) were given over the control. In dill, significant difference in seed yields were noticed when saline water beyond 4 dS/m was used. However, no significant differences in seed yields were noticed between BAW and 4 dS/m treatments. Hence, there is a scope of increasing the yield potential of mustard and safflower with the application of 3 saline water irrigations of 4 dS/m either by diluting it further or with its use in rotation with BAW. In dill, the water of sub-marginal quality with salinity of 4 dS/m can be used without any significant yield reduction when compared to BAW (Gururaja Rao *et al.*, 2001).

Mixing Ratios

The advantage from different cyclic use-modes were higher order: (2C: 1S) > (1C: 2S); canal: saline water. Irrigation of a sandy loam soil (18-26.8 per cent clay) with an alkali water increased the pH and ESP of the surface soil layers and reduced its infiltration rate to 14 per cent. The yields of rice and wheat decreased progressively with time and were 62 and 57 per cent of the potential (canal water) during the 6 years. However, when alkali water was used in cyclic modes with canal water, yields of both the crops were maintained at par with canal water, except in the CW-2AW mode (Bajwa and Josan, 1989). Cyclic use of two waters reduced the sodication of soils. Less reduction in yield as well as reduced sodication than expected by the mean composition of mixed waters points towards the advantage of using alkali and canal waters cyclically over mixing the two supplies.

Table 39.13: Water Application Modes of Canal and Saline Water on Yield Reduction (per cent) of Crops

BAW:SW	*Urd*	*Potato*	*Moong*	*Cluster bean*	*Sorghum*	*Soybean*	*Pea*	*Cowpea*	*Chillies*	*Maize*	*Onion*
2:1	13	3	43	13	9	7	11	7	4	6	7
1:1	23	5	62	25	10	28	35	18	11	10	10
1:2	28	10	85	35	12	38	51	23	33	16	18
SW	42	26	100	86	24	40	70	40	46	24	25

(EC_{iw} of saline water = 6-8 dS/m), (BAW = Best available water), (EC = 0.7-1.0 dS/m).

Seeding Practices

Failures to achieve satisfactory germination and thus the required plant population is the major factor limiting crop production in saline water irrigated areas. Crops like wheat, barley and safflower though can tolerate fairly high levels of salinity and they are very sensitive at germination and early seedling stages. Thus once the plants are established, the salinity effects can be substantially reduced. One of the alternatives to overcome this problem is using higher seed rates. However, this does not always work out as desired. Other alternatives are the practices, which would ensure suitable environment in the seeding zone during the germination and seedling emergence periods. Conventional seeding of most crops is done when optimum moisture conditions for tillage and seedbed preparation are attained following a pre-sowing irrigation. When saline water is applied for pre-sowing irrigation, seeds are exposed to soil water of higher salinity ($EC_{sw} > EC_{iw}$) as a result of movement of salts towards the surface *via* evaporative drying both up to seeding and during the periods of germination and emergence. This makes the seed germination and emergence more critical, especially for summer crops seeded under high evaporation. Under such conditions a safe technique to establish crops seems to be to go in for a post-sowing irrigation to push the salts deeper and to maintain better moisture conditions (Minhas *et al.*, 1988).

Table 39.14: Various Conjunctive Uses Modes of Saline and Canal Water on Crop Yield

Irrigation Modes	*Yield of Crops (q/ha)*								
	1981			*1982*			*1983*		
	Moong	*Urd*	*Potato*	*Moong*	*Urd*	*Potato*	*Moong*	*Urd*	*Potato*
Canal water	1.9	7.4	314	4.5	8.6	328	1.5	15.2	230
2C:1S	1.4	5.8	326	2.3	5.9	305	0.7	12.5	218
1C:1S	1.0	6.4	322	2.0	6.9	308	0.3	8.7	223
1C:2S	0.2	5.3	330	0.6	6.1	289	0.2	5.7	202
Saline water	0.0	3.5	229	0.0	6.5	187	0.0	3.7	124
CD (5%)	0.2	NS	33	1.1	NS	44	0.8	1.2	28

(Chauhan *et al.*, 1992);
C: Canal water (EC: 1 dS/m), S: Saline water (EC: 6 dS/m).

Yields of potato, moong bean and urd in general decreased with increased substitution of saline water and were lower when only saline water was used (Chauhan *et al.*, 1992). The salinity build up in the soil occurred during potato and summer moong bean crop season and most of the salts were leached out of the 0.9 m profile during the growth of the urd due to monsoon rains. In only saline water irrigation, soils attained salinity of about 12.5 dS/m in potato and 20.0 dS/m in summer moong bean at harvest. Depending upon the rain received, the amount of salt retained at the harvest of urd was 22-35 per cent as a fraction of salts present compared with at the time of moongbean harvest (Chauhan *et al.*, 1992).

Table 39.15: Seeding Methods on Indian Mustard under Saline Water Irrigation

Seedling Method	*E_{cw} (dS/m) of irrigation water*							
	Plant stand/m				*Seed yield (q/ha)*			
	3	*7*	*11*	*16*	*3*	*7*	*11*	*16*
Seeding after conventional pre-sowing irrigation	10.4	9.8	2.4	1.4	16.7	17.4	5.9	1.9
Dry seeding followed by post-sowing irrigation	10.6	10.7	9.3	6.3	15.6	16.2	15.4	12.9
½ pre- and ½ post-sowing irrigation	10.0	9.7	7.4	2.5	15.1	15.7	13.8	7.5

But the timing of this irrigation should be such as to avoid the subsequent crusting problem. Post sowing irrigation with saline water can also ensure reasonably good germination and seedling establishment in winter crops and help timely seeding in areas where it is delayed due to unassured canal water supplies or the rainfed nature of agriculture. Compared with the potential (BAW), the seed yield in post sowing irrigation with saline water following dry seeding was sustained up to 11 dS/m in Indian mustard.

Application of a heavy (10 cm) post sowing irrigation with saline water resulted in improvements in crop stand and yield when compared to a light irrigation (5 cm). Modifications in the shape of the seed beds can also reduce the salt accumulation near the seed. The double row flat topped bed configuration for saline irrigation wherein most of the salts move to the centre of the beds and the shoulders are left relatively salt free to enhance the seedling establishment (Bernstein and Francois, 1957). Seeding on sloping beds also seems better (Bernstein and Ayers, 1955). Here the seeds are placed on the sloping side and seed row is placed just above the water line and irrigation is continued until the wetting front has moved well past the seed row.

Sowing near the bottom of the furrows on both sides of the ridges (50 cm wide) in a soil having EC_e = 4.7–5.2 dS/m did not increase yield of barley as compared with flat method of sowing. However, in radish, since direct injury to leaves could be avoided, ridge-and-furrow sowing was much better than flat sowing. The other alternatives are to apply irrigation in alternate rows (Bernstein *et al.*, 1957) and to seed on the north east side of the ridges, because the salts move along with water to the drying surfaces and accumulates there. The EC_e value of south west, top and the north east sides of the ridges (60 cm apart, 25 cm high) were 35.4, 24.4 and 6.7 dS/m, respectively, though the soil initially had EC_e = 15.5 dS/m (Basin and Singh, 1966). For the larger seeded crops, the seeds can be planted in the furrows (Rai and Mehrotra, 1971). The advantage of such a practice can be that the seed will be placed in a wet and a less saline zone as in making ridges more saline surface soil goes to the ridges and a post sowing irrigation will leach the salts of furrow soil more efficiently than those of the ridge soil (Minhas *et al.*, 1988).

Soaking of seeds in 3 per cent Na_2SO_4 solution was observed to increase the yield of wheat irrigated with saline waters (Puntamaker *et al.*, 1972). Similarly, some hormones like gibberellic acid (GA) and indole acetic acid (IAA) when applied in optimum doses have been reported to increase the yields of crops under saline environment (Chippa and Lal, 1986). However, no significant response was observed to soaking the seeds of barley and wheat in solutions of NaCl and Na_2SO_4, although yield was enhanced at high EC_{iw} (Manchanda and Bhandari, 1976). Yields of wheat, sorghum and rice did not improve either by pre-soaking of seeds in salt solutions or by increasing the seed rate by 25 per cent under the saline water irrigation. In western parts of Rajasthan, saline waters contain high silica contents. When these waters are used for irrigation, soils become prone to crusting problems and subsequently affect germination and yields of crops (Jain, 1984). Additions of gypsum at the rate of 1-2 t/ha have been observed to reduce crusting, facilitate emergence of sorghum and thus increase its yields.

Organic Manures

At a given salinity level increasing the application of farm yard manures (FYM) improved the yields of all the crops tested. However, when salinity of the irrigation water was more, the per cent response was reduced when no FYM was applied. Application of press mud 12.5 t/ha or FYM 12.5 t/ha or daincha 6.25 t/ha and primarily treated distillery effluent 2.5 lakh litres/ha favoured germination and vigour of rice seedlings in nursery when compared to gypsum.

Table 39.16: Application of Farmyard Manure on Yield of Crop

EC_{iw} (dS/m)	*Yield (q/ha) at the applied FYM (t/ha)*								
	Potato			*Wheat*		*Pearl millet*		*Maize*	
	0	*3.2*	*6.4*	*0*	*50*	*0*	*50*	*0*	*50*
BAW	174	229	240	36.8	39.7	31.1	39.1	37.7	59.8
6	160	192	161	36.3	44.3	23.2	27.0	26.5	32.8
10	134	161	172	–	–	–	–	–	–

Table 39.17: Addition of Organic Materials on Yield of Crops with Saline Waters Irrigation

Organic Material	*Yield of Crop (q/ha)*				
	Chillies	*Cotton*	*Sorghum*	*Wheat*	*Chickpea*
Control	134	112	71	25.9	2.8
FYM @ 10 t/ha	138	129	77	29.5	4.6
Groundnut shells @ 5 t/ha	150	193	71	–	–
Tephrosia sp @ 5 t/ha	–	–	–	36.6	5.7

(EC_{iw} = 10 dS/m).

Nitrogen

Increased soil pH resulting from the use of alkali waters has been shown to positively influence NH_3 volatilization from urea and other ammonia sources (Singh and Bajwa, 1989). Similarly, for saline water irrigated soils, disappearance of NH_4–N was accelerated and NO_3–N losses were retarded due to decreased number of nitrifying bacteria. Thus, ammonia volatilization is extensive in salt affected soils and with use of poor quality waters which lead to low N use efficiency by crops (Garg *et al.*, 1982). A significant interaction effect was observed between salinity of irrigation water used and response to applied nitrogen in wheat. The response was maximum at EC_{iw} = 4 dS/m, started decreasing around EC_{iw} = 8 dS/m and was minimum at EC_{iw} = 16 dS/m.

Table 39.18: N and P Fertilization on Fields of Wheat

EC_{iw} (dS/m)	*N Applied (kg/ha)*			*P Applied (kg/ha)*		
	Wheat Yield (q/ha)			*Wheat Yield (kg/ha)*		
	60	*120*	*180*	*30*	*60*	*90*
BAW	26.2	36.0	39.9	32.8	34.6	34.7
6	26.2	33.6	35.8	29.8	32.6	34.3
12	24.3	25.8	26.0	21.7	25.8	28.7

Phosphorous

Increasing the level of P over the recommended dose of 60 kg P_2O_5/ha had a better effect on crop yields when saline waters were used. Application of phosphorous thus seems to mitigate the adverse effects of salinity. Type of salinity has also been observed to influence the response of crops to phosphorous application. When wheat and barley crops were irrigated with chloride dominated waters, the yield response to phosphate application was higher as compared to sulphate dominated waters (Manchanda *et al.*, 1982). Application of phosphatic fertilizers most likely will improve the threshold limits of crops to the use of chloride dominated saline water.

Table 39.19: Response of Wheat to Applied Phosphorous Under Saline Water Irrigation (6-10 dS/m)

Cl : SO_4	*Grain Yield (q/ha) at the Applied P_2O_5 (kg/ha)*		
	0	*60*	*90*
3 : 1	26.8	34.4	36.8
1 : 1	28.5	37.6	39.5
1 : 3	31.8	39.8	40.2

(Manchanda *et al.*, 1982).

Gypsum Application

The adverse effects of the long term use of alkali/sodic waters on physical and chemical properties of soils can be mitigated by the use of amendments which contain calcium (gypsum). Acids or the acid forming substances such as sulphuric acid or pyrites which on reaction with soil $CaCO_3$ release Ca^{2+} in solution can also be used for amending adverse effects of alkali waters. Another alternative is to reduce the RSC of the water itself by passing it through gypsum beds. Application of gypsum has been recommended when RSC of irrigation water exceeds 2.5 meq/litre (Bhumbla and Abrol, 1972).

Gypsum helps in maintaining the yields of the crops irrigated with alkali waters (RSC > 5 meq/litre) especially when paddy is grown in rotation and rainfall of the area being double cropped is

< 50 cm. However, no response to gypsum in wheat-fallow rotation has been reported on well drained light textured (sandy loam) soils when irrigated with waters having RSC up to 10 meq/litre (Gupta, 1980). In an already deteriorated soil (SAR_e 48.5) due to irrigation with an alkali water (EC_{iw} 2.6 dS/m, SAR_{iw} 20.5 and RSC 9.5 meq/litre), application of gypsum although did not affect the rice yields but the yield of succeeding wheat crop increased significantly (Sharma and Mondal, 1982). Application of gypsum in fallow-wheat also improved the yield of wheat. On a highly deteriorated sandy loam soil (pH = 10, ESP = 92 and infiltration rate = 2 mm/hour) due to the long term use of an alkali water (TSS 1,000 ppm, RSC 10 meq/litre), even a small dose of gypsum @ 25 per cent of the gypsum requirement (GR) increased the yields of wheat to 26.7 q/ha from almost nil (0.6 q/ha). When gypsum @ 100 per cent GR was added, it further enhanced the yields to 63.3 q/ha (Manchanda *et al.*, 1985). Similarly, sensitive crops like pearl millet and cluster bean yielded 9.7 and 25.5 q/ha when gypsum @ 100 per cent GR was added to a soil (pH 9.5, ESP 45) yielding only 0.71 q/ha and nil, respectively, due to the use of a high RSC water (Manchanda *et al.*, 1985).

Cluster bean/pearl millet-wheat crops can be successfully grown in rotation with RSC waters provided the ESP of the soils is maintained below 15 and 20 with the addition of gypsum @ 100 per cent GR of the soil. Gypsum applied at each irrigation was more effective for increasing the maize yields in maize-wheat sequence irrigated with water RSC of 8 meq/litre as compared with its single dose applied annually (Bajwa *et al.*, 1983). Gypsum improved the soil properties and significantly increased the yields of rice and wheat crops irrigated with water of RSC 6.8 meq/litre, EC_{iw} = 0.85 dS/m. Response to gypsum, either applied annually as one dose or at each irrigation was the same. With higher RSC (10.3 meq/litre) water, the improvement in wheat yields was similar for the 2 modes of gypsum application but rice responded better to gypsum applied at each irrigation. This was because more water was applied to rice which led appreciable increase in soil sodicity during the season affecting rice yields. The depth of irrigation water applied for wheat being smaller, the increase in sodium saturation was not sufficient to adversely affect the wheat yields. Gypsum to supply 2.5 and 5.0 meq/litre to alkali irrigation water for wheat and rice respectively was sufficient for the maintenance of higher yields. Pyrite has also been used for amending the deleterious effect of high RSC waters. Pyrite application once before the sowing of wheat has proved better than splits at each irrigation or mixing it with irrigation water (Chauhan *et al.*, 1986). Alkali water (EC_{iw} 1.83 dS/m, RSC 15 meq/litre) could be reclaimed by passing through a gypsum bed of assorted sized fragments (2-50 mm) (Pal and Poonia, 1979). Dissolution of gypsum is affected by factor such as size distribution of gypsum fragments, flow velocity, salt content and chemical composition of water (Singh *et al.*, 1986).

Management of Nitrate Waters

Highly saline and sodic waters limit the crop production due to reduced germination and lack of continued availability of water through increased osmotic pressure of soil solution in addition to some other detrimental soil properties. Some waters contain the toxic constituents like boron, fluorine and an excess of carbonates of sodium and nitrates. The nitrate containing waters encourage vegetative growth, prolong maturity and encourage succulence and susceptibility to insect pest attack. The use of such waters for wheat crop invariably results in its lodging. The nitrate content of irrigation water was 2.37 meq/litre.

The lodging considerably decreased with increased potash levels and was almost absent when potash was applied at 150 kg K_2O/ha. The nitrate water supplied 3.5 kg N/ha for each 2.5 cm depth irrigation. Assuming that a minimum of 5 irrigations of 5.0 cm depth were given, the crop would have received 83.0 kg N/ha. Besides the cultivars usually apply through urea without considering the quantity being supplied through irrigation waters. This induced considerable lodging during the grain formation stage of the crop (Khurana *et al.*, 1982). Application of 150 kg K_2O/ha resulted in higher grain and straw yields of wheat (Khurana and Bhaya, 1990).

Table 39.20: Potash Levels on Yield of Wheat

K_2O Level (kg/ha)	*Grain Yield (q/ha)*		*Straw Yield (q/ha)*	
	1986	*1987*	*1986*	*1987*
0	38.4	32.3	51.2	49.3
50	42.8	37.1	53.9	49.2
100	44.9	40.4	56.4	52.5
150	47.8	43.9	57.4	54.4
CD (5 per cent)	1.15	2.12	1.79	3.06

(Khurana and Bhaya, 1990).

Table 39.21: Methods of Potash Application on Wheat Yield

Methods of Application	*Grain Yield (q/ha)*		*Straw Yield (q/ha)*	
	1986	*1987*	*1986*	*1987*
As basal	42.7	37.9	54.5	50.1
In splits	47.5	40.7	57.4	52.6
CD (5 per cent)	0.95	1.52	1.46	2.15

(Khurana and Bhaya, 1990).

Application of potash in splits resulted in significantly higher grain and straw yields over its basal application (Khurana and Bhaya, 1990). This might be due to continued availability of potash to the crop all through its growth period. Further more, because of high solubility of muriate of potash its one time basal application enabled its leaching beyond the root zone every successive irrigation (Khurana *et al.*, 1983).

References

Aggarwal, M.C. and Khanna, S.S. (1983). Efficient soil and water management in Haryana. *Bull.*, HAU, Hisar, pp. 118.

Bajwa, M.S. and Josan, A.S. (1989a). *Agric. Wat. Mgmt.*, 16: 217-228.

Bajwa, M.S. *et al.* (1983). *Irri. Sci.*, 4: 191-199.

Chauhan, C.P.S. *et al.* (1992). *Annals of Arid Zone*, 31(1): 69-70.

Chauhan, R.P.S. *et al.* (1986). *Indian. J. Agric. Sci.*, 56: 717-721.

Chopra, N.K. and Chopra, N. (1997). *Indian J. Agron.*, 42(3): 468-470.

Gupta, I.C. (1980). In: *Proc. International Symposium on Salt Affected Soil*. CSSRI, Karnnal, pp. 382-388.

Gururaja Rao, *et al* (2001). Resources characterization and management options for salt affected black soils of agro-ecological region V of Gujarat State. *Technical Bulletin* 1/2001, CSSRI, Regional Station, Anand, Gujarat, p. 83.

Jain, B.L. (1984). *Indian Fmg.*, 34: 13-14.

Khurana, G.P. and Bhaya, G.P. (1990). *Indian J. Agron.*, 35(4): 429-431.

Khurana, G.P. *et al.* (1982). *Curr. Agric.*, 6: 1-11.

Khurana, G.P. *et al.* (1983). *Trans. Isdt. and Ucds.*, 8: 79-83.

Manchanda, H.R. *et al.* (1982). *J. Indian Soc. Soil Sci.*, 30: 353-360.

Manchanda, H.R. *et al.* (1985). *J. Indian Soc. Soil Sci.*, 33: 366-371.

Mass, E.V. and Hoffman, G.J. (1997). *ASCEJ Irri. Drainage Div.*, 103: 115-134.

Minhas, P.S. (1996). *Water Management*, 30: 1-24.

Minhas, P.S. and Gupta, R.K. (1990). Quality of irrigation water assessment and management. *ICAR Bull.*, ICAR, New Delhi.

Minhas, P.S. *et al.* (1990). *Irrigation Sci.*, 11: 57-62.

Minhas, P.S. *et al.* (1988). *J. Indian Soc. Soil Sci.*, 36: 584-587.

Minhas, P.S. *et al.* (1989). *J. Indian Soc. Soil Sci.*, 37: 140-146.

Nayak, A.K. *et al.* (2001). *J. Indian Soc. Soil Sci.*, 49(2): 328-331.

Rai, M. (1977). *Indian J. Agron.*, 22(4).

Rhodes, J.D. (1989). In: Water crop and soil management relating to the use of saline water. FAO, Rome, pp. 125-136.

Sharma, D.R. and Mondal, R.C. (1981). *J. Indian Soc. Soil Sci.*, 29: 270-273.

Sharma, D.R. and Mondal, R.C. (1982). In: *Proc. 12th Int. Congr. of Soil Sci.*, February 8-12, New Delhi, p. 170.

Singh, B. and Bajwa, M.S. (1989). *J. Indian Soc. Soil Sci.*, 39: 779-780.

Singh, M. *et al.* (1985). *Agric. Wat. Mgmt.*, 11: 293-301.

Subashchandra Bose, M. *et al.* (2001). In: National seminar on the use of poor quality water and sugar industrial effluents in agriculture, February 5, A.D. Agricultural College and Research Institute, Tiruchirapalli, p. 171.

40

Red Mud Pond Near NALCO Industry: A Future Death Trap for Aquatic Fauna and an Agent for Degradation of the Environment

B.N. BEURA
Department of Zoology, Aeronautics College, Sunabeda, Koraput, Orissa
ALAKA SAHU
Department of Botany, R.N. College, Dura, Ganjam, Orissa
S.K. SAHU & ASHOK K. PANIGRAHI
Environmental Science Research Centre, Department of Botany, Berhampur University, Berhampur, Orissa

ABSTRACT

The present study was designed to study the effects of the leached red mud waste/leaked effluent from red mud pond of the Alumina Industry (NALCO) situated at Damonjodi, Koraput on freshwater fish, *Anabas scandens*, Cuv. and Val. and the ecological significance of this type of contamination. Laboratory analysis indicated that the leaked chemicals from the red mud pond are almost equivalent to the laboratory prepared red mud waste extract with very little variation. The acute toxicity study revealed the following information.

Fishes were exposed to graded series of concentrations of RMWE/leached effluent (LE) for acute toxicity studies. The MAC value deduced was 4.85 per cent of RMWE/LE in 50 litres of water for 30 days. A safety concentration of 4.5 per cent of RMWE/leached effluent was selected for this study. The LC_{10}, LC_{50}, LC_{100}, determined for *Anabas* fish was found to be 6.5, 9.4 and 17.4 per cent respectively after 30 days of exposure. Exposed fishes appeared lethargic when compared to the control fish. Inappetence and ataxia was observed in the exposed fish. Exposed fishes showed loss of equilibrium, gradual onset of inactivity etc. when compared to the control *Anabas* fish. The body colour changed from yellowish white to brownish yellow during exposure period and during recovery period the exposed fish could get back its original colour. This change was probably due to the coloured toxicant, RMWE/LE of the red mud pond situated at Damonjodi. In the present set of experiment, we could not mark blindness in the early period of exposure. However, if the fishes were allowed to live for more than three months at sub-lethal concentrations even the exposed fishes became blind. The body weight of the LE (leached effluent)/RMWE exposed *Anabas* fish decreased significantly, when compared to control *Anabas* fish, where a significant increase in body weight was marked. When the exposed fish was transferred to toxicant free medium, no recovery was noted. The BSI and LSI significantly declined in the RMW exposed fish when compared to the control fish. Partial insignificant recovery was noted in BSI and HSI/LSI values of the exposed fishes, when compared to control values. The insignificant recovery during recovery period carries weight, due to non-recovery; the weight loss can be contributed to the damage caused to the exposed system.

The whole body oxygen uptake significantly depleted by 89.3 per cent in the 28 d exposed fishes, when compared to control fishes. Lower exposure period helped for better recovery in the exposed animals. The per cent depletion increased with the increase in exposure period, showing a positive correlation and the per cent recovery decreased with the increase in exposure period showing a negative correlation. Prolonged exposure of the fish might cause significant damage to the vital systems of the exposed fish, causing irreversible damage. The exposed *Anabas* fish could avoid stress due to the bimodal gaseous exchange habit. The ventilation rate decreased significantly with the increase in exposure period. Interestingly in *Tilapia* fish, an initial increase in ventilation was marked when exposed to any type of toxicant. But here, in case of *Anabas* fish such a phenomenon was not marked. The tissue slice respiration studies of brain, liver, muscle tissues and gill filaments in exposed fish, when compared to control fish, indicated toxic effects of the red mud waste of NALCO plant. The gill and liver tissue was more affected than the muscle and brain tissue. Brain tissue was least affected during 28 days of exposure. Significant depletion of total ATPase activity in the leached chemicals of the red mud waste exposed fish, affected the movements of ions across the membrane and severely affected the energy metabolism and disturbed energy budget of the exposed fish. The gill and liver was highly affected when compared to brain and muscle of the exposed fish, when compared to control fish tissues. Partial recovery in gill, liver and brain tissue was marked. Less recovery was noted in the exposed muscle tissue. Higher depletion rate in the enzyme activity during exposure period and less recovery in the enzyme activity during recovery period, decreased the metabolic activity of the whole exposed fish which was reflected in the total behaviour of the exposed fishes. Drastic depletion in potassium and magnesium ion content in the RMW exposed fish was marked when compared to the control fish. No significant variation in the sodium and calcium ion content in the exposed fish tissues was noted. Least effect was seen in gill than in muscle, brain and liver. Significant changes were noted in liver, which was highly affected. Significant changes were noted in AChE activity in brain, liver, gill and muscle of the RMW exposed *Anabas* fish, when compared to control. No recovery in the enzyme activity was noted in the brain tissue. Partial insignificant recovery was noted in muscle and gill tissue during recovery studies. Least recovery was marked in the exposed liver tissue during recovery period.

No significant variation was marked in the morphology of erythrocytes of the exposed fish. No finger like proliferations or beak like structure was seen in the red blood cells. Enlargement of red blood cell was also not marked in the exposed fishes. Shrinkage was observed in the exposed red blood cells when compared to control red blood cells. The blood haemoglobin concentration significantly depleted in the exposed fish when compared to control fish. The red blood cell count drastically reduced by 53.2 per cent in the exposed fish when compared to control fish. The haematocrit percentage depleted by 46.8 per cent in the exposed fish when compared to control fish. The oxygen carrying capacity of the blood reduced significantly by 61.8 per cent in the exposed animal when compared to control animal. The depletion in oxygen carrying capacity of blood reduced because the blood haemoglobin concentration significantly reduced. The destruction of haemoglobin content or non-synthesis of haemoglobin in the exposed animal reduced the oxygen carrying capacity of the whole blood of the exposed animal.

Introduction

In the aquatic world, biodiversity is very important in maintaining the purity of water for multiple uses. This importance was recognized as a significant means to understand the ecology of aquatic systems in and around the industrial complexes. Today the main reason of environmental pollution is the over confidence of man who thought that just because of his brain he is the master of nature. The mad rat race between the developed countries and developing countries leads to rapid progress in agriculture, industry, transportation and technology. Such activities of man had created adverse effects on all living biota in the system. And this progress in modern living system left polluted rivers, contaminated soil, depleted wild life and exhausted natural resources for us. Thus pollution is the addition of undesirable constituents to water, air or land, which affect adversely to the natural quality of the environment. In some cases, pollution may involve removal, rather than addition of the constituents from the environment. In other words pollution means the addition of any foreign material like inorganic, organic, biological or radiological or any physical change occurring in nature which may harm or affect living organisms, directly or indirectly, immediately or after a long time. According to the U.S. President's Science Advisory Committee, Environmental pollution is the unfavourable alteration of our surroundings, wholly or largely as a by-product of man's actions, through direct or indirect effects of changes in the energy pattern, radiation levels, chemical and physical constitution and abundance of organisms. It is not only the developed countries that are responsible for pollution by showing high resource consumption but also effective patterns that make them the real and bigger polluters in the world, when compared to other countries. The developing countries are also similarly responsible for this. Their higher population growths, weak environmental regulations, use of out dated technologies are also dangerously polluting the environment. Modern ecologists depict various factors for pollution, such as human population explosion, rapid industrialization, deforestation, unplanned urbanization, scientific and technological advancement etc. (Sharma and Kaur, 1993).

Industrial Pollution and India

In recent years, it was observed that most of the industries cause pollution of the natural environments. Industry is responsible for creating a fantastic array of new chemicals every year all of which eventually find their way into the environment. For most of these chemicals, not even the chemical formulae are known and much less are known about their acute, chronic or genetic effect on the living organisms. At present, the industry is the focus of attention, the world-over, as the most important source of pollution of the environment.

Chemical industry in India has grown up phenomenally since independence. There are to-day about 4000 chemical factories in India. They release large quantities of chemicals in the form of gas, in the exposed red blood cells when compared to control red blood cells. The blood haemoglobin concentration significantly depleted in the exposed fish when compared to control fish. The red blood cell count drastically reduced by 53.2 per cent in the exposed fish when compared to control fish. The haematocrit percentage depleted by 46.8 per cent in the exposed fish when compared to control fish. The oxygen carrying capacity of the blood reduced significantly by 61.8 per cent in the exposed animal when compared to control animal. The depletion in oxygen carrying capacity of blood reduced because the blood haemoglobin concentration significantly reduced. The destruction of haemoglobin content or non-synthesis of haemoglobin in the exposed animal reduced the oxygen carrying capacity of the whole blood of the exposed animal.

oduction

In the aquatic world, biodiversity is very important in maintaining the purity of water for multiple . This importance was recognized as a significant means to understand the ecology of aquatic ms in and around the industrial complexes. Today the main reason of environmental pollution over confidence of man who thought that just because of his brain he is the master of nature. nad rat race between the developed countries and developing countries leads to rapid progress riculture, industry, transportation and technology. Such activities of man had created adverse ts on all living biota in the system. And this progress in modern living system left polluted s, contaminated soil, depleted wild life and exhausted natural resources for us. Thus pollution addition of undesirable constituents to water, air or land, which affect adversely to the natural ty of the environment. In some cases, pollution may involve removal, rather than addition of nstituents from the environment. In other words pollution means the addition of any foreign rial like inorganic, organic, biological or radiological or any physical change occurring in nature n may harm or affect living organisms, directly or indirectly, immediately or after a long time. rding to the U.S. President's Science Advisory Committee, Environmental pollution is the ourable alteration of our surroundings, wholly or largely as a by-product of man's actions, gh direct or indirect effects of changes in the energy pattern, radiation levels, chemical and cal constitution and abundance of organisms. It is not only the developed countries that are nsible for pollution by showing high resource consumption but also effective patterns that them the real and bigger polluters in the world, when compared to other countries. The oping countries are also similarly responsible for this. Their higher population growths, weak onmental regulations, use of out dated technologies are also dangerously polluting the onment. Modern ecologists depict various factors for pollution, such as human population sion, rapid industrialization, deforestation, unplanned urbanization, scientific and technological cement etc. (Sharma and Kaur, 1993).

strial Pollution and India

recent years, it was observed that most of the industries cause pollution of the natural nments. Industry is responsible for creating a fantastic array of new chemicals every year all ch eventually find their way into the environment. For most of these chemicals, not even the cal formulae are known and much less are known about their acute, chronic or genetic effect living organisms. At present, the industry is the focus of attention, the world-over, as the mportant source of pollution of the environment.

nemical industry in India has grown up phenomenally since independence. There are to-day 4000 chemical factories in India. They release large quantities of chemicals in the form of gas,

Gururaja Rao, *et al.* (2001). Resources characterization and management options for salt affected black soils of agro-ecological region V of Gujarat State. *Technical Bulletin* 1/2001, CSSRI, Regional Station, Anand, Gujarat, p. 83.

Jain, B.L. (1984). *Indian Fmg.*, 34: 13-14.

Khurana, G.P. and Bhaya, G.P. (1990). *Indian J. Agron.*, 35(4): 429-431.

Khurana, G.P. *et al.* (1982). *Curr. Agric.*, 6: 1-11.

Khurana, G.P. *et al.* (1983). *Trans. Isdt. and Ucds.*, 8: 79-83.

Manchanda, H.R. *et al.* (1982). *J. Indian Soc. Soil Sci.*, 30: 353-360.

Manchanda, H.R. *et al.* (1985). *J. Indian Soc. Soil Sci.*, 33: 366-371.

Mass, E.V. and Hoffman, G.J. (1997). *ASCEJ Irri. Drainage Div.*, 103: 115-134.

Minhas, P.S. (1996). *Water Management*, 30: 1-24.

Minhas, P.S. and Gupta, R.K. (1990). Quality of irrigation water assessment and management. *ICAR Bull.*, ICAR, New Delhi.

Minhas, P.S. *et al.* (1990). *Irrigation Sci.*, 11: 57-62.

Minhas, P.S. *et al.* (1988). *J. Indian Soc. Soil Sci.*, 36: 584-587.

Minhas, P.S. *et al.* (1989). *J. Indian Soc. Soil Sci.*, 37: 140-146.

Nayak, A.K. *et al.* (2001). *J. Indian Soc. Soil Sci.*, 49(2): 328-331.

Rai, M. (1977). *Indian J. Agron.*, 22(4).

Rhodes, J.D. (1989). In: Water crop and soil management relating to the use of saline water. FAO, Rome, pp. 125-136.

Sharma, D.R. and Mondal, R.C. (1981). *J. Indian Soc. Soil Sci.*, 29: 270-273.

Sharma, D.R. and Mondal, R.C. (1982). In: *Proc. 12th Int. Congr. of Soil Sci.*, February 8-12, New Delhi, p. 170.

Singh, B. and Bajwa, M.S. (1989). *J. Indian Soc. Soil Sci.*, 39: 779-780.

Singh, M. *et al.* (1985). *Agric. Wat. Mgmt.*, 11: 293-301.

Subashchandra Bose, M. *et al.* (2001). In: National seminar on the use of poor quality water and sugar industrial effluents in agriculture, February 5, A.D. Agricultural College and Research Institute, Tiruchirapalli, p. 171.

40

Red Mud Pond Near NALCO Industry: A Future Death Trap for Aquatic Fauna and an Agent for Degradation of the Environment

B.N. BEURA
Department of Zoology, Aeronautics College, Sunabeda, Koraput, Orissa
ALAKA SAHU
Department of Botany, R.N. College, Dura, Ganjam, Orissa
S.K. SAHU & ASHOK K. PANIGRAHI
Environmental Science Research Centre, Department of Botany, Berhampur University, Berhampur, Orissa

ABSTRACT

The present study was designed to study the effects of the leached red mud waste/leaked effluent from red mud pond of the Alumina Industry (NALCO) situated at Damonjodi, Koraput on freshwater fish, *Anabas scandens*, Cuv. and Val. and the ecological significance of this type of contamination. Laboratory analysis indicated that the leaked chemicals from the red mud pond are almost equivalent to the laboratory prepared red mud waste extract with very little variation. The acute toxicity study revealed the following information.

Fishes were exposed to graded series of concentrations of RMWE/leached effluent (LE) for acute toxicity studies. The MAC value deduced was 4.85 per cent of RMWE/LE in 50 litres of

water for 30 days. A safety concentration of 4.5 per cent of RMWE/leached effluent was se for this study. The LC_{10}, LC_{50}, LC_{100}, determined for *Anabas* fish was found to be 6.5, 9.4 ar per cent respectively after 30 days of exposure. Exposed fishes appeared lethargic when con to the control fish. Inappetence and ataxia was observed in the exposed fish. Exposed showed loss of equilibrium, gradual onset of inactivity etc. when compared to the control fish. The body colour changed from yellowish white to brownish yellow during exposure and during recovery period the exposed fish could get back its original colour. This chan probably due to the coloured toxicant, RMWE/LE of the red mud pond situated at Dam In the present set of experiment, we could not mark blindness in the early period of ex However, if the fishes were allowed to live for more than three months at sub-lethal concen even the exposed fishes became blind. The body weight of the LE (leached effluent)/ exposed *Anabas* fish decreased significantly, when compared to control *Anabas* fish, significant increase in body weight was marked. When the exposed fish was transf toxicant free medium, no recovery was noted. The BSI and LSI significantly decline RMW exposed fish when compared to the control fish. Partial insignificant recovery w in BSI and HSI/LSI values of the exposed fishes, when compared to control val insignificant recovery during recovery period carries weight, due to non-recovery; th loss can be contributed to the damage caused to the exposed system.

The whole body oxygen uptake significantly depleted by 89.3 per cent in the 28 d fishes, when compared to control fishes. Lower exposure period helped for better re the exposed animals. The per cent depletion increased with the increase in exposur showing a positive correlation and the per cent recovery decreased with the increase in period showing a negative correlation. Prolonged exposure of the fish might cause s damage to the vital systems of the exposed fish, causing irreversible damage. The *Anabas* fish could avoid stress due to the bimodal gaseous exchange habit. The ventil decreased significantly with the increase in exposure period. Interestingly in *Tilap* initial increase in ventilation was marked when exposed to any type of toxicant. B case of *Anabas* fish such a phenomenon was not marked. The tissue slice respiration brain, liver, muscle tissues and gill filaments in exposed fish, when compared to c indicated toxic effects of the red mud waste of NALCO plant. The gill and liver tissue affected than the muscle and brain tissue. Brain tissue was least affected during exposure. Significant depletion of total ATPase activity in the leached chemicals of th waste exposed fish, affected the movements of ions across the membrane and severe the energy metabolism and disturbed energy budget of the exposed fish. The gill an highly affected when compared to brain and muscle of the exposed fish, when c control fish tissues. Partial recovery in gill, liver and brain tissue was marked. Le was noted in the exposed muscle tissue. Higher depletion rate in the enzyme acti exposure period and less recovery in the enzyme activity during recovery period, de metabolic activity of the whole exposed fish which was reflected in the total beha exposed fishes. Drastic depletion in potassium and magnesium ion content in the RM fish was marked when compared to the control fish. No significant variation in the calcium ion content in the exposed fish tissues was noted. Least effect was seen in muscle, brain and liver. Significant changes were noted in liver, which was hig Significant changes were noted in AChE activity in brain, liver, gill and muscle exposed *Anabas* fish, when compared to control. No recovery in the enzyme activi in the brain tissue. Partial insignificant recovery was noted in muscle and gill t recovery studies. Least recovery was marked in the exposed liver tissue during rec

No significant variation was marked in the morphology of erythrocytes of fish. No finger like proliferations or beak like structure was seen in the red Enlargement of red blood cell was also not marked in the exposed fishes. Shrinkage liquid and solid wastes, into the environment. Many of these chemicals are toxic and create pollution problem. The problem of toxic hazard has already reached alarming proportions in this country and is bound to grow with increasing industrialization. India is the largest manufacturer of pesticides in the whole of South Asia and Africa. As many as 139 organic chemicals, heavy metals like zinc, lead, chromium, copper, mercury and various other compounds are used in the manufacture of dyes only. Sundaresan *et al.* (1983) have given the growth of industries dealing with toxic chemicals and generating toxic and hazardous wastes during 1950s, 1970s and 1980s, in India.

Industries, which are known to produce potentially toxic and hazardous wastes, are pesticides, dyes and pigments, organo-chemicals, fertilizers, non-ferrous metals, steel and chlor-alkali plants. The major foci of such industries in India are Mumbai, Kolkata, Kanpur, Delhi, Chandigarh, Jamshedpur, Bokaro, Hyderabad, Vishakhapatnam, Chennai, Baroda, Cochin etc. The wastes from the industry are generally disposed by land filling or released into water bodies. Generally the toxic effluents from industry are neutralized before their discharge, but still they contain substantial amount of toxic substances that can cause pollution. At present it is believed that rivers are most severely polluted by industries followed by estuaries, lakes and ocean in declining order.

The Industry Under Study

Every underdeveloped or developing country aiming to be highly developed, explore nature and tried to find out methods and adopted industrial revolution, leaving behind the consequence of such trend. When the problems of environmental status and quality was taken into consideration it became apparent that, it was not enough to alleviate undesirable consequences of man's activities by technological means or to prevent them forbidding certain practices of them are of certain chemicals. Rapid industrialisation and exploitation of natural resources on massive scale, accumulated undesirable substances in huge quantity thus, pollute the environment. Pollution is caused due to deliberate or unavoidable discharge of industrially produced wastes. These pollutants causing health hazards and definite are undesirable problems to flora and fauna, apart from influencing the ecological balance and brings significant climatic changes.

The field under study refers to the Alumina industry specifically to Mining and Refinery complex, NALCO, Damaojodi in the district of Koraput, Orissa, India. For analysis of various factors of pollution in alumina industry it becomes imperative for us to discuss the manufacturing process from ore mining to refining in brief with reference to NALCO. The alumina is extracted in two main stages *i.e.* Mining operation and Ore refinery.

Mining Operation

The ore of Aluminium *i.e.* Bauxite mine is situated at Panchapatmali in the district of Koraput, Orissa. It is at the height of 400 ft from sea level. NALCO has marked 17 sq.km. of its area for getting deposited Bauxite. 370 million tones deposit is estimated. The mining capacity of NALCO is 24,00,000 tpa and being expanded to 48,00,000 tpa. The method used to mine Bauxite includes exploration and mine planning, removal of topsoil and over burdened and excavation of ore blasting operation. The hard cap rock at the surface is blasted to allow it to be mined together with the ore beneath.

Heavy earth moving equipment' like dozer, back-end hoe, front-end loader and dampers etc. are used for mining and transporting operations. Then the ore is crushed by single state double roll tooth crusher of 900 tph for each transportation and dispatch by various means to the refinery complex, which is done by a long cable conveyer belt systems of 14.6 km long at single fidget multi curve with 900 tph capacity at M&R complex, Damonjodi. It is the longest conveyer belt of the world. The entire distance from the mining site to refinery complex, the air of the area looks cloudy due to heavy dust loading from operation and open transportation by conveyer belt.

In the course of mining exploration and transportation, minute dust particles looked like the labourers working in that area as well as the inhabitants residing nearby inhaled few particles. This air heavily loaded with dust or suspended particles cause lung cancer and tuberculosis. A health report indicated that high percentage of the inhabitants suffer from these two diseases and also respiratory distress and bronchitis: Not only the human beings are affected but also this pollution spoils very valuable flora of the environment due to heavy deposition of red coloured dust. The dust are deposited on the green leaves of the plants, reducing the incidence and trapping of sunlight, coming from sun, the only and ultimate sources of energy. Thus, reducing the photosynthetic efficiency of plants.

It is also noted that during the transportation by conveyer belt, huge amount of crushed Bauxite powders are dispersed, mixed with air, thus, the inhabitants are suffering a lot. Many news items appeared on this issue in different Newspapers, and public reacted very badly opposing the way the area is being polluted. Due to the set up of the industry, not only 17 km^2 of Panchaputamali hills will be deforested but also 1742 acres of acquired land were subjected to deforestation, disappearance of valuable flora and disappearance of wild life. The NALCO project-plantation program in barren lands thus reclaimed by back filling and on the two sides of the newly constructed roads. But yet this did not fulfil the purpose of reappearance of the forest species and lead to degradation of biodiversity. The noise and vibration due to continuous blasting and heavy machinery equipment caused noise pollution in the mining area. 'Environment Management Plan at Alumina Refinery Complex', NALCO has taken all possible measures during construction and also taking all possible measures for monitoring the pollution control parameters so that no pollutant is allowed to be discharged from plant as process effluent or township effluent to the water body without treatment, conforming the IS norms.

The workers NALCO Damonjodi as well as the tribal inhabitants are affected by the poisonous gases and ash emitted from the chimney. This also affects the natural life in that area. It is also seen that micro particles of alumina and ash emitted during the process are deposited on the leaves of the herb and trees and cause chlorosis. The physicians suspect the increasing number of patients suffering from asthma and lung cancer may also be due to inhalation of powdered alumina and ash from the plant along with the ore dust spread over the area.

Ore Refinery

The capacity of Alumina plant at NALCO, Damonjodi is 8,00,000 tpa and being expanded to capacity of 15,75,000 tpa. The ore refining is done by a classical process called BAYER's Process in which the aluminium in the Bauxite is digested in caustic at 105°C and ultimately recovered by precipitation with the removal of impurities. After successive operation of digestion, desilication, setting and washing the precipitated hydrate *i.e.* $NaAl(OH)_4$ is filtered and calcined at 1000°C. There are two production stream of 4,00,000 tpa capacity each atmospheric pressure digestion process is carried out at 107°C with energy efficient fluidized bed calciners of 1400 tph (3×18.5 M.W.) power is co-generated by use of back pressure turbine in steam generation plant. The residual mud is separated and pumped to the red mud pond.

Red Mud Pond

In the process of bauxite refining the red mud, which is generated after digestion is the major waste product. For 8 lakhs MT alumina production 10 lakh MT of red mud is generated and is being discharged in red mud pond by 2nd stage pump in slurry form at the rate of 370 m^3/hr with 150 gps solid consistency and 2.3 gpl Na_2O the dry red mud flow in the RMP is 110 MT/hr. To store such a huge quantity of red mud slurry, the Red Mud Pond has been constructed towards the west of refinery complex.

Three earthen dams called Orissa Dam-l, 2 and 3, flank the red mud pond. Fortunately or unfortunately, natural hill ridges with high elevation surround the red mud pond. These three earthen dams connected the natural ridges. The Orissa Dam-l is towards east *i.e.* towards the refinery complex, the Orissa Dam-2 towards north, the Orissa Dam-3 is towards west *i.e.* opposite to Dam-l, this dam separated RMP from ash pond area. Hill ridges with high elevation cover South of the pond. Orissa Dam-2 has a natural hill ridge with 930 metres elevation. There is no spill way is provided in the dams to avoid spillage contamination to natural nalaha. Elevation of dams during the first phase is 930 metres the anticipated life is for 3.5 years. It is planned to expand the elevation upto 945 mts and 957 metres during second and 3rd phase operation. The anticipated life of the pond is 7 and 11 years respectively.

The initial elevation of RMP was 905 mts at different increased elevations the catchment area and the capacity of RMP are given below. At 920 rot elevation catchment area is 16.544 hectares with total capacity of 301×10^4 m^3 at present elevation *i.e.* system which needs very less area for the disposal and stacking of 930 mts the area is raised up to 108.6 hect. and 1016.17×10^4 m^3 capacity. After second phase expansion up to 945 mts; the area will be raised to 187 hects. with capacity of 3305.87×10^4 m^3. At final phase of expansion up to 957 mts elevation the total catchment area of the pond will be 212.40 hects. with total capacity of 5728.82×10^4 m^3 with 2 mts free board. After the life of the red mud pond is over, one area has been already earmarked during the feasibility study for the construction of second pond. Horizontal expansion of red mud pond is feasible to certain extent. During last 3 years it is noted that the pond level is rising abnormally due to more precipitation than evaporation at Damonjodi. In order to check the pond water level the alkaline water is returned back from the pond and is recycled for use in plant various process mainly to carry fresh generation of red mud slurry as a result alkalinity of the pond water is increasing day by day. The shore of this is extremely affected. The water of the pond is toxic to the plants.

Although the life of the dam in phase-I is over, yet the second phase expansion work has not started, rather in order to explore the long term solution to the wet tailing system, dry mud disposal system, which needs very less area for the disposal and stacking of dry mud. It does not call for construction of large earthen dams. The system multiplies alkalinity of the area of dispersal. The run-off leachates of red mud stacked near RMP are entering into the pond with rainwater. This is also an addition to the alkali strength to the RMP water instead of diluting it.

NALCO-The Industry

National Aluminium Co. Limited (NALCO) is considered to be a turning point in the 50 years old history of Indian Aluminium Industry. In a major leap forward NALCO has addressed itself to the country's need for self sufficiency in aluminium as well given the country a technology edge in making this strategic metal on the best of world standard. NALCO was set up in 1981, in the public sector, following a major investment decision of government of India in 1980, to exploit a part of large deposit of Bauxite, discovered in the east coast in 1975. Aluminium Pechiny of France provided the technology and basic engineering for bauxite mine, refinery and smelter. The project cost of Rs. 2408 crore was partly financed by Rs. 1119 crore equivalent of Euro-dollar lone extended by a consortium of international banks. The project was most expeditiously implemented, different, segments of NALCO went into production in a phased manner from November 1985 sale of Aluminium was commercial on May 1987 and export was started on September 1988. Initially 1987-88 the Aluminium production was 160,000 tones the production rate remarkably increased to 807000 tones in 1995-96, in 1996-97, 840,000 MT and in 1997-98 it was 883,000 MT and has chartered a course of international confidence in India's industrial capability. Today NALCO has emerged as the largest integrated Bauxite Alumina-Aluminium complex in Asia, enabling India to witness a quantum jump in Alumina and Aluminium production. NALCO produces calcined alumina at refinery complex, Damonjodi.

Topology and Location

The mines and refinery complex of NALCO, Damonjodi is situated Similigude block, under Potangi tahsil in the district of Koraput, Orissa state, India. Damonjodi is at a height of about 1300 metres from sea level, located at latitude 18°6′–18°58′ towards North and longitude 82°57′–83°04′ East. The area enjoys an annual rainfall of 1723 mm in 1998. The relative humidity of area ranges from a minimum of 47 per cent in April and maximum of 85 per cent in July 1998. The temperature of the area varied from a minimum of 15°C to maximum 35°C in 1997-98 and from a minimum of 9°C to maximum of 42°C in 1998-99. Southeast side is covered by Deomali hill range. The slope of the area is towards the northwest. A small river Karandia is flowing near by the Refinery complex and the Damonjodi township. It carries almost all the emissions of the plant and township. At its down stream mixed with the Kolab river at its catchment area of Upper Kolab project near Sunabeda.

Affected Area

As we know the red mud pond is situated in such a place that the pond water is restricted by high hill ridges at south, ash pond at west separated Orissa Dam-3. But at its north seepage is marked at a natural nalaha at its down stream about 100 mts away from Orissa Dam-2 of RMP. This seepage might be due to the storage of huge quantity of effluent at a high level. From the original point of leaching, the leachates gradually flow downwards mixed with natural nalaha and ultimately enters into the crop fields and freshwater ponds. These leachates from RMP gone down in its slope towards the river Karandi in the northwest direction and finally joins with the river Kolab near Sunabeda. The contaminated water in the catchment area of Upper Kolab project is being lifted for Sunabeda as well as Damonjodi township for all domestic use and human consumption. Ultimately this polluted water flows down along the river and being used by the inhabitants of Jeypore town as well as the village at the river bank.

Toxic Effects

The leachate at its initial point is highly alkaline in nature with Fe, Al, Ti, Na ions, the pH of it ranging from 9 to 10 even more. Due to the effects of toxic chemicals rice plants do not grow well. It is found that the soil colour of the crop field changed to violet red with high caustic smell. No singular instance of fish poisoning in ponds by the leached chemicals of the red mud pond was available. Even no news item was also published. One or two typical fish death as reported by the villagers were not due to the leached chemicals of the red mud pond of the NALCO plant. The deaths were mostly due to some point-spilled pesticide in a particular area. However, in future, these leached chemicals if they find an entry into any pond or river, heavy fish death cannot be ruled out, due to high toxic nature of the toxicant.

The general public opinion flashed several times in local dailies and weekly news papers that the undesirable waste emissions of NALCO mixing with water bodies, pollute the river and the reservoir, which man used. It was reported in one of the newspaper that, there is every possibility of loosing the memory power as well as reproduction ability by animals including man, if the polluted water is used for consumption. It is published in a newspaper that by using the contaminated water for drinking purpose, many people suspected their deafness and some have lost their reproductive ability. Now users report different types of skin diseases by polluted water is to be because of the contamination. On the present issue/problem many physicians have expressed their fear and also apprehension that in future, this industry causing pollution may create wide spread health hazards. Not only the human community is facing hazards but also the aquatic flora and fauna are also affected. It is expected that the toxic pollutants might be accumulated in the body of the contaminated fishes, which can be observed from the symptoms like the scale souring. Some time heavy emission lead to fatal condition of fish and aquatic organism. The abnormal deposit of red mud in the red mud pond and increasing in alkalinity of red mud pond water may become a future threat for the people near by locality in Koraput district, because of heavy leaching from the red mud pond.

Keeping in view, the leaching of chemicals from the red mud pond of the industry (NALCO, Damonjodi, Koraput, Orissa), this project was masterminded to evaluate the toxicological effects of the leached chemicals of the red mud pond on the growth and survivability of freshwater fish, *Anabas scandens*, Cuv. and Val. and its ecological significance under laboratory controlled conditions.

Materials

Selection of Test Animal

The present experiment was designed on *Anabas* fish because *Anabas scandens* exhibits bimodal gaseous exchange. *Anabas* fish can tolerate higher concentrations of the toxicants as the fish mostly depend on direct air breathing rather than gill breathing. In this fish, the gills are rudimentary, hence very little chance of entry of pollutants through the gills. In this fish most of the pollutant entry will be either through the body surface, little through the gills and some through the gastro-intestinal tracts.

Selection of the Toxicant

The soluble leached chemicals of the red mud waste of the NALCO plant of Damonjodi prepared in the laboratory has been used as the test chemical for the experimental study. The red mud collected from red mud pond is air dried, powdered and sieved. A known (2 kg.) quantity of dried powder is mixed with known amount (2 lit.) of distilled water and stirred for 15 days continuously with the help of mixer grinder. The mixture was allowed to stand for one day and the supernatant was decanted and again passed through a multi-sieve soaking and filtering system and the obtained extract is the leached chemical of the waste, which is stored for future experiments.

Maintenance of Fishes in Laboratory Aquarium

Anabas scandens, Cuv. and Val. of medium size (18-22 g) were collected from the local nursery of the Fisheries Department of Berhampur (Ganjam), Orissa. The fish were allowed to grow in the laboratory reservoirs for acclimatization at least for 15 days before starting the experiment. The fish were maintained in aquarium of 60×58 cm. containing 70 litres of water. Chlorine-free tap water was used in both control and experimental aquarium. The water was changed daily.

Air was bubbled through water of the aquarium to maintain the dissolved oxygen at 85 ± 5 per cent air saturation value. The physico-chemical quality of the water of both control and exposed aquarium were measured during the experimental period (APHA, 1974). Living earthworms from garden showing no contamination by any toxicant were collected and fed daily to both control and exposed fish initially and slowly the diet was changed to toxicant-free chopped goat liver and then to small slices of boiled eggs during holding and through out the experimental exposure period. After acclimatization, the fish were washed thoroughly with 1 per cent dilute Potassium permanganate ($KMnO_4$) solution, so as to prevent any infection. The test solution of the experimental aquarium was changed daily so as to maintain the constancy of the red mud waste extract concentration. The below table indicates the water quality of control and exposed aquariua.

Water Quality	*Control*	*Exposed*
pH	7.2 ± 0.5	7.8 ± 0.5
Temperature	26 ± 2°C	26 ± 2°C
Illumination	2200 ± 200 lux	2200 ± 200 lux
Total hardness	78.5 ± 9.8 mgl^{-1}	89.4 ± 11.6 mgl^{-1}
Specific conductivity	3.52×100 μmhos	3.64×100 μmhos
Transparency	0.01–0.025	0.04–0.065

Methodology

Respiration

Whole Animal Oxygen Uptake

The whole animal oxygen uptake rates of the test and control fish were measured using five wide mouth 2 litres capacity flasks. Each flask containing the test solution and a test fish was hermetically sealed. A reference flask was kept without fish to check any change of oxygen concentration during the experiments, due to the presence of microorganisms. Any change of oxygen concentration, caused by the microorganism was computed with the final data. After one hour, the dissolved oxygen of all experimental flasks was determined according to the modified Winkler's method (Ashby, 1973 and Panigrahi, 1980).

Tissue Slice Respiration

The tissue slice respiration, manometric reading of oxygen uptake or carbon dioxide evolution and preparation of tissue extracts were carried out by the methods reported by Patel *et al.* (1973) and Fox *et al.* (1975). All calculations were made on the basis of wet weight of slices. Under our experimental conditions, dry weight of the tissues were calculated and presented below: Brain tissue–11.41 per cent of the wet weight; Liver tissue–12.78 per cent of the wet weight; Muscle tissue–16.24 per cent of the wet tissue; Gill tissue–17.54 per cent of the wet tissue.

Ventilation Rate

Number of opercular movement per minute was counted in both control and red mud waste extract exposed fishes.

Enzyme Analysis

Activity of ATPase was determined in fish tissues by measuring the amount of inorganic phosphate produced when adenosine triphosphate was converted to adenosine di-phoshphate. Inorganic phosphate produced as a result of the cleavage of ATP to ADP was measured by the method of Fiske and Subbarao (1925) as modified by Martinek (1970). Colour development proceeded at room temperature for 30 minutes. Protein was determined by the procedure developed by Lowry *et al.* (1951), using a spectrophotometer. The ATPase activity was expressed as µmoles of inorganic phosphate liberated mg^{-1} of protein h^{-1}.

Ion Analysis

The amount Na^+, K^+, Ca^{++} in fish tissues were determined by a flame photometer taking NaCl, KCl and $CaCl_2$ as standards (Panigrahi, 1980). For magnesium determinations, the procedure followed by Orange and Rhei (1970) was adopted.

Haematology

Since unclotted blood samples are required for haematological studies and the anticoagulant should have no influence on the red cells or other cells, heparin, one of the most powerful anticoagulant was used for the collection of blood (Hurkat and Mathur, 1976). Blood was directly collected from the ventral aorta with a heparinised syringe and a drop of blood was spread on a clean slide following the usual procedure described by Dacie and Lewis (1970). The haemoglobin percentage was measured with the aid of a Spectrophotometer by acid hematin method (Sieverd, 1964). Red blood corpuscle (RBC) counts were made by the aid of a Neubauer's haemacytometer and a research microscope (Balxhall and Daisley, 1972 and 1973). The haematocrit percentage was determined following Wintrobe's method (Dacie and Lewis, 1970). Oxygen carrying capacity of blood was estimated (Johanson, 1970).

AChE Activity

The supernatant of the enzyme extracts of fish tissues were analysed for acetylcholinesterase activity using Hestrins (1949) technique as modified by Sudderuddin (1973).

Acute Toxicity of Red Mud Waste Extract

The purpose of acute toxicity test is to assess various abnormalities caused due to administration of chemicals to animals on one occasion or other and to determine the order of lethality of the chemical. Many of the schemes are available at present to test for adverse effects that may result when chemicals enter into environment. Considerable efforts have been directed towards the establishment of short-cut procedure. In aquatic toxicology, acute lethal toxicity tests with fish or invertebrates are usually intended to assess the numerical value of toxicity, to compare potencies of toxicants, to assess the effect of environmental variables on toxicity. Besides, statistical analysis of toxicity is essentially a tool in all these studies. The rapid increase in contamination of aquatic environments with toxicants in recent years has resulted in an escalation of scientific interests in the biological effects of toxicants in the environment; any deleterious effect of the toxicants on fish is likely to be reflected on the entire eco-system.

Matida *et al.* (1971) studied the effect of low concentration of methyl mercuric chloride on rainbow trouts, which showed poisoning symptoms. Macleod and Passah (1973) reported effects of temperature on mercury accumulation, toxicity and metabolic rate in rainbow trout (*Salmo garidneri*). Fathead minnows when exposed to low concentrations of methyl mercuric chloride (Olson *et al.*, 1975) showed no effect on survival at concentrations causing an acceptable residue. McKim *et al.* (1976) reported long term effects of methyl mercuric chloride (MMC) on three generations of brook trout, *Salvelinus fontinalis*. Hannerz (1968) reported accumulation and distribution of several forms of mercury in aquatic animals. Amminikuty and Rege (1977) mentioned few toxicological effects of Agallol's (Methoxy ethyl mercuric chloride, (MEMC) on widow tetra (*Gymno-corymbus ternetzi*).

Das and Misra (1982) studied the toxicity of MEMC (methoxy ethyl mercuric chloride) on *Cyprinus carpio* L. Panigrahi and Misra (1980) found severe poisoning symptoms in *Anabas scandens* exposed to inorganic mercury. The fish-pesticide Research Laboratory of the US Bureau of Sport Fisheries and Wild life tested a large number of pesticide chemicals mainly on two species of freshwater fish, rainbow trout (*Salmo gairdneri*) and blue gills (*Lepomina macrochirus*), Huges and Davies (1963) tested the toxicities of several formulations of the phenoxy herbicides, such as, 2, 4-D and 2, 4, 5-T to blue gills. Bond *et al.* (1960) made a critical study on species like Chinook (*Oncorhynchus tshawytscha*), Coho salmon (*G. kisutch*) and large mouth bass (*Micropterus salmoides*). The toxicity of some combinations of insecticides to mosquito fish was studied by Ferguson and Bingham (1966) using DDT, endrin, toxaphene and methyl parathion. Richard (1966) studied the toxicity of Thiodan in several fish species and aquatic invertebrates. Doudoroff *et al.* (1953) and Mathur (1969) demonstrated that insecticides such as aldrin, toxaphene, lindane, dieldrin, BHC and DDT were extremely toxic to fish. Amend (1974) made a comparative toxicity study of two iodophors on rainbow trout eggs. Caldwell *et al.* (1978) reported the toxicity of the fungicides, "Captan" to the dungeneas crab, *Cancer magister*. Relative toxicity of Dichlorovos (DDVP) was studied by Rath and Misra (1979a, b) on *Tilapia mossambica*, Peters belonging to 3 different age groups, Roald (1967) reported the acute toxicity of Lignosulphonates to rainbow trout. Gouda *et al.* (1981) tested the toxicity of dimecron, sevin and lindax to *Anabas scandens* and *Heteropneustees fossilis*. Choudhury (1975) reported effects of organochloro insecticides (aldrin, dieldrin and endrin) on fish and other pond organisms.

David *et al.* (1976) studied the effects of heavy metals on juvenile bay scallop, *Agropectan irradians*, Ray and David (1962) studied the fish mortality caused by precipitation of ferric ion in the river Daha at Siwan (North Bihar). Reports on temperature induced changes in acute toxicity of zinc to Atlantic Salomon (*Salmo salar*) by Hodson and Spraque (1975), and Copper toxicity to embryos and larvae of 8 species of freshwater fish by McKim *et al.* (1978) were available. Cadmium toxicity of gold fish, *Carassius auratus* by McCarty *et al.* (1978), lethal and sub-lethal effects of Binary mixtures of cyanide and hexavalent chromium, zinc or ammonia to the fathead minnow (*Pimephales promelas*) and rainbow trout (*Salmo gairdneri*) by Broderius *et al.* (1979); all these add to the knowledge of heavy metal toxicity on fishes. Hartung (1973) pointed out that much information was lacking in the due course of such changes during cadmium and other heavy metal toxicity assays. McCarty *et al.* (1978) noted the accurate assessment of water quality criteria and data pertaining to heavy metal toxicity which are derived from static bioassay experiments is of much practical concern. McCarty *et al.* (1978) studied the toxicity of cadmium to gold fish, *Carassius auratus* in hard and soft water. Panigrahi (1984) reported the toxicity of a mercury based pesticide (MEMC, Emisan-6 Brand name) on a freshwater fish *Anabas scandens*, Cuv. and Val.

Although adequate literature on acute toxicity of different killer chemicals to fish is available, no information is available regarding this toxicant, red mud waste extract (RMW Extract or leached effluent) and its acute toxicity. This study was designed to find out the lethal concentration values and maximum allowable concentration (MAC) of Red Mud Waste Extract on *Anabas scandens*, Cuv. and Val.

Materials and Methods

Test fish, *Anabas scandens* were collected, acclimatized in the laboratory as described earlier. A graded series of concentrations of Red mud Waste Extract ranging from 0.1 per cent to 20 per cent were prepared. 10 healthy fish were exposed to each concentration in 10 litre glass jars. The experiments were conducted in chlorine-free tap water at a room temperature of 28 ± 2°C.

The mortality rate of test-fish was studied following the method described by Panigrahi (1980). Observation on the toxicity of red mud waste extract was made at 24, 48, 72, 96 hours and 28 days after the experimental animals were first exposed. Individuals showing no respiratory movements, no opercular movements and no response to a tactile stimulus were recorded as dead, and were immediately removed. The test fish exposed to lower range of the toxicant were exposed for a period of five weeks to find out the maximum allowable concentration (MAC), where no mortality was noticed and this was expressed in percentage (%). Different values such as LC_0, LC_{10}, LC_{50}, LC_{100} were deduced from graphical interpolation (Finney, 1977).

Results

The result showed that the MAC value of red mud waste extract was found to be 4.85 per cent for 30 days and to be on the safer side 4.5 per cent was considered for 28 days of the exposure for sub-lethal toxicological studies. The observations showed LC_{10} values at 24 hours to be 6.5 per cent and 6.4 per cent for 48 hours. The results showed that the LC_{50} values for test fish were 20.2 per cent at 24 hours, 19.5 per cent at 48 hours of exposure. The results showed that the LC_{90} values for test fish were 24.5 per cent at 24 hours, 23.8 per cent at 48 hours of exposure. The results showed that the LC_{100} values for test fish were 26.7 per cent at 24 hours, 25.1 per cent at 48 hours of exposure. The LC_{10}, LC_{50}, LC_{90} and LC_{100}, values after 28 days were recorded to be 6.5, 9.4, 14.2 and 17.4 per cent of the red mud waste extract. No mortality was observed in the control set.

Discussion

The mortality percentage for any fixed period increased with the increasing concentration of the toxicant. There was an increase in mortality percentage with the increase in exposure period also. The proportion of these animals that died in a fixed time increased with the increase in the toxicant concentration and the median tolerance limit value decreased with the exposure period (Hodson *et al.*, 1977; Das and Misra, 1982 and Panigrahi, 1984). The close response approach to assessment of pollutant toxicity involves several assumptions. The route of entry of the pesticide is generally agreed to be via the gills (Holden, 1962; Ferguson *et al.*, 1966) and thus enters directly into the circulatory system. This caused damage to tissues as a result of which there is depression in active metabolism (Macleod and Passah, 1973). Fry (1957) demonstrated depression in active metabolism that directly reduced scope of activity of the animal. Probably this has relevance with the impaired swimming ability, erratic movements, paralysis, anorexia and ataxia, which were observed in all test fish at higher exposure periods. Initial disturbance was caused by the pesticide but latter on the fishes stabilized and performed normal activity.

Hence, with the recognition that all heavy metals, industrial wastes, effluents, red mud wastes, waste extracts and synthetic killer chemicals (pesticides) are potentially lethal to fish even at relatively low concentrations. It is now a normal practice to test all new chemicals for their toxicity to fish. It is equally important to know the amount and type of chemicals that are leaching by way of leaking from the stocking spots into the environment. These chemicals pollute the natural environments. The MAC values can be helpful to understand about the amount of the red mud waste to be discharged into the environment that would have no side effect on target and other non-target organisms like fish in aquatic environments.

Behavioural Studies

Long term sub-lethal effects resulting from chronic exposure of organisms to lower levels of toxicants, probably have the most important effects on the aquatic organisms. Killer chemicals, heavy metals, industrial wastes and pesticides in the aquatic environment was recognized a few decades back as a serious problem. A good number of workers dealt these problems to find out the impact of toxicants on aquatic flora and fauna. Panigrahi (1980) reported the behavioural changes observed in inorganic mercury exposed freshwater fishes. Panigrahy (1984) reported that fishes showed a change in behaviour when exposed to Emisan-6 (MEMC). Macleod and Passah (1973) reported the behavioural changes induced by mercury in Rainbow trout. Larson and Lewander (1973) reported metabolic effects of starvation in the eel, *Anguilla anguilla* L. Gibilin and Massaro (1973) observed behavioural changes induced by methyl mercury in Rainbow trout. Stone *et al.* (1977) reported changes in body weight, kidney and liver weight in Japanese Quail induced by lead. Bhatia *et al.* (1973) reported behavioural changes induced by dieldrin in albino rats.

During the acute toxicity studies fish were observed for behavioural changes influenced by red mud waste extract. Experimental fish were sacrificed; brain and liver were dissected out carefully and washed thoroughly with distilled water. The sacrificed fish were weighed. Brain and liver were separated and weighed carefully in a single pan electric balance (Dhona make) and the different somatic indices were calculated. The individual changes in body weight of both control and experimental fish were noted at 7-day interval and the per cent change in body weight was computed. This piece of work was designed to study the behavioural changes induced by red mud waste extract (industry leached effluent), on a fresh waster fish, *Anabas scandens*, Cuv. and Val.

Observation

All the exposed fish appeared lethargic after exposure to Red mud waste extract. The major clinical symptoms such as inappetance and ataxia appeared after 2 to 3 days exposure. At higher concentration of the toxicant, the exposed fish showed erratic movements. The other signs of toxicity such as loss of equilibrium, gradual onset of inactivity, erratic swimming with irregular collision to the inner glass walls of the aquarium was observed. Infection of eyes, exophthalmia and involusions of test fish were not observed compared to results obtained by Panigrahi and Misra (1978), Panigrahi (1980) and Panigrahy (1984). Exposed fish could not regain their pre-exposed activity after transferring the exposed fish to toxicant free water in recovery studies.

Body weight of the control *Anabas* fish increased by 12.2 per cent after 28 days of exposure and by 21.4 per cent after 56 days of exposure including 28 d of recovery. Whereas, the body weight of the exposed *Anabas* fish gradually decreased during the exposure period and a maximum of 8.2 per cent decrease was observed after 28 days of exposure (Table 40.1), when compared to control fish. Control fish did not show any signs of toxicity. The decrease in body weight slightly improved but the recovery was not significant. A maximum of 20.4 per cent total decrease in body weight was recorded in the exposed fish after 28 d of exposure (Table 40.1). No recovery in body weight was marked, when the exposed fish was transferred to toxicant free normal medium. Rather depletion by 20.7 per cent was marked after 14 days of recovery. After 28 d of recovery, 22.6 per cent decrease was recorded (Table 40.1). During recovery period, it was expected that some recovery in the body weight would be marked. But interestingly no recovery was marked in the exposed *Anabas* fish. Rather further depletion by 0.3 per cent and 2.2 per cent was recorded on 14 and 28 d of recovery. The body weight showed a linear increase showing the existence of significant positive correlation ($r = 0.971$, $P \leq 0.05$) with the exposure period in the control set. A significant negative correlation ($r = 0.974$; $P \leq 0.05$) was observed between the exposure period and body weight in the exposed *Anabas* fish. With the increase in exposure period, the body weight significantly declined.

Table 40.1: Per Cent Changes in the Body Weight of the Control and Alumina Industry Effluent Exposed Fish, *Anabas scandens* at Different Exposure and Recovery Periods

Condition/Status	*Exposure in Days*					*Recovery in Days*	
	0	*7*	*14*	*21*	*28*	*14*	*28*
Control	0.00	2.6 ± 0.4	5.9 ± 0.6	7.6 ± 0.3	12.2 ± 0.5	17.5 ± 0.6	21.4 ± 0.4
Exposed	0.00	0.8 ± 0.3	- 1.9 ± 0.2	- 4.8 ± 0.4	- 8.2 ± 0.7	- 3.2 ± 0.1	- 1.2 ± 0.4
% Decrease in body weight	0.00	- 1.8	- 7.8	- 12.4	- 20.4	- 20.7	- 22.6
% recovery in body weight during recovery period, calculated from 28d exposure value						NR (- 0.3)	NR (- 2.2)

Data are mean of 5 samples ± standard deviation.
'–' Indicate decrease in body weight, when compared to the body weight of the fish, on the fay of first exposure.

Autopsy studies revealed that the liver of RM extract exposed fish was congested, pale and tender. The liver somatic index and brain somatic index decreased with the increase in exposure period, when compared to control fish. In the control fish, the BSI ranged between 0.751 to 0.758 and the LSI ranged from 0.824 to 0.831. The BSI decreased significantly from 0.753 ± 0.032 to 0.692 ± 0.023 after 28 days of exposure. The brain somatic index did not improve in the exposed fish, when the exposed fish was transferred to toxicant free medium. Rather further depletion was marked on 14 d of recovery. However, on 28 d of recovery little improvement was noted. The LSI decreased from 0.827 ± 0.031 to 0.514 ± 0.036 after 28 days of exposure. The liver somatic index improved in the exposed fish, when the exposed fish was transferred to toxicant free medium (Table 40.2). The control fish remained clinically healthy throughout the period of experiment. The per cent decrease of BSI, increased with the increase in exposure period and a maximum decrease of 8.1 per cent was recorded on 28th day of exposure. In case of LSI, a similar trend was observed but here the index decreased maximum by 37.84 per cent, when compared to control fish, after 28 days of exposure. When the fish was transferred to toxicant free medium, partial recovery was marked on 28 d of recovery, rather per cent decrease increased up to 9.1 per cent was recorded in BSI on 14 d of recovery and on 28d of recovery, the per cent decrease was 5.9 per cent, whereas, no further depletion in LSI was

Table 40.2: Changes in the Brain Somatic Index and Hepato Somatic Index (HSI) in Control and Red Mud Waste Effluent Exposed *Anabas* Fish at Different Days of Exposure and Recovery

Parameters Studied	*Status of fish*	*Exposure in Days*					*Recovery in Days*	
		0	*7*	*14*	*21*	*28*	*14*	*28*
Brain Somatic Index	Control	0.758 ± 0.022	0.751 ± 0.045	0.756 ± 0.031	0.754 ± 0.027	0.753 ± 0.032	0.755 ± 0.023	0.756 ± 0.029
	Exposed	0.758 ± 0.022	0.749 ± 0.031	0.748 ± 0.035	0.712 ± 0.026	0.692 ± 0.023	0.686 ± 0.025	0.711 ± 0.021
Liver Somatic Index	Control	0.824 ± 0.019	0.826 ± 0.024	0.831 ± 0.014	0.828 ± 0.026	0.827 ± 0.031	0.826 ± 0.018	0.827 ± 0.024
	Exposed	0.824 ± 0.019	0.761 ± 0.022	0.702 ± 0.024	0.632 ± 0.031	0.514 ± 0.036	0.521 ± 0.033	0.554 ± 0.022

Data are mean of 5 samples ± standard deviation.

marked. The LSI showed 36.9 per cent decrease on 14th day and 33.01 per cent decrease on 28th day of recovery was marked in the exposed *Anabas* fish during recovery periods. When the exposed fish was transferred to toxicant free medium, no recovery on 14 d and 2.15 per cent recovery on 28 d of recovery were marked in BSI data, when compared to 28 d exposure value. When the exposed *Anabas* fish was transferred to toxicant free medium, 0.92 per cent and 4.83 per cent recovery was marked in LSI data, when compared to 28 d exposure value.

When the exposed fish were transferred to toxicant-free tap water, no visible significant recovery in their activity was observed.

Discussion

The rapid absorption of the red mud waste extract/leaching effluent through the gill, skin and gastro-intestinal tract of fish was well evident in the observed exposed fish. The toxicity of red mud waste extract becomes more apparent in a very shorter period in aquatic animals. Panigrahi (1980) documented the detail behavioural changes in relation to inorganic-mercury intoxication. Panigrahy (1984) reported the impact of MEMC on *Anabas scandens*, a freshwater fish and noted a substantial change in behaviour of the test fish, although he has observed swelling of eyes and consequent blindness. Such type of symptom was not observed in red mud waste extract exposed fish, probably due to absence of mercury as a base in the toxicant. Higgins (1974) and Macleod and Passah (1973) reported that loss of appetite, loss of weight, nervousness, dizziness, loss of equilibrium, erratic swimming and gradual onset of inactivity were the sub-clinical effects of inorganic mercury intoxication. Panigrahi (1984) also reported similar symptoms relating to a mercury-based fungicide. Neurological damage in inorgano-mercury intoxicated fish relating to behavioural studies was also reported (Panigrahi, 1980), and he opined that the damage was caused only by inorganic mercury. Sarangi (2002) observed that *Tilapia* fish showed significant variation in body weight change and significant variation in BSI and HSI in the exposed animals when compared to control fish. In the present investigation, *Anabas* fish could tolerate higher amount of leached effluent than *Tilapia* fish. The higher level of tolerance can be linked to bimodal gaseous exchange habit of *Anabas* fish, when compared to *Tilapia* fish. The tolerance was related to the rate of absorption of toxicants by the surviving aquatic animals in the aquatic contaminated ecosystems. The present experimental animals could withstand higher concentration of the toxicants, because this *Anabas* fish in contaminated environments mostly depend on the atmospheric oxygen rather than the dissolved oxygen of contaminated water. The ventilation rate significantly declined in the exposed fish leading to decrease in toxicant uptake by the exposed animal.

The observed depression in active metabolism in exposed fish were indicative of damage to nervous tissues, inhibition of enzymes or a vital system, was totally in agreement with Panigrahi (1980). Fry (1957) demonstrated the depression in active metabolism directly reduced "scope for activity". This in turn may result in decreased growth, impaired swimming ability with erratic out burst at times. Loss in weight due to starvation was reported by Larsen and Lewander (1973). But the loss due to starvation was very less. Panigrahi and Misra (1978, 1980) reported the loss of body weight due to mercury intoxication and confirmed that this loss in body weight was only due to mercury stress. In addition, expose fish maintained their feeding habit only after a short span of time. Hence, the weight loss cannot be correlated with starvation but can only be related with toxicant intoxication. Whereas, the control fish showed increase in weight. The differences observed between control fish and exposed fish in terms of body weight is only due to the administration of red mud waste extract in exposed aquarium. Borg *et al.* (1970) reported inappetance, muscular weakness, ataxia and loss of body weight as the main clinical symptoms of mercury poisoning. On autopsy, they found muscular atrophy, which can be correlated with the weight loss in the goshawks. Hanko *et al.* (1970) reported loss of appetite, weakness of the extremities, excitation in the animal and loss of body weight due to mercury poisoning in chickens. A decline in liver somatic index and brain somatic index in exposed fish could be correlated with the degeneration of cells and decrease in other macromolecular variables or decrease in moisture content in both liver and brain tissues.

Concisely, at this stage it is not possible to establish the cause of loss in body weight and decrease in liver and brain somatic index, except a generalised comment on the behaviour of the exposed fish to stress that the observed symptoms were most probably be due to the toxicant poisoning. Bhatia *et al.* (1973) reported that change in behaviour in test fish had some relationship with insecticide intoxication. Rath and Misra (1980) reported the change in behavioural activity of fish exposed to dichlorvos (DDVP) and confirmed the idea that the change in activity was related to pesticide intoxication, which is totally agreeable with our observation in this study. The present study and the findings reported by Mishra (2002) and Sarangi (2002) clearly indicated that the leached effluent of the red mud pond, the red mud waste and the red mud waste extract are strictly poisonous to aquatic flora and fauna.

Results

The whole body oxygen uptake of the control fish was within the range of 0.187 ± 0.015 to 0.193 ± 0.009 mg of O_2 g^{-1} h^{-1} during the entire period of experimentation. However, the whole body oxygen uptake decreased from 0.187 ± 0.010 to 0.081 ± 0.006 mg of O_2 g^{-1} h^{-1} after 7 days of exposure, showing a maximum decrease in oxygen uptake by 56.68 per cent over the control value. After 7 days of exposure, the whole body oxygen uptake declined significantly to 0.022 ± 0.009 mg of O_2 g^{-1} h^{-1} after 28 days of exposure (Table 40.3), where, a maximum decrease by 88.29 per cent, was recorded, over the control value. When the red mud waste extract/waste effluent exposed fish was transferred to toxicant free tap water medium, partial recovery in whole body oxygen uptake was marked. During recovery period, the whole body oxygen uptake declined from 0.191 ± 0.008 to 0.065 ± 0.011 mg of O_2 g^{-1} h^{-1}, on 28th day of recovery. During recovery studies the whole body oxygen uptake decreased by 75.66 per cent and 65.96 per cent was recorded on 14th and 28th day of recovery (Table 40.3). Instead of showing any recovery, the whole body oxygen uptake further declined, indicating the acute lethal value of the waste leached effluent. Partial recovery by 12.63 per cent and 22.32 per cent was recorded on 14th and 28th day of recovery, respectively when compared to 28th day exposure value.

For each exposure period, the exposed fish was allowed to recover for the same period of exposure to find out the possibility of 100 per cent recovery to its pre-test level. The 7 day exposed fish was allowed to recover for 7 days. The whole body O_2 uptake increased from 0.081 to 0.126 mg of O_2 g^{-1} h^{-1}. During this period only 32.62 per cent decrease in oxygen uptake was recorded, but the fish could recover by 24.06 per cent. The 14 d exposed fish was allowed to recover for 14 days. The whole body O_2 uptake increased from 0.049 to 0.094 mg of O_2 g^{-1} h^{-1}. During this period only 51.29 per cent decrease in oxygen uptake was recorded, but the fish could recover by 23.32 per cent. The 21 day exposed fish was allowed to recover for 21 days. The whole body O_2 uptake of exposed fish increased from 0.032 to 0.059 mg of O_2 g^{-1} h^{-1}. During this period only 68.94 per cent decrease in oxygen uptake was recorded, but the fish could recover only by 14.21 per cent. The 28 day exposed fish was allowed to recover for 28 days. The whole body O_2 uptake decreased from 0.022 to 0.048 mg of O_2 g^{-1} h^{-1}.

During this period only 74.46 per cent decrease in oxygen uptake was recorded, but the exposed fish could not recover but further depletion in oxygen uptake by 13.82 per cent was recorded. The data clearly indicated that the exposed fish could recover partly during smaller periods of exposure. But prolonged exposure caused significant damage, where the fish could not recover to its pre-test activity or level. The per cent recovery decreased with the increase in exposure period and the per cent decrease in oxygen uptake increased with the increase in exposure period. The correlation coefficient analysis indicated that the control set did not show any correlation with the exposure period (P = NS). Whereas, in the exposed set a negative significant correlation existed between the exposure period and whole body oxygen uptake ($r = -0.988$, $P \leq 0.01$). With the increase in exposure period, the oxygen uptake significantly declined in exposed fishes.

Table 40.3: Changes in Respiration Rate (mg of O_2 g^{-1} hr^{-1}) of Both Control and Alumina Industry Effluent Exposed Fish, *Anabas scandens* at Different Exposure and Recovery Periods

Status of Fish	*Exposure in Days*					*Recovery in Days*	
	0	*7*	*14*	*21*	*28*	*14*	*28*
Control	0.192 ± 0.005	0.187 ± 0.010	0.193 ± 0.009	0.190 ± 0.008	0.188 ± 0.006	0.189 ± 0.004	0.191 ± 0.008
Exposed	0.192 ± 0.005	0.081 ± 0.006	0.049 ± 0.005	0.032 ± 0.011	0.022 ± 0.009	0.046 ± 0.008	0.065 ± 0.011
% decrease	0.00	-56.68	-74.61	-83.15	-88.29	-75.66	-65.96
% recovery in respiration rate when compared to 28 d exposure value						+12.63	+22.32
P≤	NS	0.01	0.01	0.001	0.001	0.01	0.01

Data are the mean of 5 samples ± standard deviation.
Per cent decrease in exposed fish during experimental period calculated from the mean of the samples over 28 d exposure value.

The ventilation rate ranged between 66 ± 11 to 68 ± 15 strokes/minute in the control fish set during the entire period of experimentation. The ventilation rate of the red mud waste extract exposed fish decreased from 68 ± 15 to 25 ± 12 strokes/minute on 28th day of exposure (Table 40.4). The ventilation rate declined significantly at higher exposure period. When the exposed fish was transferred to toxicant free medium, little increment in the ventilation rate was observed. After 28 d of recovery the ventilation rate decreased from 68 to 36 strokes/minute and on 28 d of recovery a maximum of 47.1 per cent decrease in ventilation rate was marked. The ventilation rate decreased with the increase in exposure period. During this period the exposed *Anabas* fish depended mostly on atmospheric oxygen rather than the dissolved oxygen of the experimental aquarium water. The opercular movement of the fish has been intended either for collection of food from the environment or for absorbing the dissolved oxygen available in the medium. When the medium is contaminated with the toxicant, the exposed fish tries to depend on the oxygen from air. The exposed fish engulfs air and absorbs the oxygen through the air breathing apparatus positioned in the head. Hence, the ventilation rate decreased significantly during the period of stress. In the present set, the ventilation rate declined significantly with the increase in exposure period, in the exposed fish when compared to control fish and a maximum of 63.2 per cent decrease in the ventilation rate were noted and a maximum of 47.1 per cent decrease was noted on 28 d of recovery. A maximum of 5 per cent and 16.1 per cent recovery was noted on 14th and 28th day of recovery, indicating non-significant recovery during recovery period. The ventilation rate of the exposed fish did not show any higher value, when compared to the control fish up to 14 days of exposure, like other exposed fishes. On 28th day of exposure, the ventilation rate declined by 63.2 per cent (Table 40.4) over the control value. When the exposed fish was transferred to toxicant free medium, the ventilation rate instead of showing any further decrease, showed increase by 5 per cent over the 28th day exposure value, on 14th day of recovery and 16.1 per cent decrease over the 28th day of exposure value on 28th day of recovery (Table 40.4). The correlation coefficient analysis between exposure period and the ventilation rate of the control fishes did not show the existence of any significant correlation (P = NS). It seems the ventilation rate has slowed down when the fish was kept in stagnant water for a prolonged period, with the same type of food throughout the experimental period. Rather during this holding period, little depression in the rate was observed. Whereas, in the toxicant exposed set, the ventilation rate initially increased insignificantly up to 2 days of exposure, than the ventilation rate declined significantly, but this value was less than the control value on 28th day of exposure. The correlation coefficient analysis between days of exposure and ventilation rate in the exposed fish, indicated, the existence of a significant positive correlation ($r = 0.986, P \leq 0.05$). This indicated that with the increase in exposure period, the ventilation rate decreased. The initial increase in ventilation rate might be due to the stress. The fish probably out of suffocation, tried to pump more amount of water through the gills, either to trap more oxygen or to wash the gills.

Table 40.4: Changes in Ventilation Rate (Strokes/Minute) of Both Control and Alumina Industry Effluent Exposed Fish at Different Exposure and Recovery Periods

Status of Fish	*Exposure in Days*					*Recovery in Days*	
	0	*7*	*14*	*21*	*28*	*14*	*28*
Control	68 ± 15	66 ± 11	68 ± 13	67 ± 12	68 ± 14	67 ± 13	68 ± 11
Exposed	68 ± 15	48 ± 10	42 ± 9	31 ± 8	25 ± 12	28 ± 11	36 ± 8
% decrease	0.00	-27.3	-38.2	-53.7	-63.2	-58.2	-47.1
% recovery, when compared to 28 d exp. Value						+5.0	+16.1

Data are mean of 5 samples ± standard deviation.
Per cent change values were calculated from the mean of the samples, when compared to respective control values.
Per cent recovery was calculated from 28 d exposure value.

Table 40.5 indicated the changes in oxygen uptake (µl of O_2 g^{-1} h^{-1}) by brain, liver, muscle tissue slices and gill filaments of control and red mud waste extract (waste effluent) exposed fish. In case of brain tissue slices, significant decrease in oxygen uptake was marked on 28th day of exposure. The O_2 uptake declined from 586.4 ± 19.1 to 484.2 ± 11.8 µl of O_2 g^{-1} h^{-1}, where a maximum of 17.4 per cent decrease was marked over the control value. When the exposed fish was transferred to red mud waste extract free medium, no recovery in the oxygen uptake after 14 days and no recovery after 28 day of recovery was marked, respectively. In case of brain, the oxygen uptake declined from 584.2 ± 8.2 to 434.2 ± 12.1 µl of O_2 g^{-1} h^{-1}, where a maximum of 25.7 per cent decrease was marked over the control value, after 28 days of recovery. No recovery was marked on 28th day of recovery. The correlation coefficient analysis between days of exposure and the oxygen uptake of brain tissue slices of the control set indicated the existence of a significant correlation ($r = -0.299$, P = NS), but the brain tissue slices of the toxicant exposed fish, showed the existence of a significant ($r = -0.973$, $P \leq 0.05$) negative correlation with the exposure period.

The liver tissue slices showed the maximum decrease in oxygen uptake. The values declined from 614.2 ± 16.6 to 314.5 ± 19.3 µl of O_2 g^{-1} h^{-1} on 28th day of exposure, showing a maximum decline by 48.8 per cent (Table 40.5). With the increase in exposure period, the oxygen uptake steadily and significantly declined during the entire period of experimentation. When the exposed fish was transferred to red mud waste extract free medium, the oxygen uptake declined from 616.7 ± 19.2 to 354.2 ± 24.8 µl of O_2 g^{-1} h^{-1} on 14 d of recovery showing a maximum decline by 42.6 per cent. When the exposed fish was transferred to red mud waste extract free medium, the oxygen uptake declined from 616.2 ± 20.4 to 392.2 ± 14.2 µl of O_2 g^{-1} h^{-1} on 28 d of recovery showing a maximum decline by 36.4 per cent. A partial recovery by 6.2 per cent and 12.4 per cent was marked after 14 and 28 days of recovery, when the exposed fish was transferred to toxicant free medium. The correlation coefficient analysis between the exposure period and the oxygen uptake of liver tissue slices of the control fish indicated the existence of non-significant correlation. Whereas, the liver slices of exposed fish showed the existence of a significant negative correlation ($r = -0.989$, $P \leq 0.01$) between days of exposure and oxygen uptake of liver tissue slices.

Table 40.5: Changes in Oxygen Uptake (µl of O_2 g^{-1} hr^{-1}) by Brain and Liver of Control and Alumina Industrial Effluent Exposed Fish, *Anabas scandens* at Different Days of Exposure and Recovery

Tissue	*Condition*	*Exposure in Days*					*Recovery in Days*	
		0	*7*	*14*	*21*	*28*	*14*	*28*
Brain	Control	582.2 ± 12.4	585.2 ± 11.8	586.2 ± 18.4	584.5 ± 17.4	586.4 ± 19.1	582.8 ± 10.2	584.2 ± 8.2
	Exposed	582.2 ± 12.4	551.4 ± 12.4	532.2 ± 16.2	502.2 ± 18.2	484.2 ± 11.8	421.6 ± 14.6	434.2 ± 12.1
Liver	Control	614.2 ± 14.2	612.5 ± 11.5	614.4 ± 16.4	616.8 ± 18.2	614.2 ± 16.6	616.7 ± 19.2	616.2 ± 20.4
	Exposed	614.2 ± 14.2	504.2 ± 12.2	454.2 ± 18.8	398.6 ± 26.4	314.5 ± 19.3	354.2 ± 24.8	392.2 ± 14.2

Data are mean of 5 samples ± standard deviation.

Table 40.5 indicated the changes in oxygen uptake (µl of O_2 g^{-1} h^{-1}) by muscle tissue slices of control and red mud waste extract exposed fish. In case of muscle tissue slices, significant decrease in oxygen uptake was marked on 28th day of exposure. The O_2 uptake declined from 569.4 ± 14.8 to 404.8 ± 12.5 µl of O_2 g^{-1} h^{-1}, where a maximum of 28.9 per cent decrease was marked, on 28 d of exposure when compared to the control value. When the exposed fish was transferred to red mud waste extract free medium, the oxygen uptake improved showing 3.2 per cent and 12 per cent after 14 and 28 days of recovery, respectively. In case of muscle, the oxygen uptake declined from 568.2 ± 12.2 to 472.4 ± 12.4 µl of O_2 g^{-1} h^{-1} where a maximum of 16.9 per cent decrease was marked over the control value, after 28 days of recovery. A partial recovery by 12.0 per cent was marked on 28th day of recovery. The muscle tissue slices in the exposed fish showed gradual decrease in oxygen uptake with the increase in exposure period, showing an inverse relationship. The oxygen uptake in muscle tissues declined significantly and a maximum of 28.9 per cent decrease was marked on 28th day, when compared to its control value. The exposed fish could recover partly after 14 day and 28 day of recovery, respectively. The correlation coefficient analysis conducted between days of exposure versus the oxygen uptake of muscle tissue slices of the control fish indicated the existence of a negative, non-significant ($r = -0.385$, P = NS) correlation, whereas, a negative but highly significant ($r = -0.991$, $P \leq 0.01$) correlation existed between days of exposure and the oxygen uptake rate of the exposed fish. The regression analysis between days of exposure versus the per cent decrease in oxygen uptake by muscle tissue indicated a significant ($P \leq 0.01$), positive correlation ($r = 0.988$). The existence of an inverse relationship indicated the effect of stress on the animal.

Table 40.6 indicated the changes in oxygen uptake (µl of O_2 g^{-1} h^{-1}) by gill filaments of control and red mud waste extract (leached effluent) exposed fish. In case of gill filaments, significant decrease in oxygen uptake was marked on 28th day of exposure. In case of gill, the O_2 uptake declined from 682.6 ± 12.4 to 302.8 ± 14.6 µl of O_2 g^{-1} h^{-1} where a maximum of 55.6 per cent decrease was marked over the control value. When the exposed fish was transferred to leached effluent free medium, the oxygen uptake improved showing 8.8 per cent and 16.3 per cent after 14 and 28 days of recovery respectively. In case of muscle, the oxygen uptake declined from 680.2 ± 28.6 to 412.6 ± 24.4 µl of O_2 g^{-1} h^{-1} where a maximum of 39.3 per cent decrease was marked over the control value, after 28 days of recovery. A partial recovery by 16.3 per cent was marked on 28th day of recovery. The gill filaments in the exposed fish showed gradual decrease in oxygen uptake with the increase in exposure period, showing an inverse relationship. The oxygen uptake in gill filaments declined significantly and a maximum of 55.6 per cent decrease was marked on 28th day, when compared to its control value. The exposed fish could recover partly after 14 day and 28 day of recovery respectively.

Table 40.6: Changes in Oxygen Uptake (µl of O_2 g^{-1} hr^{-1}) by Muscle and Gill of Control and Alumina Industrial Effluent Exposed Fish, *Anabas scandens* at Different Days of Exposure and Recovery

Tissue	*Condition*	*Exposure in Days*					*Recovery in Days*	
		0	*7*	*14*	*21*	*28*	*14*	*28*
Muscle	Control	568.6 ± 18.4	568.4 ± 12.6	568.2 ± 5.7	568.6 ± 12.2	569.4 ± 14.8	568.2 ± 14.3	568.2 ± 12.2
	Exposed	568.6 ± 18.4	534.2 ± 12.2	484.2 ± 9.4	428.4 ± 11.1	404.8 ± 12.5	422.2 ± 18.6	472.4 ± 12.4
Gill	Control	688.2 ± 12.2	688.2 ± 14.8	686.4 ± 12.6	686.8 ± 22.7	682.6 ± 12.4	682.5 ± 16.5	680.2 ± 28.6
	Exposed	688.2 ± 12.2	524.6 ± 10.4	444.2 ± 11.2	364.4 ± 16.8	302.8 ± 14.6	362.5 ± 14.2	412.6 ± 24.4

Data are mean of 5 samples ± standard deviation.

The changes in total ATPase activity in brain, liver, muscle and gill of control and red mud waste extract (leached effluent) exposed fish at different days of exposure and recovery and it's per cent changes were shown in Table 40.7. The brain tissue showed a maximum decrease by 65.2 per cent and the enzyme activity declined from 32.2 ± 2.3 to 11.2 ± 3.1 µmole of ip liberated mg^{-1} of protein hr^{-1} on 28th day of exposure. When the exposed fish was transferred to toxicant free medium, the ATPase activity declined from 32.1 ± 1.5 to 11.8 ± 2.4 µmole of ip liberated mg^{-1} of protein hr^{-1} on 14th day of recovery and from 31.5 ± 2.4 to 19.2 ± 2.2 µmole of ip liberated mg^{-1} of protein hr^{-1} on 28th day of recovery. This enzyme activity decreased by 63.2 per cent on 14th day recovery and decreased by 39.1 per cent over the 28th day exposure value, after 28 days of recovery, indicating a partial recovery. The exposed fish brain enzyme activity could recover partly by 2.0 per cent and 26.1 per cent on 14th and 28th day of recovery (Table 40.12). No significant recovery was marked on 14th day of recovery in the 28 d exposed fish, when transferred to toxicant free medium. The liver tissue showed a maximum decrease by 71.1 per cent and the enzyme activity declined from 28.4 ± 3.2 to 8.2 ± 3.4 of µmole of ip liberated mg^{-1} of protein hr^{-1} on 28th day of exposure. When the exposed fish was transferred to toxicant free medium, the ATPase activity declined from 28.4 ± 3.8 to 11.2 ± 2.4 µmole of ip liberated mg^{-1} of protein hr^{-1} on 14th day of recovery and from 28.6 ± 2.2 to 16.2 ± 3.2 µmole of ip liberated mg^{-1} of protein hr^{-1} on 28th day of recovery. This enzyme activity decreased by 60.6 per cent on 14th day recovery and decreased by 43.4 per cent over the 28th day exposure value, after 28 days of recovery, indicating a partial recovery. The exposed fish brain enzyme activity could recover partly by 10.5 per cent and 27.7 per cent on 14th and 28th day of recovery. The muscle tissue showed a maximum decrease by 46.7 per cent of the enzyme activity on 28th day of exposure, when compared to control and the enzyme activity declined from 22.9 ± 1.2 to 12.2 ± 3.2 μmole of ip liberated mg^{-1} of protein hr^{-1} on 28th day of exposure. When the exposed fish was transferred to toxicant free medium, the ATPase activity of muscle declined from 22.4 ± 2.3 to 15.2 ± 2.4 μmole of ip liberated mg^{-1} of protein hr^{-1} on 14th day of recovery and from 22.8 ± 2.1 to 16.2 ± 3.2 μmole of ip liberated mg^{-1} of protein hr^{-1} on 28th day of recovery. This enzyme activity decreased by 31.7 per cent on 14th day recovery and decreased by 28.9 per cent over the 28th day exposure value, after 28 days of recovery, indicating a partial recovery. The exposed fish muscle enzyme activity could recover partly by 15 per cent and 17.8 per cent on 14th and 28th day of recovery. The gill filament showed a maximum decrease by 73.3 per cent of the enzyme activity on 28th day of exposure, when compared to control and the enzyme activity declined from 20.6 ± 1.6 to 5.5 ± 2.2 μmole of ip liberated mg^{-1} of protein hr^{-1} on 28th day of exposure. When the exposed fish was transferred to toxicant free medium, the ATPase activity of gill declined from 20.7 ± 1.4 to 8.5 ± 1.7 μmole of ip liberated mg^{-1} of protein hr^{-1} on 14th day of recovery and from 21.1 ± 2.2 to 11.2 ± 1.4 μmole of ip liberated mg^{-1} of protein hr^{-1} on 28th day of recovery. This enzyme activity decreased by 58.9 per cent on 14th day recovery and decreased by 46.7 per cent over the 28th day exposure value, after 28 days of recoveiy, indicating a partial recovery. The exposed fish gill enzyme activity could recover partly by 14.4 per cent and 26.6 per cent on 14th and 28th day of recovery. The correlation coefficient analysis between days of exposure and total ATPase activity in control fish brain showed a positive but non-significant correlation, whereas, the exposed fish brain showed a significant negative correlation ($r = -0.986$, $P \leq 0.01$). The control set showed positive but insignificant correlation. The per cent change in the enzyme activity in the exposed fish muscle when compared to control fish muscle showed the existence of a negative significant correlation ($r = -0.984$, $P \leq 0.01$) with the exposure period. The correlation coefficient analysis between days of exposure and the total ATPase of liver slices of the control fish did not show any significant correlation, whereas the exposed fish liver showed a negative and significant correlation ($r = -0.989$, $P \leq 0.01$) with the exposure period. The per cent change in ATPase activity showed the existence of a negative, significant correlation ($r = -0.988$, $P \leq 0.01$) with the exposure period. The control muscle did not show any significant correlation between exposure

Table 40.7: Changes in Total ATPase Activity (μ moles of ip liberated mg^{-1} protein hr^{-1}) in Control and Alumina Industrial Effluent Exposed *Anabas* Fish Brain, Liver, Muscle and Gill at Different Days of Exposure and Recovery

Tissue	*Condition*	*Exposure in Days*					*Recovery in Days*	
		0	*7*	*14*	*21*	*28*	*14*	*28*
Brain	Control	30.7 ± 2.1	31.4 ± 1.2	31.4 ± 2.2	31.3 ± 1.6	32.2 ± 2.3	32.1 ± 1.5	31.5 ± 2.4
	Exposed	30.7 ± 2.1	28.5 ± 1.2	21.2 ± 2.4	16.2 ± 1.3	11.2 ± 3.1	11.8 ± 2.4	19.2 ± 2.2
Liver	Control	28.2 ± 2.4	28.6 ± 2.2	28.4 ± 4.1	28.6 ± 2.1	28.4 ± 3.2	28.4 ± 3.8	28.6 ± 2.2
	Exposed	28.2 ± 2.4	18.2 ± 3.2	14.2 ± 1.5	10.2 ± 2.2	8.2 ± 3.4	11.2 ± 2.4	16.2 ± 3.2
Muscle	Control	22.5 ± 1.1	22.2 ± 1.2	22.6 ± 2.1	22.8 ± 1.5	22.9 ± 1.2	22.4 ± 2.3	22.8 ± 2.1
	Exposed	22.5 ± 1.1	20.2 ± 2.2	16.4 ± 4.2	15.2 ± 2.5	12.2 ± 3.2	15.2 ± 2.4	16.2 ± 3.2
Gill	Control	20.6 ± 1.7	19.5 ± 0.8	20.5 ± 1.4	21.4 ± 1.2	20.6 ± 1.6	20.7 ± 1.4	21.1 ± 2.2
	Exposed	20.6 ± 1.7	12.6 ± 1.3	9.8 ± 1.6	7.4 ± 1.2	5.5 ± 2.2	8.5 ± 1.7	11.2 ± 1.4

Data are mean of 5 samples ± standard deviation.

period and enzyme activity. However, the muscle of the exposed fish showed negative and significant correlation ($r = -0.981$, $P \leq 0.05$) with the exposure period. In case of gill tissue a significant correlation existed between the exposure period and the per cent decrease in the enzyme activity. With the increase in exposure period, the enzyme activity significantly declined and the per cent decrease significantly increased. The per cent change in the enzyme activity showed negative and significant correlation with the exposure period. The exposed fishes were when transferred to red mud waste leached effluent free medium, the enzyme activity showed good recovery but not statistically significant. Exposed liver showed the highest recovery (27.7 per cent), exposed muscle showed the lowest recovery by 17.8 per cent, gill showed 26.6 per cent recovery and brain showed recovery by 26.1 per cent in the exposed fishes, after 28 days of recovery.

Table 40.8 indicates the changes in sodium ion, potassium ion, calcium ion and magnesium ion of both control and red mud waste extract exposed fish brain, liver, muscle and gill tissues after 28 days of exposure. The sodium ion content declined from 565.4 ± 16.5 to 514.5 ± 21.5 mg g^{-1} in exposed fish brain, from 570.6 ± 31.6 to 449.5 ± 17.8 mg g^{-1} in exposed fish liver, from 530.2 ± 13.6 to 422.8 ± 15.5 mg g^{-1} in exposed fish muscle and from 625.7 ± 17.3 to 584.6 ± 25.4 mg g^{-1} in exposed fish gill, after 28 days of exposure showing maximum decrease by 9.0 per cent, 21.2 per cent, 20.2 per cent and 6.6 per cent in brain, liver, muscle and gill tissues of the exposed fishes, respectively. The exposed liver tissue showed the highest per cent decrease, when compared to muscle, brain and gill tissues of the exposed fish. The potassium ion content declined from 565.2 ± 17.6 to 128.6 ± 23.4 mg g^{-1} in exposed fish brain, from 582.2 ± 18.5 to 316.4 ± 15.8 mg g^{-1} in exposed fish liver, from 575.8 ± 28.5 to 453.7 ± 13.5 mg g^{-1} in exposed fish muscle and from 415.6 ± 18.5 to 362.4 ± 15.6 mg g^{-1} in exposed fish gill tissue, respectively, after 28 days of exposure. Brain, liver, muscle and gill tissues of the exposed fish showed 77.3 per cent, 45.6 per cent, 21.2 per cent and 12.8 per cent decrease in potassium ion content, when compared to the control value, where brain was highly affected showing the maximum depletion. Significant ($P \leq 0.05$) decrease in potassium ion content in brain, liver, muscle and gill of exposed fish, compared to the control fish, was marked. Significant variation was marked in the calcium ion content in different tissues of the exposed fish, when compared to control fish tissues. The calcium ion content, decreased from 205.6 ± 14.5 to 172.4 ± 16.8 mg g^{-1} in exposed brain, from 225.8 ± 15.5 to 214.2 ± 14.6 mg g^{-1} in exposed liver, from 181.5 ± 12.5 to 152.2 ± 18.5 mg g^{-1} in exposed muscle and from 358.5 ± 14.6 to 285.2 ± 28.4 mg g^{-1} in exposed gill tissue, when compared to control tissues after 28 days of exposure. Maximum depletion by 20.4 per cent in gill tissue, 16.2 per cent in brain tissues, 16.1 per cent in muscle tissues and least depletion by 5.1 per cent in liver tissues of the exposed fish, when compared to control fish tissues was recorded. Least significant differences were observed in the exposed fish tissues, when compared to control fish tissues. The magnesium ion content, decreased from 265.7 ± 17.4 to 172.6 ± 19.7 mg g^{-1} in exposed fish brain, from 325.4 ± 20.2 to 107.6 ± 18.5 mg g^{-1} in exposed fish liver, from 271.4 ± 14.2 to 181.2 ± 14.6 mg g^{-1} in exposed fish muscle and from 265.8 ± 11.5 to 182.4 ± 14.2 mg g^{-1} in exposed fish gill tissue of exposed fish, when compared to control after 28 days of exposure. Maximum depletion by 67.0 per cent in liver tissue, 35.0 per cent in brain tissues, 33.2 per cent in muscle tissue and least depletion by 31.4 per cent in gill tissues, of the exposed fish, when compared to control fish tissues was recorded. Out of the four ions studied in the exposed fish tissues, non-significant decrease was recorded in sodium ion and calcium ion content and highly significant differences were observed in potassium ion and magnesium ion content in all the tissues studied. Out of the four tissues studied the liver and muscle was the most affected in some ions and than the most affected was brain for some ions and for the rest of the ions the gill was least affected. The analysis of variance ratio test for ions like Na^{+}, K^{+}, Ca^{++} and Mg^{++} in brain, liver, muscle and gill of the control and red mud waste leached extract (leached effluent) exposed fish showed significant difference between rows and non-significant difference between columns. Significant differences in ion content of different tissues of the exposed, when compared to control fish clearly indicate the nature of the toxicant and effects caused by the toxicant (Red mud waste extract/waste leached effluent) on the nervous tissue, synaptic transmission, nerve impulse generation, ionic balance and membrane transport system of the exposed fish.

Table 40.8: Changes Na^+, K^+, Ca^{++} and Mg^+ Ion Concentration in Control and Alumina Industrial Effluent Exposed Fish Brain, Liver, Muscle and Gill After 28d of Exposure

Tissue	*Condition*	*Ion Concentration*			
		Na^+	*K^+*	*Ca^{++}*	*M^{++}*
Brain	Control	565.4 ± 16.5	565.2 ± 17.6	205.6 ± 14.5	265.7 ± 17.4
	Exposed	514.5 ± 21.5	128.6 ± 23.4	172.4 ± 16.8	172.6 ± 19.7
Liver	Control	570.6 ± 31.6	582.2 ± 18.5	225.8 ± 15.5	325.4 ± 20.2
	Exposed	449.5 ± 17.8	316.4 ± 15.8	214.2 ± 14.6	107.6 ± 18.5
Muscle	Control	530.2 ± 13.6	575.8 ± 18.5	181.5 ± 12.5	271.4 ± 14.2
	Exposed	422.8 ± 15.5	453.7 ± 13.5	152.2 ± 18.5	181.2 ± 14.6
Gills	Control	625.7 ± 17.3	415.6 ± 18.5	358.5 ± 14.6	265.8 ± 11.5
	Exposed	584.6 ± 25.4	362.4 ± 15.6	285.2 ± 28.4	182.4 ± 14.2

Data are mean of 5 samples ± standard deviation.

The changes in AChE activity in brain, liver, gill and muscle of control and red mud extract exposed fish at different days of exposure and recovery and it's per cent changes were shown in Table 40.9. The brain tissue showed a maximum decrease by 23.0 per cent and the enzyme activity declined from 0.538 ± 0.039 to 0.414 ± 0.032 μmole of ACh hydrolysed hr^{-1} mg^{-1} of tissue on 28th day of exposure. When the exposed fish was transferred to toxicant free medium, the AChE activity declined from 0.540 ± 0.041 to 0.392 ± 0.037 of μmole of ACh hydrolysed hr^{-1} mg^{-1} of tissue on 14th day of recovery and from 0.540 ± 0.045 to 0.409 ± 0.039 of μmole of ACh hydrolysed hr^{-1} mg^{-1} of tissue on 28th day of recovery. During exposure period, a maximum 23 per cent decrease in the enzyme activity was recorded. This enzyme activity decreased by 27.4 per cent on 14th day recovery and decreased by 24.3 per cent when compared to control value, after 28 days of recovery, indicating no recovery at all. The per cent decrease increased with the increase in exposure period from 7th day to 28th day. Initially the per cent decrease was 1.1 per cent on 7th day, 7.7 per cent on 14th day, 8.0 per cent on 21st day and the highest of 23 per cent being on 28th day of exposure. The per cent decrease increased significantly with the increase in exposure period in exposed fish exposed to RMWE. The exposed fish brain enzyme activity could not recover even partly on 14th and 28th day of recovery showing permanent damage caused to the system. The liver tissue showed a maximum decrease by 54.9 per cent and the enzyme activity declined from 0.479 ± 0.018 to 0.216 ± 0.034 of μmole of ACh hydrolysed hr^{-1} mg^{-1} of tissue on 28th day of exposure. When the exposed fish was transferred to toxicant free medium, the AChE activity declined from 0.481 ± 0.044 to 0.236 ± 0.031 of μmole of ACh hydrolysed hr^{-1} mg^{-1} of tissue on 14th day of recovery and from 0.480 ± 0.041 to 0.254 ± 0.028 of μmole of ACh hydrolysed hr^{-1} mg^{-1} of tissue on 28th day of recovery. This enzyme activity decreased by 50.9 per cent on 14th day recovery and decreased by 47.1 per cent over the

28th day exposure value, after 28 days of recovery, indicating partial insignificant recovery .The per cent decrease increased with the increase in exposure period from 7th day to 28th day. Initially the per cent decrease was 15.5 per cent on 7th day, 21.9 per cent on 14th day, 34.9 per cent on 21st day and the highest of 54.9 per cent being on 28th day of exposure. The per cent decrease increased significantly with the increase in exposure period in exposed fish exposed to RMWE. The enzyme activity decreased by 50.9 per cent and 47.1 per cent on 14th and 28th day of recovery, when the exposed fish was transferred to toxicant free medium. The exposed fish liver enzyme activity could recover partly 3.96 per cent on 14th and 7.8 per cent recovery on 28th day of recovery showing drastic damage caused to the system. The muscle tissue showed a maximum decrease by 26.6 per cent and the enzyme activity declined from 0.383 ± 0.048 to 0.281 ± 0.026 of µmole of ACh hydrolysed hr^{-1} mg^{-1} of tissue on 28th day of exposure. When the exposed fish was transferred to toxicant free medium, the AChE activity declined from 0.384 ± 0.058 to 0.285 ± 0.037 of µmole of ACh hydrolysed hr^{-1} mg^{-1} of tissue on 14th day of recovery and from 0.385 ± 0.064 to 0.294 ± 0.049 of µmole of ACh hydrolysed hr^{-1} mg^{-1} of tissue on 28th day of recovery. This enzyme activity decreased by 25.8 per cent on 14th day recovery and decreased by 23.6 per cent over the 28th day exposure value, after 28 days of recovery, indicating partial recovery. The per cent decrease increased with the increase in exposure period from 7th day to 28th day. Initially the per cent decrease was 12.1 per cent on 7thday, 20.7 per cent on 14th day, 22.9 per cent on 21st day and the highest of 26.6 per cent being on 28th day of exposure. The per cent decrease increased significantly with the increase in exposure period in exposed fish exposed to RMWE. The exposed fish muscle enzyme activity could recover only partly by 0.81 per cent on 14th and by 2.96 per cent on 28th day of recovery showing damage caused to the system. Out of the three tissues studied, the exposed brain showed the highest damage than liver and muscle. The gill tissue showed a maximum decrease by 46.9 per cent and the enzyme activity declined from 0.586 ± 0.036 to 0.311 ± 0.064 of µmole of ACh hydrolysed hr^{-1} mg^{-1} of tissue on 28th day of exposure. When the exposed fish was transferred to toxicant free medium, the AChE activity in gill filaments declined from 0.588 ± 0.041 to 0.346 ± 0.032 of µmole of ACh hydrolysed hr^{-1} mg^{-1} of tissue on 14th day of recovery and from 0.590 ± 0.042 to 0.385 ± 0.028 of µmole of ACh hydrolysed hr^{-1} mg^{-1} of tissue on 28th day of recovery. This enzyme activity decreased by 41.1 per cent on 14th day recovery and decreased by 34.8 per cent over the 28th day exposure value, after 28 days of recovery, indicating partial recovery. The per cent decrease increased with the increase in exposure period from 7th day to 28th day. Initially the per cent decrease was 12.9 per cent on 7th day, 26.6 per cent on 14th day, 33.9 per cent on 21st day and the highest of 46.9 per cent being on 28th day of exposure. The enzyme activity decreased by 41.1 per cent and 34.8 per cent on 14th and 28th day of recovery when compared to its respective control value. The per cent decrease increased significantly with the increase in exposure period in exposed fish exposed to RMWE. The exposed fish muscle enzyme activity could recover only partly by 5.74 per cent on 14th and by 12.2 per cent on 28th day of recovery showing damage caused to the system. Out of the four tissues studied, the exposed liver and gill showed the highest damage than brain and muscle. The correlation coefficient analysis between days of exposure and AChE activity in control fish brain showed a positive but non-significant correlation, whereas, the exposed fish brain showed a significant negative correlation ($r = -0.986$, $P \leq 0.01$). The control set did not show any positive significant correlation. The per cent change in the enzyme activity in the exposed fish brain when compared to control fish brain showed the existence of a negative significant correlation ($r = -0.986$, $P \leq 0.01$) with the exposure period. The correlation coefficient analysis between days of exposure and the AChE activity of liver slices of the control fish did not show any significant correlation, whereas the exposed fish liver showed a negative and significant correlation ($r = -0.985$, $P \leq 0.01$) with the exposure period. The per cent change in

AChE activity showed the existence of a negative, significant correlation ($r = -0.988$, $P \leq 0.01$) with the exposure period. The control muscle did not show any significant correlation between exposure period and enzyme activity. However, the muscle of the exposed fish showed negative and significant correlation ($r = -0.954$, $P \leq 0.05$) with the exposure period. The per cent change in the enzyme activity showed negative and significant correlation with the exposure period. The exposed fishes were when transferred to red mud extract free medium, the enzyme activity showed recovery in liver and muscle tissues but not statistically significant. Liver showed the highest damage (54.9 per cent), brain showed the no recovery during recovery period and gill showed maximum recovery by 12.2 per cent in the recovery fishes, after 28 days of recovery.

Table 40.9: Changes in Total AChE Activity (μ moles of ACh hydrolyzed mg^{-1} hr^{-1}) in Control and Alumina Industrial Effluent Exposed *Anabas* Fish Brain, Liver, Muscle and Gill, at Different Days of Exposure and Recovery

Tissue	*Condition*	*Exposure in Days*					*Recovery in Days*	
		0	*7*	*14*	*21*	*28*	*14*	*28*
Brain	Control	0.546±0.045	0.541±0.048	0.543±0.042	0.535±0.036	0.538±0.039	0.540±0.041	0.540±0.045
	Exposed	0.546±0.045	0.535±0.018	0.501±0.036	0.492±0.044	0.414±0.032	0.392±0.037	0.409±0.039
Liver	Control	0.478±0.032	0.476±0.024	0.479±0.031	0.478±0.040	0.479±0.018	0.481±0.044	0.480±0.041
	Exposed	0.478±0.032	0.402±0.036	0.374±0.039	0.311±0.038	0.216±0.034	0.236±0.031	0.254±0.028
Muscle	Control	0.385±0.095	0.388±0.036	0.381±0.074	0.384±0.054	0.383±0.048	0.384±0.058	0.385±0.064
	Exposed	0.385±0.095	0.341±0.044	0.302±0.051	0.296±0.022	0.281±0.026	0.285±0.037	0.294±0.049
Gill	Control	0.586±0.042	0.588±0.036	0.589±0.028	0.587±0.014	0.586±0.036	0.588±0.041	0.590±0.042
	Exposed	0.586±0.042	0.512±0.044	0.432±0.041	0.388±0.036	0.311±0.064	0.346±0.032	0.385±0.028

Data presented are mean of samples ± standard deviation.

Changes in few haematological parameters in control and red mud waste extract exposed fish have been presented in Table 40.10. The haemoglobin content decreased at all exposure periods. The haemoglobin content decreased from 13.56 ± 0.28 to 5.18 ± 0.28 g/100ml after 28 days of exposure. When the exposed fish was transferred to red mud waste extract free medium, on 28th day of recovery, the haemoglobin content decreased from 13.66 ± 0.28 to 7.85 ± 0.26 g/100ml (Table 40.10). The haemoglobin content decreased steadily and the per cent decrease increased with the increase in exposure period. On 28th day of exposure, a maximum of 61.8 per cent decrease was recorded. The per cent decrease increased from 31.6 per cent to 61.8 per cent on 28th day of exposure. A partial recovery was recorded on 28th day of recovery, when the exposed fish was transferred to toxicant free medium. On 28th day of recovery, 42.5 per cent decrease, when compared to control was marked. But this decreased value was more than the 28th day exposure value, indicating recovery by 9.96 per cent and 19.3 per cent on 14th and 28th day of recovery. The RBC count decreased in the exposed fish from 1.22 ± 0.18 to 0.57 ± 0.08 × 10^6 cells/cm^3 when compared to control fish after 28 days of exposure. When the exposed fish was transferred to leached effluent free medium, on 28th day of recovery, the RBC count decreased from 1.25 ± 0.14 to 0.78 ± 0.16 x 10^6 cells/cm^3 (Table 40.10). The RBC count decreased steadily and the per cent decrease increased with the increase in exposure period. On 28th day of exposure, a maximum of 53.2 per cent decrease was recorded. The per cent decrease increased from 3.4 per cent to 53.2 per cent on 28th day of exposure. Partial recovery was recorded on 28th day of recovery, when the exposed fish was transferred to toxicant free medium. On 14th day of recovery, 50.0 per cent decrease and on 28th day of recovery 37.6 per cent decrease, when compared to control was marked. But this decreased value was more than the 28th day exposure value, indicating partial recovery by 3.2 per cent and 15.6 per cent on 14th and 28th day of recovery, respectively. The haematocrit value decreased from 34.26 ± 0.17 to 18.24 ± 0.14 per cent in the exposed fish, when compared to control fish after 28 days of exposure (Table 40.10). When the exposed fish was transferred to red mud waste extract free medium, on 28th day of recovery, the haematocrit value decreased from 34.22 ± 0.16 to 18.25 ± 0.22 per cent. The haematocrit value decreased steadily and the per cent decrease increased with the increase in exposure period. On 28th day of exposure, a maximum of 46.8 per cent decrease was recorded. The per cent decrease increased from 3.1 per cent to 46.8 per cent on 28th day of exposure. A partial insignificant recovery was recorded on 28th day of recovery, when the exposed fish was transferred to toxicant free medium. On 28th day of recovery, 46.6 per cent decrease, when compared to control value was marked. But this decreased value was less than the 28th day exposure value, indicating partial recovery by 0.2 per cent on 28th day of recovery. The oxygen carrying capacity of blood of the exposed fish decreased with the increase in exposure period. On 28th day of exposure, the oxygen carrying capacity decreased from 16.95 to 6.48–volume per cent. When the exposed fish was transferred to toxicant free medium, the oxygen carrying capacity slightly increased, when compared to 28th day exposure value. The oxygen carrying capacity declined from 17.08 to 9.81–volume per cent on 28th day of recovery (Table 40.11). The per cent decrease of oxygen carrying capacity increased from 31.6 per cent on 7th day to 42.3 per cent on 14th day to 53.7 per cent on 21st day and to 61.8 per cent on 28th day of exposure. During recovery studies, the per cent decrease value instead of increasing decreased indicating partial recovery. On 14th and 28th day of recovery 51.8 per cent and 42.6 per cent decrease in oxygen carrying capacity was noted. On 14th day of recovery 10.0 per cent recovery and on 28th day of recovery 19.2 per cent recovery was calculated, when compared to 28th day exposure value (Table 40.11).

Table 40.10: Changes in Haematological Parameters of Control and Alumina Industrial Effluent Exposed Fish at Different Days of Exposure and Recovery

Parameters Studied		*Exposure in Days*					*Recovery in Days*	
		0	*7*	*14*	*21*	*28*	*14*	*28*
Haemoglobin (g/100 ml)	Control	13.60±0.32	13.50±0.28	13.65±0.35	13.55±0.42	13.56±0.28	13.62±0.26	13.66±0.28
	P ≤	NS	0.01	0.01	0.001	0.001	0.001	0.01
	Exposed	13.60±0.32	9.24±0.20	7.88±0.31	6.28±0.32	5.18±0.28	6.56±0.12	7.85±0.26
RBC Count (10^6 Cells/cm^3)	Control	1.26±0.14	1.22±0.20	1.23±0.21	1.24±0.15	1.22±0.18	1.24±0.11	1.25±0.14
	P ≤	NS	NS	0.05	0.01	0.001	0.001	0.01
	Exposed	1.26±0.14	1.18±0.11	1.02±0.13	0.84±0.12	0.57±0.08	0.62±0.14	0.78±0.16
Haematocrit (per cent)	Control	34.10±0.21	34.21±0.12	34.26±0.14	34.24±0.11	34.26±0.17	34.22±0.12	34.22±0.16
	P ≤	NS	NS	0.05	0.01	0.01	0.01	0.01
	Exposed	34.10±0.21	33.14±0.21	26.15±0.23	21.22±0.15	18.24±0.14	16.98±0.29	18.25±0.22

Data are mean of 5 samples ± standard deviation.
P = Levels of significance, NS = Not significant.

Table 40.11: Computed Oxygen Carrying Capacity (vol. %) and per cent Decrease in Oxygen Carrying Capacity of Control and Alumina Industrial Effluent Exposed *Anabas* Fish Blood at Different Days of Exposure and Recovery

Tissue	*Condition*	*Exposure in Days*					*Recovery in Days*	
		0	*7*	*14*	*21*	*28*	*14*	*28*
Fish blood	Control	17.00	16.98	17.06	16.94	16.95	17.03	17.08
	Exposed	17.00	11.55	9.85	7.85	6.48	8.20	9.81
	(% Decrease)	0.00	-31.6	-42.3	-53.7	-61.8	-51.8	-42.6
Per cent recovery during recovery period When compared to 28 d exposed value							+10.0	+19.2

Data computed from the mean of samples.

Basing on the tables and all statistical data, it can be inferred that the toxicant, red mud waste extract/leached waste effluent is toxic to fish in prolonged exposure for the entire period of exposure. When the exposed fish was transferred to toxicant free medium, no recovery was marked in many parameters studied. A partial insignificant recovery in very few negligible parameters indicated nothing. No inference can be drawn from such type insignificant recovery. This study fills a gap in the environmentally available toxicant research on aquatic animals. This type of study on all types toxicant is essential to understand the environment and environmental deterioration caused by industrial wastes in general. Most important point needs further discussion relates to careless handling of industrial wastes, safe disposal of wastes and fate of these waste chemicals in the environment, needs attention.

Discussion

The excessive discharge of red mud leached chemicals in recent years tends to contaminate the environment. Controversy has centered on both the mode of action of red mud leached chemical in effecting animal metabolism and the extent of its toxicity resulting from penetration into the system. The literature on residual accumulation of red mud leached chemical and its effects on fish is scanty. Towards last part of the present century the agricultural contamination of red mud leached chemical and its entry through food chain, have not only drawn the attention of the scientists of all over the world, but also a lot of research work in this line have already been undertaken. The eco-toxicology of chemicals is an important aspect of environmental management and the management of pollution. The eco-toxicological assessments of toxic chemicals depend on two basic items of information. First, the environmental available concentrations of the chemical which occur as a result of discharge and distribution in the natural environment and secondly, the toxicological properties of the chemical at that concentration and its possible impact on the biota including humans. Environmental concentrations and toxicological properties also include the concentrations of by-products or transformation products as well. Different methods and procedures are used to draw all the available eco-toxicological information together to arrive at an assessment. The basic requirements for such procedures are the pattern of use of chemicals; their disposal and release; physico-chemical characteristics; degradation and bioaccumulation; and biological factors such as growth and development inhibition, toxicity testing etc.

The industrial wastes once thrown into the environment, these wastes spread all over and affect the flora and fauna. For example, the red mud waste discharged from the NALCO plant was stored in a natural pond surrounded by hills from all sides except one side, where the industry people have constructed three dams. Unfortunately, they have never visualized probably that the leached chemicals/chemicals from the red mud pond, will create future problems. The data obtained in this piece of investigation revealed that the extract of the red mud is deadly toxic and can seriously affect the animals, particularly fish. Samantaray (2000) had selected a resistant variety, still then, the extract effected the crop plant seriously, particularly to the production/yield of the crop plant. The morphological changes were clearly visible and effects were more pronounced. Hence, care should be taken to check leaching of the chemicals from the red mud pond or they must find out ways and means for safe disposal of the waste. We are not conservative enough to request/direct the closure of the industry but some safe disposal methods are ought to be adopted to safe guard the interest of the public in general and environment in particular. However, literature explaining the following aspects in details is lacking:

1. Toxicant dynamics of leached chemicals of red mud wastes in the environment.
2. Red mud waste dynamics in fish in relation to fish biology.
3. Amount of red mud leached chemical transfer to aquatic system, aquatic animals, particularly freshwater fish etc.
4. Effect of leached chemicals of the red mud pond on fish toxicology, fish behaviour studies, fish physiology, changes induced in the biochemical pattern, effects caused to the fish enzyme system, fish viability, fish tolerance etc.
5. Effect and bio-concentration of the leached chemicals in the trophic levels, and trophic level transfer leading to bio-magnification etc.

In the present investigations the red mud waste dynamics in fish and fish biology during growth and development was studied.

The disturbances in growth and growth substances are always related with metabolic changes inside the plant body (Levitt, 1972; Poljakoff-Mayber and Gale, 1975) animal body. Biochemical changes proceed any change in growth because growth is the outcome of many biochemical processes; and physiological and biochemical effects, are the underlying cause of measurable whole organism effects (Mehrle and Mayer, 1980).

The route of entry of the toxicant is generally agreed to be via the gills (Holden, 1962; Ferguson *et al.*, 1966) and thus gets directly into the vascular system. This cause damage to tissues and as a result of which there is depression in active metabolism (MacLeod and Pessah, 1973). Fry (1957) demonstrated depression in active metabolism directly reduced scope of activity in the experimental animals. Probably this has relevance with the impaired swimming ability, erratic movements, which were observed in all test fish. High percentage mortality of fish due to action of the leached chemicals of the red mud waste might be due to the pathological changes as Mathur (1969) pointed out that fish subjected to insecticide such as lindane, dieldrin, DDT and BHC die, due to pathological disorders. Samant (1989) showed that phenyl mercuric acetate was highly toxic to fish even at very low concentrations. Ray and David (1962) concluded that deficiency of dissolved oxygen concentration caused by decomposition of algae or bottom deposits was mostly the cause of fish kill in pond and impounded water. Excess discharge of any industrial wastes, leakage of red mud wastes from red mud ponds as in the present case might cause severe damage to the ecological system. Inorganic ions can equally prove fatal by limiting oxygen intake causing respiratory and circulatory failure influencing smooth osmotic exchange of gases caused by flocculation of iron and other waste chemicals discharged from the red mud pond on gill filaments. There are reports that a thick film of red mud waste is deposited on the gill filaments of fish causing severe disruption in gaseous exchange. The leached chemicals of the red mud waste increased the turbidity of the water and transparency decreased significantly. Which might have played an important role to decrease the dissolved oxygen present in the exposed aquarium.

David *et al.* (1976) showed that of the metal tested sliver and mercury were the most toxic *i.e.* silver was 95 per cent toxic at 0.086 ppm and mercury at 0.150 while LD_{50} values were 0.333 ppm for silver and 0.089 ppm for mercury. During the experimental period; temperature, hardness, time of exposure, size of the test fish and all such factors will alter the MAC values. Smaller fishes can tolerate lower concentrations. Bigger fishes can resist up to a concentration where all small fishes die. With the increase in exposure period toxic tests becomes more pronounced in small fishes, when compared to bigger fishes. Probably the toxicity varies and corresponds to age of the fish. The solid waste of the chlor-alkali industry contained mainly mercury as toxicant in different forms at different concentrations. In this piece of investigation it was presumed that all types of chemical compounds with other soluble chemicals would be discharged out as leachates. The extract prepared in the laboratory contains the leachates of the red mud pond, which was observed in the spot at Damonjodi, Koraput. Sahu and Panigrahi (1999) reported loss of mercury by volatilization and adsorption which has not often been reported by investigators and hence suggested that mercury concentrations given in the literature should be considered less than those reported, unless test solution measurements were included. Mercury loss from the water in renewed static tests occurred by both biological activity and physical effects (adsorption and volatilization), while some loss also occurred in flow through systems; concentrations remained fairly stable throughout the test period, because the test solution was not changed during the entire period of experimentation (Pattanaik, 2002). In the present investigation, the medium of the experimental aquarium was changed periodically to ensure the constancy of the toxicant.

The rapid absorption of (inorganic and organic) mercury through the gill, skin and gastro-intestinal tract of fish is well evident (Gibilin and Massaro, 1973, Panigrahi, 1984 Samant, 1989 and Pattanaik, 2002). The toxicity of mercury becomes apparent in a shorter period in aquatic animals. Panigrahi (1980) documented the detail behavioural changes in relation to inorganic mercury intoxication. Panigrahi (1984) reported the impact of MEMC on *Anabas scandens*, a freshwater fish and noted substantial changes in behaviour of the test fish. Although he has observed swelling of eyes and consequent blindness, such types of symptoms was not observed in the exposed fish in the present investigation, probably due to the absence of mercury as a base in the toxicant (Panigrahi, 1985). In the present investigation, reddening of eyes was marked. This might be due to other toxicants present in the red mud waste. Swelling of eyes and reddening of eye can be attributable to mercury poisoning and consequent early blindness, might be due to the combined effect of the chemicals present in the leachate of the solid waste (Pattanaik, 2002). Higgins (1974), MacLeod and Pessah (1973), Panigrahi (1980) and Samant (1989) reported that loss of appetite, loss of weight, nervousness, dizziness, loss of equilibrium, erratic swimming and gradual onset of inactivity were the sub-clinical effects of inorganic organic mercury intoxication. Identical observations were marked here in exposed *Anabas* fish. Panigrahi (1984) also reported similar symptoms relating to a mercury-based fungicide. MacLeod and Pessah (1973) and Panigrahi (1980) reported neurological damage caused by mercury on fish, basing on the behavioural studies. In the present investigation though we found similar results, but it will be difficult to say, what really caused the changes or the effects are due to the combined effect of the red mud waste.

In the preliminary study the animals showed all regular features of mercury poisoning; such as excitation, irritation and restlessness. Towards the end of the experimental period *i.e.* prior to death the locomotion of fish almost ceased and remain suspended vertically in the water medium, indicated loss of equilibrium. In addition, periodic and erratic paralytic movements were observed in exposed fish. The early signs of poisoning were probably due to the effect of combined chemicals of the extract, which showed probably synergistic effects on fish. Control fish remained clinically healthy, throughout the experimental period. The observed depression in active metabolism in red mud waste extract exposed fish are indicative of damage to nervous tissues, inhibition of enzymes or vital system was totally in agreement with MacLeod and Pessah (1973), Panigrahi (1980) and Panigrahi (1984). Fry (1957) demonstrated that depression in active metabolism directly reduce "scope for activity". This in turn may result in decreased growth, impaired swimming ability with erratic outbrust at times. It is possible that even a small reduction in the normal active metabolic rate may be indicative of stress in the animals (MacLeod and Pessah, 1973). Loss in weight due to starvation was noted by Larson and Lewander (1973). But the loss due to starvation was far less. Panigrahi and Misra (1978 and 1980) reported the loss of body weight due to mercury intoxication and confirmed that this loss in body weight was only due to mercury stress. In addition, exposed fish maintained their feeding habit only after a short span of time. Hence the weight loss cannot be correlated with starvation but can be related with the red mud waste intoxication. A significant correlation exists between body weight and days of the exposure. The effects of mercury on the brain and optic nerves presumably caused blindness and exophthalmia in fish. Brain tissue should therefore be carefully examined for carcinogenic or histopathological signs. Gopalkrishnan (1961) reported the different stages of eye infection, the second stage of infection of eye being the reddening of the eyeball and whitening of the corneas. Similar eye infection was observed with a prior swelling of the eyes. This was probably due to the destruction of brain cells and nerve cells, which caused loss in consistency in the whole tissues.

Yamamura *et al.* (1987) reported no change in body weight in short-term exposure in both control and MMC exposed rats. However, a decrease in body weight was reported after 8 weeks Me Hg-exposed rats. Stevenson *et al.* (1977) reported no change in body weight gain between control and lead exposed rats up to 8 weeks exposure. Growth has often been used to assess the effect of toxicants on fish. However, biochemical changes during toxicosis should precede reductions in growth because growth is the culmination of many biochemical processes, and physiological and biochemical effects are the underlying cause of measurable whole organism effects (Mehrle and Mayer, 1980). A decline in liver somatic index and brain somatic index in exposed fish could be correlated with degeneration of cells and decrease in other biochemical variables. Stone *et al.* (1977) reported no significant effect on body weight or size of kidneys, liver as related to body weight in lead exposed Japanese Quail. Concisely, at this stage it is not possible to establish the cause of blindness, loss in body weight and decrease in liver and brain somatic index, except a generalized comment on the behaviour of the exposed fish to stress that the observed symptoms were most probably be due to the leached chemicals of the red mud waste. Bhatia *et al.* (1973) reported that changes in behaviour in test fish had some relationship with insecticide intoxication. Rath and Misra (1980) reported the change in behaviour activity of exposed fish to dichlorovous (DDVP) and confirmed the idea that the change in activity was related to pesticide intoxication, which is totally agreeable with our observation in this study. Panigrahi (1980), Panigrahi (1984) and Samant (1989) reported similar decline in the parameters. However, in this present piece of investigation, the extent of damage was probably more acute and pronounced, might be due to the synergistic effect of the chemicals present in the leachate of red mud pond.

In pollution monitoring programmes, efforts have been directed towards obtaining information on the body burdens of common or commercially important species. Besides studies of bioaccumulation, the physiological responses of the animals must also be followed since the whole body response is critical to an animal's survival (Goldberg *et al.*, 1978 and Micallef and Tyler, 1990). The effect of inorganic mercury and organic mercury on the whole animal oxygen uptake the respiratory metabolism was as expected and presented earlier by Panigrahi (1980) and Samant (1989). A high concentration toxicant inhibition of metabolism was almost complete. Oxygen consumption decreased with the increase in exposure period. An inverse correlation reported between oxygen consumption and exposure period. Similar inverse relationship of oxygen consumption is noticed with the concentration of the leachate of the red mud waste in the medium. The ventilation rates of the exposed *Tilapia* fish increased significantly when compared to control fish. With the increase in exposure period the whole animal oxygen uptake of *Anabas* decreased significantly. Regression coefficients are all negative confirming the increase in inhibition at higher concentrations and also with the exposure period. A depression in the filtration rates of bivalves has been shown to take place following exposure to a variety of metal ions other than mercury. The results for the effects of mercury on the filtration rates showed that mercury rapidly decreased the filtration rate in muscles (Micallef and Tyler, 1990). Panigrahi (1980) reported that with the increase in pollutant the whole animal oxygen uptake and respiratory metabolism declined sharply both in *in vitro* and *in vivo*. Rath and Mishra (1979) reported a striking decline in whole animal oxygen uptake and respiratory metabolism declined sharply both *in vitro* and *in vivo*. Rath and Mishra (1979) reported a striking decline in whole animal oxygen uptake and respiratory metabolism when exposed to dichlorvos an organophosphorous compound. Panigrahi (1984) reported an identical data showing agreement with the fmdings of Panigrahi (1980). Incidentally the observation of this study tallied with the findings of Panigrahi (1980) and of Rath and Misra (1979). Rath and Misra (1980) suggested an inverse relationship in active metabolism and pollutant concentration. The decline in whole body oxygen uptake can be attributable to several factors. Some important points are first, might be due to coagulation of mucus with the gill lamellae induced by the toxicant (Samant, 1989), second, due to degeneration of epithelial cells, from gill lamellae by the toxicant, third, disruption of gases exchanges at gills due to differential erythrocyte behaviour and haemolysis induced by mercury (Panigrahi, 1980), fourth, loss of appetite and ataxia due to mercury intoxication etc. The idea is being well supported by Panigrahi (1984) and Samant (1989). The above authors indicated that residual mercury concentration was probably responsible for the decline in oxygen uptake by the exposed fish when compared to control fish. At this stage, it is very difficult to assess the exact cause of the decrease in the parameters. However, the role of mercury can never be ignored at this stage of discussion. But in the red mud waste extract, mercury or any other heavy metal was not found. At lower concentration of the red mud waste an increase in oxygen consumption by tissues slices was observed. Dom (1974) reported the alteration of ionic distribution and osmo-regulatory activity of the animal (Passow *et al.*, 1961) and accounted for the increased respiratory rate in *Congeria*. Fox *et al.* (1975) reported that inorganic mercury has no effect on the oxygen consumption by the brain tissue slices up to 0.5 mM concentration of mercury nitrate in contrast to organic mercury. Organomercurials inhibit the CO_2 production at lower concentrations even (Fox *et al.*, 1975) in brain slices of guinea pig. Panigrahi (1980) reported an initial increase in oxygen consumption at lower concentration in liver, brain and muscle slices and gill filaments, and suggested that this was probably due to the fact that at much lower concentrations mercury stimulated an increase in oxygen consumption. Such a generalization in this stage would at best be prematured. Panigrahi (1984) reported a similar trend in freshwater fishes when exposed to mercury-based fungicide. In contrast, the observation of this study reveals no such increase at lower concentration of the toxicant. However, a decline in active metabolism, which related to respiratory metabolism, was marked with the increase in exposure period and toxicant concentration. These observations totally agree with the findings of Rath and Misra (1981) and Samant (1989) and do not coincide with the trend of Panigrahi (1984). The difference may be due to the fact that at much lower concentration mercury induces an increase in oxygen consumption (Panigrahi, 1980). In the present investigation, such type of increase in the oxygen consumption rate of the whole fish or different tissue slices was not observed. In all exposure cases, significant decrease in the parameters were marked in the red mud waste extract studies. The gill surfaces of the freshwater teleosts are in intimate contact with the water are particularly susceptible to aquatic contaminants, and changes in respiratory metabolism (Schaumburg *et al.*, 1967). Since the respiratory and circulatory systems are intimately connected, any significant changes in one system would reflect in the other. The gills are the primary site for the active absorption and respiratory exchange of gases. Because of the insufficiency of low solubility, the gills in obtaining the required oxygen for survival, low solubility of oxygen and increased toxicity of water, large amount of water must be passed over the gills to meet the oxygen demands of the fish (O'Conner and Fromm, 1975). At the first introduction of the toxic chemical to the aquatic systems, the ventilation rate increased over the control value. When the fish slowly acclimatized to the concentration of the toxicant, different organs of the fish like in brain, liver, muscle and gill, the physiological activity decreased with the increase in ventilation rate. With prolonged exposure to a particular concentration, the fish ultimately faces death due to high accumulation of the toxicant inside the body tissue. The ventilation rate increased probably because the body demands for oxygen increased. The opercular movement, to make up the oxygen demand by the body, probably pumped more amount of water. Lunn *et al.* (1976) reported the increased respiration rates and decreased heart rates in fish exposed to DDT in rainbow trout. Brown and Newell (1972) reported a 50 per cent depression in respiration rate of the whole animal by copper. Vernberg and Vernberg (1972) reported a decrease in respiration rate with the exposure of fiddler crabs, *Uca pugilator* by inorganic mercury ($HgCl_2$). Macinnes and Thurberg (1973) found that silver depressed whole animal oxygen consumption in the mud snail, *Nassarius obsoletus*. Dorn (1974) reported the alteration of ionic distribution and osmo-regulatory activity of the animal (Passow *et al.*, 1961) and accounted for the increased respiratory rate in *Congeria*. The decline in exposed gill respiration can be correlated with the residual mercury concentration (Pattanaik, 2002). The gill filaments coagulate with the mucus forming a coating over the gill epithelial layer, which restrict and reduce the area of gaseous exchanges. The gill of the fish is most important, because gill is the only site of O_2 absorption (Panigrahi, 1980 and Rath and Misra, 1980a). Due to the decrease in the area of gaseous exchanges, the body demand for oxygen increased significantly. As a result, increase in ventilation rate was seen in the exposed fish (Rath and Misra, 1980a). The interpretation of metal induced changes in respiration is complicated by the fact that such alterations differs from metal to metal, from species to species and from one experimental condition to another (Thurberg *et al.*, 1974 and Panigrahi, 1980). Presence of mercury in the environment has received considerable attention. The rate of accumulation of mercury is dependent on the type of species (Matsunaga, 1978), and its metabolism (Norstrom *et al.*, 1976 and Cutshall *et al.*, 1978). Panigrahi (1980) reported that mercury in fish tissue causes enzyme alteration even at very low concentration. Upon continued exposure toxicity leads finally to structural alterations (Hossain and Dutta, 1986). The maintenance of certain concentration gradient in different tissues as a matter of fact depends on the integrity of the Na^+,K^+, Mg^+-ATPase activity. ATPase is specifically required for the transport of ions against concentration gradient as well as across the membrane. Yagi *et al.* (1976) reported the existence of a strong inter-relationship between Na^+, K^+, Mg^+–ATPase activity and Na^+, K^+, Ca^{++} and

Mg^{++} content of the tissues. ATPase is specifically required for the transport of ions against concentration gradation and across membranes. Ahuja and Subrahmanyam (1978) concluded that ATPase plays a central role in synaptic transmission and nerve impulse generation. The decrease in brain ATPase activity may be due to over excitation of the fish to the stress, Ahuja and Subrahamanyam (1978). The same author reported, decrease in brain Na^+, K^+–ATPase activity in the trained rats may be due to their nervous over excitation during training. A drastic depletion in this enzyme activity in liver, gill and muscle was observed in the exposed fish. Panigrahi (1984) reported a similar trend in mercury based fungicide exposed fishes. The data of this investigation coincides with the findings of Rath and Misra (1980), Mishra (2002) and Sarangi (2002) in different vital organs of a freshwater fish exposed to dichlorovos. Since liver is the active site of all metabolisms, depletion in ATPase activity in the liver caused diminishing of enzymes activity at 10 mg/kg and 20 mg/kg MMC in rats. The same author reported a slight activation of the enzymes activity with an injection of Hg. Panigrahi and Misra (1978) reported depression in ATPase activity by chronic Chlordance intoxication in freshwater fishes and suggested that the mechanism of the action of the pesticides on ATPase system may be due to the uncoupling of oxidative phosphorylation which causes a depletion in the phosphorylation product–ATP. Panigrahi (1980) noted that the reduction in available free phosphate would be directly proportional to the reduction of total ATP produced. Panigrahi (1980) reported that a significant difference in the response of the ATPase activity to inorganic mercury was observed between *in vivo* and *in vitro*. In the exposed fish a gradual decrease of Na^+, K^+, Mg^{++}–dependent ATPase activity with the exposure period was marked in all the tissues. In contrast, in *in vitro* studies, mercury contained in the solid waste at low concentrations stimulates total ATPase activity. However, on further increase in solid waste concentration, the total ATPase activity decreased significantly. In the present investigation, the difference in action of the enzyme activity was due to the red mud waste extract. Stimulation occurs at low concentration of the stress and the fish behaviour was more erratic. The extreme behaviour of the exposed fish, exposed to red mud waste, now can be correlated with the enzyme activity. There exists a strong correlation among the ions concentration in the tissue, ATPase activity, AChE activity and metabolic rate (Panigrahi, 1980; Webb, 1966) with the behaviour of the fish. Any change in the above said parameters due to the toxicant, will change the behaviour of the fish directly or indirectly. The above investigation reported by Pattanaik (2002) strongly indicated that the enzyme activities were related to residual mercury concentration in the tissues. We can conclude here that probably the causative agent for all such observed changes in this study is the soluble chemicals present in the red mud waste extract of the NALCO plant, which leak from the cemented dams as leached effluent at Damanjodi.

Life depends on a complex network of chemical reactions brought about by specific enzymes. The enzymes are the primary instruments for the expression of gene action. Enzymes are of biological importance in metabolic functions, which have the most highly specialized class of proteins. The change in metabolism can well be correlated with the disturbance in behaviour and action of the enzymes (Panigrahi, 1980). The reduction in active metabolism has been correlated with the toxic effective different types of toxicants. The disturbance in enzymatic action can be related to the specific action of certain toxicants by inhibiting the enzymes to act, bind the active sites of the enzyme and enzymes become non-functional due to disintegration. Inhibition of enzymes by chemical agents may be reversible or irreversible, competitive or non-competitive. There are many different mechanisms through which thiol groups of enzymes can be acted upon by heavy metals. The importance of enzymatic studies in the system lies in the functional interpretation of the pollutant causing different types of disorders. Membranes play a role in the structure and function of the cell and its organelles. One of the way by which membranes can affect cellular metabolism is by interacting with enzymes. Some membrane function as the organizing matrix for enzyme systems, the micro-environment in all these cases at or near the membrane surface may modulate the activities of the enzymes located in the region. No information is available on the effects of red mud wastes of aluminium industry on different biota. Few preliminary reports and news publications are only available on pollution problems of NALCO plant. The purpose of acute toxicity test is to assess various abnormalities caused due to administration of chemicals to animals on one occasion or other and to determine the order of lethality of the chemical. Many of the schemes are available at present to test for adverse effects that may result when chemicals enter into environment. Considerable efforts have been directed towards the establishment of short-cut procedure. In aquatic toxicology, acute lethal toxicity tests with fish or invertebrates are usually intended to assess the numerical value of toxicity, to compare potencies of toxicants, to assess the effect of environmental variables on toxicity. Besides, statistical analysis of toxicity is essentially a tool in all these studies. The rapid increase in contamination of aquatic environments with toxicants in recent years has resulted in an escalation of scientific interests in the biological effects of toxicants in the environment; any deleterious effect of the toxicants on fish is likely to be reflected on the entire eco-system. This piece of study is the initial study in this line, where there is scope to study in greater details before any suggestion or any recommendation is made for pollution control and abatement.

The ions studied play a vital role in the circulatory system of the animals. The plasma ion concentration plays a significant role in different metabolism. Any change in ion concentration will definitely reflect in the metabolic activity and physiological activity of the organism. Dawson (1979) reported significant decrease in different haematological parameters with the decrease in plasma calcium and increase in plasma sodium of winter flounder exposed to mercury. Panigrahi and Misra (1979) reported changes in blood morphology in mercury exposed fishes. The same author correlated the finger like projections, beak like proliferations and vacuolization in the erythrocytes of mercury-exposed fish with that of change in ion concentrations and drastic depletion in different enzyme activities. Similar decline in enzyme activities and disturbance in ion concentrations were marked in solid waste exposed fish. Hence, it can be concluded that the changes in different enzymes activities and ion concentrations were related to nervous over excitation, which leads to impaired swimming and erratic movements of the exposed fish. The decline in Na^{+}, K^{+} and Mg^{++} concentration in gill tissues indicates the depression in gaseous exchange in the exposed fish (Pattanaik, 2002 and Sarangi, 2002). The results obtained in this investigation do not agree with the findings of Misra (1985), as they showed decrease in (loss of DNA and RNA) polynucleotides and protein and a concomitant increase in total free amino acid content inside the seedlings. The previous workers of this laboratory confirmed the above possibilities and similar results were also observed by us.

Most of the results of different parameters were found to be statistically significant, it does mean that, the trend or changes observed in the present piece of work are to be taken into confidence. In this connection here it can be stressed that as the study was undertaken under laboratory-controlled conditions hence most of the parameters did confirm to the statistical 'significance' test, as observed by other workers. From the above observations and interpretations, erecting a model becomes a difficult task in absence of residual level data. However, many interferences and predictions raised during discussion, provides gaps or fields unexplored yet. In this piece of investigation, though we felt to a link a gap but after completion of this piece, we find many questions remain unanswered. In this piece of investigation, we tried to find out the effects of red mud leached chemical/leached effluent from the red mud pond, on a freshwater fish, *Anabas scandens*, Cuv. & Val. and its eco-toxicological significance. From the results obtained in present study, it is clearly evident that, red mud leached chemical, which is now well known for its interference in biochemical metabolism, becomes toxic when used in lower and moderate concentrations. Thus, we can conclude that even though many advantages have been derived from the industry but we must not neglect the environment by way of constructing poison ponds like red mud ponds. Many similarities exist in biological process among all forms of life, and investigative effects of this nature are not only important to plants, fish and wildlife, but to all life forms, including man. However, as long as these chemicals are thrown into the environment without proper check, this type of leaching will always be there. These leached chemicals will always enter into crop fields, aquatic ecosystems like pond, small canals, rivers, dams, contaminate groundwater and water bodies. There is always the possibility in future that unexpected and undesirable complication may arise even though minimum immediate hazards are occurring at present. Safe disposal of these wastes and proper care to avoid leaching of these chemicals can protect the target and non-target organisms.

The decrease in activity of the fish, the decrease in oxygen uptake by the fish, the decrease in tissue slices respiration, the lethargic movement by the fish may be well correlated to the decrease in ATPase activity. Although differences were well marked between *in vivo* and *in vitro* results. At lower concentrations, there appears to be sufficient evidence from those studies to indicate that ATPase inhibition was probably the site of attack of inorganic mercury in fish tissues. The decrease in the enzyme activity in the present investigation might be only due to the soluble chemicals of the red mud waste. Similar pattern results were observed by Panigrahi (1985), Sarangi (2002) and Mishra (2002). Hence, basing on the present data, we cannot assume that the observed changes were only due to mercury contained in the toxicant, as the present toxicant, red mud waste extract/leached effluent does not contain mercury even at very low concentrations. Hence, the observed depletion was probably due to the combined effect all the chemicals present in the waste effluent rather than a single chemical. We are not inclined to criticize the obtained data earlier by other workers in the field but a clear message can be given by saying that the enzyme activity can be influenced by any of the toxicant(s) available in the environment. Hence, the measurement of this enzyme can be a single parameter indicator for measuring and monitoring pollution of the aquatic environments.

A significant depression of Na^+, K^+-ATPase activity was observed with the excessive absorption of the toxicant from the environment. The decrease in total ATPase activity in gill *in vivo* of the exposed fishes lead to the decrease in ion exchange and also, in addition, gaseous exchange. The depletion in total ATPase activity with the exposure period in brain, liver, and muscle caused the depression in the activity of the fish was most probably due to mercury present in the solid waste, as reported by Pattanaik (2002). In the present investigation, it was found that this toxicant (red mud waste extract, RMWE) can depress the enzyme activity and hence the active metabolism. The neuro-toxic action of mercurial compounds as reported earlier may be due to a secondary response of the fish to the stress (Panigrahi, 1980). It may be quite possible that mercury and metabolites in the tissues, jointly responsible for the observed toxic action of mercury *in vivo* and at higher concentration *in vitro*. Further more, Yagi *et al.* (1976) correlated the decrease in Na^+, K^+, Mg^{++} content in different tissues of PCB poisoned rat with Na^+, K^+, Mg^{++} dependent ATPase activity. Similarly a strong correlation was observed between Na^+, K^+ and Mg^{++} ion content of different tissues with the Na^+, K^+, Mg^{++} dependent ATPase activity in respective tissues. A direct positive correlation exists between the minerals and the activity of Na^+, K^+, Mg^{++}-dependent ATPase. The fact that there is no clear parallel between ATPase inhibition and per cent mortality, deserves attention. Desaiah *et al.* (1975) suggested that such a case is possible due to detoxifying mechanism operating *in vivo* preventing development of a graded dosage inhibition type of response even though such graded response has been reported *in vitro*. But in contrast, in our recovery experiments, no significant recovery of total ATPase activity in brain, liver and muscle was noticed. Hence, the hypothesis that detoxifying mechanism decreases the dosage inhibition type of response was not agreeable, as no significant recovery was noticed in the fishes. The amount of recovery may be mostly due to the excretion of toxicants from the body. The cation exchange system has direct relation with the ATPase enzyme activity. The activity of this particular enzyme depends on several factors such as concentration of substrate (ATP), products (inorganic phosphate, ADP) and metal accelerators (Na^+, K^+ and Mg^{++}). Hence, any sort of change in one of these above mentioned factors would lead to the inhibition of enzyme activity or it might be due to the resultant of all the above factors. The decrease in Na^+-ATPase, K^+-ATPase and Mg^{++}-ATPase might be due to the disturbance of Na^+, K^+ and Mg^{++} ion concentration in different tissues of the fish. The response of ATPase to inorganic and organic mercury varied from fish to fish, tissue to tissue and also within the same tissue. The toxicant causes permanent damage to cell and other vital systems. Due to decline in enzyme activity the metabolic activity declined significantly. The decline in enzymatic activity, periodic failure of nerve impulse generation, depletion in metabolic activity can be correlated to red mud waste toxicity which gets reflected in the behaviour of the organism in the form of impaired swimming, periodic outburst, imbalance, ataxia, inappetence, lethargicity etc. Although quantifying these effects and changes, will be a very difficult task in the present set up of conditions, still then an approximation can be made from the structural and behavioural changes. The change in behaviour in exposed organisms was always compared with that of control organisms. Both the sets were kept under identical conditions and the only difference being the exposed set gets the pollutant stress. So the difference in both the sets might be due to the toxicant stress. In both control and exposed *Tilapia* fish, a great variation in total ATPase was marked among brain, liver, and muscle tissues (Sarangi, 2002). It has been observed that at very low concentrations of the waste no activation of the enzyme activity was marked in all the three types of tissues as observed by Samant (1989). This was probably due to the total effect of all chemicals present in the leachate acted as a whole to produce such a higher toxic effect. Panigrahi (1984) reported reduction in Na^+, K^+, Mg^{++} and Ca^{++} ion concentration in brain, liver, muscle and gill tissues of MEMC exposed fish, when compared to control fish. These ions are very important because they control the permeability of the cell membrane. Katz (1963) have found that the Hg*(II)* ions are bound predominantly to the pyrimnidine and purine bases in pure DNA, preference being shown in AT-base pair (Wong *et al.*, 1979) even though a reaction with phosphate group cannot be excluded. Neurotransmission in animals were controlled by these metallic ions. The change in ion concentration leads to disruption of neurotransmission. The present findings totally agree with the findings of Panigrahi (1984) and Samant (1989).

In the fishes a declining trend in minerals has been noticed. The probable decrease as predicted by Yagi *et al.* (1976) may be impairment in intestinal absorption of these minerals. Due to decrease in these minerals the ionic regulation of different tissues decreased (Prossor, 1973). Hurkat and Mathur (1976) reported that these minerals are highly essential for the maintenance of osmotic fragility and mechanical resistance of blood cells and proper function of all types of tissues. Reports that Ca^{++} plays a vital role in blood coagulation can be correlated to the fact that the pollutant induces decline in ion content of the exposed fish. The existence of activator ions of alkali metal series for alkaline earth metal series were highly essential for the activity of most of the enzymes which required ions for their optimum action, as a result of which the decrease in ion concentrations in different tissues were likely to influence any of the biological functions of the enzymes. Mercury, acting as a protein denaturant, binds to membranes, altering ionic distribution and osmo-regulatory activity of the animal (Passow *et al.*, 1961). Prossor (1973) reported that the decrease in the concentration of minerals tend to decrease the ionic regulations of the different tissues and also added that the depletion in Na^+ and K^+ level tends to decrease in the active membrane transport. No significant variation in case of exposed liver and brain was marked when compared to control fish, so far as Ca^{++} studies are taken into account. However, a significant depletion in Ca^{++} content was observed in gill tissues. It is feasible that the requirement of calcium in liver and brain tissues increased in mercury intoxicated fishes and the decrease of calcium content of gills did not promote a homeostatic adaptation as inferred by Yagi *et al.* (1976) and Samant (1989). Hence, mercury plays an important role in the regulation system of calcium metabolism. These ions play a vital role in the circulatory system of the animals. The plasma ion concentration plays a significant role in different metabolism. Any change in ion concentration will definitely reflect in the metabolic activity and physiological activity of the organism. Dawson (1979) reported significant decrease in different haematological parameters with the decrease in plasma calcium and increase in plasma sodium of winter flounder exposed to mercury. Panigrahi and Misra (1979) reported changes in blood morphology in mercury-exposed fishes. The same author correlated the finger like projections, beak like proliferations and vacuolization in the erythrocytes of mercury-exposed fish with that of change in ion concentrations and drastic depletion in different enzyme activities. Similar decline in enzyme activities and disturbance in ion concentrations were marked in solid waste exposed fish. Hence, it can be concluded that the changes in different enzymes activities and ion concentrations were related to nervous over excitation, which leads to impaired swimming and erratic movements of the exposed fish. The decline in Na^+, K^+ and Mg^{++} concentration in gill tissues indicates the depression in gaseous exchange in the exposed fish (Pattanaik, 2002).

Haematological studies in fishes are promising tools for investigating physiological stress and related diseases in fish. The haemoglobin concentration decreased significantly with the increase in exposure period and a drastic depletion in the number of red blood cells was noticed with the increasing exposure period. When the exposed fish was transferred to toxicant free medium no recovery was marked. It seems the industrial effluent induced irreversible damage to the exposed fish. The decrease in red blood cells is probably due to (a) reduction of erythrocyte life span due to effluent poisoning. The effluent might be causing osmotic resistance and mechanical fragility of the red blood cells. Reports of Vincent and Blackburn, (1958) and Joyce *et al.* (1959) on the defects in membrane permeability such as to derange the normal monovalent cation exchange, are clearly evident. The decrease in red blood cells may be related to the destructive or suppressive effects of erythrogenic tissues, which is expressed in a failure in red cell production (de Bruin, 1976). The decrease in red blood cells content is probably due to injury to peripheral red cells or a defect in erythrogenic tissues, which is expressed in a failure in red cell production (de Bruin, 1976). The decrease in red blood cell content is probably due to injury to peripheral red cells or a defect in erythropoiesis. Reports of Webb (1966) and Panigrahi and Misra (1978) suggest that the decrease in red blood cells is due to haemolysis. Haemolysis causes haemolytic anemia. The anemia's associated with chronic or acute poisoning by certain toxicants is by no means entirely accountable to direct detrimental action upon the red cell membrane. The toxicant causes haemolysis by interacting with legands, constituents of the erythrocyte lipoprotein membrane.

Fish haemoglobins are unstable tending to form a variety of derivatives which lead to alterations in electrophoretic patterns (Yamanaka *et al.*, 1965). With the exposure, the blood oxygen carrying capacity decreases which plays a central role in oxygen demand and subsequent oxygen consumption by the fish (Houston *et al.*, 1976). The decrease in haematocrit value is probably due to either shrinkage of erythrocytes or destruction of red blood cells. The effluent of the industry probably due to change in osmotic resistance induced the shrinkage of the erythrocytes. Under reduced oxygen tension, swelling/shrinkage of red cells has been reported to take place, but is reversible in O_2 equilibrated condition (Sovio and Nyholm, 1974 and Panigrahi and Misra, 1978). No recovery was marked in this study. This indicated that such type of phenomenon has not occurred in this investigation. Interactions of several pollutants and or natural factors can also alter the character or degree of a pollutant effect (Thurberg *et al.*, 1974 and Gray *et al.*, 1974). Dubale (1963) concluded that air-breathing fishes possessed a higher oxygen holding capacity than the non-migratory water breathing fishes. Lenfant and Johansen (1972) have also shown that there is a tendency for haemoglobin concentration to be higher in those fishes that get their oxygen from air. Ramaswamy and Gopalkrishna Reddy (1978) reported higher haemoglobin concentration and higher oxygen holding capacity in *Anabas*. The *in vivo* oxygen haemoglobin equilibrium is often remarkably sensitive to the metabolic needs of the animal (Riggs, 1970). Krogh and Leitch (1919) reported the higher oxygen capacity in air breathing fishes to reflect the degree of oxygen deficiency of the ambient medium. The decrease in haematocrit value was due to shrinkage of the red blood cells as reported by Sovio and Nyholm (1974) seemed probable. At very high exposure period, instead of further shrinkage of the red blood cells has undergone haemolysis. Finger like proliferations, beak like structures and vacuoles were not marked in the red blood cells of the exposed fishes exposed to industrial effluent of the alumina industry, as observed by Panigrahi and Misra (1978, 1979 and 1980) and Panigrahi (1980) in mercury exposure. In the present investigation only shrinkage followed by hemolysis was noted. The haematological responses of teleosts are usually related to environmental situations influencing metabolic oxygen demand or oxygen availability. The depletion in haemoglobin content and also the red blood cells in exposed fishes warrant attention. All blood parameters showed a negative correlation with the exposure period. The decrease in haemoglobin content might be due to inhibition in haemoglobin biosynthesis or destruction of haemoglobin and non-synthesis of haemoglobin due to the toxicant. The decrease in red blood cells in exposed fishes may be mostly due to the decrease in erythrocyte life span. The effluent of the industry probably disturbs the osmotic resistance and mechanical fragility of red blood cells.

Most of the results of different parameters were found to be statistically significant; it does mean that, the trend or changes observed in the present piece of work are to be taken into confidence. In this connection here it can be stressed that as the study was undertaken under laboratory-controlled conditions hence most of the parameters did confirm to the statistical 'significance' test, as observed by other workers. From the above observations and interpretations, erecting an eco-toxicological model becomes a difficult task in absence of residual level data. However, many interferences and predictions raised during discussion, provides gaps or fields unexplored yet. In this piece of investigation, though we felt to a link a gap but after completion of this piece, we find many questions remain unanswered. In this piece of investigation, we tried to find out the toxicological effects of leached alumina industrial effluent, on a freshwater fish, *Anabas scandens*, Cuv. and Val. and its ecological significance.

From the results obtained in present study, it is clearly evident that, the leached effluent of the Alumina industry, which is now well known for its interference in biochemical metabolism, becomes toxic when used in lower and moderate concentrations. Thus, we can conclude that even though many advantages have been derived from the industry but we must not neglect the environment by way of constructing poison ponds like red mud ponds near NALCO (Damonjodi) and discharging liquid effluents into the RM ponds for storage and consequent evaporation.

Many similarities exist in biological process among all forms of life, and investigative effects of this nature are not only important to plants, fish and wildlife, but to all life forms, including man.

However, as long as these chemicals are thrown into the environment without proper check, this type of leaching will always be there. These leached chemicals will always enter into crop fields, aquatic ecosystems like pond, small canals, rivers, dams, contaminate groundwater and water bodies. There is always the possibility in future that unexpected and undesirable complication may arise even though minimum immediate hazards are occurring at present. Safe disposal of these wastes and proper care to avoid leaching of these chemicals can protect the target and non-target organisms. Due to careless handling of chemicals, unplanned disposal of wastes, future and fate of the waste chemicals in the environment, non-adoption of safety measures, non-maintenance of ecological balance, improper adoption of plantation, non-selection of pollution reducing plantation, will significantly harm the man made agricultural ecosystem in particular and the ecosystem in general.

References

Ahuja, A.K. and D. Subrahmaniyam (1978). *Proc. Ind. Acad. Sci.*, 87:189.

Amend, D.F. (1974). *Trans. Am. Fish. Soc.*, 103(1): 73.

Amend, D.F., W.T. Yasutake and R. Morgan (1969). *Trans. Amer. Fish. Soc.*, 98: 419.

Amminikutty, C.K. and M.S. Rege (1977). *J. Expt. Biol.*, 15(3): 197.

APHA (American Public Health Association) (1974). *Standard Methods for the Examination of Water and Wastewater*. Environmental Protection Agency, USA.

Ashby, M. (1973). *Introduction to Plant Ecology*. Macmillan and ELBS, London, p. 271.

Balxhali, P.C. and Daisley, K.W. (1972). *J. Fish. Biol.*, 4: 593.

Bhatia, S.C., Sharma, S.C. and Venkitosubramanian, T.A. (1973). *Toxicol. Appl. Pharmacol.*, 24: 216.

Bond, C.C., Lewis, R.H., and Frayer, J.L. (1960). *Trans. Ind. Seminar on Biol. Problems in Wat. Poll.* R.A. Tar. Sanit. Eng. Centre, 95.

Borg, K., Erne, K., Hanko, E. and Wanntorp, H. (1970). *Environ. Pollut.* 1: 91.

Broderius, S.I. and Lioyd, L.G. (1979). *J. Fish. Res. Bd.*, Canada, 7(2): 164.

Brown, B. and R. Newell (1972). *Mar. Biol.*, 16: 108.

Brubaker, P.E., G.W. Laicer and R. Klein (1971). *Biochem. Biophys. Res. Commum.*, 44: 1552.

Bruin, A. de (1976). Nucleic and Microbolism. In: *Biochemical Toxicology of Environmental Agents*. Elsevier, North Holland Biomedical Press, The Netherlands, p. 612.

Caldwell, P.C., E.L. Mackor and C. Hinshelwood (1950). *J. Chem. Soc.*, 31: 51.

Choudhury, H. (1975). *J. Ind. Fish. Soc.* India, 7: 189.

Cutshall, N.H., J.R. Naidu and W.G. Peary (1978). *Science*, 200: 1489.

Das, A.K. and Misra, B.N. (1982). *J. Environ. Biol.*, 3(2): 19.

David, A.N., A. Calabrese, B.A. Nelson, J.R. MacInnes and D.R. Nenzloff (1976). *Bull. Environ. Contam. Toxicol.*, 16(3): 275.

Dawson, M.A. (1979). Haematological effects of Long-term, mercury exposure and subsequent period of recovery on the winter Flounder. *Pseudopleuronactus americanus*, In: *Marine Pollution: Functional Responses*, p. 171.

Decie, I.V. and S.M. Lewis (1970). *Practical haematology*, 4th. Edn. J. and A. Churchill, London, p. 19.

Desaiah, D. and Koch, R.B. (1975). *Biochem. Biophys. Res. Comm.*, 64(1): 13.

Desaiah, D. and Koch, R.B. (1975). *Bull. Environ. Contman. Toxicol.* 13(2): 238.

Desaiah, D., Cutkomp, L.K.R.B. and Garvinean, A. (1975). *Arch. of Environ. Contam. Toxicol.*, 3(2): 132.

Dorn, P. (1976). *Bull. Environ. Contam. Toxicol.*, 15: 714.

Doudoroff, P., Katz, M. and Tarrwell, C.M. (1953). *Sewage and Industrial Wastes*. 25(7): 840.

Dove, W.F. and T. Yamane (1960). *Biochem. Biophys. Res. Commun.*, 3: 608.

Dubale, M.S. (1963). *Aerial Respiration in Fishes: A Review*. Proc. First Summer School of Zoology (Simla, 1961), p. 328.

Ferguson, D.E. and Bingham, C.R. (1966a). *Bull. Environ. Contam. Toxicol.*, 1: 97.

Ferguson, D.E., Ludke, J.L. and Murphy, G.G. (1966). *Trans. Am. Fish. Soc.*, 95: 335.

Finney, D.T. (1977). *Probit Analysis*. Cambridge Univ. Press, Cambridge.

Fiske, C. and Subbarow, Y. (1925). The colourimetric determination of phosphorous. *J. Biol. Chem.*, 66: 375.

Fox, J.H., Patel-Mandlik, K. and Cohen, M.M. (1975). *Jr. of Neurochem.*, 14: 757.

Goldberg, E.D. (1978). In: *Advances in Oceanography*. Joint Oceanographic Assembly (Edited by H. Charmoch and G. Deacon), Scotland, 49.

Gopalakrishnan, V. (1961). *Indian J. of Fish.*, 8: 222.

Gouda, R.K. Tripathy, N.K. and Das, C.C. (1981). *Comp. Phyiol. Ecol.*, 6(3): 170.

Gray, J.S. (1974). In: *Pollution and Physiology of Marine Organisms*. (Eds.) F.J. Vernberg and W.B. Vernberg, Academic Press, New York, p. 465.

Hanko, E., Erne, K., Wanntorp, H. and Borg, K. (1970). *Acta. Vet. Scand.*, 11:268-282.

Hannerz, L. (1968). *Resp. Insti. Freshwater Res. Swed.*, 48A: 120.

Hartung, R. (1973). Biological effects of some heavy metal pollutant in water. In: *Metal Ions in Some Biological System*. (Ed.) S.K. Dhar, Plenum Press, New York, p. 161.

Hodson, P.V. and Spraque, J.B. (1975). *Jour. Fish. Res. Bd.* Canada., 32: 1.

Hodson, P.V., Blunt, B.R., Spray, D.J. and Austen, K. (1977). *J. Fish. Res. Bd.* Canada, 34: 501.

Holden, A.V. (1962). *Ann. Appl. Biol.*, 50: 467.

Hossain, A. and H.M. Dutta (1986). *Bull. Environ. Contam. Toxicol.*, 36: 460.

Houston, A.H., K.M. Mearow and J.S. Smeda (1976). *Comp. Biochem. Physiol.*, 54(A): 267.

Hughes, J.S. and Davis, J.T. (1963). *Weeds*, 11: 50.

Hurkat, P.C. and Mathur, P.N. (1976). *Text Book of Animal Physiology*. S. Chand and Co. (P) Ltd., New Delhi, p. 172.

Inagaki, T. (1940). *J. Biochem.*, Tokyo, 32: 57.

Johanson, K. (1970). *Fish Physiology*. (Ed.) W.H. Hoar and D.J. Randall, Academic Press, New York, 4: 367.

Jonasson, I.R. (1970). *Mercury in the Natural Environment: A Review of Work*. Geological Survey of Canada, Paper, 70.

Joyce, C.R.B., H.A. Moore and M. Weatherale (1959). *Brit. J. Pharmacol., Chemother.*, 9: 463.

Kastori, R., Popovic. M., Gasic, O., Petrovic, N., Stajner, D. (1993). *Zemljiste Biljka*, 42: 177-184.

Katz, S. (1952). *J. Am. Chem. Soc.*, 74: 2238.

Katz, S. (1962). *Nature*, 194: 569.

Katz, S. (1963). *Biochem. Biophys. Acta.*, 68: 240.

Katz, S. and V. Santilli (1962). *Biochem. Biophys. Acta.*, 55: 621.

Krough, A. and A. Leitch (1919). *J. Physiol.*, London, 52: 288.

Larson, A. and Lewander, K. (1973). *Comp. Biochem. Physiol.*, 44A: 367.

Lenfant, C. and K. Johansen (1972). *Respir. Physiol.*, 14: 211.

Levitt, J. (1972). *Response of Plants to Environmental Stresses*. Academic Press, Inc. N.Y.

Lowry, C.H., Rosebrough, J.N., Farr, A.L. and Randall, R.J. (1951). *J. Biol. Chem.*, 193: 265.

Lunn, C.R., D.P. Toews and D.J. Pree (1976). *Can. J. Zool.*, 54: 214.

Mac Innes, J. and F. Thurberg (1973). *Mar. Poll. Bull.*, 4: 185.

MacLeod, J.C. and Passah, E. (1973). *J. Fish. Res. Bd.* Canada, 30: 485.

Mathur, D.S. (1969). Toxicity of Aldrin to certain fish. *Sci. and Cult.*, 35: 1.

Matida, Y., Kumada, H., Kimura. S., Nose, T., Yokote, M. and Kowateu, H. (1971). *Bull. Fresh Wat. Fish. Res. Lab.*, Japan, 21: 197.

Matsunga, K.(1978). *Bull. Fac. Fish.*, Hokkaido Univ., 29: 70.

McCarty, L.C., J.A.C. Herg and A.H. Houston (1978). *J. Fish. Res. Bd.* Canada, 35(1): 35.

McKim, J.M., Eaton, G.J. and Holcombe, G.W. (1978). *Bull. Environ. Contam. Toxicol.*, 19: 608.

McKim, J.M., Olson, G.F., Holcombe, G.W. and Hunt, E.P. (1976). *J. Fish. Res. Bd.* Canada, 33: 2726.

Mehrle, P.M. and F.L. Mayer (1980). Clinical tests in aquatic toxicology: State of the art. *Environ Health Perspect*, 34: 139.

Misra, M. (2002). *Toxicological Effects of Red Mud Waste Extract on a Freshwater Fish and its Ecological Implications*. Ph.D. thesis, Berhampur University.

Mulvihill, J. (1972). *Science*, 176: 132.

Norstron, R.J., A.E. Makinnon and A.S.W. de Freitas (1976). *J. Fish. Res. Bd. Can.*, 33: 248.

Nystrom, T. and Kjelleberg, S.(1987). *FEMS Microbiol. Ecol.*, 45:143-153.

Nystrom, T., Albertson, N., Flardh, K., and Kjelleberg, S. (1990). *Microbiol. Ecol.*, 74: 129-140.

O'Connor, D.V. and P.O. Fromm (1975). *Bull. Environ. Contam. Toxicol.*, 13: 406.

Olson, G.F., Mount, D.I., Snarski, V.M. and Throaland, T.W. (1975). *Bull. Environ. Contam. Toxicol.*, 14: 129.

Orange and Rhein (1970): In: *Manual of Clinical Laboratory Procedures* (Eds. Faulkner, W.R. and King, J.W.). The Chemical Rubber Co., U.S.A., p. 59.

Panigrahi, A.K. (1980). Eco-physiological effects of mercurial compounds on some freshwater fish (Toxicological effects of inorganic mercury [$Hg(NO_3)_2$] on two freshwater fishes, *Tilapia mossambica*, Peters and *Anabas scandens*. Cuv. & Val. and its ecological implications. Ph.D. thesis, Berhampur University.

Panigrahi, A.K. and Mishra, B.N. (1978). *Environ. Pollut.*, 16(1): 31.

Panigrahi, A.K. and Mishra, B.N. (1979). *Bull. Environ. Contamn. Toxicol.*, 23(6): 784.

Panigrahi, A.K. and Mishra, B.N. (1980). *Arch. Toxicol.*, 44: 269.

Panigrahi, K. (1984). Toxicological effects of Emisan-6 (Methoxy Ethyl Mercuric Chloride) on a freshwater fish. *Anabas scandens*. Cuv. and Val. Ph.D. thesis, Berhampur University.

Passow, H., A. Rothstein and T.W. Clarkson (1961). *Pharmacol. Rev.*, 13: 185.

Patel, K.K., Hurtmann, J.F. and Cohen, M.M. (1973). *Stroke*, 4: 221.

Pattnaik, J. (2002). Intensity of mercury pollution in and around a Chlor-alkali industry and its eco-toxicological significance. Ph.D. thesis, Berhampur University.

Poljakoff-Mayber, A. and J. Gale (1975). *Plants in Saline Environments* (Ecological studies 15). Springer-Verlag, Berlin.

Prosser, C.L. (1973). Inorganic ions. In: *Comparative Animal Physiology*. (Ed.) C.L. Prosser, W.B. Saunders Co., p. 79.

Ramaswamy, M. and T. Gopalkrishna Reddy (1978). *Proc. Indian Acad. Sci.*, 87(B): 331.

Ramel, C. (1969). *Hereditas*, 61: 208.

Rath, S. and Misra, B.N. (1979). *Exp. Geront.* 14: 307.

Rath, S. and Misra, B.N. (1980). *Environ. Poll.* (Series A), 23: 95.

Rath, S. and Misra, B.N. (1980). *Indian J. Fish.*, 27 (1 and 2): 76.

Rath, S. and Misra, B.N. (1981). *Toxicology*, 19: 123.

Ray, P. and David, A. (1962). *Ind. Jr. Fish.*, 9: 117.

Richard, A.S. (1966). *Fish. Wild.*, 35: 3.

Riggs, A. (1970). *Fish Physiology*. (Eds.) W.S. Hoar and D.J. Randall. Academic Press, New York, 4: 367.

Roald, S.D. (1967). *Aquacultures*, 12: 327.

Samant, D.P. (1987). Toxicological effects of a mercury based fungicide on a freshwater fish and its ecological implications. Ph.D. thesis, Berhampur University.

Samantaray, A.C. (1999). Eco-toxicological effects of leached wastes of an Aluminium factory on a crop plant. Ph.D. thesis, Berhampur University.

Sarangi, P.K. (2002). Eco-toxicological effects of leached chemicals of the Waste of the Alumian industry on a freshwater fish. Ph.D. thesis, Berhampur University.

Schaumberg, F.D., T.E. Howard and C.C. Walden (1967). *Water Res.*, 1: 731.

Sharma, B.K. and H. Kaur (1993). *Environmental Chemistry*. Goel Publishing House, Krishna Prakashan Media (P) Ltd., Meerut.

Shaw, B.P. (1987). Eco-physiological studies of Industrial effluents on Biosystems. Ph.D. thesis, Berhampur University, Orissa.

Shaw, B.P. (1987). Eco-physiological studies of industrial, effluent of a chlor-alkali factory on biosystems. Ph.D. thesis, Berhampur University, Orissa, India.

Soivio, A. and K. Nyholm (1974). *Aquaculture*, 3: 395.

Stevenson, A.J., S. Kacew and R.L. Singhal (1977). *Toxicol. Appl. Pharmacol.*, 40: 161.

Stone, C.L., Foy, M.R.S., Jones, A.L. and Mahaffey, K.R. (1977). *Poultry Science*, 56: 174-181.

Sudderudin, K.I. (1973). *Comp. Gen. Pharmacol.*, 4: 219.

Sunderessan, B.B., P.V.R. Subrahmanyam and A.D. Bhinde (1983). An overview of toxic and hazardous waste in Indian industry and Environment. Special issue (1983): In: *The State of Indian Environment, 1984-85*. The second citizen's report, p. 196.

Thurberg, F.P., A. Calabrese and M.A. Dawson (1974). Effect of silver on oxygen consumption of bivalves at various salinities. In: *Pollution and Physiology of Marine Organisms* (Eds. F. J. Vernerg and W. B. Vernberg), Academic Press, N.Y., p. 67.

Trim, A.R. (1959). *Biochem. J.*, 73: 928.

Vernberg, W.B. and F.J. Vernberg (1972). *Fish. Bull.*, 70: 415.

Vincent, P.C. and C.R.B. Blackburn (1958). *Aust. J. Exp. Bio. Med Sci.*, 36: 471

Webb, J.L. (1966). *Enzymes and Metabolic Inhibitors*, 1st ed. Academic Press, New York, 2: 729.

Wong, P.T.S., G. Burmison and Y.K. Chau (1979). *Bull. Environ. Contam. Toxicol.*, 23: 487.

Yagi, N., M. Kimura and Y. Itokawa (1976). *Bull. Environ. Contam. Toxicol.*, 16: 516.

Yamamura, K., N. Machara, H. Ohno, N. Veno, A. Kohyama, T. Satoh, A. Shimoda and R. Kishi (1987). *Bull. Environ. Contam. Toxicol.*, 38: 985.

Yamanaka, H., K. Yamaguchi and F. Matsumura (1965). *Bull. Jap. Soc. Scient. Fish.*, 31: 833.

Index